5th International Conference on Fisheries and Marine Sciences (INCOFIMS 2022)

IOP Conference Series: Earth and Environmental Science
Volume 1273

Online
28 September 2022

Editors:

Akhmad Taufik Mukti
A. Shofy Mubarak
Eka Saputra

ISBN: 978-1-7138-8742-3
ISSN: 1755-1307

Printed from e-media with permission by:

Curran Associates, Inc.
57 Morehouse Lane
Red Hook, NY 12571

Some format issues inherent in the e-media version may also appear in this print version.

Institute of Physics
Dirac House, Temple Back
Bristol BS1 6BE UK

Phone: 44 1 17 929 7481
Fax: 44 1 17 920 0979

techtracking@iop.org

Additional copies of this publication are available from:

Curran Associates, Inc.
57 Morehouse Lane
Red Hook, NY 12571 USA
Phone: 845-758-0400
Fax: 845-758-2633
Email: curran@proceedings.com
Web: www.proceedings.com

TABLE OF CONTENTS

The 5th International Conference on Fisheries and Marine Sciences (INCOFIMS) Surabaya Indonesia, 28 September 2022

International conference on fisheries and marine sciences (INCOFIMS) is an annual conference organized by Faculty of Fisheries and Marine Universitas Airlangga, Surabaya, Indonesia. The main aim is to provide a sharing platform that enables researchers, academics and practitioners from all over the world to share their most recent findings as well as to propose the best strategies to address issues and challenges which we have been currently facing in aquaculture and fisheries practices worldwide. The 1st INCOFIMS was held successfully offline in Surabaya in 2018, the second in 2019, the third in 2020 and the 4th in 2021.

The 5th INCOFIMS was previously scheduled offline in Surabaya on 28th September 2022. However, due to the Covid-19 pandemic and travel restriction for foreigners come into Indonesia as well as traveling within the Indonesian islands, we had the 5th INCOFIMS in a virtual format with ZOOM on 28 September 2022, and hosted from Faculty of Fisheries and marine, Universitas Airlangga, Surabaya Indonesia.

The theme in the 5th INCOFIMS was "Optimizing the Utilization of Aquatic Resources Through the Blue Economy Program Towards Achieving the SDGs Program". Technically, we had the conference divided into 2 (two) sessions in general: (1) keynote speaker session and (2) guest speaker session. In the keynote session, we had 3 (three) keynote speakers delivering a speech which were Prof. Dr. Orhan Tufan Eroldogan, Prof. Lee Baumgartner and Dr. Ian Hendy. Each keynote speaker had 1.5 hours for giving a presentation using ZOOM and 30 minutes for discussion in one virtual room. After the keynote speaker session, we proceeded to the guest speaker session in which all participants were divided into 7 (seven) rooms according to our subtopics for oral and poster presentations:

Room 1: Aquaculture technology
Room 2: Fish Nutrition
Room 3: Fish Diseases
Room 4: Marine and Aquatic Sciences
Room 5: Estuarine and Coastal Ecosystems
Room 6: Fisheries Management
Room 7: Fisheries Socio-economics

In this session, every speaker had 15 minutes for presentation and 5 minutes for discussion. Total participants joined in this conference was 200 participants from at least 6 different countries (Australia, Japan, Austria, Malaysia, Taiwan and Indonesia).

The conference was in general quite successful, acknowledging the number and enthusiasms of participants during the discussion sessions in both the keynote speaker session and guest speaker's session. We thank all participants and organizing committee for their support to this conference and see you in the 6th INCOFIMS 2023.

Chairman
Ayulana Nafisah, Ph.D

IOP Conf. Series: Earth and Environmental Science 1273 (2023) 011001 doi:10.1088/1755-1315/1273/1/011001

ORGANIZING COMMITTEE

Chair Person
Ayulana Nafisah Ph.D (Universitas Airlangga, Indonesia)

Reviewer
Prof. Dr. Sri Subekti, drh.,DEA (Universitas Airlangga, Faculty of Fisheries and Marine, Indonesia)
Assoc. Prof. Sukree Hajisamae (Prince of Songkla University, Faculty of Science and Technology, Hatyai, Thailand)
Assoc. Prof. Wanmimol Klaypradit (Kasetsart University, Department of Fishery Products, Bangkok, Thailand)
Assoc. Prof. Prapansak Srisapoome (Kasetsart University, Department of Aquaculture, Bangkok, Thailand)
Prof. Ir. Moch. Amin Alamsjah, M.Si., Ph.D (Universitas Airlangga, Faculty of Fisheries and Marine, Indonesia)
Dr. Laksmi Sulmartiwi, S.Pi., M.P (Universitas Airlangga, Faculty of Fisheries and Marine, Indonesia)
Prof. Dr. Ir. Gunanti Mahasri, M.Si (Universitas Airlangga, Faculty of Fisheries and Marine, Indonesia)
Dr. RR. Juni Triastuti, S.Pi., M.Si (Universitas Airlangga, Faculty of Fisheries and Marine, Indonesia)
Dr. Woro Hastuti Satyantini, Ir., M.Si. (Universitas Airlangga, Faculty of Fisheries and Marine, Indonesia)
Wahju Tjahjaningsih, Ir., M.Si (Universitas Airlangga, Faculty of Fisheries and Marine, Indonesia)

Editor
Dr. Akhmad Taufik Mukti, S.Pi., M.Si (Universitas Airlangga, Faculty of Fisheries and Marine Surabaya, Indonesia)
Dr. A. Shofy Mubarak, S.Pi., M.Si (Universitas Airlangga, Faculty of Fisheries and Marine Surabaya, Indonesia)
Eka Saputra, S.Pi., M.Si (Universitas Airlangga, Faculty of Fisheries and Marine Surabaya, Indonesia)

Steering Commitee
Prof. Moch. Amin Alamsjah, Ph.D (Universitas Airlangga, Faculty of Fisheries and Marine Surabaya, Indonesia)
Dr. Woro Hastuti Satyantini, Ir., M.Si. (Universitas Airlangga, Faculty of Fisheries and Marine Surabaya, Indonesia)
Dr. Laksmi Sulmartiwi, S.Pi., M.P (Universitas Airlangga, Faculty of Fisheries and Marine Surabaya, Indonesia)
Dr. Eng. Sapto Andriyono, S.Pi., MT. (Universitas Airlangga, Faculty of Fisheries and Marine Surabaya, Indonesia)
Dr M. Hery Riyadi Alauddin, S.Pi., M.Si. (Director of Marine and Fisheries Polytechnic)
Dr. Sri Rahmaningsih, S.Pi., MP. (Faculty Fisheries and Marine Science, University of Ronggolawe)

Secretary
Dwi Yuli Pujiastuti, S.Pi., M.P., M.Sc (Universitas Airlangga, Faculty of Fisheries and Marine Surabaya, Indonesia)
Abdul Manan, S.Pi., M.Si., Ph.D (Universitas Airlangga, Faculty of Fisheries and Marine Surabaya, Indonesia)

Treasury
Lailatul Lutfiyah, S.Pi., M.Si (Universitas Airlangga, Faculty of Fisheries and Marine Surabaya, Indonesia)
Dwitha Nirmala, S.Pi., M.Si (Universitas Airlangga, Faculty of Fisheries and Marine Surabaya, Indonesia)

Human Relation Division
HapsariKenconojati, S.Si., M.Si (Universitas Airlangga, Faculty of Fisheries and Marine Surabaya, Indonesia)
Mohammad Faizal Ulkhaq, S.Pi., M.Si (Universitas Airlangga, Faculty of Fisheries and Marine Surabaya, Indonesia)
Arif Habib Fasya, S.Pi., M.P (Universitas Airlangga, Faculty of Fisheries and Marine Surabaya, Indonesia)
Darmawan Setia Budi, S.Pi., M.Si (Universitas Airlangga, Faculty of Fisheries and Marine Surabaya, Indonesia)

Organizer Division
Dr. Eng. Patmawati, S.Pi., M.Si (Universitas Airlangga, Faculty of Fisheries and Marine Surabaya, Indonesia)
Muhammad Amin, S.Pi., M.Sc., Ph.D (Universitas Airlangga, Faculty of Fisheries and Marine Surabaya, Indonesia)
Dr. Adriana Monica Sahidu, Ir., M.Kes (Universitas Airlangga, Faculty of Fisheries and Marine, Indonesia)
Dr. Anna Fauziah, Indah Puspitasari, S.Si., M.Sc. (Marine and Fisheries Polytechnic)
Atika Marisa Halim, S.Pi., MP., M.Sc .(Marine and Fisheries Polytechnic)
Muhammad Zainuddin, S.Pi., M.Si (Faculty Fisheries and Marine Science, University of Ronggolawe)
Dr. Marita Ikajoysidawati, ST., M.Si. (Faculty Fisheries and Marine Science, University of Ronggolawe)

Secretariat
Luthfiana Aprilianita Sari, S.Pi., M.Si (Universitas Airlangga, Faculty of Fisheries and Marine Surabaya, Indonesia)
Wahyu Isroni, S.Pi., MP. (Universitas Airlangga, Faculty of Fisheries and Marine Surabaya, Indonesia)
Daruti Dinda Nindarwi, MP (Universitas Airlangga, Faculty of Fisheries and Marine Surabaya, Indonesia)
Syifania Hanifah Samara, S.Pi., M.Sc (Universitas Airlangga, Faculty of Fisheries and Marine Surabaya, Indonesia)

IOP Conf. Series: Earth and Environmental Science 1273 (2023) 011002 doi:10.1088/1755-1315/1273/1/011002

Peer Review Statement

All papers published in this volume have been reviewed through processes administered by the Editors. Reviews were conducted by expert referees to the professional and scientific standards expected of a proceedings journal published by IOP Publishing.

- **Type of peer review:** Double Anonymous
- **Conference submission management system:** Morressier
- **Number of submissions received:** 120
- **Number of submissions sent for review:** 120
- **Number of submissions accepted:** 97
- **Acceptance Rate (Submissions Accepted / Submissions Received × 100):** 80.8
- **Average number of reviews per paper:** 2
- **Total number of reviewers involved:** 24
- **Contact person for queries:**
  Name: Dwi Yuli Pujiastuti
  Email: dwiyp@fpk.unair.ac.id
  Affiliation: Department of Marine, Faculty of Fisheries and Marine, Universitas Airlangga

The effect of the innards silage of fish and coconut pulp fermentation on the biomass and population of silkworm (*Tubifex tubifex*)

D Puspitasari[1], **L Lutfiyah**[2,3], **L Sulmartiwi**[2]

[1]Program study of Aquaculture, Faculty Fisheries and Marine, Universitas Airlangga, Surabaya Indonesia
[2]Department of Marine, Faculty Fisheries and Marine, Universitas Airlangga, Surabaya Indonesia

[3]Corresponding author: lailatullutfiyah@fpk.unair.ac.id

Abstract. Silkworm is a type of worm that can be found in waterways that contain organic matter. Silkworms are difficult to find especially during the rainy season and silkworms can carry parasites because their habitat is in dirty places. The purpose of this study was to determine the effect and effective dose of a combination of the innards silage of fish and coconut pulp fermentation on the biomass and population of silkworms. The research method used was the experimental method, data analysis in this study used Analysis of Variance (ANOVA) and continued with Duncan's Multiple Distance test. The result showed application combination of the innards silage of fish and coconut pulp fermentation had a significant effect ($p<0.05$) on the biomass and population of silkworms. The highest biomass with a value of 16,296 g was found in treatment P3 (rice field mud+25% innards silage of fish +75% fermentation coconut pulp), while the highest population was obtained in treatment P3 with a total population of 1463 individuals.

1. Introduction

Feed must be considered in terms of quantity and quality. Silkworms are natural food for fish seeds because they have good quality, but the availability of silkworms is very little. The weakness of silkworms is that their availability is decreasing all the time, especially during the rainy and flood seasons [1].

Silkworm habitat is in ditches or ditches, silkworms can carry diseases or parasites from their habitat [2]. Therefore it takes an effort to cultivate silkworms. Cultivation media that can be used are fish offal and fermented coconut dregs.

2. Material and methods

2.1 Maintenance media

Maintenance media using fish offal and fermented coconut dregs. These two materials are combined with rice field mud according to the treatment used.

2.2 Research treatment

This study used four treatments and five replications to determine the effective dose on silkworm biomass and population, the dose determination used in this study 100% rice field mud (control) Rice field mud + combination result (75% fish offal silage and 25% coconut dregs fermentation), Rice field mud + combination result (50% fish offal silage and 50% coconut dregs fermentation), Rice field mud + combination result (25% fish offal silage and 75% coconut dregs fermentation).

3. Result and discussion

3.1 Silkworms biomass

In observation of absolute biomass of Silkworm the highest average biomass was found in P3 treatment with a value of 16.29 while the lowest yield was found in P0 treatment of 5.14.

Table 1. Silkworm Absolute Biomass Average Value

Treatment	Initial biomass	Final biomass	Absolut biomass ±SD
P0	10	15,14	$5{,}14 \pm 1{,}15^{c}$
P1	10	16,82	$6{,}82 \pm 1{,}04^{c}$
P2	10	19,56	$9{,}56 \pm 1{,}51^{b}$
P3	10	26,29	$16{,}29 \pm 1{,}85^{a}$

Information: Different superscript lowercase notations in the same column showed significant differences ($p>0.05$). P0 (100% paddy mud), P1 (paddy mud + fish innards silage 75% + coconut dregs fermentation 25%), P2 ((paddy mud + fish innards silage 50% + 50% coconut dregs fermentation), P3 (paddy mud + fish offal silage 25% + coconut dregs fermentation 75%).

Provision of silage of fish offal and fermentation of dregs coconut affects silkworm biomass. Lack of nutrients found in cultivation media will affect the lack of food intake so that causing low biomass and nutrient content of silkworms. Growth silk worm biomass is influenced by the availability of organic matter that can used by silk worms [3]. The decomposition of organic matter into organic sources by bacteria causing the nutrient content of the maintenance media [4]. In addition to protein, carbohydrates also play a role in increase in biomass, carbohydrates are elements composed of C-Organic needed by microorganisms to produce energy for processes metabolism, thereby accelerating bacterial growth and decomposition can be digested by silk worms [5].

3.2 Silkworms population

In observation of absolute population of silkworm the highest average population was found in P3 treatment with a value of 1463 ind.while the lowest yield was found in P0 treatment of 856 ind. The highest population was obtained because silk worms were able to grow with utilizing cultivation media rich in organic matter [6]. Crude protein content in fermentation coconut pulp by 8.11%, crude fiber 20.89%. The protein content is useful for population growth and biomass of silkworms which require nutrients to reproduce [7].

Table 2. Average Value of Silk worm Population

Treatment	Initial population	Final population	Absolute biomass ±SD
P0	5480	6336	856 ±154,60^{b}
P1	5480	6393	913 ±84,47^{b}
P2	5480	6532	1052 ±151,08^{b}
P3	5480	6943	1463 ±292,54^{a}

Information: Different superscript lowercase notations in the same column showed significant differences ($p>0.05$). P0 (100% paddy mud), P1 (paddy mud + fish innards silage 75% + coconut dregs fermentation 25%), P2 ((paddy mud + fish innards silage 50% + 50% coconut dregs fermentation), P3 (paddy mud + fish offal silage 25% + coconut dregs fermentation 75%).

4. Conclusion

The combination of silage of fish offal and fermentation of coconut dregs has an effect on the biomass and population of silkworms. The resulting biomass ranged from 5.14 g to 16.296 g with an initial stocking of 10 g. The silkworm population produced ranged from 856 ind to 1463 ind. The effective dose of silage of fish offal and fermentation of coconut dregs is P3 treatment (paddy mud + (25% fish offal silage + 75% coconut dregs fermentation) with the highest biomass yield of 16.296 g and population of 1463 ind.

5. References

[1] Akhril M, Muskita W H, Idris M 2019 *Med Akua.* 4(3) 125-132.
[2] Nuraini, S Nasution, A Tanjung, H Syawal 2019 *J of Rural Urb Com Enpow.* 1(1).
[3] Pardiansyah D, Eddy S, Daniel D 2014 *J Akua Ind.*13(1) 2 8-35.
[4] Shafrudin D, W Efiyanti, Widanarni 2005 *J Akua Ind.* 4(2) 97-102.
[5] Astuti K, Tri C, Lutfhi D W 2017 *J Istek.* 10(1).
[6] Widyawati N F, Suminto, J Hutabarat 2014 *JAMT.* 3(4).
[7] Biyatmoko D, Syarifuddin dan Lilis H 2018 *ZIRA'AH.* 43(3).

6. Acknowledgement

This research was conducted jointly with the team at the Faculty of Fisheries and Marine, Airlangga University.

Chemical characteristics of red swimming crab (*Thalamita spinimana*) in Dompak Waters, Tanjungpinang City

Y Oktavia[1] and W Muzammil[2,3,4]

[1]Aquatic Product Technology Department, Faculty of Marine Science and Fisheries, Raja Ali Haji Maritime University, Tanjungpinang City, Indonesia
[2]Aquatic Resources Management Department, Faculty of Marine Science and Fisheries, Raja Ali Haji Maritime University, Tanjungpinang City, Indonesia
[3]Marine Biotechnology Laboratory, Faculty of Marine Science and Fisheries, Raja Ali Haji Maritime University, Tanjungpinang City, Indonesia

[4]Corresponding author: wahyu.muzammil@umrah.ac.id

Abstract. Red swimming crab (*Thalamita spinimana*) is one of the fishery commodities caught by fishermen in Senggarang Village, Tanjungpinang City. Dompak Waters is one of the red swimming crab fishing ground for Senggarang fishermen. Utilization of red swimming crab have not used optimally as raw material for food and industry. This study aimed to determine the nutritional composition, fatty acids, amino acids, and heavy metals in red swimming crab. The sample used is fresh red swimming crab meat contained in the carapace and claws. The results showed that the nutritional composition of fresh red swimming crab meat contained 81.13% moister, 15.48% protein, 0.59% fat, 2.14% ash, 0.67% carbohydrate, and 0.53% crude fiber content. The highest amino acid content in red swimming crab is glutamic acid at 16,043.78 mg/kg. Red swimming crab has 0.38% unsaturated fatty acids and 0.20% saturated fatty acids. Heavy metals Hg, Pb, and Sn contained in red swimming crab are still below the based on Indonesian National Standard for heavy metal maximum threshold contained in food. Red swimming crab caught by fishermen in Senggarang Village from Dompak Waters fishing ground area are safe to use as food and nutritious raw materials.

1. Introduction

Riau Islands is a cluster of small islands that have high natural resource potential, especially in aquatic resources [1,2,3,4,5,6], one of which is Bintan Island. Bintan Island has a long been known as one of the bauxite-producing islands in Indonesia [7,8,9,10]. Since bauxite mining activities have been banned, the fisheries sector has become one of the leading players in addition to the tourism sector. Senggarang is one of the coastal areas on Bintan Island, precisely in Tanjungpinang City, Riau Islands Province where in general the people make a living as fishermen. One of the marine life of the economically valuable crustacean species caught by Senggarang fishermen apart from crabs is the red swimming crab (*Thalamita spinimana*).

Utilization of red swimming crab in Senggarang is generally still done on a small-scale fisheries, similar in common crab fisheries in Indonesia [11,12,13,14]. Fishermen use trap fishing gear (local people called *bubu*) to catch red swimming crab. The fishing boat <5GT used is also simple with the distance to the fishing area relatively close and categories as daily fishing fisherman. One of fishing ground area of fishermen from Senggarang are Dompak Waters (near Senggarang coastal area).

The potential for capture fisheries in Dompak waters, Tanjungpinang City such as red swimming crab (*T. spinimana*), blue swimming crab (*Portunus pelagicus*), penaeid shrimp, and various types of crustaceans that have important economic value for Senggarang fisherman. The catch of fisherman in Dompak waters from red swimming crab commodities has a selling price that is still relatively low when compared to blue swimming crab. In terms of meat taste and texture, it is not much different from blue swimming crab, but it is only sold locally, the price of red swimming crab is not as expensive as the price of blue swimming crab. This is also possible because blue swimming crab is Indonesia's leading export crab commodity to the United States [10,15,16,17], so the price follows the market price using the USD exchange rate.

Some research about red swimming crab such as maturity level of female red swimming crab [5], morphometric characteristics of red swimming crab [18]. The lack of research on chemical characteristics contained of red swimming crab still unknown, even though this resource has economic value for Senggarang fisherman. Chemical characteristics research related to the potential nutritional composition, fatty acids, amino acids, and heavy metals in red swimming crab is very important to do to see the safety food of red swimming crab uses as food and nutritious raw materials, especially from Dompak Waters habitat, Tanjungpinang City.

2. Material and methods

2.1 Material

Red swimming crab meat (*Thalamita spinimana*) obtained from Sengarang fishermen caught in Dompak Island Waters, Tanjungpinang City, Riau Islands Province. Samples were collected randomly regardless of age, sex, and size.

2.2 Methods

Red swimming crab meat preparation was carried out on live crabs. Red swimming crab meat was then carried out proximate analysis using the SNI 01-2891-1992 method, amino acid profile analysis (ultra-performance liquid chroatography-photodiode array), fatty acids (gass chromatography-flame ionization detector), and heavy metals (inductively coupled plasma mass-spectrometry). Each test was repeated twice and the average value was taken.

3. Result and discussion

3.1 Proximate

Crab as well as fish is an important food ingredient because it contains both macronutrients (protein and fat) and micronutrients (vitamins and minerals) [19]. The nutritional content of crab can be used to select crab according to their nutrition and function. This provides a basis for consumers to choose crab according to their needs. The chemical composition (proximate) of crab includes determination of water content, protein. Fat and ash are 98-99% of fish body compositions [20]. Red swimming crab (*Thalamita spinimana*) as one of the aquatic biota has a chemical composition as a compiler of its body. The chemical composition of the red swimming crab (*Thalamita spinimana*) is shown in Table 1.

Table 1. Proximate test result of red swimming crab (*Thalamita spinimana*)

Nutritional	Chemical Composition (%)
Ash	2,14 ± 0,02
Water	81,13 ± 0,11
Fat	0,59 ± 0,01
Protein	15,48 ± 0,10
Carbohydrate	0,67 ± 0,01
Fiber	0,53 ± 0,03

The chemical content of fish including crab are varies depending on several factors. In general, the protein content of fish (including crab) meat ranges from 17-22%. The fat content of fish fillets is grouped into low-fat fish (fat content below 3%), medium (3-7% fat), and high-fat fish (fat content more than 7%). Fish and seafood have the potential to provide calories between 199-200 cal per 100 g, which is associated with the protein and fat content of fish. Fish oil is a great source of vitamins A, D, and E. Mostly in fish liver oil should be concentrated vitamin A. The total minerals (macro-elements) content in marine organism such as calcium, sodium, potassium, phosphorus, and magnesium. Micro-elements such as selenium, iodine, cobalt, fluorine, molybdenum, and manganese in the muscles and invertebrates of raw marine fish is in the range of 0.6 - 1.5% on the wet weight [21].

Ash content shows the mineral content contained in food. The ash content of the red swimming crab (*Thalamita spinimana*) in this study was 2.14%. The low ash content of red swimming crab is because the samples used are only fresh meat. This ash content is lower than blue swimming crab/*Portunus pelagicus* (13%) [22] and *Thalamita* sp. (13.02%) [23].

Water content is the largest component of living organism. The water content of red swimming crab in this study was 81.13%. Moisture content greatly affects the shelf life of a food ingredient. Foods with high water content will spoil faster.

The fat content of the red swimming crab (*Thalamita spinimana*) is 0.59%. The results of this study are lower when compared to research conducted by Ayaz and Ozogul. [24] on blue crab. The results of research conducted by Ayaz and Ozogul. [24] showed that the fat content of blue crab (*Callinectes sapidus*) was low (0.96% for females and 1.11% for males).

The protein content of red swimming crab (*Thalamita spinimana*) is 15.48%. The results of this study are lower when compared to research conducted by Ayaz and Ozogul. [24] on blue crab. The results of research conducted by Ayaz and Ozogul. [24] showed that the protein content in blue crab (*Callinectes sapidus*) was relatively high (21.40% for males-22.45% for females). The results of another study on blue swimming crab (*Portunus pelagicus*) species showed protein levels ranging from 20.75% - 23.44%, lipids from 1.13% - 1.43%, and water from 73.31% - 76.04%, total minerals (TMS) from 1.81% - 2.10% [25].

3.2 Amino acid

Amino acid are components of protein that are needed by the body [26]. Red swimming crab (*Thalamita spinimana*) contains amino acids, (essential and non-essential). Amino acid testing in the future could be used as a reference in the development of functional and pharmaceutical foods [27]. The essential amino acid composition of red swimming crab (*Thalamita spinimana*) is shown in Table 2 and the non-essential amino acid composition of red swimming crab (*Thalamita spinimana*) is shown in Table 3.

Table 2. Essential amino acid composition of red swimming crab (*Thalamita spinimana*)

Essential Amino Acid	**Average (mg/kg)**
Histidine	4,110.42
Arginine	9,908.43
Threonine	5,902.47
Valin	5,567.37
Isoleucine	5,229.34
Leucine	8,453.20
Phenylalanine	5,374.13
Lysine	8,879.51
Total	53,424.85

Table 3. Non-essential amino acid composition of red swimming crab (*Thalamita spinimana*)

Non-Essential Amino Acid	Average (mg/kg)
Aspartic acid	9,709.03
Glutamic acid	16,043.76
Serine	4,745.47
Glycine	13,540.14
Alanine	6,838.69
Proline	3,937.48
Tyrosine	4,008.77
Total	58,823.33

Red swimming crab (*Thalamita spinimana*) contains essential amino acids, including: histidine, arginine, threoin, valine, isoleucine, leucine, phenylalanine, lysine. In this study, arginine was the highest value of essential amino acid content (9,908.43 mg/kg), while the lowest was histidine (4,110.42 mg/kg). The results of research conducted on crustaceans in Egypt and Saudi Arabia showed that the most abundant amino acid was arginine. This amino acid is found in higher amounts because it is an important component in the osmotic pressure regulation process in crustacean muscles [27]. Amino acid such as histidine (His) plays an important role in the antioxidant capacity of aquatic organism [28].

Red swimming crab (*Thalamita spinimana*) contains not only essential amino acids, but also non-essential amino acids such as glutamic acid, aspartic acid, serine, glycine, proline, alanine, and tyrosine. Glutamic acid was the highest non-essential amino acid that found 16.043,76 mg/kg, while proline was the lowest non-essential amino acid was that found 3,937.48 mg/kg. Profile of amino acids (protein) in crab plays an important role not only in various physiological and biological activities, but also in maintaining human health. Aspartic acid, glycine, and glutamic acid as amino acids promote wound healing [29]. Glutamic acid, glycine, and aspartic acid are responsible for the appearance of flavors in marine fish [30]. Tyrosine is an amino acid that has a role in self-protection, wound healing, and skin hardening. Hydroxylation of tyrosine leads to the formation of dihydroxylphenylalanine (DOPA) [31].

3.3 Fatty acid

Fatty acids are the components that compiler of fat. Fatty acids could be divided into saturated and unsaturated fatty acids [32]. The fatty acids found in red swimming crab (*Thalamita spinimana*) could be seen in the Table 4.

Table 4. Fatty acid content in red swimming crab (*Thalamita spinimana*)

Fatty Acid	**Fatty Acid Content (%)**
Linoleic acid	0.03 ± 0.00
Oleic acid	0.13 ± 0.00
Arachidonic acid	0.07 ± 0.00
Omega 6 fatty acid	0.10 ± 0.00
DHA	0.08 ± 0.00
Omega 3 fatty acid	0.15 ± 0.00
PUFA	0.25 ± 0.00
Stearic acid	0.03 ± 0.04
Palmitic acid	0.14 ± 0.00
Unsaturated fatty acid	0.38 ± 0.00
Omega 9 fatty acid	0.13 ± 0.00
EPA	0.07 ± 0.00
Mono-unsaturated fat	0.13 ± 0.00
Saturated fat	0.20 ± 0.00

The results showed that the highest fatty acids found in red swimming crab (*Thalamita spinimana*) were unsaturated fatty acids value was 0.38% when compared to saturated fatty acids value was 0.20%. The content of Polyunsaturated Fatty Acid (PUFA) has a value of 0.25%. This shows Red swimming crab contains good unsaturated fats. This is in line with the results of research by Balzano *et al.* [33] which states that crabs are a source of food rich in n-3 PUFA (omega-3 polyunsaturated fatty acids), with a multiple higher content of eicosapentaenoic acid, with 131.6 ± 31.6 mg/100 g edible portion than mantis shrimp and caramote shrimp.

3.4 Heavy metal

Heavy metal are required for a variety of biological processes, but become dangerous when present in excessive concentrations [34,35]. In particular, heavy metals induce oxidative stress by reducing antioxidant and generating free radicals levels [36]. Heavy metals not only alter the confirmation of DNA and protein, but also inhibit their function [37]. The results of testing for heavy metals in red swimming crab (*Thalamita spinimana*) could be seen in the Table 5.

Table 5. Heavy metal contains in red swimming crab (*Thalamita spinimana*)

Heavy Metals	**Contains (mg/kg)**	**SNI Threshold (mg/kg)***
Mercury (Hg)	0.06	<1.0
Lead (Pb)	-	<0.5
Tin (Sn)	-	-

Description: *Indonesian National Standard (SNI) 7387: 2009

The results of heavy metal testing on red swimming crab (*Thalamita spinimana*) showed that the detected heavy metal content was mercury with a concentration of 0.06 mg/kg. The heavy metal contained in the red swimming crab (*Thalamita spinimana*) is still below the allowable threshold. Muscle tissue of cuttlefish (*Sepia officinalis*), squid (*Loligo vulgaris*), and shrimp (*Penaeus semisulcatus*) showed that the content tested for lead content showed a value of <0.003–1.60 mg/kg [38]. Based on Liu *et al.* [39] research, shows that heavy metal content varies between fish species level. The metal contains studied in aquatic resources were found the highest concentrations in crustaceans, molluscs, and fish [40]. The metals cadmium, arsenic, copper, and zinc in the marine organisms tissues are relatively high concentrations, while mercury, chromium, and lead are identified as having low concentrations. Most elements that exposure doses are safe for coastal communities, except for arsenic, which indicates a cancer risk and high hazard value [41].

4. Conclusion

Red swimming crab from Dompak waters has a protein content of 15%, water 81%, ash 2,14%, fat 0,59%, carbohydrate 0,67% and contains essential amino acids, non-essential amino acids and fatty acids and does not contain Pb and Sn.

5. References

[1] Apriadi T, Muzammil W, Melani WR, Safitri A 2020 *DEPIK J Ilmu-Ilmu Perairan, Pesisir dan Perikanan.* 9(1) 119-130.
[2] Aprilia R, Susiana S, Muzammil W 2021 *J Kelaut Indones J Mar Sci Technol.* 14(2) 111-119
[3] Muzammil W, Prihatin N, Melani WR. 2021 *J Pengelolaan Perikan Trop.* 5(1) 20-28.
[4] Muzammil W, Zahra A, Oktavia Y 2021 *Panrita Abdi-Jurnal Pengabdi pada Masyarakat* 5(3) 356-364.
[5] Novitri S, Susiana S, Muzammil W, Kurniawan D. 2021 *Akuatikisle J Akuakultur, Pesisir dan Pulau-Pulau Kecil.* 5(2) 35-38.
[6] Yanto F, Susiana, Muzammil W 2020 *J Trop Fish Manag.* 4(2) 1-9.
[7] Apriadi T, Pratama G, Putra RD, Jumsurizal, Jaya YV, Firdaus M, Arpas HD, Suryanti A 2018 *Biodiversitas*, 19 967-973.
[8] Azizah D, Hamidy R, Mubarak M, Efriyeldi E, Muzammil W, Lestari F, Murtini S 2021 *E3S Web Conf.* 324 01008.
[9] Melani WR, Zulfikar A, Apriadi T, Muzammil W, Sabriyati D 2022 E3S *E3S Web Conf.* 339 02010.
[10] Muzammil W, Apriadi T, Melani WR, Damora A 2021. *J Trop Fish Manag.* 5(2) 65-72.
[11] Fikri IA, Darmono OP, Tetelepta JMS, Damora A, Muzammil W 2018 *IOP Conf Ser Earth Environ Sci.* 216 012038.
[12] Hamid A, Wardiatno Y, Irawati N 2019 *AACL Bioflux.* 12(4) 1335–1348.
[13] Muzammil W, Wardiatno Y, Butet NA 2015 *J Ilmu Pertan Indones.* 20(1) 78-84.
[14] Muzammil W, Apriadi T, Melani WR, Handayani KD 2020 *Aceh J Anim Sci.* 5(1) 18-25.
[15] APRI [Asosiasi Pengelola Rajungan Indonesia]. 2014. *Stock assessment, fisheries and environment parameters for BSC (Portunus pelagicus) in the Java Sea.* Research Report. Jakarta.
[16] Ghofar A, Redjeki S, Madduppa H, Abbey M, Tasunar N 2018 *IOP Conf Ser Earth Environ Sci.* 116 012068.
[17] Prince J, Creech S, Madduppa H, Hordyk A 2020 *Reg Stud Mar Sci.* 36 101309 1-13.
[18] Triana A, Susiana S, Muzammil W, Kurniawan D 2022 *Akuatikisle J Akuakultur, Pesisir dan Pulau-Pulau Kecil.* 6(1) 75-78.
[19] Hadijah S, and Lokapirnasari W P 2021 *IOP Conf Ser Earth Environ Sci.* 679(1) 012051.
[20] Ahmed I, Jan K, Fatma S, Dawood MAO 2022 *J Anim Physiol Anim Nutr.* 106 690–719.
[21] Venugopal V. 2006. *Seafood processing : adding value through quick freezing, retortable packaging, cook-chilling, and. Boca Raton*: CRC Press.
[22] Premarathna AD, Rajapakse RPVJ, Pathirana E, Senaratne VP, Karunarathna SC, Jayasooriya AP 2015 *Int J Sci Technol Res.* 4(7)197-201.
[23] Luthfiyana N, Bija S, Irawati H, Awaludin, Ramadani A 2021 *J Pengolah Has Perikan Indones.* 24(2) 188-199.
[24] Ayaz D and Ozogul Y 2011 *J Fish Sci.* 5(3) 262-269.
[25] Ayaz D and Ozogul Y 2011 *J Fish Sci.* 5(4) 308-316.

[26] Ma'ruf A, Widjaja NMR, Hidajati N, and Damayanti R 2018 *Philipp J Vet Med.* *55*(Special issue) 79-84.
[27] Abdel-Salam HA 2014 *Am J Biosci.* 2(2) 70-78.
[28] Huang Z, Aweya JJ, Zhu C, Tran NT, Hong Y, Li S, Yao D, Zhang Y 2020 *Front. Immunol.* 11 574721.
[29] Chyun JH, Griminger P 1984 *J. Nutr.* 114 1697–1704.
[30] Ryu B, Shin KH, Kim SK 2021 *Fish. Mar. Drugs*, 19 377.
[31] Kim J, Marshall MR, Wei C. 2000. *Polyphenoloxidase.* Di dalam : Haard NF, Simpson BK, editor. *Seafood Enzymes: Utilization and influence on postharvest seafood quality.* New York: Marcel Dekker, Inc. 271-316p.
[32] Nopita W, Lamid M, and Agustono 2020 *World's Vet Journal.* 10(4) 653-657.
[33] Balzano M, Pacetti D, Lucci P, Fiorini D, Frega NG 2017 *J Food Compos Anal.* 59 88-94.
[34] Handayani KS and Soegianto A 2019 *Ecol Environ Conserv.* 25 S46-S49.
[35] Nurhayati D, and Putri DA 2018 *J Aquac Sci.* 3(1) 38-45.
[36] Widiastuti IM, Hertika AMS, Musa M, Arfiati D, and Soegianto A 2019 *Pollut Res.* 38 S49-S53.
[37] Kim JJ, Kim YS, Kumar V. 2019 *J Trace Elem Med Biol.* 54 226-231.
[38] Korkmaz C, Ay Ö, Çolakfakıoğlu, C 2019 Thalassas. 35 65–71.
[39] Liu Q, Liao Y, Shou L 2018 *Mar Pollut Bull.* 131 356-364.
[40] Soegianto A, Wahyuni HI, Yulianto B, and A Manaf 2022 *Environ Toxicol Pharmacol.* 90 103810.
[41] Soegianto A, Nurfiyanti PE, Saputri RNR, Affandi M, and Payus CM 2022 *Mar Pollut Bull.* 177 113573.

6. Acknowledgement

This research was funded by the Ministry of Education, Culture, Research, and Technology through the 2022 Beginner Lecturer Research (Penelitian Dosen Pemula) Grant scheme according to Decree Number: 033/E5/PG.02.00/2022 with Master Contract Number: 116/E5.PG.02.00.PT/2022 dated 10 May 2022 and Derivative Contract Number: 08/DRTPM/PDP/I/2022 dated 11 May 2022. Thank you to Mr. Roni as a fisherman who has helped researchers in sampling in the field.

IOP Conf. Series: Earth and Environmental Science 1273 (2023) 012003 doi:10.1088/1755-1315/1273/1/012003

Stress level of vaname shrimp (*Litopenaeus Vannamei*) with infested ciliata pathogens after open wet transportation

A F Apriyadi[1], G Mahasri[2,4], L Sulmartiwi[3]

[1]Aquaculture,Faculty of Fisheries and Marine, Universitas Airlangga, Surabaya, Indonesia
[2]Departement of Aquaculture, Faculty of Fisheries and Marine, Universitas Airlangga, Surabaya, Indonesia
[3]Department of Marine, Faculty of Fisheries and Marine, Universitas Airlangga, Surabaya, Indonesia.

[4]Corresponding author: mahasritot@gmail.com

Abstract. Vaname shrimp (*Litopenaeus vannamei*) is one of the leading fishery commodities in Indonesia. The unbalanced interaction between the environment, host, and pathogen causes the shrimp's body defense to decrease and it is easy for parasites to infest and cause disease. This study aimed to determine the correlation between stress levels and the degree of ectoparasite infestation in vaname shrimp which was infested with pathogenic ciliates prior to open wet system transportation. This research uses survey method. Sampling of vaname shrimp at IBAP Lamongan. The parameters observed were blood glucose levels and the degree of infestation. The results of the ectoparasite examination showed that the ectoparasites that infested vaname shrimp had the degree of infestation in the category of very mild to moderate infestation. The number of vaname shrimp with a very light category is 4, the category of light infestation degree is 22, the number of vaname shrimp with a moderate level of infestation category is 24. There is a correlation between blood glucose levels and the level of infestation of ectoparasites which have a very high power. Thus, an increase in blood glucose levels will be followed by an increase in the rate of ectoparasite infestation.

1. Introduction

The volume of shrimp exports in Indonesia shows a significant pattern of increasing from around 207,704 tons in 2019 to 239,282 tons in 2020 [1]. Good transportation procedures are needed to ensure that the seeds survive to the cultivation pond. In an open wet system, fish are transported in tanks but continuously aerated to provide sufficient oxygen during transportation, usually this method of transportation is carried out only for short distances and short time [2].

Vaname shrimp when transported has an increase in metabolic rate up to three times that of routine metabolism, an increase in metabolic rate will affect the decrease in water quality and will adversely affect the health and survival of seeds in transportation activities [3]. Inappropriate interactions between the environment, host and disease-causing bodies can cause stress in shrimp and decrease the immune system so that they are more susceptible to disease [4]. In intensive ponds and traditional ponds in Gresik Regency, it has been proven that there are several ectoparasites that infest, namely *Zoothamnium, Epistylis and Vorticella* [5].

2. Material and methods

The research was conducted in January – April 2022. This study used a survey method. Sampling used a purposive sampling method of 50 fish from the total population with shrimp measuring 7 – 8 cm in the Brackish Water Cultivation Installation (IBAP) Lamongan, East Java in a pond with HDPE plastic base.

Examination of blood glucose levels and ectoparasite infestation in white shrimp was carried out at the Anatomy and Aquaculture Laboratory, Faculty of Fisheries and Marine Affairs, Airlangga University. Measurement of blood glucose levels is carried out by taking blood in the ventral section, namely on the second segment [6]. Next, a minimum of 0.1 ml of blood is taken, then remove the needle at the tip of the syringe and the blood is dripped on the Easy Touch branded test script, the blood glucose level will be displayed on the screen, and after that the data is read [7]. Examination of ectoparasites was carried out using the native method [8]. To determine the correlation between blood glucose levels and ectoparasite infestation using simple correlation analysis [9].

3. Results and discussion

3.1 Results

3.1.1 Blood glucose level measurement results

The results of measuring blood glucose levels of vaname shrimp after transportation showed normal to high values. The results of measuring blood glucose levels of 50 white vaname shrimp can be seen in Table 1.

Table 1 Results of Measurement of Blood Glucose Levels

∑ Shrimp sample	Blood Glucose Levels (mg/dL)	Category
0	<10	Under normal
27	10-34	Normal
23	>34	High

3.1.2 Ectoparasite examination results on vaname shrimp

The results of the examination of ectoparasites on 50 vaname shrimp after transportation, the ectoparasites found to be infesting vaname shrimp were from the Protozoa group, including *Zoothamnium, Vorticella, and Epistylis*. These ectoparasites are found in swimming legs, walking legs, and tails. The results of the examination of ectoparasite infestation can be seen in Table 2

Table 2. Examination Results of Ectoparasite Infestation

∑ Infested Shrimp	∑ Ectoparasites (Zooid)	Infestation Degree
4	0	Very light
22	11-25	Light
24	28-48	Average

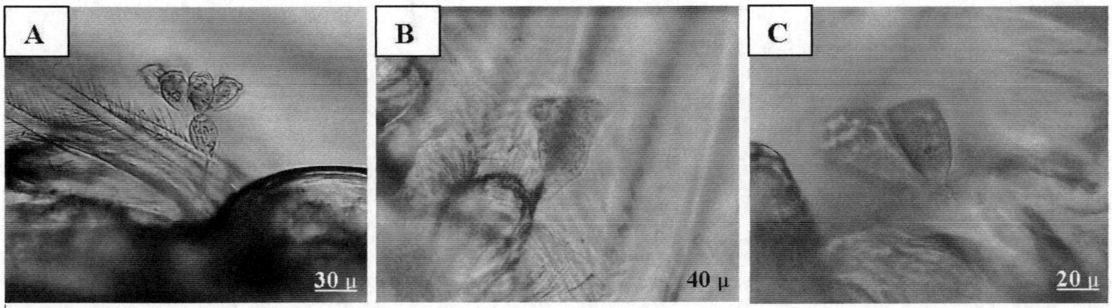

Figure 1. Ectoparasites found to infest vaname shrimp (A). *Zoothamnium* sp. at the foot of the road magnification 100x (B). *Vorticella* sp. at the tail of 100x magnification (C). *Epistylis* sp. on the body surface magnification 100x.

3.1.3 Results of regression analysis and correlation between blood glucose levels and ectoparasite infestation

The results of the regression and correlation analysis showed that there was a very close relationship between blood glucose levels and the level of ectoparasite infestation in vaname shrimp, which was indicated by a positive correlation coefficient of $r = +0.927$ ($p<0.01$) meaning that it had a very close positive relationship with significance very real.

3.1.4 Water quality measurement results

Water quality measurements were carried out once, namely after transportation. Water quality parameters measured include temperature, salinity, DO, pH, ammonia, nitrite and nitrate. The results of water quality measurements can be seen in Table 3.

Table 3. Results of water quality measurements

Parameter	Measurements Results	Criteria	Standard
Temperature (°C)	24	Normal	23-24
Salinity (ppt)	15	Normal	15-25
DO (mg/L)	11.2	Normal	> 3.0
pH	7	Normal	7-8
Amonia (mg/L)	0,5	Abnormal	< 0.01
Nitrite (mg/L)	0	Normal	< 0.01
Nitrate (mg/L)	10	Normal	20

3.2 Discussion

The results of the examination of ectoparasites after transportation, the types of ectoparasites found were three species from the phylum protozoa, namely *Zoothmanium* sp, *Episylis* sp. and *Voticella* sp. The most common ectoparasites were *Zoothmanium* sp. and *Episylis*, sp. while *Voticella* sp. only slightly, this is because *Zoothamnium* sp. are ciliates that live normally in low-quality waters, so even though the water quality is good, this parasite can still grow [10]. In addition, there are more *Zoothamnium* and *Episylis* than *Vorticella* because *Zoothamnium* and *Episylis* live in colonies so that they reproduce faster than *Vorticella* which live solitary [11].

Fegan *et al.* [12] stated that there are several characteristics based on the number of individuals. Very light infestation if the number of parasitic individuals < 5 zooids, mild infestation if the number of parasitic individuals 5-25 zooids, moderate infestation if the number of parasitic individuals is 26-50 zooids, and heavy infestations if the number of parasitic individuals > 50 zooids. Referring to these characteristics, it shows that the ectoparasites that infest vaname shrimp have the degree of infestation in the category of very light to average infestation. The number of vaname shrimp with a very light category is 4, the category of light infestation degree is 22, the number of vaname shrimp with a average level of infestation category is 24.

The correlation between blood glucose levels and ectoparasite infestation results obtained is a high blood glucose level accompanied by an increase in the number of ectoparasite infestations, so it can be said that the higher the blood glucose level, the higher the level of ectoparasite infestation found in vaname shrimp. Kilawati *et al.* [4] stated that shrimp in a state of stress resulted in a decrease in immunity so that it was easy for parasites to be infested. The condition of stressed shrimp is characterized by an increase in blood glucose levels which can stimulate parasite infestation in shrimp. Parasite infestation can occur due to unbalanced interactions between the environment, biota and disease-causing agents [13].

The results of measuring blood glucose levels of vaname shrimp after transportation showed normal to high values. The number of shrimp with normal blood glucose levels (10-34 mg/dL) after transportation was 27, while the number of shrimp with high blood glucose levels (>34 mg/dL) was 23. Lornzon *et al.* [14] stated that the lowest normal glucose level in shrimp was 10 mg/dL and the highest normal glucose level in shrimp was 34 mg/dL.

The survival rate of vaname shrimp in open wet system transportation activities is 97%. The results of the survival rate are quite good. Saputra *et al.* [15] stated that the survival rate of fry is still above 90% which means it is still good in terms of pressure resistance during travel. The high value of survival is due to transportation activities carried out according to transportation standards, measurement of water quality after transportation on water temperature parameters is 24 °C. Afriansyah *et al.* [16] stated the ideal temperature range to support transport activities in fish is 23-24 °C. The ideal dissolved oxygen (DO) for shrimp growth is more than 3 mg/L [17]. Meanwhile, the DO measurement results in the transport water were 11.2 mg/L.

The pH value at the time of transportation was 7, the pH value was still in the optimal range of shrimp life, Berka *et al.* [18] stated that the optimum pH for transportation was 7-8. Salinity at the time of transportation is 15 ppt. The ideal salinity for shrimp growth ranges from 15-25 ppt. The value of nitrate at the time of transportation is 10 mg/L, this value is still considered feasible for shrimp farming because it is below the threshold value [17]. The nitrate threshold value for vannamei shrimp culture was 20 mg/l [19].

The results of the ammonia measurement showed the value of ammonia was 0.5 mg/L and nitrite was 0 mg/L. The value of ammonia is less than 0.01 mg/L and nitrite is less than 0.01 mg/L [17]. The increase in ammonia levels is due to the large amount of leftover feed, and the metabolic products of fish that settle to the bottom. The condition of the waters that contain a lot of ammonia causes fish to become stressed, which is characterized by a weak body, inactive movement, silence, decreased appetite, endurance. the body decreases so that it is susceptible to disease, which can eventually lead to death [20]. The decrease in water quality during transportation was responded to by the seeds, ranging from a mild response involving the hormonal system, a moderate response indicated by increased blood glucose levels and the most severe response causing seed death [21].

The simple correlation analysis (r) the correlation coefficient between ectoparasite infestation and blood glucose level (r) is 0.927, the basis for making the decision is based on the value of r which is greater than the confidence interval of 0.05. The results can be seen because the significant value is 0.927>0.05. This shows that there is a correlation between ectoparasite infestation and blood glucose levels. Meanwhile, in the analysis of the degree of correlation between ectoparasite infestation and blood glucose levels, the decision-making was based on a significant value of 0.927 which was between 0.81 - 1.00 which showed a perfect correlation [9].

4. Conclusion

The blood glucose level of vaname shrimp (*Litopenaeus vannamei*) which was infested with pathogenic ciliate before transportation in the wet system was classified as normal to high. Ectoparasite infestation on vaname shrimp (*Litopenaeus vannamei*) before transportation was categorized as very light to average.There is a very close positive correlation (r=0.927) between the blood glucose level of white vaname shrimp and ectoparasite infestation, which means that the higher the blood glucose level of white vaname shrimp, the higher the protozoan ectoparasite infestation of white vaname shrimp

5. References

[1] Ministry of Maritime Affairs and Fisheries 2021 *Export Statistics of Fishery Products for 2016-2020 Directorate General of Strengthening the Competitiveness of Marine and Fishery Products*. P 25-38.
[2] Muammar M, Syahputra F, and Muhazzir S 2021 *Jurnal TILAPIA* 2(1) 15-22.
[3] Paulo C F C, Pedro H S K, Elaine A, Correia dan Bernardo B 2009 *Neotrop Ichthyol.* 7(2) 283-288.
[4] Kilawati, Y. dan Y. Maimunah 2015 *J. Life Sci. Res.* 2 (1) 50-59.
[5] Mahasri G, Suprapto H, and Farras A 2017 *J Ilm Perikan Kelaut*, 9 (2) 118-126.
[6] Braak V D 2002 Thesis Wageningen Institute of Animal Sciences Netherlands 168 p.
[7] Irdalisa I, Safrida S, Khairil K, Abdullah A, dan Sabri M 2015 *J Edubio Trop*. 3(1) 1-50.
[8] Rohmin M F T, G Mahasri dan F A Rantam 2017 *J Biosains Pascasarj*. 19 (2) 143-157.
[9] Al-Arief, M. A. 2016. *Experimental design. Lantern Jaya Madina*. Surabaya. p 33-40.
[10] Nurlaila, Dewiyanti I dan Wijaya S 2016 *Jurnal Ilmiah Mahasiswa Kelautan dan Perikanan Universitas Syiah Kuala*. 1(3) 338- 396.
[11] Li L, Song W, Warren A, Shin M K, Chen Z, Jia D and Sun P 2008 *J Eukaryot Microbiol.* 55(5) 448-456.
[12] Fegan D, Nieles F, Flagel T, Rossuwan S, and Waiyakaruitata M 1993 *The Development of a Method for Determining the Quality of Post Larva of Penaeus monodon Fab* Asian Fisheries Society Conferences. 23 p.
[13] Austin B, and D A Austin 2007 *Bacterial Fish Pathogens. Diseasein Farmed and Wild Fish. Fourth Edition. Ellis Horword Limited* Chichester pp 663-692.
[14] Lorenzon, S, Edomi, P, Giulianini P G, Mettulio R, & Ferrero E A 2004 *J Exp Biol.* 207(24) 4205-4213.
[15] Saputra H K H 2021 *J Sains Terap*. 11(1) 9-19.
[16] Afriansyah P, Rosmawati R, & Mumpuni F S 2016 *J Mina Sains*. 2(1) 39-44.
[17] Indonesian National Standard 2006 *The production of vannamei shrimp (Litopenaeus vannamei) is ponded with intensive technology. National Standardization Agency.* SNI 01-7246-2006 13 p.
[18] Berka R 1986 *The transport of live fish: a review Rome: Food and Agriculture Organization of the United Nations* 52 p.
[19] Government Regulation of the Republic of Indonesia Number 82 year 2001 concerning Management of Water Quality and Control of Water Pollution. State Secretariat of the Republic of Indonesia Jakarta 28 pages.
[20] Arifin M Y 2017 *J Ilm Univ Batanghari Jambi.* 16(1) 159-166.
[21] Weber E S 2011 *Vet Clin North Am - Exot Anim Pract.* 14(1) 21–32.

6. Acknowledgement

The author would like to thank Faculty of Fisheries and Marine, Airlangga University, for conducting the seminar.

Study on the preference of rosella mangrove syrup with different concentrations of sucrose

S N Ramadhani[1], E Saputra[2], S Andriyono[2,3]

[1]Undergraduated Program of Fisheries Processing Technology, Faculty of Fisheries and Marine, Airlangga University, Campus C Universitas Airlangga Mulyorejo Surabaya, East Java, Indonesia 60115
[2]Departement of Marine, Faculty of Fisheries and Marine, Airlangga University, Campus C Universitas Airlangga Mulyorejo Surabaya, East Java, Indonesia 60115

[3]Corresponding Author: sapto.andriyono@fpk.unair.ac.id

Abstract. Syrup products are preferred considering the tropical climate of Indonesia. One of the mangrove fruits that can be processed into syrup is pedada (*Sonneratia caseolaris*) which is combined with the natural colour of roselle flower petals. With the addition of roselle, the resulting syrup colour becomes attractive with a pink color. The manufacture of roselle mangrove syrup requires the addition of the right sugar. In this study, the manufacture of 1 kg of mangrove fruit was added with 3 concentrations of sucrose sugar (1; 1.5, and 2 kg). Tests were carried out by measuring antioxidant activity, the value of sucrose content, viscosity, and hedonic test. The test results showed that the value of sucrose content increased along with the addition of sucrose sugar to the syrup, namely 65.61 to 71.46%. The value of this sucrose content has met the SNI for syrup. The highest antioxidant value in the 2 kg sugar treatment was 79.58. In the study of the viscosity, the highest value at treatment with the composition of 2 kg sugar additional. However, the panelists preferred syrup with a sugar composition of 1.5 kg with the highest score marked on the hedonic test.

1. Introduction

Mangrove forest is one of the unique and distinctive forest ecosystems in tidal and coastal areas, and beaches, and is a very potent natural resource. Mangrove forests have high economic and ecological value but are vulnerable to damage if they are not wise in maintaining, preserving, and managing them [1]. One type of plant that grows in mangrove waters is *Sonneratia caseolaris* or known as pedada. This plant produces fruit, namely pedada fruit which can be used as a food source [2]. Pedada fruit (*Sonneratia caseolaris*) has been widely processed to be made into several food products such as jenang, lunkhead, jam, and syrup.

Rosella (Hibiscus sabdariffa) is a type of outdoor ornamental plant from the hibiscus family that has many benefits, one of which can be used to prevent disease. Flowers and seeds can be used as ingredients to make healthy drinks. Rosella plants contain chemical compounds including gossypetin, anthocyanins and glucosides. Anthocyanins are included in flavonoid compounds. Anthocyanins act as antioxidants and are believed to cure degenerative diseases by reducing them, meaning that these compound molecules can react with free radicals and counteract them [3]. Anthocyanin compounds in rosella flowers can be extracted using the maceration technique [4].

One of the processed forms that are commonly found is in the form of syrup. The syrup is a drink in the form of a concentrated solution with various flavours [5]. Fruit syrup is a drink made from raw fruit materials, and sugar plays a very large role in the beverage industry [6]. Sugar can be used as a source of nutrition, to form texture and flavour through the browning reaction [7]. Reducing sugars are sugars (carbohydrates) that reduce electron-accepting compounds, such as glucose and fructose. Reducing sugars are sugars that contain completely free aldehyde or ketone groups, except for sucrose and starch (polysaccharides), all monosaccharides (glucose, fructose, galactose) and disaccharides (lactose, maltose) are included as reducing sugars [8]. The main ingredients commonly used in making syrup include fruit juice, sugar as a sweetener, and stabilizers [9].

The sweetener commonly used in the syrup-making process is the type of sugar sucrose. Sucrose sugar has a high-calorie content of 400 calories in 100 grams of material as a sweetener and is the most common type of sugar obtained from sugarcane stem extract [10]. Based on SNI 01-3544-1994, the syrup quality requirement with the test criteria for total sugar (calculated as saccharose) is at least 65%. In this study, the addition of different sugar concentrations in the manufacture of rosella mangrove syrup needs to be done, because to determine the total sugar content and level of preference for rosella mangrove syrup.

2. Material and methods

2.1 Place and time of research

This research was conducted from March to May 2022 at the Food Processing Laboratory and the Chemistry and Analysis Laboratory, Faculty of Fisheries and Marine Affairs, Airlangga University, Surabaya.

2.2 Production of pedada fruit syrup

The pedada fruit is taken first, then the pedada fruit is washed until clean and peeled from the skin using a stainless-steel knife. Then weighed with a ratio of 1: 2 (1 kg of pedada fruit: 2 litres of water). Stirring is done manually on the pedada fruit by adding 1 litre of water until the fruit becomes pedada fruit pulp. The next process is the first filtering using a coarse sieve to separate the seeds so that the juice and dirty seeds can be obtained. Dirty seeds that have been filtered are added with 1 litre of water again for a second filtering process to obtain clean fruit juice and seeds. The juice from the first and second filtering is taken as raw material for making syrup, while the clean seeds are separated. The fruit juice was filtered again using a calico cloth to obtain pure fruit juice filtrate. Furthermore, the additional granulated sugar (according to the treatment with various concentrations) was dissolved at a warm temperature (50-60°C) followed by a stirring process using a stirring rod until the sugar was completely dissolved. Then cooled and packed in sterilized bottles.

2.3 Roselle extraction

Rosella extraction was carried out to obtain anthocyanin and phenolic compounds from rosella through the maceration extraction method. Rosella extraction was carried out using the maceration method which is a modification of the [4] method. The solvent used in the extraction of rosella is using water. The rosella petals are first dried and ground using a chopper to become a powder. Then the rosella powder was sifted using a filter. Rosella powder that has been filtered, added hot water that had previously been boiled to 100°C with a ratio of 1:10 (w/w) in an Erlenmeyer flask. Then stir using a magnetic stirrer for 5-10 minutes with a closed container. The extract was filtered to obtain pure rosella extract and then put in a dark glass bottle with a closed state so that the flavonoid content that produces antioxidants in the rosella extract is not damaged during the storage process. Pure rosella extract was taken as much as 15% to be poured into bottles containing pedada fruit syrup according to treatment. Then manual homogenization was carried out to obtain rosella mangrove syrup.

2.4 Addition of sugar concentration

The process of adding sugar concentration to syrup in this study refers to [11]. The additional material used in this study is granulated sugar which gives a sweet taste and is a natural preservative for the syrup produced. Add sugar with rosella mangrove syrup according to the specified treatment of G1, G2, and G3 (G1 = Rosella mangrove syrup with the addition of 1 kg of sugar; G2 = Rosella mangrove syrup with the addition of 1.5 kg of sugar; G3 = Rosella mangrove syrup with the addition of 2 kg of sugar). After the granulated sugar was mixed, a manual stirring was carried out for each treatment until the mangrove rosella syrup solution was homogeneous and thickened, so that mangrove rosella syrup with different sugar concentrations could be obtained.

2.5 Rosella mangrove syrup testing

First, the testing is sugar content. The sample has been tested used by the Luff Schoorl method. This method is to determine the sugar content and samples. This method is used to determine moderate carbohydrate levels and is the best method because it has an error of 10% for measuring carbohydrate levels, and is more practical and inexpensive [12]. Each syrup with the formulation of each treatment was weighed to take 2 grams of sample and put into a 250 ml volumetric flask and added water and then shaken. Then 5 ml of half-base bp-acetate was added and shaken. After the solution is homogeneous, 1 drop of 10% $(NH_4)_2HPO_4$ solution is dripped (if a white precipitate appears, then half-base Pb-Acetate is sufficient). After that, 15 ml of 10% $(NH_4)_2HPO_4$ solution was added to test that the half-base Pb-Acetate had been completely precipitated, adding 1-2 drops of 10% $(NH_4)_2HPO_4$. If no precipitate appears, the 10% $(NH_4)_2HPO_4$ additional is considered sufficient. Then homogenization was carried out by shaking and determining the contents of the volumetric flask to mark the line with aquadest, shake it 12 times, and allow it to be filtered. Then the filter results were taken using a pipette of as much as 50 ml to determine reducing sugar in a 100 ml volumetric flask. Next, add 25 ml of 25% HCl, attach a thermometer and perform hydrolysis in a water bath. When the temperature reaches 68-70°C, it is maintained for 10 minutes. When the process is complete, remove and rinse the thermometer with water. Then the test solution was added with 30% NaOH until neutral (pink colour) with phenolphthalein indicator. Then the Erlenmeyer is connected to an upright cooler and heated on a hot plate so that within 3 minutes it should start to boil. Continue to heat for up to 10 minutes. After that lift and immediately cool in a bath containing ice. After the solution has cooled, 10 ml of 20% potassium iodine solution and 25 ml of 25% H_2SO_4 are added. Then titrate using 0.1 N $Na_2S_2O_3$ with 0.5% starch solution as an indicator (the indicator drops are not dark blue) to obtain the titration volume (V1 ml). Then make a test blank by replacing solution 1 with 10 ml of distilled water with the same procedure so that the titration volume of the test blank (V2 ml) is obtained. Calculating reducing sugar content can use the formula:

$$\text{Table Value (AT)}=(\text{V2-V1})\times\frac{\text{Normal } Na_2S_2O_3 \text{ Standardized}}{0,1}$$

$$\text{Reductase sugar (\%)}=\frac{(\text{AT} \times \text{dilution factor})}{\text{bottle of samples (mg)}}\times 100\%$$

Second, the antioxidant activity of pedada fruit syrup with rosella flower extract was tested quantitatively by the modified method [13,14,15]. Testing of antioxidant activity can be divided into several stages: (1) The syrup was cooled to 3-5°C for 24 hours and centrifuged at 3000 rpm for 15 minutes to separate the supernatant. The supernatant obtained was used as a sample solution; (2) 1.97 mg of DPPH was dissolved in 100 ml of methanol in a flask to give a solution at a concentration of 50 M [16]; (3) A total of 20 mg of the syrup sample that had been prepared previously was dissolved in 20 ml of methanol to a concentration of 1000 ppm. The sample was kept in a test tube for up to 4 ml, and 1 ml of DPPH solution was added. The sample mixture and DPPH solution were then incubated in a shading chamber for 30 minutes by coating the test tube with aluminium foil. Inhibition in both samples was expressed as % inhibition and was calculated according to the equation:

$$\%\text{ Inhibition} = \left(1-\frac{\text{As}}{\text{Ak}}\right)\text{x } 100$$

The third test is the viscosity Test. Viscosity testing on mangrove rosella syrup was carried out to determine the viscosity level of rosella mangrove syrup by the addition of different sugar concentrations. The viscosity test method is the Ostwald method, namely the Ostwald pipe using the Ostwald viscometer. A viscometer is a device for measuring the viscosity of a fluid. Commonly used viscometers are falling bullet viscometers, tube/capillary/ostwald viscometers, and rotation systems [17]. The concept of the Ostwald pipe method is the flow velocity of a fluid in a tube pipe. The next test is to measure the flow time of aquadest using 10 ml of distilled water which is inserted into the Ostwald pipe and sucked up to the limit listed on the top. The time to drop the aquadest to the limit indicated at the bottom is calculated using a stopwatch. Furthermore, testing the measurement of syrup sample flow time using the same method. The formula used in the calculation of viscosity is as follows:

$$\eta = \eta 0 \frac{t \times \pi}{t0 \times \pi 0}$$

Description: η = sample liquid viscosity; $\eta 0$ = viscosity comparison fluid; t = sample liquid flow time; t0 = comparison fluid flow time; π = the density of sample liquid; $\pi 0$ = the density of the comparison fluid

The last test is the organoleptic Test. In organoleptic testing of pedada mangrove syrup, it was carried out following SNI 2346: 2015 regarding sensory testing on fishery products using hedonic tests or the level of preference for scoring methods with a minimum number of untrained panelists of 30 people. Panelists will be given 3 types of rosella mangrove syrup with a code of 3 syrups with different sugar concentrations. Then the panelists will rate the level of preference on 3 assessment criteria which include appearance, aroma, and taste.

2.6 Data analysis.

Data from the results of this study were obtained using SPSS Analysis of Variance (ANOVA) 1-factor software to determine the effect of the treatment given. Treatments that show significant results can be continued with the Smallest Significant Different (BNT) further test.

3. Results and discussion

3.1 Luff schoorl method test result

Based on the result of the table above, it showed the mean results of the analysis of sucrose levels using the luff schoorl method on rosella mangrove syrup with different values in each treatment. The analysis results of the best sucrose content were rosella mangrove syrup in the G3 treatment (2kg sugar concentration) with 71.46±0.59 Table 1.

Table 1. The results of the test sucrose level

Treatment	Mean of Sucrose Level (%) ± SD	Reference (SNI, 1994)
G1(1kg)	$65{,}61^{a}\pm0{,}58$	
G2 (1.5kg)	$66{,}63^{b}\pm0{,}58$	Min. 65%
G3 (2kg)	$71{,}46^{c}\pm0{,}59$	

Note: Different superscript lowercase notations in the same column show significant and significant differences (p>0.05).

Of the three samples of rosella mangrove syrup with different sugar concentrations, treatment III with sample code G3 (2kg sugar concentration) was the best result and met the syrup quality requirements based on SNI 01-3544-1994. The sucrose content in the syrup increases as sugar is added. This is because the concentration of sugar can affect the level of sucrose produced. The more sugar concentration is added, the more sucrose content will increase [18]. The level of sweetness in simple sugars varies. Among the simple sugars most often used as a sweetener additive in food or beverage products is sucrose [19]. This is explained by [20], sucrose as a sweetener has a high-calorie content of 400 calories in 100 grams of material.

3.2 Antioxidant activity results

Table 2. Antioxidant activity test results

Treatment	Mean of Antioxidant Level (%) ± SD
G1 (1kg)	$78{,}33^{a}\pm8{,}32$
G2 (1.5kg)	$67{,}50^{a}\pm7{,}60$
G3 (2kg)	$79{,}58^{a}\pm4{,}02$

Note: The same superscript lowercase notation in the column shows a non-significant and significant difference (p>0.05)

In treatment G3 (2kg sugar concentration) the percent inhibition of antioxidant activity was 79.58%, higher than in G1 (1kg sugar concentration) and G2 (1.5kg sugar concentration) treatment. This is because the Maillard reaction of reducing sugars and proteins in the form of melanoidin (the brown pigment in sugar as a product of the Maillard reaction which has antioxidants) from the heating process of sugar during syrup-making only has a small effect on free radical scavenging. The difference in the percentage of inhibition in each sample of rosella mangrove syrup is related to pedada fruit juice which has a role to influence antioxidant activity because pedada fruit is one type of plant that has potential as an antioxidant because this plant contains several compounds such as alkaloids, flavonoids, glycosides, saponins, and phenols [21].

3.3 Viscosity test result

Table 3. Viscosity test results

Treatment	Mean of Viscosity (dPas)± SD
G1 (1kg)	7,606^{a}±2,60
G2 (1.5kg)	11,826^{b}±4,09
G3 (2kg)	15,354^{c}±5.27

Note: The same superscript lowercase notation in the column shows a significant and significant difference ($p>0.05$)

The viscosity of a liquid with added sugar depends on the length of the heating time [22]. If the heating process is carried out too long, the syrup produced will be thicker, because the higher the heating temperature, the higher the solubility of sugar. Processing using a mixture of sugar and water can cause sugar and water molecules to bind together and unite. The longer the heating process, the thicker the syrup will be [23].

3.4 Organoleptic test result

Table 4. Hedonic test results

Treatment	Appearance	Aroma	Taste
G1	5,4	5,4	5,63
G2	7,23	6,57	7,47
G3	6,23	5,03	5,43

The different sugar concentrations added when making the syrup did not affect the colour change. The results of organoleptic tests on rosella mangrove syrup with appearance parameters ranged from 5.4 to 7.23. Panelists most liked rosella mangrove syrup with a sugar concentration of 1.5 kg with a value of 7.23 (Like). The colour produced from rosella mangrove syrup is a red colour from natural dyes called anthocyanins in rosella flowers [24]. The constituent components of the aroma consist of volatile compounds that easily evaporate at high temperatures so that there will be the formation of a distinctive aroma from the food itself [25]. That additional sugar to a food product does not influence the aroma because sugar does not have a strong and prominent aroma [11]. The addition of sugar concentration can improve the taste of rosella mangrove syrup because sucrose is the sugar that gives the product a sweet taste. The higher the sucrose concentration in a food product, the higher the level of sweetness [26]. Sugar has a function as a sweetener, perfecting the sour taste, and other flavours and also giving a filling taste because it can improve viscosity [18].

4. Conclusion

Based on the results of research on the rosella mangrove syrup in each concentration of sucrose, it can be concluded that adding sugar in rosella mangrove syrup with different sugar concentrations on sucrose levels ranged from 65.61% to 71.46%. In contrast, the antioxidant activity value in rosella mangrove syrup with various sugar concentrations ranged from 67.65 to 79.58. In the viscosity test, the highest value was in the treatment with a composition of 2 kg of added sugar. Based on the results of the preference level test, the rosella mangrove syrup with the addition of 1.5 kg of sugar concentration had the highest appearance value of 7.23, the aroma of 6.57, and the taste of 7.47 in all tests.

5. References

[1] Susanto N, Kusmana C, Sudarma D, dan Sukmadi R 2008 *KKMN.*
[2] Manalu R D E 2011 Thesis. Bogor: Bogor Agricultural Institute.
[3] Djaeni M, Ariani N, Hidayat R, Utari F D 2017 *JATP*. 6(3) 148 – 151.
[4] Cisse M, Vaillant F, Acosta O, Mayer C D dan Dornier M 2009 *JAFC*. 57(14) 6285–6291.
[5] Satuhu S 1994 *Fruit Handling and Processing* (Jakarta: Penerbit Swadaya).
[6] Matute A I R, Soria A C, Sanz M L, Castro I M 2010 *JFCA*. 23(3) 260-263.
[7] Sularjo 2010 *Magistra*. 74 39-48.
[8] Lehninger A L 1982 *Fundamentals of Biochemistry Volume 1* (Jakarta: Erlangga).
[9] Darwin P 2013 *Enjoy Sugar Without Fear* (Yogyakarta : Sinar Ilmu) 24-32 hal..
[10] Koswara S 2008 *Sugary Foods and Tooth Decay (Online Ebook Store for Food, Nutrition and Argoindustry) 3 87-96 p.*
[11] Hadiwijaya H 2013 *Jurnal THP* Universitas Andalas. Padang.
[12] Underwood 2014 *JFI.* (2) 92 – 101.
[13] Latief M, Tafzi F, Saputra A 2013. *Prosiding Semirata* FMIPA Universitas Lampung..
[14] Maesaroh K, Kurnia D, Anshori J A 2018 *Natura Acta*. 6(2):93–100.
[15] Molyneux P 2004 *The use of the stable free radikal diphenyl picrylhydrazyl (DPPH) for estimating antioxidant activity.*
[16] Ridwan R, Wiseno E, dan Suwargo P G 2012 Thesis. Depok: Gunadarma University.
[17] Luthony T L 1990 *Sweetener Source Plants* (Jakarta: Penebar Swadaya).
[18] Kusnandar F 2019 *Macro Component Food Chemistry* (Jakarta: Bumi Aksara).
[19] Syafutri M I, Lidiasari E, dan Indawan H 2010 *J. Gizi dan Pangan.* 5(2) 78- 86.
[20] Avenido P, dan Serrano A E 2012 *Europe JEB*.
[21] Buckle K A, Edwars R A, Fleet G H, dan Wooton *Food Science* (Jakarta: Universitas Indonesia).
[22] Suyitno, Haryadi, dan Supriyanto 1989 *Food Engineering* (Yogyakarta: Universitas Gajah Mada)..
[23] Mastuti E, Sari N P, Simangunsong R A 2013 *Ekuilibrium*. 12(1) 43 – 47.
[24] Soekarto S T, dan Hubeis M 1985 *Organoleptic Research Methodology (Bogor: Bogor Agricultural University).*
[25] Safitri D E 2012 Jurnal *IPB*. 15(1) 11-20.

6. Acknowledgements

The author gives thanks for this work being supported by Faculty of Fisheries and Marine, Universitas Airlangga.

Blood glucose level and infestation degree by *Zeylanicobdella* sp. in cantang grouper before and after closed wet transportation system

G Mahasri[1,5], R M Khatami[2], A A Romadhon[2], N Musa[3], and Kusnoto[4]

[1]Department of Fish Health Management and Aquaculture, Faculty of Fisheries and Marine, Universitas Airlangga, Surabaya, Indonesia

[2]Aquaculture Study Program, Faculty of Fisheries and Marine, Universitas Airlangga, Surabaya, Indonesia.

[3]Department of Aquaculture, Faculty of Fisheries and Food Science, Universiti Malaysia Terengganu, Terengganu, Malaysia.

[4]Departement of Parasitology, Faculty of Veterinary Medicine, Universitas Airlangga, Surabaya, Indonesia.

[5]Corresponding author: mahasritot@gmail.com

Abstract. The cantang grouper is one of the high-value fisheries commodities. Transportation is an integral activity in aquaculture, notably in transporting cantang grouper. Among the methods is the closed wet transportation system. The disrupted environment in the closed wet transportation system, exhaustion, and deficit of the immune system can cause the cantang grouper to be infested by marine leech or *Zeylanicobdella*. In this study, *Zeylanicobdella* sp. infestation levels and cantang grouper blood glucose levels are compared before and after a closed wet transportation system. This research used an experimental method. The 35 individuals' sample was examined for blood glucose levels and the infestation degree of *Zeylanicobdella*. Subsequently, they were transported from Situbondo, East Java, to the Faculty of Fisheries and Marine Universitas Airlangga Surabaya and re-examined. The results of measuring blood glucose levels of cantang grouper were in the below-normal category, normal, and above-normal. There was no significant difference in the infestation degree by *Zeylanicobdella* sp. between before and after transportation. This study concludes that closed wet transportation affected blood glucose levels and did not affect the degree of infestation by *Zeylanicobdella* sp. in cantang grouper.

1. Introduction

The cantang grouper is one of the fisheries commodities with the advantage of growing twice as quickly as the community's preferred tiger grouper. This grouper originated from female tiger grouper (Epinephelus fuscoguttatus) and male kertang grouper. Both fish were crossed (hybridization) to produce the cantang grouper (Epinephelus lanceolatus). The cantang grouper's benefits include its ability to catch valuable commodities [1]. According to statistical data, cantang grouper production is increasing yearly [2].

Transportation is an important part of commercial fish marketing activities to supply aquaculture needs, especially cantang grouper seeds [3]. One method is to use a closed wet transport system. Fish fry is shipped using Styrofoam plastic bags, with limited oxygen being calculated according to the needs during transportation [4]. During transportation, fish are in an enclosed environment, and their metabolic wastes affect water quality parameters such as the pH and the concentrations of ammonia, nitrite, and carbon dioxide. The accumulation of total ammonia–nitrogen and nitrite–nitrogen in water can lead to transportation stress [5]. A closed transport system may expose fish to multiple stressors, including capture, handling, overcrowding, abrupt changes in temperature, and physical trauma [6]. A decline in water quality is caused by the accumulation of ammonia and fluctuations in dissolved oxygen and pH [7].

This disturbance causes fish to experience decreased immunity, and disease is easy to attack fish [8]. The disease will attack fish if there is an interaction between the host, pathogen, and the environment, especially parasite infestation [9]. The environment can significantly influence every important parasite species, especially in wild or cultured marine fish [10]. This problem can lead to pathological changes in host species, decrease fitness, and reduce live fish's market value [11].

The purpose of this study was to determine the average blood glucose levels and the degree of infestation of *Zeylanicobdella* sp. in cantang grouper before and after a closed wet transportation system and to determine differences in blood glucose levels and the degree of infestation of Zeylanicobdella sp. in cantang grouper before and after closed wet transportation system. Moreover, this research is expected to be preliminary research in the development, handling, and reference in transporting sick fish.

2. Materials and methods

This research was conducted in January – March 2022. A sampling of cantang grouper was carried out at a predetermined location, and samples of cantang grouper were 5 - 10% of the population and obtained a total of 35 individuals. The sample of cantang grouper was taken by purposive sampling, totaling 35 fish from the total population. A sampling of 5-10% of the total population [12]. Sampling, blood glucose levels, and the degree of *Zeylanicobdella* infestation before transportation were carried out in private aquaculture with the commodity of cantang grouper cultivated in Krajan Village, Bungatan District, Situbondo, East Java. Examination of blood glucose levels and the degree of ectoparasite infestation after transportation was carried out at the Anatomy and Aquaculture Laboratory, Faculty of Fisheries and Marine, Airlangga University, Surabaya.

Blood glucose levels and the degree of infestation of *Zeylanicobdella* sp. in each cantang grouper were measured and observed before and after transportation. 0.1 ml of grouper blood was taken using a 1 ml syringe plus EDTA solution as an anticoagulant. Examine *Zeylanicobdella* sp. in cantang grouper before transportation by observing and counting the number of these parasites in each fish. After that, the cantang grouper was transferred to a plastic bag filled with water and oxygen for transportation. The cantang grouper is packaged using double plastic bags measuring 120x53 cm [13] and then filled with 10 liters of seawater. According to [14], 10 liters of water can contain cantang grouper for a size of 10-14 cm, as many as 500 cantang groupers. Next, pure oxygen is added to the water medium in a ratio of 1:3 [3]. After that, the plastic bag is stored in a Styrofoam box to which 1.5 kg of ice has been added. Data analysis of blood glucose levels and the degree of infestation of *Zeylanicobdella* sp. on cantang grouper was tested using Dependent T-Test (Pairs). Grouper blood glucose levels were measured with reference [15], and the degree of infestation of *Zeylanicobdella* sp. was calculated by reference [16]

3. Results and discussion

3.1. Results

The results of measuring blood glucose levels of cantang grouper before and after transportation showed values below normal, normal, and above normal, ranging from 24 - 68 mg/dL. The results of measuring blood glucose levels in 35 cantang groupers can be seen in Table 1.

Table 1. The results of the blood glucose level average on cantang grouper before and after closed wet transportation system

Blood glucose level of cantang grouper						**Blood glucose level category (Martínez-Porch*et al.*, 2009)**
Before transported (mg/dL)	**Amount of fish (individual)**	**Category**	**After transported (mg/dL)**	**Amount of fish (individual)**	**category**	
24-27	9	Low	25-27	6	Low	Low (<28,8 mg/dL)
28-34	24	Normal	28-33	20	Normal	Normal (28,8-34,2 mg/dL)
43-49	2	Above normal	39-68	9	Above normal	Above normal (>34,2 mg/dL)

The results of statistical tests stated that there was a significant difference between blood glucose levels of cantang grouper before and after transportation in the normal and above normal categories ($p < 0.05$). Meanwhile, in the below-normal category, the results showed no significant difference ($p > 0.05$). However, even though there was a significant difference or not, the average value of blood glucose levels before and after transportation was still in the normal range.

The examination results showed that the cantang grouper was infested with Zeylanicobdella sp. are healthy, light, moderate, and heavy categories. The results of observations of the degree of infestation of *Zeylanicobdella* sp. in cantang grouper can be seen in Table 2.

Table 2. The results of the blood glucose level average on cantang grouper before and after closed wet transportation system

Degree of Infestation by *Zeylanicobdella* sp. In cantang grouper				**Degree of Infestation Category [16]**
∑ Cantang grouper before transported	**∑ *Zeylanicobdella* before transported**	**∑ Cantang grouper after transported**	**∑ *Zeylanicobdella* after transported**	
25	0 (normal)	25	0 (normal)	0 (normal)
5	2 – 5 (mild)	5	2 – 5 (mild)	1 – 5 (mild)
3	6 – 7 (moderate)	3	6 – 7 (moderate)	6 – 10 (moderate)
2	13 – 18 (heavy)	2	12 – 19 (heavy)	>11 (heavy)

The results of the examination carried out on 35 samples of cantang grouper were found to be infested with Zeylanicobdella sp. On the surface of the body and fins are most commonly found on the dorsal, ventral, and pectoral fins. The results of observations of Zeylanicobdella sp. can be seen in Figure 1.

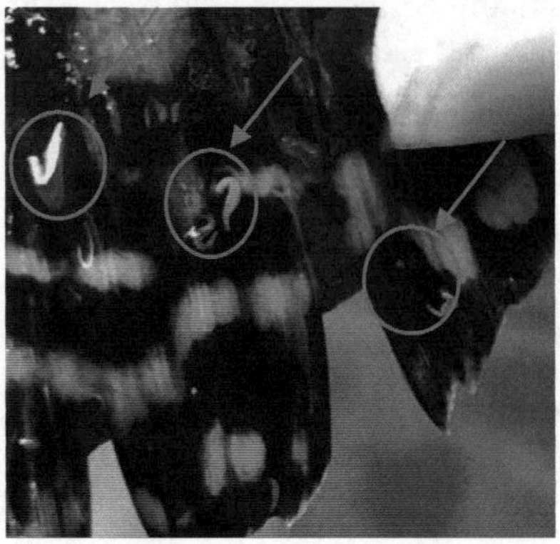

Figure 1. Infestation of *Zeylanicobdella* sp. on the pectoral fins of the cantang grouper is indicated by red arrows

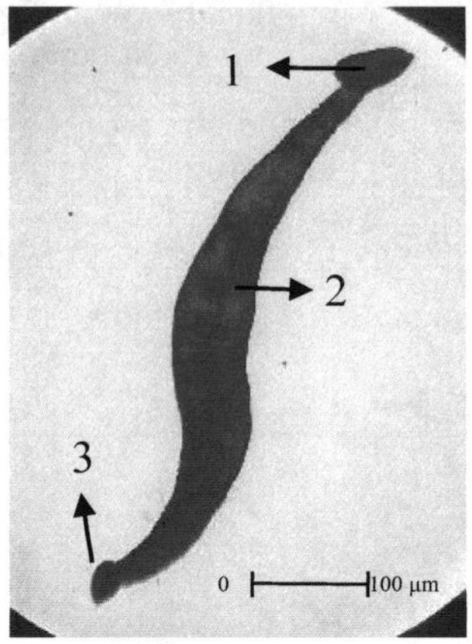

in Figure 2. Identification of *Zeylanicobdella* sp. With an optical microscope (magnified by 100μm), (a) *Anterior sucker*; (b) *Pulsating vesicles*; (c) *Posterior sucker*.

3.2 Discussion

Measurement of blood glucose levels of cantang grouper before and after transportation was below normal, normal, and above normal. The number of cantang groupers before transportation is 9, and after transportation is 6, which have low blood glucose levels (24-27 mg/dL), for the number of cantang groupers with normal blood glucose levels (28-34 mg/dL).) before transportation was 24 and after transportation was 20, and for the number of cantang groupers with a high blood glucose level (39 – 68 mg/dL) before transportation was 2 and after transportation was 9.

Statistical tests performed using the Dependent T-Test showed a significant difference between the blood glucose levels of cantang grouper before and after transportation in the normal and above normal categories ($p < 0.05$). Meanwhile, in the below-normal category, the results showed no significant difference ($p > 0.05$). Changes in the level of high and low blood glucose levels in cantang grouper are caused by the level of stress caused by the poor quality of the surrounding water beyond the limits of physical ability [17]. Measurement of blood glucose levels can be measured to diagnose whether the fish is simply experiencing stress caused by various stressors such as shocks and decreased water quality [18].

The degree of infestation of *Zeylanicobdella* sp. The 35 samples of cantang grouper before and after transportation were included in the healthy, light, medium, and heavy conditions. The number of cantang grouper before transportation is 25 and after transportation is 25, which has a healthy infestation degree category (0 parasites), for the number of cantang grouper, which has a light infestation degree category (1-5 parasites) at the time before transportation is 5 fish and after transportation, namely 5 tails while the number of cantang groupers which have a moderate level of infestation category (6-10 parasites) at the time before transportation is 3 and after transportation is 3 fish and for the number of cantang groupers which have a category of severe infestation degree (> 11 Parasites) at the time before transportation, namely 2 tails and after transportation, namely 2 tails.

There is a difference in the degree of infestation in this cantang grouper, which is caused by factors of water quality and poor water conditions. Also, the water carried, namely water that has the parasite *Zeylanicobdella* sp. This water condition follows the statement [19] that the number of samples of *Zeylanicobdella* sp. infest cantang grouper by several factors of water quality and poor water conditions.

Based on the statistical tests carried out using the Dependent T-Test showed no significant difference between the degree of infestation of *Zeylanicobdella* sp. in cantang grouper before and after transportation in the heavy category ($p > 0.05$). However, in table 3, there is an increase in *Zeylanicobdella* sp. in the heavy category. The increasing parasite number is because the water used during transportation already contains living *Zeylanicobdella* sp. The *Zeylanicobdella* sp. can tolerate a temperature of 24°C – 40°C.

4. Conclusion

Blood glucose levels of cantang grouper were categorized below normal, normal, and above normal. The degree of infestation of *Zeylanicobdella* sp. in cantang grouper is in the normal to heavy category. There was a significant difference between the blood glucose levels of the cantang grouper before and after transportation. In contrast, there is no significant difference in the degree of infestation of Zeylanicobdella sp. in the cantang grouper.

5. References

[1] Shen G M, Shi C Y, Fan C, Jia D, Wang S Q, Xie G S, Li G Y, Mo Z L, and Huang J 2017 *J Fish Dis.*;40(10) 1351–62.
[2] Ismi S, Asih Y S, Nasukha A, Astuti N W W 2020 *JFMR-Journal Fish Mar Res* 4(3) 339–44.
[3] BKIPM 2020 BKIPM Statistik: Grouper Live Fishery Product Export Volume Year 2020 [Internet]. Available from: Retrieved from www.bkipm.kkp.go.id, 24 August 2020.
[4] Cerreta A J, Lewbart G A, Harrison T M 2020 *J Surv Fish Sci.* 7(1) 105–12.
[5] Zheng T, Song Z, Qiang J, Tao Y, Zhu H, Ma, and Xu P 2021 *Front Immunol.* 12(October) 1–14.
[6] Masud N, Ellison A, Cable J 2019 *Dis Aquat Organ.*134(1) 25–32.
[7] Sampaio F D F, Freire C A 2016 *Fish Fish.* 17(4) 1055–72.
[8] Bunlipatanon P, U-taynapun K 2017 *Aquac Res.* 48(4) 1711–23.
[9] Afifah D D, Mahasri G, Satyantini W H. T 2021 *IOP Conf Ser Earth Environ Sci.* 679(1).
[10] Noga E J. *Fish Disease. 1 ed. Fish Disease.* Iowa: Wiley-Blackwell;. 519 pp.
[11] Ihwan M Z, Shaharom-Harrisson F, Marina H, Wahidaelw 2013 *J Sustain Sci Manag.* 8(1) 121–5.
[12] Cameron A. Survey Toolbox for Aquatic Animal Diseases. *A Practical Manual and Software Package* [Internet]. Canberra: Australian Centre for International Agricultural Research; 2002. 375 pp. Tersedia pada: http://aciar.gov.au/files/node/2268/mn94.pdf
[13] Miniartini M, Sujarwani S, Alat B 2020 *Bul Tek Litkayasa Akuakultur.* 18(2) 119–22.
[14] Ismi S, Kusumawati D, Asih Y.N 2016 *J. Ilmu dan Teknologi Kelautan Tropis.* 8(2) 625-632.
[15] Martínez-Porchas M, Martínez-Córdova L R, Ramos-Enriquez R. *Panam J Aquat Sci.* 4(2) 158–78.
[16] Ernest H. Williams J, Bunkley-Williams L. *Parasites Of Offshore Big Game Fishes Of Puerto Rico And The Western Atlantic.* Vol. 1, Puerto Rico Department of Natural and Environmental Resources, San Juan, PR, and the University of Puerto Rico, Mayaguez. Puerto Rico: Puerto Rico Department of Natural and Environmental Resources, San Juan, PR, and the University of Puerto Rico, Mayaguez; 1996. 382pp.
[17] Paulo CFC, Pedro HSK, Elaine A, Correia S & Bernardo B 2009. *J. Neo Ich,* 7(2) 283-288.
[18] Sulmartiwi L, S Harweni, A T Mukti, dan Rr Triastuti 2013 *Jur Ilm Per Kel.* 5 (1) 19-20.

[19] Slamet A, Ahmad T, P T Imanto, Muchari, A Bashari, P Sunyoto, Operational Technical Guidelines for Growing Grouper Fish in Floating Net Cages. Maros Aquaculture Research Institute. 2008. p 6-7.
[20] Kua B C, Choong F C, Leaw Y Y 2014 *J Fish Dis.* 37(3) 201–7.

6. Acknowledgements

The author is grateful to the Dean of the Faculty of Fisheries and Marine, Airlangga University, that has supported this research with RKAT-fund.

IOP Conf. Series: Earth and Environmental Science 1273 (2023) 012006 doi:10.1088/1755-1315/1273/1/012006

The effects of *Chaetoceros* sp. meal as a feed supplement on color expression, growth performance and survival rate of discus (*Symphysodon discus*)

A Fauziah[1], M Z Arifin[1], A Widodo[1], A B Cahyanurani[1,2], A M Halim[1], and A A Aonullah[1]

[1]Politeknik Kelautan dan Perikanan Sidoarjo, Department of Aquaculture Techniques, Sidoarjo 61253, East Java, Indonesia

[2]Corresonding author: annisacahyanurani@gmail.com

Abstract. Discus fish is one of the ornamental fish that has high economic value based on its color characteristic. To increase fish body color, its need carotenoids obtained from feed. *Chaetoceros* sp. is a natural food rich in carotenoids, especially fucoxanthin. This study aims to evaluate the effect of diets consisting of 0% (control), 5%, 10% and 15% *Chaetoceros* sp. powder on color expression (brightness value and color density), growth performance and survival rates of *Symphysodon discus. Chaetoceros* sp. powder was added into the basal diet. One hundred of discus fish with an average length of 0.5 cm were cultured for 56 days. The result showed that the supplementation of 5-15% *Chaetoceros* sp. powder had no significant effects on the survival rate and the brightness value of discus fish. The increasing of *Chaetoceros* sp. powder supplementation showed a lower rate of growth performance compared to control ($p<0.05$), however the color density was observed to increase significantly at 15% *Chaetoceros* sp. treatment ($p<0.05$). As the conclusion, this study showed that dietary *Chaetoceros* sp. could improve the color expression especially the color density of discus fish and the supplementation level was suggested to be 10% *Chaetoceros* sp powder.

1. Introduction

Discus fish, *Symphysodon discus* belongs to the family Cichlidae originating from calm waters of Amazon River and also known as the King of Aquariums. Discus fish has a potential to be developed in order to meet domestic and export market demands due to its economic value [1,2]. As an ornamental fish, discus fish have many variations of body color including both structural colors and pigmentary. It also known for its beautiful shape and unique variety of patterns [3,4].

Discus fish are considered difficult to cultivate. Discus fish cultivation require special attention in their cultivation techniques because they require good environmental conditions and suitable feed [5]. This fish also very sensitive to changes in the aquatic environment. If the environment is not suitable, the fish become less appetite which will eventually affect growth and can lead to death [6,7]. Beside that, compared to similar species in their natural habitat, cultured fish run into color quality degradation due to stress because inappropriate feeding, handling, and environmental factors [2,8].

An important characteristic in evaluating ornamental fish is the color of the body. In most of ornamental fish, the color that appears comes from the deposition of carotenoids in the tissue and by the presence of chromatophores containing these pigments [9]. Carotenoids are the main source of color in fish skin. Nevertheless, the synthesis of carotenoids cannot be directly performed by fish [10-12]. Fish must obtain optimal levels of carotenoids in the feed to improve their skin color [10,13,14]. The method commonly used by ornamental fish cultivators is to provide natural food such as Tubifex worms, Bloodworms, and *Moina* sp to accelerate growth and brighten colors [15]. Along with the development of fish feed manufacturing technology, sources of carotenoids that were previously only provided in the form of raw materials, now these raw materials can be incorporated into feed [16]. Microalgae are well-known as organisms that contain high-value ingredients such as various nutritional content including polyunsaturated fatty acids (PUFA), pigments such as carotenoids and also other active compounds [17,18]. Microalgae have been widely used to improve color in ornamental fish because they are rich in carotenoids [19]. Application of diets containing 1.52% *Spirullina platensis* enriched with carotenoids and 1% of *Haematococcus pluvialis* for 3 weeks showed an improvement of the color intensity on red velvet sword tail (*Xiphophorus helleri*), rainbow fish (*Pseudomugil furcatus*) and topaz cichlids (*Cichlasoma myrnae* [20]. In discus fish, application of spirulina and astaxanthin was observed to increase survival as well as improve the color intensity of the fish body [21,22]. *Chaetoceros* sp. including marine diatoms that are commonly cultivated and rich in carotenoids, especially fucoxanthin where fucoxanthin production will increase when entering the stationary phase [23-25].

According to [26], the addition of alga oil from *Chaetoceros* sp. on diets compared to other types of algae on Zebra fish *Daniorerio rerio* has the ability to increase total carotenoids and also potentially for fish color enhanchment. In view of the above, this study was conducted to evaluate the effect of diets containing *Chaetoceros* sp. powder on color expression (brightness value and color density), growth performance and survival rates of *S. discus*.

2. Materials and methods

2.1 Fish

Discus fish, *S. discus* were obtained from Research Center for Ornamental Fish Cultivation, Marine and Fisheries Ministry, Depok, Indonesia. *S. discus* with 1,27 cm (0,5 inch) in length were used as many as 360 fish. Fish were acclimatized for a week in an aquarium with environmental conditions similiar to laboratory conditions and fed with a commercial diet.

2.2 Feed Preparation

Basal feed as control consisted of semi-pure raw materials, namely 60% beef heart, 15% ground fish, 15% shrimp, 5% carrot, 2% vitamin premix, 3% spinach. Three diets were designed and *Chaetoceros* sp. powder was added as much as 5%, 10% and 15%, reducing the share of the basal feed. Raw maerials were ground and put into the mold to form 'beef heart hamburger'. The feed is then stored in the freezer.

2.3. Feeding Experiment

The fish were reared in 12 experimental aquariums (50 x 20 x 40 cm^3) that are equipped with aeration. The test feed and control feed were given by applying ad satiation method twice a day (8:00 am and 15:00 pm). Each aquarium was carried out each containing 30 fish. The monitored dissolved oxygen ranged from 6-7 mg L^{-1} through stable aeration, temperature ranged from 28-30 °C, pH 6.5-7.5. The water in the aquarium is siphoned daily in the morning before the first feeding by replacing 1/3 of the existing water. The siphoned water is removed to remove the remaining feed particles. The experiment was carried out at 56 days. The animals were measured the height at the beginning and at the end of study (days 56) and analyzed for color expression on days 56 (the end of the study).

2.4. Color Expression Analysis

Discus color expression was measured on the body surface of the total fish from each aquarium at the end of the study. Color measurement was carried out using a CR-400 colorimeter (Konica Minolta, Osaka, Japan), as one of the measuring instruments considered objective in assessing color quality. Color assessment with a colorimeter produces color quality characteristics, namely lightness or brightness, which characterizes light intensity (bright or dark) and chroma or density, which is an expression of the amount of color-forming substance [27]. Discus fish in this study have two types of colors, namely yellow and silver. In this study only focus on the yellow color, therefore the brighter and higher the concentration value, the better the color quality of discus fish.

2.5. Statistical Analysis

Data from this study were analyzed by one way ANOVA followed by post hoc Tukey test to identify significant differences among treatments. Data are expressed as mean standard deviation and all the analysis was conducted by using SPSS 16.0.

3. Results and discussion

3.1. Survival Rate

The survival rate of discus fish during the rearing period ranged from 81.11±8.39 to 97.22±4.81% (Figure 1). The results of the dose ratio of *Chaetoceros* sp. powder showed no effect on survival rate of discus fish ($p>0.05$). In general, the deaths that occurred in all treatments were thought to be due to the nature of discus fish that were sensitive to a disturbance, both changes in water quality and external disturbances [28], but the water quality during maintenance showed a range of values that were still feasible for discus fish. This is due to the efforts that made to improve water quality during the study, such as water changes as much as 65% of the total volume of the aquarium and maintenance carried out in the aquarium container as well as the installation of light bulbs that are turned on continuously so that the temperature remains stable and a recirculation filter is given. This can be expected to minimize toxic metabolic waste and increase oxygen solubility, so that high mortality does not occur until the end of maintenance. In addition, fish deaths are suspected to be due to disturbances outside of treatment or during sampling. This can cause fish to experience stress which ends in death. Discus fish are fish that are very sensitive to disturbances that can startle the fish, be it noise, human traffic, or sounds and movements that cause stress to the fish. In the treatment with a concentration ratio of 15% and 10%, there were some discus fish that were aloof and had no appetite, were not aggressive and were black all over their bodies. This indicates that the fish are under stress. [29] states that changes in food composition will result in disruption of physiological processes and fish behavior which in turn can reduce fish health and physiological conditions resulting in decreased food utilization, growth, and survival. Seen in the highest percentage ratio, *Chaetoceros* content tends to have more deaths than other treatments.

3.2. Growth Performance

The length growth of discus fish during the rearing period ranged from 0.8±0.08 to 0.96 ±0.03 cm (Figure 2). The results of the ratio of the composition of the *Chaetoceros* sp. powder affected the length growth of discus fish ($p>0.05$). Based on the results of further tests showed that treatment D (15% concentration was significantly different from the control, while the 10% and 5% treatments were not significantly different from the other treatments). Growth is a change in size, both weight and length in a certain period or time [30]. The growth of discus fish slowed down along with the increase in the percentage ratio of *Chaetoceros* sp. powder content. The average increase in fish length for each treatment ranged from 0.80 to 0.96 cm. Based on visual observations, discus fish are voracious fish for food, that is, when the fish are fed it looks very aggressive. However, because fish have a territorial nature (maintain territory), the opportunities for fish to get food become uneven and affect the physiological condition of fish. Discus fish are fish that are difficult to cultivate because they are vulnerable to an environment that is less conducive to life which causes fish to become stressed and can reduce or even inhibit fish growth [31]. In addition, discus fish also have a level slow growth or relatively long fish growth. In general, it takes ±3 months to reach the market size (2 inches) [32]. Therefore, one of the factors that affect the growth rate of fish is the ability of the fish to utilize the feed.

3.3. Color Expression

The results of color quantification using a colorimeter (Table 1) show that the color brightness value of discus fish ranges from 47.90-50.69% and the density value is between 8.68-16.42%. The brightness value did not have a significant effect on the color of discus fish, but the addition of *Chaetoceros* sp. powder with different doses in the feed gave a significant effect on the color density value (P<0.05). The addition of *Chaetoceros* sp. powder feed as a source of carotenoids in the diet affects the color of discus fish compared to the feed that is not added with Chaetoceros both qualitatively and quantitatively, which increases the concentration value by 41-89% (Table 1). This shows that Chaetoceros in the feed is capable or effective to improve the color quality of discus fish. The development of color in ornamental fish is greatly influenced by the coloring pigments produced by carotenoids. According to [26], the highest carotenoid concentration of 0.70 mg/ml was noted in the animal they are fed with the meal mixed with oil obtained from *Chaetoceros* sp., this study claimed that diet could enhanced the coloration in ornamental fishes and become source of carotenoids. Similar to previous reports where the addition of carotenoids in the feed can affect the color of male red cherry shrimp *Neocaridina heteropods* [33], even [34] using synthetic material Carophyll pink with astaxanthin content of 8% as a source of carotenoids in feed was able to improve the color performance of *Lutjanus sebae* red snapper. Another study showed that the supplementation of astaxanthin in the diet of 100-300 mg kg^{-1} and 50-400 mg kg^{-1} could increase color intensity of the sword platy *Xiphophorus helleri* and body pigmentation on discus *Symphysodon* spp respectively [35].

Color quality is related to the amount of pigment in the skin [36]. The quality and presence of color is due to the presence of color cells in the skin called chromatophores. Chromatophores are pigment cells that have a spherical shape and are located spread throughout the epidermal cell layer of fish skin [37]. According to [19] astaxanthin added to fish feed is one of the dominant and effective carotenoids to increase fish color brightness and form red pigment cells. Astaxanthin (red xanthophyll) is applied in aquaculture, pharmaceutical and food industries [38]. so it is suspected that the carotenoid content in *Chaetoceros* sp. powder can influence discus color expression.

Table 1. Color quantification of discus fish feed with different doses of *Chaetoceros* sp. powder

Parameter	Doses of *Chaetoceros* sp. Powder (%)			
	0	5	10	15
Brightness value (%)	50.69±1.33 ^{a}	47.90±1.57 ^{a}	49.34±.51 ^{a}	48.34±1.11 ^{a}
Color density (%)	8.68±0.58 ^{a}	12.26±0.47 ^{b}	15.16±1.17^{bc}	16.42±1.30 ^{c}

Note: Data are mean ± SD. Different letters indicate significant differences (P < 0,05)

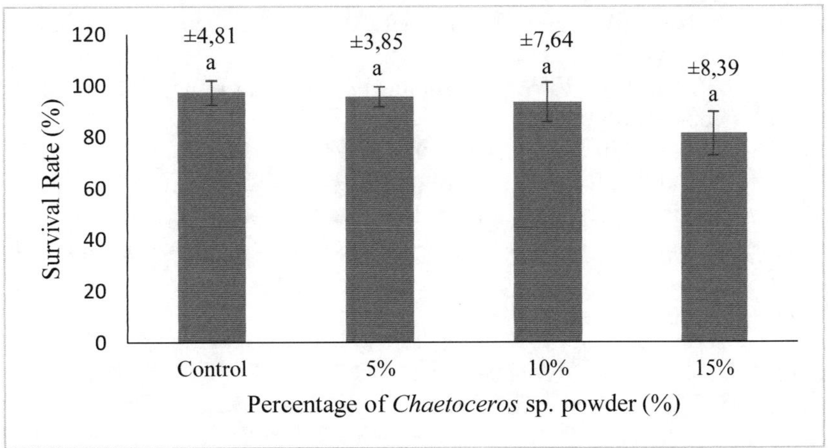

Figure 1. Effects of *Chaetoceros* sp. powder as on survival rate of discus (*Symphysodon discus*) during rearing period. Indicates significant difference (P < 0.05).

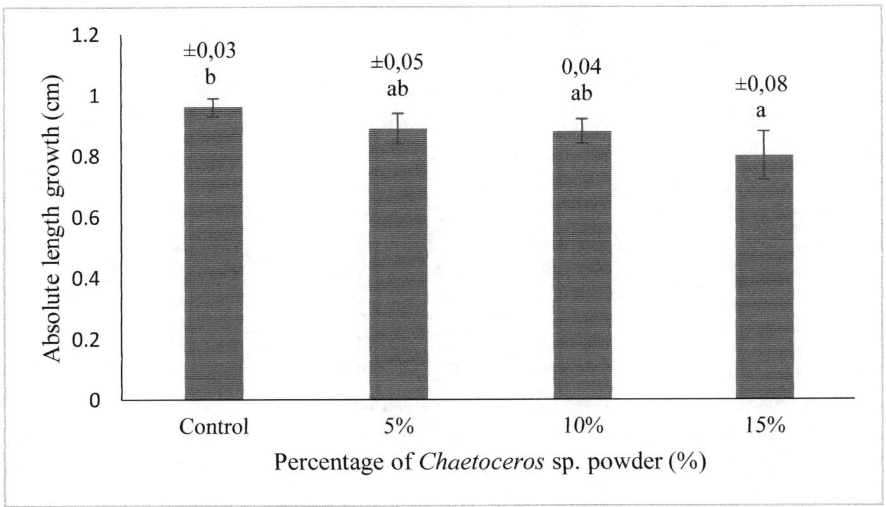

Figure 2. Effects of *Chaetoceros* sp. powder as on growth performance of discus (*Symphysodon discus*) during rearing period.

4. Conclusion

The supplementation of 5-15% *Chaetoceros* sp. powder had no significant effects on the survival rate and the brightness value of discus fish. The increasing of Chaetoceros sp. powder supplementation showed a lower rate of growth performance compared to control (p<0.05), however the color density significantly increased, 15% *Chaetoceros* s sp. showed highest color density compared to the other treatments (p<0.05). Dietary *Chaetoceros* sp. could improve the color expression especially the color density of discus fish and the supplementation level was suggested to be 10% *Chaetoceros* sp powder.

5. References

[1] Wan A K Y 2014 *Drivers in the demand for the ornamental trade of discus (Genus Symphysodon) between international markets* mres (University of Kent,)
[2] Kusrini E and Priono B 2011 *Media Akuakultur* 6 (1) 32–5
[3] Sidi F P and Sungkowo A B 2022 *Pros Work Pap Ser Manag.* 14 340–57.

[4] Yang B-T, Wen B, Ji Y, Wang Q, Zhang H-R, Zhang Y, Gao J-Z and Chen Z-Z 2021 *J. Proteomics.* 233 104085.
[5] Lesmana D S and Daelami D 2009 A complete guide to popular freshwater ornamental fish (Jakarta: Penebar Swadaya).
[6] Lambert D 2009 Cultivation of freshwater ornamental fish (Jakarta: Gramedia Pustaka Utama).
[7] Pirhonen J, Aaltonen S and Järvenpää H 2014 *Aquac Res.* 45 5 940–3.
[8] Khairunnisa K, Waspodo S and Setyono B D H 2020 *J Perikan Unram.* 10 77–83.
[9] Chatzifotis S, Pavlidis M, Jimeno C D, George Vardanis, Sterioti A and Divanach P 2005 *Aquac Res.* 36 15 1517-1525.
[10] Sinha A and Asimi O A 2007 *Aquac Res.* 38 11 1123-1128.
[11] Wallat G, Lazur A and Chapman F 2005 *North Am J Aquac.* 67 42–51.
[12] Gupta S K, Jha A K, Pal A K and Venkateshwarlu G 2007 *NPR.* 6 (1) 46-49.
[13] Sudirman, D A, Arief, M, & Fasya, A H 2020 *IOP Conf Seri Ear and Envi Scie. 441* 1 012086.
[14] Hudaidah, S, Putri, B, Samara, S H, & Adiputra, Y T 2019 *IOP Confe Seri Ear and Envi Scie. 236* 1 012073.
[15] Marginata E, Rusliadi R and Putra I 2015 *J Online Mhs Fak Perikan Dan Ilmu Kelaut Univ Riau.* 2 2 1–8.
[16] Prastari C and Harefa A 2020 *J Penelit Terap Perikan Dan Kelaut.* 2 1 1–5.
[17] Chu W-L 2012 *IeJSME.* 6 S24–37.
[18] Kusnanda A J, Perdana B A, Dharma A and Chaidir Z 2021 *J Kim dan Kem.* 43 1 38-43.
[19] Gupta S, Jha A, Pal A and Gudipati V 2014 *Ind J Nat Prod Resour.* 6 46–9.
[20] Ako H, Tamaru C S, Asano L, Yuen B, and Yamamoto L 2000 *UNJR Tec Rep.* 28 1-4.
[21] Song X, Wang L, Li X, Chen Z, Liang G and Leng X 2017 *Aquac Res.* 48 1359–67.
[22] Tu N P C, Ha N N, Linh N T T and Tri N N 2022 *Aqua* 553 738048.
[23] Sedjati S, Supriyantini E, Ridlo A, Yudiati E and Prasetyo L D 2019 *J Kelaut Trop.* 22 173–80.
[24] Prasetyo L D, Supriyantini E and Sedjati S 2022 *Bul Osea Mar.* 11 59–70.
[25] Wang LJ, Fan Y, Parsons RL, Hu GR, Zhang PY and Li FL 2018 *Mar Drugs.* 16 1 1-13.
[26] G B, Chellappan A, Citarasu T and Babu M 2017 *J Aquac Res Dev.* 8 (4) 1-9.
[27] Guillaume J, , Kaushik S, Bergot P and Metailler R 2001 *Nutrition and Feeding of Fish and Crustaceans* (Springer Science & Business Media).
[28] Daelami D 2001 *Freshwater ornamental fish hatchery business* (Jakarta: Penebar Swadaya)
[29] Wedemeyer G A 1996 *Physiology of fish in intensive culture systems* (Chapman and Hall).
[30] Alamsjah M A, Kusumaningrum G A and Masithah E D 2014 *J Ilm Perikan Dan Kelaut.* 6 25–30.
[31] Kadarini T, Sholichah L and Gladiyakti M 2018 *Pros FORUM Inov Teknol AKUA* 409–16.
[32] Indarta D 2002 *Maintaining and Cultivating Superior Discus* (Depok: Penerbit Agromedia)
[33] Amin M I, Rosidah, Lili W, Sukarman and Priyadi A 2012 *J Ilm Perikan Dan Kelaut.* 3 4 243-252.
[34] Aslianti T, Afifah and Priyono A 2011 Penel. Hayati Edisi Khusus 4B (51–57).
[35] Rachmawati D, Samidjan I and Pinandoyo P 2016 *Pen AkuaJ Ilm Perikan Dan Kelaut.* 13 1 58-67.
[36] Jegan K 2008 *J Vet Anim Sci.* 4.
[37] Subamia I W, Meilisza N and Permana A 2016 *J Ris Akua.* 8 429–38.
[38] Rodríguez-Sáiz M, de la Fuente J L and Barredo J L 2010 *Appl Microbiol Biotechnol.* 88 645–58.

6. Acknowledgments

We thank the editor and reviewers for their insightful comments on the manuscript.

Fluctuating asymmetry patterns of common carp (*Cyprinus carpio*) in fish hatchery center of Kabat, Banyuwangi, Indonesia

L Lutfiyah[1,3], D S Budi[2]

[1]Department of Marine, Faculty of Fisheries and Marine, Universitas Airlangga, Surabaya 60115 Indonesia
[2]Study Program of Aquaculture Banyuwangi Campus, Sekolah Ilmu Kesehatan dan Ilmu Alam, Universitas Airlangga, Indonesia

[3]Corresponding author: lailatullutfiyah@fpk.unair.ac.id

Abstract. A decrease in common carp (*Cyprinus carpio*) fry quality has occurred in many hatcheries indicated by fry abnormality increase, which causes low survival and growth rates. This research aims to obtain quantitative information of asymmetry level on common carp fry in the Fish Hatchery Centre (FHC) of Kabat as a reference for the breeding program. A total of 200 common carp fries with a total length of 5-7 cm were obtained from FHC of Kabat. The observed magnitude (FAm) and number (FAn) are fluctuating asymmetry. To calculate both of them; bilateral meristic characteristics were measured are pectoral fins radii, ventral fins radii, and scales on the lineal lateralis number. We found that there was an incidence of asymmetric fluctuations in FHC of Kabat with the highest FAm and FAn were obtained in lateral line scales. This implies inbreeding which causes a decrease in hatchery productivity, so the introduction of brood stock from other areas is required.

1. Introduction

Common carp (*Cyprinus carpio*), the five top aquaculture commodity, is an important animal protein source in the human diets in many parts of the developing world [1]. Many breeding programs have been carried out to increase the productivity and quality of common carp fry so that many varieties of fish have varied body colours and morphology [2,3]. Quality seeds will ensure the sustainability of fish farming.

Fish Hatchery Centre of Kabat (located in Banyuwangi Regency, East Java Province, Indonesia) is a Unit of Technical Executive of Food and Maritime Department of Banyuwangi Regency Government supplies common carp seeds for farmer in Kabat district and the surrounding area. A decrease in common carp fry quality has occurred in this hatchery centre indicated by fry abnormality increase which causes low survival and growth rates. Many fish hatcheries have the same problem of quality deteriorating of their seed produced such as high abnormality and lower viability and survival that causing by inbreeding [4].

The fish quality strongly depends on genetic variability [5]. Fry abnormalities may obtained because of high pressure in cross breeding due to inbreeding and limited number of brood stocks causing low genetic variation [6]. Depression of inbreeding more often occurred in life-history stages than morphological traits [7]. However, some studies were also found inbreeding depression evidence on morphological characteristics developmental stability [8].

In aquaculture, poor genetic quality has a negative impact on critical qualities such as growth and survival. Individual development stability was reduced genetically, which resulted in an increase in individual abnormality and asymmetry [9].

The stability of paired organs in animals development called fluctuating asymmetry related to genetic diversity [10] that measured by the average of asymmetry characteristics number per fish and magnitude asymmetry number. The developmental stability measurement with these two numbers on each fish to obtain the average of fluctuating asymmetry (the overall asymmetry) can be used to forecast genetic diversity reduction causing by inbreeding [8].Fluctuating asymmetry show a variation in development stability although relatively simple [11]. The differences of phenotypic in individuals for bilateral and meristic characteristics show fluctuating asymmetry that is the difference between the right and left body characters because of the individual's disability to be growing normally [12].

Previous research on fluctuating asymmetry of common carp [13], Nile tilapia (*Oreochromis niloticus*) [5,14], Chinook Salmon (*Oncorhynchus tshawytscha*) [15] and cantang hybrid grouper (*Epinephelus fuscoguttatus* x *Epinephelus lanceolatus*) [16] has been obtained to determine the fry quality as basic knowledge for the breeding programs. This research purposes to obtain qualitative representation about the degree of asymmetry on common carp fry in the Fish Hatchery Center of Kabat as a consideration for the breeding program. This information shows the status of this hatchery which can be used as a basis for evaluating the development of breeding programs both locally and nationally by building a good hatchery network through brood stock exchange programs to prevent inbreeding.

2. Material and methods

2.1. Time and place

This study was conducted in October 2019, the sample was obtained from Fish Hatchery Center of Kabat (Kabat District, Banyuwangi, East Java, Indonesia) and the sample was measured at Anatomy laboratory - Universitas Airlangga, Banyuwangi Campus (Giri District, Banyuwangi, East Java, Indonesia) Figure 1.

2.2. Fish origin and measurement of meristic

A total of 200 common carp fry with total length 5-7 cm were obtained from Fish Hatchery Centre of Kabat. A total of 120 good random samples (not damaged in fins and scales due to handling) were chosen. The stability development observation of individuals was done by the asymmetry percentages and fluctuating asymmetry approach, i.e., comparing the number of bilateral meristic traits of each individual on the right and left side.

The bilateral meristic traits were observed are the pectoral fins radii, ventral fins radii, and scales on the linea lateralis number. Bilateral meristic traits were measured following Saanin (1984). Observed data of all traits were computed for calculation of magnitude and number fluctuating asymmetry. Fluctuating asymmetry of magnitude (FAm) derived from the total of characteristics number difference observed in the right and left divided by the total samples. Fluctuating asymmetry of number (FAn) is the number of asymmetric individuals was found in the measurements divided by the total samples. Fluctuating asymmetry was calculated on the following formula [18]:

$$FAm = \frac{\sum(L-R)}{n}, \; FAn = \frac{\sum(Z)}{n}$$

Information:

L : left side organs number
R : right side organs number
Z : asymmetric individuals for specific meristic traits number

FAm : Fluctuating Asymmetry magnitude
FAn : Fluctuating Asymmetry number
n : total samples

Overall fluctuating asymmetry was calculated from the sum of fluctuating asymmetry total of each bilateral meristic traits. The data were descriptively analysed and graphical presented.

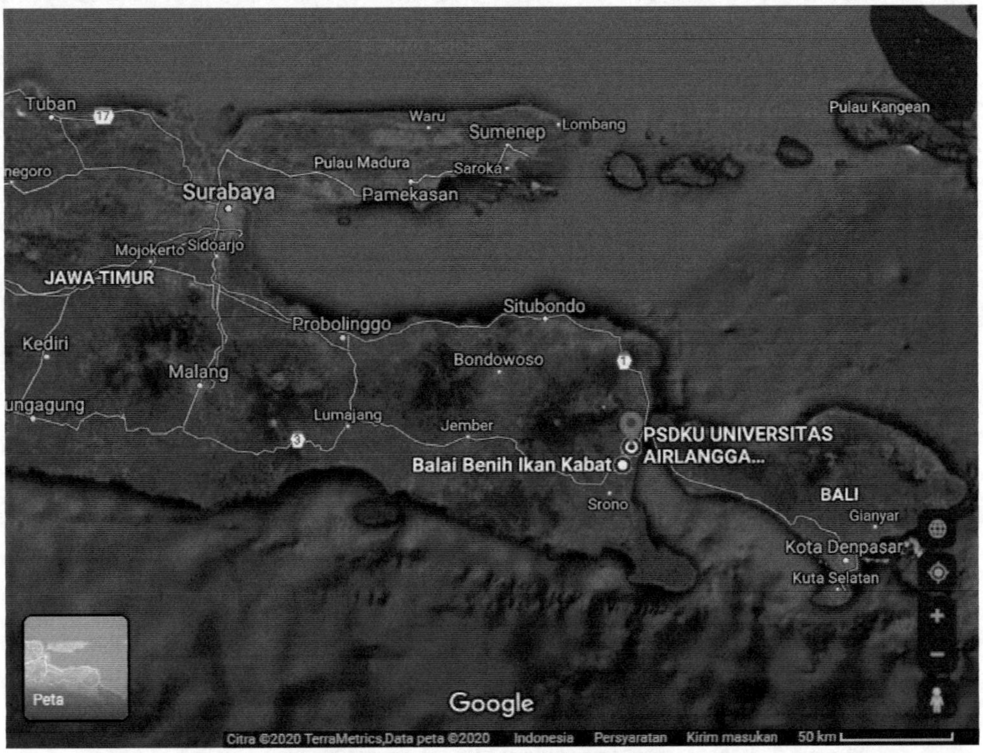

Figure 1. Map of research locations.

3. Results and discussion

Observed parameters including magnitude (FAm) and number (FAn) fluctuating asymmetry can be seen in Figure 2. and 3. The highest FAm and FAn are obtained in scales number on lateral line. The highest FAm value is 1.10 that means from 100 measured fish, the difference of meristic characteristics (the number of scales) on the right and left body sides are 110 pieces. The highest FAn value is 0.55 that means from 100 measured fish, there were 55 asymmetrical individuals. Furthermore, there was same value of FAm in pectoral and ventral namely 0.5; and there was same value of FAn also in pectoral and ventral namely 0.4. Overall value of FAm (2.1) higher than FAn (1.35), see Figure 4.

The highest FAm and FAn in lateral line scales also occurred in previous study in Nile tilapia (*Oreochromis niloticus*) [5,14] and Cantang hybrid grouper (*Epinephelus fuscoguttatus* & *Epinephelus lanceolatus*) [16]. In species, fluctuating asymmetry refers to non-directional variations from ideal bilateral structure symmetry [19,20]. Right and left paired character asymmetry has been related during ontogeny to genotypic and environmental variation [19,21] and is seen as a form of noise from development [20]. The lower energy priority for its growth than other pair organs causes the large fluctuating asymmetry value of bilateral meristic features on a paired organ [5]. The present study shows the highest value of fluctuating asymmetry (FAm and FAn) were occurred in scales on linea lateralis. This is predicted because the scales number of lateral line higher than the radii number of ventral and pectoral fins, so the energy used for lateral line scales development is higher.

The fluctuating asymmetry overall value (FAm and FAn) shows a lower value than other species such as Nile tilapia [5,14] and Cantang hybrid grouper [16]. However, the value of fluctuating asymmetry overall in present study are relatively high. It is predicted cause by low genetic variability related with inbreeding occurrence that increases the homozygosity. Even, inbreeding occurrences in a low number of brood stocks can be causing homozygous offspring [22]. Inbreeding may impact the developmental instability of morphological traits because of enzymatic diversity lack in homozygotes can reduce developmental homeostasis processes normality [4]. Homozygosity and instability of individual development as observed by fluctuating asymmetry have positive relationship. High heterozygosity in individuals causing high developmental stability and high homeostatic meaning as phenotype traits approaching normal population [18].

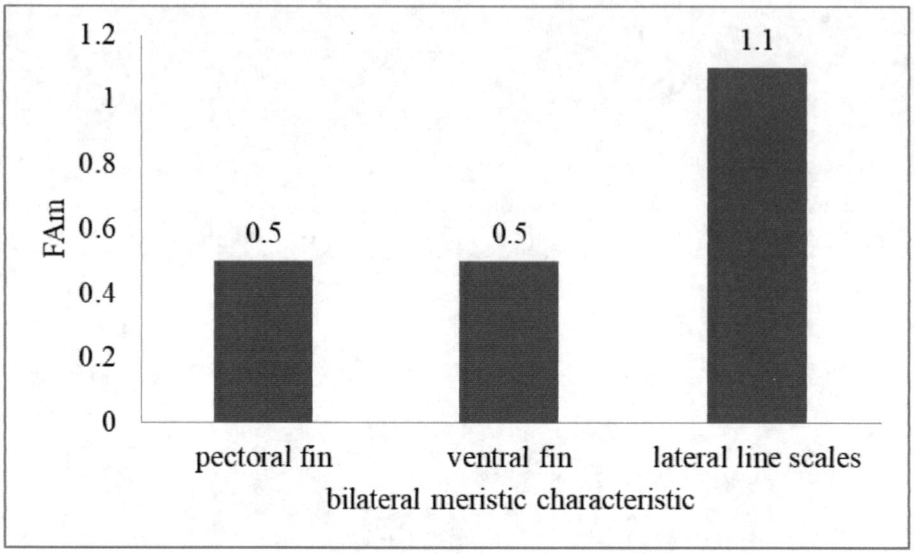

Figure 2. Fluctuating asymmetry of magnitude (FAm) of common carp (*Cyprinus carpio*) fry in Fish Hatchery Center of Kabat.

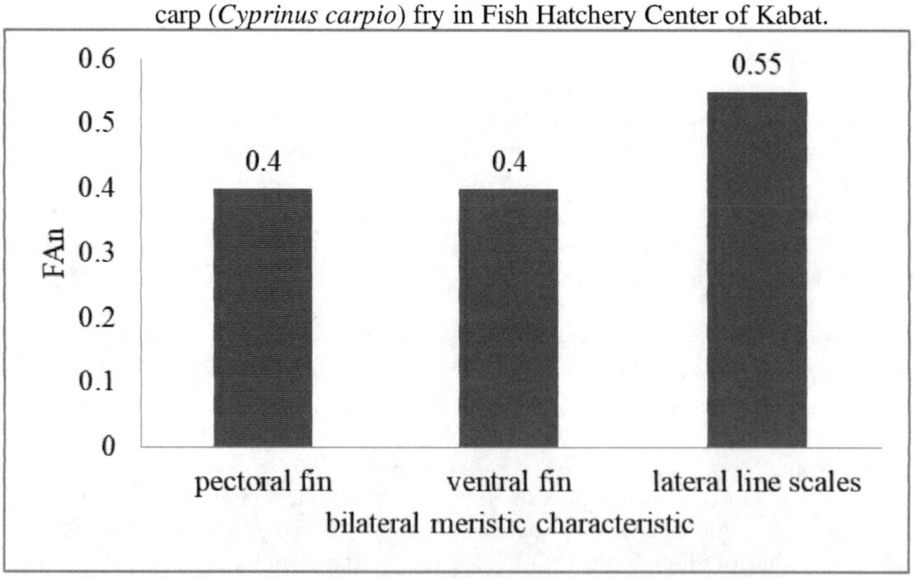

Figure 3. Fluctuating asymmetry of number (FAn) of common carp (*Cyprinus carpio*) fry in Fish Hatchery Center of Kabat.

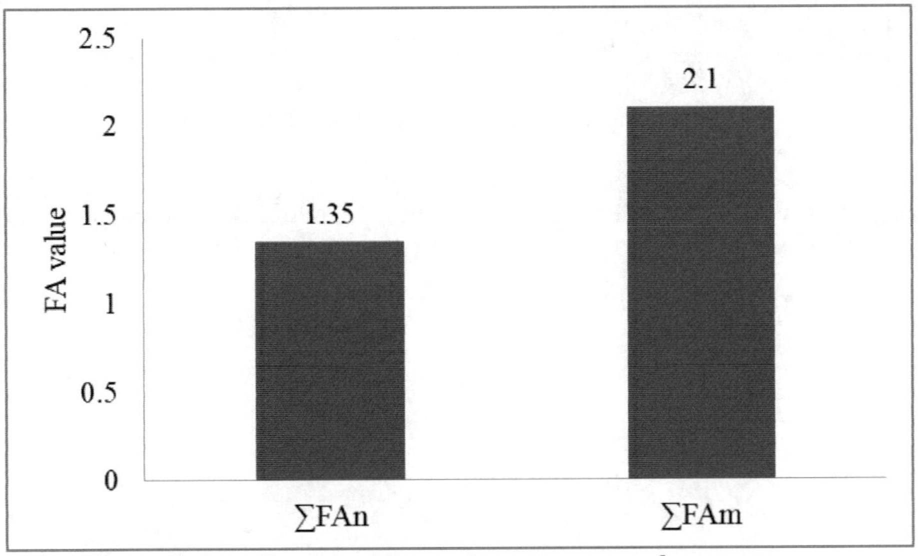

Figure 4. Overall value of fluctuating asymmetry of common carp (*Cyprinus carpio*) fry in Fish Hatchery Center of Kabat.

4. Conclusion

The highest fluctuating asymmetry of magnitude (FAm) and number (FAn) of common carp (*Cyprinus carpio*) in Fish Hatchery Center of Kabat were obtained in lateral line scales. The overall value of fluctuating asymmetry (FAm and FAn) are lower than other species. The incidence of asymmetric fluctuations implies inbreeding which causes a decrease in hatchery productivity, so the introduction of brood stock from other areas is required.

5. References

[1] Hulata R T, Hulata G, and David L 2019 *Heredity.* 123 565–78.
[2] Vandeputte M 2003 *Aquat. Living Resour.* 16 399–407.
[3] Hulata G 1995 *Aquaculture* 129. 143–55.
[4] Fessehaye Y, Komen H, Rezk M A, van Arendonk J A M, and Bovenhuis H 2007 *Aquaculture.* 264 27–35.
[5] Budi D S, Lutfiyah L, and Triastuti R J 2017 *Omni-Akuatika.* 13 1–4.
[6] Sheridan L, and Pomiankowski A 1997 *Heredity.* 79 515–23.
[7] DeRose M A, and Roff D A 1999 *Evolution.* 53 1288–1292.
[8] Alibert P, Renaud S, Dod B, Bonhomme F, Auffray C, Alibert P, Renaud S, Dod B, Bonhomme F, and Auffray J 1994 *Biol Sci.* 258 53–59.
[9] Vøllestad L A, Hindar K, and Møller A P 1999 *Heredity.* 83 206–18.
[10] Clarke G 1992 *Acta Zool. Fenn.* 191 31–5.
[11] Eriksen M S, Espmark Å M, Poppe T, Braastad B O, Salte R, and Bakken M 2008 *Environ. Biol. Fishes.* 81 87–99.
[12] Clarke G M 1995 *Conserv. Biol.* 9 18–24.
[13] Bongers A B J, Ben-Ayed M Z, Doulabi B Z, Komen J, and Richter C J J 1997 *J. Exp. Zool.* 277 72–9.
[14] Budi D S, and Lutfiyah L 2017 *J. Aquac. Sci.* 2 72–80.
[15] Koeberle A L, Arismendi I, Crittenden W, Leer D, and Noakes D L G 2020 *Aquat. Ecol.* 54 431–46.
[16] Lutfiyah L, and Budi D S 2019 *J. Aquacuture Sci.* 4 21–6.
[17] Saanin H 1984 Taxonomy and fish identification keys (Bandung: Binacipta) p 508.
[18] Leary R F, Allendorf F W, and Knudsen K L 1985 *Trans. Am. Fish. Soc.* 114 230–5.

[19] Merila J, and Bjorklund M 1995 *Syst. Biol.* 44 97–101.
[20] Palmer A R, and Strobecke C 1986 *Annu. Rev. Ecol. Syst.* 17 391–421.
[21] Valen L Van 1962 *Evolution.* 16 125–42.
[22] Widiyati A, and Sumantadinata K 2007 *J. Ris. Akuakultur.* 2 395–8

6. Acknowledgments

The authors gratefully thank all colleagues in Universitas Airlangga Banyuwangi campus for their help and technical assistance. The research was carried out independently without funding support from any agency.

IOP Conf. Series: Earth and Environmental Science 1273 (2023) 012008 doi:10.1088/1755-1315/1273/1/012008

Use of sea urchin (*Diadema setosum*) shells as antibacterial

E Saputra[1,2], D Y Pujiastuti[2], A A Abdillah[3]

[1]Marine Department, Faculty Fisheries and Marine, Universitas Airlangga Surabaya. Indonesia

[2]Corresponding author: ekasaputra@fpk.unair.ac.id

Abstract. Sea urchin (*Diadema setosum*) is one type of aquatic biota originating from the echinoderm phylum whose distribution is seen in almost all waters. The toxin produced by sea urchins can be used in various fields, one of which is a potential treatment as antibacterial because it contains bioactive compounds. *Staphylococcus aureus* (*S. aureus*) is a small portion of microorganisms that live in the oral cavity that can cause disease.

1. Introduction

Sea urchin (*Diadema setosum*) is one type of aquatic biota originating from the echinoderm phylum. The distribution of sea urchins is seen in almost all water zones; there are as many as 950 species of sea urchins scattered throughout the world. *Diadema setosum* is one type of sea urchin that is spread throughout the coral reef zone, including in the sand zone, algae growth zone, seagrass zone the edge area. Toxins produced by organisms, one of which is sea urchins, can be used in various fields, one of which is the field of medicine which has the potential to be used as a new type of antibiotic to be developed in the pharmaceutical field because it contains bioactive compounds. Extract of *Diadema setosum* sea urchin shells with methanol as a solvent showed the ability as an antibacterial. *Staphylococcus aureus* (*S. aureus*) is a small portion of all microorganisms that live in the oral cavity. There are about 10 to 1000 colonies of *S. aureus* per liter of saliva. *S. aureus* can turn into a pathogen if there is trauma or abrasion on the mucosal surface. Currently, many toothpaste manufacturers produce toothpaste by utilizing natural or herbal ingredients such as betel leaf toothpaste, areca nut toothpaste, and gambir toothpaste. However, no one has exploited aquatic resources such as sea urchins which have the potential as natural antibacterials.
Sea urchins are one type of aquatic biota originating from the echinodermata phylum. The spread of sea urchins is seen in almost all water zones and there are 950 species of sea urchins spread throughout the world. The spread of sea urchins is in the waters of Indonesia, Malaysia, the Philippines and North Australia around 316 species, while around 84 species consisting of 48 clans and 21 tribes are found in Indonesian waters themselves. *Diadema Setosum* (D. Setosum) is one type of sea urchins that spread throughout the coral reef zone, among others in the sand zone, algae growth zone, seagrass zone to the tubir area [1]. The population of species is more commonly found in coral areas whose condition has been damaged. D. Setosum, because his life is below the lowest receding limit, the food source comes from various types of algae and organic/detritus particles.

2. Material and Methods

2.1. Time and Place

The present research is performed in the Laboratory of Aquaculture, Faculty of Fisheries and Marine, Airlangga University.

2.2. Materials and Equipment

Research materials needed are Wagner's reagent, Meyer's reagent, magnesium powder (Mg), concentrated hydrochloric acid (HCL), chloroform (CHCl3), acetic anhydride (C4H6O3), concentrated sulfuric acid (H2SO4), ferric chloride (FeCl3) 3% , distilled water (H2O), sodium hydroxide (NaOH), sodium carbonate (Na2CO3) 7.5%, iron (III) chloride (FeCl3), spiritus and aluminum foil. 96% alcohol, distilled water, sea urchin extract, S. aureus bacterial culture, physiological NaCl, 70% ethanol, and NA medium (Nutrient Agar).

2.3. Research Methods

In this study using an experimental method with a quantitative and qualitative approach. This study was conducted to determine the causal relationship of one or all of the variables or treatment given and connect with without treatment or control. The method used to find elements, characteristics, properties of a phenomenon. This method starts by collecting data, analyzing data and interpreting them. Conducted with case studies, comparative studies, documentary analysis.

3. Results And Discussions

Average yield of inhibition zone against S. aureus bacteria (mm) ± SD, the results obtained are listed in Table 1.

Table 1. Results of analysis of proximate levels

	shell extract 0%	shell extract 70%	shell extract 80%	shell extract 90%	shell extract 100%
The average inhibition zone (mm) ± sd	2,5^{a}±0,707107	4^{a}±1,08012	6,375^{b} ±2,86865	4,125^{a} ±1,49304	3,625^{a} ±0,85391

Based on the results of the antibacterial activity test in the table above, it was found that commercial toothpaste solutions with the addition of sea urchin shell extract at different concentrations were known to inhibit S. aureus bacteria. Antibacterial activity test using a disk test or paper disk which is one of the tests by measuring the inhibition area or clear zone formed around the paper disk containing a solution of antibacterial material according to the concentration of the treatment. At a concentration of 70%, the highest inhibition zone was 4.5 mm with an average of 4 mm. Concentration of 80% obtained the highest inhibition zone is 9 mm with an average of 6.375 mm. At 90% concentration, the highest inhibition zone was 6 mm with an average of 4.125 mm and at 100% concentration, the highest inhibition zone was 4.5 mm with an average of 3.62 mm. While in the positive control, namely toothpaste without the addition of extract, the highest inhibition zone was 3.5 mm with an average of 2.5 mm. In the graph above, it can be seen that each treatment has a different significance and the treatment with a concentration of 80% has the highest inhibition zone with a value of 9 mm which means it is the most effective concentration to inhibit the growth of S. aureus bacteria [2].

The addition of sea sea urchin shell extract used in this study (70%, 80%, 90%, and 100%) with the best effectiveness in P2 with a concentration of 80%can inhibit the growth of S. Aureus bacteria. This is because the bioactive component which plays a role in inhibiting the general bacterial activity originating from the triterpenoid and alkaloid groups. This is in line with [3], a bioactive component that plays a role in inhibiting the activity of bacteria from saponin compounds, steroids/triterpenoids, alkaloids, and flavonoids with the mechanism of work of bacteriostatic [4].

4. Conclusions

Bioactive components or secondary metabolites from sea urchin shells of D. setosum which act as antibacterials that can inhibit S. aureus bacteria are alkaloids and triterpenoids. Sea urchin shell extract D. setosum which has good effectiveness in influencing S. aureus bacteria is in P2 treatment with an additional concentration of extract as much as 80%.

5. References

[1] Hadinoto S, Sukaryono I D, Siahay Y 2017 The Nutritional Content of Sea Urchins (Diadema Setosum) and the Potential of Their Shells as Antibacterials. Ambon, Research and Community Service Institute, Lambung Mangkurat University, pp. 260-265.
[2] Gonzalez-Gutierrez J, Partal P, Garcia-Morales M, Gallegos C 2010 Bior tech 101 (6) 2007-2013.
[3] Ningrum N A 2018 Identification of Antibacterial Compounds of Sea Urchin (Diadema setosum) Gonad Extract With N-Hexane Solvent Against Escherichia coli and Staphylococcus aureus. Thesis. Fisheries Product Technology Study Program, Department of Aquatic Resources Management, Faculty of Fisheries and Marine Sciences, Universitas Brawijaya
[4] Uma B, Parvathavarthini R 2010 Intern J Phar tec Res 2 (3) 1677 - 1680.

6. Acknowledgement

The author expresses many thanks to the supervisors who have given a lot of knowledge and energy. The author also thanks the editors and reviewers for their suggestions and input on our manuscript.

IOP Conf. Series: Earth and Environmental Science 1273 (2023) 012009 doi:10.1088/1755-1315/1273/1/012009

Effect of feeding a combination of silage of fish offal and fermented tofu dregs with different doses on the population and biomass of silkworms (*Tubifex tubifex*)

V N Sari[1], L Lutfiyah[2,3] and L Sulmartiwi[2]

[1]Study Program of Aquaculture, Faculty of Fisheries and Marine, Airlangga University, 60115 Surabaya, East Java, Indonesia.

[2] Department of Marine, Faculty of Fisheries and Marine, Airlangga University, 60115 Surabaya, East Java, Indonesia.

[3]Corresponding author: lailatullutfiyah@fpk.unair.ac.id

Abstract. Silkworm (*Tubifex tubifex*) is one of the natural food that is often used in fish farming. Determining the success of silkworm cultivation is strongly influenced by the culture media, which is generally rice mud. Rice field mud is a living medium for silk worms in nature, but the low nutrient content requires the addition of silage of fish offal and fermentation of tofu dregs. This study used a maintenance medium dose of 100% rice field mud (control), 75% fish offal silage + 25% tofu dregs fermentation (P1), 50% fish offal silage + 50% tofu dregs fermentation (P2), and 25% of tofu silage. fish + 75% tofu dregs fermentation (P3). The main parameters in this study were the growth of biomass and population of silkworms during 21 days of rearing. While the supporting parameters of this research are water quality parameters which include temperature, pH, and DO. Analysis of the data in this study using Analysis of Variance (ANOVA) 1 factors. The results of the calculation of optimal biomass and population were obtained in 25% silage of fish offal + 75% fermented tofu dregs with an absolute biomass value of 18.6 grams and an absolute population value of 8,698 individuals.

1. Introduction

Aquaculture requires feed with sufficient nutritional content to produce healthy and fresh fish. In fish farming activities, the availability of feed sources is an important factor for the hatchery process [2]. One of the natural food used in fish farming is silk worms. Silkworms (*Tubifex tubifex*) are favored by many fish because of their high nutritional content. Silkworm population growth is influenced by C-organic, which plays an important role as energy [8].

One of the determinants of the success of silkworm cultivation is strongly influenced by the culture media used. The general media used in the cultivation of silk worms is rice field mud which is a living

IOP Conf. Series: Earth and Environmental Science 1273 (2023) 012009 doi:10.1088/1755-1315/1273/1/012009

medium for silk worms in nature [7]. Fish offal silage contains 14.01% protein, 60.62% water, 4.75% ash and 20% fat [5] In addition to silage of fish offal, the media used in this cultivation is fermented tofu dregs. Tofu pulp contains 18.67% crude protein, 24.43% crude fiber, 9.43% crude fat, 3.42% ash and 41.97% BETN [6]. Looking at the composition, silage of fish offal and fermented tofu dregs has a high enough protein content so that it can meet the nutritional needs of silk worms which can support the increase in population and biomass of silk worms. So in this case, a research was conducted on the effect of feeding a combination of silage of fish offal and fermented tofu dregs with different doses on the population and biomass of silk worms.

2. Material and methods

2.1. Materials

The main ingredients used include the innards of Siamese catfish obtained from PT. Dimaz Reiza Officer, molasses, tofu dregs, *Effective Microorganisms* (EM4), rice field mud, water, silkworms. Meanwhile, the tools used in this study were plastic containers, culture racks, water pumps, paralon pipes, PVC glue, filter cotton, 1 mm mesh filter, buckets, tubs for fermentation, water barrels, digital scales, thermometers, magnifying glass, hand counter, pH pen, DO meter, label paper, tissue and stationery.

2.2. Methods

This study used a completely randomized design (CRD) experiment with four treatments and five repetitions, the samples were coded P0, P1, P2, and P3. The treatment used was the addition of feed in the form of a combination feed of fish offal silage and fermented tofu dregs in rice field mud culture media. Data from the results of the study were obtained using Analysis of Variance (ANOVA). If there is an effect of treatment, it is continued with Duncan's Multiple Distance test.

3. Result and discussion

3.1. Result

Calculation of absolute biomass, population and water quality of silkworms (*Tubifex tubifex*) was carried out to determine the increase in biomass during 21 days of rearing.

Table 1. Average Value of Initial Biomass, Final Biomass and Absolute Biomass (g).

Treatment	Initial Biomass	Final Biomass	Absolute Biomass
P0	10	17,8	$7{,}8\pm0{,}13^{d}$
P1	10	22,5	$12{,}5\pm0{,}76^{c}$
P2	10	25	$15\pm0{,}31^{b}$
P3	10	28,6	$18{,}6\pm0{,}59^{a}$

Information: Notations with different superscript letters showed significantly different values ($P<0.05$) according to Duncan's Multiple Region test.

Table 2. Initial Population Mean Value, Final Population and Absolute Population (ind).

Treatment	Initial Population	Final Population	Absolute Population
P0	5.320	8.757	$3.437\pm29{,}62^{d}$
P1	5.320	10.933	$5.613\pm312{,}61^{c}$
P2	5.320	12.130	$6.810\pm131{,}76^{b}$
P3	5.320	14.018	$8.698\pm213{,}89^{a}$

Information: Notations with different superscript letters showed significantly different values ($P<0.05$) according to Duncan's Multiple Region test.

Table 3. Average Results of Water Quality Parameters.

Water Quality	P0	P1	P2	P3
Temperatur (°C)	28-30 °C	26-30 °C	27-30 °C	26-28 °C
pH	5-6,2	4,5-5	5,3-5,8	6,1-7,3
DO (mg/L)	3,67-3,96	3,5-3,87	3,25-3,5	3,51-3,84

Information: Notations with different superscript letters showed significantly different values ($P<0.05$) according to Duncan's Multiple Region test.

3.2. Discussion

The provision of enrichment of culture media using tofu dregs organic matter with a higher dose in treatment 3 was able to provide the nutritional needs of silk worms to grow so that the absolute biomass growth of silk worms was higher than other treatments. This is presumably because the protein and carbohydrate content of tofu dregs is higher than the protein and carbohydrate content found in fish offal silage. Fish offal silage contains 14.01% protein and 13.61% carbohydrates. While the nutritional content of tofu dregs is 18.67% protein, 24.43% crude fiber, 9.43% crude fat, 3.42% ash and 41.97% BETN [6].

Silkworm biomass did not get maximum results in the control treatment (P0) with a value of 7.8 grams, this is presumably because in treatment 0 without the addition of growing media so that the food source for silkworms was insufficient for growth because they only expected leftover food that was inside mud and sand. Rice field mud contains 2-5% organic matter [12]. According to the criteria for assessing C and N nutrient status described in the research of [3]. the organic matter content of rice field mud is relatively low, so it is necessary to add alternative materials that have high protein and carbohydrate content to support the increase in organic C and N values in the maintenance media.

Population is the number of individuals in a group of silk worms. The highest population is in P3 which is 8,698 individuals, while the lowest absolute population is in P0 (control) which is 3,437 individuals. The low population is due to the dose of feed given having a smaller nutritional value than the nutritional value in treatment 3. This is in accordance with the research of [4]. The lowest population was obtained in treatment A, which was 124244 individuals/m2 with a worm weight of about 2.8 mg/head. This is thought to be due to the low protein content of the silkworm feed. According to [10]. protein utilization is influenced by energy content and amino acid content in feed. The number of populations is related to the reproductive process, in addition to the quantity of food available, the quality of food must also be considered, so that it can meet the needs for both growth and reproduction [1].

The results of measurements of water quality showed that the temperature values ranged from 26-30 °C. Temperature can affect the physical and chemical properties of water and can accelerate biochemical processes [9]. That if the water temperature increases, the metabolic rate and the need for oxygen also increase, as well as the toxicity of pollutants. The temperature range during maintenance is considered feasible for the maintenance of silkworms because the allowable range is between 25-28 °C [11]. According to [11] the use of plastic material as a container is suspected to be the cause of the high temperature range, besides the small size of the container also helps increase the temperature quickly.

4. Conclusion

Feeding a combination of silage of fish offal and fermented tofu dregs has an effect on the increase in biomass and silkworm population. Silkworm biomass ranged from 7.8±0.13d to 18.6±0.59a. As for the population of silk worms, the results obtained in the range of 3.437±29.62d to 8.698±213.89a.

The optimal dose of combined feed in increasing biomass and silkworm population was found in treatment 3 with a combination of 25% fish offal silage and 75% tofu dregs fermentation which produced an absolute biomass of 18.6 grams and an absolute population of 8,698 individuals.

5. References

[1] E W Cahyono, J Hutabarat and V E Herawati 2015 *JAMT*. 4(4):127–135.
[2] W Detiben, H Pangkey, Jeffrie F M *Budidaya Perairan Mei 2018, 6* (2), 25-31.
[3] Eviati and Sulaeman 2009 Technical Instructions: Chemical Analysis of Soil, Plants, Water and Fertilizers. 2nd ed vol. 2 Bogor.
[4] Y Hadiroseyani, Nurjanah dan D Wahjuningrum 2007 *J. Akuakultur* 6 (1) : 79-78.
[5] Hossain and Alam 2015 *Agric Sci.*13(2): 13-25.
[6] H Imam, H Rahmat and Mansyur 2005 *JIT*. 5(2): 94-99.
[7] Safrina, Berta P and H Wijayanti 2015 Prosiding Seminar Nasional Politeknik Negeri Lampung 520-525.
[8] Santi F, Dadang S and E Supriyono 2020 *JIPM, 2* (3), 429-434.
[9] Spotte S H 1970 Fish and invertebrata culture Wiley Interscience a Divisions of John Wiley Inc. New York.
[10] M A Suprayudi, G Edriani and J Ekasari 2012 *JAI*, 11 (1): 1-10.
[11] D Suryadin, Senny H and Rustadi 2017 *J Fis Sci.* 19 (2): 97-10.
[12] A Tangketasik, N M Wikarniti and N N Soniari 2012 *Agrotop*. 2(2):101-107.

6. Acknowledgement

This research was conducted jointly with the team at the Faculty of Fisheries and Marine, Airlangga University.

IOP Conf. Series: Earth and Environmental Science 1273 (2023) 012010 doi:10.1088/1755-1315/1273/1/012010

Toxicity of brilliant blue dye on development organ and survival rate of tilapia larvae (*Oreochromis niloticus*)

F A Wicaksono [1], J Triastuti[2,4], L Lutfiyah[2], and N N Dewi[3]

[1]Study Program of Aquaculture, Faculty of Fisheries and Marine, Airlangga University, 60115 Surabaya, East Java, Indonesia.
[2]Department of Marine, Faculty of Fisheries and Marine, Airlangga University, 60115 Surabaya, East Java, Indonesia.
[3]Department of Aquaculture, Faculty of Fisheries and Marine, Airlangga University, 60115 Surabaya, East Java, Indonesia.

[4]Corresponding author: juni.triastuti@fpk.unair.ac.id

Abstract. One of the most widely used textile dyes is brilliant blue. However, brilliant blue has a bad influence on the waters, namely the effect of carcinogenic, polluted air quality, cannot be degraded or non-biodegradable and affects fish life in the waters. The purpose of this research was to determine the effect of the toxicity of brilliant blue dye on organ development and survival rate of tilapia (Oreochromis niloticus) larvae. The stage of this research was a completely randomized design with the provision of brilliant blue dye to the larvae of tilapia (Oreochromis niloticus). The results showed that the highest concentration treatment (40 ppm) experienced the fastest organ development, the smallest yolk volume (0.14 mm^3), the highest abnormality (20%), and the lowest larval survival rate (65%).

1. Introduction

Untreated waste from the textile industry induces critical water pollution by degenerate quality of water, making it ineffective for ecosystems of aquatic [12]. The most common dyes found in textile waste are azo dyes with usage reaching 80% [10]. One of the organic pollutants which has an azo chromophore group (-N=N-), is classified as a synthetic dye and the most reactive dye is brilliant blue dye [9].

The increasing content of brilliant blue textile dye pollution in the waters will disturb aquatic organisms, especially tilapia larvae [7]. Brilliant blue textile dye can cause malformations in the development of fish larvae such as swim bladder defects, yolk sac edema, and curved spine [9].

2. Material and methods

2.1. Sample collection

Larvae (*Oreochromis niloticus*) samples were collected from UPT. Laboratory Fish Health and Environment, Umbulan, Pasuruan.

2.2. *Research methods*

The research method used is a laboratory experiment. Experimental research was carried out by means of certain treatments on the object of research under controlled conditions to determine the effect of exposure to blue textile dye on organ development and survival rate of tilapia larvae. This research is divided into 2 parameters, namely the primary parameters and secondary parameters. The primary parameters in this research are organ development (yolk volume and larval abnormalities) and larval survival rate. The secondary parameters observed are air quality parameters (DO, pH, and temperature).

Data from research on egg volume, abnormalities, and survival rates of larvae were obtained using Analysis of Variance (ANOVA), if the results showed significant differences, then followed by a significant difference test. The data obtained from the results method of Duncan's Multiple Range Test (DMRT).

3. Result and discussion

3.1. *Organ and yolk sac volume development*

Observation of organ development identified 400 individuals of tilapia larvae *O. niloticus*.

Table 1. Organ development of tilapia larvae (*Oreochromis niloticus*) incubated at different dye concentrations. dph = day post hatching

Organ Development	*Time of Development (dph)*					(Morrison, 2001)
	A (0 ppm)	B (14 ppm)	C (20 ppm)	D (28 ppm)	E (40 ppm)	
Eyespots	1-2	1-2	1-2	1-2	1-2	1-2
Body pigmentation	2-3	2-3	2-3	2-3	2-3	2-3
Anal	3-4	3-4	3-4	3-4	3-4	3-4
Caudal fin	4-5	4-5	4-5	4-5	4-5	4-5
Pectoral fin	5-6	5-6	5-6	5-6	5-6	5-6
Dorsal and ventral fin	5-6	5-6	5-6	5-6	5-6	5-6
Swim bubbles	6-7	6-7	6-7	6-7	6-7	6-7
Mouth opening	6	6	7	7	7	6
Depleted egg yolk	7	7	7	6	6	7

Based on observation organ development of tilapia larvae exposed to brilliant blue dye there is a difference in the mouth opening and run out of egg yolk so it will result in organ development and larval survival rate.

Table 2. Yolk sac volume of tilapia larvae (*Oreochromis niloticus*) incubated at different dye concentrations

Treatment	1 (dph) Yolk sac volume ± SD ($mm^{3)}$)	7 (dph) Yolk sac volume ± SD ($mm^{3)}$)
A (0 ppm)	$9{,}26{\pm}0{,}21^{a}$	$0{,}45{\pm}0{,}70^{a}$
B (14 ppm)	$8{,}27{\pm}0{,}13^{b}$	$0{,}41{\pm}0{,}29^{b}$
C (20 ppm)	$8{,}39{\pm}0{,}24^{b}$	$0{,}34{\pm}0{,}43^{ab}$
D (28 ppm)	$8{,}34{\pm}0{,}08^{b}$	$0{,}45{\pm}0{,}40^{c}$
E (40 ppm)	$8{,}42{\pm}0{,}20^{b}$	$0{,}14{\pm}0{,}23^{c}$

The difference in letters above the bar indicates a significant difference ($P<0.05$) based on Duncan's Multiple Range Test (DMRT).

Brilliant blue dye affects organ development, volume egg yolk, abnormalities, and larval survival rate. Fresh water environment is hypotonic so that water containing brilliant blue dye diffuses into the blood vessels of the larval yolk and affects the disability yolk sac so that it can be detrimental in providing nutrition for fish larvae [6]. The yolk sac is a source of nutrition (endogenous feeding) in fish larvae for organ development processes [8]. Lateness absorption of egg yolks can occur due to decreased body metabolism of negative effect of brilliant blue dye on egg yolk [8]. Using inefficient nutrition of egg yolks can reduce available energy for cell proliferation and inhibit the process of organ development [1].

3.2. Survival rate of tilapia larvae

Pressure difference osmosis and the entry of water through the pores of the yolk between the body and environment due to brilliant blue dye resulting in more energy metabolism needed to carry out osmoregulation in an effort to fish larvae adaptation and survival rate.

Table 3. The results of survival rate of tilapia larvae (*Oreochromis niloticus*) incubated at different dye concentrations of brilliant blue

Treatment (ppm)	Survival Rate ± SD (%)
0	20 ± 0,00
14	17 ± 0,95
20	16 ± 0,81
28	15 ± 0,81
40	13 ± 0,81

The results of survival rate of tilapia larvae treated with 40 ppm brilliant blue dye showed that 13 live tilapia larvae.

The difference in osmotic pressure and the entry of water through the pores of the yolk between the body and the environment due to brilliant blue dye resulting in more metabolic energy needed to carry out osmoregulation as an effort to adapt and survival rate of fish larvae [11]. Based on the results of the study, the highest percentage of survival rate in treatment A (0 ppm/control) followed by treatment B (14 ppm), treatment C (20 ppm), and treatment D (28 ppm), while the lowest larval survival rate was in treatment E (40 ppm). The lowest survival rate in treatment E (40 ppm) caused the organ cannot develop properly, this has an impact on the low larval survival rate. Fish larvae need more energy higher than adult fish, because it is used for growth and organ development [3].

Table 4. Water Quality Data Temperature, pH, and DO of tilapia larvae (*Oreochromis niloticus*) incubated at different dye concentrations of brilliant blue

Water Quality	Concentration (ppm)				
	ppm	14 ppm	20 ppm	28 ppm	40 ppm
Temperature (°C)	8,0 -28,8	28,2 -29,1	28,0 -29,9	28,3 -29,9	28,2 -28,9
pH	,0 - 8,3	7,7 - 8.1	7,5 - 8,1	7,63 - 8,1	7,6 - 8,1
DO (mg/l)	,0 - 8,7	7,1 - 9.8	7,6 - 9,4	7,30 - 9,0	7,3 - 9,6

The water quality data in this research were in accordance with the standard for tilapia larvae cultivation.

The value of dissolved oxygen DO content in the tilapia aquarium ranges from between 7-12 mg/L, still within the tolerance limit of tilapia life or in optimal conditions. The state of unbalanced dissolved oxygen will cause stress on the larvae because the brain does not get enough oxygen supply and perish due to deficiency of oxygen caused by body tissues unable to bind oxygen dissolved in the blood [4]. The temperature content in the aquarium during the research ranged from 27-29°C. The optimal temperature for tilapia is 28-32°C. When the temperature is below 25°C, the movement activity and appetite of the fish decreases. Temperature conditions above 35°C, fish aquaculture will experience stress and difficulty breathing because fish oxygen consumption increases, while the solubility of oxygen in water decreases [5].

The results showed that the pH value in the tilapia aquarium ranged from between 7.03-8.35 is within the tolerance limit of tilapia life or at optimal conditions. Tilapia has a tolerance value of water acidity for optimal life between 5-8.5 [2]. Acidity levels (pH) that are not optimal cause fish to be stressed, susceptible to disease, and low growth [4].

4. Conclusion

The highest concentration of brilliant blue dye (40 ppm) had the fastest organ development, the smallest yolk volume (0.14 mm^3), the highest abnormality (20%), and the lowest larval survival rate (65%).

5. References

[1] Adhvaryu J 2017 10.13140/RG.2.2.23122.56005
[2] Andriani Y, Kamil T I, dan Iskandar I 2018 *Aqua Fish Sci.* 7 (3): 25-33
[3] Andriani R and Hartini 2017 *JHS.* 1(2) 32-39.
[4] Dahril 2017 *The Terubuk Fish. Period.* 45: 21-29.
[5] Irianto A 2003 Gadjah Mada University Press
[6] Li J, Liang Y, Zhang X, Lu J, Zhang J, Ruan T, Zhou Q, Jiang G 2011 *Env Sci Tech.* 45: 9750- 9757
[7] Maghfiroh, Astuti S P, Kurdi M 2018 *Env Eng J.* 4(2) 18.
[8] Malinda 2019 University of Islamic Malang
[9] Oliveira G A R, Lapuente J, Teixid ó, E, Porredón C, Borrąs M, and Oliveira D P 2016 *EnvToxic Chem.* 35(2) 29-34.
[10] Sudha M 2014 *J Cur Micro. App. Sci.* 3(2) 670-690.
[11] Sumarmin R 2012 *Sci Tech.* 4 (2): 94-100.
[12] Verma Y 2011. *Toxic Ind Health.* 27(1) 41-49.

6. Acknowledgement

I would like to thank Mrs. Juni Triastuti for her support and encouragement. I also would like to thank Mrs. Lailatul Lutfiyah and Mrs. Nina Nurmalia Dewi for their support, comments, and suggestions. This research was conducted for the anatomy and aquaculture laboratory at Fisheries and Marine UNAIR and UPT. Fish health and environment laboratory, Pasuruan.

Effect of antibacterial activity and analysis of active compounds of *Hyrtios erecta* sponge ethanol extract using maceration method

R Wardany[1], Patmawati[2] and D Nirmala[2,3]

[1]Bachelor Programme in Aquaculture, Faculty of Fisheries and Marine, Airlangga University, Surabaya 60115, East Java, Indonesia.
[2]Department of Marine, Faculty of Fisheries and Marine, Universitas Airlangga, Surabaya 60115 Indonesia

[3] Corresponding author: dwitha.nirmala@fpk.unair.ac.id

Abstract. This type of sponge *Hyrtios erecta* has not been widely studied as an antibacterial agent. The method used to extract the *Hyrtios erecta* sponge is using maceration. The use of this maceration method aims to compare the use of extraction in producing yields. This study used tests such as water content test using the Loss on Drying method, yield test, phytochemical test using the tube method, and antibacterial activity test using the disc diffusion method (Kirby Bauer). The research design used was RAL. Antibacterial activity data using ANOVA. While the data from the other tests were analyzed by descriptive method. The results showed that the use of different extraction methods and extract concentrations resulted in significantly different diameters of the inhibition zones. The results showed that the best treatment in producing the diameter of the inhibition zone was the maceration method with a concentration of 30000 ppm/disk which was 7.08±1.81 for Escherichia coli and 2.83±1.21 for Staphylococcus aureus. *Hyrtios erecta* extract was more sensitive to gram negative than gram positive. The ethanol extract of the *Hyrtios erecta* sponge which has been tested in the form of phytochemicals shows the presence of bioactive compounds of alkaloids, flavonoids, and saponins.

1. Introduction

The growth of bacteria that cause infection needs to be inhibited, one of which is the use of antibacterials. Antibacterial is a substance that can inhibit the growth of bacteria and can kill pathogenic bacteria [1] Antibacterial from natural ingredients is now widely developed to treat diseases in humans and animals because it is environmentally friendly. In addition, natural ingredients also do not cause harmful side effects, are easy to find in the surrounding environment and do not require large costs to get them [2] One of the natural materials that are currently being researched are sponges. Sponges are one of the constituents of coral reef ecosystems from the phylum porifera which have a different structure compared to other metazoans [3,4] Sponges produce several classes of secondary metabolites, including alkaloids, steroids, flavonoids, phenols, saponins, and peptides [5]

The secondary metabolite content of sponges has potential as antibacterial, antiviral, anti-leukemic, anti-inflammatory, anti-cancer, insecticide, and anthelmintic [6].

Hyrtios erecta contains several active compounds such as spongistatin as an anticancer compound, 15-oxopupeheonol compound as an inhibitor of cancer and malaria cells, sesterpen 1 which inhibits leukemia cells and dipupehedion as an inhibitor of cancer cells. In addition, research conducted by Hernindya *et al* [7] showed that the *Hyrtios erecta* sponge contains alkaloids, steroids, and polyphenols. Until now there has been no research on the antibacterial activity of the *Hyrtios erecta* sponge from the aquaculture of Talengan Bay, Sangihe Islands, North Sulawesi.

Bioactive compounds from sponges can be obtained by extraction. Differences in extraction methods can affect the type and quantity of compounds extracted [8] Using two extraction methods, namely maceration and ultrasonic. Maceration is an extraction method that is often used because the process is easy, which is done by immersing the sample using an organic solvent at room temperature [9] However, maceration requires a longer time to extract the active substance in the sample [10]While the Ultrasonic-Assisted Extraction (UAE) method is an extraction method using ultrasonic waves which can damage cell wall permeability, cause spontaneous waves (cavitation) and increase cell damage. The advantage of the ultrasonic method is that the process does not require high costs and does not require a long time so that the energy expended is not too large [11]

The choice of solvent in the extraction process can also affect the content of the extracted compounds. This study uses ethanol as a solvent because it has the advantage of being cheap, easy to obtain, and being able to mix with water at various concentrations which makes it easier to adjust the polarity of the solvent to optimize the extraction of secondary metabolites [12] In addition, ethanol is able to dissolve almost all substances, both polar, semipolar, and nonpolar [13,14]

Antibacterial activity of the ethanolic extract of the *Hyrtios erecta* sponge against Escherichia coli and Staphylococcus aureus bacteria, what groups of compounds contained in the ethanolic extract of the *Hyrtios erecta* sponge have the potential as antibacterial, and whether the different extraction methods and extract concentrations affect the antibacterial activity. The aims of this study were to determine the ethanolic extract of the *Hyrtios erecta* sponge against the antibacterial activity of Escherichia coli and Staphylococcus aureus, to analyze the active compounds contained in the ethanolic extract of the *Hyrtios erecta* sponge, to determine the effect of different methods and concentrations of sponge ethanol extract in yielding yields and the bioactivity of compounds in inhibiting Escherichia coli and Staphylococcus aureus bacteria.

2. Material and methods

2.1. Sample collection

Bahan yang digunakan adalah spons *Hyrtios erecta* dari Perairan Talengan, Sangihe, Sulawesi Utara, etanol 96%, chloramphenicol C30 Oxoid, aquades, paper dish blank, es batu, alkohol, media Nutrient Broth (NB), media Mueller Hinton Agar (MHA), NaCl, serbuk magnesium, asam klorida (HCl), reagen mayer, reagen wagner, reagen dragondorff, kloroform, dan asam asetat.

2.2 Antibacterial Activity Test

The method used in this study is the agar diffusion method. In this antimicrobial activity test, the paper disk used is 6 mm in size with an absorption capacity of 50 L per disc.

2.3 Bioactive Component Test (Phytochemical Screening)

Phytochemical test is a qualitative test conducted to determine the presence or absence of bioactive components in *Hyrtios erecta* sponge extract. Phytochemical tests in this study included tests for alkaloids, flavonoids, saponins, and steroids.

2.4 Water Content Measurement

The moisture content of the *Hyrtios erecta* sponge was measured using the Loss on Drying (LOD) method by weighing a sample of 0.75 grams, then measuring the moisture content using a moisture analyzer.

3. Result and discussion

3.1 Hyrtios erecta Sponge Moisture Content

The measurement of water content was carried out in a dry sample state and in powder form. The moisture content of the sponge in this study was measured using a moisture analyzer with the Loss on Drying method. The results show that the moisture content of the *Hyrtios erecta* sponge is 10.10%.

The sponge sample used in this study was a cultivated dry *Hyrtios erecta* sponge from Talengan Bay, Sangihe Islands, North Sulawesi. The *Hyrtios erecta* sponge before being extracted was cut into small pieces and then pounded with a pestle mortar until it became powder. This aims to increase the surface area of the sponge sample so that during the extraction process the solvent can be absorbed maximally into the sample and the extraction results obtained will be optimal. In this study, *Hyrtios erecta* sponge powder was extracted using two methods, namely maceration and Ultrasonic-Assisted Extraction (UAE).

3.2 Hyrtios erecta Sponge Extract Yield

In the early stages of the study, the *Hyrtios erecta* sponge sample was extracted using the maceration method and Ultrasonic-assisted Extraction (UAE) using 96% ethanol as solvent. The extracts from both methods were then dried in an oven at 40oC and weighed for each yield calculation. The results of the dry extract obtained were then tested for antibacterial activity against pathogenic bacteria, namely Escherichia coli and Staphylococcus aureus.

Yield is the total weight of secondary metabolites extracted from a sample. Yield results are needed in the extraction process because it is to determine the amount of extract obtained during the extraction process from a sample. According to Sayuti [15] the yield has a relationship with the amount of active compound content of a sample so that the more yields produced, the more active compound content. The yield of the highest percentage of *Hyrtios erecta* sponge ethanol extract was obtained by the ethanol extract using the ultrasonic method, which was 4.5% while the yield from the maceration method was only 3.4% as shown in Table 2. The choice of using the extraction method can affect the yield. According to [16] the use of ultrasonic methods in the extraction process can increase the amount of crude yield. Research conducted by [17] also showed that the yield of mangosteen peel extract extracted using ultrasonic produced higher yields than the yield extracted using maceration.

Tabel 1. Yield Extract on Spons *Hyrtios erecta*

Extraction Method	Initial Weight (g)	Dried Extract (g)	Yield (%)
Maceration	2	0,068	3,4 %
Ultrasonic-assisted Extraction (UAE)	2	0,09	4,5%

3.3 Antibacterial Activity of Sponge Ethanol Extract

The antibacterial activity test of the ethanolic extract of the *Hyrtios erecta* sponge from maceration and ultrasonic methods was carried out using the disc diffusion method. The results of the activity test of *Hyrtios erecta* sponge ethanol extract against the growth of Escherichia coli and Staphylococcus aureus bacteria after an incubation period of 1x24 hours based on the extraction results and different concentrations are as follows:

IOP Conf. Series: Earth and Environmental Science 1273 (2023) 012011 doi:10.1088/1755-1315/1273/1/012011

Table 2. Results of Antibacterial Activity Test Against Escherichia coli Based on Different Extractions and Concentrations

Extraction Results (method)	Concentration (µg/disk)	Test 1	Test 2	Test 3	Average Inhibitory Zone Diameter (mm)
Maceration (A1)	75 (B1)	1,8	1,05	0,3	1,50^{c}±0,75
	150 (B2)	2,3	1,85	1,05	1,73^{c}±0,63
	300 (B3)	8,65	5,1	7,5	7,08^{ab}±1,81
Chloramphenicol	30	21,45	20,9	18,1	20,15±1,8
Ultrasonic (A2)	75 (B1)	3	1	1,1	1,7^{c}±1,13
	150 (B2)	5,7	6,15	6,3	6,05^{b}±0,31
	300 (B3)	10,25	7,8	9,6	9,22^{a}±1,27
Chloramphenicol	30	20,75	20,4	19	20,05±0,93

Explanation: a, b, c : Numbers followed by the same letter show that the results are not significantly different ($p < 0.05$)
A1, A2 : Extraction Method
B1, B2, B3 : Extract concentration of both methods

Table 3. Results of Antibacterial Activity Test Against Staphylococcus aureus Based on Different Extractions and Concentrations

Extraction Results (method)	Concentration (µg/disk)	Test 1	Test 2	Test 3	Average Inhibitory Zone Diameter (mm)
Maceration (A1)	75 (B1)	1,1	0,5	1,1	0,9^{c}±0,35
	150 (B2)	1,2	0,7	1,5	1,13^{bc}±0,40
	300 (B3)	1,9	2,4	4,2	2,83±^{b}1,21
Chloramphenicol	30	18.05	18,3	19,05	18,47±0,52
Ultrasonic (A2)	75 (B1)	1,05	1,4	1	1,15^{bc}±0,22
	150 (B2)	2	1,7	1,2	1,63^{bc}±0,40
	300 (B3)	6,55	8,4	7,5	7,48^{a}±0,93
Chloramphenicol	30	18,7	19,65	18	18,78±0,83

Explanation: a, b, c : Numbers followed by the same letter show that the results are not significantly different ($p < 0.05$)
A1, A2 : Extraction Method
B1, B2, B3 : Extract concentration of both methods

3.4 Bioactive Compound Test (Phytochemical Screening)

Phytochemical testing was carried out to identify the content of active compounds contained in an extract of simplicia. In this study, testing of bioactive compounds included tests for flavonoids, alkaloids, saponins

Table 4. Phytochemical Test Results of *Hyrtios erecta* Sponge Ethanol Extract

Phytochemical Test	Maceration Extract	Ultrasonic Extract	Information
Alkaloid			
-Mayer	++	++	Maceration: there is a change in color to yellowish white
-Wagner	+	++	Ultrasonic: color changes to white (cloudy)
-Dragendorff	+	+	Maceration: color changes to brown

IOP Conf. Series: Earth and Environmental Science 1273 (2023) 012011 doi:10.1088/1755-1315/1273/1/012011

Flavonoid	+	+	Maceration: the color changes to orange Ultrasonic: color changes to reddish black
Steroid	_	_	Maceration: no color change to blue or green Ultrasonic: no color change to blue or green
Saponin	+	+	Maceration: stable foam is formed Ultrasonic: stable foam is formed

Information:
(+) : There is a color change
(++): There is a significant color change
(-) : No color change

The results of the phytochemical screening in Table 5 show that the *Hyrtios erecta* sponge was positive for alkaloids extracted from both maceration and ultrasonic methods. Positive results for alkaloids were indicated by a change in color to white in the Mayer reagent. It is thought that this change is due to the reaction between the alkaloids and the metal ion K+ of potassium tetraiodomercurate (II). In Dragendorff's reagent, the color changes to yellow. These changes are thought to occur due to the reaction of alkaloids with potassium tetraiodobismutate (III) [18] While thealkaloid test with Wagner reagent showed a color change to brown. This change is due to the K+ metalion forming a coordinate covalent bond with nitrogen in the alkaloid to form a precipitated potassium- alkaloid complex [19] When compared, there are differences in color intensity resulting from ultrasonic and maceration methods. The color formed is more dominant in the ultrasonic method extract. This is most likely due to differences in the working principle of the extraction process of the two methods so that the alkaloid compounds in the ultrasonic method are more perfectly extractedthan maceration. Alkaloids have basic groups which when in contact with bacteria will react with amino acid compounds that make up the bacterial cell wall and bacterial DNA which is the main constituent of the cell nucleus. Changes in the composition of amino acids will change the genetic balance of DNA acids so that bacterial DNA is damaged and encourages lysis in the cell nucleus. This results in bacterial cells not being able to metabolize [20]

The flavonoid test showed positive results from both methods indicated by a color change to orange or red. The purpose of adding Mg and HCl is to reduce the benzopyron core contained in the flavonoid structure so that red or orange flavilium salts are formed [20]. Flavonoid compounds have the ability to inhibit bacterial growth by causing damage to the permeability of the bacterial wall. The hydroxyl groups found in flavonoids are able to change organic components and transport nutrients which ultimately lead to toxic effects on bacteria [21]. While the steroid test showed negative results because the sample did not change color to green or blue or it was possible that the *Hyrtios erecta* sponge contained small amounts of steroid compounds so that it was not detected if qualitative testing was carried out.

The saponin test showed a positive result which was indicated by the formation of foam. The foam is a glycoside which is hydrolyzed into glucose and other compounds and forms foam [21] The mechanism of inhibition of saponin compounds as antibacterial is by lowering the surface tension of the cell wall. Saponins will bind to lipopolysaccharides on bacterial cell walls which can cause increased cell wall permeability and decrease cell wall surface tension so that cell walls will break or undergo lysis and make antibacterial substances enter cells easily and interfere with metabolism and ultimately cell death [22].

4. Conclusion

Hyrtios erecta sponge from Talengan Waters, Sangihe Islands, North Sulawesi contains bioactive components including alkaloids, flavonoids, and saponins. The antibacterial activity produced by the ethanolic extract of the sponge *Hyrtios erecta* has a greater inhibition zone against Escherichia coli bacteria than Staphylococcus aureus. *Hyrtios erecta* sponge was more sensitive in inhibiting gram-negative bacteria Escherichia coli than gram-positive bacteria Staphylococcus aureus. Different extraction methods used affect the yield of bioactive components, yield, and antibacterial activity. The extraction result from the Ultrasonic-Assisted Extraction method with a concentration of 300 g/disk was the best treatment in inhibiting the growth of Escherichia coli and Staphylococcus aureus and the diameter of the inhibition zone increased with high concentration.

5. References

[1] Paju N, Yamlean P V, Kojong N 2013 *Pharmac* 2 (1) 51–61.
[2] Karlina C Y Y 2013 *Berk Ilmi Biol* 2 (1) 87-93.
[3] Katili S S, Wewengkang D S dan Rotinsulu H 2020 *Pharma* 9 (1) 100-107.
[4] Hardiningtyas S D 2009 Antibacterial activity of fragmented and unfragmented soft coral extracts of Sarcophyton sp in the waters of Pramuka Island, Thousand Islands. Department of Aquatic Product Technology, Faculty of Fisheries and Marine Sciences, Bogor AgriculturalUniversity.
[5] Nugraha W A, Mubarak F, Husaini E dan Evendi H 2020 *J Ilmi Perik dan Kela* 12 (1) 131-139.
[6] Kim S K 2012 Marine Pharmacognosy: Trends and Applications. New York: CRC Press.
[7] Hernindya A, Swantara M D dan Suaniti N M 2014 *Indo E-Jour of App Chem.* 2 (1) 25-30.
[8] Dewi I P, Maisaroh S dan Verawaty V 2020 *Jurn Farma Hige* 12 (1) 48-54.
[9] Handayani P A dan Nurcahyanti H 2015 *Jurn Bah Al Terba* 3 (1) 1-7.
[10] Susanty S dan Bachmid F 2016 *Jurn Konv* 5 (2) 87-92.
[11] Verdiana M, Widarta I W R dan Permana I D G M 2018 *J Ilm dan Tekno Pangan.* 7 (4) 213-222.
[12] Sari D I dan Triyasmono L 2017 *J Pharm* 4 (1).
[13] Sulastri E, Cristadeolia O, Yusriadi 2015 *J Pharma* 2 (2) 2.
[14] Tjahyaningsih W, Alamsjah M A dan Abdillah A A 2013 *J Ilmi Perik dan Kel.* 5 (2) 123-128.
[15] Sayuti M 2017 *Techno Sci and Engi Jour* 1 (3).
[16] Handayani H, Sriherfyna F H dan Yunianta Y 2016 *J Pangan dan Agro.* 4 (1).
[17] Sholihah M A, Ahmad U dan Budiastra I W 2017 *J ketekn perta* 5 (2).
[18] Ergina E, Nuryanti S dan Pursitasari I D 2014 *J Akadem Kim.* 3 (3) 165-172.
[19] Ikalinus R, Widyastuti S K dan Setiasih N L E 2015 *Indo Medi Veter.* 4 (1) 71-79.
[20] Siregar A F, Sabdono A dan Pringgenies D 2012 *J of mar rese.* 1 (2) 152-160.
[21] Purwati S, Lumowa S V dan Samsurianto S 2017 *In Prosiding Seminar Kimia.* 153- 158.
[22] Dwicahyani T, Sumardianto S dan Rianingsih L 2018 *J Pengol dan Biotek Has Perik.* 7 (1) 15-24.

6. Acknowledgement

Recognition to the Head of Laboratory of the Faculty of Fisheries and Marine Universitas Airlangga, Surabaya for giving permission for the research site and Head of Chemical and Analysis Laboratory and Microbiology Laboratory for research samples.

IOP Conf. Series: Earth and Environmental Science 1273 (2023) 012012 doi:10.1088/1755-1315/1273/1/012012

Stress response and oxygen consumption level of vaname shrimp (*Litopenaeus vannamei*) infested by Ciliata pathogens on culture system with high stocking densities

M Marista[1], G Mahasri[2,3], Kusnoto[2]

[1]Aquaculture,Faculty of Fisheries and Marine, Universitas Airlangga, Surabaya, Indonesia
[2]Departement of Aquaculture, Faculty of Fisheries and Marine, Universitas Airlangga, Surabaya, Indonesia
[3]Departement of Parasitology Veteriner, Faculty of Veterinary Medicine, Universitas Airlangga, Surabaya, Indonesia

[3] Corresponding: mahasritot@gmail.com

Abstract. One of the technology used to increase the production of vaname shrimp is by culture system with high stocking densities. Disease cases that are often reported to attack vaname shrimp in rearing with high stocking densities are derived from ectoparasites Ciliata pathogens such as *Vorticella* sp., *Zoothamnium* sp. and *Epistylis* sp. This study aimed to determine the effect of high stocking density, rearing time and the interaction between high stocking density and rearing time on the stress response (blood glucose level) and oxygen consumption level of vaname shrimp (*Litopenaeus vannamei*) infested by Ciliata pathogens. This study used an experimental method using a Completely Randomized Design (CRD) witha 3x5 factorial pattern consisting of 3 treatments and 6 replications. Factor A is high stocking density (10 prawns/10L, 15 prawns/10L and 20 prawns/10L) and factor B is rearing time (day- 0, 2, 4, 6 and 8). The results of this study showed that high stocking density and rearing time had a significant effect ($p<0.05$) and there was a significant interaction ($p<0,01$) with a high stocking density of 20 prawns/10L and rearing time on day-8 of increasing blood glucose level and oxygen consumption level of vaname shrimp infested by Ciliata pathogens.

1. Introduction

Vaname shrimp culture with high stocking density can cause a decrease in water quality as a medium for shrimp maintenance so that shrimp are susceptible to disease, one of which is caused by ectoparasites. The ectoparasites that are often reported to attack vaname shrimp are from the genus *Zoothamnium* sp., *Epistylis* sp. and *Vorticell*a sp. [1]. Maintenance of high stocking density causes competition in the use of space which causes direct contact between shrimp bodies so that they can transmit ectoparasites easily and quickly. Increased intensity of pathogenic ciliate attacks can have an impact on the physiological response of vaname shrimp in terms of stress response [2].

Stress can be interpreted as a form of response to living things that are unable to regulate normal physiological conditions due to disturbances or pressures both from within the body and influences from outside the body [3]. Indicators of shrimp experiencing stress can be characterized by an increase in blood glucose levels or *hyperglycemia*. Blood glucose levels increase due to vaname shrimp experiencing homeostatic instability in the shrimp body utilizing energy from the glycolysis process (glycogen is converted into glucose) which is regulated by *Crustacean hyperglycemic Hormone* (CHH) in the shrimp hepatopancreas. Oxygen consumption level is the amount of oxygen needed by an organism. An increase in the level of oxygen consumption due to the breakdown of glucose into *Adenosine Triphosphate* (ATP) in aerobic respiration always requires oxygen (O_2) as a gas component and a vital element, so that the body's metabolism increases which causes of oxygen consumption level to be high [4].

2. Material and methods

The study was conducted in February - March 2022. The research method used is an experimental method using a 3x5 factorial completely randomized design (CRD) consisting of factor A and factor B. Factor A is the number of high stocking density 10 prawns/10L, 15 prawns/ 10L and 20 prawns/10L while factor B was the time of rearing on days 0, 2, 4, 6, and 8. Samples of vaname shrimp infested by Ciliata Pathogens with age PL 30-40 were taken from the Brackish Water Cultivation Installation (IBAP) Lamongan, East Java as many as 500 shrimp. Then for the initial examination of ectoparasites, 50 vaname shrimp were taken.

Examination of ectoparasites Ciliata Pathogens on 50 vaname shrimp at the Laboratory of Anatomy and Culture, Faculty of Fisheries and Marine, Airlangga University, Surabaya. Examination of Ciliata Pathogens on the body of vaname shrimp using a light microscope with a magnification of 100-400x. Measurements of water quality, blood glucose levels and oxygen consumption levels were carried out behind Building A of the Faculty of Fisheries and Marine, Airlangga University, Surabaya. Blood glucose levels were measured using the Blood Glucose Monitoring System while the level of oxygen consumption according to [5], can be determined by the following formula:

$$oc = (DOo - DOt)\ x \frac{V}{t\ x\ w}$$

OC = oxygen consumption expressed in milligrams of oxygen per hour per gram of shrimp (mg O_2/g/hour)
DO_0 = initial dissolved oxygen (mg O_2/L)
DO_t = dissolved oxygen after t hours (mg O_2/L)
V = volume of water (L)
t = time elapsed for the shrimp consuming oxygen (hours)
W = weight of test shrimp (g)

The measurement data of blood glucose levels and the level of oxygen consumption of vaname shrimp infested with Ciliata pathogens were analyzed using IBM SPSS (Statistical Package for the Social Sciences) Statistics Version 24. The statistical test used was Factorial ANOVA (Univariate Analysis of Variance) to determine the effect of different treatments. If there is a significant difference, it will be continued with Duncan's Multiple Range Test with a level of 1% to find out the most influential treatment.

3. Results and discussion

3.1 Ectoparasite examination results on vaname shrimp

Early examination of ectoparasites in samples of 50 vaname shrimp infested by Ciliata pathogens reared at IBAP Lamongan, East Java found ectoparasites from the genera *Vorticella* sp., *Epistylis* sp. and *Zoothamnium* sp. The three genera of ectoparasites infest many parts of the shrimp body such as

IOP Conf. Series: Earth and Environmental Science 1273 (2023) 012012 doi:10.1088/1755-1315/1273/1/012012

walking legs, swimming legs, body surface, tail and carapace. Ectoparasite *Vorticella* sp., *Epistylis* sp. and *Zoothamnium* sp. can be seen in Figure 1.

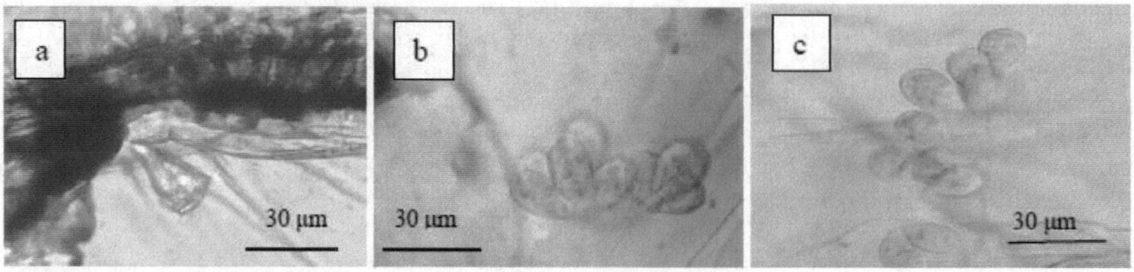

Figure 1. Ectoparasite Ciliata pathogens (a) *Vorticella* sp. at the foot of the road magnification 400x (b) *Epistylis* sp. on the cephalothorax magnification 400x and (c) *Zoothamnnium* sp. on the body surface of vaname shrimp in magnification 400x.

3.2 Blood Glucose Level Measurement Results
Vaname shrimp infested with Ciliata pathogens were reared at high stocking densities for 8 days with measurements every 2 days. The results of measuring blood glucose levels can be seen in Table 1.

Table 1. Average blood glucose levels of vaname shrimp during treatment.

High Stock density/ 10L	Blood Glucose Level (mg/dL) ± SD				
	Day-0 (B1)	Day-2 (B2)	Day-4 (B3)	Day-6 (B4)	Day-8 (B5)
(A1)	$24,67^{h}\pm2,81$	$26,83^{gh}\pm2,23$	$26,83^{gh}\pm2,23$	$32,5^{e}\pm1,76$	$33^{e}\pm1,27$
(A2)	$24,67^{h}\pm2,94$	$28^{fgh}\pm3,09$	$30,67^{efg}\pm2,50$	$37^{d}\pm4,19$	$41,67^{c}\pm3,2$
(A3)	$25,83^{h}\pm3,81$	$31,67^{ef}\pm2,6$	$32,67^{e}\pm2,50$	$47,83^{b}\pm6,46$	$52,5^{a}\pm5,96$

Analysis using ANOVA statistics showed that high stocking densities and rearing time had a significant effect ($p<0.05$) on blood glucose levels of vaname shrimp infested with Ciliata pathogens. The results of statistical tests showed that there was a significant interaction effect ($p<0.01$) between high stocking density and rearing time on blood glucose levels of vaname shrimp infested with Ciliata pathogens. It can be seen from the data in Table 1 that the average blood glucose levels of vaname shrimp infested with Ciliata pathogens continue to increase with increasing rearing time.

3.3 Oxygen consumption level measurement results
Vaname shrimp infested by Ciliata pathogens were then taken for measurement of oxygen consumption level using a DO meter and calculated according to the formula [5]. The results of measuring the level of oxygen consumption can be seen in Table 2.

Table 2. Average Oxygen Consumption Level of vaname shrimp during treatment.

High Stock Density/ 10L	Oxygen Consumption Level ($mgO_2/g/hour$)±SD				
	Day-0 (B1)	Day-2 (B2)	Day-4 (B3)	Day-6 (B4)	Day-8 (B5)
(A1)	$0,048^{j}$±0,014	$0,06^{ij}$±0,011	$0,08^{ghi}$±0,014	$0,098^{fg}$±0,017	$0,131^{e}$±0,014
(A2)	$0,068^{hi}$±0,011	$0,08^{gh}$±0,007	$0,14^{de}$±0,017	$0,161^{c}$±0,019	$0,171^{c}$±0,019
(A3)	$0,076^{hi}$±0,010	$0,111^{f}$±0,011	$0,15^{cd}$±0,018	$0,195^{b}$±0,018	$0,236^{a}$±0,013

Analysis using ANOVA statistics and further tested using DMRT (*Duncan's Multiple Range Test*) showed that high stocking densities and rearing time had a significant effect ($p<0.05$) on oxygen consumption level of vaname shrimp infested with Ciliata pathogens. The results of statistical tests showed that there was a significant interaction effect ($p<0.01$) between high stocking density and rearing time on oxygen consumption level of vaname shrimp infested with Ciliata pathogens. It can be seen from the data in Table 2 that the average oxygen consumption level of vaname shrimp infested with Ciliata pathogens continue to increase with increasing rearing time.

3.4 Water quality

Water quality measurements were carried out during rearing using a thermometer, pH paper, Dissolved Oxygen (DO) and ammonia, nitrat and nitrit test kits. The results of water quality measurements in ponds could be seen in Table 3.

Table 3. Water quality measurement results.

Water Quality		Values of Water Quality Parameters in Treatment			Optimum Values (SNI,2014)
		A1	A2	A3	
Temperature(°C)	Morning	26-27	26-27	26-27	28-33°C
	Afternoon	26-28	26-28	26-28	
pH		8	8	8	7,5-8,5
Salinity (ppt)		12	12	12	4-32 ppt
Dissolved Oxygen (mg/l)		8,3 – 9,6	8,5-10	8,5 – 10	>4,0 mg/l
Amonia (mg/l)		0,03 – 0,27	0,03 – 0,27	0,03 – 0,53	<0,1 mg/l
Nitrit (mg/l)		0,5	0,5-1	0,5-1	<0,1 mg/l
Nitrat (mg/l)		10	10-25	10 -25	< 1 mg/l

Based on the results of the initial examination of ectoparasites Ciliata Pathogens, all of them were infested by ectoparasites with zooids found in each sample of 5-25 zooids. According to [6] infestations degree of zooids 5-25 included in the light category, 26-50 zooids in the medium category and more than 50 zooids in the heavy category. The ectoparasites found to attack vaname shrimp (*L. vannamei*), namely *Vorticella* sp. with a body shape like an inverted bell equipped with cilia and peduncle and live solitary, *Epistylis* sp. with an elongated body shape equipped with cilia and peduncle and living in colonies with a number of 2-5 zooids per colony, *Zoothamnium* sp. with a globular body shape equipped with cilia and peduncle and live in colonies with the number of 2-10 zooids per colony.

Vaname shrimp (*L. vannamei*) infested by Ciliata pathogens on culture system with high stocking density had a significant effect ($p<0.05$) on blood glucose levels. Blood glucose level increased (*hyperglycemia*) is due to the influence of high stocking density and longer rearing times which can cause an increase in feed residues and feces so that organic matter in the water increases, so that vaname shrimp need a lot of energy to adapt to their environment [7]. High stocking density also causes competition in the use of space which causes direct contact between individuals so thatectoparasite infestations can increase easily and quickly. Some of these factors can cause vaname shrimp increase in the degree of ectoparasite infestation easily and quickly [8].

Vaname shrimp under stress causes hepatopancreas to secrete *Crustacean Hyperglycemic Hormone* (CHH) which functions in increasing blood glucose levels in haemolymph and lactate concentration in mobilizing intracellular glycogen reserves. CHH that comes out of the hepatopancreas then encourages the process of glycogenolysis and glycolysis to produce energy in the form of ATP. When the production of glucose by CHH exceeds the needs of normal shrimp, the glucose will be released byhepatopancreas cells and return to the blood (*haemolymph*) causing an increase in blood glucose levels(hyperglycemia) [9]. The lowest blood glucose level was in treatment A1 (stocking density 10 prawns/10L) with rearing time of day-0 (24.67 mg/dL), This is due to the absence of high stocking density and the absence of competition in the use of space and food so that blood glucose levels are still in the normal range.

Oxygen consumption level of vaname shrimp infested by Ciliata pathogens had increases along with the increase of high stocking densities and rearing time. Significant increase of oxygen consumption level occurs in stock density 20 prawns/10L (A3). So that, vaname shrimp (*L. vannamei*) infested by Ciliata pathogens on culture system with high stocking density had a significant effect ($p<0.05$) on oxygen consumption level.There is an increase in oxygen consumption level because when the shrimp undergoes a drastic change in homeostasis in their body, it causes the immune response to decrease, so vaname shrimp need oxygen which plays a role it is important in the breakdown of glucose into energy in the glycolysis process which is later converted in the form of Adenosine Triphosphate (ATP) in aerobic respiration [4].

According to [10] any are several factors that affect of oxygen consumption level, including internal factors such as length, body weight, age, and body activity as well as external factors such as the influence of feed and poor aquatic environment. Meanwhile, according to [11] an increase in organic matter due to an increase in shrimp biomass in maintenance with high stocking densities can affect the increase in shrimp metabolism so that the oxygen demand required is higher for therespiration process.

Based on the results of water quality measurements, there are several water quality parameters that exceed the optimum limit for shrimp life, namely ammonia (NH_3), nitrite (NO_2) and nitrate (NO_3). Ammonia content during treatment increased by 0.5 mg/l, while according [12] the optimal ammonia content in shrimp culture was <0.1 mg/l. Conditions like this can cause the waters to become toxic and support stress on white vaname shrimp.

The value of nitrite (NO_2) during treatment was around 0.5-1 mg/l, while according to [12] the optimal nitrite content in shrimp rearing was <0.1 mg/l. The high content of nitrite in a waters affects the stress response of vaname shrimp. This is because the blood cells in shrimp (haemocyanins) which function in binding oxygen are inhibited by nitrite compounds due to competition between nitrite ions and Chloride ions (Cl) for the oxygen transport system [13]. The values of Nitrate (NO_3) during treatment range from 10-25 mg/l, while according to [12] the optimal nitrate content in shrimp rearing is <1 mg/l. An increase in nitrate (NO_2) as much as 0.3 mg/l in a waters can cause the process of eutrophication or the emergence of excessive nutrients into the aquatic ecosystem resulting in an explosion of plankton population (blooming plankton) so that it can inhibit the process of vaname shrimp respiration so that oxygen consumption level also increased [13].

Based on the research of [14] and [15] rearing vaname shrimp infested with pathogenic ciliates with high stocking density on the 8th day, there was an increase of *Zoothamnium, Epistylis* and *Vorticella* with a moderate level of infestation degree with 25-50 zooid per individual. This increase inectoparasites Ciliata pathogens occurred due to the influence of stocking density and rearing time. If the vaname shrimp infested with Ciliata pathogens is maintained with high stocking density and longertime, then the pathogenic ciliate will develop continuously and form new colonies to affect the physiological condition of white vaname shrimp.

Treatments of vaname shrimp infested by Ciliata pathogens with high stocking density and different rearing times has a very significant interaction effect ($p<0.01$) on the stress response (blood glucose levels) and the level of oxygen consumption. The longer of rearing time and the higher of stocking densities, the lower the carrying capacity of the shrimp's living environment and makes it easier for ectoparasites to grow and develop rapidly and increased of infestation degree that can cause physiological disturbances (stress) of vaname shrimp.

4. Conclusion

High stocking density and rearing time had a significant effect ($p<0.05$) on stress response (blood glucose level) and oxygen consumption level of vaname shrimp (*Litopenaeus vannamei*) infested by Ciliata pathogens. There was a significant interaction effect ($p<0.01$) between high stocking density and rearing time on the stress response (blood glucose level) and oxygen consumption level of vaname shrimp infested by Ciliata pathogens at a stocking density of 20 prawns/10L and rearing time on the 8th day with the value of blood glucose level is 52.5 mg/dL and an oxygen consumption level of 0.236 mgO2/g/hour.

5. References

[1] Inkasari D 2019 Skripsi. Medan: Universitas Sumatera Utara. 1- 47 hal.
[2] Widodo A F, Pantjara B, Adhiyudanto N B dan Rchmansyah 2011 *JRA*. 6(2) 225-241.
[3] Donsu J D T 2017 *Nursing Psychology (Yogyakarta: Pustaka Baru Press) 66-77 pp.*
[4] Faturrohman K 2017 Tesis. Bogor: Institut Pertanian Bogor. 1-80 hal.
[5] Kawamoto N 1977 *Fish Physiology*. (Tokyo: Koseisha-Koseikaku In Japanese) 355 hal.
[6] Fegan D, Nieles F, Flagel T, Rossuwan S, and Waiyakaruitata M 1993 Asian Fisheries Society Conferences. 23 p.
[7] Supono 2018 *Water Quality Management for Shrimp Farming. CV. Main Grace Award. Bandar Lampung*. 132 hal.
[8] Eren A., Koesdarto S, dan Mahasri G 2020 *JSV*. 38(2) 1-10.
[9] Qyli M., Aliko V, Sgarra R, Giulianini P G, and Manfrin C 2021 *Turk J Zool*. 45 25-32.
[10] Domenici P, Herbert N A, Lefrancois C, Steffensen J F and McKenzie D J 2013 *Springer*. 129 159.
[11] Luong T C, Lemonnier H, Hochard S, Royer F, and Letourneur Y 2016. *JAR*. 10 (1) 1-12.
[12] SNI 8008 2014 *Intensive Vaname Shrimp (Litopenaeus vannamei Boone, 1931) Production in Lining Ponds. National Standardization Agency. 11 p.*
[13] Izzati M 2011 *BIOMA*. 13(2) 80-84.
[14] Trinendyah F V 2022 Skripsi. Surabaya: Universitas Airlangga. 1- 85 hal.
[15] Putri J K K N 2022 Skripsi. Surabaya: Universitas Airlangga. 1- 84 hal.

6. Acknowledgement

Authors wishing to acknowledge assistance or encouragement from colleagues, special work by technical staff or financial support from organizations should do so in an unnumbered Acknowledgments section immediately following the last numbered section of the paper.

IOP Conf. Series: Earth and Environmental Science 1273 (2023) 012013 doi:10.1088/1755-1315/1273/1/012013

Exposure metanil yellow as textile dye

J Triastuti [1,3], AP Rahayu [2]

[1] Department of marine, Faculty of Fisheries and marine, University Airlangga Campus C Mulyorejo , Surabaya 60115, Indonesia
[2] Fisheries Science Program, Faculty of Fisheries and marine, University Airlangga Campus C Mulyorejo , Surabaya 60115, Indonesia

[3]Corresponding authors: juni.triastuti@fpk.unair.ac.id

Abstract. The textile industry is one of the fastest growing industries today. The more the textile industry, the more dye waste is generated. Waste that is not managed properly and is wasted in the waters can cause pollution. Fish that are commonly found in freshwater are the Cichlid group, one of which is tilapia (*Oreochromis niloticus*) which can be exposed to textile dye waste. Exposure to chemical-based textile dye waste can affect fish physiology, including the formation of fish bones when fish are growing. This can be seen from the results of research on the description of malformation of tilapia exposed to metanil yellow textile dye. This research is an experimental study on juveniles conducted in the laboratory for one month. The results showed that there were malformation in the cephal, truncus and caudal tilapia.

1. Introduction

Textile industry is one of the fastest growing industries today. This industry contributes to aquatic waste due to the use of textile dyes [1]. The textile dye will be a hazard if it is not treated properly because it contains chromophore and auxochrome groups. The effect given to aquatic biota is reduced oxygen and is able to cause malformations in the body of aquatic biota, especially fish [2].

Tilapia is one of superior consumption fish in Indonesia that lives in rivers. However, its quality and growth need to be considered [3]. The growth of tilapia that must be considered starts from the embryonic stage, larvae to the early juvenile stages [4].

The stages of embryogenesis and larval development are a developmental phase in the life process of fish. This is a very critical phase that must be understood in cultivation activities. Embryogenesis or embryonic development, namely the whole process that includes the development process after fertilization (fertilization) to organogenesis before the fish hatch. While the larval stage in fish is post hatching after the egg stage whose organs are not yet complete such as the size of the child and its mother and play an important role in recruitment. Morphologically, the development of fish larvae stadia is divided into two phases, namely pro-larvae and post-larvae [5]. However, in this study, what will be considered is the development of larvae.

Based on the exposure of the dangers of textile dye waste and the importance of considering to the development of tilapia larvae, a study of textile dye exposure on tilapia larvae was conducted to observe the malformed organs. The textile dye used in this study is metanil yellow.

2. Material and methods

2.1 Selection Parent

Parent selection is the initial stage in each study to determine which fish are good for use in the study. This stage will determine the success of the hatchery so it must be carried out carefully and accurately based on predetermined criteria. Parent selection serves to obtain quality fish. Tilapia broodstock used in this study were obtained from UPT. Fish and Environmental Health, Umbulan, Pasuruan.

2.2 Acclimatization

Acclimatization is an effort of physiological adjustment or adaptation of fish to a new environment that they will enter. Acclimatization of fish is done so that fish do not experience stress. The acclimatization process was conducted for 3-7 days. The acclimatization process was carried out in tilapia rearing ponds in the new plot.

2.3 Spawning

Spawning is a mating process between male and female fish that releases egg cells from the female, sperm cells from the male and occurs outside the body of the fish (external). Spawning uses a male: female parent ratio = 1: 3, namely 3 males paired with 9 females.

2.4 Embryo Treatment

The embryonic development phase was observed under a microscope by taking the embryo using a pipette and then placing it on a concave glass slide. Time and pictures of each stage of embryonic development are recorded and documented, especially in certain phases such as cell division, morula, blastula, gastrula, organogenesis and hatching. Observation of embryos was conducted out for 96 hours.

2.5 Larvae Treatment

The observation stages of tilapia larvae were conducted for 7 days in the aquarium with 3 L of water. Observations were made in order to observe the eye pigmentation, body pigmentation, fin development, mouth opening and egg yolk.

2.6 Measurement Water Quality

Water measurements were carried out to maintain water quality. Water quality measurements are carried out once a day. The indicators measured are temperature, pH, and DO. Water quality measurements were conducted continuously carried out from the beginning of maintenance to the end.

3. Result and Discussion

3.1. Organ Development

Table 1. Development of Tilapia Larvae Organs

Organ Development	*Time of Development* (dpf)					Fujimura and Okada (2007)
	A (0 ppm)	B (125 ppm)	C (156 ppm)	D (196 ppm)	E (245 ppm)	
Eyespot	1-2	1-2	1-2	1-2	1-2	1-2
Body pigmentation	2-3	2-3	2-3	2-3	2-3	2-3

An al fin	3-4	3-4	3-4	3-4	3-4	3-4
Caudal fin	4-5	4-5	4-5	4-5	4-5	4-5
Pectoral fin	5-6	5-6	5-6	5-6	5-6	5-6
Dorsal fin	5-6	5-6	5-6	5-6	5-6	5-6
Ventral fin	5-6	5-6	5-6	5-6	5-6	5-6
Swim bubbles	6-7	6-7	6-7	6-7	6-7	6-7
Mouth opening	5	6	6	7	7	6 -7
Depleted egg yolk	11	12	13	13	13	7

Note: dpf = day post fertilization

Based on the results of the study, it shows that the development time of tilapia larval organs was still in the normal time range according to [4]. However, egg yolks that run out have a longer time compared to research by [4] and research by [9]. Organ malformations due to exposure to metanil yellow dye can be seen in Figure 1 and Figure 2. Figure 1 shows malformations in the cephalic and truncus. The cephalic malformation of the mouth opening and operculum. The truncus has malformations in the fins and digestive tract. The last malformation occurred in the caudal part, it was seen that there was a bending of the spine and bleeding in the caudal fin. This can happen because the fish organs are continuously exposed to metanil yellow dye [6].

Malformations in fish can affect the growth and survival of tilapia larvae. Morphological malformations can affect the way fish swim so that it can reduce growth and increase mortality. The relatively high malformation rate during the larval stage suggests that developmental and environmental factors such as diet, light, aquarium size, density and water quality play an important role in malformation [7].

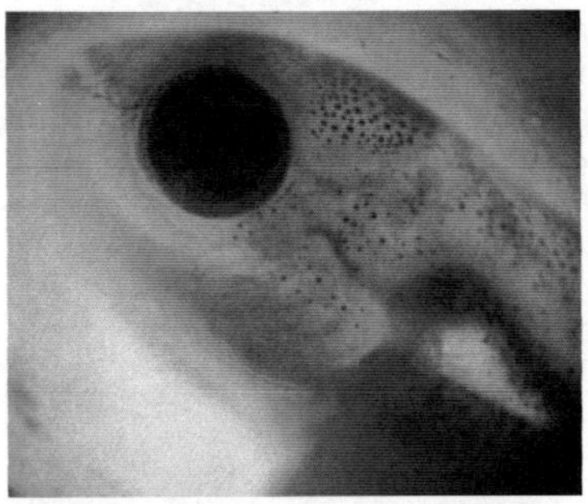

Figure 1. Malformation of the cephal

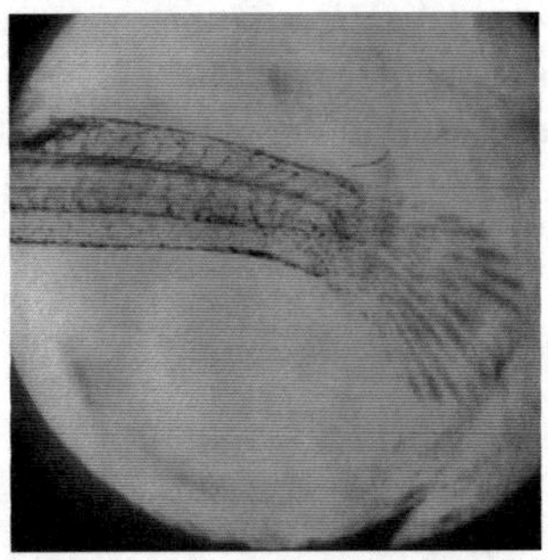

Figure 2. Hemorrhage of the caudal

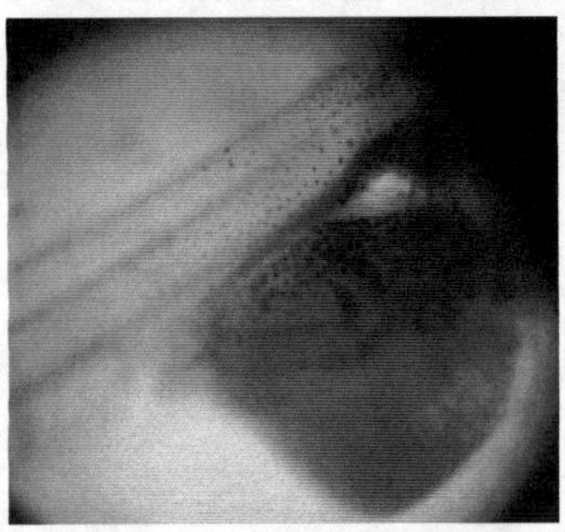

Figure 3. Edema in egg yolk

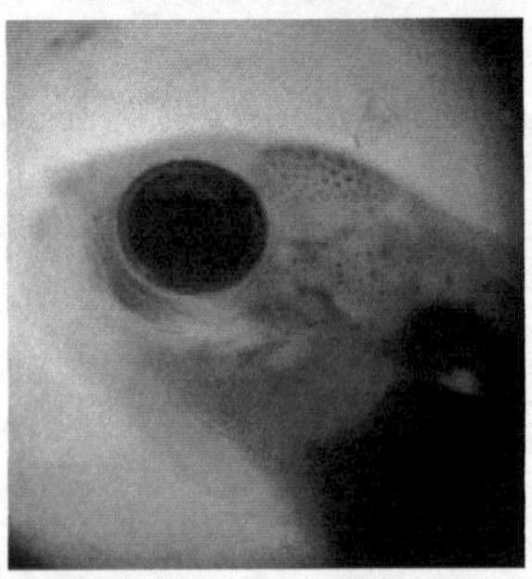

Figure 4. Malformation in mouth

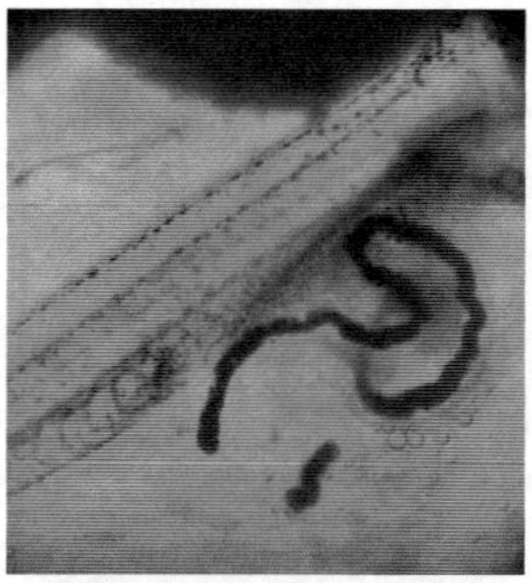

Figure 5. Egg yolk lost

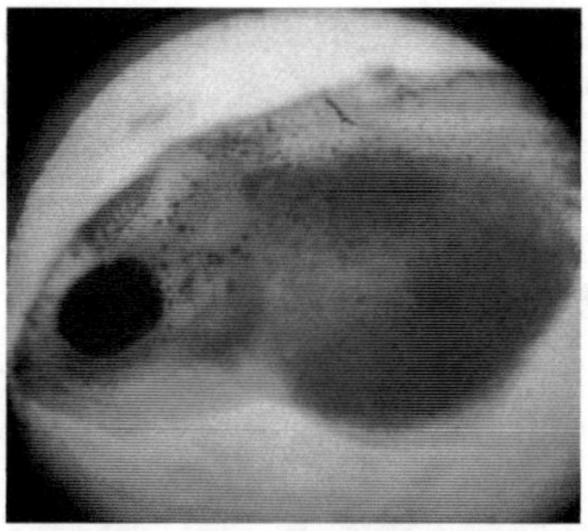

Figure 6. Malformation in chepal and egg yolk

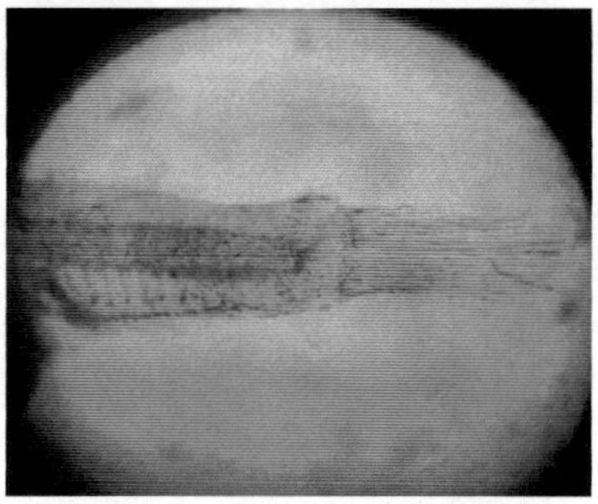

Figure 7. Caudal narrow

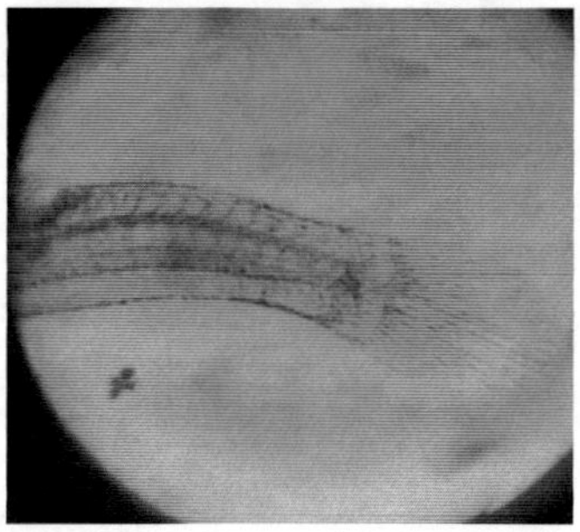

Figure 8. Malformation vertebrae

Table 2 . Larvae Treatment Water Quality

Water Quality	Concentration				
	A (0 ppm)	B (125 ppm)	C (156 ppm)	D (196 ppm)	E (245 ppm)
Temperature (°C)	28.1 - 28.8	28.3 - 28.8	27.0 – 28.3	27.2 – 28.6	28.2 - 28.9
pH	7.0 - 7.7	7. 1 - 7.9	7.1 - 7.9	7.1 - 7.9	7.1 - 7.9
DO (mg/l)	6.3 - 8.7	6.7 - 8.5	7.4 - 8.8	6.5 - 8.3	6.4 - 8.4

Water quality is one of the indicators that must be considered in the rearing of fish larvae. Based on the results of water quality shows that the value of water quality in each treatment is not much different. This means that there is no significant environmental effect when tilapia larvae are reared. The maximum temperature for freshwater aquaculture is 28 – 32 and the maximum pH ranges from 7 – 8. The solubility of oxygen (DO) is affected by temperature, besides that the solubility of oxygen in water is also influenced by dissolved particles [8].

4. Conclusion

Based on the results of the study showed that there was a malformation in the organ development of tilapia exposed to metanil yellow dye. Malformations occur in the truncus, cephalic, caudal and yolk development of fish. Egg yolk which normally develops for 7 days to 13 days at the highest concentration.

IOP Conf. Series: Earth and Environmental Science 1273 (2023) 012013 doi:10.1088/1755-1315/1273/1/012013

5. References

[1] Carney A B 2021 *Ecotox and Environ Saf* 207.
[2] Sharma K 2022 *Chemosphere* 287 132103.
[3] Iskandar O, and Mustofa B 2016 *Bulletin of Aquac Rese Techn* 14 (2) 83.
[4] Fujimura K and Okada N 2007 *Develop Grow and Differen* 49 (4) 301–324.
[5] Herjayanto M, Carman O and Soelistyowati D T 2017 *Aquat Indon* 2 (1) 1.
[6] Aydin 2022 *Aquacu* 550.
[7] Witzmann F 2021 *Internat Jour of Paleop* 33 30–42.
[8] Azhari D, Mose N I and Tomasoa A M 2018 *Scient Tinda* 4 (1) 23–26.
[9] Morrison 2001 *Morpholo* 247 172-195.

6. Acknowledgment

The reseachers thank UPT. Fish and Environmental Health, Umbulan, Pasuruan who have provided a place to carry out the study.

Study of underwater sound effects relation to current velocity and *Siganus* sp. population in seaweed cultivation area

M Kasim[1,4], M Musaruddin[2], LK Mansur, RD Palupi[1], W Jalil[3] , A Takwir[1]

[1]Faculty of Fishery and Marine Sciences, Halu Oleo University, Indonesia. Kampus Bumi Tridarma UHO. Andounohu, Kendari, Southeast Sulawesi, 93231, Indonesia.
[2]Faculty of Engineering, Halu Oleo University, Kendari, Indonesia
[3]Faculty of Fishery, Dayanu Iksanuddin University, Baubau, Southeast Sulawesi, Indonesia.

[4]Corresponding Authors: marufkasim@uho.ac.id

Abstract. Underwater sound is one of the interesting phenomena in the ocean. Underwater sounds can impact marine organisms, especially fish and mammals. One of the organisms that can be influenced by underwater sound is the Siganus fish. This study aims to determine sound propagation with a certain level of fish pest repellents about the surface currents velocity and Siganus canaliculatus. population in seaweed cultivation areas. This research was conducted in a seaweed cultivation area in Baubau City, Southeast Sulawesi, Indonesia, in May-July 2022. The tool used is a seaweed pest repellent Ver.04 (Usirikan Ver 03). Underwater sound, at a strength level of 76.19-83.09 dB, can propagate quickly at 60 m in diameter of the sea area. The farthest distance multiplied at a strength of 62.41-69.30 dB can reach 150 m in diameter of the sea area. The fish population (S. canaliculatus)was not found in the seaweed cultivation area, with a sound level of 62.41 dB. The population of Siganus fish is around 2-4 ind/m2 at a sound level of 48.62 dB. Underwater sound with a sound level above 50 dB can impact the population of S. canaliculatus. Seaweed fish repellent can significantly affect behavior. of S. canaliculatus

1. Introduction

The sound propagation waves in the ocean are the speed and scope of sound propagation that can travel very far and fast. The speed of sound waves in air is 333 m/s to 340 m/s, which is different from the sound speed waves in the water column, about 4 times faster compared to the sound speed air. The sound speed waves in the water column range from 1500 m/s to 1520 m/s. Propagation of sound occurs because of particles' up and down pressure in a medium. In the ocean, the deeper the depth of the sea, the greater the pressure. Particles of high-pressure water were compressed-decompressed so that it continues to propagate sound in the water column without losing much energy. In addition, the density of water is greater than the density of air. The sound to have a greater speed and can travel far in water. However, the speed of sound waves in the ocean will always be different. The value of underwater sound propagation varies based on the weather in the morning and evening, current pattern and speed, depth conditions, salinity, and temperature. The water temperature could affect the sound speed in the water column.

Underwater sound is illustrated as a special path that travels from the main sound source, and its propagation depends on the structure of the underwater sound speed itself and also on the oceanographic environment. Combining the water column and the basic topography is an important overview of changing or directing sound and its source [1]. Oceanographic factors produce variations in the speed of sound with depth, which strongly influences long-distance propagation [2]. Ocean currents are one of the most robust conductors for dispersing underwater sound waves [3]. The sound spread at sea level is very strong, caused by surface currents [4]. In shallow areas, with a certain topography, the sound source will lead to more comprehensive.

One example The main noise source in nearshore locations is the waves, while open ocean noise dominates the open beachfront [5]. The sounds that appear in shallow areas are also heavily influenced by natural sources of sound, such as the sound of air bubbles while breaking a beach wave [6]. High-level underwater noise, mainly due to human activities near the coast, can disrupt the behavior of water animals [7, 8, 9]. Studies have been conducted on several organisms, such as seals and sea turtles [10]. The impact of underwater sound on organisms is still being discussed [11]. The behavior of organisms to avoid sound sources is temporary and has no long-term effecton certain organisms. Conduct to avoid noisy sound sources is a short-term response of organisms, especially when foraging activities or carrying out the reproductive process [7, 10]. The impact of loud underwater sounds also occurs in adult fish and mammals. The immediate impact is auditory closure, which causes cochlear damage and behaviour of population recruitment, and would furtheraffect marine ecosystems' health function and services [12]. Using noise at an optimum level is very likely to affect fish populations in certain areas. This study tried to determine the relationship between underwater sound and ocean physical factors, especially current patterns and speeds. The present, as a driver for the spread of sound, is thought to have a strong enough influence on the spread of sound. In addition, this study also uses an underwater sound source with a high noise level to dispel Siganus sp fish that eat seaweed in the cultivation area.

2. Material and methods

2.1. Sample Collection

This research was conducted from May to August 2022 in a seaweed cultivation area in Southeast Sulawesi, Indonesia (Figure 1). This research was conducted by testing the underwater sound to control seaweed pests. The underwater sound is produced by a seaweed repellent (Patent No: SID201906036) (Picture 2). The underwater sound thrower has a frequency of 3 Hz - 2 kHz and a power of 40-100 dB. An underwater sound ejector is a device designed to produce sounds not liked by fish, especially the Siganus sp and turtle species (13,). The tool dimensions are 15 cm in diameter and 30 cm in height. This tool is waterproof and can be operated with a remote system. This tool floats on the water's surface but leaves only the top of the device. At the bottom, as much as the part is below the water's surface.

Figure 1. Underwater sound thrower ver03 (UsirikanVer.03).

The underwater sound system is operated at five different points. At each central point where the tool is placed, measurements of the sound power and frequency of the sound produced are carried out. Furthermore, the measurement of sound strength and frequency of sound waves in the direction and distance away from the center of the sound point is carried out. The measurement direction is carried out on the east, west, north, and south every 5 m distance until the minimum sound power is obtained. At each observation point with a distance of 5 m, the condition of the fish population of Siganus sp. Recording pattern by modifying the fish population measurement method [14]. The recording is done with an underwater Canon camera rotated slowly in 360°.

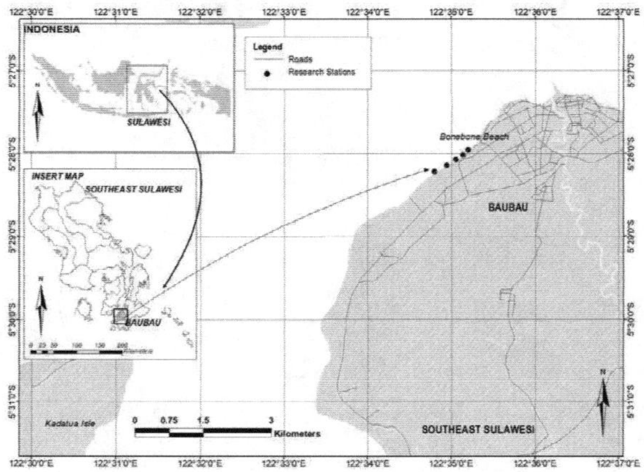

Figure 2. Study sites during the research

The method of determining the location of the sample points using GPS and purposive sampling [15]. In each measurement of sound strength and frequency of sound waves, current measurements were made using a current meter. Measurement of water currents was carried out for a period of 24 hours by dipping the tool every 1 hour at 5 station points. Current measurements carried out have covered one cycle of daily tidal periods. At the same point, the activity and number of individual populations of Siganus sp fish were recorded using an underwater camera.

2.2. Statistical Analysis
Current analysis and modeling are carried out using MIKE21 software; a hydrodynamic model will be created using Mike21 with the Flow Model FM module [16].

The data from the model using MIKE21 software will then be validated. Validation is carried out to know the extent to which the model is close to the actual conditions in the field. The validation method used is Root Mean Square Error (RMSE) [17]. RMSE is used as a standard statistical metric inorder to determine model performance [18].

3. Results and discussion
3.1. Water physical condition
The study results explain that the current pattern moves towards the northeast and south at the high tide in the morning and again moves towards the west when the tide approaches in the afternoon. The backflow pattern occurs during high and low tides, especially in real seaweed cultivation. The average current speed reaches 0.12 – 0.16 m/s in the morning before high tide and changes rapidly to 0.56-0.60 during low tide (Figure 3) and near low tide in the afternoon (Figure 4). This current movement pattern continues during the west monsoon in July-September.

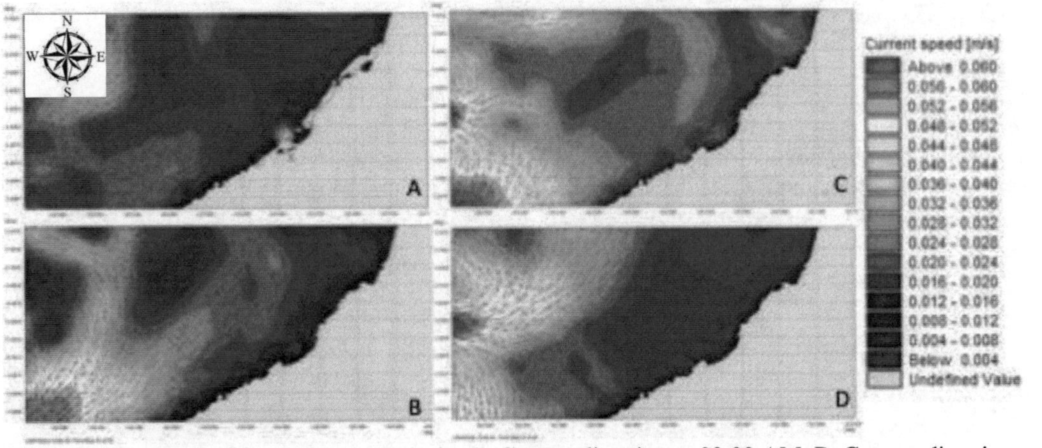

Figure 3. Current direction at the research site A. Current direction at 09.00 AM, B. Current direction at 10.00 AM, C. Current direction at 11.00 AM, D. current direction at 12.00 AM.

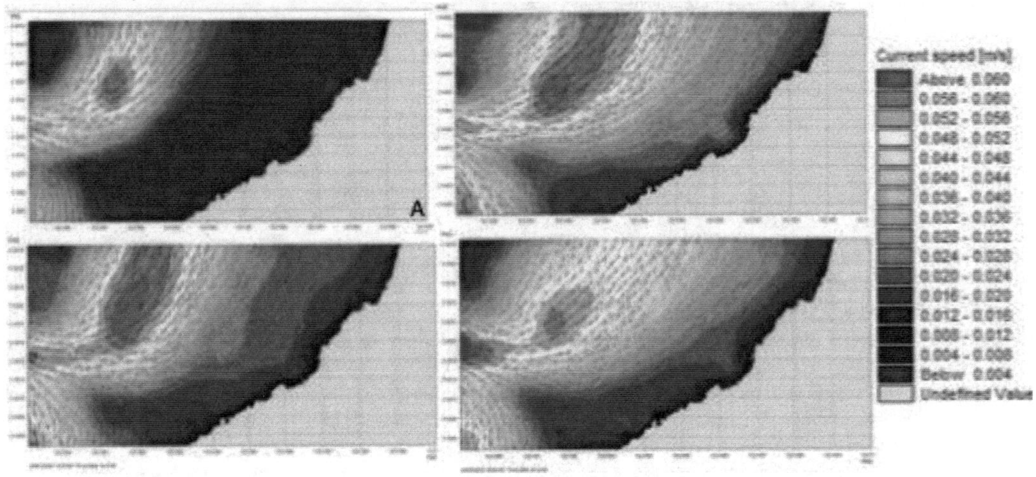

Figure 4. Current direction at the research location A. Current direction at 13.00 PM, B. Current direction at 14.00 PM, C. Current direction at 15.00 PM, D. current direction at 16.00 PM.

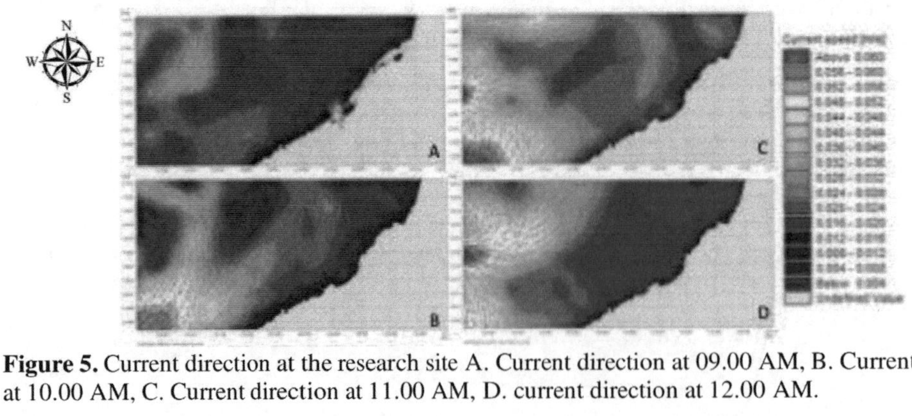

Figure 5. Current direction at the research site A. Current direction at 09.00 AM, B. Current direction at 10.00 AM, C. Current direction at 11.00 AM, D. current direction at 12.00 AM.

IOP Conf. Series: Earth and Environmental Science 1273 (2023) 012014 doi:10.1088/1755-1315/1273/1/012014

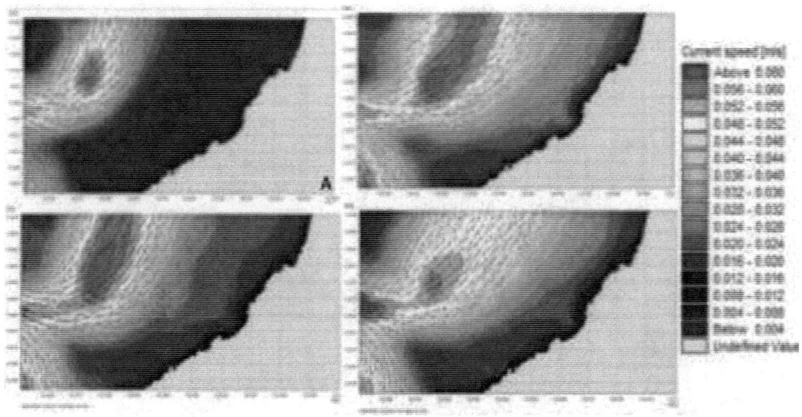

Figure 6. Current direction at the research location A. Current direction at 13.00 PM, B. Current direction at 14.00 PM, C. Current direction at 15.00 PM, D. current direction at 16.00 PM.

Changes in the current pattern and current velocity at the research location occurred at the optimum value during the afternoon towards the afternoon. This pattern is optimal when the water mass changes from east to west. The optimum speed is also seen in the afternoon before low tide. The tidal current pattern occurs until the sea level rises to 2.05 m when the water deflects pattern towards the east. At low tide, it reaches a height of 0.7 m with an average speed of 0.04 m/s.

The current pattern that occurs before the tide is seen as the vector rotating along with the tide due to the addition of water volume, this condition resembles the current pattern that occurs during high tide. The reversal of the current pattern occurs when the tide is approaching to form the opposite pattern when the tide occurs. The speed of the current increases when there is a high tide and slows down when there is a low tide.

The tidal model simulation shows a current pattern that tends to be elliptical and spreads out (Figure 5) During the west monsoon, which occurs in March-May and during the rainy season, there is a concentration of current elliptical patterns cantered towards the middle during the west monsoon which occurs in July-September.

The results of secondary data analysis show that the tidal pattern occurs twice in 24 hours, leading to an increase in current velocity and constriction. Tidal amplitude has the same wave height and distance on each side of the rising tide. During the study, there was a tidal pattern with the highest tide at the height of 205 cm (2.05 m) and the lowest ebb at a height of 7 cm (0.07 m). At the time of the first reversal, the elevation occurred at a distance of 90 cm, and the second elevation occurred at 30 cm high tide (Figure 6).

3.2. Underwater sound propagation

Pattern Sound propagation occurs in a circle at a distance of up to 75 m throughout the level evenly so that the sound propagation has a diameter of 150 m for an average range of 76.19-83.09 dB. This strength is reduced and reaches a diameter distance of up to 240 m at a minimum power of 55.51-62.41 dB. Sound propagation still forms a circle up to a distance of 150 m in diameter and breaks irregularly at a distance of more than 160 m in diameter. This condition is strongly influenced by currents that occur because currents strongly influence the refraction pattern of underwater sound. The results of measurements at five different stations with uniform conditions showed a similar pattern that described the sound distribution pattern, which was influenced by the current pattern (Figure 7).

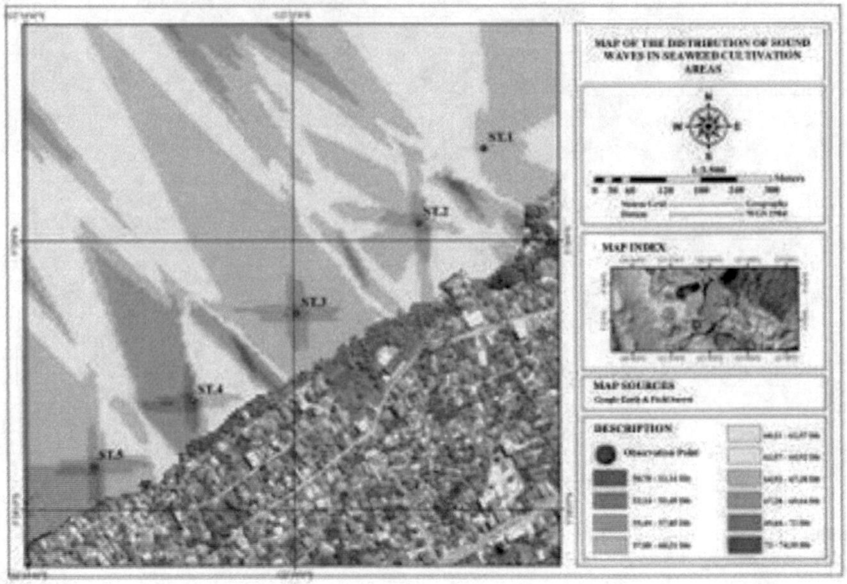

Figure 7. The pattern of underwater sound propagation that occurs at the research location

The pattern and speed of the current also have a fairly good influence on the sound distribution.The distance of the spread from the point of the sound source is seen in the direction of the current pattern and the increase in ocean currents. At stations 1,3 and 4, the sound propagates towards the northwest which is in the direction of the tidal current pattern that occurs during the day. And the area moves to narrow again and moves towards the northeast when the tide pattern occurs. The tide patterns that occur during the day and evening provide an overview of changes in sound propagation that occur during the study. An interesting thing happened at stations 3 and 4 which had sound propagation away and consistent with the strength of 55.51-62-41 at a distance of up to 200 m fromthe center of the sound. At the center of sound origin with a power of up to 76.19-83.09, there was alsoa significant widening at station 4.

The fluctuation of the underwater sound strength graph moves in tandem with the sound strength at all observation stations. The highest power up to above 80 dB occurs at the optimum point of current velocity above 0.09 m/s. This current velocity is quite high during the research. The lowest strength of sound occurs at 0.02 m/s. However, especially at station 3, there is a stagnant current speed which does not affect the fluctuation of sound propagation at the existing observation points.

3.3. Sound propagation effect on the fish population of Siganus sp.

The population of Siganus canaliculatus fish in the seaweed cultivation area is quite large with a population of 8-12 individuals/m^2. This salting is done with an underwater camera that sees the activity of fish eating seaweed that occurs in July-August every year. Many populations of siganus fishare often accompanied by fish enlargement after the reproductive period that occurs in early February after the end of the west season (November-January). The number of fish is also seen in the relatively small size of 8-11 cm long.

Sound propagation effect on the fish population of Siganus sp. The population of Siganus canaliculatus fish in the seaweed cultivation area is quite large, with a population of 8-12 individuals/m^2. This salting is done with an underwater camera that sees the activity of fish-eating seaweed that occurs in July-August every year. Many populations of siganus fish are often accompanied by fish enlargement after the reproductive period that occurs in early February after the end of the west season (November-January). The number of fish is also seen in the relatively small size of 8-11 cm.

Underwater sound is produced by various sources propagating following ocean waves' current pattern and direction. From the study results, it can be seen that the sound propagation coming from the seaweed repellent moves quickly and tends to point in the current direction. The current reversal pattern also affects the sound spread during the study. The ocean acoustic properties, such as the path that sound travels from a local source, depend primarily on ocean's sound velocity structure, that would be turn based on the oceanographic environment [19]. The distribution of underwater sound in the water column is influenced by various physical parameters, although by looking at its relation to the frequency of the sound produced from the source. A relevant example is in shallow water; tidal conditions affect the spread and level of sound. This is because the depth of the water limits the propagation of low-frequency sound [20,21].

The water column combine to basic properties synergizing with sound propagation pathways can also provide information on most propagation phenomena in the oceans. It should be noted that underwater sound travels faster than air. Different media will impact the transmittance of the sound itself [eg. The speed of sound in water (about 1500 ms-1) is much greater than in air (about 330 ms-1)][22,23]. The spread of sound that spreads in the water column will decrease with the distance the sound travels at a certain distance.

The reduction in sound level power occurs because the sound wave radiating spherically from the source contains only a finite amount of energy. Thus, the sound wave energy will be distributed over a wider area at a longer distance. However, the transmission loss is much more complex in shallow waters where the seabed and the water surface can reflect sound [24]. In this study, the sound level strength was reduced when measurements were made at a distance of more than 50 m in diameter. Thereduced power level of the sound is more associated with the mass of seawater which is the determining factor for the reduction in the sound level. Although it is seen that the sound speed is faster than the air medium due to the density of the object, it provides a wider spread opportunity than on land. Prediction of the level of sound reception at a certain distance can be done by reducing the dispersion that disappears or sound absorption in the aquatic environment from the sound source. For example, changes in salinity and ocean temperature in the ocean will produce changes in the speed of sound that can cause complex reflecting boundary layers that transmit sound [25]. In the water, sound can be responded to well by various marine organisms. The response occurs mainly as part of the workof the senses, which appear synergistically with the senses, such as sight, touch, smell, and other reactions. The response is also an action of the existing transmission signal strength. Various marine animals, such as some invertebrates, fish and marine mammals, have developed specialized mechanisms for emitting and detecting underwater sounds [26,27]. Some types of marine fish often develop communication with sound. This communication is also associated with the environmental conditions around the fish, especially in introducing them to the environment [28]. This sound communication, by fish, is transmitted at frequencies below 1 kHz, with sound power levels below
120 dB. Some fish will show reactions and hearing mechanisms through sensitivity to particle movement. This pattern is the presence of particle motion through the 'generalist' so that one can hear the existing movement, especially at lower frequencies below 1 kHz [29,30]. Fish can listen to sound using the inner ear, which is sensitive to the moving components of the sound field particles. Sound pressure is only detected in fish with gas-filled organs, such as swim bladders. Most fish can respond well to sound. Fish can generally detect the sound in the field by extending the detected frequency range and increasing sensitivity to sounds underwater.

Thus, hearing range and sensitivity to sound will substantially vary, particularly between fish species. The most crucial difference in this study is between species that can detect sound pressure andfish that only see particle motion (31, 32). Fish with swim bladders generally have better hearing abilities, which is done to respond to sound than less sensitive species. The response that occurs in fishwill depend on the sound level, the source of the ambient sound, the activity of the fish while listening to the sound, and previous exposure to the same sound and other sounds. Another condition that fish will be able to respond to underwater sounds is a condition of self-rescue or carrying out life and safety activities [33,34]. All fish species can detect particle motion directly from sound, which is thought to be necessary for this ability. The motion and direction of underwater sound particles may differ at each location and affect the power of the fish. The range of sound is usually quite complex and information-rich for marine organisms such as fish that have several dimensions that humans do not have [35].

The presence of sound in the underwater environment, especially in fish, which is very worrying is the bad impact on several types of fish close to the sound source. Sound can have a direct effect on some marine fish. Sounds with high threshold levels can potentially damage several tissues and organsof fish, including non-auditory tissues such as (swimming bladder and muscle tissue in fish, increasingthe number of gas bubbles in marine mammals and also possibly resulting in traumatic brain injury/neurotrauma [36,37]. However, the damage that occurs is highly dependent on the location of the sound source, the distance from the sound source to the organism, the high sound level, and the strength of the sound frequency [38]. The death of fish is the most significant damage that often occurs in the presence of sound at the threshold of organism tolerance. With a high enough sound entry, damage to the tissues and organs of fish can occur so that it can cause death sometime after the animal leaves the sound source area [39,40,41]. The sound propagation moves from the source in the ocean, and the excellent level decreases and disappears at a certain distance. The middle area of the sound source is called the noise zone. The noise zone can be seen from noise propagation modelling. As sound propagates, in principle, in an omnidirectional manner from the source, the noise influence zone, given as the distance from the start, represents a radius rather than a straight line. This model in seawater only gives a very rough estimate of the zone of influence because sound in the ocean is always in three-dimensional. Reflection, interference and refraction patterns in the propagation of sound that water in the sea will cause a much more complex sound field, especially under certain topographic conditions. Complexity arises from effects such as the increased sound energy received bythe fish and with distance, especially when multiple sound sources are used simultaneously and simultaneously. This condition can cause stress on fish and other indirect effects [42]. The interesting thing that also happens to the auditory system in fish is that fish in a culture location that is close to thesound source and cannot move or move away from the sound source can expose fish to high noise levels. At certain times, fish will be able to adapt to sound exposure [39]. Fish can search for prey using unidirectional hearing or avoid predation, detecting other special sound signals. Fish will generally come to or avoid certain underwater sounds. Some fish can also distinguish different directions and sound sources, including the distance from which sound comes [40]. Fish generally gaina significant sensory advantage by hearing underwater sounds. This is because fish hearing can usuallyprovide extensive information compared to other senses [41]. Fish will be able to determine the direction and distance from the sound source in the ocean with their hearing ability. This condition also provides the ability to respond quickly and precisely when the fish is interested or not with a sound that appears. The fish's ability to locate the source of underwater sounds may be essential because in the ocean, light levels may be lower, and vision at a distance may be impaired. It is possibleto detect approaching or moving away from the sound source, which is a special advantage for all fish.Another thing is the ability to detect signals that may be needed for direction and space, and orientation of locations around fish. This condition can often be done with the sense of hearing when the other senses are not working optimally. [42]. All fish will be able to detect sounds at frequencies below 50 Hz to 1000 Hz. This detection depends on the species. Certain fish species that can see underwater sounds at frequencies above about 300 Hz are generally marine or freshwater fish. However, fish can also hear sounds up to 3000 Hz or higher. In this regard, there should be a lot of variation in some species regarding sensitivity to underwater sound. Because this also reflects the high level of fish diversity in the ocean [43].

4. Conclusion

The distribution of underwater sound is closely related to the distribution pattern of surface currents. The propagation of sound travels rapidly in water and places a limit on the propagation of sound. Siganus fish populations were not found at sound levels above 60 dB. The distribution of underwater sound with a level of up to 90 dB can spread widely up to a radius of 70 m from the sound source. Thediameter radius of underwater sound distribution reaches 150 m.

5. Reference

[1] Kuperman WA. and Roux, P. 2014. *Und Acou.* 5
[2] Frosch, 1964. R.A. Frosch 1964 *Science.* pp 889-894.
[3] Godin, O A 1992 *J de Phys.* 4 2 C1.
[4] Kim, S H Choi, B K Kim, B N 2020 *J of Mar Scie and Engine.* 9 1.
[5] Wilson, O B, Wolf, S N, and Ingenito, F 1985 *J Acoust Soc Am.* 78 190–195.
[6] Didenkulov, I N, Korotkov, A S 1993 *In Kerman eds Nat Phys Sourc of Under Sou.*

[7] Southall B L, Bowles A E, Ellison W T, Finneran J J, Gentry R L, Greene C R Jr, Kastak D, Ketten D R, Miller J H, Nachtigall P E, Richardson W J, Thomas J A, & Tyack P L 2007 *Aqua Mam.* 33 4 pp 411 - 521.
[8] Hastings, M C 2008 *Acou Tod.* 4 2 22-34.
[9] Popper, A N, Hawkins, A D, Fay, R R, Mann, D, Bartol, S, Carlson, T, Coombs, S, Ellison, W T, Gentry, R, Halvorsen, M B, L kkeborg, S, Rogers, P, Southall, B L, Zeddies, D, and Tavolga, W N 2014 *Sound Exposure Guidelines for Fishes and Sea Turtles: A Technical Report* Springer and ASA Press Cham Switzerland.
[10] Hall, M V 2013 *Proce of Aco.* 13-20.
[11] Beale C M 2007 *Inter J of Comp Psyc.* 20 pp 111-120.
[12] Peng C, Zhao X and Liu G, 2015 *Int J Environ Res.* 12304-12323.
[13] Kasim 2021 *Egyp J of Aqua Res.* 47 179-184.
[14] English, S, Wilkinson, C and Baker, V 1997 *Town Aust Inst of Mar Scie.*
[15] Sugiyono 2015 Educational Research Methods Quantitative, Qualitative and Approaches *R&D 47* 6 AlfaBeta 451 15.
[16] DHI. 2017. *MIKE 21 & MIKE 3 Flow Model FM - Hydrodynamic and Transport Module.*
[17] Shiddekh, M A I, dan Suryani, E 2018 *J Tek ITS.* 7 1 138–142.
[18] Chai, T, and Draxler, R R 2014 *Geosc Mod Devel,* 7 3 1247–1250.
[19] Roux P, Kuperman W A, 2014 *Under Acou.* 32-39.
[20] Yasin, M, Apsari, R, Ningsih, K, Yhuwana, Y G Y, Zulkarnaen, M, Djamal, M, Meilano, I, Evita, M, Aminah, N S, Srigutomo, W, Setiawan, A, Darmawan, H, Suyanto, I, Wahyudi, Suryanto, W, Putra, H, Irzaman, Perdinan, & Harun, S W 2022 *J of Optoe and Adv Mate. 24* 5 201-205.
[21] Idris, N F, Lokman, M Q, Rahim, H R A, Saidin, N, Yasin, M, & Harun, S W *Micro and Opt Tech Let. 59* 9 2196-2201.
[22] Ain, K, Kurniadi, D, Suprijanto, S, & Santoso, O 2017 *J of Elec Bio. 8* 1 3-10.
[23] Jensen, F B, Kuperman, W A, Porter, M B, and Schmidt, H 2011 *Computational Ocean Acoustics, second edition* (Springer, New York).
[24] Gotz T, Hastie G, Hatch, L T, Raustein, O, Southall B L, Tasker, M, Thomsen, T, 2009 *Osp Comm.* 33-37.
[25] Richardson, W J, Malme, C I, Green, C R jr and D H Thomson 1995 *Marine Mammals and Noise* Academic Press San Diego CA 576 pp.
[26] Berta, A, Sumich, J L and Kovacs, K M 2006 *Marine mammals evolutionary biology 2nd edition* Elsevier and Academic Press San Diego 547 pp.
[27] Ladich, F and M L Fine 2006 *Sound generating mechanisms in fishes: A unique diversity in vertebrates* Science Publishers Enfield NH pp 3-43.
[28] Ladich, F and A A Myrberg 2006 *Agonistic behaviour and acoustic communication* Science Publishers Enfield NH pp 121-148.
[29] Popper A N, Hawkins A D, Sand O, Sisneros J A 2019 *Journal of the Acoust Soci of Ame.* 146 948 – 955.
[30] Fay, R, 2011 *Direct Hear in Fish.*
[31] Richardson W J, Greene C R Jr, Malme C I, Thomson D H 1995 *Marine mammals and noise* Academic Press San Diego CA.
[32] Hastings, M C, & Popper, A N 2005 *Effects of sound on fish* California Department of Transportation Contract 43A0139 Task Order 1.
[33] Madsen P T, Wahlberg M, Tougaard J, Lucke K, Tyack P 2006 *Mar Ecol Prog Ser.* 309 279–295.
[34] Thomsen, F, Lüdemann, K, Kafemann, R, & Piper, W 2006 *Biola.*

[35] Wright, A J, Soto, N A, Baldwin, A L, Bateson, M, Beale, C M, Clark, C, Deak, T, Edwards, E F, Fernández, A, & Godinho, A 2007 *Int J of Comp Psycho*. 20 2 250-273.
[36] Filiciotto, F, Giacalone, V M, Fazio, F, Buffa, G, Piccione, G, Maccarrone, V, Di Stefano, V, Mazzola, S, & Buscaino, G 2013 *Aqua. 414-415* 36-45.
[37] Hawkins, A D, & Popper, A N 2018a *J of the Acou Soc of Amer. 144* 6 3329-3350.
[38] Fay, R R 2000 *J Assoc Res Otola*. 1 120–128.
[39] Fay, R R, and Popper, A N 2000 *Hear Res.* 149 1–10.
[40] Putri, N A, Sulmartiwi, L, & Pursetyo, K A 2020 *AACL Biof. 13* 1 250-260.
[41] Avrilia, D, Suprapto, H, & Rahardja, B S 2022 *Worl Veter J. 12* 1 87-94.
[42] Hawkins, A. D, & Popper, A N 2018b *Effects of man-made sound on fishes* New York Springer Nature pp 145-177.
[43] Saade, E, Usman, U, Sulaeman, H A and Jannah, N M 2021 *J of Aqua and Fis Hea*. 11 1 10–20.

6. Acknowledgment

Grateful to the Ministry of Research and Technology/BRIN for Supporting the full Research Grant. Our appreciation goes to LPPM-UHO for the facilitation provided to make this research a success. Appreciation also goes to the Laboratory of the Faculty of Fisheries and Marine – UHO for the assistance in sample analysis provided.

Morphological and molecular identification of java barb (*Barbonymus gonionotus*) from East Java Province

M Rahmawati[1], A T Mukti[2], S Andriyono[3,4]

[1]Magister Program Biotechnology of Fisheries and Marine, Faculty of Fisheries and Marine, Universitas Airlangga, Campus C Mulyorejo, Surabaya 60115, Indonesia
[2]Departement of Fish Health Management and Aquaculture, Faculty of Fisheries and Marine, Universitas Airlangga, Campus C Mulyorejo, Surabaya 60115, Indonesia
[3]Departement of Marine, Faculty of Fisheries and Marine, Universitas Airlangga, Campus C Mulyorejo, Surabaya 60115, Indonesia

[4]Corresponding Author: sapto.andriyono@fpk.unair.ac.id

Abstract. This study aims is compare the identification of Java barb fish (*Barbonymus gonionotus*) in East Java Province by morphometric and molecular identification. Molecular identification was conducted using mitochondrial DNA based on the Cytochrome Oxidase Sub Unit I (COI) region. Java barb fish samples were collected from various regions in East Java Province including Gresik, Mojokerto, Madiun, Malang, and Banyuwangi. The identification results by morphological and molecular methods showed the same results, namely *Barbonymus gonionotus*. Based on the morphological identification, *Barbonymus gonionotus* has a long and flat body shape with a raised back, terminal mouth position, silver body colour with a darker dorsal area, dorsal fin shape with weakly hardened rays, and homocercal caudal fin shape. The morphometric measurements of Java barb fish using the one-way ANOVA test showed that the five locations differed significantly. The results of the molecular research showed that all java barb fish sequences were successfully amplified with the COI gene (648 bp). The BLASTN results show the percentage identity ranges from 99.85-100%.

1. Introduction

Java Barb (*B. gonionotus*) is a local endemic fish native to Indonesian freshwater found in several areas such as Sumatra, Java, and Kalimantan [1]. Java Barb (*B. gonionotus*) is one of the local endemic fish in East Java. Java Barb (*B. gonionotus*) has high economic value as a fish consumption and has the potential as a source of protein. Java Barb (*B. gonionotus*) protein content reaches 60.25% [2]. The economic potential of Java Barb (*B. gonionotus*) results in overfishing, which impacts decreasing Java Barb production. Therefore, it is necessary to carry out conservation activities to prevent extinction and to be able to explain the current status of the freshwater fish population [3]. This population study will also have to deal with the identification process.

Species identification can be made by morphological and molecular methods [4]. One technique that is used as a reference for knowing the morphology is morphometrics [5]. Morphometric measurement is one of the important studies in fisheries that must be carried out widely. This is considered important because of morphometric variations of a population in different environmental conditions [6]. Molecular identification is identical to DNA barcoding, which is an identification method that refers to the arrangement of nucleotide bases belonging to a species [7]. DNA barcoding research in fish often uses mitochondrial DNA bases on the Cytochrome Oxidase Sub Unit I (COI) region. The use of COI genes has been recognized globally in species identification [8]. Research on morphological and molecular identification *B.gonionotus* in East Java Province is still relatively lacking. Therefore, it is necessary to research the identification *B.gonionotus* from East Java Province. This study compares the identification results *B.gonionotus* using morphological and molecular methods.

2. Materials and methods

The instruments used in this study were a digital scale, ruler, tube, incubator, frezeer, centrifuge, vortex, and PCR thermalcycler. The materials used were *B. gonionotus*, gSYNC^{TM} DNA Extraction kit, and primers FishF1 (5' TCA ACC AAC CAC AAA GAC ATT OGC AC 3') and FishR1 (5'TAG ACT TCT GGG TGG CCA AAG AAT CA 3').

2.1. Data collection

This research was conducted in March-June 2021. Java Barb samples were caught from fishermen. This research was conducted in five locations in East Java Province, including Gresik, Mojokerto, Madiun, Malang, and Banyuwangi (Figure 1). The samples were found in the Lamong Gresik River, the Rolag Songo River in Mojokerto, the Bening Reservoir in Madiun, the Karangkates Reservoir in Malang, and the Setail River in Banyuwangi (Table 1). Each location contributes five fish.

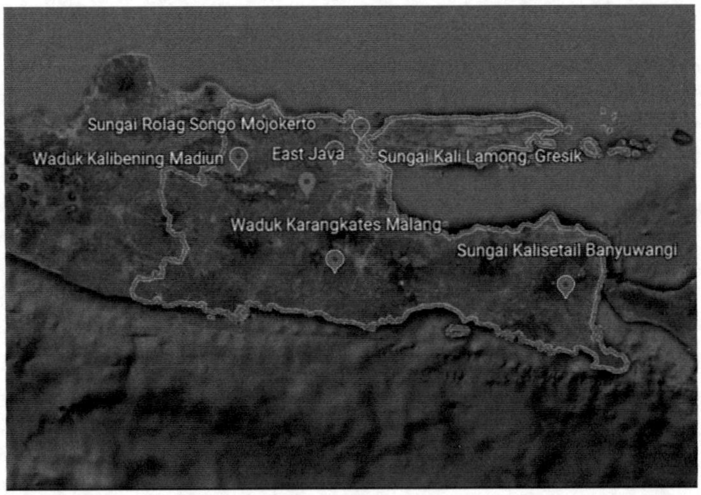

Figure 1. Java Barb Fish Sampling Location Map (https://earth.google.com/web)

Table 1. Location and Coordinate Point of Java Barb Sampling

No.	Location	Coordinate Point
1.	Lamong River, Gresik	7°12'02"S 112°40'09"E
2.	Rolag Songo River, Mojokerto	7°26'44"S 112°28'01"E
3.	Bening Reservoir, Madiun	7°32'42"S 111°47'44"E
4.	Karangkates Reservoir, Malang	8°09'38"S 112°26'55"E
5.	Setail River, Banyuwangi	8°17'32"S 114°14'28"E

2.2. Identification of Java Barb (*B. gonionotus*)

2.2.1 Morphological identification

Morphological identification of Java Barb is undertaken visually on external characteristics including body shape, colour, mouth shape, and fin shape [9]. Analysis of morphological data refers to www.

Fishbase.se and various supporting references from other researchers. Observations and measurements in morphology identification also use morphometric methods [10]. Samples measurements using a ruler as a measuring tool. Measurements of the measured morphometric characteristics including several characters can be seen in Figure 2, and description characters can be seen in Table 2.

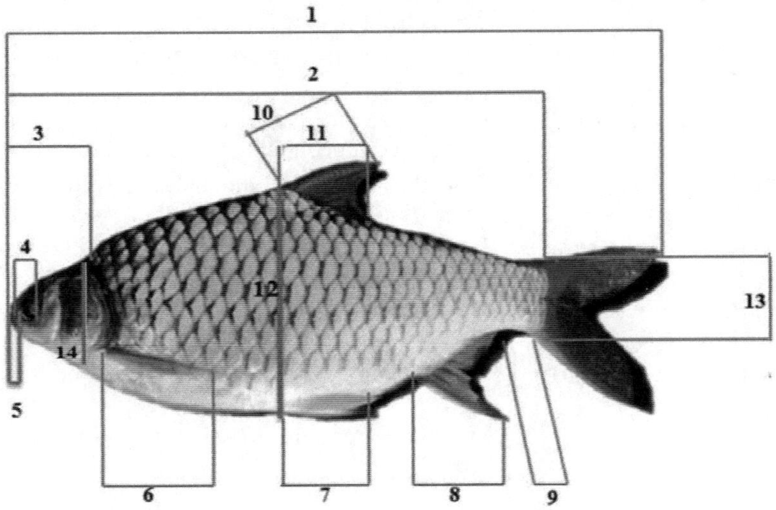

Figure 2 Morphometric Characters Measurement [10]

2.2.2 Molecular Identification

Molecular identification was carried out using DNA barcode techniques. The steps carried out in the DNA barcoding technique include DNA extraction, PCR, DNA visualization by electrophoresis, and DNA sequencing. Sampling was carried out by cutting the caudal peduncle of Java Barb by approximately 25 mg. Furthermore, preservation is carried out with 90% alcohol and stored in a -20C freezer for further analysis [11]. DNA isolation was carried out using the gSYNC™ DNA Extraction kit [12]. The quality of DNA isolation was observed on 1.2% agarose gel using electrophoresis, then stored in a -20 C freezer. The next step is the PCR step, which aims to multiply the target DNA by using primers FishF1 (5' TCA ACC AAC CAC AAA GAC ATT OGC AC 3') and FishR1 (5'TAG ACT TCT GGG TGG CCA AAG AAT CA 3') with the target gene Java Barb fish 648 bp [13,14].

PCR has three stages, namely, initial denaturation at 95 C for 2 minutes, denaturation at 94 C for 30 seconds, annealing at 54 C for 30 seconds, elongation at 72 C for 1 minute and final extension at 72 C for 10 minutes. The number of cycles of the denaturation-elongation stage is 35 times. The PCR results were visualized on 1.2% agarose gel employing electrophoresis, and the photos were photographed with a UV transilluminator. The PCR product is then sent to PT Genetics Science Indonesia for the sequencing process [14].

Table 2 Description of Java Barb Morphometric Characters Measurement

No	Morphometric Characters	Description
1.	Total Length (TL)	The distance between the tip of the head (anterior) to the posterior of the caudal fin
2	Standard Length (SL)	The distance from the anterior of the mouth to the caudal base
3	Head Length (HL)	The distance from the anterior of the mouth to the posterior of the head
4	Eye Diameter (ED)	The length of the diameter of the eye socket is measured from the anterior to the posterior side of the eyeball
5	Snout Length (SnL)	The distance from the beginning of the mouth to the end

No	Morphometric Characters	Description
		of the mouth
6	Pectoral Fin Length (PFL)	The distance from the base of the fin to the longest tip of the pectoral fin
7	Ventral Fin Length (VFL)	The distance from the base of the fin to the longest tip of the ventral fin
8	Anal Fin Length (AFL)	The distance from the base of the fin to the longest tip of the anal fin
9	Caudal Penducle Length (CPL)	The distance measured from the posterior of the dorsal fin to the caudal peduncle
10	Dorsal Depth (DD)	Base distance from the radius to the tip of the longest dorsal fin
11.	Dorsal Base Length (DBL)	The distance measured is a straight line from the anterior of the dorsal fin base to the posterior of the dorsal fin base
12.	Body Depth (BD)	Measured at the highest ventral between the dorsal and ventral
13.	Caudal Penducle Depth (CPD)	Measured at the stem of the caudal peduncle at the lowest place
14.	Head Depth (HD)	Measured from the top of the head to the bottom of the head

2.3 Data Analysis

Morphological analysis was carried out using the one-way ANOVA test by calculating the F value for each morphometric parameter tested, referring to Ayyubi [9]. Molecular analysis was carried out in two stages, first, the sequencing result was presented as a chromatogram edited using Chromas 2.6 Software. Second, the edited sequence results are stored in a fasta file, followed by BLASTN analysis on the GenBank database at NCBI (http://blast.ncbi.nlm.nih.gov/) [15,14].

3. Results and discussion

3.1. Results

3.1.1 Morphological Identification

Based on the result of morphological identification observation at the five locations, the fish samples were identified as Barbonymus gonionotus. The morphological characteristics of the five locations looked similar, namely terminal mouth position, round eye shape, long and flat body shape with a raised back, dorsal fin shape of weakly hardened fin rays, homocercal caudal fin shape, and silver body color with dorsal area darker. The characteristics shown are the following www.fishbase.se and references [10]. The result of the morphological identification observations can be seen in Figure 3.

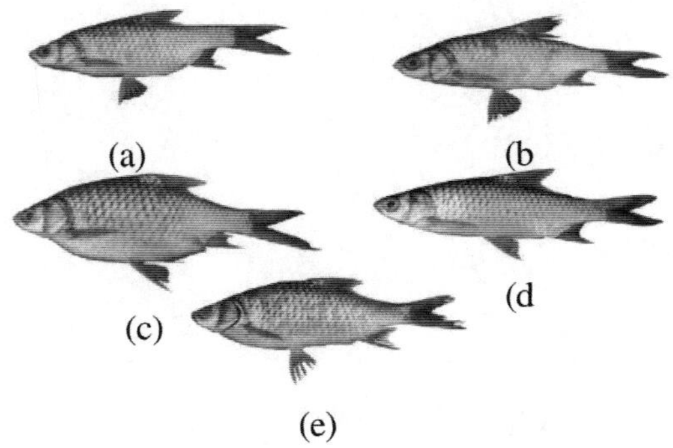

Figure 3. Morphological Identification Observation Results, information: *Barbonymus gonionotus*comes from Gresik (a), Mojokerto (b), Madiun (c), Banyuwangi (d), and Malang (e).

Morphometric analysis was performed using the one-way ANOVA test. The number of morphometric parameters calculated is 14 parameters. Several characteristics can distinguish the species B.

Table 3. Characteristics Morphometrics of Java Barb (*Barbonymus gonionotus*)

No	Characteristics morphometric	Gresik	Mojokerto	Malang	Madiun	Banyuwangi	F Value
		Mean ±SD	Mean ±SD	Mean ±SD	Mean ±SD	Mean ±SD	
1	TL	15.66 ± 0.531	18.1 ± 2.096	21.34 ± 0.868	19.76 ± 1.719	17.16 ± 1.753	10.703
2	SL	11.38 ± 0.634	13.8 ± 1.242	16.42 ± 0.606	15.24 ± 1.316	13.06 ± 1.457	15.377
3	HL	2.20 ± 0.393	3.02 ± 0.432	3.78 ± 0.286	3.44± 0.251	2.96 ± 0.428	13.172
4	ED	0.80 ± 0.122	1.06 ± 0.134	1.16 ± 0.114	1.08 ± 0.083	0.94 ± 0.089	8.082
5	SnL	0.50 ± 0.122	0.84 ± 0.151	0.90 ± 0.100	0.94 ± 0.089	0.66 ± 0.207	8.566
6	PFL	2.50 ± 0.122	2.88 ± 0.258	3.70 ± 0.234	3.68 ± 0.192	2.78 ± 0.303	28.357
7	VFL	2.42 ± 0.164	2.72 ± 0.303	3.52 ± 0.238	3.48 ± 0.109	2.56 ± 0.320	23.436
8	AFL	2.22 ± 0.164	2.64 ± 0.270	3.16 ± 0.114	3.06 ± 0.151	2.18 ± 0.363	19.451
9	CPL	0.38 ± 0.083	0.54 ± 0.114	0.70 ± 0.031	0.54 ± 0.089	0.44 ± 0.054	10.278
10	DD	3.10 ± 0.100	3.56 ± 0.305	4.14 ± 0.304	4.10 ± 0.509	3.32 ± 0.432	8.378
11	DBL	3.14 ± 0.207	3.36 ± 0.251	4.30 ± 0.331	4.08 ± 0.178	3.52 ±0.334	16.736
12	BD	4.74 ± 0.279	6.42 ± 0.736	8.00 ± 0.667	5.90 ± 0.254	5.70 ± 0.595	24.104
13	CPD	1.30 ± 0.212	1.98 ± 0.303	2.64 ± 0.152	2.06 ± 0.378	1.68 ± 0.476	11.591
14	HD	2.26 ± 0.320	2.58 ± 0.311	3.24 ± 0.181	2.86 ± 0.296	2.50 ± 0.339	8.092

3.1.2 Molecular Identification

The results of the molecular research showed that all Java Barb fish sequences were successfully amplified with the COI gene of 648 bp. The average length of DNA sequence is the result of trimming obtanined from the results of the sequences that have been cut at the right and left ends of the chromatogram using Chromas 2.6 Software. Based on BLASTN samples that have been compared with the GenBank database show that the five samples identified as *Barbonymus gonionotus*, with a ratio of 99.85-100% (Table 4).

Table 4. BLASTN Results of Java Barb (*Barbonymus gonionotus*)

No.	Sample Origin	Max Score	Total Score	Query Cover	E Value	Percentage Identity	GenBank Accession No.
1.	*B. gonionotus* from Gresik	1205	1205	100%	0.0	100%	KJ936769
2.	*B. gonionotus* from Madiun	1214	1214	100%	0.0	100%	KJ936769
3.	*B. gonionotus* from Mojokerto	1214	1214	100%	0.0	100%	KJ936769
4.	*B. gonionotus* from Malang	1216	1216	100%	0.0	99.85%	KJ936769
5.	*B. gonionotus* from Banyuwangi	1214	1214	100%	0.0	100%	KJ936769

3.2. Discussion

Morphological identification is carried out as the first step in identifying fish species. The results of the analysis of morphological identification in the five locations have identified Barbonymus gonionotus, which has characteristics including a round eye shape, terminal mouth position, dorsal fin shape with weakly hard fin rays, homocercal caudal fin shape, long and flat body shape with an elevated back, silver body colour with the darker dorsal area. The morphological characteristics of Java Barb are also the same as shown in the study of Ayyubi [10]. The result of morphometric measurements of Java Barb based on one-way ANOVA analysis, of 14 morphometrics characters, showed that there was a significant difference based on the calculated F value that was greater than the F table ($P < 0.05$). Morphometric measurements of 14 characters, three distinguishing characters with the largest F value showed on PFL, BD, and VFL characters (Table 3).

The results of morphological identification seen from environmental aspects show that Java Barb from rivers (Gresik, Mojokerto, and Banyuwangi) tend to have a slimmer body than fish obtained from the reservoir (Madiun and Malang). The results of BD character showed that the average fish origin from Gresik was 4.74 cm, Mojokerto 6.42 cm, Banyuwangi 5.70, Madiun 8 cm, and Malang 5.9 cm (Table 3). The difference in body dept (BD) is related to environmental adaptation. The Java Barb that lives in rivers has a slim body shape because the fish tend to go against the current and move more actively, compared to fish that live in reservoirs [16]. Differences in morphology and morphometrics are caused by several factors, such as differences in environmental conditions, topography and habitats [17].

The result of molecular identification at the five locations also showed that all samples were identified as *Barbonymus gonionotus* with a query cover of 100%, an E value of 0.0, and a percentage identity between 99.85-100% (Table 4). References from Triandizal and Madduppa [8] state that species are said to be similar if the max score and total score are the same, the query cover values are close to 100%, the E value is close to 0.0 and the percentage identity is close to 100%. The similarity of DNA results obtained with the GenBank database accession No KJ936769 indicates that molecular identification can be used as an alternative method of identifying a species. DNA barcoding provides fast and accurate identification due to the constant nature of DNA compared to morphology [4]. The COI gene has been widely used for identification. The COI gene is relatively stable, and not easy to change when compared to other genes, and the number of copy numbers that are widely used makes it easy for amplified identification. The COI genes can be used to search for species, and study diversity, and kinship relationships between intraspecies and interspecies [18].

4. Conclusion

Morphological and molecular identification showed that the Java Barb fish samples obtained from five locations in East Java Province were *Barbonymus gonionotus*. Both methods are suitably performed for fish identification, but molecular identification supports of genetic information for future studies.

5. References

[1] Radona D, Soelistyowaty D T, Carman O, dan Gustino R 2016 *Jur Ikht Ind*16 (3) 259-268.
[2] Hadisusanto, S, dan Suryaningsih S 2011 *Bio* 16 (2): 214-220.
[3] Akbar N, dan Labenua R 2018 *Dep Jurn* 7 (2) 164-176.
[4] Putri A, dan Madduppa H 2020 *Jurn Kela* 13(3) 168-175.
[5] Fadhil R, Muchlisin Z A, dan Sari W 2016 *Jurn Ilmi Mahas Kela Peri Uns* 1(1) 146-159.
[6] Tzeng T D, Chiu C S and Yeh S Y 2001 *Fishe Rese* 53 (3) 211-217.
[7] Triandiza T, dan Madduppa H 2018 *Jur Sumber Akua Indopas* 2 (2) 81-90.
[8] Dawnay N, Ogden R R, McEwing, Carvalho G C and Thorpe R S 2007 *Foren Scien Intern* 173 1-6.
[9] Duong T, Tran L V D, Nguyen N T, Jamaluddin J A F, and Azizah M N S 2020 *Tropi Zool* 33 63-76.
[10] Ayyubi H, Budiharjo A, dan Sugiyarto 2018 *Jurnal Iktiologi Indonesia*, 19(1): 65-78.
[11] Andriyono S, and Suciyono S, 2020 *Egyptian Jour of Aqua Biol and Fishe* 24, 427-439.
[12] Nuryanto A, Baghawati D, Kusbiyanto K, Sastranegara M H, and Rachmawati F 2020 *Jurn Ilmi Biol* 8 145-156.
[13] Ward R D, Zemlak T S, Innes B H, Last P R, and Hebert P D 2005 *Biolo Scien* 360 (1462) 1847 – 1857.
[14] Panprommin D, Soontornprasit K, Tuncharoen S, Pithakpol S, and Keereelang J 2019 *Jour of Asi-Paci Biod* 12 382-389.
[15] Kumar S, Stecher G, M Li Knyaz C, and Tamura K 2018 *J Mole bio evo* 35 1547.
[16] Santos A B, Terra B, and Araújo F G 2010 *Zoo* 27(5) 732–740.
[17] Baur H, and Leuenberger C 2011 *System Biolo* 60(6) 813-825.
[18] Hajibabei M, Singer G A C, Hebert P D N, and Hickey A. 2007 *Tren in Gene* 23(4) 167-172.

6. Acknowledgements

The writer is thankful to the supervisor who has devoted much knowledge, time, patience, enthusiasm, and guidance. The writer is also grateful for direction, advice, input, and corrections from the supervisors. We also thank the editors and reviewers for their suggestions to improve our manuscript.

IOP Conf. Series: Earth and Environmental Science 1273 (2023) 012016 doi:10.1088/1755-1315/1273/1/012016

The effect of red ginger extract (*Zingiber officinale* var. *rubrum*) on infestation of *Argulus japonicus* in goldfish (*Carassius auratus*)

S Laurensia[1], **Kismiyati**[2,3], **M Amin**[2]

[1]Undergraduate Program of Aquaculture, Faculty of Fisheries and Marine, Universitas Airlangga, Surabaya
[2]Department of Aquaculture, Faculty of Fisheries and Marine, Universitas Airlangga, Surabaya

[3]Corresponding author: kismiyati@fpk.unair.ac.id

Abstract. Goldfish (*Carassius auratus*) is one of the freshwater ornamental fish that is widely known to the public. The activity of cultivation goldfish can not be separate from pathogen infection. One of the parasites that often infests goldfish is *Argulus japonicus*. Red ginger extract Red is one of the natural ingredients can be an alternative to reduce *Argulus japonicus* infestation due to the presence of antiparasitic compounds. The purpose of this research to know the effect and determine the optimal concentration of red ginger extract for controlling *Argulus japonicus* in goldfish. The research method used is the laboratory experimental. The fish were infected with *Argulus japonicus* by artificial infestation. This study, treatment by giving red ginger extract to fish infested with *Argulus japonicus* and finding for the best concentration. The result showed a significant effect on reducing *Argulus japonicus* infestations in goldfish p(<0,05). The best treatment in this study was treatment C with concentration of red ginger extract 350 ppm with an average release of 61.11%. In this study, red ginger extract has an effect in reducing *Argulus japonicus* infestations and can be an alternative material for controlling *Argulus japonicus*.

1. Introduction

Indonesia has the potential to has abundant natural resources in an abundance of ornamental fish. One of the freshwater ornamental fish that is quite popularly cultivate in Indonesia is the goldfish. The Directorate General of Aquaculture, Ministry of Maritime and Fisheries targets goldfish production in 2019 to increase from the previous year. In 2018 the goldfish production target was 81,284.1 fish and in 2019 it increased to 102,000 fish [1]. In this case, it can be said that goldfish cultivation activities have increase and become an economic driver for ornamental fish cultivation.

The maintenance of goldfish is inseparable from the risk of disease attacks, especially diseases caused by parasites [2]. One of the parasitic species that often attacks goldfish is *Argulus japonicus*. Severe infestations by Argulus cause changes in feeding behavior, pathological changes, reduced host growth, and in most cases, death. This parasitic infection can also trigger secondary infections by bacteria and fungi [3]. Parasite control can use chemicals or antibiotics, but in long-term use, it can result in contamination of the aquatic environment, toxicity to animals and humans, and other negative impacts [4].

One of the natural product that can be used to control *Argulus japonicus* is red ginger. Red ginger extract contains secondary metabolites including tannins, flavonoids, alkaloids, terpenoids, and saponins [5]. The bioactive compounds contained in the red ginger extract are thought to reduce *Argulus japonicus* infestations. The presence of active compounds in red ginger prompted research on the effect of red ginger extract (*Zingiber officinale* var. *rubrum*) in reducing *Argulus japonicus* infestations that attack goldfish (*Carassius auratus*).

2. Material and methods

2.1 Place and time

The research was conducted in January – March 2022 at the Anatomy and Aquaculture Laboratory, Faculty of Fisheries and Marine, Universitas Airlangga, Surabaya.

2.2 Research tools and materials

The tools used in this study consisted of 20 aquariums with size 15x15x30 cm^3, aerator, water quality test (thermometer, DO meter, pH meter), plastic clip, blender, beaker glass, petri dish, analytical scales, filter paper, aluminum foil, measuring glass, buchner funnel, stirrer, erlenmeyer flask, petri dish, rotary evaporator vacuum. The materials used in this research include of 40 goldfish (*Carassius auratus*) size 7-10 cm, 200 *Argulus japonicus*, red ginger, ethanol 96%, aquades, and fish feed (pellets).

2.3 Research method

The research method used is a laboratory experimental method. The data presentation using descriptive method. The experimental method used in this study was a completely randomized design as the experimental design.

2.4 Research procedure

Identification species of *Argulus* on experiment were examined under a microscope to ensure that the species obtained was *Argulus japonicus*. The key to identifying morphology of *Argulus japonicus* species based on [6] is observing shape of the respiratory area and swimming leg. The research procedure *Argulus japonicus* to infested in goldfish was carried out according to the artificial infestation procedure [7]. *Argulus japonicus* was fasted by separating it from the host for 24 hours so that the artificial infestation process was faster attached to the fish. Then add the goldfish and *Argulus japonicus* (1 fish infested 3 *Argulus*), and wait about 15 minutes until the *Argulus* sticks to the fish. The red ginger extraction process refers to [8] using the maceration method for 3x24 hours with ethanol 96% as solvent. Extraction of red ginger using 96% ethanol solvent with a ratio of 1:5. The results of the red ginger macerate were evaporating with a rotary evaporator at a temperature of 40° and a rotating speed of 60 rpm to obtain a thick red ginger extract. The concentrations of red ginger extract used in this study were: 0 ppm, 250 ppm, 350 ppm, 450 ppm, and 550 ppm. Goldfish that had been infested by *Argulus* were put into the treatment aquarium which had been filled with red ginger extract as many as 2 fish per aquarium. Observations were made for 60 hours. Observations of water quality were carried out before and after treatment.

3. Result and discussion

3.1 Identification of Argulus japonicus

Based on the identification key [6], the ectoparasites used in this study were the genus *Argulus* and included the species *Argulus japonicus*. *Argulus japonicus* has a flattened dorsoventral. On the cephalothorax, there are a pair of antennae I and antennae II and a pair of eyes. The thorax has a proboscis, a pair of maxilla I, a pair of maxilla II which on the basal plate has three protrusions, a respiratory area, and a swimming leg. The abdomen contains the testes in males and spermathecae in females. (Fig.1)

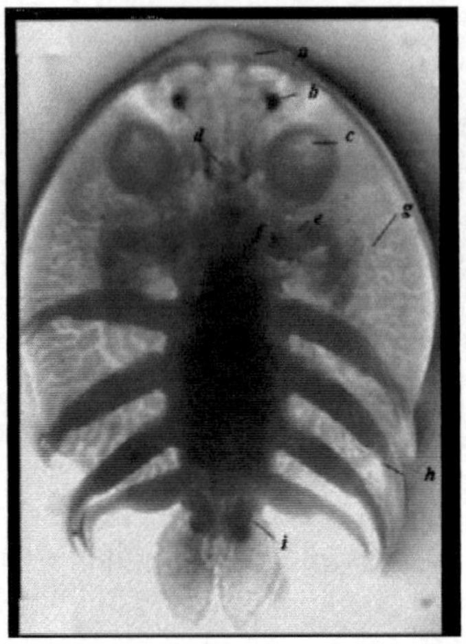

Figure 1. Morphology of *Argulus japonicus*; a, antenna; b, eyes; c, maxilla I; d, proboscis; e, maxilla II; f, basal plate; g, respiratory area; h, legs (pinnate setae); i, spermathecae (magnification 40x)

Specific characteristics of *Argulus japonicus* were observed, namely in the maxilla I contained 6-7 supporting rods, small anterior respiratory area and large posterior respiratory area, the third swimming leg has protrusions, and spines on the basal plate maxilla II (Fig.2).The difference *Argulus japonicus* with others seen in size and shape of carapas. The size of a adult *Argulus japonicus* and *Argulus foliaceus* is almost the same at 6-8mm, while the size of *Argulus coregoni* reaches 12mm [9,10]. Based on the statement of [11], *Argulus japonicus* has a carapace shape that is more pointed than *Argulus foliaceus* and has a border that extends beyond the abdomen. According [12] the posterior end of *Argulus coregoni* abdomen is pointed and sharp, the posterior end of *Argulus foliaceus* is sharply triangular, while the posterior end of *Argulus japonicus* is rounded.

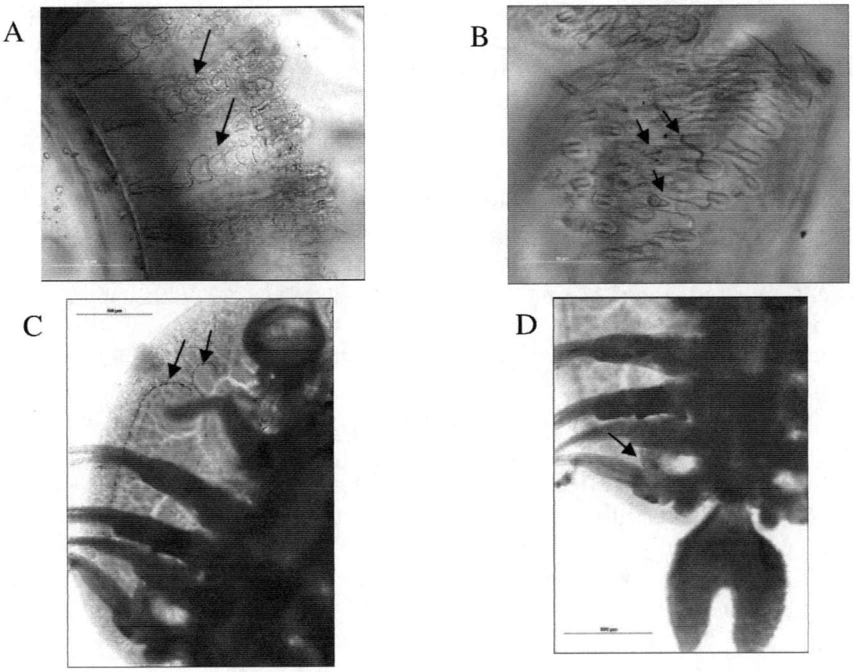

Figure 2 Identification of *Argulus japonicus*, A. Supporting rods on maxilla I (scale bar=50μm), B. Spina on basal plate of maxilla II (scale bar=50μm), C. Respiratory area (scale bar=500μm), D. Leg (scale bar=500μm)

3.2 Release rate of Argulus japonicus in goldfish

The results showed that red ginger extract had a significant effect on reducing *Argulus japonicus* infestation ($p<0.05$) in goldfish. The highest percentage of release was in treatment E (550 ppm) with an average release of 72.22%, then followed by treatment D (450 ppm) at 66.67%, treatment C (350 ppm) at 61.11%, treatment B (250 ppm) was 38.89%, and treatment A (control) was 0%. However, the best treatment for giving red ginger extract to *Argulus japonicus* in this study was treatment C (350 ppm) with an average release of 61.11% because in treatment D and treatment E there was death of goldfish. The graph of the release rate of *Argulus japonicus* is shown in Fig.3.

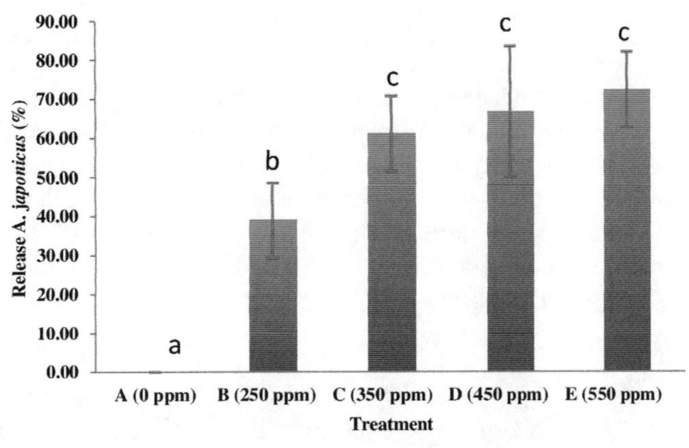

Figure 3. Graph of Release Rate of *Argulus japonicus*

The results of this study indicate that red ginger extract affects the release rate of *Argulus japonicus*. This is because red ginger extract has compounds that can weaken and even cause the death of *Argulus japonicus.* According to [13], phytochemical compounds contained in red ginger include flavonoids, saponins, tannins, and alkaloids that act as antimicrobials.
Flavonoid compounds can be antiparasitic because they can damage cell membranes by denaturing cell proteins so that cell membrane permeability is disrupted which causes leakage of cell contents [14]. Saponin compounds interfere with the work of enzymes and damage the cytoplasmic membrane which results in inhibition of nutrient transport [15]. Saponin compounds can cause hemolysis of erythrocytes [3]. Alkaloid compounds can inhibit the work of enzymes to synthesize parasitic proteins. Inhibition of this enzyme can result in disturbed parasite metabolism [16]. Tannin compounds can interfere with the work of enzymes, adhesins, and protein transport which results in decreased cell wall permeability and eventually cells will die [15]. Red ginger also contains gingerol which is a phenol derivative compound that causes protein coagulation so that cells become lysed [17].

Goldfish as a host in the study experienced death, namely in treatment D as many as 1 fish and in treatment E as many as 2 fish. This can occur due to the different immune systems of fish in adapting to environmental conditions and also the effect of the content of the red ginger extract. According to [3], the coumpound of saponin and alkaloids in the ethanolic extract and with phenolic compound might be the reason for fish mortality. Goldfish that are still alive after soaking red ginger extract is transferred to a rearing container filled with clean fresh water and aerated so that oxygen intake in the water is met, thereby reducing stress on fish.

3.3 Water quality
Water quality is a supporting factor that must be considered during the research. The water quality parameters observed were DO, pH, and temperature (Table 1.)

Table 1. Water quality data for each treatment

Treatment	Temperature (°C)	DO (mg/L)	pH
A (Control)	26-28,8	4,6-6,3	7,3-7,5
B (250 ppm)	27-29	4,4-6	7,3-7,6
C (350 ppm)	26,9-29,3	4,6-6,1	7,3-7,6
D (450 ppm)	27,5-29,1	4,7-6	7,3-7,7
E (550 ppm)	26,8-28,6	4,5-6,5	7,3-7,6

The water quality parameters measured in this study were temperature, dissolved oxygen (DO), and acidity (pH). The result of observations of the average water quality before and after treatment increased and decreased. Changes in water quality that occur have no significant effect and was still in the optimal range. The result of temperature observation was obtained between 26-29°C, DO between 4-6 mg/L, and pH between 7.3-7.7. These results indicate that the water quality condition is classified as optimal. Based on the [18], the optimal water quality values for goldfish rearing are temperatures in the range of 25-32°C, pH between 5.5-9, and DO between 5-7 mg/L.

4. Conclusion
Red ginger extract (*Zingiber officinale* var. *rubrum*) has an effect in reducing *Argulus japonicus* infestations in goldfish (*Carassius auratus*). The optimal concentration of red ginger extract that can reduce *Argulus japonicus* infestation in goldfish in this study was 350 ppm with a release percentage of 61.11%.

IOP Conf. Series: Earth and Environmental Science 1273 (2023) 012016 doi:10.1088/1755-1315/1273/1/012016

5. References

[1] Director General of Aquaculture 2019 Quarterly I Performance Report 2019 (Ministry of Maritime Affairs and Fisheries) p 21.

[2] Anggraini R and Gultom E S 2017 *J.Biosains* 3 86-89.

[3] Kumari P, Kumar S, Rames M , Shameena S, Deo A D, Rajendran K V, Raman R P 2019 *Aquaculture* 551: 1-7.

[4] Leal J F, Santos E B H, and Esteves V I 2018 *Reviews in Aquaculture* 1–19.

[5] Munadi R 2020 *Cokroaminoto J of Chem Sci* 2 1-6.

[6] Rushton-Mellor S K 1994 *Systematic Parasit* 28 51-63.

[7] Kismiyati. 2009 Ectoparasitic infestation (*Argulus japonicus*) on goldfish (*Carassius auratus*) and efforts of fish control with fish of Sumatra (Surabaya: Airlangga University).

[8] Indrawati I, Miranti M, and Mayfi I R 2017 *Nusa Bio* 9 92-96.

[9] Rushton-Mellor S K and G A Boxshall 1994 *J Nat Hist* 28 763-785.

[10] Gurney R 1948 *Proc Zool Soc London* 118 553-558.

[11] Mousavi H E, Behtash F, Rostami-Bashman M, Mirzargar S S, Shayan P, Hooman 2011 *HVM Bioflux* 3 198-204.

[12] Zoysa M D, Ryu S, Kim H, Park B K 2017 *Pakistan J Zool* 49 727-732.

[13] Iheanacho S C, Ogunji J O, Oguenji E O, Nwuba L A, Nnatuanya I O, Ochang S N, Mbah C E, Usman I B, Haruna M 2017 *J of Pharma and Phytoc* 6 761-767.

[14] Rohyani I C, Aryanti E, and Suripto 2015 *Pros Sem Nas Biodis Indon* 1 288-391.

[15] Wila H, Yusro F, and Mariani Y 2018 *J Tengkawang* 8 38-49.

[16] Latief A 2009 *Traditional Medicinal Plants (Jakarta: EGC Medical Book).*

[17] Handrianto P 2016 *J of Res and Tech* 2 1-4.

[18] Indonesian National Standard 2013 Production of goldfish (*Carrasius auratus)* (SNI-7872:2013).

6. Acknowledgements

The author expresses many thanks to the supervisors who have given a lot of knowledge and energy. The author also thanks the editors and reviewers for their suggestions and input on our manuscript.

IOP Conf. Series: Earth and Environmental Science 1273 (2023) 012017 doi:10.1088/1755-1315/1273/1/012017

Study of the population *Vibrio* Sp. on the health of vanname shrimp (*Litopenaeus vannamei*) in Cv. Reksa Bumi Situbondo

TA Setyastuti[1,2], DN Suseno[1], IGP Rumayasa[1], N Hakimah[1], FM Hilmy[1]

[1] *Sidoarjo Marine and Fisheries Polytechnic, Sidoarjo, Indonesia*

[2]Corresponding author *:* *t*asetyastuti@gmail.com

Abstract. Vannamei shrimp is one of the commodities of excellence in aquaculture. The production system is carried out mass with high stocking density causes vannamei shrimp culture susceptible to disease. One type of disease that can cause mass mortality in cultured shrimp is vibriosis. This activity aims to study the abundance of bacterial populations *Vibrio sp*. and its effect on the health of Vanname Shrimp (*Litopenaeus vannamei*) at CV. Reksa Bumi Situbondo East Java. Identification of Vibrio sp. performed on TCBS agar media. Clinical symptoms of shrimp that are attacked by vibriosis. Besides the attack vibrio sp. produce clinical symptoms such as body flushing, red swimming legs and red tail. The abundance of Vibrio sp. that exceeds the threshold causes WFD disease attacks and IMNV. The high abundance of Vibrio sp. caused by conditions unfavorable aquatic environment due to entering the transition season resulting in sudden changes in water quality and increase the abundance of bacteria.

1. Introduction

Cultivation is one of the activities in increasing production fishery. Vanname shrimp (*Litopenaeus vannamei*) is one of the leading commodity in aquaculture. This apart from the price is competitive, the production system can also be done independently mass with high stocking densities [1].

Disease is a major obstacle in business development efforts cultivation because it can cause elatively high mortality. Consequence disease attack on shrimp farming, shrimp farming production decreased from 409,590 tons in 2008 to 338,060 tons in 2009 [2]. Occurrence of related diseases closely related to pond water quality. Pond water quality is one location with other locations are different. Pond water quality is affected by the layout of the pond and the influence of the surrounding environment such as agriculture, settlements, mangroves and estuaries. This causes the pond have different water quality. When the water quality in the pond not good, it will cause disease in biota cultivation [3,4,5].

One type of disease that can cause mass death in cultured shrimp is vibriosis. This disease is caused by bacteria Vibrio genera such as *V. harveyi*, *V. alginolyticus*, *V. Parahaemolyticus* and *V. penaeicida* [6,7, 8]. Clinical symptoms of infected shrimp vibriosis, i.e., brownish red gills and hepatopancreas, the telson, uropod and abdominal parts are red, and swim slow [9,10,11]. One of the determining factors incidence of disease is low environmental quality, for example high organic matter content in the water. Therefore one way to prevent disease is to keep environment, namely maintaining good water quality [12].

2. Material and Methods

The research was conducted from March to July 2022 at CV. Bumi Reksa located in Buduan Village, Suboh District, Situbondo Regency, East Java Province. The testing process for *Vibrio sp*. carried out once a week planted on TCBS agar media using the spread method and calculating the amount bacterial colonies using the petri dish count method or so-called Total Plate Counts. Shrimp health monitoring is carried out regularly macroscopically and microscopically once every 10 days at the same time by sampling the growth of shrimp.

3. Results and Discusion

3.1. Abundance of Vibrio sp.

Tests on the abundance of Vibrio bacteria were measured starting from DOC 27 up to DOC 76 with values ranging from 4.9 x 10^3 – 2.8 x 10^4 CFU/ml. The highest number of colonies was found in DOC 69 and the number of colonies lowest at DOC 27. Based on the results of testing the abundance of bacteria Vibrio sp. showed that the abundance of bacteria exceeded the threshold minimal bacteria in the water. According to SNI (2016), the minimum thresholdthe presence of Vibrio sp. in water according to is 10^3 CFU/ml. The results of testing the abundance of Vibrio sp. can be seen on Figure 1.

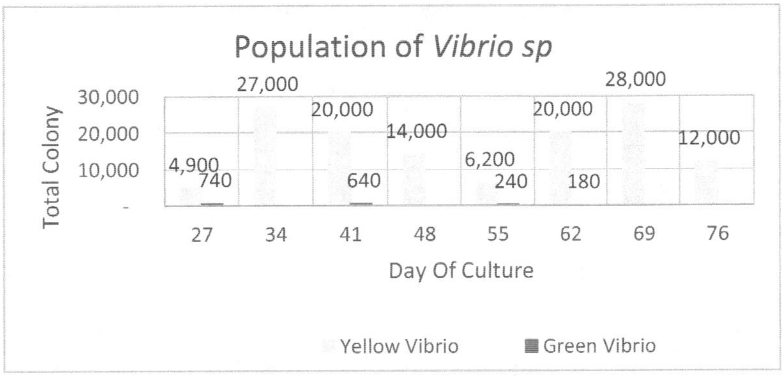

Figure 1. Population of *Vibrio sp*

The abundance of *Vibrio sp*. DOC 34 and 69 are suspected because the value of increase in salinity and organic matter (TOM). Thing This is in accordance with the opinion of Cui et al., [13] *Vibrio sp* on intensive shrimp farming activities will continue to increase due to the presence of high accumulation of organic matter, as well as salinity levels saturated. Organic matter has a static effect on biodiversity and bacterial community structure in aquatic ecosystems. Organic matter that increases with increasing feed input will create a community structure of microorganisms in pond waters increasingly diverse [14].
Efforts were made to suppress the growth of Vibrio sp. In CV. Mutual Earth by siphoning regularly, adding water and Bacillus probiotic administration. As mentioned by Kuebutornye et.al [15], as a probiotic, Bacillus has the following characteristics: exceeds other probiotics, namely its ability to produce spores and secondary metabolites that are effective against many microbial pathogens. Thing This causes a decrease in the number of Vibrio in water media maintenance after application.

3.2. Macroscopic Shrimp Health Monitoring

Shrimp that are attacked by vibriosis have decreased until they lose their appetite eat so that the intestines are empty and cause growth retardation shrimp. In addition, the attack of Vibrio sp. give rise to symptoms clinical features such as flushed body, red swimming legs and red tail. This is in accordance with Sarjito et al.,[16] that the clinical symptoms of shrimp attacked by vibriosis is the body (carapace) reddened, melanosis in the skin, necrosis of the tail, swimming and walking legs reddened The reddened hepatopancreas tends to be dark in color. Clinical symptoms of *Vibrio sp.* on the shrimp can be seen in Figure 2 below.

Figure 2. Clinical symptoms of Vibrio sp. on shrimp (a) red body (b) empty intestine (c) red tail
Source: Primary Data (2022)

Shrimp sampling is carried out every 10 days starting from DOC 40. The results of the shrimp sampling can be seen in Figure 3.

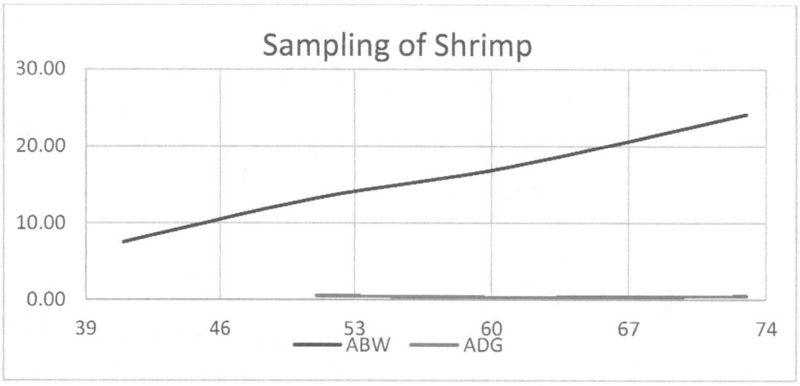

Figure 3. Sampling of shrimp for ABW and ADG

Based on the results of routine sampling, it can be seen that the value of ABW shrimp is always increasing steadily. On the plot with a low ABW value is thought to be caused by a decrease in the ADG. value caused by WFD disease so that shrimp experience a decrease in appetite eat.

3.3. Microscopic Shrimp Health Monitoring

According to Xu et al.,[17] as a vital organ for storage nutrition in vannamei shrimp (*Litopenaeus vannamei*), hepatopancreas has various functions, including for energy metabolism, absorption of nutrients, synthesis of digestive enzymes and also for immunity body and functions as a detoxifier. The

hepatopancreas is the center from the digestion of shrimp and located in the head, in conditions normally brown. Through visual observation of hepatopancreas can be identified the condition and quality of shrimp that related to appetite. If the hepatopancreas organ has damage will disrupt the metabolic system in the shrimp body : (a) Wrinkle. Wrinkles are the number of microphiles in the hepatopancreas, microfili is the outer skin that coats the lipids of the hepatopancreas related to the function of absorption of nutrients from food. wrinkle count based on wrinkles in the hepatopancreas, (b) Empty. The condition of vacancy in the hepatopancreas is calculated from the number of empty. The more empty gaps indicate that the shrimp's appetite low, (c) Necrosis. Necrosis is acute cellular damage and can be focal or massive, resulting in the network is not formed intact anymore complete shrinkage or shrinkage of the nucleus. Cell necrosis caused by biological agents such as viruses, bacteria, fungi, and parasites [18].
Based on the results of the assessment of the shrimp hepatopancreas, the results obtained were on pond E3 : Wrinkles ranged from 45%-55%. While the blanks ranged from between 45%-55%. There was not necrosis found as in the assessment hepatopancreas can be seen in table 1 below.

Table 1. Shrimp Hepatopancreas Assessment Results

DOC	**Kerut (%)**	**Kosong (%)**	**Nekrosis**
32	45	50	-
40	45	45	-
51	50	50	-
60	50	50	-
72	55	55	-

Source : Data Primer (2022)

Based on the results obtained, the longer the shrimp culture, the the higher the value of wrinkle, empty and also necrosis in the shrimp hepatopancreas. This is due to the condition of water quality which gets worse along with the growth of shrimp resulting in decreased appetite so that shrimp are more susceptible to disease. According to Annisa et al.,[19] that the hepatopancreas of infected shrimp *Vibrio sp.* shows severe necrosis and damage network structure. Based on the assessment of hepatopancreas experiencing increase in wrinkle, blank, and necrosis as the an increase in the abundance of *Vibrio sp.* on shrimp. As for wrinkles and necrosis of the hepatopancreas can be seen in Figure 4.

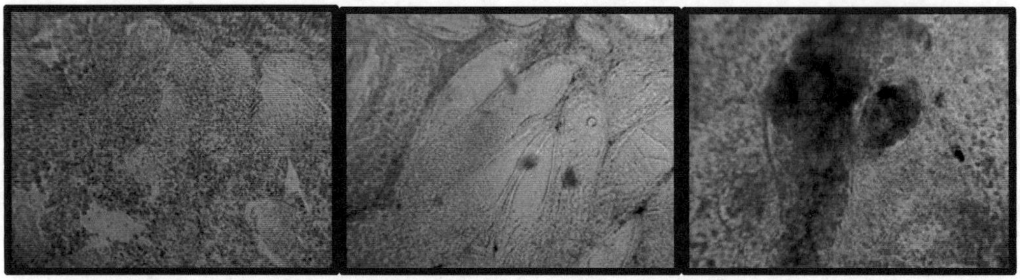

Figure 4. Wringkle and Necrosis of Hepatopancreas
(a) healthy hepatopankreas (b) wringkle 70% (c) Mild Necrosis

3.4 Water Quality Monitoring

Optimal pond water quality will provide living space so that shrimp can live properly and will grow optimally. When the environment can provide adequate water quality according to needs shrimp, the survival rate will be high and the growth of shrimp will optimally so that the production target is achieved as expected. Water quality determined by the constituent variables. Water quality must monitored regularly to determine the environmental conditions of cultivation whether the water quality is in accordance with the optimal growth of vanname shrimp.

Monitoring of water quality is carried out on a daily basis and weekly includes physical, chemical and biological factors. Quality Parameters Water physics consists of temperature and brightness. Water quality parameters Chemical composition consists of pH, salinity, DO, NH3, and TOM. Biological parameters consists of a total of bacteria.

4. Conclusion

Based on the results of the analysis obtained in this study on CV. Reksa Bumi concluded that the abundance value of vibrio sp. in the vanname shrimp rearing pond ranges from 4.9×10^3 - 2.8×10^4 CFU/ml. The abundance of bacteria exceeds the minimum bacterial threshold in waters SNI minimum threshold for the presence of Vibrio bacteria sp. in water is 10^3 CFU/ml. High abundance of vibrio bacteria sp. in pond waters, shrimp enlargement causes shrimp to be attacked WFD disease or white stool, besides that it also causes weakening of shrimp immunity so that shrimp are attacked by IMNV disease. Follow-up and prevention efforts carried out in CV. Reksa Bumi Situbondo, namely by siphoning regularly, adding water, scale water changes and administration of probiotics *Bacillus sp*.

5. Reference

[1] Mangampa, M dan Suwoyo, HS, 2016 *J Ris Akua. 5* 3 pp 351-361.
[2] Ministry of Maritime Affairs and Fisheries. 2013. Maritime Affairs and Fisheries in Numbers. Jakarta. p 118 .
[3] Manan, A., & Kharisma, A. 2012. *J Ilm Perikan Kelaut*, *4*(2), 128-134.
[4] Galang, D. P., Ashari, A. K., Sulmatiwi, L., Mahasri, G., & Sari, L. A. 2019 *IOP Conf. Ser.: Earth Environ. Sci.* Vol. 236(1) 012014.

[5] E D Masithah, D D Nindarwi, A L A Suyoso and D Husin. IOP Conf. Ser.: Earth Environ. Sci. 236 012018
[6] Helmi, A. M., Mukti, A. T., Soegianto, A., and Effendi, M. H. 2020. *Sys Rev Pharm, 2020; 11 (8): 51, 58.*
[7] Haifa-Haryani, W. O., Amatul-Samahah, M. A., Azzam-Sayuti, M., Chin, Y. K., Zamri-Saad, M., Natrah, I., and Ina-Salwany, M. Y. 2022 *Microorganisms*, *10*(9), 1851.
[8] Asplund, M E 2013 *Ecological Aspects of Marine Vibrio Bacteria*. University of Gothenburg Kristineberg 48 p.
[9] Ramesh, K, Natarajan, M, Sridhar, H, and S Umamaheswari 2014 *Glob J of Bio Scie and Biotec.* 3 1 109-114.
[10] Kharisma, A., Tjahjaningsih, W., and Sigit, S. 2020. IOP Conf. Ser.: Earth Environ. Sci. 441(1) 012012).
[11] Mahasri, G., Sulmatiwi, L., Pamenang, G. D., & Harifa, A. I. 2019. *Indian Vet. J. 96*(5), 21-23.
[12] Murdjani. M. 2002. Identification and pathology of Vibrio alginolyticus in rat grouper (*Cromyleptes altivelis*) Dissertation of Postgraduate Program, University of Brawijaya Malang.
[13] Ariadi, H, Fadjar, M, Mahmudi, M, & Supriatna 2019 *Aquac Aqua Cons & Legis*. 12 6 2103-2116.
[14] Kuebutornye, F K A, Abarike, E D, Lu, Y, Hlordzi, V, Sakyi, M E, Afriyie, G, Wang, Z, Li, Y, Xie, C X, 2020 *Fis Physiol Biochem*. 1–23.
[15] Sarjito, M Apriliani, D Afriani, dan A H Condro Haditomo 2015 *J Kel Trop*. 18 3 189-196.
[16] Xu, C, E Li, Y Liu, X Wang, J G Qin and L Chen 2017 *J of Prote*. 1-10.
[17] Putri, A M, Prayitn, S B, & Sartijo 2015 *J of Aqua Manag and Tech*. 4(4), 141-149.
[18] Annisa, N, Sarjito, dan Slamet Budi Prayitno 2015 *Skrip*. Hal 54-60.
[19] SNI (SNI 01-7246-2006) 2006

6. Acknowledgments

Thank to all the staff of CV Bumi Reksa Situbondo, IOP reviewers and Sidoarjo Marine and Fisheries Polytechnic for all the opportunities that have been given.

IOP Conf. Series: Earth and Environmental Science 1273 (2023) 012018 doi:10.1088/1755-1315/1273/1/012018

Effect of Probiotics in Recirculating Aquaculture System (RAS) on the Concentration of Ammonia, Nitrite, and Nitrate in the Aquaculture of Catfish (*Clarias sp.*)

A Nurhasanah[1], B S Rahardja[2,3], and Prayogo[2]

[1]Aquaculture Study Program, Faculty of Fisheries and Marine Affairs, Airlangga University, Surabaya 60115, Indonesia
[2]Department Aquaculture, Faculty of Fisheries and Marine, Airlangga University, Surabaya 60115, Indonesia

[3]Corresponding author: bs_rahardja@fpk.unair.ac.id

Abstract. The purpose of this research is to determine the effect of probiotic administration on ammonia, nitrite, and nitrate levels in catfish cultivation *(Clarias sp.)* with the RAS system. The research was conducted in February – March at the Anatomy Laboratory and chemical laboratory of the Faculty of Fisheries and Marine Airlangga University. This study was conducted experimentally using the RAL method consisting of 5 treatments four replays, namely control treatment (P0), probiotic dose 0.5 mL/L (P1), probiotic dose 1 mL/L (P2), probiotic dose 1.5 mL/L (P3), and probiotic dose 2 mL/L (P4). Data analysis is processed using ANOVA and DMRT. The result of this study was the administration of probiotics with the bacterium *Nitrosomonas* sp. And *Nitrobacter sp.* with different doses has an effect ($p<0.05$) on the concentration of ammonia, nitrite, and nitrates.

1. Introduction

Catfish farming can be done intensively using high density, artificial feeding, and conducting regular water turnover. Catfish cultivation can intensively use a high solid spread that reaches 1500 fish /m^3 [8]. Catfish production is inseparable from waste leftover feed and the metabolism of fish containing ammonia. The rest of the feed and the build-up of feces can increase ammonia, nitrites, and nitrates. Average ammonia increase of 0.022-0.030 mg/L either by artificial feeding or by commercial feeding [5]

RAS system (*Recirculating Aquaculture System)* is a closed water system that reprocesses water used using filters so that the water quality becomes better. Filters in the RAS system serve to filter water to be clearer and neutralize the content of ammonia through the process of nitrification. The filter consists of a physics filter that acts as a filter of solids content in water, and a physics filter can separate suspended particles efficiently. Chemical filters serve to filter the content of chlorine and organic impurities using activated carbon. Biological filters consist of decomposing microorganisms that help in improving water quality [4]

The addition of Probiotics with bacterial content *Nitrosomonas sp* and *Nitrobacter sp.* in RAS system (*Recirculating Aquaculture System)* can help optimize the work of biological filters in degrading organic matter to degrade ammonia compounds in the waters [10]. Bacteria *Nitrosomonas sp.* and *Nitrobacter sp.* is a bacterium that can help lessen ammonia into nitrates through the nitrification process. *Nitrosomonas sp* will help ammonium interact with oxygen so nitrite, and *Nitrobacter sp.* will help the oxidation process of nitrites into nitrates [13]

2. Methodology

2.1. Place and Time

The research was conducted on February 22 - March 23, 2021, at the anatomy Laboratory and Chemical Laboratory of the Faculty of Fisheries and Marine Airlangga University, Surabaya.

2.2 Materials

2.2.1 Research Equipment

The research equipment used is plastic buckets as many as 20 pieces of volume 15 L and 40 pieces of volume 10 L, aerators, hoses and aeration stones, PVC pipes, streaming wire, water pumps, fishing nets, digital scales, and water quality measuring instruments, spectrophotometers.

2.2.2 Research Materials

The research material used is catfish length 10.58 ± 0.65 cm and weight 7.45 ± 1.68 gr from IBAT Mojokerto as much as 400 heads. Probiotic commercial bacterial *content* Nitrosomonas sp. and Nitrobacter sp. density 1.0 x 10^8 CFU/mL *as much as* 1 L. Filtration media consists of dacron, coral stone, and bioball. The feed used is HI-PRO-VIT 781-1 with 29.97% protein. Ammonia, nitrate, and nitrite test materials are Phenol, NaCl, Sodium nitroprusside, Trinatrium Citrate, NaOH, Natrum Hypochlorite, Aqueous, KNO, and HCl.

2.3 Research Methods

The research method used in this study is experimental.

2.3.1 Research Plans

The research design used is The Complete RandomIzed Design (RAL). This study used five treatments namely P_0, P_1, P_2, P_3 and P_4 with each there are four times the replay consists of: P_0 = 10 L water + 20 fish (control), P_1 = 10 L water + 20 fish + probiotics 0.5 ml / liter, P_2 = 10 L water + 20 fish + probiotics 1 ml / liter, P_3 = 10 L water + 20 fish + probiotics 1.5 ml / liter and P_4 = 10 L water + 20 fish + probiotics 2 ml / liter. The observed variables include, variabel free: the addition of commercial probiotics with the content of *Nitrosomonas* sp. and *Nitrobacter* with different doses, variabel terikat: ammonia, nitrite, nitrate, and variable control: catfish: type, solid amount spread, fish size, feed requirement, RAS installation size and filter media.

2.3.2 Working Procedures

a) preparation of maintenance media and seed dispersal: cleaning equipment in the form of plastic buckets as a place of maintenance of fish, filtration media, and tubs of stock washed using soap, rinsed, and dried. Applying water for one night before use for cultivation purposes. Water filling is done after a dry maintenance tub of 10 liters. Catfish seed is then put in a maintenance tub with a density of 2 fish/ L. b) made RAS installation: starting with a hole in the inlet and outlet in the plastic bucket, stringing PVC pipes so that it takes the form of the recirculation system. The filters used are dacron, coral stone, and bioball. c) catfish maintained for 30 days. Sampling measurements of average weight and initial biomass of fish. The amount of feed given daily is as much as 5% of the total weight of fish. Feeding is done with frequency three times a day, and adding probiotics are done once a week.

2.3.3 Research Parameters

a) the main parameters observed were ammonia, nitrite, nitrate, and b) parameter supporting observed is water quality including DO, temperature, Ph

2.3.4 Data Analysis

The data were analyzed using *analysis of Variance* (ANOVA)tests to determine the effect of the treatment given. If it shows the results of a noticeable difference, it will continue with a double comparison test Ducan (*Ducan's Multiple Range Test*), with a fundamental level of 5%.

3. Results and Discussion

3.1. Ammonia Concentration

Table 1. Average Concentration of Ammonia in Catfish Maintenance

Treatment	Ammonia concentration (mg/l) ± SD)			
	Week 1	Week 2	Week 3	Week 4
P0	$0.219^{a} \pm 0.007$	$0.301^{bc} \pm 0.0174$	$0.313^{bc} \pm 0.031$	$0.316^{c} \pm 0.033$
P1	$0.265^{b} \pm 0.038$	$0.278^{ab} \pm 0.0281$	$0.285^{ab} \pm 0.009$	$0.272^{b} \pm 0.016$
P2	$0.278^{b} \pm 0.023$	$0.264^{a} \pm 0.0072$	$0.263^{a} \pm 0.018$	$0.227^{a} \pm 0.009$
P3	$0.278^{b} \pm 0.016$	$0.321^{c} \pm 0.0141$	$0.341^{c} \pm 0.038$	$0.339^{c} \pm 0.029$
P4	$0.259^{ab} \pm 0.039$	$0.304^{bc} \pm 0.0312$	$0.313^{bc} \pm 0.009$	$0.335^{c} \pm 0.020$

Description: Different superscript values show real different results (P<0.05).

Table 1. The results of the analysis of variants (ANOVA) showed a noticeable influence of probiotic administration on ammonia concentrations ($p<0.05$) in each treatment. *Duncan's Multiple Range Test* results in week 1 showed that P0 is significantly different from P1, P2, and P3. Week 2 shows P2 is significantly different from P0, P3, and P4. Week 3 shows that P2 is significantly different from P0, P3, and P4. Week 4 shows that P2 differs significantly from P0, P1, P2, P3, and P4. The ammonia fluctuation graph can be seen in Figure 1.

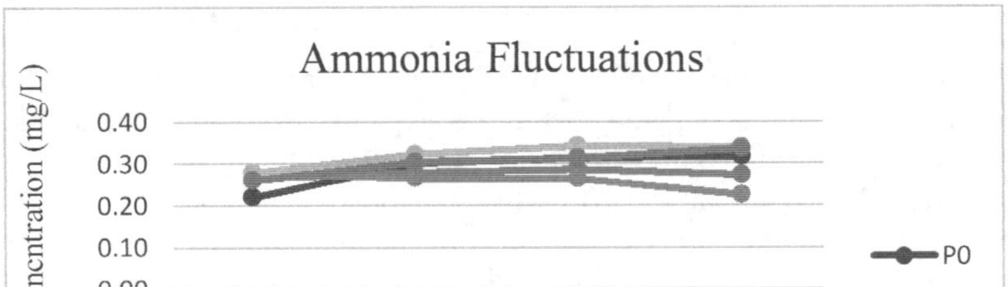

	1	2	3	4
P0	0.219	0.301	0.312	0.316
P1	0.265	0.278	0.285	0.272
P2	0.278	0.264	0.263	0.223
P3	0.278	0.321	0.341	0.338
P4	0.259	0.304	0.313	0.335

Figure 1. Graph of ammonia fluctuations

ANOVA calculation of P2 treatment obtained the best ammonia concentration results from other treatments with probiotics containing *Nitrosomonas* and *Nitrobacter* by 1 ml/L. Administration of probiotics with such doses shows that probiotics work effectively to reduce the content of ammonia in aquaculture water. Administration of probiotics with *Nitrosomonas* bacteria content in the cultivation maintenance media can help decompose organic matter by decomposing ammonia into nitrites so that the concentration of ammonia decreases [10]. Probiotics in the maintenance medium will suppress the increase in ammonia by decomposing organic matter in the waters [2]

Ammonia is one of the most important parameters of water quality; ammonia is the final product of the protein catabolism process that can become a water toxin that can cause damage to fish gills, difficulty breathing, until death. Water quality is also one of the factors that influence the growth of bacteria. The pH value for bacterial growth nitrification is 7-8.5 at 20-30cC. *Nitrosomonas sp.* and *Nitrobacter sp.* bacteria require oxygen to degrade ammonia. When the water's oxygen content is low, it can cause the ammonia degradation process to be imperfect. The administration of high doses of probiotics causes competition in obtaining oxygen and nutrients between bacteria, causing a reduction in ammonia that is not maximal [1]. Overdosing on probiotics will result in changes in maintenance media conditions that impact low fish growth.

3.2 Nitrite Concentration

Table 2. Average Nitrite Concentration in Catfish Maintenance

Treatment	Nitrite Concentration (mg/l) ± SD			
	Week 1	Week 2	Week 3	Week 4
P0	$0.515^{b} \pm 0.014$	$0.441^{ab} \pm 0.096$	$0.415^{bc} \pm 0.066$	$0.350^{a} \pm 0.013$
P1	$0.508^{b} \pm 0.041$	$0.509^{abc} \pm 0.069$	$0.367^{ab} \pm 0.053$	$0.323^{a} \pm 0.010$
P2	$0.441^{a} \pm 0.027$	$0.417^{a} \pm 0.047$	$0.342^{a} \pm 0.031$	$0.323^{a} \pm 0.008$
P3	$0.522^{b} \pm 0.038$	$0.541^{c} \pm 0.029$	$0.465^{c} \pm 0.025$	$0.413^{b} \pm 0.039$
P4	$0.548^{b} \pm 0.007$	$0.529^{bc} \pm 0.027$	$0.413^{bc} \pm 0.017$	$0.400^{b} \pm 0.035$

Description: Different superscript values show real different results ($P<0.05$).

Table 2. The results *of the analysis of variants* (ANOVA) showed a noticeable influence of probiotic administration on the concentration of nitrites ($p<0.05$) on each treatment. *Duncan's Multiple Range Test* results in week 1 show that P2 is significantly different from P0, P1, P3, P4. Week 2 shows that P2 is significantly different from P3 and P4. Week 3 shows that P2 is significantly different from P0, P3, and P4. Week 4 shows that P0, P1, and P2 are significantly different from P3 and P4. Graphs of nitrite fluctuations can be seen in Figure 2.

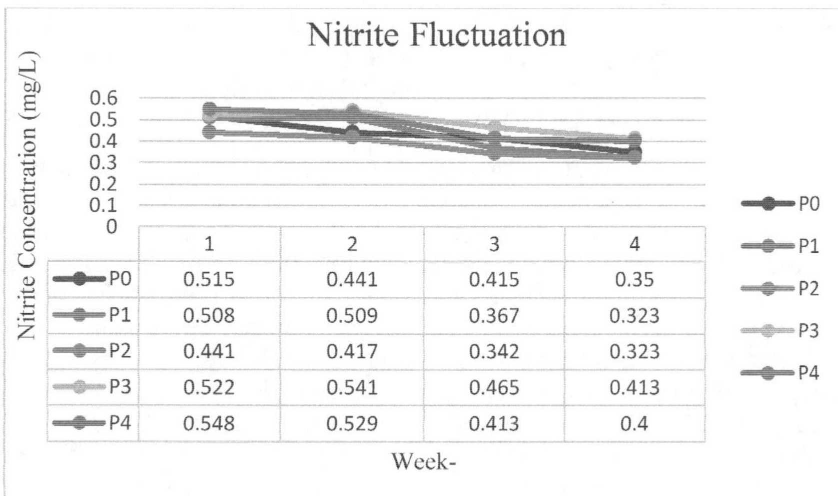

Figure 2. Graph of nitrite fluctuations

The results of the calculation of ANOVA nitrite concentration is best found in P2 treatment compared to other treatments with the addition of probiotics with the content of *Nitrosomonas sp* and *Nitrobacter sp.* as much as 1 ml/L. Low nitrate concentrations in maintenance media can be caused by *Nitrobacter*, which works effectively in remodeling nitrites into nitrates in the nitrification process. Nitrite is an intermediate product performed by bacteria in the nitrification process performed by *Nitrosomonas* bacteria by remodeling ammonia into nitrites. Nitrosomonas bacteria carry out the oxidation of ammonia into nitrites in the process of nitrification, and Nitrobacter bacteria carry out the oxidation of nitrites into nitrates in aerobic conditions [14]. Nitrite is one form of inorganic nitrogen in the water, the nitrite content in catfish culture is not more than 1.0 mg/l [11]

The process of converting nitrites into nitrates is carried out by *Nitrobacter* bacteria affected by oxygen. If oxygen is insufficient will cause the degradation process. Nitrification bacteria need oxygen to remodel ammonia into nitrites and nitrates for the nitrification process to run correctly. The speed of degradation of nitrites into nitrates will be achieved in aerobic conditions and supplemented by an abundance of nitrification bacteria [8]. Nitrobacter bacteria convert nitrites into nitrates; these bacteria require enough oxygen as a source of energy to support metabolism in the process of nitrate [15]. The recommended number of nitrites for catfish cultivation is less than 0.5 mg/L [16]. Excessive nitrate compounds in the water can decrease the blood ability of aquatic organisms in the process of oxygen binding (*Methemoglobin*) because nitrites will act stronger with hemoglobin, causing tissue damage, discoloration of the data to brown, and high mortality rates in aquatic organisms [12].

3.3 Nitrate concentration

Table 3. Average Nitrate Concentration in Catfish Maintenance

Treatment	Nitrate Concentration (mg/l) ± SD)			
	Week 1	Week 2	Week 3	Week 4
P0	$1.877^{a} \pm 0.138$	$2.877^{b} \pm 0.307$	$3.057^{b} \pm 0.234$	$3.134^{ab} \pm 0.272$
P1	$2.698^{b} \pm 0.259$	$2.892^{b} \pm 0.403$	$2.986^{b} \pm 0.144$	$3.084^{ab} \pm 0.154$
P2	$2.709^{b} \pm 0.415$	$2.803^{b} \pm 0.433$	$3.011^{b} \pm 0.551$	$3.288^{b} \pm 0.129$
P3	$2.203^{a} \pm 0.092$	$2.226^{a} \pm 0.087$	$2.189^{a} \pm 0.128$	$3.044^{ab} \pm 0.421$
P4	$2.190^{a} \pm 0.357$	$2.573^{ab} \pm 0.110$	$2.462^{a} \pm 0.412$	$2.715^{a} \pm 0.337$

Description: Different superscript values show real different results ($P<0.05$).

Table 3. The results *of the analysis of variants* (ANOVA) showed a discernible influence of probiotic administration on nitrate concentrations ($p<0.05$) in each treatment. *Duncan's Multiple Range Test* results in week 1 showed that P0 is significantly different from P1 and P2. Week 2 shows that P3 is significantly different from P0, P1, and P2. Week 3 shows that P3 is significantly different from P0, P1, and P2. Week 4 shows that P4 is significantly different from P2. A graph of nitrite fluctuations can be seen in Figure 3.

The results of ANOVA calculation of the best nitrate concentration are found in P2 treatment compared to other treatments by administering a dose of probiotics with *Nitrosomonas sp.* and *Nitrobacter sp.* content of 1 ml/L. This is thought to be due to the nitrification process that works optimally. The nitrification process works well, indicated by the opposite result between nitrite and nitrate when a low nitrite concentration eats a high nitrate concentration [6]. Nitrification causes ammonia in the maintenance medium to be reduced and causes the concentration of nitrites to increase. Still, when the nitrification bacteria work optimally, high oxygen nitrite will be immediately converted into nitrates (nitrates) so that the nitrate concentration becomes high [1]. Nitrates are compounds produced from the perfect oxidation of nitrogen compounds by *Nitrobacter* bacteria in water and are highly soluble in water.

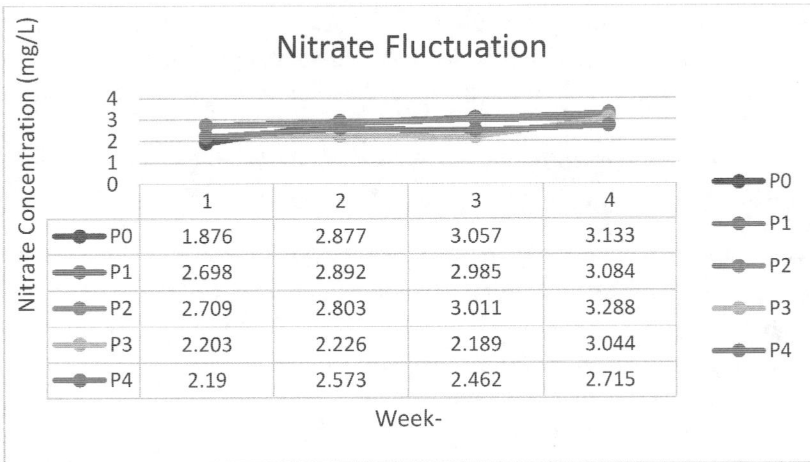

	1	2	3	4
P0	1.876	2.877	3.057	3.133
P1	2.698	2.892	2.985	3.084
P2	2.709	2.803	3.011	3.288
P3	2.203	2.226	2.189	3.044
P4	2.19	2.573	2.462	2.715

Figure 3. Graph of Nitrate Fluctuation

The nitrate content in the maintenance medium added with probiotics with the best dosage is 0.023-3.52 mg/L [9]. Nitrate (NO3) is a nutrient that, at excessive levels, can cause a decrease in the water quality of a body of water that eventually causes water pollution and results in eutrophication [14]. Water quality also affects the performance of bacteria in the nitrification process to remodel ammonia into nitrates. Nitrification bacteria grow at a temperature of 20-30cC with a DO content of more than 3 mg/L. *Nitrosomonas* bacteria can grow optimally at 5 – 30°C, pH 5.8 – 8.5 habitats in seawater, freshwater, and soil, *Nitrobacter* good at effective nitrification process is 7.5-8.5 [3]. *Nitrobacter* bacteria require oxygen to oxidize nitrites into nitrates so that sufficient dissolved oxygen content is needed in the waters to optimize the nitration process. It is supported that the poor performance of nitrification bacteria and supported low oxygen then causes the nitration process to be imperfect, resulting in nitrate concentrations [1].

4. Conclusion

Based on the results of research that has been done, the conclusion can be drawn: a) Administration of probiotic bacteria *Nitrosomonas* sp. And *Nitrobacter sp.* with different doses has an effect ($p<0,05$) on the concentration of ammonia with the best ammonia production at P2 (dose 1 ml / L). b) Administration of probiotic bacteria *Nitrososmonas sp.* And *Nitrobacter sp.* with different doses has an effect ($p<0.05$) on the concentration of ammonia with the best nitrite production at P2 (dose 1 ml / L). c) Administration of probiotic bacteria *Nitrososmonas sp.* And *Nitrobacter sp.* with different doses has an effect ($p<0.05$) on the concentration of ammonia with the best nitrate production at P2 (dose 1 ml / L).

5. References

[1] Andriani, Y., Kamil, T. I., & Iskandar, I. 2018 *Depik J Ilmu-Ilmu Perairan, Pesisir Dan Perikan.* 7(3), 209-217.
[2] Aquarista, F., Iskandar, U. Subhan. 2012 *J Perikan dan Kelaut.* 3(4): 133–140
[3] Bhunia, P. 2014 *Fundam Biol Treat.* 47–73.
[4] Darmayanti., E. I. Raharjo., Farida. 2018 *J Ruaya J Penelit Dan Kaji Ilmu Perikan Dan Kelautan* 6(02):1-8
[5] Dhiba, A. A. F., H. Syam., Ernawati. 2019 *J Penddikan Teknol Pertan.* 5(1):131-144
[6] Fujiana, F., Setyowati, D. N. A., & Setyono, B. D. H. 2020 *J Perikan* 10(2), 148-157.
[7] Hastuti, Y. P. 2011 *J Akuakultur Indones.* 10(1), 89-98.
[8] Hermawan, T. E. S. A., Sudaryono, A., & Prayitno, S. B. 2014 *J Aquac Manag Technol.* 3(3), 35-42.

[9] Julianti, V., Salamah., dan Muliani 2016 *Aquat Sci J.* 3(2):66-74
[10] Lisna Dan Insulistyowati. 2015 *J Penelit Univ Jambi Seri Sains.* 17 (2): 18 – 25.
[11] Prayogo., T.A. Putri., B.S Rahardja. 2021 *IOP Conf Ser Earth Environ Sci.* 718 (2021) 012049
[12] Setijaningsih, L., and Suryaningrum, L. H. 2015 *Ber Biol.* 14(3), 287-293.
[13] Uemoto, H., Shoji, T., & Uchida, S. 2014 *Life Sci Sp Res.* 1, 89-95.
[14] Wantasen, S. 2015 *J Bumi Lestari.* 15(2):176-183
[15] Widayat, W. 2015 *J Air Indones.* *8*(1):48-64
[16] Žibienė, G., & Žibas, A. 2017. Water Quality and Waste Contamination Study in Catfish Rearing Recirculating Aquaculture System. In International Scientific Conference Rural Development 2017 (Pp. 850-856).

IOP Conf. Series: Earth and Environmental Science 1273 (2023) 012019 doi:10.1088/1755-1315/1273/1/012019

Morphology of Female Tilapia (*Oreochromis* sp.) Broodstock at Hatchery Centers in South Sulawesi

Amrullah[1,2], Ardiansyah[1], Dahlia[1], Wahidah[1], I Dwinhoven[1]

[1]Department of Aquaculture, Pangkep State Polytechnic of Agriculture, Pangkep 90655, South Sulawesi, Indonesia.

[2]Corresponding author: ulla_285@yahoo.com

Abstract. This study aimed to evaluate the morphology of female tilapia (*Oreochromis* sp.) broodstock used in 3 hatcheries in South Sulawesi. A total of 92 female tilapia broodstock were collected from 3 hatcheries in South Sulawesi: Gowa Regency (30 fish), Bantaeng (30 fish), and Soppeng (32 fish). Morphological measurements were taken using a ruler on eight characters including the TL, SL, HL, SnL,CPL, BH, CPH and HH. Comparison of values among morphometric characters SL, HL, SnL, CPL, BH, CPH, and HH to TL was analyzed for variance (Anova) and continued with Duncan's comparison test. The results of the analysis showed that the SL/TL characters were not significantly different ($P > 0.05$) between hatchery locations, while the HL/TL, SnL/TL, CPL/TL, BH/TL, CPH/TL, and HH/TL characters were significantly different ($P < 0.05$) between locations. Most of the morphological characters in female broodstock originating from Gowa and Soppeng contributed greatly to the formation of the first main axis (F1), while the rest were scattered on the second (F2) and third (F3) axes These results indicate that, in general, in terms of morphometric variations, the female broodstock at the three tilapia hatcheries have different morphometric characteristics.

1. Introduction

Tilapia is an important commercial freshwater fish species that plays a major role in aquaculture activities [1,2,3]. This fish is ideal for cultivation because of its fast growth, high fecundity, and tolerance to poor water quality [4,5,6], making it the second most widely cultivated fish species in the world after carp [7,8,9].

Research on tilapia is generally related to aquaculture activities. Research about hatchery activities is also needed, including the use of broodstock in hatcheries, especially regarding the morphological characters of the female broodstock so that they could produce quality seeds [10,11]. Morphological evaluation, especially morphometric evaluation, has been carried out by [12,13,14]. Morphometrics is a field that studies variations and changes in shape, namely the size and shape of an organism [15]. Morphometric evaluation of fish reveals the relationships between body parts such as the length of the head, snout, eyes, body, fins, and tail. This method describes the measurement or distance between different points or landmarks [16].

The problem in tilapia seeds is a decrease in the quality of superior strains due to broodstock mismanagement, thus reducing the growth performance of the resulting seeds. The absence of an inventory of the broodstock in hatchery activities is due to the lack of broodstock management. To overcome this problem, it is necessary to manage the female broodstock in hatchery activities based on morphological character information. This study aimed to evaluate the morphology of female tilapia broodstock at three hatcheries in South Sulawesi.

2. Material and methods

Evaluation of 92 female tilapia broodstock obtained from 3 hatcheries in South Sulawesi, Gowa Regency (30 fish), Bantaeng (30 fish), and Soppeng (32 fish). Morphological measurements were taken using a ruler of eight characters: total length (TL), standard length (SL), head length (HL), snout length (SnL), caudal peduncle length (CPL), body height (BH), caudal peduncle height (CPH), and head height (HH). The ratio of values between the morphometric characters SL, HL, SnL, CPL, BH, CPH, and HH to TL was analyzed for variance (Anova) and followed up with Duncan's comparison test if differences in the characters between locations were found. Morphological characters that characterized each location underwent multivariate analysis using PCA (Principal Component Analysis).

3. Results and Discussion

This study evaluated the morphology of female tilapia from three hatcheries. Most of the ratio values of tilapia morphometric characters from the three hatcheries were significantly different ($P<0.05$). The difference in the age of the tilapia evaluated was not considered in the analysis because there were similarities in the total length and standard length. According to [17], fish size is more important than age, mainly because some factors in taxonomy, ecology, and physiology depend more on size than age.

The results of the analysis of the ratio of morphological characters to TL are listed in Table 1. The ratio of SL/TL characters was not significantly different ($P > 0.05$) between hatchery locations. This shows that, in general, the SL of the broodstock used in the three hatcheries have almost the same size ratio with a population range of Gowa (0.69 – 0.86), Bantaeng (0.78 - 0.84), and Soppeng (0.74 - 0.86).

Table 1. The ratio of morphological characters to tl in female tilapia from Gowa, Bantaeng, and Soppeng

Location	Value	Character Ratio						
		SL/TL	HL/TL	SnL/TL	CPL/TL	BH/TL	CPH/TL	HH/TL
Gowa	Minimum	0.69	0.15	0.05	0.08	0.26	0.13	0.16
	Maximum	0.86	0.23	0.08	0.12	0.33	0.22	0.26
	Average	0.80	0.18	0.07	0.10	0.30	0.17	0.21
Bantaeng	Minimum	0.78	0.16	0.06	0.08	0.29	0.14	0.17
	Maximum	0.84	0.22	0.10	0.14	0.35	0.24	0.27
	Average	0.81	0.19	0.08	0.11	0.32	0.18	0.21
Soppeng	Minimum	0.74	0.15	0.05	0.09	0.26	0.10	0.20
	Maximum	0.86	0.24	0.09	0.14	0.35	0.20	0.28
	Average	0.81	0.17	0.07	0.12	0.31	0.16	0.24

The ratios of HL/TL, SnL/TL, CPL/TL, BH/TL, CPH/TL, and HH/TL characters were significantly different ($P < 0.05$) between locations (Table 1). The results of Duncan's comparison test showed that the range of morphological characters of HL/TL in Soppeng (0.15 - 0.24) was not significantly different ($P > 0.05$) from Gowa (0.15 - 0.23) but was significantly different ($P < 0.05$) from Bantaeng

(0.16 - 0.22) which was not significantly different (P > 0.05) from Gowa. The range of morphological characters of SnL/TL in Soppeng (0.05 - 0.09) was not significantly different (P > 0.05) from Gowa (0.05 - 0.08), but both were significantly different (P < 0.05) from Bantaeng (0.06 - 0.10). Bantaeng had the highest SnL/TL ratio range. The range of morphological characters of CPL/TL in Gowa (0.08 - 0.12) was not significantly different (P > 0.05) from Bantaeng (0.08 - 0.14), but both were significantly different (P < 0.05) from Soppeng (0.14 - 0.12). Soppeng had the highest CPL/TL ratio range value. The range of the ratio of the morphological characters BH/TL in Gowa (0.26 - 0.33) was not significantly different (P > 0.05) from Soppeng (0.26 - 0.35) but was significantly different (P < 0.05) from Bantaeng (0.29 - 0.35) which was not significantly different (P > 0.05) from Soppeng. Bantaeng had the highest BH/TL ratio range value. The range of morphological characters of CPH/TL in Gowa (0.13 - 0.22) was not significantly different (P > 0.05) from Bantaeng (0.14 - 0.24), but both were significantly different (P < 0.05) from Soppeng (0.10 - 0.20). Soppeng had the lowest CPH/TL ratio range value. The range of morphological characters of HH in Gowa (0.16 - 0.26) and Bantaeng (0.17 - 0.27) were not significantly different (P > 0.05), but both were significantly different (P < 0.05)from Soppeng (0.20 - 0.28). Soppeng had the highest HH/TL ratio range value.

The Bantaeng hatchery had the highest SnL/TL and BH/TL character ratios, while the Soppeng hatchery had the highest CPL/TL and HH/TL character ratios. The lowest ratio of CPL/TL was found at Soppeng.

Based on this evaluation, female Tilapia from Soppeng were significantly higher (P<0.05) in the tail and head morphological characters than in the Gowa and Bantaeng populations. This is in accordance with [18] who stated that differences between populations of tilapia species are mostly reflected in head size.

The morphological characters that varied between locations indicated that each site consisted of a phenotypically heterogeneous group. Variations are related to the environment. According to [19], variations of several measured morphological characters (height, caudal peduncle height, snout length, pectoral fin length) will be phenotypic due to the environment. The impact of environmental variations on fish morphometric and meristic differences has been reported [20,21]. Furthermore, [12] stated that high morphological plasticity in fish can be caused by environmental conditions, wherein environmental variability is found in many freshwater fish species [23].

Most of the ratios of morphological characters in female broodstock originating from Gowa and Soppeng contributed greatly to the formation of the first main axis (F1) and the rest were scattered on the second (F2) and third (F3) axes (Table 2). The amount of variation described in the first three main axes (F1-F3) reached 63.60% with characteristic root values of 1.7541 (F1), 1.4711 (F2), and 1.2269 (F3), respectively.

Table 2. Contribution of factorial axis and correlation of variables to axis factorial in female tilapia from Gowa, Bantaeng, and Soppeng

Eigenvalues:	F1	F2	F3	F4	F5	F6	F7
Eigenvalue	1.7541	1.4711	1.2269	0.8819	0.6630	0.5615	0.4416
Variability (%)	25.0582	21.0161	17.5266	12.5980	9.4717	8.0211	6.3084
Cumulative %	25.0582	46.0743	63.6009	76.1989	85.6706	93.6916	100.0000

The morphological character variables CPL/TL, CPH/TL, and HH/TL contributed greatly to the formation of the first major axis (F1), HL/TL and BH/TL contributed to the formation of the second major axis (F2), and SL/TL contributed to the formation of the third principal axis (F3), while SnL/TL contributed to the formation of the fourth principal axis (F4) (Table 3)

Table 3. Squared cosines of the female tilapia from Gowa, Bantaeng, and Soppeng variables

The ratio of the Morphological Character to TL	F1	F2	F3	F4	F5	F6	F7
SL	0.0133	0.2457	**0.4712**	0.1187	0.0255	0.0331	0.0924
HL	0.0000	**0.4060**	0.2459	0.0260	0.3004	0.0168	0.0049
SnL	0.0299	0.1088	0.3079	**0.4732**	0.0783	0.0018	0.0000
CPL	**0.5889**	0.0024	0.0490	0.0204	0.0718	0.2368	0.0307
BH	0.1826	**0.4121**	0.0533	0.0899	0.1502	0.0200	0.0920
CPH	**0.3630**	0.2951	0.0005	0.0381	0.0207	0.2529	0.0296
HH	**0.5763**	0.0010	0.0991	0.1156	0.0162	0.0000	0.1919

Values in bold correspond for each variable to the factor for which the squared cosine is the largest.

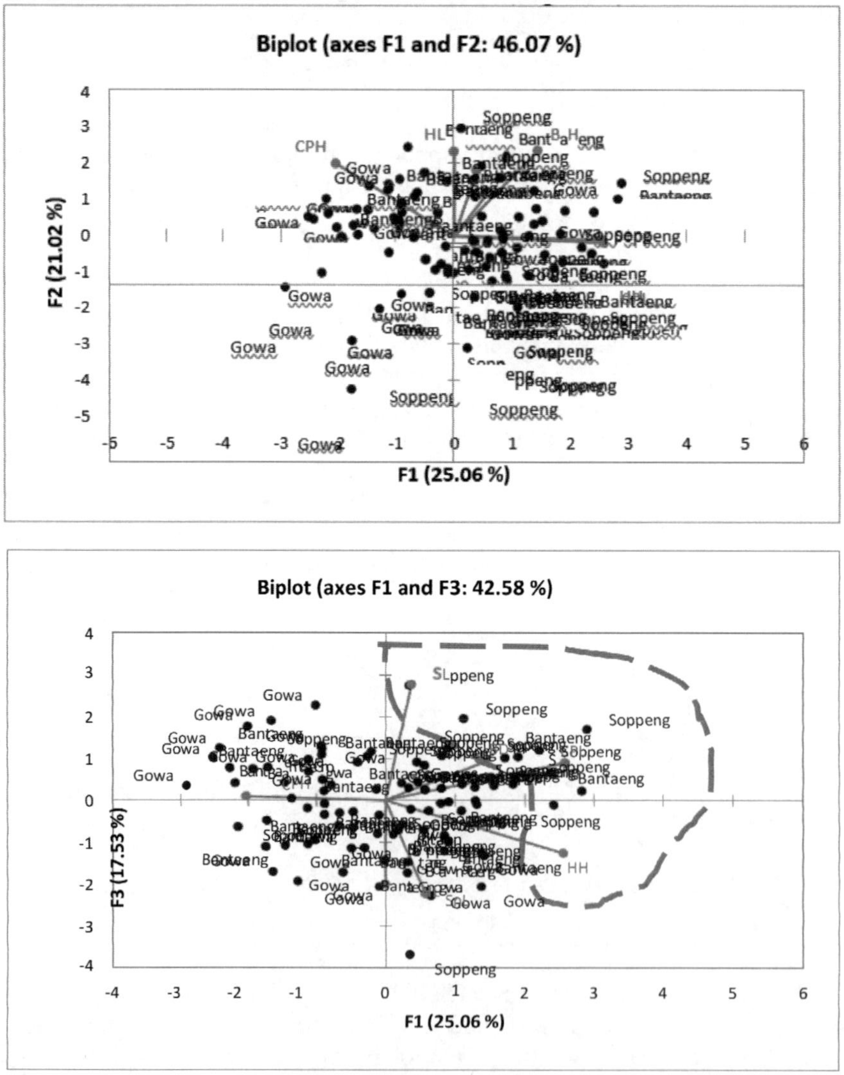

Figure 1. Profile of observation and morphological character variables distribution of female tilapia from Gowa, Bantaeng, and Soppeng.

The Gowa observation group was characterized by a relatively high ratio of CPH/TL and low HH/TL morphological characters. The Bantaeng observation group was characterized by high HL/TL, high SnL/TL, and low CPH/TL. The Soppeng observation group was characterized by high SL/TL, CPL/TL, BH/TL, and HH/TL variables, and low CPH/TL (Figure 1).

This descriptor illustrates the observed differences between locations. These morphological character ratio traits may reflect some ecological or behavioral characteristics of the fish. Among the most distinguishing ratios of the morphometric characters is CPL/TL.

The difference in the ratio of morphological characters in female tilapia broodstock may be due to the mating system used to produce the next broodstock and adaptations to different environmental conditions. These results indicate that, in general, in terms of morphometric variations, the female broodstock at the three tilapia hatcheries have differing morphometric characteristics.

4. Conclusion

In general, the morphometric variations at the three tilapia hatcheries showed that the female broodstock had different morphometric characters, with significantly different ratios of HL/TL, SnL/TL, CPL/TL, BH/TL, CPH/TL, and HH/TL characters ($P < 0.05$) between hatchery locations.The CPL/TL character ratio was the discriminator between the three sites.

5. References

[1] Mert R, and Cicek E 2010 *J Ani Vet Adv* 9 1753-6.
[2] Kurniawan D, and Fasya A H 2021 *Ear and Environ Scie* 679(1) 012039.
[3] Sinaga V O B, and Mukti R C 2022 *Jou of Aquac and Fis Heal* 11 (1) 90-96.
[4] Hasan V, Mukti A T, and Putranto T W C 2019 *Ecol Enviro and Conser* 25 187-189.
[5] Hasan V, and Tamam M B 2019 *Chec Lis* 15(1) 225-227.
[6] Gunanti M, Wulansari P D, and Kinzella K 2019 *Eart and Environ Scie* 236 (1) 12089.
[7] Zhou L, Dan, Li Y, Li X, Chen X, Li Y, and Gan L 2015 *Irani Jour of Fish Scien* 14 (4) 997-1009.
[8] Kurniawan D and Fasya A H 2021 *Eart and Environ Scie* 679 (1) 12039).
[9] Samara S H, andFathurrozi A W 2020 *Eart and Environ Scie* 441 (1) 012006.
[10] Putri T A, Maya S, and Santanumurti M B 2020 *Eart and Environ Scie* 441 (1) 12052.
[11] Cahyani R, Satyantini W H, Nindarwi D D, and Cahyoko Y 2021 *Eart and Environ Scie* 679 (1) 12042.
[12] Asmamaw B, and Tessema M 2021 *Jour of Aquac and Fis Heal* 10 (3) 341-356.
.
[13] Hassanien H A, Kamel E A, Salem M A, and Dorgham A S 2011 *Jour of the Arab Aquac Soci* 6(2) 237-250.
[14] Yoboue A N, Boussou K C, Konan K S, Konan K F, Adepo-Gourene A 2019 *Jour of Anim and Pla Scie* 41 (3) 7029-7041.
[14] Wahidah, Rusli A, Rajamuddin M A L, dan Amrullah 2021 *Biof* 14 (3) 1811-1817.
[15] Webster, M. 2006. Introduction to geometric morphometrics. Department of the Geophysical Sciences, University of Chicago.
[16] Simon K D, Bakar Y, Temple S E, and Mazlan A G 2010 *J Zhejiang Univ Sci B* 11 871-9.
[17] Yakubu A, and Okunsebor S A, 2011 *Intern Jour of Morpho* 21(4) 1429-1434.
[18] Turan C, Oral M, Öztürk B, and Düzgünes E 2006 *Fishe Resea* 79 (1-2) 139-147.
[19] Dyne J, Magnan P, Bernatchez L, and Rodriguez M A 1999 *Jour of Fis Biol* 54 955-972.
[20] Akinrotimi O A, Ukwe O I K, and Amadioha F 2018 *Intern Jour of Poul and Fishe Sci* 2(1):1-8.
[21] González M A, Rodriguez J M, Angón E, Martínez A, Garcia A, and Peña F 2016 *Arch Anim Bree* 59(4) 435–444.
[22] Eyo J E 2003 *The Zoolo* 2(1) 1-17.
[23] Ezeafulukwe C F, Njoku D C, Ekeledo C B and Adaka G S, 2015 *Intern Jour of Veter Scie* 4(3) 131-135.

6. Acknowledgments

We thank the Directorate of Resources, Directorate General of Vocational Education, Ministry of Education, Culture, Research and Technology for the financial support for this research through Penelitian Produk Vokasi Unggulan Perguruan Tinggi (PPVUPT) scheme, based on the Contract Number 096/PL.22.7.1/SP-PG/2022.

Effectiveness of depuration with zeolite stone filter to reduce levels of heavy metal lead (Pb) in the feather shell (*Anadara antiquata*)

A I Shaleh[1], A S Mubarak[1,2], W Tjahjaningsih[1]

[1]Faculty of Fisheries and Marine, Universitas Airlangga, Surabaya, Indonesia

[2]Corresponding author: shofy.ua@gmail.com ; mubarak.as@fpk.unair.ac.id

Abstract. Lead toxicity in humans causes neurological disorders, bone metabolism disorders, kidney damage, and impaired liver function. Feather shells that live in marine waters with high Pb heavy metal content can be exposed to these heavy metals, posing a risk to humans who eat shellfish. Depuration is one method that can reduce the Pb content in the body of the feather shell. The depuration process is influenced by several factors, including the length of the depuration and the filter used. This study aims to determine the effective depuration time to reduce the Pb content in feather shells (*Anadara antiquata*) using zeolite as a filter. This study applied the Latin Square Design (RSBL) method as a research design consisting of four treatments with four replications. The duration of the depuration treatment was 0 hours (P0), 24 hours (P1), 48 hours (P2), and 72 hours (P3). The effectiveness of the depuration of shells using a zeolite filter can be seen in the reduced Pb content in shellfish. The depuration process for 24 hours showed an effective depuration time with a reduction rate of Pb of 1.095 ± 0.05 mg/kg.

1. Introduction

Sedati Waters is one of the shellfish fishing areas with a high lead (Pb) content. Hassan *et al.* [1] stated that the Pb content in Sedati waters was more than the threshold of 0.44 ppm. This is not in accordance with the water threshold set according to Minister of Environment Decree No 51 of 2004 of 0.008 ppm. Yona *et al.* [2] proved that the Madura sediment contains more Pb than Cu, Cd, and Zn. Activities that cause heavy metal pollution can contribute to heavy metal accumulation over time [3].

Shellfish from waters with high Pb content will be dangerous if consumed by humans. According to Yona *et al.* [2], sea shellfish are a biological indicator of heavy metal pollution in waters and also a source of protein for humans. Based on the Regulation of the Head of BPOM RI No. 23 of 2017, the maximum limit for Pb content in processed shellfish products is 0.2 ppm. Yona *et al.* [2] argue that Pb and Cd are non-essential metals whose biological function is not yet known, while Cu and Zn are essential metals required for metabolism.

Lead is a heavy metal that is cumulative, so it can cause acute and chronic toxicity in humans. Children are more susceptible to lead poisoning than adults [4]. Based on the impact of consuming Pb-contaminated shellfish, a solution is needed to reduce the Pb content in feather shells. Depuration is a way to reduce the Pb content in the body of the feather shell [5].

The working principle of the depuration process is to condition the shellfish in seawater media with low heavy metal content and the water is still flowing [5]. Flowing water causes diffusion so that the dissolved oxygen content becomes high and can accelerate the detoxification process. Detoxification of organisms occurs in the liver, which consists of two stages. The first stage is an oxidation process and the second stage is a heavy metal binding reaction [6].

The Pb content in the waters can affect the decrease in the excretion of feather shells, this situation causes the excretion of Pb to take place slowly so that it will accumulate back into the body. Based on these conditions, it is necessary to add a filter to accelerate the absorption of Pb in the waters [7]. The addition of materials that have adsorption, precipitation, filtration, osmosis and electrodialysis capabilities can increase the excretion of Pb contained in the body of feather shellfish [8]. Zeolite is one of the materials capable of adsorbing Pb. This is because zeolites contain a hydrate crystallinaluminosilicate compound with high cation exchange properties capable of exchanging ions [9]. Zeolite has been widely used in the waste treatment process. The zeolite structure is hollow and has a negative charge that can be used to absorb unwanted cations in a water affected by waste. The use of zeolite as a biofilter media has the ability to reduce organic pollutants higher, which is ± 90% compared to 70% without media. Zeolite can act as a medium for microorganisms to form a biological layer (biofilm) that can help decompose organic matter [10].

Arifin [11] proved that depuration with a zeolite filter affected the reduction of lead and cadmium. However, from this study it is still unknown how long it takes to effectively reduce lead levels, therefore a study was conducted to obtain this information.

2. Materials and methods

2.1. Materials

This study used ±2.7 kg zeolite, sea water 1200 liters which was first deposited for 24 hours. The feather shells (*A. antiquata*) used measuring 3.0 – 4.0 cm were 16 kg [5] caught by fishermen at TPI Sedati Sidoarjo. 500 liter capacity depuration tank with water/aeration pump and flow meter.

2.2. Methods

2.2.1. Research Design

This study used the Latin Square Design (RBSL) method with four treatments and four replications. Depuration time is a treatment consisting of: 0 hours (P0), 24 hours (P1), 48 hours (P2), and 72 hours (P3).

2.2.2. Feather Shell Depuration Process

The depuration process used 300 grams of zeolite for each replication placed in the filter bath. The filter tub is placed outside the depuration tub which is directly connected to the outlet hole of the depuration tub. The shellfish depuration process was carried out according to the treatment. During the process, seawater is flowed into a filter tank containing zeolite stones. The seawater is then pumped from the filter tank to the depuration tank. Each replicate contained 1 kg of test shellfish placed in a basket.

2.2.3. Test of Pb Content with Atomic Absorption Spectrohotometry (AAS)

Determination of heavy metal content using the Atomic Absorption Spectrophotometer (AAS) method [12]. The Pb levels before and after treatment on shellfish meat, zeolite and sea water were tested in a laboratory under the Faculty of Public Health, Universitas Airlangga, Surabaya.

2.2.4. Research Parameters

The main parameter of this study was the percentage of Pb reduction in mussel meat. The supporting parameters are the Pb content in water and zeolite.

Percentage of heavy metal reduction [13] = $\frac{C_{before} - C_{after}}{C_{before}} x\ 100\%$

Description:

C_{before} : heavy metal levels before treatment (ppm)

C_{after} : heavy metal levels after treatment (ppm)

2.2.5. Water Quality

Water quality parameters were observed every day during the study. Parameters of temperature, pH, salinity, and dissolved oxygen (DO) were measured using a thermometer, pH meter, refractometer and DO meter.

2.2.6. Analysis of Data

Data were analyzed using ANOVA (Analysis of Variance). If the treatment has a significant effect on the results, Duncan's test is performed [14].

3. Result and discussion

3.1. Result

The depuration process caused a decrease in the Pb content in the feather shells (Table 1).

Table 1. The rate of decrease in Pb in feather shells

Depuration(hours)	Rate of Decrease ± SD (mg/kg)
0	-
24	0.030 ± 0.003^{a}
48	0.016 ± 0.002^{c}
72	0.015 ± 0.001^{b}

Description: Different letter superscripts in the same column show significant differences ($p < 0.05$).

The Pb content in the zeolite before depuration was 0.000 mg/kg, after 24 hours depuration (P1) was 0.214 mg/kg, 48-hour depuration (P2) was 0.257 mg/kg and 72-hour depuration (P3) was 0.339 mg/kg. This indicates an increase in the absorption of Pb by the zeolite after the depuration process was used (Figure 1).

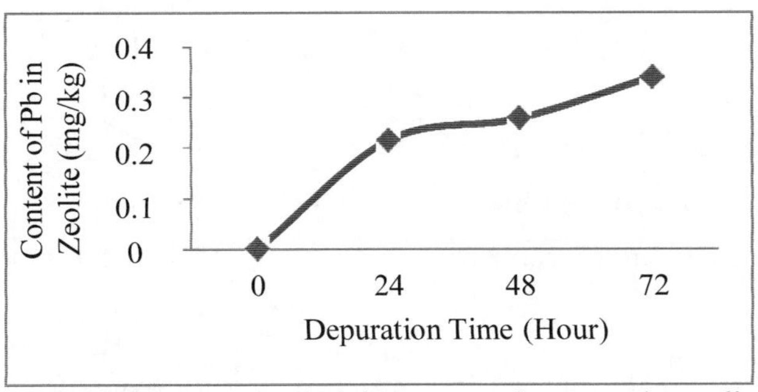

Figure 1. Graph of Pb heavy metal content in zeolite rock filter

The Pb content in seawater used for the depuration process has increased. Before depuration, the Pb content of seawater was 0.015 mg/L. After a 24-hour depuration process (P1) was 0.279 mg/L, a 48-hour depuration (P2) was 0.262 mg/L and a 72-hour depuration (P3) was 0.265 mg/L (Figure 2).

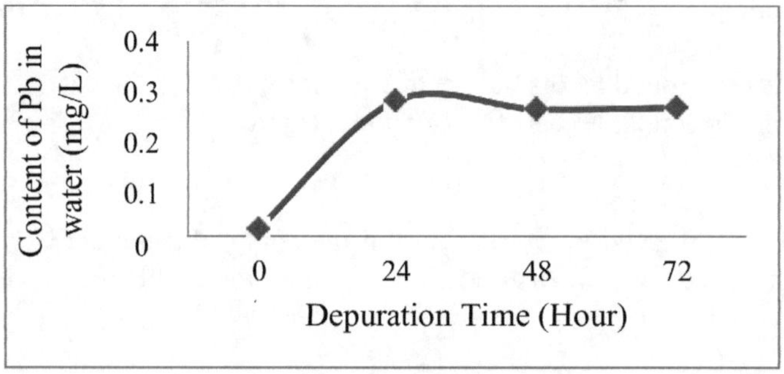

Figure 2. Graph of Pb heavy metal content in water

The Pb content in shellfish decreased, followed by an increase in the Pb content in the water and zeolite. The regression equation for the relationship between Pb content in the water media and shellfish is y = -4.1735x + 1.8386 with a coefficient of determination $R^2 = 0.7293$ (Figure 3A). While the relationship between Pb content in zeolite and shellfish is described through the regression equation y = -4.2418x + 1.842 with a coefficient of determination $R^2 = 0.9761$ (Figure 3B).

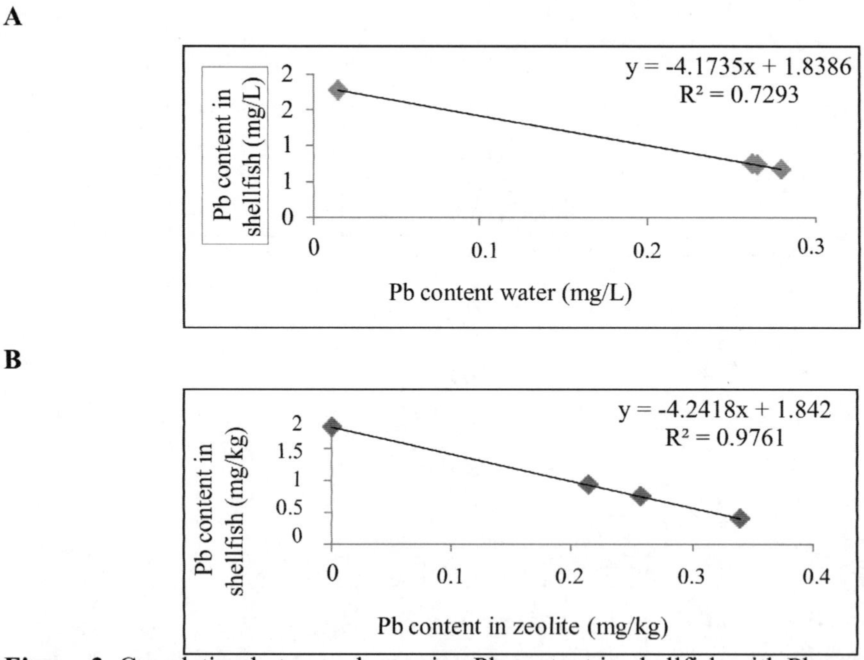

Figure 3. Correlation between decreasing Pb content in shellfish with Pb content in water (**A**), and Pb content in zeolite (**B**).

The Pb content of seawater for the depuration process is 0.0024 mg/L. The water temperature range during the depuration process was 26.8-28.6oC, DO 4.66-5.97 mg/L, and pH 6.8-7.8. While the seawater salinity used is 30 ppt.

3.2. Discussion
The Pb content in feather shells after going through the depuration process was lower than before the depuration process. The Pb levels in mussels decreased during the depuration process followed by

increased Pb levels in water and zeolite rock media. This is evident from the coefficient of determination (R^2) which is close to the value of 1 (Figure 3). According to Shaobin and Yuelian [15], zeolite is a hydrated alumino silicate that has an open three-dimensional framework structure, which is built by tetrahedral-tetrahedral [SiO4]4- and [AlO4]5- with O atoms as a link between Si atoms and Al atoms. forming regular cavities and channels occupied by water molecules and cations. Cations in the zeolite cavity can move freely so as to allow ion exchange without damaging the zeolite structure.

The absorption of Pb by zeolite stone continued to increase starting from 24 hours, 48 hours to 72 hours. The negative charge of the zeolite is able to bind heavy metals so that it can reduce heavy metals from shellfish excretion. Elysabeth *et al.* [16] have proven that the longer the contact time of the zeolite with wastewater, the less the amount of Fe and Pb contained in the wastewater. According to Chmielewská [17], the adsorption of pollutants onto porous solids begins with the transport of pollutants from the solution to the surface of the adsorbent, adsorption on the surface particles, then ends with transportation in the adsorbent.

The shellfish used in this study had uniform sizes, so there was no significant difference in the initial Pb content in the body of the shellfish between treatments. The difference in the size of the shellfish can affect the difference in the Pb content in the shellfish [18].

Media water temperature ranged from 26.9-28.6°C during the depuration process. The optimal temperature range for mussels is between 25-32°C [19]. The use of an aerator during the depuration process serves to maintain optimal DO for shellfish metabolism. The DO value of media water ranged from 4.66 to 5.97 mg/L. Respiration with abundant oxygen will make the shellfish metabolic process go well [6]. The range of media water pH values is between 6.8 – 7.8. During the depuration process, the media water salinity was 30 ppt. According to Yanti *et al.* [20], shellfish have a very high survival rate at a salinity of 30 ppt.

4. Conclusion

The rate of reduction of Pb in feather shells is affected by the duration of the depuration process. The depuration process for 24 hours proved to be more efficient than 48 hours and 72 hours with a Pb reduction rate of 1.095 ± 0.05 mg/kg.

5. References

[1] Hasan A F, Suprapto H, Rahardja B S, and Pursetyo K 2017 *JMCS.* **6**(3), 148-156 [in Indonesian].
[2] Yona D, Sartimbul A, Rahman M A, Sari S H J, Mondal P, Hamid A, and Humairoh T 2021 *JIPK.* **13**(1), 20-28.
[3] Yona D, Vernandes D, Kasitowati R D, and Sari S H J 2020 *JIPK.* **12**(1), 140-148.
[4] Prilly D, Akili R H, and Maddusa S 2018 *Jurnal Kesmas.* **7**(4), 1-8 [in Indonesian].
[5] Lee R, Lovatelli A, and Ababouch L 2008 *Bivalve Depuration: Fundamental and Practical Aspects* (Rome: Food and Agriculture Organization).
[6] Rose R L and Hodgson E 2004 *Metabolism of Toxicants.* In E. Hodgson, *A Textbook of Modern Toxicology*Third Edition (New Jersey: John Wiley & Sons Inc).
[7] Eshmat M E, Mahasri G, and Rahardja B S 2014 *JIPK.* **6**(1), 101-108.
[8] Minceva M, Markovska L, and Meshko V 2007 Maced. J. Chem. Eng. **26**(2), 125-134.
[9] Sulistyowati N, Sriyanti, and Darmawan A 2018 *JKSA*, **21**(2), 102-106 [in Indonesian].
[10] Budijono, Hasbi M and Ahmali 2010 *Jurnal Ilmu Perairan* **8**(2), 64-70 [in Indonesian].
[11] Arifin W 2016 *Depuration of Pb and Cd Heavy Metal Content in Mussels (Anadara antiquata) with Different Filters. Thesis. (Surabaya: Faculty of Fisheries and Maritime Affairs. Airlangga University) [in Indonesian].*
[12] Al-Kazaghly R F, Hamid M, and Ighwela K A 2021 *JIPK.* **13**(1), 81-86.
[13] Suligundi B T 2013 *Jurnal Teknik Sipil Untan.* **13**(1), 29-44 [in Indonesian].

[14] Kusriningrum R S 2008 *Experiment Design* (Surabaya: Airlangga University Press) [in Indonesian].
[15] Shaobin W, and Yuelian P 2009 *J. Chem. Eng.* **156**, 11-24.
[16] Elysabeth T, Jufrodi and Hudaeni 2015 *J Chemtech.* **1**(1), 26-28 [in Indonesian].
[17] Chmielewská E 2019 Natural Zeolite: *Alternative Adsorbent in Purification or Post-Treatment of Waters* Chapter 4 In Mariano Mercurio, Binoy Sarkar, Alessio Langella (Eds) *Modified Clay and Zeolite Nanocomposite Materials: Environmental and Pharmaceutical Applications* (pp 87-112) (Amsterdam: Elsevier Inc).
[18] Amriani A, Hendrarto B, and Hadiyarto A 2011 *Jurnal Ilmu Lingkungan* **9**(2), 45-50 [in Indonesian].
[19] Broom M J 1985 *The Biology and Culture of Marine Bivalve Molluscs of the Genus Anadar*a (Manila, Philippines: International Center of Living Aquatic Resources Management).
[20] Yanti H, Muliani, and Khalil M 2017 *Acta Aquatica* **4**(2), 53-58 [in Indonesian].

6. Acknowledgment

The authors are grateful to the Faculty of Fisheries and Marine, Universitas Airlangga which had provided the space and laboratory to conduct this research.

IOP Conf. Series: Earth and Environmental Science 1273 (2023) 012021 doi:10.1088/1755-1315/1273/1/012021

Iodine content and acceptability of seaweed flour from various types of seaweed, drying methods, and types of fishy odor removal

F Retiaty[1,2,3] **and N Nurjanah**[1]

[1] National Research and Innovation Agency, Cibinong Science Center, Jl. Raya Jakarta-Bogor, Pakansari, Kecamatan Cibinong, Kabupaten Bogor, Jawa Barat 16915, Indonesia
[2] Department Food Science and Technology, Faculty of Agricultural Technology, IPB University, Indonesia

[3]Corresponding author: fifiretiaty@gmail.com

Abstract. This study aimed to develop seaweed flour from various types of seaweed (*Euchemma cottonii, Gracillaria verrucosa*) with various drying methods (sun, vacuum oven, cabinet drying, drum drying) and types of fishy odor removal [solution of $Ca(OH)_2$ and lime each 5%, 0.25% star anise, 0.33% clove powder], analyze iodine content and acceptability. Iodine content and acceptability were analyzed by ICP-OES and multiple comparison test, respectively. The results showed that the types of fishy odor removal significantly affected the organoleptic acceptance of seaweed flour with the best acceptance being 5% $Ca(OH)_2$ solution ($p<0.05$). Iodine content between drying methods and types of seaweed was significantly different ($p<0,05$). Seaweed flour with drum drying significantly had the highest iodine content, 89.96 (*Euchemma cottonii*) and 564.15 mg/kg dw (*Gracillaria verrucose*). Seaweed flour from *Gracillaria verrucosa* had significantly higher iodine content than *Euchemma cottonii*, 205.26 and 41.58 mg/kg dw, respectively ($p<0.05$). The study concluded that types of fishy odor removal with the best panelist acceptance, types of seaweed, and drying methods that produce seaweed flour with the highest iodine are 5% $Ca(OH)_2$ solution, *Gracillaria verrucosa* and drum drying, respectively. Seaweed flour had the potential as an ingredient in the development of nutrient-rich products, especially iodine.

1. Introduction

One of Indonesia's marine biological natural resources whose iodine content is high and is considered very potential in overcoming the problem of iodine deficiency is seaweed. The iodine content in seaweed ranges from 0.1 - 0.8% depending on the type of seaweed. Seaweed is a macro alga included in the *Thallophyta* division, which is a plant that has a skeleton structure consisting of stems/thallus and has no leaves and roots[1,2]. Types of seaweed that are widely found in Indonesian waters with a fairly high iodine content are read seaweed, species of *Rhodophyta* i.e. *Gracillaria* sp. and *Euchemma sp.* [3,4,5].

Seaweed production in Indonesia in 2018 was 10.32 million tons, consisting of *Euchema cottoni* of 8.62 million tons and *Gracilaria verrucosa* of 1.12 million tons. The benefits of seaweed as a food ingredient have long been known. In Indonesia, seaweed has long been used by coastal residents for vegetables, vegetables, pickles, cakes, puddings, and sweets. Seaweed will have a higher economic value if it is processed, among other things, into food products ready for consumption. Currently, seaweed is encouraged to be processed into processed food products whose acceptability in the community is high so that it can increase post-harvest processing of seaweed [6].

The effort to create seaweed products as food ready for consumption is to make seaweed flour which in addition to being able to obtain nutrients can also be used as a functional food. The expected form of the product is fortificant for sprinkling on foods consumed in households, especially in endemic areas of iodine deficiency disorders. In connection with the characteristics of iodine, which is a water-soluble mineral and prone to heating treatments and fishy flavor, it is necessary to research how to process and how to use seaweed flour at the household level by sprinkling it in several types of hot and cold foods such as rice, porridge, sweet iced tea and orange ice. Some studies state that the problem in processing seaweed-based products is a fishy flavour that can reduce the acceptability of food products. This method of processing and using seaweed flour is expected to increase the acceptability of seaweed flour for consumption in addition to preventing the retention of iodine contained in it [7,8,9,10].

Seaweed is known for its smell that is not liked by the general public. In the results of organoleptic tests of food products sourced from seaweed, generally, the panelists did not like the fishy flavor and taste of the product. Several studies have reported that the addition of *Euchemma cottonii* seaweed flour concentrations to various products such as rice and analog rice, extruded rice, kwetiau, boba balls, and dodol significantly decreased the acceptance of the color, taste, texture and flavor of these products [11,12,13,14,15]. The specific smell that is not liked will arise when the seaweed is exposed to air. In seaweed of the red algae type, the specific smell is due to the content of dimethyl sulfide. The use of spices that have a dense odor is thought to reduce the fishy smell in the seaweed [8,9], namely cloves (*Syzygium aromaticum*), star anise (*Illicium verum*), lime (*Citrus acida*) and $Ca(OH)_2$ solution [16].

Research on the use of lime peel in overcoming fishy odors in catfish obtained the results of fried catfish that were not fishy with the addition of dried lime peels [17]. In addition, $Ca(OH)_2$ is also used in fishy odor additives in rabbit meatballs that obtain good flavor resulting in rabbit meatballs [18]. Star anise contain of the anethole compounds are responsible for its characteristic taste (85-90%) and essential oil that acts as a flavoring agent, i.e. γ *-terpineol,* β-terpineol, acetaldehyde, methyl p-anisate, 4-ethyl benzaldehyde, benzyl alcohol, bisabolene, 4-methoxy cinnamaldehyde, 9,12-octadecadienoic acid (Z,Z)-, methyl ester/methyl linoleate, acetic acid geranyl ester, nopol, trans-nerolidol, geranyl isobutyrate and p-Allylphenol/Chavicol (odorant) [19].

This study aimed to make iodized food products in the form of sprinkles based on seaweed *Euchemma cottonii* and *Gracillaria verrucosa* using various types of drying methods, and types of fishy odor removal. The resulting product is expected to be easy to use at the household level according to individual needs and is organoleptically well received by the community so that iodine consumption can be fulfilled and overcome the problem of disorders due to iodine deficiency.

2. Materials and methods

Seaweed types *Euchemma cottonii* and *Gracillaria verrucosa* were obtained from Lontar Village, Tirtayasa District, Banten Regency, Banten Province. The fishy odor removal consisting of $Ca(OH)_2$, lime (*Citrus acida*), star anise (*Illicium verum*), and clove (*Syzygium aromaticum*) powder were purchased from local market (Bogor, Indonesia).

2.1. Processing of seaweed flour

Seaweed washed and cleaned with water. Seaweed was soaked for 24 hours with various types of fishy odor removal, i.e. a solution of $Ca(OH)_2$ and lime each 5%, 0.25% star anise, and 0.33% clove powder. Seaweed is drained, then a reduction in size is carried out by cutting it $\pm$ 1 cm and re-washing with water for seaweed with sun-dried, vacuum oven, and cabinet dryer. Meanwhile, the seaweed to be dried with a drum dryer is blended with a ratio of water and seaweed 1: 2 to become seaweed pulp. Seaweed pieces are dried by different drying methods, i.e sun drying, vacuum oven, and cabinet drier for 6, 4 and 5 days at 80°C until 15% of moisture (maximum). Seaweed pulp was dried by drum dryer at 90-100°C, pressure 4 Bar. Dried seaweed is mashed and sifted 100 mesh to obtain seaweed flour.

2.2. *Acepptability of seaweed flour*

Acceptability tests are carried out to get the best types of fishy odor removal. The acceptability test is carried out by the method of sensory analysis of multiple comparison test [20]. The tests were conducted on 31-32 semiexpert panelists. Panelists were asked to determine the level of comparison and the magnitude of the level of difference in sensory acceptability of color and odor in seaweed flour that had been soaked with various types of fishy odor removal and dried in sunlight compared to seaweed flour without fishy odor removal (control) and the taste and flavor of rice that had been sprinkled with seaweed flour from various types of fishy odor removal compared to seaweed flour without fishy odor removal (control). The sprinkling concentration of seaweed flour is 0.15 g of seaweed flour in 100 g of rice. The comparison scale used are 1= very very badly than the control; 2= very worse than control; 3= worse than the control, 4= somewhat worse than the control, 5= as good as the control, 6= Somewhat better than the controls, 7= Better than the control, 8= Very good from the controls, 9= very very good from the control. The scale of the magnitude of the level of difference used are 1= none, 2= few, 3= medium, 4= many, and 5= very many [20].

The panelists were also asked descriptively to choose one type of fishy odor removal that was most sensory preferred overall both based on the color and smell of seaweed flour, as well as the taste and flavor of rice that had been sprinkled with seaweed flour. Ethical approval was obtained from The National Institute of Health Research and Development, Ministry of Health, Indonesia (exempted ethical approval document number: LB.02.01/5.2/KE 398/2016).

2.3. *Determination of iodine content*

Iodine analysis was performed in a commercial test laboratory (SGS Indonesia). The sample used for iodine analysis is seaweed flour made using the best types of fishy odor removal, i.e. a solution of $Ca(OH)_2$ 5% concentration and dried by various drying methods. Iodine content was analyzed for 2 replications for each sample with Inductively Coupled Plasma- Optical Emission Spectrometry (ICP-OES). A total of 0.5 samples were reacted with concentrated HNO_3, 30% H_2O_2 in vessels and vortexes. The sample is digested in microwave digestion at 105°C. The sample is cooled in a water bath for 30-45 minutes, and filtered with Whatman 41 paper. The sample was transferred to a 50mL flask and diluted until the tera mark with Mili-Q water, then the sample was injected in ICP-OES (AOAC 994.10 and 993.14:2005; SOP LBFD-07445:2007) [21].

2.4. *Statistic analysis*

Iodine content and sensory acceptability (the level of comparison and magnitude of the level of difference) of seaweed flour from various drying methods, and types of fishy odor removal was analyzed by Analysis of Variance (ANOVA) [22]. Iodine content from various types of seaweed was compared by t-test [23].

3. Results and discussion

3.1. *Acceptibillity of seaweed flour*

The acceptability test is carried out using the multiple comparison test method, which includes testing the level of comparison and the magnitude of the level of difference of sensory acceptance of seaweed flour with various types of fishy odor removal compared to seaweed flour without fishy odor removal (control). The level of comparison and the magnitude of the level of difference of sensory acceptance are presented in table 1 and 2 (*Euchemma cottonii* seaweed flour), and table 3 and 4 (*Gracillaria verrucose* seaweed flour), respectively.

This test is used to find out whether there is a difference between one or more samples and the standard example (control) and to estimate the magnitude of the existing difference [20]. This test was carried out which aimed to get the best types of fishy odor removal that could be accepted by the panelists.

Tables 1 and 3 showed that the type of fishy removal odor significantly affected the degree of comparison of sensory acceptance of all sensory parameters, except smell for *Euchemma cottonii* seaweed flour and only color for *Gracillaria verrucose* flour ($p<0.05$). However, the type of fishy odor removal significantly only affected the magnitude of the level of differences of sensory acceptance of color *Euchemma cottonii* seaweed flour, color and smell *Gracillaria verrucose* flour ($p<0.05$) (table 2 and 4).

Table 1. The level of comparison of sensory acceptance of *Euchemma cottonii* seaweed flour with various types of fishy odor removal against flour without types of fishy odor removal soaking treatment.

Sensory attributes	n	Types of fishy odor removal			
		Clove powder	Star anise powder	Lime	$Ca(OH)_2$
Color	31	4,44±1,243 [a]	4,66±1,066 [a]	5,19±1,378 [b]	6,38±1,408 [c]
Flavor	31	5,19±1,330 [a]	4,69±1,120 [a]	5,72±1,486 [b]	5,84±1,526 [b]
Taste	31	4,56 ±1,458 [a]	5,09±1,376 [a]	5,88±1,212 [b]	5,72±1,611 [b]
Smell	31	5,00±1,244 [a]	4,91±1,254 [a]	5,47±1,606 [a]	5,50±1,437 [a]

Note: The values within the same row followed by different superscript letters are significantly different (*p*<0.05)

Table 2. The the magnitude of the level of difference of sensory acceptance of *Euchemma cottonii* seaweed flour with various types of fishy odor removal against flour without types of fishy odor removal soaking treatment.

Sensory attributes	n	Types of fishy odor removal			
		Clove powder	Star anise powder	Lime	$Ca(OH)_2$
Color	31	2,31±0,998 [a]	2,31 ±0,965 [a]	2,16±0,723 [a]	2,91±0,928 [b]
Flavor	31	2,91±0,818 [a]	2,63±1,070 [a]	2,41±0,979 [a]	2,75±1,047 [a]
Taste	31	2,69±0,896 [a]	2,47 ±0,950 [a]	2,66 ±0,971 [a]	2,91 ±0,893 [a]
Smell	31	2,47±1,135 [a]	2,34±0,902 [a]	2,56±0,982 [a]	2,53±1,016 [a]

Note: The values within the same row followed by different superscript letters are significantly different (*p*<0.05)

Table 3. The level of comparison of sensory acceptance of *Gracillaria verrucosa* seaweed flour with various types of fishy odor removal against flour without types of fishy odor removal soaking treatment.

Sensory attributes	n	Types of fishy odor removal			
		Clove powder	Star anise powder	Lime	$Ca(OH)_2$
Color	32	5,65±1,226 [a]	5,61 ±1,383 [a]	6,48±1,363 [b]	6,68±1,720 [b]
Flavor	32	5,58±0,992 [a]	5,77±1,203 [a]	5,65±1,404 [a]	5,61±1,230 [a]
Taste	32	5,48 ±0,996 [a]	5,65±1,199 [a]	5,32±1,249 [a]	5,23±1,257 [a]
Smell	32	5,90 ±1,221[a]	5,81±1,223 [a]	5,65±1,743 [a]	5,42±1,608 [a]

Note: The values within the same row followed by different superscript letters are significantly different (p<0.05)

Table 4. The the magnitude of the level of difference of sensory acceptance of *Gracillaria verrucosa* seaweed flour with various types of fishy odor removal against flour without types of fishy odor removal soaking treatment.

Sensory attributes	n	Types of fishy odor removal			
		Clove powder	Star anise powder	Lime	$Ca(OH)_2$
Color	32	2,81±0,946 [a]	2,71±1,039 [a]	2,94±0,929 [a]	3,77±1,056 [b]
Flavor	32	2,23±1,023 [a]	2,29±1,071 [a]	2,68±1,222 [a]	2,52±1.208 [a]
Taste	32	2,13±0,991 [a]	2,29±1,039 [a]	2,55±0,961 [a]	2,48±1,092 [a]
Smell	32	2,42±1,025 [a]	2,55±1,150 [a]	3,10±1,136 [b]	2,71±1,101 [b]

Note: The values within the same row followed by different superscript letters are significantly different (p<0.05)

The soaking of *Euchemma cottonii* and *Gracillaria verrucose* seaweed with $Ca(OH)_2$ significantly resulted in the best seaweed flour color compared to other types of a fishy odor, with an average rating score of 6.38-6.68 (somewhat better to better than seaweed without the addition of fishy odor removal-control), followed by lime with the greatest of the level of difference between slightly to moderately (2.91) on *Euchemma cottonii* and medium to many (3.77) on *Gracillaria verrucose* (table 1-4). Ca(OH)2 gives the same good color as lime on *Gracillaria verrucose* seaweed flour with the same scoring score, which is somewhat better to better than seaweed flour without the addition of fishy odor removal (6.48-6.68). This happens because Ca(OH)2) affects pigment damage, resulting in whiter seaweed flour than other types of fishy odor removal soaking [17]. Seaweed flour of *Euchemma cottonii* and *Gracillaria gigas* has a degree of whiteness of 59,80% [11] and 47.78% [24], respectively.

Seaweed juice soaked for 36 hours in a solution of $Ca(OH)_2$ concentration of 1% produces an flavor that is not significantly different from soaking using 5% solution of lime. As for the taste and texture of soaking with $Ca(OH)_2$ is better than lime [25]. The same study using lime as an ingredient fishy odor removal on fish petis also produced brightly colored flour [26].

Cloves and star anise powder produce seaweed flour both *Euchemma cottonii* and *Gracillaria verrucose* with the lowest color comparison assessment, which is somewhat worse to somewhat better than the control (4.44 – 5.65) with a slight to moderate difference level (2.31-2.81) (table 1-4). Soaking with cloves and star anise powder produces a rather dark flour due to the influence of the black color of both ingredients.

$Ca(OH)_2$ and lime provided the aroma and taste of seaweed flour *Euchemma cottonii* significantly the best compared to other types of fishy odor removal, but have no effect on *Gracillaria verrucose*, with the same good to somewhat better comparison level score from the control, i.e. 5.72-5.84 (flavor) and 5.72-5.88 (taste) with a slight to moderate difference, i.e. 2.41-2.75 (aroma) and 2.66-2-91 (taste). This is in line with research on soaking *Euchemma cottonii* with Ca(OH)2 to extract its carrageenan and produce flour that gives a flavor close to neutral [27].

Research on the use of lime in processing mackerel skins showed that soaking with solution lime for 60 minutes resulted in mackerel skin chip products most preferred by panelists. This suggests that lime can be used to eliminate the fishy odor on fish skin [26].

Lime contains essential oils are a mixture of volatile components such as terpenes and oxygenated compounds, which are pleasant in smell and rich in flavor, i.e. o-cymene (16.62%) α-cedrene (10.57%), decadienal (8.04%), bisabolene (5.07%) and β-humelene (4.14%). Citronellyl acetate (2.37%), linalool acetate (2.37%), carvone (1.81%), decanone (1.47%), isopulegol acetate (1.30%), farnesol (1.25%), 40-methoxyacetophenone (1.21%), and D-carene (1.07%), α-terpineole (0.61%), dihydroxylinalool acetate (0.65%), cis-nerone (0.57%), caryophyllene oxide (0.43%), and 2,2-dimethyl-3,4-octadienal (0.38%) [28]. It is used as a flavoring agent in food, beverages, liquors, and other food products [29]. The use of lime essential oil is often limited because of low water solubility, making it immiscible with water. This is because its essential oil is hydrophobic and consists of a mixture of varying lipophilic flavor compounds including lipophilic terpenoids, phenylpropanoids, and short-chain aliphatic hydrocarbon derivatives [30]. Unsaturated compounds such as monoterpene hydrocarbons as 95% of the volatile fraction in lime are highly unstable and easily decompose when directly exposed to heat, humidity, light, or oxygen, thus affecting the function and activity of essential oil in food applications [31].

The addition of types of fishy odor does not significantly affect the level of comparison and difference of flavor of *Gracillaria verrucosa* seaweed flour (table 3 and 4). This is because the type of seaweed *Gracillaria verrucosa* is not too fishy like *Euchemma cottonii.*

The soaking of seaweed in various types of fishy odor removal significantly does not affect the level of comparison of the smell of seaweed flour produced in both *Euchemma cottonii* and *Gracillaria verrucose*. However, panelists identified a difference in the smell of *Gracillaria verrucose* seaweed flour soaked with various types of fishy odor removal compared to controls. *Gracillaria verrucose* seaweed flour soaked with $Ca(OH)_2$ and lime has the highest level of difference in smell compared to cloves and star anise powder, with a moderate to large difference score (2.71 – 3.10) (table 4).

3.2. Selected types of fishy odor removal are most preferred

The frequency distribution of panelist acceptance to sensory preferred overall color and smell of seaweed flour and the taste and flavor of rice sprinkled with seaweed flour is shown in table 5.

Table 5. The frequency distribution of panelist acceptance to sensory preferred overall on seaweed flour (color and smell) and rice sprinkled with seaweed flour (taste and flavor).

Types of fishy odor removal	*Euchemma cottonii*		*Gracillaria verrucosa*	
	n	Percentage (%)	n	Percentage (%)
Star anise powder	2	6,45	2	6,2
Clove powder	3	9,67	3	9,4
Lime	6	19,35	6	18,8
$Ca(OH)_2$	20	65,51	21	65,6

Table 5 showed that the seaweed flour assessed to have the highest acceptance of the overall sensory attributes is seaweed flour with soaking treatment in a 5% $Ca(OH)_2$ solution, for both *Euchemma cottonii* and *Gracillaria verrucosa*. Soaking with $Ca(OH)_2$ produces a relatively whiter color, taste, flavor, and smell are less fishy compared to soaking treatments in other types of fishy odor removal.

Several studies on the use of $Ca(OH)_2$ as fishy odor removal provide good acceptance results in processed seaweed products [25,32,33]. Soaking seaweed in a solution of 0.5 % $Ca(OH)_2$ for 15 minutes produces a seaweed stick product with good acceptability [30]. The use of 1% $Ca(OH)_2$ with 1 day of soaking resulted in seaweed dodol that was preferred by the panelists with a maximum addition of seaweed of 40% [33]. Ca(OH)2 is reported used by seaweed farmers in processing seaweed into seaweed flour at a seaweed cultivation center in West Nusa Tenggara, Indonesia [34].
Thus, the best type of fishy odor removal selected is a 5% $Ca(OH)_2$ solution. The solution of 5% $Ca(OH)_2$ is further used in the production of seaweed flour by various drying methods.

3.3. Iodine content of seaweed flour

The iodine content of seaweed flour derived from 2 types of seaweed and dried by 4 drying methods in dry weight (dw) is presented in table 6. The types of seaweed and drying methods have a significant effect on the average iodine content of seaweed flour ($p<0.05$). Table 6 showed that seaweed flour derived from *Gracillaria verrucosa* significantly has an average iodine content 5 times higher than Euchemma cottonii ($p<0.05$) with an iodine content of 205.26 and 41.58 mg/kg dw, respectively.
Seaweed is the natural resource richest in minerals and traces elements due to its cell-wall polysaccharides, proteins, sulfate, anionic carboxyl, and phosphate groups [35]. Seaweed flour is a source of minerals, especially iodine [5,36].

Table 6. Iodium content of seaweed flour (mg/kg dw)

Types of seaweed	Drying methods				Average
	Sun drying	Vacuum oven drying	Cabinet drying	Drum drying	
Euchemma cottonii	25,27±0,20	24,62±1,01	26,46±5,72	89,96±13,04	41,58±30,35^{a}
Gracillaria verrucosa	79,47±19,20	88,97±14,55	88,46±10,58	564,15±61,81	205,26±223,01^{b}
Average	52,37±33,35^{a}	56,80±38,09^{a}	57,46±36,46^{a}	327,06±276,19^{b}	

Note: The values within the same row and the same column followed by different superscript letters are significantly different (*p*<0.05)

The iodine content in the seaweed flour produced is relatively higher than seaweed flour and dried seaweed from other studies (table 6). *Euchemma cottonii* seaweed flour was reported to contain iodine of 19.4 μg/g dw [3,11]. However, the resulting *Euchemma cottonii* seaweed flour has a lower iodine content and *Gracilaria verucossa* seaweed flour has a higher iodine content than other types of dried seaweed of similar species. The seaweed used is a species of *Rhodophyta* (red seaweed). Research on iodine content in five red seaweed (*Chondracanthus teedei* var. lusitanicus, *Chondrus crispus*, *Gracilaria gracilis*, *Grateloupia turuturu*, *Osmundea pinnatifida*) showed that iodine content between species varied. *Grateloupia turuturu* contains the lowest iodine concentration with 47 μg/g dw. *Gracilaria gracilis* had iodine content 94 μg/g dw. *Chondracanthus teedei* var. lusitanicus, *Chondrus crispus*, and *Osmundea pinnatifida* showed higher levels of iodine, i.e. 126, 206, up to 302 μg/g dw, respectively [36]. The other red seaweed, i.e. *Palmaria palmata* had iodine content 293 μg/g dw [5].

The variations of iodine content found in the results for the same species are related to the different harvesting conditions and geographical origin of the species, mechanism of iodine uptake (physiological adaptation) [36]. The composition of seaweed mineral content will depend on the various external and internal factors such as an elemental abundance of ocean water, location with sunlight exposure, pH, salinity, temperature, light, availability of natural nitrogen content, climatic conditions, aging of seaweeds, different metabolic processes and others [37], species and varieties [36], locations and biomass source [5].

In addition to iodine *Gracillaria verrucosa* seaweed contains other minerals. *Gracillaria verrucosa* seaweed is reported to have a maximum content of K, Ca, Fe, Mn, Zn, P, Co, and Bo with 3.65 g/kg, 0.39 g/kg 49.55, 11.44, 5.45, 2.45, 2.11 and 0.97 mg/kg of dry weight biomass, respectively [38]. The total mineral or ash content of *Euchemma cottonii* seaweed flour was reported to be lower compared to Gracillaria. *Euchemma cottonii* seaweed flour dried with a dryer cabinet and drum dryer has a successive ash content of 14.81, and 5.11% dw, respectively [11,39], while the ash content of *Gracillaria gigas* seaweed flour was 16.18% dw [24]. *Euchemma cottonii* dried seaweed on Karimunjawa Island treated by salty drying, fresh drying, and alkali has a higher ash content of 46.95, 18.59, and 20.15% dw respectively [40].

The drying method of drum drying produces seaweed flour with a significant average iodine content 6 times higher (327.06 mg/kg dw) than sun drying, vacuum oven, and cabinet drying ($p<0.05$). The iodine content of seaweed flour with a drum dryer is 89.96 mg/kg dw (*Euchemma cottonii*) and 564.15 mg/kg dw (*Gracillaria verrucosa*). Sun-dried seaweed flour, vacuum oven, and cabinet dryer have not much different iodine content ($p>0.05$), which is 52.37, 56.80, and 57.46 mg/kg dw, respectively (table 5).

The drum drying method, in addition to being proven to produce seaweed flour with an average iodine content significantly higher than the other drying methods, was reported to be able to produce higher yields of *Euchemma cottonii* seaweed flour than oven drying [12]. Drying with a drum dryer is carried out at a temperature of ± 90-100°C, a pressure of 4 Bar in a short time. Although the drying temperature of the drum dryer is relatively higher than the other drying methods, because it is done in a short time (< 1 hour) it is possible to be able to maintain the iodine content from damage. Drying seaweed with the sun, vacuum oven, and cabinet dryer takes quite a long time, 6, 4, and 5 days, respectively although it is carried out at a lower temperature than the drum dryer. It is possible that iodine in seaweed flour under such drying conditions undergoes oxidation.

Research on the stability of iodine in salt in 12 developing countries reports that iodine can be highly unstable influenced by various factors, such as high humidity, packaging materials woven high-density polyethylene bags, salt impurities, storage conditions, oxidation, temperature, food processing, washing, and cooking processes in the household. Storage of salt samples at 40°C, 100% relative humidity resulted in the greatest iodine loss. Exposure to sunlight on the packaging material will increase iodine loss by as much as 25%, because the storage temperature of the product can increase to 60°C [41].

Iodine loss increases significantly with an increase in processing temperature. Iodine loss is as much as 2-3.8% at 80°C processing temperature and increases to 4% at 100°C. The maximum iodine loss is 18.5% at a processing temperature of 350°C [42].

Potassium iodide can be oxidized to elemental iodine by oxygen or other oxidizing agents, especially in the presence of catalysts, such as metal ions, and moisture. Elemental iodine readily sublimes and is then rapidly lost to the atmosphere through diffusion [39]. Iodine is unstable in an acidic solution and will volatilize in the presence of light and air. Iodine is capable of volatilization at all pH values <10.5, especially the most volatile in the pH range of 0–1 [43].

Seaweed flour with sun drying, vacuum oven, and cabinet drying has 2 times more washing intensity compared to seaweed dried by drum drying[44,45]. This possibly leads to the solubility of iodine in washing water. *Eucheuma cottonii* seaweed dried seaweed that received repeated washing treatment until the seaweed became fresh before drying was reported to have a lower ash content of 2.5 times compared to directly dried seaweed [40].

Seaweed flour can be an alternative food ingredient to supply the need for iodine in the daily diet[46,47,48]. The recommended dietary allowance (RDA) of iodine for Indonesians aged 13-80 years is 150 μg/day [49]. The iodine content in *Euchemma cottonii* and *Gracillaria verrucosa* seaweed flour was dried with a drum dryer of 89.96 mg/kg dw (76.46 mg/kg ww) and 564.15 mg/kg (499.98 mg/kg ww), respectively (table 6). Thus, the consumption of *Euchemma cottonii* and *Gracillaria verrucosa* seaweed flour dried by drum dryer of 1.96 g and 0.30 g per day, respectively can already supply the daily needs of iodine.

4. Conclusion

The types of fishy odor removal significantly affected the level of comparison of of sensory acceptance of all sensory parameters, except smell for seaweed *Euchemma cottonii* flour and only color for *Gracillaria verrucose* flour ($p<0.05$). the type of fishy odor removal significantly only affected the magnitude of the level of differences of sensory acceptance of color *Euchemma cottonii* seaweed flour, color and smell *Gracillaria verrucose* flour ($p<0.05$). The solution of Ca $(OH)_2$ 5% is the best type of fishy odor removal which produces color, flavor, taste, and smell as good to better than seaweed without soaking with fishy odor removal (5.23-6.68) with the level of the difference a slight to large (2.52 -3.77) and overall preference for sensory by 6.51-65.6% of the panelists. The types of seaweed and the drying methods significantly affected the average iodine content of seaweed flour. Seaweed flour from *Gracillaria verrucosa* significantly had an average iodine content of 5 times higher (205.26 mg/kg dw) than *Euchemma cottonii* (41.58 mg/kg dw). The drum drying method produced seaweed flour with an average iodine content of 6 times significantly higher (327.06 mg/kg dw) than sun drying, vacuum oven and cabinet drying ($p<0.05$). Sun drying, vacuum oven and cabinet drying resulted the same average iodine content of seaweed flour ($p>0.05$). Research on the shelf life and physical properties of *Euchemma cottonii* and *Gracillaria verrucose* flours need to be done.

5. References

[1] Sutinah, Riniwati H, Sahidu A M, and Suryani 2020 *Syst. Rev. In Parm.* 11(2) 44-50.
[2] Firdarini I R, Kismiyati, and Manan A 2019 *IOP EES*. 236(1) 01 2126.
[3] Astawan M, Koswara S, and Herdiani F 2004 *J. Teknol. dan Ind. Pangan.* 15 61–9.
[4] Hoer M A 2018 Skirpsi. Malang: Universitas Brawijaya.
[5] Roleda M Y, Skjermo J, Marfaing H, Jónsdóttir R, Rebours C, Gietl A, Stengel D B, and Nitschke U 2018 *Food Chem.* 254 333–9.
[6] Soetjipto *et al* 2019 *Peluang Usaha dan Investasi Rumput Laut* ed I N Sjarif (Jakarta: Directorate of Business and Investment, Directorate General of Strengthening Competitiveness of Marine and Fishery Products, Ministry of Maritime Affairs and Fisheries of the Republic of Indonesia).
[7] Fadhlullah M, Soeprijadi, Ratnaningtyas S and Mukhaimin 2022 *Iran. J. Fish. Sci.* 21 2022.
[8] Keyimu X G, and Abdullah A 2014 *AIP Conf. Proc.* 1614 399–403.
[9] Kurniawan R, Nurjanah, M.Jacoeb A, Abdullah A and Pertiwi R M 2019 *Ulva lactuca JPHPI.* 22 573–80.
[10] Frøst M B, Hartmann A, Petersen M A, Duelund L, and Mouritsen O G 2021 *Int. J. Gastron. Food Sci.* 25 100363.
[11] Nurbaity S, and Apriani K 2014 *J. Pascapanen dan Bioteknol. -JPB Perikan.* 9 1–10.
[12] Muslikatin 2012 Skripsi. Bogor: Pertanian, Institut Pertanian Bogor.
[13] Gunita E R 2020 *Pros. Pendidik. Tek. Boga Busana.* 15 1–9.
[14] Barokah D 2022 Skripsi. Malang: Universitas Brawijaya.
[15] Lukito M S, Giyarto G, and Jayus J 2017 *J. Agroteknologi.* 11 82–95.
[16] Hakim L 2015 *Spices and Herbs in Community Home Yards (Yogyakarta: Diandra Creative).*
[17] Puspitarini T, and Pratjojo W 2014 *Indo. J. Chem. Sci.* 3 113–8.
[18] Dahlan M, Anggaeni E D, and Aspriati D W 2021 *Int. J. Anim. Sci.* 2 6–14.
[19] Patra J K, Banerjee S, Das G, Bose S, Rodriguez-torres M P, and Shin H-S 2020 *Phyther. Res.* 34 1248–67.
[20] Setyaningsih D, Apriyantono A, and Sari M P 2010 *Sensory Analysis for the Food and Agro Industry (IPB Press).*
[21] [AOAC] Associatian of Official Analytical Chemist 2005 *Official Methods of Analytical of The Association of Official Analytical Chemist* (Washington DC: AOAC).
[22] Walpole R E 1995 *Pengantar Statistika* ed B Sumantri (Jakarta: Gramedia Pustaka Utama).
[23] Steel, RGD, Torrie JH D D 1996 *Principles and Procedures of Statistics* (New York: McGraw-Hill Inc).

[24] Yakhin L A, Wijaya K, and Santoso J 2013 *JPHPI-Jurnal Pengolah. Has. Perikan. Indones.* 16 177–82.
[25] Yohanista M, and Rume M I 2021 *Jurnal Ilmu Kelaut. dan Perikan.* 3 1-5.
[26] Purnomo, and Suhanda J 2017 *Fish Sci.* 7 83–99.
[27] Eveline, Santoso J, and Widjaja I 2011 *J. Pengolah. Has. Perikan. Indones.* XIV 98–105.
[28] Mahmud S, Saleem M, Siddique S, Ahmed R, Khanum R and Perveen Z 2009 *J. Saudi Chem. Soc.* 13 195–8.
[29] Ferhat M A, Meklati B Y, and Chemat F 2007 *Flavour Fragr. J.* 22 494–504.
[30] Kubeczka K H, and Ullmann I 1981 *Phytochemistry* . 20 828–30.
[31] Perez-Cacho P R, and Rouseff R 2008 *J. Agric. Food Chem* . 56 9785–96.
[32] Badarudin M I 2019 *J. Ris. Perikan. dan Kelaut.* 1 14–25.
[33] Karina I, and Desrizal 2021*Teknol. Pangan Media Inf. dan Komun. Ilm. Teknol. Pertan.* 12 220–230.
[34] Salman S, Triadi A A, Sulistyowati E D, Zainuri A, and Padang Y A 2019 *J. Karya Pengabdi.* 1 49–56.
[35] Ródenas de la Rocha S, Sánchez-Muniz F J, Gómez-Juaristi M and Marín M T L 2009 *J. Food Compos. Anal.* 22 330–6.
[36] Milinovic J, Rodrigues C, Diniz M, and Noronha J P 2021 *Algal Res.* 53 1–7.
[37] Sanchez-Rodriguez I, Huerta-Diaz M A, Choumiline E, Holguin-Quinones O, and Zertuche-Gonzalez J A 2001 *Environ. Pollut.* 114 145–60.
[38] Subramanian G, Nagaraj A, Sasikala J, Ravi P, and Sona P 2020 *Int. J. Anal. Exp. modal Anal.* XII 1721–7.
[39] Wresdiyati T, Hartanta A B, and Astawan M 2011 *J. Vet.* 12 126–35.
[40] Wisnu, Restiana D R 2009 *Pena Akuatika.* 1–11.
[41] Diosady L L, Alberti J O, Mannar M G V, and FitzGerald S 1998 *Food Nutr. Bull.* 19 5–12.
[42] Bhatnagar A, Maharda N S, Ambardar V K, Dham D N, Magdum M, and Sankar R 1997 *Indian J. Pediatr.* 64 883–5.
[43] Cook M K, Dial A R, and Hendy I L 2022 *Mar. Chem.* 245.
[44] Fauziyah N, Fermanto, Amin M N G, Alamsjah M A, Subekti S, and Pujiastuti D Y 2021 *J. of Appl. Phy.* 33(2) 1209-1216.
[45] Yolanda N T, and Agustono 2018 *JMCS.*7(3) 127-138.
[46] Alamsjah M A, Sulmartiwi L, Pursetyo K T, Amin M N G, Wardani K A K, and Arifianto M D 2017 *J. of the Indian Academy of Wood Sci.* 14(1) 32-45.
[47] Charles A L, Sridhar K, nad Alamsjah M A 2020 *J. of Appl phyc.* 32(1) 563-572.
[48] Savitri N A, Masithah E D, and Tjahjaningsih W 2022 *IOP Conf. Series : EES.* 1036(1) 012008.
[49] Ministry of Health 2019 Regulation of the Minister of Health of the Republic of Indonesia Number 28 of 2019 Concerning Recommended Nutrition Adequacy Rates for Indonesian People (Jakarta: Ministry of Health, Republic of Indonesia).

6. Acknowledgment

The authors would like to acknowledge the financial support from National Institute of Health Research and Development, Ministry of Health, Indonesia.

Inhibition of *Lactobacillus acidophillus, Lactobacillus, Bacteria plantarum* and *Bifidobacterium spp.* against *Escherichia coli* and *Salmonella sp.* in the digestive tract in vitro

B Agustono[1], R Apriliawati[2], A E wijaya[2],N husna[2], F N inta[2], M R maulana[2], M R habibi[2], P risqiah[2], Isnaini[2], M Y yunita[3], M G A yuliani[4]

[1]Division of Animal Husbandry, Veterinary Medicine, School of Health and Natural Sciences, Jl. Mulyorejo, Kampus C, Universitas Airlangga, Surabaya, Indonesia
[2]Veterinary Medicine, School of Health and Natural Sciences, Jl. Mulyorejo, Kampus C, Universitas Airlangga, Surabaya, Indonesia
[3]Division of Pathology Veteriner, Veterinary Medicine, School of Health and Natural Sciences, Jl. Mulyorejo, Kampus C, Universitas Airlangga, Surabaya, Indonesia
[4] Division of Basic Veteriner Medicine, Veterinary Medicine, School of Health and Natural Sciences, Jl. Mulyorejo, Kampus C, Universitas Airlangga, Surabaya, Indonesia

[1]Corresponding author: bodhiagustono@fkh.unair.ac.id

Abstract. The bacteria *Lactobacillus acidophillus*, *Lactobacillus plantarum* and *Bifidobacterium spp.* is a lactic acid bacteria that is used as a probiotic. The research was conducted with the aim of knowing the inhibition of the pribiotic bacteria *Lactobacillus acidophillus, Lactobacillus plantarum* and *Bifidobacterium spp.* against *Escherichia coli* and *Salmonella sp.* Supporting tests such as motility tests, catalase tests, and fermentation tests were also carried out. The media used in this study were: Mueller Hinton Agar (MHA) with the Kirby-Bauer method and Nutrient Broth (NB) media, TSIA, SIM. The control groups used in this study were P1 : *Lactobacillus acidophillus*, P2 : *Lactobacillus plantarum*, P3 : *Bifidobacterium*, P4 : *Lactobacillus acidophilus* + *Bifidobacterium*, P5 : *Lactobacillus plantarum* + *Bifidobacterium*, P6 : *Lactobacillus acidophilus* + *Lactobacillus plantarum* + *Bifidobacterium*, and P0 as control. The greatest inhibitory power on *Escherichia coli* bacteria was at P3, P4, P6 at 0.30 mm and *Salmonella sp.* namely at P4 by 15 mm. While the smallest inhibition on Escherichia coli bacteria is at P5, P7 is 0.2 mm and *Salmonella sp.* namely at P3, P5, P6 by 0.25 mm.

1. Introduction

E Coli is a Gram negative bacterium, short rod-shaped, facultative anaerobic, non-sporing, and widely found in our environment. *Escherichia coli* bacteria are species with natural habitats in the digestive tract of humans and animals. *Escherichia coli* bacteria can form colonies in the digestive tract of humans and animals within a few hours after birth. *Escherichia coli* if it enters the digestive tract in large quantities can cause health [1]. Although *Escherichia coli* is part of the normal microbe of the digestive tract, it has now been proven that certain strains are capable of causing moderate to severe gastroenteritis in humans and animals. There are three types of *Escherichia coli*, namely: Enterotoxigenic *Escherichia coli* (Enterotoxigenic Escherichia coli (ETEC)). Enterotoxin production by *Escherichia coli* was discovered in the 1970s from strains associated with diarrheal disease. Several strains of *Eschericia coli* cause infections in humans, such as urinary tract infections, meningitis infections in neonates, and intestinal infections (gastroenteritis). Most of the diseases caused by infection with *Escherichia coli* are transmitted through uncooked food by direct contact and usually occur in places that have less sanitation and a less clean environment.

Salmonella is a gram-negative bacterium and belongs to the Enterobacteriaceae family. *Salmonella* is an enteric pathogenic bacterium and the main cause of foodborne disease. (Klotchko, 2011). There are more than 2500 Salmonella serotypes that can infect humans. However, the serotypes that are often the main cause of infection in humans are the following, namely *Salmonella paratyphi* A (serogroup A), *Salmonella paratyphi* B (serogroup B), *Salmonella cholerasius* (serogroup C1) and *Salmonella typhi* (serogroup D). There are more than 2500 *Salmonella* serotypes that can infect humans. However, the serotypes that are often the main cause of infection in humans are the following, namely *Salmonella paratyphi* A (serogroup A), *Salmonella paratyphi* B (serogroup B), *Salmonella cholerasius* (serogroup C1) and *Salmonella typhi* (serogroup D) [2].

In the treatment of diseases caused by bacterial infections, it is often done by consuming antibacterial agents (antibiotics). The application of antibiotics is widely used in preventing and treating infectious agents, as well as as additional feed to spur improved livestock performance [3]. Journalists et al., [4] stated that inappropriate and continuous use of antibiotics can cause bacterial residues, so it is found that E. coli and Salmonella sp. which are resistant to several antibacterials [5].

There are several ways to prevent infections caused by *E. coli* and *Salmonella sp*. bacteria. One of them is using probiotics that can inhibit the growth of pathogenic bacteria by competing for a number of substrates for fermentation [6]. Many types of bacteria are currently often used as probiotics, but not much is known with certainty about their activity against pathogenic bacteria. Therefore, this research will be conducted to test the probiotic activity of *Lactobacillus acidophilus*, *Lactobacillus plantarum*, and *Bifidobacterium sp.* The types of probiotics used are *Lactobacillus* and *Bifidobacterium* bacteria which can improve health by stimulating the immune response and inhibiting pathogens [4].

2. Research Methods

This study used several methods with treatment P0: control, P1: *Lactobacillus acidophillus*, P2: *Lactobacillus plantarum*, P3: *Bifidobacterium*, P4: *Lactobacillus acidophillus+Bifidobacterium*, P5: *Lactobacillus plantarum+Bifidobacterium*, P6: *Lactobacillus acidophillus+Lactobacillus plantarum+Bifidobacterium* .

2.1 Tools and materials

2.1.1 Tool

The tools used in this practicum include Petri dishes, Incubator, Analytical Balance, Ose, Aluminum Foil, Cotton, Tweezers, Bunsen, Matches, Label Paper, Tissue, Micropipette, Yellow tip, Stirrer, Test Tube, Glass Measuring, Dropper, Volumetric Pipette, Bulb, Paper disk, Spreader, Caliper

2.1.2 Ingredient

The materials used include isolates of *Eschericia Coli* bacteria and *Salmonella sp. Lactobacillus Acidophillus* Probiotic Bacteria Solution, *Lactobacillus Plantarum, Bifidobacterium,* Nutrient Broth (NB), Mueller Hinton Agar (MHA) Media, Aquadest, Spiritus.

2.2 Research Procedure

2.2.1 Motility Test of Escherichia coli and Salmonella sp.

Bacterial isolates were put into a semi-solid SIM medium in a tube reaction using a sterile ossicle needle. Then incubated for 24 hours at 37ºC. A positive test is indicated by the growth of bacteria that spreads, then the bacteria mobile (motile) and if bacterial growth does not spread in only one pathway, then these bacteria are immobile (non motile) [8].

2.2.2 Catalase Test

At first, the cloudy medium was dripped onto the slide, then H2O2 was added. Catalase activity on microbes can be determined by the formation of O2 gas bubbles. A positive test is indicated by the formation of oxygen bubbles which indicates that the organism produces the enzyme catalase which converts hydrogen peroxide into water and oxygen.

2.2.3 Fermentation Test

The fermentation test used TSIA media which was made the same as the H_2S test. The media will be sterilized at 121°C for 15 minutes. Then the media will be poured into each test tube with a dose of 7 ml and then tilted until the media becomes solid. The media that has become solid will then be inoculated with bacterial isolates aseptically using a loop needle inserted in the middle until it reaches 3/4 of the surface of the media. The ose needle that was inserted will be scratched on the inclined plane of the media and will then be incubated for 1x24 hours at 37°C [9].

2.2.4 Test of Rejuvenation of Escherichia coli and Salmonella sp.

The test bacteria used must be rejuvenated first by scratching using a sterile loop and then inoculated at 37°C for 24 hours. After that, the test isolate was made by means of the rejuvenated *Eschericia coli* bacterial colonies suspended in a Nacl solution with a ratio of 1:9 where 1 ml of bacterial suspension was dissolved with 9 ml of Nacl and then homogenized to obtain the final bacterial result with a concentration of 108 CFU/ml which was then compared to a standard 1 ml McFarland tube. The process of rejuvenation of *Salmonella sp.* This was done by scratching on SSA (Salmonella Shigella Agar) media. SSA (Salmonella Shigella Agar) media is a selective medium that can grow *Salmonella sp.* [10]. Then inoculated at 37°C for 24 hours.

2.2.5 Antibacterial Activity Test Against Escherichia coli and Salmonella sp.

One of the methods used in the antibacterial activity test against bacteria is the disc diffusion method (Kirby-Bauer test). Before carrying out the bacterial activity test using the disc diffusion method, the two test bacteria used (*Escherichia coli and Salmonella sp.*) were adjusted to the Mac Farland standard of 1 ml. Using a micropipette, take 100 . of bacterial suspensionl into The surface of the media used is Hueller Hinton Agar (MHA) by spreading it evenly using a spreader. Prepare a probiotic solution, including *Lactobacillus acidophillus, Lactobacillus plantarum, Bifidobacterium*, a combination of *Lactobacillus acidophillus* and *Bifidobacterium, Lactobacillus plantarum* and *Bifidobacterium, Lactobacillus acidophillus* and *Lactobacillus plantarum*, and a combination of the three probiotics. Drops of paper dish (6 mm diameter) with each probiotic solution as much as 50lin a baking glass to infuse for 10 minutes. Then the paper dish is placed on the media containing *Salmonella sp.* and media containing 2-3 paper dishes of *Escherichia coli* each, press the paper dish slowly until it sticks to the surface of the media. Then incubated at 37ºC for 24 hours. Antibacterial activity was declared positive if an inhibition zone was formed in the form of a clear zone around the paper disk. [11]. disc diffusion method (Kirby-Bauer test).

2.2.6 Antibacterial Inhibition Test

The media that had been incubated for 24 hours was then observed the inhibition zone formed around the paper dish. The inhibition zone formed indicates the strength of probiotic bacteria against the test bacteria *Eschrechia coli* and *Salmonella sp.* The diameter of the inhibition zone formed. After the formation of the inhibition zone around the paper dish, the diameter of the inhibition zone formed was measured using a caliper.

IOP Conf. Series: Earth and Environmental Science 1273 (2023) 012022 doi:10.1088/1755-1315/1273/1/012022

3. Results and Discussion

3.1. Motility Test of Escherichia coli and Salmonella sp.

Based on their motility, bacteria are classified into motile and non-motile groups. Motile bacteria are bacteria that have locomotion and are generally in the form of flagella so that it is possible to move, while non-motile bacteria are bacteria that do not have locomotion. Based on the arrangement and location of the flagella, bacteria are divided into monotric, amphitric, lophotric and peritrichous. Monotrik ie bacteria that have one flagellum on one side of the cell body; amphitrik i.e. bacteria that have one flagellum on two sides of the cell body; lophotrik, namely bacteria that have many flagella but only in one part of the cell body; and peritrichous bacteria that have flagella on all sides of the cell body [12].

Motility test was carried out to determine the motility of EE bacterial isolates. Bacteria were inoculated into tubes containing the medium aseptically with the stab technique. The tubes were then incubated for 24 and 48 hours. The results of 24-hour observations have shown that EE bacterial isolates grow only on the stab line, do not grow around the stab line, and there is also a change in color on the stab line only [13]. 48 hours of observation was not observed.

3.2 Catalase Test

Based on the research conducted, the results showed that the bacteria *Salmonella sp., Escherichia coli, Bifidobacterium,* and *Lactobacillus acidofhilus* showed positive results. While the bacteria *Lactobacilus plantarum* showed negative results. From these data, only four bacterial isolates produced the enzyme catalase. According to Lay, [14], to determine the presence of catalase enzyme in bacterial isolates using the catalase test. Catalase is an enzyme that can catalyze the decomposition of Hydrogen Peroxide (H2O2) into water and O2. Hydrogen peroxide is toxic to bacterial cells because this material is able to deactivate enzymes in cells and is very harmful to bacterial cells themselves. This test is important to do to determine the nature of a bacterium to oxygen demand.

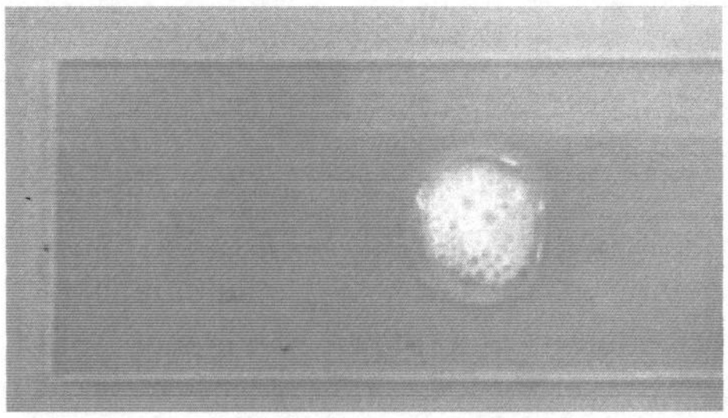

Figure 1. Catalase test positive of *Escherichia coli* and *Salomonella sp.*

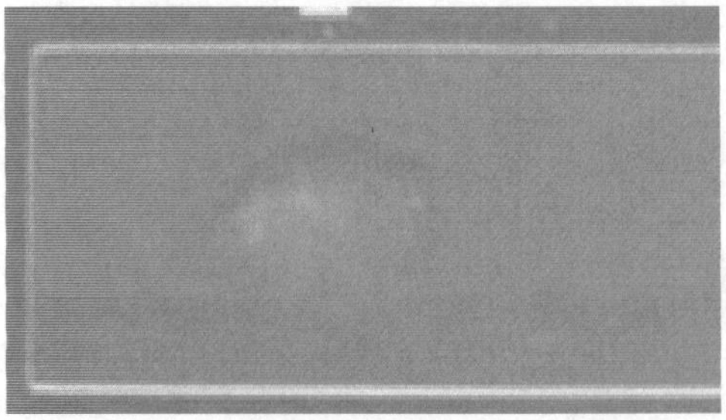

Figure 2. Catalase test negative of *Escherichia coli* and *Salmonella sp.*

3.3 Fermentation Test

Based on the tests carried out, the results showed that *E.coli* and *Salmoella sp.* is facultative, *Bifidobacterium* is anaerobic, *L. acidophillus* and *L. plantarum* are known to not have bacterial growth. The fermentation test is carried out to determine the ability of a bacterium to ferment sugar to produce acid or gas [9]. The results of the tests carried out are listed in table 1.

Table 1. Result of the test motility, catalase, fermentation

Isolat	Motility	Catalase	Fermentation	H_2S
Escherichia coli	+	+	b/f	-
Salmonella sp.	+	+	b/f	+
L. acidophillus	-	+	t/t	-
L. plantarum	-	-	t/t	-
Bifidobacterium	-	+	b/a	-

Note : + : Positive;- : Negative; b/f : Facultative; b/a : Anaerobic; t/t : Not growing

3.4 Antibacterial Inhibition Test

Based on the results of this study, the probiotics *Lactobacillus acidophilus, Lactobacillus plantarum, Bifidobacteriun sp.* and several combinations, namely Lactobacillus acidophilus with *Lactobacillus plantarum, Lactobacillus acidophilus* with *Bifidobacteriun sp., Lactobacillus acidophilus* with *Lactobacillus plantarum* with *Bifidobacteriun sp.* inhibit the growth of *Escherichia coli* and *Salmonella* bacteria. The inhibition of *Lactobacillus acidophilus* against *E.coli* is in the weak category, which is 0.25 mm. The inhibition zone is classified as weak for *Salmonella*, which is 2.42 mm.

This research was conducted with the aim of knowing the ability of probiotics such as *Lactobacillus acidophilus, Lactoacilus plantarum*, and *Bifidobacterium sp.* This probiotic study by comparing the zone of inhibition on the bacteria *E. coli* and *Staphylococcus sp.* Using a control without treatment which showed no inhibition zone formed. The zone of inhibition is a reflection that probiotic bacteria have the ability to inhibit the growth of pathogenic bacteria. *L acidophilus* and *Bifidobacterium* have the ability to inhibit the growth of pathogenic bacteria, where the inhibitory power against H pylori is the highest. These results are in accordance with the research of Fuller, [15] and Fukushima, [16], *Lactobacillus acidophilus* or *Bifidobacterium* is able to produce natural antibiotics, namely bifidin, while *L. acidophilus* has the ability to produce bacteriocin bacteriocin. Fuller, [15] stated that *L. acidophilus* has the ability to stimulate local bacteria and is able to survive in the digestive tract. Furthermore, *L acidophilus* produces many antibiotics such as Acidolin, Acidophillin, Lactobacillin, and Lactocidin. and *Bifidobacterium* produce Bifidin [17].

In the treatment of probiotics, there were differences in results with control m, namely the formation of an inhibition zone. In each treatment, the results were quite varied as shown in the table. One of the inhibiting factors of the bacteria *Lactobacillus acidophilus, Lactobacillus plantarum, Bifidobacteriun sp.* and several combinations, namely *Lactobacillus acidophilus* with *Lactobacillus plantarum, Lactobacillus acidophilus* with *Bifidobacteriun sp., Lactobacillus acidophilus* with *Lactobacillus plantarum* with *Bifidobacteriun* sp. against *Escherichia coli* and *salmonella sp.* is the storage of bacteria at room temperature for too long so that the resulting effect is less than optimal. The overall anti-bacterial activity was declared positive because there was an inhibition zone formed around the parer disk.

From the test results obtained the results of the formation of inhibition zones in all treatments. This proves that the bacteria *Lactobacillus acidophilus*, *Lactobacillus plantarum, Bifidobacteriun sp.* and the combination can inhibit the growth of *E.coli* and *Staphylococcus sp* bacteria by producing special substances. According to Rachmawati, [18] *Lactobacillus achidophilus* bacteria can inhibit the growth of *E. colli* and *Staphylococcus sp.* by producing Lactin B and acidocin. The inhibition zone obtained from the treatment of *L. plantarum* was caused by the primary metabolites of *L. plantarum* in the form of lactic acid, ethanol, diacetyl, and carbon dioxide as well as secondary metabolites in the form of bacterosine and hydrogen peroxide compounds [19].

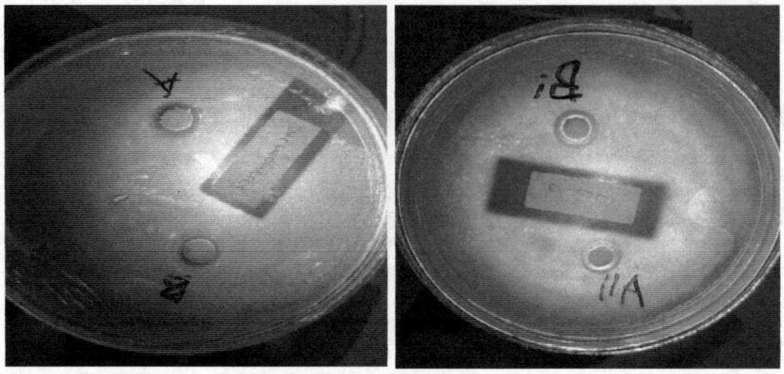

Figure 3. picture of inhibitory test results from probiotic bacteria.

Table 2. Inhibition of the probiotik against *E.coli* and *Salmonella sp.*

Sample	Label	Zone of Inhibition (mm) / SD
Escherichia coli	P0	0 ± 0.00000
	P1	0.25 ± 0.01000
	P2	0.25 ± 0.01000
	P3	0.3 ± 0.01000
	P4	0.3 ± 0.01000
	P5	0.2 ± 0.01000
	P6	0.3 ± 0.01000
	P7	0.2 ± 0.01528
Salmonella sp.	P0	0 ± 0.00000
	P1	0.5 ± 0.01000
	P2	0.35 ± 0.01000
	P3	0.25 ± 0.01000
	P4	15 ± 1.00000
	P5	0.25 ± 0.01000
	P6	0.25 ± 0.01000
	P7	0.35± 0.01000

Note: P1 : *Lactobacillus acidophilus;* P2 : *Lactobacillus plantarum*; P3 : *Bifidobacterium*; P4 : *Lactobacillus acidophilus+Bifidobacterium*; P5 : *Lactobacillus plantarum+Bifidobacterium*; P6 : *Lactobacillus acidophilus+Lactobacillus plantarum+Bifidobacterium*

4. Conclusion

Inhibition of the pribiotic bacteria *Lactobacillus acidophillus, Lactobacillus plantarum* and *Bifidobacterium spp*. against *Escherichia coli* and *Salmonella sp*.

5. Reference

[1] Suwito, and Andriani W. 2018. *J of Trop Biol.*
[2] Klotchko, a. 2011. salmonellosis. Available from:https://emedicine.medscape.com/article/228174- overview#a4 .
[3] Brooks, g.f., butel j.s & morse s.a. 2004. Jawetz, Melnick & Adelberg's medical microbiology twenty second edition Lange Medical Books/Mcgraw-hill. medical publishing division.
[4] Haryati, T. U. T. I. 2011. *Wartazoa*, 21(3), 125-132.
[5] Journalist, Y.D., Yorva S., Aslinar. 2009. *Andalas Med Mag*. 01(33).
[6] Widianingsih, M. 2018. *JST*, *7*(2), 178-187.
[7] Sumarsih, S., Sulistiyanto, B., Sutrisno, C. I., and Rahayu, E. S. 2012. *J Lit Prov Jateng*, *10*(1), 1-9.

[8] Sudarsono A. 2008. Isolation and Characterization of Bacteria in Marine Fish in Gindara Fish Species (Lepidocibium flavobronneum). Thesis. Bogor: Bogor Agricultural University
[9] Kosasi, C., Lolo, Widya A., Sedewi, Sri. 2019. J. Agardh and Biochemical Ident. 8(2).
[10] Ningrum. 2014. Analysis of Salmonella sp and Formalin Content in Milkfish Brain Foods (Chanos chanos Forsk) sold in Gresik City Gift Shops as a High School Biology Learning Resource, Thesis. University of Muhammadiyah Malang, Malang.
[11] Kaseng, E. S., Muhlishah, N., and Irawan, S. 2016. *Bionature*, 17(1).
[12] Atlas, R. M. 2014. Handbook of Microbiological Media, 4th ed. ASM Press, Washington DC: 2037 hlm.
[13] Talaro, K. P. 2005. Immunization and Immune Assays In: Foundations in Microbiology Basic Principles. Nev York, McGraw Hill, 490-491.
[14] W. Lay, Bibiana. 1994. Microbial Analysis in the Laboratory. Jakarta: Raja Grafindo Persada.
[15] Fuller, R. 1992. History and development of probiotics. Springer, Dordrecht. In probiotics (pp 1-8).
[16] Fukushima, Y., Kawata, Y., Hara, H., Terada, A., & Mitsuoka, T. 1998. *Int. J. food microbiol*, *42*(1-2), 39-44.
[17] Chaitow, L., and Trenev, N. 1990. Pro Biotics: The revolutionary "friendly bacteria" Way to Vital Health and Well-Being. Thorsons Publishing Group.
[18] Rachmawati, I., Suranto, S., dan Setyaningsih, R. 2005. *Asian J Trop Biotechnol*, *2*(2), 43-48.
[19] Maurya, A. P., & Thakur, R. L. 2012. *Current Science*, *103*(12), 1405-1407.

IOP Conf. Series: Earth and Environmental Science 1273 (2023) 012023 doi:10.1088/1755-1315/1273/1/012023

Iodium fortification on traditional salt with spray different potassium iodate

H F Salsabila[1], M A Alamsjah[2,3], D Y Pujiastuti[2]

[1]Study Program of Fisheries Product Technology, Faculty of Fisheries and Marine, Universitas Airlangga, Campus C UNAIR, Jl. Mulyorejo, Surabaya, Indonesia
[2]Department of Marine, Faculty of Fisheries and Marine, Universitas Airlangga, Campus C UNAIR, Jl. Mulyorejo, Surabaya, Indonesia

[3]Corresponding author: alamsjah@fpk.unair.ac.id

Abstract. The purpose of this study was to increase the iodine content of traditional salt after fortification of iodine with potassium iodate spray. In addition, this study also aims to determine the NaCl content and water content of traditional salt after potassium iodate fortification is carried out. This research is also expected to provide an alternative solution to increase the iodine content of salt by fortification of potassium iodate. And hopefully this research can help the government in handling cases of Iodine Deficiency Disorder in Indonesia. The treatments used were the addition of KIO_3 solution with different weight of traditional salt P1 (50 gram traditional salt : 3 ml KIO_3 solution), P2 (33,33 gram traditional salt : 3 ml KIO_3 solution) dan P3 (25 gram traditional salt : 3 ml KIO_3 solution). Potassium iodate fortification in traditional salt affects iodine and NaCl levels. Fortification of 3 ml potassium iodate with a salt weight of 25 grams (P3) produced thehighest iodine content with an average of 64.74 ppm, an average NaCl content of 86.14 %, and an average water content of 3.7%.

1. Introduction

Iodine is an important compound that is generally contained in salt. Iodine itselfhas many benefits for the human body, it can even be called a very essential substance because it is a composition of the thyroxine hormone. Iodine is also needed on an ongoing basis to maintain the body's metabolism. To increase brain intelligence, iodine intake is also needed. Iodine deficiency can cause various diseases such as goiter, growth disorders (dwarf), and physical disorders such as muteness and deafness.

Based on WHO data in 2008 a third of the world's population still suffers from iodine deficiency, with a total of 260 million school-age children and 80% of cases are mild to moderate iodine deficiency. According to Agus [1], iodine deficiency in Indonesia has been widely discussed since 1927 and has been found in almost all partsof Indonesia, from the northern part of Aceh to the Papua area. The problem of iodine deficiency generally appears among the lower-middle class, but recently developed countries such as Australia, England, and New Zealand have been identified as havingmild-scale iodine deficiency disorders.

Iodine mineral substances are usually found in table salt that is sold freely in themarket, but not all types and brands of table salt contain iodine. The results of the 2005 iodized salt survey showed that

13% of the salt used by households did not contain iodine. As much as 16% of the salt used by households in rural areas and 9% in urbanareas does not contain iodine. Salt without iodine is still circulating in the market because there are still direct sales from salt farmers or small traders.

According to data from Badan Pusat Statistik (BPS) the realization of Indonesian salt imports throughout 2020 reached 2.61 million tons with a value of US$ 94.55 million, an increase in volume compared to the realization of imports in 2019 and 2018. As for during January-February 2021, Indonesia is recorded as still importing salt with a volume of 80.2 thousand tons or equivalent to US$ 2.61 million.

This realization is greater than the realization of imports from January to February 2020 which reached 123.76 thousand tons. One of the things that causes Indonesia to still import salt is not the inability of the farmers to provide salt in large quantities. However, the quality of the salt itself varies. The salt produced by Indonesian farmersis classified as not suitable for industrial salt. Because if the industrial sector continuesto impose the use of local salt, it will automatically change the taste of the product andhave the potential to have a negative impact on the industrial sector.

Therefore, the scientists of this country must show more concern for the little people in the form of research in the field of science. In the field of fishery product technology, it is possible to study efforts to increase iodine in salt. One of the ways fortification of iodine in salt by fortification of potassium iodate. Potassium iodate itselfis a compound that can be added to salt to increase the level of iodine in salt. One of the iodine fortification techniques is the spray technique.

2. Material and method

2.1 Research settings

This research was conducted from March to April 2022 at the Chemistry and Analysis Laboratory, Faculty of Fisheries and Marine Affairs, Airlangga University, Surabaya.

2.2 Research material

The main ingredients used include traditional salt, potassium iodate solution, HCL, 5% potassium chromate, sulfuric acid, silver nitrate, 10% KI, 0.005 N sodium thiosulfate, 0.1 N potassium dichromate, 1% starch and aquadest.

The tools needed in this study were 10 ml and 50 ml (pyrex) burettes, oven, 250 ml erlenmeyer (pyrex), 10 ml measuring cups (pyrex), 100 ml and 250 ml volumetric flasks (pyrex), microburette 5 ml (pyrex), analytical balance (mettle toledo), oven, mixer, blender, 0.5 ml, 50 ml and 20 ml volume pipettes (pyrex), sprayer and dropper.

2.3 Research design

This study used a completely randomized design (CRD), with 3 treatmentsand 6 replications referring to the formula (t-1)(n-1)≥15.

Table 1. Research design

Treatment	KIO_3 Solution (ml)	Traditional Salt (g)
P1	3	50
P2	3	33.33
P3	3	25

2.4 Research procedures

2.4.1 Refining stage

Before the salt is fortified. Traditional salt has to go through a refining stage. Traditional salt is ground using a mortar and pestle. The purpose of this stage is to reduce the salt particles so that they are suitable for consumption salt.

IOP Conf. Series: Earth and Environmental Science 1273 (2023) 012023 doi:10.1088/1755-1315/1273/1/012023

2.4.2 Weighing stage

After passing through the washing stage, the traditional salt is then weighed usingan analytical balance. Weighing is done to determine the weight of salt to be fortified.

2.4.3 Fortification KIO_3

Make a solution by mixing aquadest with powder in a ratio of 10:1, then the solutionis put into a 3 ml sprayer and then sprayed according to the weight of each treatment. After spraying, the salt is stirred evenly and then tested for iodine, NaCl, and water content.

3. Results and discussion

Characterization of traditional salt before fortification needs to be done to findout the initial data and also as a consideration in the analysis of this research. The characterizations analyzed in this study were the traditional salt water content, NaCl content, and the main analysis was the iodine content. The results of the analysis of variance of traditional salt that have been fortified can be seen in table 1.

Table 2. Results analysis traditional salt of fortified KIO3

Treatment	Iodine Content (ppm)	NaCl Content (%)	Water Content (%)
P1	40.28 ^{a}±0.33	77.88 ^{a} ±0.11	4.53 ^{a}±0.07
P2	52.22 ^{b}±0.18	80.54 ^{b}±0.46	4.23 ^{bb}±0.06
P3	64.74 ^{c}±0.33	86.14 ^{c}±0.08	3.70 ^{bb}±0.49

Results Analysis of iodine levels is the main data in this study. Iodine levels in salt can be calculated by various methods such as using AAS or iodometric titration. In this study, the iodine level test was carried out using the iodometer titration methodbecause the cost used was more affordable and required a relatively shorter time than the AAS method. The drawback of the iodometric titration method is that the results ofthe analysis are not as accurate as if the measurements were carried out with AAS. Iodometric titration is carried out with more or less the same method as in general titration. Iodine data for traditional salt after fortification ranged from 40.17 to 64.91 ppm. Where this is in accordance with the SNI for consumption salt 3556:2000. The results of traditional salt characterization, namely the iodine content of traditional saltthat had been fortified potassium iodate analyzed using ANOVA showed significantlydifferent results indicated by the P-value (<0.05). which had been fortified P1, P2, and P3 were significantly different. the analysis of traditional salt characterization, namely the NaCl content in traditional salt that has been fortified, analyzed using ANOVA showed significantly different results indicated by the P-value (<0.05) as well as the analysis of the water content results.

Fortification is an effort made to add micronutrients such as vitamins and minerals to food to improve the quality of nutrients that are useful for public health. Food fortification with micronutrients is one of the main strategies that can be used toimprove the micronutrient status of food. According to Budiarto and Rini [2], there are several things that must be considered in the food fortification process, including fortified food ingredients that are safe for consumption and there is a guarantee against possible side effects. Fortification is carried out with potassium iodate because it is more stable than potassium iodide in salt fortification. According to Permatasari [3] KI is more easily lost if iodized salt is exposed to sunlight, heat, exposure to high temperatures, oxidizing agents, high humidity, high acidity, and there are impurities (impurities) in the salt. The choice of fortification with potassium iodate in traditionalsalt is also done because it is relatively easier.

The increase was quite significant from P1, P2, and P3 due to spraying with thesame amount carried out on samples with different weights. Where the heavier the sample weight, the lower the iodine content. On the other hand, where the iodine content of salt with a lighter weight will result in higher iodine yields. At the stage of fortification of potassium iodate of traditional salt, there are several factors that allow the level of iodine in traditional salt to fluctuate unevenly. These factors, among others, according to Permatasari [3] are homogeneity, increased water content, increased pH, and the nature of potassium iodate itself, which is volatile. In this research, the main factor that might influence the inequality in the fortification results in each sample P1, P2, and P3 was homogeneity. This is due to the lack of homogeneity when mixing traditional salt with potassium iodate. The possibility of a lack of homogeneity is verylarge, considering the tools used are simple tools, namely a spatula and a stirring rod.

Another factor that may influence the uneven fluctuation is the nature of the potassium iodate itself. Potassium iodate is an unstable form because of its high solubility in water, the iodine content is easily lost under extreme conditions, such as light, heat, and humidity. According to Karwiti [4], substances that are reducing and hygroscopic in salt have the most effect on the loss of iodine through the oxidation- reduction (redox) process in an acidic environment. These substances are formed simultaneously with the process of forming salt.

In this research based on the results of ANOVA analysis showed the higher thelevel of potassium iodate spraying, the higher of NaCl level. This is because accordingto Pakaya *et al.*, [5] statement, the ecological cycle of iodine in nature begins with the form of seawater vapor (which contains iodine) which is carried by wind and clouds toland areas. Seawater vapor containing iodine will then fall as rainwater and replace theiodine-deficient soil layer. It can be interpreted that potassium iodate itself has similarproperties to seawater, where purification and washing of salt with old water (sedimentated seawater) can increase NaCl levels significantly. Based on the results of the analysis of the water content of traditional salt that has been fortified, it is in accordance with the SNI standard for consumption salt of 3556:2000, which is $<7\%$. The results of the ANOVA analysis of the water content of traditional salt that had beenfortified P1 had a lower concentration than P2 and P3. It can be said that theconcentration of water content in this study is inversely proportional to the content of iodine and NaCl levels. Where the greater the water content, the lower the levels of iodine and NaCl.

4. Conclusion

Based on the results of the research that has been done, the followingconclusions can be drawn that iodine fortification of traditional salt with potassium iodate spray can increase the level of iodine in traditional salt with the highest average iodine being found in P3. Furthermore, iodine fortification of traditional salt also affects changes in NaCl content and watercontent. There was an increase in NaCl levels at P1, P2, and P3 and a decrease in water content at P1, P2, and P3. It is necessary to re- analyze the chemical reaction that occurs between potassium iodate and NaCl, and it isnecessary to review the mixing method of mixing potassium iodate with NaCl so that fluctuations in iodine in salt need to be reviewed and traditional fortification can be increased homogeneously.

5. References

[1] Agus, Z. 2007 *J Kesehat Masy Andalas*. 1(2): 59-64.
[2] Budiarto, H. and Rini, D. A. S. 2019 *Indones J Mar Sci Technol*. 12(2): 104-111
[3] Permatasari, S. M. 2017 *Darussalam Nutr J*. 1(1): 8-15.
[4] Karwiti, W., Basa, I. H., Asrori, A. and Silvia, V. 2018 *J Kesehat Poltekkes Palembang*. 13(2): 98-110.
[5] Pakaya, N. K., Sulistijowati, R. and Dali, F. A. 2015 *NIKe J*. 3(1): 52-65.

Testing the antibacterial effectiveness of mangosteen peel extract (*Garcinia mangostana* L) against *Aeromonas hydrophila* Bacterial infection in dumbo catfish (*Clarias gariepinus*)

B Cahya[1], D Nirmala[2,3], B Triantoro[1]

[1] Bachelor of Fisheries and Marine Sciences Faculty of Brawijaya University
[2]Department of Marine, Faculty of Fisheries and Marine, Universitas Airlangga, Surabaya 60115 Indonesia

[3]Corresponding author: dwitha.nirmala@fpk.unair.ac.id

Abstract. Plants belonging to the Clusiaceae species are Garcinia mangostana Linn (*mangosteen fruit*). The skin layer of this fruit has been used as a traditional medicine for wound healing, topical infections. Mangosteen rind contains compounds including mangostin, mangostenol, mangostinon A, mangostenon B, trapezifolixanthone, tovofillin B, alpha mangostin, beta mangostin, garsinon B, mangostanol, flavonoids, epicatechin. Gartanin, gamma mangostin, garsinon E, epicatechin. Farming African catfish (*Clarias Gariepinus*) when facing serious problems such as fish are susceptible to disease, mainly due to bacteria. One type of bacteria that causes the disease is Aeromonas hydrophila. This type of bacteria can cause high fish mortality in African catfish (Clarias Gariepinus) of various sizes. The purpose of this study was to determine the effect of mangosteen rind extraction on the growth of Aeromonas hydrophila bacteria in vitro, as well as the 96-hour Lethal Dose (LD50) value of mangosteen rind extract after exposure to African catfish (Clarias Gariepinus). In this study, the antibacterial activity of the material was tested in vitro with concentrations of 0%, 1.56%, 3.125%, 6.25%, 12.5%, 25% and 50%.

1. Introduction

The first time catfish came to Indonesia by a private company around 1986, then developed almost throughout Indonesia. Catfish cultivators have high economic value because they are one of the potential fishery product commodities for export. Catfish (*Clarias* sp) is in third place after carp (*Cyprinus carpio*) and catfish (*Clarias Gariepinus*) in the production of freshwater fish farming in Indonesia [1]. With the ever-increasing demand for catfish, aquaculture business opportunities are finally opening up. Intensive catfish farming has been carried out a lot, such as using superior seeds, providing quality feed, and controlled maintenance management. However, intensive cultivation does not guarantee that catfish will avoid diseases, such as diseases caused by *Streptococcus* sp.

Streptococcus sp bacteria is a bacteria that can cause Streptococciasis which attacks seawater and fresh water fish and can lead to fish death. Many ways have been done to deal with disease attacks in cultivated fish, such as administering vaccines, disinfectants, antibacterials and antibiotics. These efforts are made to increase the body's immunity (immunity) in fish from disease.

Lusiastuti *et al.* [2] stated that giving antibiotics as a treatment for fish can trigger microbial resistance to antibiotics and the danger of antibiotic residues which have a negative impact on society causing several types of broad spectrum antibiotics to be prohibited from circulating and used for fish medicine.

Mangosteen skin shown to have several compounds that beneficial for health. Research by Puspitasari *et al.,* [3] explained that 95% ethanol extract of mangosteen rind contains alkaloids, tannins, flavonoids, triterpenoids, polyphenols, saponins [3]. Previous studies have shown mangosteen peel potential to inhibit various types of bacteria such as *Staphylococcus sanguis, Streptococcus sallivarius, Streptococcus oralis, Streptococcus mutans, Lactobacillus acidophilus* [4] and has an inhibitory effect on *S. aureus*, *Salmonella* sp and *E. Coli* [5] and effectively inhibits bacteria The cause of pneumonia is *klebsiella pneumonia* [6], *Streptococcus mutans* [7].

In an effort to increase immunity in fish, there are various herbal plants that can be used, such as mangosteen rind extract (*Garcinia mangostana* L).). In vitro testing of xanthone natural ingredients from mangosteen rind extraction is expected to occur between natural ingredients in bacterial cell receptors, which is then followed by a transpeptidase reaction and inhibition of peptidoglycan synthesis in the bacterial cell wall. The mechanism ends with the termination of the activity of the autolytic enzyme inhibitor on the cell wall. In an isotonic environment, lysis occurs in a hypertonic environment, so that water moves out of the cell. In a hypertonic environment, osmotic pressure causes water to flow out of the cell. If enough water is removed in this way, the cytoplasm will have such a low concentration of water that the cell will no longer function. If peptidoglycan synthesis is inhibited so that no peptidoglycan is formed, water entering the cell from the environment will cause the bacterial cell to undergo lysis. The lysis of the bacterial cell is caused by the absence of a structure that is able to withstand the osmotic pressure caused by incoming water forming a change in the bacterial cell wall into a protoplast (no cell wall) or spheroflas which is only covered by a fragile cell membrane. If the utilization of mangosteen rind extract synthesizes peptidoglycan in the cell wall of *Streptococcus iniae* bacteria, then the active ingredient has the potential as an anti-bacterial agent [8].

2. Material and methods

2.1. Material

The research was conducted at the Educational Laboratory of the Faculty of Fisheries and Maritime Affairs, Airlangga University, the Bacteriology Laboratory of Class I Fish Quarantine Center (BKI) Juanda Surabaya, the Laboratory for Control and Quality Testing of Fishery Products (LPPMHP) Surabaya and the Testing Service Unit, Phytochemical Laboratory, Faculty of Pharmacy, Airlangga University.

2.2. Methods

The research was conducted in three stages. The first stage of the research was to test the antibacterial sensitivity of the liquid dilution method (broth dilution assay) to determine the Minimum Inhibitory Concentration (MIC) and Minimum Bactericidal Concentration (MBC) in vitro with concentrations of 1.56%, 3.125%, 6.25%, 12, 5%, 25% and 50%. In phase I testing, the statistical analysis used in determining the extract fraction that affected the MIC and MBC values was the ANOVA (Analysis of Variance) test with a Randomized Block Design (RBD) because this study had two sources of variation, namely treatment and experimental media or materials. random influence. The groups as replicates for RAK, between groups of media or experimental materials are considered uniform. The concentrations used were 6 treatments in the repetition calculation RAK with error-free degrees RAK > 15 is 3 repetitions.

3. Result and discussion

3.1 Lethal Dosage 50 (LD50) Bacteria S. Iniae

Based on the data in Table 1, it can be seen that the percentage of closest deaths is below 50%, namely at a bacterial density of 10^7 with a mortality percentage of 45%, while the closest mortality percentage is above 50%, namely at a bacterial density of 108 and 75%, so that the LD50 value is between the bacterial densities 107 and 108. Determination of the LD50 value using the method of Reed and Muench [9]. The LD50 value that will be used in the main test is bacteria with a density of 10^7 CFU/mL.

Table 1. LD50 Test Value of Pathogenicity of *S. Iniae* Bacteria in Dumbo Catfish (*Clarias gariepinus*)

Dilution *S. iniae*	Number of Fish	Number of Deaths	Total Deaths	% Death
10^{10}	20	20	20/20	100
10^9	20	15	15/20	75
10^8	20	13	13/20	65
10^7	20	9	9/20	45
10^6	20	5	5/20	25
10^5	20	3	3/20	15
10^4	20	1	1/20	5
10^3	20	0	0/20	0
10^2	20	0	0/20	0
10^1	20	0	0/20	0
Kontrol	20	0	0/20	0

3.2 Testing of MIC and MBC Stratified Extraction of Mangosteen (G. mangostana L) Peel against S. iniae in vitro

Based on MBC observations, it is known that mangosteen rind extract (*G. mangostana* L) extracted using ethanol as a solvent can kill S. iniae bacteria in vitro. The resulting MIC was grown on TSA agar media, then incubation for 24 hours at 27°C. MBC testing was carried out from a concentration of 1.56% to a concentration of 50%. The minimum concentration that can inhibit the growth of S. iniae is 10^7 CFU/ml at a concentration of 12.5%.

MBC test results can be seen in Table 2. According to Katzung (1989) *in* Naiborhu [10] explained that the mechanism of action of antimicrobial compounds begins with inhibition of cell wall synthesis, changes in cell membrane permeability or active transport through the cell membrane, inhibition of protein synthesis, namely inhibition of translation and transcription of genetic material and inhibition of nucleic acid synthesis. Damage to the cell membrane causes the transport of compounds and ions to not take place into the bacterial cell so that the bacteria experience a lack of nutrients needed for growth and eventually die. The natural substance xanthones from the extraction of mangosteen rind forms a bond between the natural ingredients on the bacterial cell receptor, which is then followed by a transpeptidase reaction and inhibition of peptidoglycan synthesis in the bacterial cell wall. Putra [11] stated that the mechanism of xanthone antibacterial activity occurs due to the reaction of the carbonyl group in xanthones to amino acid residues in cell membrane proteins, extracellular enzymes and cell wall proteins, which causes the protein to lose its function.

Table 2. MBC observations of mangosteen rind extract (*G. mangostana* L) against *S. Iniae*

No.	Solvent (I)	Solvent (J)	Average Solvent (I – J) ± Std. Error
1	Etanol	Etil-asetat	$1{,}07^b \pm 0{,}44$
		Heksane	$1{,}07^b \pm 0{,}44$
2	Etil-asetat	Etanol	$-1{,}07^b \pm 0{,}44$
		Heksane	$0{,}00^a \pm 0{,}44$
3	Heksane	Etanol	$-1{,}07^b \pm 0{,}44$
		Etil-asetat	$0{,}00^a \pm 0{,}44$

MBC test results can be seen in Table 2. According to Katzung (1989) *in* Naiborhu [10] explained that the mechanism of action of antimicrobial compounds begins with inhibition of cell wall synthesis, changes in cell membrane permeability or active transport through the cell membrane, inhibition of protein synthesis, namely inhibition of translation and transcription of genetic material and inhibition of nucleic acid synthesis. Damage to the cell membrane causes the transport of compounds and ions to not take place into the bacterial cell so that the bacteria experience a lack of nutrients needed for growth and eventually die. The natural substance xanthones from the extraction of mangosteen rind forms a bond between the natural ingredients on the bacterial cell receptor, which is then followed by a transpeptidase reaction and inhibition of peptidoglycan synthesis in the bacterial cell wall. Putra [11] stated that the mechanism of xanthone antibacterial activity occurs due to the reaction of the carbonyl group in xanthones to amino acid residues in cell membrane proteins, extracellular enzymes and cell wall proteins, which causes the protein to lose its function.

3.3 Toxicity Test of Ethanol Extract of Mangosteen Peel (G. mangostana Linn) on dumbo catfish (Clarias gariepinus)

In Table 3 of the acute toxicological test of the substance, at a concentration of 100,000 ppm there were several deaths of test animals with a mortality rate below 50%, while at dipping concentrations of 0 ppm, 10 ppm, 100 ppm, 1000 ppm and 10000 ppm, no deaths occurred (Mortality). The results of the Probit Analysis statistical test obtained an LD50 value of 96 hours at a concentration of 104,343 ppm.

Table 3. Percentage of survival rate of test results ethanol extract of mangosteen rind (G. mangostana Linn) on dumbo catfish (*Clarias gariepinus*)

Treatment	Initial Amount Fish	Final Amount Fish	Average ± std dev
P_0 (0 ppm)	17	17	$9{,}00^a \pm 0{,}00$
P_1 (10 ppm)	17	17	$9{,}00^a \pm 0{,}00$
P_2 (100 ppm)	17	17	$9{,}00^a \pm 0{,}00$
P_3 (1000 ppm)	17	17	$9{,}00^a \pm 0{,}00$
P_4 (10000 ppm)	17	17	$9{,}00^a \pm 0{,}00$
P_5 (100000 ppm)	17	10	$8{,}70^b \pm 0{,}62$

Graph of concentration log Mortality toxicity test percentage survival rate (Survival Rate) test results of the ethanol extract of mangosteen rind (*G. mangostana Linn*) on dumbo catfish (*Clarias gariepinus*) occurred in P5 treatment. This is due to the active ingredient in the ethanol fraction of the rind Mangosteen begins to damage the catfish's endurance and can kill it.

4. Conclusion

Giving the extract of the ethanol fraction of mangosteen rind extract (*Garcinia mangostana* L.) has an effect on the growth of pathogenic bacteria *Streptococcus iniae* with a density of 10^7CFU/ml in vitro.

5. Reference

[1] Kordik, M. Ghufran. 2013. *Superior Tilapia Cultivation*. Agromedia Library. Printing 1. ISBN 979-006-443-8. Jakarta. 148pp
[2] Lusiastuti A.M, Purwaningsih U, and T Sumiati. 2010 *Aquac Res J*. 5(2): 237-243.
[3] Puspitasari, L., Swastini, D. A., & Arisanti, C. I. A. 2013 *J Farm Udayana*. *2*(3), 1-4.
[4] Janardhanan, S., Mahendra J., Girija, S., Mahendra, L., Priyadharsini V., 2017 *J Clin Diagnostic Res*. 11(1),19–22.
[5] Sukasih, E., Sukarti, T. and Broto, W. 2011 *J. Pascapanen*. 8(1), 32–38
[6] Melkianus, B., Fatimawali, F., Sudewi, S., 2019 *Pharmacon J Ilm Farm*. 8(1), 88–93

[7] Widayat, M., Purwanto., Dewi, A., 2016 *Pustaka Kesehat.* 4(3), 514–518.
[8] Husen, S., Kalqutny, S., Ansori, A., Susilo, R., Alymahdy, A., Winarni,D., 2017 *Inf Serv Netw.* 14(4), 1238–1245.
[9] Reed, L.J., Muench, H., 2012 *Am J Hyg.* 27:493-497.
[10] Naiborhu PE. 2018. *Extraction and benefits of mangrove extracts (Sonneratia alba and Sonneratia caselaris) as antibacterial natural ingredients on the tiger shrimp pathogen Penaeus Monodon, Vibrio harveyi.* [Thesis]. Aquatic Science Study Program. Bogor Agricultural Institute.
[11] Putra I.N.K. 2010 *J. Teknol. Dan Ind Pangan*, *21*, 1-5.

6. Acknowledgement

Award to the Assistant of the Educational Laboratory of the Faculty of Fisheries and Maritime Affairs, Airlangga University, to the Head of the Bacteriology Laboratory of Class I Fish Quarantine Center (BKI) Juanda Surabaya, to the Head of the Surabaya Fishery Product Quality Control and Testing Laboratory (LPPMHP) and Service Testing Unit, Phytochemical Laboratory for providing permission to practice to get results.

Characterization of *Bacillus subtilis* (BR 610) isolated from the intestines of Rabbitfish, *Siganus guttatus,* maintained in Floating Net at Awerange Bay, Barru

B R Tampangallo[1,3], H Anshary[2], S Wulan[2], and R Syah[3]

[1]Doctoral Study Program in Fisheries Science, Faculty of Marine Science and Fisheries, Hasanuddin University, Makassar, South Sulawesi, Indonesia
[2]Faculty of Marine Science and Fisheries, Hasanuddin University, Makassar, South Sulawesi, Indonesia
[3]Research Center for Fishery, National Research and Innovation Agency Republic of Indonesia

[4]Email: bungatampangallo@yahoo.com.

Abstract. The use of some species of Bacillus as a probiotic has been widely reported. In general, using probiotics in the exact isolation location will improve the performance of these probiotics. This study aims to characterize the *Bacillus subtilis* (BR610) isolate, successfully isolated from the intestines of Rabbitfish previously cultivated in the floating net of Awerange bay, Barru, South Sulawesi. The research includes morphological characterization of BR610 isolates, activity tests of amylase, cellulose, and proteinase enzyme. BR610 isolate inhibition test against several types of pathogenic bacterial isolates (*Vibrio harveyi, V. parahaemolyticus, V. alginolitycus*, and Streptococcus) as well as BR610 isolate resistance test against various kinds of antibiotics. The results showed that the *Bacillus subtilis* (BR610) has a spherical morphology, wavy edges, flat elevation, and is cream-white in color. BR610 can produce the enzyme amylase 20.36±17.8mm, proteinase 14.2±6.7mm, and Cellulose 20.9±2.7mm. These isolates can also inhibit the growth of *V. harveyi* (13.0±0.3mm), *V. alginolyticus* (15.1±2.8mm), *V. parahaemolyticus* (14.7±2.7mm), and Streptococcus (17.6±1.6mm). *Bacillus subtilis* BR610 is still worthy of being used as an environmentally friendly probiotic because it is still sensitive to several types of antibiotics tested.

1. Introduction

The use of isolated isolates of probiotic bacteria, such as Bacillus, in aquaculture, has been widely reported. Several types of Bacillus have been reported to increase the immune response of vannamei shrimp [1–3], tiger prawns [4], giant prawns [5], mangrove crab [6], rainbow trout [7], tilapia [8,9], carp [10], *Trachinotus ovatus* [11], and increase the content of Gracillaria seaweed agar [12].

This is due to the favorable properties possessed by Bacillus. *Bacillus subtilis*, for example, can inhibit the growth of pathogenic bacteria, immun stimulation, maintenance of intestinal homeostasis and prevention of intestinal inflammation, maintenance of normal intestinal flora, regular bowel movements, and survival in the gastrointestinal tract [13]. Bacillus can produce lytic enzymes such as protease, chitinase, cellulose, and β-1,3-glucanase that lysis the cell wall of pathogenic microbes [14,15]. Another characteristic of Bacillus is the provision of nutrients and digestive enzymes that can promote growth through secretion [16].

Bacillus can be isolated from the soil [17], sediment [18,19], water [20,21], shrimp intestines [22], fish intestines [23–25], and seaweed [12,26]. The discovery of this type of bacteria in almost all places suggests that Bacillus has a reasonably wide range of adaptability. This adaptability is because Bacillus can produce spores when its environmental conditions are less favorable.

Based on the description above, research on the composition of bacteria in the intestines of Rabbitfish, *Siganus guttatus*, is considered necessary. Rabbitfish has a reasonably good economic value, and the food preferences of this fish tend to consume seaweed (herbivores), so it requires bacteria-producing enzymes-digestive enzymes to symbiotic in their intestines. In our previous research, we isolated bacteria from the intestines of Rabbitfish obtained from the cultivation results in floating net, Awerange bay, Barru, South Sulawesi. *Bacillus subtilis* BR610 is one of the types of isolates found. Therefore, this study aims to characterize phenotypes, biochemistry, and the ability to produce amylase enzymes, protease enzymes, cellulose enzymes, antibacterials, and the sensitivity of BR610 isolates to several types of antibiotics.

2. Materials and Methods

2.1. Isolate Rejuvenation

This study used the isolate *Bacillus subtilis* BR610, which had been successfully isolated from the intestines of the Rabbitfish. This Rabbitfish was obtained from cultivation in the floating net bay Awerange, Barru. The *B. subtilis* BR610 isolates collected in skim milk media were first rejuvenated on Tryptic Soy Agar (TSA) media obliquely before being used for subsequent tests.

2.2. Morphological characterization and enzyme activity test

Morphological characterization is carried out by observing colony shape and size, edge shape, elevation, and color. Next, bacterial isolates are subjected to gram staining [27]. Catalase activity was assessed using hydrogen peroxide (H2O2) [28]. The test of the activity of the enzymes protease (15 g bacto agar, 5 g bacto peptone, 1 g yeast extract, 1% skim milk) and amylase (5 g yeast extract, 2 g soluble starch, 20 g bacto agar, 1 L Borigin Medium) refers to the method [29]. Test the activity of the cellulose enzyme using agar media added with a CMC of 1% [30]. Indications of positive activity will be indicated by forming a halo zone around the colony of bacteria. In the test media, amylase will appear after being dripped with a Lugol solution (10 g KI, 5 g I2, 5 g sodium acetate, 100 mL aquades) while in the cellulose test using 0.1% congo red solution. The diameter of the colony and the clear zone formed are then measured using a caliper. The amylolytic, proteolytic and cellulolytic indices are calculated using the formula:

$$Enzyme\ Activity\ \ Index = \frac{(Clear\ zone\ Diameter\ \ - Diameter\ of\ bacterial\ colony)}{Diameter\ of\ bacterial\ colony}$$

2.3. Antimicrobial activity

Antimicrobial activity tests using several types of pathogenic bacteria, namely *Vibrio alginolitycus, Vibrio parahaemolitycus, Vibrio harveyii* isolated from shrimp and *Streptococcus* sp regarding the scratch method [31–33]. This test used Muller Hinton media (3.8 g in 100 mL of aqueous autoclave for 15 min at a temperature of 121oC, pressure 1 atm). The BR610 isolate inhibition test begins with rejuvenating the pathogen isolates used in nutrient broth media. Incubation of pathogenic isolates is carried out for 24 hours on a swaying shaker. After incubation, each pathogenic bacterial inoculum is diluted using a physiological solution (10^{-2}), then spread on a pre-prepared MH plate medium. This plate is dredged for 5 minutes, after which the BR610 isolate is scratched. The inhibitory activity of BR610 isolates was observed after the plate was incubated for 48 hours. The diameter of the formed clear zone is then measured using a caliper.

2.4. Sensitivity against antibiotics test
The BR610 isolate sensitivity test for several antibiotics aims to determine whether a bacterial isolate is still sensitive or has resistance to an antibiotic. BR610 bacterial sensitivity test against antibiotics using the Kirby-Bauer paper disk method [34]. The antibiotics used in this study were Rifampicin, oxytetracycline, gentamicin, ampicillin, kanamycin, hygromycin, cefotaxime, Polymyxin, and Streptomycin with a concentration of 15iu / paper disk each. The antibiotic sensitivity test of BR610 isolates starts from the propagation of bacteria in nutrient broth media. After 24 hours of incubation on a shaker incubator, this BR610 isolate was taken as much as 100 µL and flattened over the muller Hinton plate media. This plate is dredged, and the pressed paper dripped with antibiotics is affixed. This plate is then incubated for 24 hours. The clear zone formed after incubation shows that the isolates are not resistant to antibiotics. Measurement of the formed clear zone using a caliper.

2.5. Data analysis
The data obtained are analyzed descriptively using the help of drawings, graphs, and table.

3. Results and Discussion
3.1. Morphological Characteristics and activity of enzyme B. subtilis BR610
The results of the observation of the phenotype characteristics of the BR610 isolate after rejuvenation on the TSA plate media are round in shape, with a jagged edge, flat elevation, and white in color. The gram-staining results show that the BR610 isolate is rod-shaped and gram-positive. BR610 isolates have positive catalase activity and can ferment aerobically or anaerobically (facultative aerobic). Along with Bergeys' statement, the genus Bacillus is gram-positive, rod-shaped, and has a wide physiological diversity. The morphological form and gram staining results of BRr610 isolates can be seen in Figure 1.

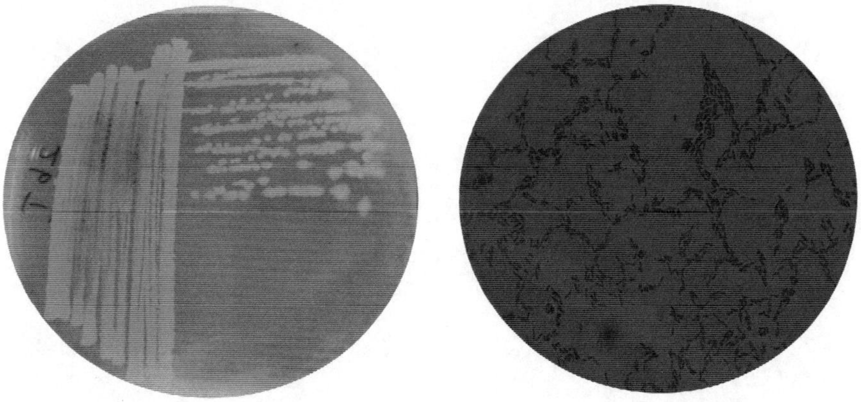

Figure 1. Phenotype characteristics (left) and gram staining of *B. subtilis* strain BR610 (right)

The *B. subtilis* BR610 isolate has the positive activity of the enzyme's amylase, protease, and cellulose. The activity index of enzyme amylase ranges from 2.4 ± 1.8, proteinase 1.4 ± 0.7, and cellulose

2.1±0.3 (Table 1). The cellulose activity index of BR610 is still lower than isolates isolated from *Macrotermes gilvus* [30] but still higher than isolates from catfish [35].

Table 1. Enzyme activity index of the Bacillus subtilis BR610

Enzyme	Diameter of colony (mm)	Diameter of halo zone (mm)	Enzyme activity index
Amylase	9,2 ± 8,6	20,9 ± 7,2	2,4 ± 1,8
Proteinase	10,8 ± 4,7	24,2 ± 7,3	1,4 ± 0,7
Cellulose	11,9 ± 0,6	36,8 ± 1,4	2,1 ± 0,3

Bacillus produces several enzymes and has been utilized for pharmaceuticals, biotechnology, and sewage treatment [35–37]. The ability to produce this enzyme is expected to help the digestive process of Rabbitfish, which is herbivorous. Enzyme cellulose is expected to help decompose the plants/seaweed consumed by Rabbitfish. Although it is known that almost all enzymes are present in all fish and do not depend on the type of food [38], the dominance of certain bacterial communities in the gut is expected to help speed up the digestive process.

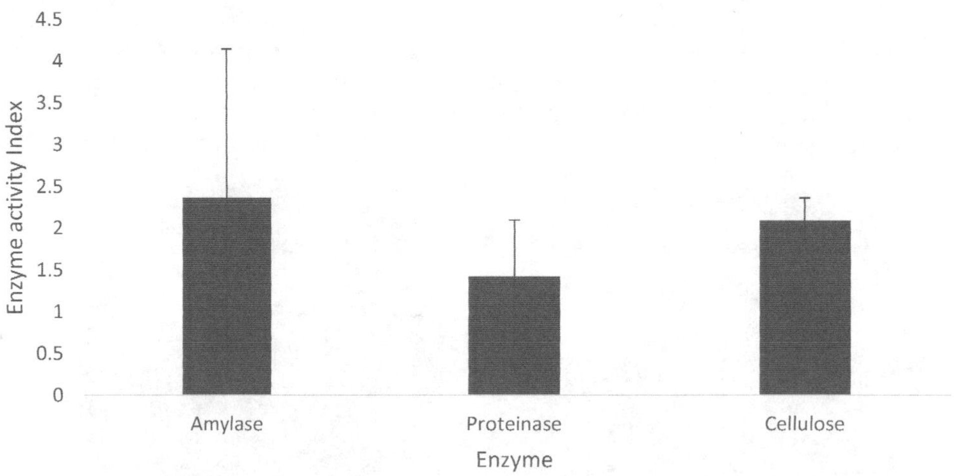

Figure 2. Activity Index of amylase, proteinase and cellulose enzymes of the B. subtilis BR610 isolate

3.2. Antimicrobial activity

BR610 isolates can inhibit several types of pathogenic bacteria tested. The highest activity was obtained in the inhibition of Streptococcus isolates, namely 17.6 ±1.6 mm, followed by *Vibrio alginolyticus* 15.1±2.8 mm, then *V. parahaemolyticus* 14.7 ± 2.7, and the lowest against *V. harveyi* 13 ± 0.3 mm. This antibacterial activity falls into the high category because it is more than 12 mm.

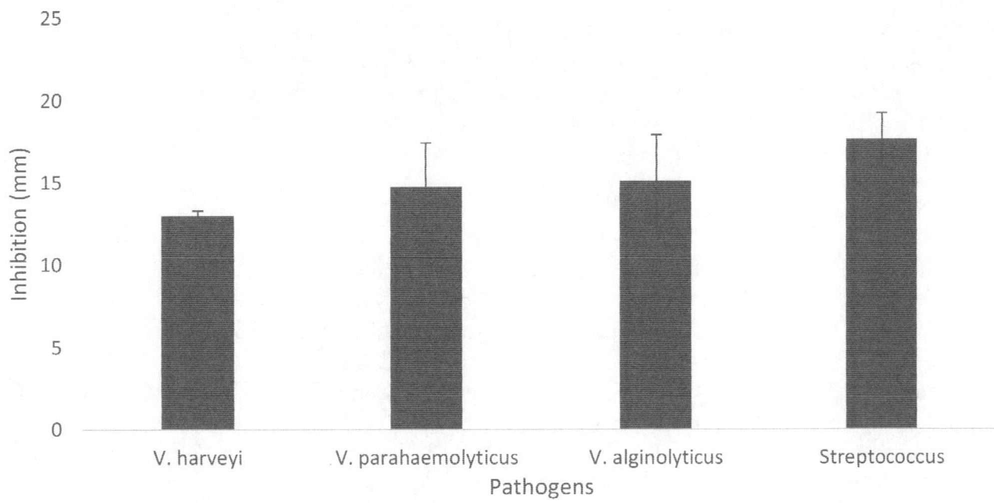

Figure 3. Inhibition activity of *B. subtilis* BR610 isolates against several types of pathogenic bacteria.

The antagonistic properties of bacteria can be categorized as bacteriocin and antimicrobial [39]. Furthermore, it is reported that some types of lactic acid bacteria have both properties, as with Bacillus [15]. Bacillus can also inhibit bacteria and has been widely applied in aquaculture [40–42].

3.3. Sensitivity BR610 to antibiotics

BR610 isolates are sensitive to all antibiotics tested at a 15 iu/disc paper dose. The highest sensitivity is to Rifampicin, cefotaxime, gentamycin, ampicillin, kanamycin, polymyxin, streptomycin, and hygromycin, and the lowest against oxytetramicyn. This shows that the BR610 isolate is not resistant to antibiotics and is still worthy of being used as a probiotic candidate.

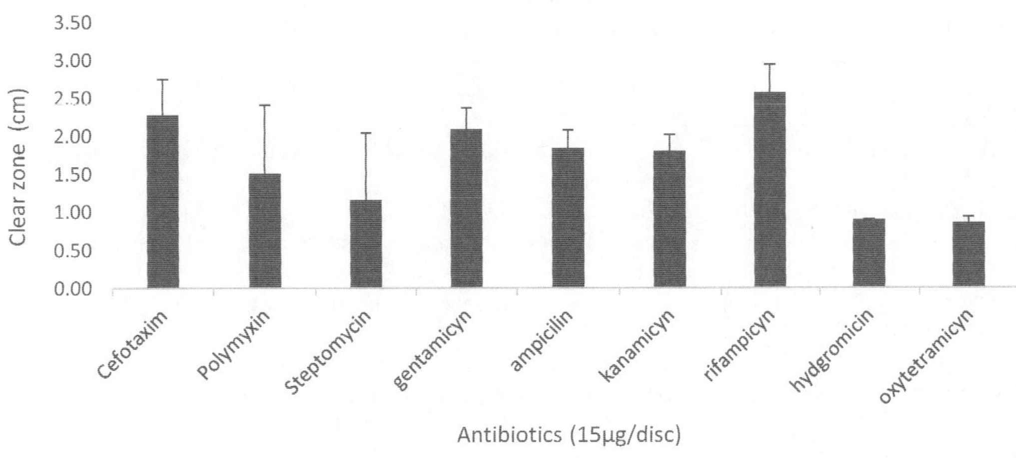

Figure 4. Clear zone diameter on the sensitivity test of B. subtilis BR610 isolates against several types of antibiotics

4. Conclusion

The characterization of the isolate *Bacillus subtilis* BR610 isolated from the intestines of the Rabbitfish is spherical in shape, with a flat elevation and jagged edges, and the color is white. BR610 isolates are rod-shaped, gram-positive, catalase-positive, and facultative, producing amylase, proteinase, and cellulose. Isolate BR610 can also produce antimicrobial and is not resistant to antibiotics so it can be used as a probiotic candidate for Rabbitfish farming.

5. References

[1] Widanarni W, Rahmi D, Gustilatov M, Sukenda S and Utami D A S 2020 *J. Akuakultur Indones.* **19**
[2] García-Medel D I, Angulo C, Escamilla-Montes R, Fierro-Coronado J A, Diarte-Plata G, Gámez-Jiménez C and Luna-González A 2020 *Aquac. Int.* **28** 2303–18
[3] Zokaeifar H, Balcázar J L, Saad C R, Kamarudin M S, Sijam K, Arshad A and Nejat N 2012 *Fish Shellfish Immunol.* **33** 683–9
[4] Lante S, Tampangallo B R and Tenriulo A 2021 *IOP Conference Series: Earth and Environmental Science* vol 860 pp 1–8
[5] Hindu S V, Thanigaivel S, Vijayakumar S, Chandrasekaran N, Mukherjee A and Thomas J 2018 *Fish Shellfish Immunol.* **73** 112–20
[6] Gunarto G, Tampangalo B R and Muliani M 2021 *Thalassas* 465–75
[7] Ghafarifarsani H, Rashidian G, Bagheri T, Hoseinifar S H, Doan H Van, Resources N, Fisheries O and Resources N 2021 *Annals of Animal Science*, *21*(2), 543-559.
[8] Mulyadin A, Yuhana M and Wahjuningrum D 2021 *J Akua Indonesia*, *20*(1), 34-46.
[9] Pan X, Liu Q, Wen L, Huang Y, Ma H, Lin Y, Chen Z, Qin J and Du X 2021 *Aquac. Reports* **20**
[10] Chang X, Chen Y, Feng J, Huang M and Zhang J 2021 *Aquac. Reports*, *20*, 100678
[11] Zhang Q, Yu H, Tong T, Tong W, Dong L, Xu M and Wang Z 2014 *Fish Shellfish Immunol.* **38** 7–14
[12] Suvega T and Arunkumar K 2019 *Biocatal. Agric. Biotechnol.* **19** 101136
[13] Suva M, Sureja V and Kheni D 2016 *J. Curr. Res. Sci. Med.* **2** 65
[14] Biziulevieius G A and Pukaite V 2021 Fermosorb **20** 12–5
[15] Urdaci M C and Pinchuk I 2022 *Norfolk, UK: Horizon Bioscience*, 171-82.
[16] Hai N Van 2015 *Fish Shellfish Immunol.* **45** 592–7
[17] Dunlap C A, Schisler D A, Perry E B, Connor N, Cohan F M and Rooney A P 2017 *Int j sys evol microbiol*, *67*(8), 2720-2725.
[18] Pepi M, Borra M, Tamburrino S, Saggiomo M, Viola A, Biffali E, Balestra C, Sprovieri M and Casotti R 2016 *Sci. Total Environ.* **562** 588–95
[19] Nakamura K and Silver S 1994 *Appl. Environ. Microbiol.* **60** 4596–9
[20] Maugeri T L, Gugliandolo C, Caccamo D, Panico A, Lama L, Gambacorta A and Nicolaus B 2002 *Biotechnol. Lett.* **24** 515–9
[21] Mohapatra S, Mohanta P R, Sarkar B, Daware A, Kumar C and Samantaray D P 2017 *Proc. Natl. Acad. Sci. India Sect. B - Biol. Sci.* **87** 459–66
[22] Sriwulan, Azwar A, Rantetondok A and Anshary H 2019 *AACL Bioflux* **12** 1866–81
[23] Kavitha M, Raja M and Perumal P 2018 *Aquac. Reports* **11** 59–69
[24] Monzón L, Jimena A, Silvia B, Daniel T, Ana M and González F 2021 aquaculture. *Probiotics Antimicrob*, 1-9.
[25] Kuebutornye F K A, Lu Y, Abarike E D and Wang Z 2020 *Probiotics Antimicrob*, *12*, 412-424.
[26] Tampangallo B R, Atmomarsono M and Muliani 2013 *Pros. Forum Inov. Teknoligi Akuakulture*

825–32
[27] Hucker G J, Conn H J, Mellen C R, Smith G A E, Pyrke C B A, Mulford A J B, Boshart C F, Rouse I, Bradley F M, Teneyck P G, Smith M H, Mellen C R, Cook E J, Thatcher R W, Agr D, Mann R, Collison R C and Wheeler W P 1923 Methods of Gram Staining *Technical Bulletin* p 37 pp
[28] Reiner K 2016 Catalase Test Protocol *ASM* pp 1–9
[29] Xu L, Yuan J, Chen X, Zhang S, Xie M, Chen C and Wu Z *Aquaculture* **540** 736–48
[30] Ferbiyanto A, Rusmana I and Raffiudin R 2015 *HAYATI J. Biosci.* **22** 197–200
[31] Taskan M, Karaman I, Yuce H and Kara G 2019 *Ann. Med. Res.* **26** 2733
[32] Kourmouli A, Valenti M, van Rijn E, Beaumont H J E, Kalantzi O I, Schmidt-Ott A and Biskos G 2018 *J. Nanoparticle Res.* **20** 2–7
[33] Emam A M and Dunlap C A 2020 *Int. J. Gen. Mol. Microbiol.* **113** 2041–52
[34] Jorgensen, J. H., & Turnidge J D 2016 Kirby-Bauer disk diffusion susceptibility test protocol *ASM* pp 1–14
[35] Yang J K, Shih I L, Tzeng Y M and Wang S L 2000 *Technol.* **26** 406–13
[36] Haque M A, Cho K M, Barman D N, Kim M K and Yun H D 2015 *Process Biochem.* **1** 1–11
[37] Ebrahimpour A and Kariminik A 2015 Iran *Int. J. Life Sci.* **9** 39–42
[38] Chakrabarti I, Gani M A, Chaki K K, Sur R and Misra K K 1995 *Comp. Biochem. Physiol. -- Part A Physiol.* **112** 167–77
[39] Ringø E, Schillinger U and Holzapfel W 2005 *Biol. Grow. Anim* vol 2 pp 418–53
[40] Abun A, Rusmana D, Widjastuti T, Haetami K and Rusmana D 2021 *J. Appl. Anim. Res*, *49*(1), 295-303.
[41] Boyd C E and Gross A 2014 *Advances in shrimp biotechnology*, 101-105.
[42] Li L T, Yan B L, Li S H, Xu J T and An X H 2016 *Ann. Microbiol.* **66** 317–28

6. Acknowledgments

The author says thanks to the team of DPPU Pertamina Hasanuddin Makassar; researchers, technicians, and laborant of: the fish and environmental health team, and the Feed and Fish Nutrition team of Research Institute for Brackishwater Aquaculture and Fisheries Extension, KKP have helped the authors to conduct this research.

Identification, prevalence and intensity of ectoparasite in sand lobster (*Panulirus homarus*) cultivated in floating cages (*floating net cage*) and bottom cage (*bottom cage*)

G Septian [1], M F Ulkhaq [2], Kismiyati [3,4]

[1]Study program of Aquaculture, Faculty of Fisheries and Marine, Universitas Airlangga, Surabaya Indonesia
[2]Study program of Aquaculture, School of Health and Natural Sciences, Universitas Airlangga, Banyuwangi Indonesia
[3]Department of Aquaculture, Faculty of Fisheries and Marine, Universitas Airlangga, Surabaya

[4]Corresponding author: kismiyati@fpk.unair.ac.id

Abstract. Lobster (*Panulirus homarus*) is one of the marine fishery commodities that is gaining popularity and has been widely cultivated in Indonesia. This is because lobster has great business prospects and is one of the important commodities and has high economic value. Lobster cultivation can now be done in various systems including floating net cages, bottom cages and ponds. The problem that is often faced by cultivators is the emergence of diseases in the lobster cultivation cycle caused by an unbalanced interaction between lobster conditions, the environment, and pathogens. This is one of the main problems in lobster cultivation that can interfere with the survival of lobsters and can also cause death. Research on ectoparasite infection in Sand Lobster (*Panulirus homarus*) cultured in floating net cages and bottom cages aims to identify the types of ectoparasites and calculate the prevalence and intensity of ectoparasites infecting sand lobster (*Panulirus homarus*). The method used in this study is a survey method. The sampling method was carried out by random sampling method and the method of analysis and presentation of data using descriptive methods. The main parameters observed in this study were the type of ectoparasites, prevalence and intensity values of sand lobster (*Panulirus homarus*) infected with ectoparasites, while the supporting parameter was lobster size. The ectoparasite found to infest the sand lobster (*Panulirus homarus*) was *Octolamis angulata*. The highest prevalence value was obtained in cultivation using floating net cages of 53,3% which belonged to the Frequently category (*commonly*) and the highest intensity value was obtained in cultivation using floating net cages of 30,1 which belonged to the Medium category (*currently*). It is necessary to conduct further studies on *Octolasmis* species based on the DNA chain arrangement and description of tissue damage due to *Octolasmis* infestation

1. Introduction

Sand lobster (*Panulirus homarus*) is one of the marine fishery commodities that is gaining popularity and has been widely cultivated in Indonesia. This is because lobster has great business prospects and is one of the important commodities and has high economic value. In 2020 the demand for lobster in the international market reaches 2,500 tons per year, but this demand is not matched by the production of aquaculture, the volume of lobster production in 2020 is only 1,337 tons and the production target in 2024 is 7,220 tons [1]. The higher demand for lobsters causes over-fishing which will damage the balance of the ecosystem because the number of lobster broodstocks in nature will decrease. This can be reduced through lobster rearing cultivation activities so that they do not depend on stocks in nature [2]. Lobster cultivation can now be done in various systems including floating net cages, bottom cages and ponds or ponds. Floating net cages are one of the cultivation media that are on the surface of the waters supported by a framework and buoys that function as supports to stay on the surface of the waters.

The use of nets must be adjusted to the size of the lobsters stocked to avoid lobsters escaping into the waters and predatory attacks [3]. Diseases that usually infest lobsters are caused by bacteria, viruses, fungi and parasites [4]. This is one of the main problems in lobster cultivation that can interfere with the survival of lobsters and can also cause death. The emergence of disease in the lobster cultivation cycle is caused by an unbalanced interaction between lobster conditions, the environment, and pathogens [5].

Based on the description above, it is necessary to conduct this research in order to assess the identification and prevalence value of ectoparasites in sand lobster (*Panulirus homarus*). The results of this study are expected to provide basic information about ectoparasite infection in sand lobsters reared in floating net cages and bottom cages.

2. Material and methods

Sampling was carried out in the Floating net cage of the "Pemuda Ketapang Bahari" Fish Cultivation Group (POKDAKAN) and the bottom cage (Bottom cage) of the Lobster Cultivation Pilot Unit (*Panulirus* spp.) POKDAKAN) "Marine Enchantment" Banyuwangi. Parasite examination was carried out under a microscope natively at 40x and 100x magnification. The gills that were infested with ectoparasites were then taken using tweezers and placed on an object glass that had been dripped with physiological NaCl. The ectoparasites found were observed with the aid of a microscope and documented using a lucida camera to be identified based on the specific morphological characteristics that appeared when observing the shape and size according to the identification key, then the total number of infested samples and the prevalence and intensity values were calculated.

2.1 *Tools and Materials*

The tools used in this study include a microscope, object glass, cover glass, styrofoam, digital scales, ruler, scissors, petri dish, sample bottles, label paper, and tissue.

2.2 *Procedures and data analysis*

The sampling location was determined by random sampling method. Sampling of sand lobster was carried out in the cultivation of sand lobster with floating net cages and bottom cages. The lobster samples observed were live lobsters. The number of samples taken from each location of floating net cages and bottom cages is 10% of the total population. The sample size criteria used in this study were to have a total length of 15 3.5 cm, weight 100-150 grams/head.

The sample of the sand lobster was measured for the length and weight of the lobster, then all parts of the body were examined and observed first, then the mucus and carapace were scraped. The organs examined were the carapace, walking legs, antennae and gills. Parasite examination was carried out under a microscope natively at 40x and 100x magnification. The gills that were infested with ectoparasites were then taken using tweezers and placed on an object glass that had been dripped with physiological NaCl. The ectoparasites found were observed with the aid of a microscope and documented using a lucida camera to be identified based on special morphological characteristics that appeared when observing the shape and size according to the key of identification. The analysis of the data used the descriptive method.

3. Result and discussion

3.1 *Identification species of ectoparasite*

The results of the examination of 60 sand lobsters cultured in floating net cages and bottom cages found an ectoparasite infestation in the sand lobster, namely *Octolasmis angulata* from the Crustaceae class. This species is found to infest the surface of the body and gills of sand lobster.

3.2. *Octolasmis angulata*

Based on the observations made, *Octolasmis angulata* is shaped like a sprout, has a body size of 6.34± 0.42 mm, its body parts are composed of tergum, scutum, carina and peduncle. The capitulum has an oval shape protected by a calcareous stone shell, has a dotted stalk called a peduncle. *Octolasmis* live in colonies on the surface of the infested lobster body. The morphology of *Octolasmis angulata* is almost the same as other species of the genus *Octolasmis*, which is like a swan's neck. The difference between this species and the others is that there is no visible tergum in the capitulum. Another organ found in the capitulum is the carina, which is shaped like an L with a slightly pointed tip and a pair of scutum resembling an inverted L.

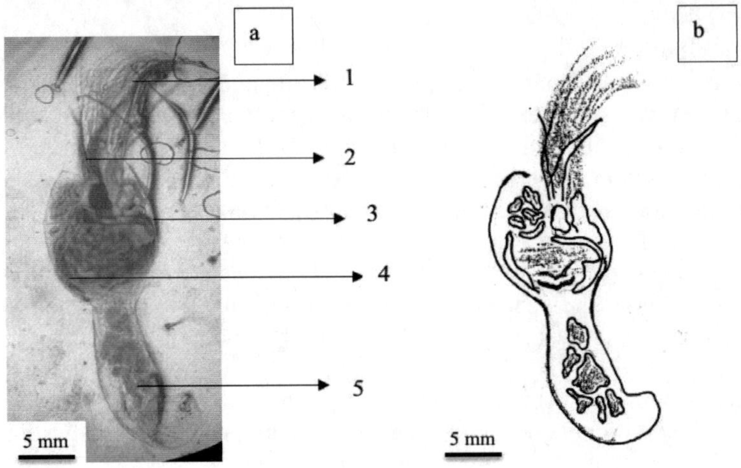

Figure 1. *Octolasmis angulata* (a) Image of *Octolasmis angulata* photographed using a 40x magnification stereo microscope; and (b) Images of *Octolasmis angulata* photographed using a stereo microscope equipped with a lucida camera. 1. *Cirrus*, 2. *Penis*, 3. *Scutum*, 4. *Carina,* 5. *Peduncle*

3.3 *Prevalence and Intensity of Sand Lobster (Panulirus homarus) Infested with Ectoparasite Octolasmis angulata*

The results of ectoparasite examination carried out on 60 sand lobsters (*Panulirus homarus*) which were cultured using a floating net cage and bottom cage system showed that the ectoparasites found were the species *Octolasmis angulata.* The highest level of ectoparasite infestation occurred in cultivation using a floating net cage system with as many as 16 infested animals *Octolasmis angulata*.

The calculation of the highest prevalence value of ectoparasites in sand lobster (*Panulirus homarus*) occurred in aquaculture using a floating net cage system of 53,3%, while the calculation of prevalence values in sand lobster (*Panulirus homarus*) in basic cage cultivation was 26,7%. The results of the calculation of the intensity value of *Octolasmis angulata* ectoparasites that infest sand lobster (*Panulirus homarus*) which are cultured using floating net cages and bottom cages. floating net cage system of 30.1.

Cultivation System	Total of Samples	Amount of infested samples	Prevalence (%)	Category by (Williams and Williams, 1996)	Total of parasites	Intensity	Category by (Williams and Williams, 1996)
Floating net cage	30	16	53,3%	*Frequently*	482	30,1	*Medium*
Bottom cage	30	8	26,7%	*Often*	79	9,8	*Medium*

Table 1. Prevalence and Intensity of Sand Lobster (*Panulirus homarus*) Infested with Ectoparasite *Octolasmis angulata*

The results of the calculation of the prevalence of ectoparasites in sand lobster cultured using a floating net cage system is 53,3%. The prevalence value of ectoparasites in sand lobster using a floating net cage system is in the Frequently category [6]. This value is higher than the calculation of the prevalence value of ectoparasites in sand lobster cultivated using a bottom cage system of 26,7% which belongs to the Often category, the high prevalence value in aquaculture using a floating net cage system is caused by the influence of aquaculture locations adjacent to the ship crossing port, thus affecting the condition of the aquaculture waters. The difference in prevalence values between the two cultivation systems can be influenced by several factors, namely the selection of cultivation sites, suitable hosts, adequate oxygen and food and safety from predators.

Environmental conditions that support the development of parasites, parasites will reproduce more quickly, so that it can have a negative impact on the host. The carrying capacity of the environment also affects ectoparasites in reproducing as well as competition between species in obtaining suitable hosts. There is an indication that a high intensity value occurs due to the increasing size and age of the lobster, the greater the opportunity for *Octolasmis* to reproduce and enlarge the colony on its host, this is supported by the statement that the older the host, the longer the time. that the host has to make contact with the parasite, so that the prevalence and intensity of the parasite can increase according to the age of the host [7]. Water quality conditions affect the presence of ectoparasites in the cultivation of sand lobster (*Panulirus homarus*), see the results of measuring water quality parameters at the location of sand lobster cultivation (*Panulirus homarus*) using floating net cages and bottom cages. The level of ectoparasite infestation was still in normal conditions and had no significant effect on the continuity of cultivation and no ectoparasites from other genera such as *Vorticella* and *Epistylis* were found, this was due to environmental conditions that did not support the survival of ectoparasites from the genus *Vorticella* and *Epistylis*. Vorticella ectoparasites are found in areas with substrates, waters that have high organic matter content and low concentrations of DO (*Dissolved Oxygen*) can increase the growth of *Vorticella* [8]. Low DO (*Dissolved Oxygen*) conditions became one of the parameters for *Epistylis* ectoparasites to increase growth and reproduce [9]. *Epistylis* sp. can reproduce optimally in waters with a temperature of 10-25°C. pH 6.5-7 and salinity 15-31 ppt [10].

Based on the data obtained from the results of this study, the level of ectoparasite infestation was still in normal conditions and had no significant effect on the continuity of the cultivation of sand lobster (*Panulirus homarus*) in the Banyuwangi waters area either by using a floating net cage system or bottom cages.

4. Conclusion

The ectoparasite species that infests the sand lobster (*Panulirus homarus*) which is cultured in floating net cages and bottom cages is *Octolasmis angulata*. The highest prevalence value of ectoparasites infesting sand lobster *Panulirus homarus*, which is cultivated in floating net cages, is 53,3% which is included in the Frequently category. Intensity value which is cultivated in floating net cages of 30,1 and included in the category of Medium.

5. References

[1] Kementerian Kelautan and Perikanan, 2020, Permen KP Nomor 12/Permen-KP/2020 Tentang Pengelolaan Lobster (*Panulirus spp.*), Kepiting (*Scylla spp.*), and Rajungan (*Portunus spp.*).

[2] Dhewantara Y L, Rahmatia F, and Nainggolan A 2021 *Jurn Akua Raw Indo* 9 (2) 163-172

[3] Lestari D P, Nurliah, Damayanti A A, and Larasati C E 2018 Water Quality Parameters in Supporting Aquaculture Activities in the Teluk Jor Area, East Lombok Regency Prosiding PKM-CSR 1 723-730 hal.

[4] Irvansyah M Y, Aldulgani N, and Mahasri G. 2012 *Sai Sen ITS* 1(1) 1-5.

[5] Indriyani, Nur and Yusnaini 2018 *AACL Biof* 9(1) 108-117

[6] Williams E H and Williams L B 1996 *Parasits of Offshore Big Gami Fishes of Puerio Rico and The Western Atlantic*. Puerio Rico Department of Natural and Environmental Resources. San Juan, PR. University of Puerio Rico. 383 pp.

[7] Bawia R H A, Tuiyo R and Mulis 2014 *Scien Jour of Fish and Mar* 2(3) 60-65.

[8] Nicolau A, Martins M J, Mota M and Lima N 2005 Chem 58 605-614

[9] Shields J D 2011 *Jour of Inver Pathol* 106 79–91

[10] Ma H, and Overstreet R M 2006 *Jour Euca Micro* 53 (2): 85-95.

6. Acknowledgements

We would like to thank those who have been involved in this research and the Study program of Aquaculture, School of Health and Natural Sciences, Universitas Airlangga, for the use facilities during the research.

Effectiveness of cultivation *Chaetoceros calcitrans* with liquid organic fertilizer *Gracilaria* sp. against growth rate and nutritional content of *Artemia* sp.

R N Kusuma[1], M A Alamsjah[2,4], Agustono[3]

[1]Study Program of Aquaculture, Faculty of Fisheries and Marine, Universitas Airlangga, Campus C UNAIR, Jl. Mulyorejo, Surabaya, Indonesia
[2]Department of Marine, Faculty of Fisheries and Marine, Universitas Airlangga, Campus C UNAIR, Jl. Mulyorejo, Surabaya, Indonesia
[3]Department of Aquaculture, Faculty of Fisheries and Marine, Universitas Airlangga, Campus C UNAIR, Jl. Mulyorejo, Surabaya, Indonesia

[4]Corresponding author: alamsjah@fpk.unair.ac.id

Abstract. This research aims to determine the effectiveness of the cultivation of *Chaetoceros calcitrans* with liquid organic fertilizer *Gracilaria* sp. on the growth rate and nutritional content of *Artemia* sp. This research used a completely randomized design (CRD) consisting of 5 treatments, namely P0 (pro analysis fertilizer 16 mg/ml), P1 (liquid organic fertilizer *Gracilaria* sp. 8 mg/ml), P2 (liquid organic fertilizer *Gracilaria* sp. 16 mg/ml), P3 (liquid organic fertilizer *Gracilaria* sp. 24 mg/ml), P4 (liquid organic fertilizer *Gracilaria* sp. 32 mg/ml) with 4 replications each. The data statistically analyzed using ANOVAvtest and continued with Duncan's Multiple Range Test to determine the effective dose as feed for *Artemia* sp. Furthermore, the research was carried out using the Independent Sample T-Test test which consisted of 2 treatments, namely P0 and P1 with 3 replications each. The results showed that the use of liquid organic fertilizer *Gracilaria* sp. on the cultivation of *Chaetoceros calcitrans* did not give a significant difference ($p \geq 0.05$) to the growth rate of length and weight in *Artemia* sp. Proximate analysis of *Artemia* sp. *Chaetoceros calcitrans* feed with liquid organic fertilizer *Gracilaria* sp. the results obtained are 53.62% protein, 14.07% carbohydrate, and 25.9% fat.

1. Introduction
Feed is one of the factors that can affect growth, especially in fish larvae. The feed that can be given to fish larvae must be by the larval mouth opening and according to needs, namely with natural food. *Artemia* sp. is one of the zooplankton with high digestibility and protein content reaches 60% which is influenced by the quality and quantity of nutrients during maintenance [1]. The nutritional content needed by *Artemia* sp. to support the growth rate is a feed containing 30% protein and 5% fat [2].

Nutrients that can be given to *Artemia* sp. one of them is the microalgae *Chaetoceros calcitrans*. The protein content of *Chaetoceros calcitrans* based on the research of Jati *et a*l,. [3] can reach 34.03% with the provision of pro-analytical fertilizers during the cultivation period. It is known that the price of pro-analytical fertilizer can reach Rp.1000/ml, so an alternative is needed with an economical price and contains nutrients that can meet the needs of microalgae. The pro- analytical fertilizer contains content according to the needs of microalgae, namely nitrogen, phosphorus, and potassium derived from KNO_3, NaH_2PO_4, $FeCl_3$, and Na_2 EDTA.

The alternative that can be used is through the processing of liquid organic fertilizer with macroalgae, *Gracilaria* sp. Based on Nugroho's research [4] the use of liquid organic fertilizer *Gracilaria* sp. can increase the growth rate of *Nannochloropsis* sp. The concentration of nitrogen, phosphorus, and potassium in liquid organic fertilizer *Gracilaria* sp. can meet the needs of microalgae. According to Eyster (1978) in Nugroho [4] the nitrogen needs of microalgae are 0.14-0.7 g/l) and phosphorus (0.015-0.62 g/l). Based on this background, a study was conducted on the effectiveness of the cultivation of *Chaetoceros calcitrans* with liquid organic fertilizer *Gracilaria* sp. on the growth rate and nutritional content of *Artemia* sp.

This research aims to determine the optimum dose of liquid organic fertilizer *Gracilaria* sp. on the cultivation of *Chaetoceros calcitrans* and determine the effect of using liquid organic fertilizer *Gracilaria* sp. on the cultivation of *Chaetoceros calcitrans* on the growth rate and nutritional content of *Artemia* sp. The benefit of this research is that it can provide information regarding the use of liquid organic fertilizer *Gracilaria* sp. as an alternative to pro-analytical fertilizers for microalgae cultivation, especially *Chaetoceros calcitrans*. This study can also provide information regarding the effect of using *Chaetoceros calcitrans* as feed for *Artemia* sp. on growth rate and nutrient content.

2. Material and method

2.1 Place and time

This Research was held from the month October 2021 to January 2022 in the Laboratory anatomy and cultivation, Faculty of Fisheries and Marine, Airlangga University, Surabaya. Analysis content of nitrogen, phosphorus, and potassium in liquid organic fertilizers *Gracilaria* sp. and analysis proximate to *Chaetoceros calcitrans* and *Artemia* sp. in the laboratory Nutrition, Faculty of Public Health, Airlangga University, Surabaya.

2.2 Tools and materials

Tools used in this research for making liquid organic fertilizer *Gracilaria* sp. are glass measure, bottle 1,5-liter plastic volume, knife, scales, and stirrer. Tools used during the cultivation of *Chaetoceros calcitrans* and maintenance of *Artemia* sp. are plastic bottle 1.5-liter volume, plastic bottle volume 600 ml, carboy volume 10 liters, pump aeration, hose aeration, aeration stone, *beaker glass*, syringe, dropper, 16 watt TL lamp, *hemocytometer*, *hand tally counter*, microscope, refractometer, thermometer, pH meter, and filter.

Ingredients used for making liquid organic fertilizer *Gracilaria* sp. are grass sea *Gracilaria* sp. obtained from Kupang Village, District Jabon, Sidoarjo Regency, distilled water, sugar solution, EM4, and fish meal. Ingredients used for cultivation of *Chaetoceros calcitrans* and *Artemia* sp. is *Artemia* sp. nauplii, *starter* culture *Chaetoceros calcitrans* pure obtained from Hall Fishery Situbondo Brackish Water Cultivation (BPBAP), seawater, Natrium Thiosulfate, alcohol, KNO_3, Na_2 EDTA, $FeCl_3$, NaH_2PO_4, NaOH and chlorine.

2.3 Research methods

Research methods used are experimental methods with Design Random Complete (RAL) which consists of 5 treatments and 4 replications to determine the dose use of liquid organic fertilizer *Gracilaria* sp. Treatment consist from use different doses in cultivation *Chaetoceros calcitrans*, P0 (pro-analysis fertilizer 16 mg/ml), P1 (liquid organic fertilizer *Gracilaria* sp. 8 mg/ml), P2 (liquid organic fertilizer *Gracilaria* sp. 16 mg/ml), P3 (liquid organic fertilizer *Gracilaria* sp. 24 mg/ml), P4 (liquid organic fertilizer *Gracilaria* sp. 32 mg/ml). The observed parameters are the number of density cells and will analyze using the *Analysis of Variance* (ANOVA) test, then continued with Duncan's Multiple Distance test.

The main parameters observed next are rate growth long and weight body of *Artemia* sp. This research used experimental methods with 2 treatments and 3 replications to feed given to *Artemia* sp., including P0 (*Chaetoceros calcitrans* with pro-analysis fertilizer) and P1 (*Chaetoceros calcitrans* with liquid organic fertilizer *Gracilaria* sp.) which is will analyze using the *Independent Sample T-Test test.*

2.4 Research procedure

2.4.1 Research preparation

Research preparation is done by sterilizing the tools and materials. Tool sterilization is carried out by washing the tools to be used with detergent, then rinsing with water until clean. Material sterilization is carried out by sterilizing seawater which will be used as a medium for maintaining *Chaetoceros calcitrans* and *Artemia* sp. with filtering. After the seawater was filtered, then chlorine was added at a dose of 60 ppm and allowed to stand for ± 24 hours, then sodium thiosulfate was added at a dose of 30 ppm. After the sterilization process, a Chlorine Test is carried out to ensure the water is neutral.

2.4.2 Pro analysis fertilizer

The process of making pro analysis fertilizer by boiling water until it boils to use the media as a solvent in a sterile state. After the distilled water boils, the first ingredient to be added is Na2EDTA and homogenized. After Na2EDTA is dissolved, it is continued by homogenizing KNO3, NAH2PO4, and FeCl3 into the solution. The mixed materials were then put into a 1000 ml Erlenmeyer and covered with aluminum foil and closed again with plastic and then given rubber to prevent contamination. Then proceed with the sterilization process by inserting an Erlenmeyer containing fertilizer into an autoclave for 15 minutes at a temperature of 121°C and a pressure of 1 atm, then the fertilizer can be stored in the refrigerator so that the quality is maintained. The following composition is seen in Table 1.

Table 1. Composition of Pro Analysis Fertilizer in BPBAP Situbondo

Material	Dose
KNO3	75 gr
NaH2PO4	5 gr
Na2EDTA	5 gr
FeCl3	3,15 gr
Silicate	30 ml
Aquades	1000 ml
B1	100 mg
B12	5 mg

2.4.3 Liquid organic fertilizer Gracilaria sp.

Manufacture of liquid organic fertilizer *Gracilaria* sp. by washing the seaweed ± 2 kg first. Next, *Gracilaria* sp. is chopped and put in a plastic bottle. Then aquadest was added as a solvent, EM4 to speed up the fermentation process, and a solution of sugar and fish meal to increase the nutrient content of the fertilizer produced [5]. The ingredients were homogenized and fermented for ±9 days [6]. The following composition of liquid organic fertilizer *Gracilaria* sp. can be seen in Table 2.

Table 2. Composition of liquid organic fertilizer *Gracilaria* sp.

Material	Dose
Gracilaria sp.	400 gr
EM4 5%	20 ml
Solution of sugar 5%	20 ml
Aquadest 250%	1000 ml
Fish meal 25%	100 gr

2.4.4 Chaetoceros calcitrans cultivation

Cultivation was carried out with a total volume of 500 ml, 80% of maintenance media, and 20% of *Chaetoceros calcitrans* seedlings. Neutral seawater as a maintenance medium was put into the culture container and aerated. Then added fertilizer as nutrients during cultivation according to the treatment, pro- analysis fertilizer at a dose of 16 mg/ml, and liquid organic fertilizer *Gracilaria* sp. with a concentration of 8 mg/ml, 16 mg/ml, 24 mg/ml, and 32 mg/ml then add the seeds of *Chaetoceros calcitrans*.

2.4.5 Density calculation of Chaetoceros calcitrans

The first step in starting the calculation is to clean the hemocytometer first and then dry it with tissue paper. Then the sample was dripped on the cross-section of the trench until it was full and then covered with a cover glass, and ensured that there were no air bubbles. Furthermore, the hemocytometer was observed under a microscope and counted the cells using a hand tally counter to make counting easier. Calculation of the density of *Chaetoceros calcitrans* was carried out every day for 8 days after initial stocking and was carried out to determine the amount to be given as feed for *Artemia* sp.

2.4.6 Preparation of Artemia sp.

Artemia cysts are hatched through a decapsulation process by inserting *Artemia* cysts into hatching containers which are then hydrated with fresh water with strong aeration or stirring for 10-15 minutes. Filtered using gauze with a diameter of 120 microns and washed. Cysts that have been washed are put into a chlorine solution with a concentration of 40-60% and aerated. This process is carried out until the color changes to orange, which is then washed with sodium thiosulfate to remove the chlorine odor. Then the cyst was hatched for 24 hours with the help of aeration.

2.4.7 Treatment of Artemia sp.

Treatment of Artemia sp. through feeding *Chaetoceros calcitrans* according to Firmansyah *et al.*, [1] during the maintenance period as many as 22500 cells/head were carried out at 10.00 every day with an inlet-outlet system and were counted using a hemocytometer and a hand tally counter. The maintenance period of *Artemia* sp. was carried out for 14 days and carried out the replacement of water media periodically to maintain water quality in optimal condition and remove organic matter in the range of 10 – 20% of the total volume of the media.

2.4.8 Calculation of growth rate Artemia sp.

The growth rate of *Artemia* was determined by calculating the absolute length and absolute weight during the cultivation period after 24 hours of feeding. Length measurement on *Artemia* sp. was carried out using a micrometer and the aid of a microscope with a light magnification of 40 times. Measurement of biomass weight in Artemia sp. was carried out by weighing the wet weight carried out at the beginning and end of the research. Samples were taken from each treatment as much as 100 ml which were then poured onto the mesh and calculated. Before the mesh is filled with 100 ml of sample, it is weighed first. The mesh that has been filled with 100 ml of sample is placed at the mouth of the glass with the aim that the water can drip out, then the blotting paper is placed on the bottom when there is no water dripping, and note that if there are stains on the blotting paper then the mesh is weighed.

2.5 Nutritional Content Analysis
Analysis of nutrient content was carried out on *Chaetoceros calcitrans* and *Artemia* sp. including measurements of water, ash, protein, fat, and carbohydrate content. Tests for proximate content were carried out at the Nutrition Laboratory, Faculty of Public Health, Airlangga University, Surabaya.

2.6 Water Quality Measurement
Water quality parameters that need to be considered in the cultural activities of *Chaetoceros calcitrans* and the cultivation of *Artemia* sp. include measurements of temperature using a thermometer, salinity using a refractometer, and pH using a pH meter. In measuring water quality for *Artemia* sp. The oxygen content was also measured using a DO meter.

2.7 Data Analysis
The data obtained were analyzed using Analysis of Variance (ANOVA) with a completely randomized design (CRD) with the help of SPSS software. If the data obtained shows that the final results have a significantly different effect or are very significantly different, then the data will be compared with the treatment that produces the best results with Duncan's test [7]. The data obtained were also tested for normality and then the Independent Sample T-Test was tested with the help of SPSS software.

3. Result and discussion
Test the nutritional content of liquid organic fertilizer *Gracilaria* sp. which has been fermented for 8 days with the addition of EM4 bio activator, sugar solution, and fish meal which contains nitrogen, phosphorus, and potassium. The following results of the nutritional content test can be seen in Table 3.

Table 3. Nutrient Content of Liquid Organic Fertilizer *Gracilaria* sp.

Parameters	Result (mg/ml)
Nitrogen	14.50
Phosphorus	14.85
Potassium	0.30

In this research, the manufacture of liquid organic fertilizer *Gracilaria* sp. through a fermentation process for ±9 days and using the addition of bio activator EM4, fish meal, and sugar solution to speed up the fermentation process and increase the nutritional content. According to Alamsjah et al. (2011), the addition of proteolytic fermentation bacteria can hydrolyze protein into amino acids and increase nitrogen content. According to Sundari et al. (2014), the addition of fish meals can increase the phosphorus content because it contains protein.

Density calculations were carried out during the 8-day cultivation period and analyzed by Analysis of Variance and Duncan's Multiple Range Test. The following data on the density of *Chaetoceros calcitrans* can be seen in Table 4.

Table 4. Density of *Chaetoceros calcitrans*

Days to-	Density of *Chaetoceros calcitrans* (10^4 cell/ml)				
	P0	P1	P2	P3	P4
1	$175.00^{d}\pm0.16$	$183.75^{cd}\pm0.09$	$206.25^{bc}\pm0.16$	$233.75^{a}\pm0.23$	$231.25^{ab}\pm0.20$
2	$363.75^{c}\pm0.13$	$350.00^{c}\pm0.26$	$443.75^{a}\pm0.15$	$397.50^{b}\pm0.12$	$421.25^{ab}\pm0.11$
3	$396.25^{b}\pm0.16$	$412.50^{b}\pm0.33$	$496.25^{a}\pm0.14$	$466.00^{a}\pm0.18$	$467.50^{a}\pm0.19$
4	$522.50^{a}\pm0.06$	$496.25^{a}\pm0.14$	$522.50^{a}\pm0.10$	$527.50^{a}\pm0.16$	$506.25^{a}\pm0.38$
5	$545.00^{b}\pm0.16$	$380.00^{d}\pm0.12$	$641.25^{a}\pm0.75$	$492.50^{bc}\pm0.13$	$471.25^{c}\pm0.09$
6	$586.25^{a}\pm0.19$	$336.25^{c}\pm0.05$	$557.50^{a}\pm0.39$	$461.25^{b}\pm0.09$	$430.00^{b}\pm0.15$
7	$466.25^{a}\pm0.17$	$268.75^{c}\pm0.11$	$481.25^{a}\pm0.09$	$417.50^{b}\pm0.06$	$406.25^{b}\pm0.13$
8	$427.50^{a}\pm0.10$	$237.50^{d}\pm0.03$	$400.00^{b}\pm0.09$	$382.50^{c}\pm0.06$	$367.50^{d}\pm0.06$

Note: Different superscripts in one line show a significant difference ($p<0.05$) P0: Pro Analysis fertilizer 16 mg/ml. P1: Liquid organic fertilizer *Gracilaria* sp. 8 mg/ml P2: Liquid organic fertilizer *Gracilaria* sp. 16 mg/ml P3: Liquid organic fertilizer *Gracilaria* sp. 24 mg/ml P4: Liquid organic fertilizer *Gracilaria* sp. 32 mg/ml

Based on Table 2. it is known that P2 with liquid organic fertilizer *Gracilaria* sp. 16 mg/ml had the highest total density and based on the Analysis of Variance and Duncan's Multiple Range Test there were significant differences between each treatment. The difference in the amount of density in each treatment was influenced by the nutritional content during the cultivation period. The main nutrients needed by *Chaetoceros calcitrans* include nitrogen, phosphorus, and potassium for the photosynthesis process. On the first day of counting, it has entered a logarithmic phase which is marked by an increase in the number of cells [8]. The growth phases in *Chaetoceros calcitrans* include the lag phase, logarithmic phase, stationary phase, and declination phase [9]. The following graph shows the growth rate of *Chaetoceros calcitrans* in Figure 1.

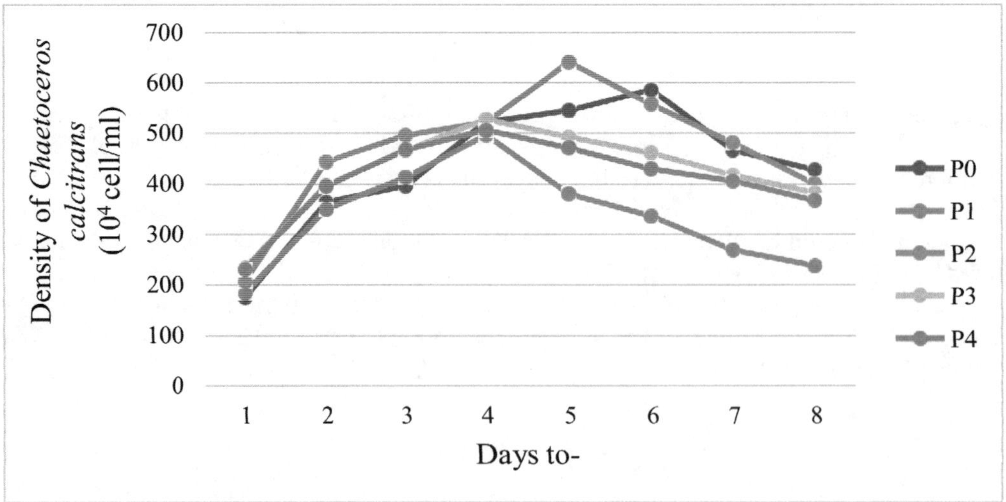

Figure 1. Growth Rate Chart of *Chaetoceros calcitrans*

Based on Figure 1, it is known that *Chaetoceros calcitrans* experienced a logarithmic phase from the first day of counting in each treatment. At P0 experienced a logarithmic phase until the sixth day of counting. In contrast to P1, P3, and P4 which experienced a logarithmic phase until the fourth day and P2 until the fifth day. After reaching the logarithmic phase with high density, *Chaetoceros calcitrans* in each treatment experienced a death phase which was characterized by a decrease in the density.

The results of observations on research on the nutritional content of *Chaetoceros calcitrans* consisted of control treatment and treatment with the highest density, namely P2 (liquid organic fertilizer *Gracilaria* sp. 16mg/ml) representing all replications. An analysis of the nutrient content based on wet weight was then converted to dry weight value. Data analysis of the nutritional content of *Chaetoceros calcitrans* in dry weight can be seen in Table 5.

Table 5. Data Analysis of Nutritional Content of *Chaetoceros calcitrans* in Dry Weight

Treatment	Protein (%)	Fat (%)	Ash (%)	Carbohydrate (%)	∑ Energy (Kcal/kg)*
P0	28.57	6.86	6.29	58.28	4613.66
P2	19.07	2.54	6.36	72.03	4252.29

Information: P0: Pro Analysis Fertilizer; P2: Liquid Organic Fertilizer *Gracilaria* sp. 16 mg/ml; *Total energy where 1 g protein = 5.6 kcal, 1 g fat = 9.1 kcal, 1 g carbohydrate = 4.1 kcal (Jauncy and Ross, 1982)

The difference in the results was due to the phase difference in each treatment when the test was carried out. The test was carried out on the sixth day when *Chaetoceros calcitrans* with pro-analysis fertilizer was in the logarithmic phase with the highest density, in contrast to the use of liquid organic fertilizer *Gracilaria* sp. who have entered the death phase. According to Jati *et al.*, [3] after entering the stationary phase, the nutritional content may change due to the lack of nitrate content in the cultivation media, which causes carbohydrates to increase and protein content tends to decrease.

The length growth rate is known by calculating the difference in length at the beginning and end of maintenance and then analyzed by Independent Sample T-Test. The following are the results of the Independent Sample T-Test in Table 6.

Table 6. The Result of Homogeneity Test and Independent Sample T-Test

		Levene's Test for Equality of Variances		t-test for Equality of Means				
		F	Sig.	T	df	Sig. (2-tailed)	Mean Difference	Std. Error Difference
Length Data of *Artemia* sp.	Equal variances assumed	.020	.894	-1.948	4	.123	-.03267	.01677
	Equal variances not assumed			-1.948	3.972	.124	-.03267	.01677

The basis for making decisions on the Independent Sample T-Test is the value of sig. (2-tailed) < 0.05 then H1 is accepted. Based on Table 4. it is known that the value of sig. (2-tailed) > 0.05, which is 0.123, which means that there is no effect of feeding *Chaetoceros calcitrans* with different nutrients on the length growth rate of *Artemia* sp. According to Toonen [10], sufficient protein content can support the body length growth of *Artemia* sp. In addition to the influence of nutritional content, according to Akhsin *et al* , [11], the length of time the process of digesting food also affects the increase in body length because during the digestion process energy is needed which is obtained through protein.

The rate of weight growth was determined by calculating the difference in weight at the beginning and end of maintenance and then analyzed using the Independent Sample T-Test. The following are the results of the Independent Sample T-Test in Table 7.

Table 7. The Result of Homogeneity Test and ndeplendent Sample T-Test

		Levene's Test for Equality of Variances		t-test for Equality of Means				
		F	Sig.	T	df	Sig. (2-tailed)	Mean Difference	Std. Error Difference
Weight Data of *Artemia* sp.	Equal variances assumed	6.291	.066	-1.986	4	.118	-.008000	.004028
	Equal variances not assumed			-1.986	2.201	.174	-.008000	.004028

The basis for making decisions on the Independent Sample T-Test is the value of sig. (2-tailed) < 0.05 then H1 is accepted. Based on Table 5.7 it is known that the value of sig. (2-tailed) > 0.05, which is 0.118, which means that there is no effect of feeding *Chaetoceros calcitrans* with different nutrients on the weight growth rate of *Artemia* sp. According to Akhsin *et al* , [11] increased weight of *Artemia* sp. is supported by the carbohydrate and fat content of the feed given. Carbohydrates are a source of energy in the metabolic process, and if the body has excess it will be deposited as body fat.

The results of observations in research on nutritional content consisted of the treatment of *Artemia* sp. *Chaetoceros calcitrans* feed with pro analysis fertilizer and *Artemia* sp. *Chaetoceros calcitrans* feed with liquid organic fertilizer *Gracilaria* sp. on the nutrient content based on the wet weight which is then converted into the dry weight value. Data analysis of nutrient content in *Artemia* sp. in dry weight can be seen in Table 8.

Table 8. Data Analysis of Nutritional Content of *Artemia* sp. in Dry Weight

Treatment	Protein (%)	Fat (%)	Ash (%)	Carbohydrate (%)	∑ Energy (Kcal/kg)*
P0	53.34	27.06	5.13	14.46	6042.36
P1	53.62	25.9	6.41	14.07	5936.49

Information: P0: Pro Analysis Fertilizer; P1: Liquid Organic Fertilizer *Gracilaria* sp.; *Total energy where 1 g protein = 5.6 kcal, 1 g fat = 9.1 kcal, 1 g carbohydrate = 4.1 kcal (Jauncy and Ross, 1982)

Differences in the results of the analysis of nutritional content can be influenced by the nutritional content of the feed given. In this study, it can be seen that *Artemia* sp. utilize carbohydrates contained in *Chaetoceros calcitrans* as a source of metabolic energy, and save the use of protein and fat to support growth so that they are not used as energy producers. According to Yanto *et al.*, [12] Carbohydrates have the function to meet the body's metabolic energy needs, if in excess amounts they will be stored in the form of glycogen as an energy reserve and for protein and fat synthesis which aims to increase growth.

4. Conclusions
The use of liquid organic fertilizer *Gracilaria* sp. as a nutrient *Chaetoceros calcitrans* did not give a significant difference to the increase in growth in length and weight. Based on the proximate content test, the results were 53.62% protein, 25.9% fat, and 14.07% carbohydrate. Based on the results of the research conducted, the use of liquid organic fertilizer as an alternative source of nutrition in microalgae cultivation with a nitrogen content that is following the needs of microalgae, namely 16 mg/ml to increase the density and nutrient content as feed for *Artemia* sp. This is because liquid organic fertilizer *Gracilaria* sp. has a more economical price than the use of pro-analytical fertilizers as microalgae nutrition

5. References
[1] Firmansyah, M. Y., R. Kusdarwati dan Y. Cahyoko. 2013 *J Ilm Perikan dan Kelaut*. 5(1):105-111.
[2] Le, T. H., N. V. Hoa, P. Sergeloos, and G. Van Stappen. 2018 *Rev Aquac*. 11 (4): 1169-1175.
[3] Jati, F. J. Hutabarat, dan V. E. Herawati. 2012 *J Aquac Manag Technol*. 1 (1): 221-235.
[4] Nugroho, A. 2019. Evaluation of the Utilization of Gracilaria sp. Seaweed Liquid Organic Fertilizer. Against the Growth of Nannochloropsis sp. Thesis. Airlangga University. Surabaya. p. 59.
[5] Sundari, I., W. F. Maruf, dan E. N. Dewi. 2014 *J Pengolah dan Bioteknol Has Perikanan*. 3(3): 88-94.
[6] Ratrinia, P. W., Uju dan P. Suptijah. 2016 *J Pengolah Has Perikan Indones*. 19 (3): 309- 320.
[7] Sasmitha, K. R. 2008. Experiment Design. Universitas Airlangga Press. Surabaya. p 44-86.
[8] Yanuaris, L. M., R. Kusdarwati dan Kismiyati. 2012 *J Ilm Perikan dan Kelaut*. 4 (1): 21-26.
[9] Muliani, E. Ayuzar, dan M. C. Amri. 2018 *Acta Aquat*. 5 (1): 30-35.
[10] Toonen R. 2004. *Brine Shrimp: Nutritional Value od Live Foods for The Coral Reef Aquarium Part 2. Aquarium Invertebrates*, 3 (2). 3p.
[11] Akhsin, M. H., Irwani, dan N. Taufiq. 2014 *J Mar Res*. 3 (4): 456-461.
[12] Yanto, H., A. E. Setiadi dan D. Kurniasih. 2019 *J Ruaya*. 7 (2): 39-46.

The effect of old water making time and heating temperature on moisture content and NaCl on traditional salt production

E Susilowati[1], MA Alamsjah[2,3], DY Pujiastuti[2]

[1]Study Program of Fisheries Product Technology, Faculty of Fisheries and Marine, Universitas Airlangga, Campus C UNAIR, Jl. Mulyorejo, Surabaya, Indonesia
[2]Department of Marine, Faculty of Fisheries and Marine, Universitas Airlangga, Campus C UNAIR, Jl. Mulyorejo, Surabaya, Indonesia

[3]Corresponding author: alamsjah@fpk.unair.ac.id

Abstract. The manufacture of salt from old water is only done in salt ponds by geomembrane methods and prisms with the help of evaporation from sunlight. This study uses experimental methods. Working procedures in this study include sampling, followed by characterizing water samples including pH, NaCl, salinity and clarity tests. Then continued with the manufacture of salt samples . After that, each sample was tested on the water content and NaCl levels. The data to be obtained will be analyzed using the Duncan and Tukey Test tests with the SPSS 25.0. While the sensory test using Krustal Wallis if there is a noticeable difference ($p<0.05$) it will be continued with the Mann Whitney test and compared to the quality standard of SNI 3556:2010 iodine consumption salts. Sensorically covering the color, taste and texture of salt with the method of making using hotplate has met the quality 1 and 2. The chemical characteristics of the moisture content parameters have met the standard SNI 3556: 2010 salt consumption of iodine. Some salt samples do not meet the SNI's standards on NaCl levels. Salt quality is affected by the NaCl levels of old water, impurities and the process of making salt.

1. Introduction

Salt is one of the essential chemicals needed in the food industry, chemistry, pharmaceuticals, to everyday life. Based on the Ministry of Industry of the Republic of Indonesia (2021) the number of national industrial salt needs in 2021 is estimated to reach 4.6 million tons. But until now Indonesia still has to import salt from other countries to meet its national salt needs. This is due to the low quality of national salt, it can be seen from the large amount of salt processed by salt farmers and salt processing plants that have not met the standards when viewed from their NaCl levels [1].

The area of salt ponds in Indonesia is about 30,786 ha and is located in various regions, the largest salt ponds are found on the island of Java which is an area of 25,541 ha. In addition to Java Island, salt is also produced in West Nusa Tenggara Province with an area of 1,155 ha, South Sulawesi Province 2,205 ha,

Sumatra Island and other areas covering an area of 1,885 ha [2].

The problem that exists in people's salt production today is the lack of quality and quantity of national salt needs along with the increasing population and the rapid development of industry towards salt needs, there are several main problems that need to be solved jointly by agencies related to national salt production, while these problems include technology and technical production. Salt farmers in the process of making salt use a very simple way that evaporates seawater in the salting plot with sunlight without any technological touch, so that although the raw materials are abundant but salinity and dissolved pollutants are very diverse. Seawater that has undergone evaporation and crystallization processes produces salts commonly referred to as coarse salts (krosok) that contain an average sodium chloride (NaCl) content of 85% and contain impurities such as magnesium sulfate ($MgSO_4$), calcium sulfate ($CaSO_4$), magnesium chloride ($MgCl_2$), Potassium chloride (KCl) and soil impurities (Sumada et al., 2016). According to Martina *et al.,* [3] A good salt category ranges if it contains NaCl levels >95%, both NaCl levels of 90-95%, and moderate NaCl levels between 80-90% but the priority is the salt content above 95%. The people's cultivation system until now uses total crystallization so that productivity and quality are still lacking or in general the NaCl level is less than 90% and contains many impurities. In general, making salt from old water is only done in salt ponds by geomembrane methods and prisms with the help of evaporation from sunlight, there have not been many studies related to making salt from old water with small-scale methods. So this study is expected to provide information related to the length of time of making old water with different temperature variations to the NaCl content and water content of traditional salt production.

Natural salts contain compounds of magnesium chloride, magnesium sulfate, magnesium bromide, and other compounds. Salt is a collection of chemical compounds with the largest constituents being sodium chloride (NaCl) and impurities namely calcium sulfate ($CaSO_4$), magnesium sulfate ($MgSO_4$), and magnesium chloride ($MgCl_2$). So far salt in Indonesia is produced by State-Owned Enterprises (SOEs) in this case PT. Salt (Persero) and salt farmers or known as the people's cultivation. The dissolved content in salt is < 20% consisting of $MgCl_2$ 0.3 - 1%, $MgSO_4$ 0.2 - 0.6%, and $CaSO_4$ 0.5 -1% (Sedivy, 2009). The chemical composition of Sodium Chloride (NaCl) consists of 39.3% Sodium (Na) and 60.7% Chlorine (Cl). Sodium Chloride is crystalline or white powder with a cube-shaped ismeric system. The salt structure (NaCl) includes the middle anion and the cation occupies the octahedral cavity as in Figure 1 [4].

Salt characteristics are divided into physical characteristics that include appearance, taste, smell, and color, as well as chemical characteristics that include water content, NaCl levels, iodine levels, and impurity content. Impurities or impurities in salt greatly affect the quality of the salt. Impurities in salt can be $CaSO_4$, $MgSO_4$, $MgCl_2$, as well as metal components [5]. Impurities of the element calcium (Ca) are usually casts and carbonates. Cast crystals are very fine and settle very slowly, so that at the crystallization stage of NaCl the cast will crystallize. This causes salt obtained by evaporation using solar power to be lower in purity compared to artificial evaporation. In addition to calcium compounds (Ca), there are also magnesium (Mg) compounds contained in the parent solution, namely the remaining deposition of NaCl. Mg compounds are attached to the outside of salt crystals and cause hygoscopic properties to be large and the taste to be bitter [6].

Sodium chloride is an ionic bonding substance. The cohesion force in ionic molecules comes from the electrostatic pull between ions and ions. Such materials are thought to be easy to form solids, because ions can directly attract ions so as to build a molecular structure. NaCl has an fcc (face center cubic) lattice structure so NaCl has a strong bond. Because the source of the bond is electrostatic, so the more negative the ions surrounding the positive ion, the more stable and strong the solid substance.

The method used for the NaCl test is Mohr Argentometry. Actually Argentometry there are 4 kinds, namely Argentometry Mohr, Argentometry Volhard, Argentometri Fajans and Argentometry Liebig. But a method often used for chloride levels is Mohr Argentometry. Mohr argentometry is used for the determination of halogenide salts with a standard solution of AgNO3. The indicator used is a solution of K2CrO4 and the titration endpoint is characterized by the formation of a pink deposit of pink. It should be noted that titration should be carried out in a slightly alkaline neutral state. Tritrasi should not be done in a state that is too alkaline or in an acidic state. During titration, the solution must be shaken firmly so that the titration endpoint is clearly visible [7].

Old water is a term of saturated NaCl content. The term starts from salt farmers, which define saturated water as "old water", the average sea water has 0-3oBe then evaporated so that it becomes old water with a salt content above 20°Be. Be is a salinity level of seawater and is used as a raw material for salt making. The tool used to measure the degree of Be is the Be (baume) meter. Old water that has reached the Be standard for production needs is then flowed on the salt table until salt crystals are produced that are ready for harvest [8].

The basis of argentometry titration is a precipitation reaction in which the substance to be determined is ended by a raw solution of silver nitrate (AgNO3) and chromate indicators. Such substances include halogenide salts (Cl, Br, I), cyanide, thiocyanide and phosphate. The titration endpoint is indicated by the presence of colored deposits [9]. Argentometry titration has 3 common methods: mohr method, Fajans method, and Volhard method. The mohr method is a method used in the measurement of chloride and bromide levels in a neutral atmosphere with a standard solution of silver nitrate (AgNO3) and the addition of potassium chromate (K2CrO4) as an indicator. Titration in an acidic atmosphere causes silver chromate to dissolve because it is formed dichromate and in a alkaline atmosphere will form a precipitate of silver hydroxide. When the chloride or bromide ions have been deposited by silver ions (Ag+), the chromatic ions will react with excess silver (Ag) to form a brown/red brick silver deposit (Ag2CrO4) as the titration endpoint [10].

Water content is the percentage content of a material that can be expressed based on wet weight (wet base) or based on dry weight (dry base). Wet heavy water content has a theoretical maximum limit of 100 percent, while water content based on dry weight can be more than 100 percent [11} Increased water content in salt can affect the content of NaCl, the higher the water content in salt, the concentration of other compounds will also decrease including NaCl compounds [12]

The determination of moisture content carried out refers to the methods that have been used by Deglas and Yosefa, [13}. Determination of moisture content can basically be done by several methods, namely: by drying method (thermogravimery), distillation method (thermovolumetry), physical method and chemical method (Karl Fischer Method). Of these methods, determination of the moisture content of dry materials can be done using the drying method in the oven or the thermogravimetry method. The thermogravimetry method to be carried out in this study refers to SNI 3556-2010.

2. Material and method

2.1 method

This research was conducted at the Laboratory of Chemistry and Analysis, Faculty of Fisheries and Marine Affairs, Airlangga University. The main ingredient in this study is old water with a variation of the length of deposition of 4, 5 and 6 months obtained directly from salt farmers in Sedayu Lawas Village, Brondong District, Lamongan Regency. Other materials used for testing include $AgNO_3$ SAP brand, K_2CrO_4 SAP brand, H_2SO_4 SAP brand, Merck brand NaOH, $Na_2S_2O_3$. $5H_2O$ brand Merck, KI brand SAP, $K_2Cr_2O_7$ brand SAP, HCl brand SAP, NaCl brand SAP, amilum brand smart and aquades technical brand. The materials used in this study consisted of the main ingredients and analytical materials.

The equipment used in this study included hot plates, desikators, analytical balance sheets, watch glass, trays, stirrer rods, 2000 ml glass beakers, 200 ml glass beakers, 200 ml glass beakers, 50 ml measuring cups, measuring cups 100 ml, spoons, measuring pumpkins 100 ml, measuring pumpkins 250 ml, measuring pumpkins 300 ml, pipettes volume 5 ml, pipette volume 10 ml, burette 50 ml, statif, clamps, drip pipettes, buchner funnels, trays, 10 cm jars and baumeters.

The research method to be used in this study is an experimental method with a complete random design (RAL). Experimental method is a method used to obtain data carried out by laboratory experiments and direct, systematic observation of the events of the object being studied. RAL is used when media and experimental materials are uniform or can be considered uniform. This design has only one source of diversity: treatment in addition to random influences [14]. The design of the study to determine the effect of temperature and time of making old water on water content and NaCl in traditional salts is Anava 2 Factors. This research design can also find out the influence of several free variables on bound variables, not just one free variable, besides that a combination of several variables can also be seen the influence or the existence of interactions.

In factorial tests, each level of one factor is connected (interacting) with every other level of factors, even the main purpose of using factorial tests is to look for the influence of interactions that are most strong in influencing the results of research. The study used two free variables: temperature and time. The temperatures used are 150°C, 200°C and 250°C. While the time of making old water used is 4 months (23 Be), 5 months (25 Be) and 6 months (26 Be).

Old water obtained from ponds with various variations in the length of deposition time in the pond (4 months of deposition, 5 months of deposition, and 6 months of deposition). Prepare the old water then measure its °Be. Put 1000 ml of old water in the beaker glass and evaporate with a hot plate at each temperature treatment of 150°C, 200°C and 250°C until salt crystals are produced and perform the same treatment for other different old water samples.

The method used to analyze sodium chloride levels is the argentometry titration of the Mohr method. The Mohr method is used to determine chloride levels with a raw solution of silver nitrate ($AgNO_3$) and an indicator of potassium chromate (K_2CrO_4). $AgNO_3$ solution used as a raw solution is first standardized using sodium chloride (NaCl) solution. This is done to find out the normality value of $AgNO_3$ solution [15]. Testing nacl levels based on SNI 8207: 2016, sodium chloride (NaCl) test is the formation of AgCl deposits using $AgNO_3$ with potassium chromate instructions in a neutral or slightly alkaline solution. First, make various reagents in the titration of Mohr. Dissolving 8.5 g of $AgNO_3$ in 500 mL aquades, makes the potassium chromate indicator, K_2CrO_4 5% by dissolving 5 grams of K_2CrO_4 in 100 mL aquades. Make a solution of sulfuric acid, H_2SO_4 1 N by pouring little by little 7 mL of concentrated H_2SO_4 (H_2SO_4 95-97%) into a 300 ml beaker glass containing 100 mL of aquades, then transfer in a 250 mL measuring flask, and dilute until the line mark. Make a solution of sodium hydroxide, NaOH 4 N by dissolving 16 g of NaOH with aquades in a 100 mL beaker glass, then tangle into a 100 mL measuring flask and dilute with aquades until the line marks and shake until homogeneous.

Second, Determination of Standardization of $AgNO_3$ solution 0.01 N by making 250 mL of primary standard solution NaCl 0.01 N. The first step is to weigh NaCl 0.0692 grams in watch glass, then transfer in a 250 mL measuring flask, dissolved with aquades to the boundary mark and shake slowly until evenly mixed. After making the primary solution of NaCl followed by standardizing the silver nitrate solution ($AgNO_3$) namely by inserting a 25 mL solution of NaCl 0.01 N into a 100 mL erlenmeyer glass, then added 5 drops of K_2CrO_4 indicator solution 5% and stirred. Titration of NaCl 0.01 N solution that has been indicatorized with AgNO3 solution until a white deposit occurs [16]. Record the volume of $AgNO_3$ solution and then calculated the normality of $AgNO_3$ raw solution with the formula:

$$\text{N } AgNO_3 = \frac{\text{V NaCl x N NaCl}}{\text{V AgNO3}}$$

Water content analysis is a test method in the food industry used to determine the quality and security of foodstuffs against possible damage [17]. The method used in the moisture content test is the thermogravimetry method or drying method using a moisture analyser tool. Water content testing according to Jolly *et al.,* [18], which prepares an aluminum base which is then placed on the base to determine the weight of the tare. A sample of 2 grams is placed evenly on an aluminum base and testing begins heating the tool to 145°C. As long as the sample is heated, the tool continues to read the sample mass and transmit it to the microprocessor until a constant mass of material is obtained and the final result will be printed on the monitor.

2.2 Data analysis

The parameters tested are the value of moisture content and NaCl on traditional salt production. Data from the results of the water content test and nacl test were obtained using the Analysis of Variance (ANOVA) of the Duncan and Tukey Tests to determine the effect of the treatment given. Treatment that shows significant results can be continued with advanced trials of Tukey's Smallest Real Difference (LSD) at 95% accuracy. As for sensory test data analysis using Kruskal Wallis Test and Mann Whitney test using SPSS 25.0 ($P<0.05$) software.

3. Results and discussion

3.1 The results of analysis water content (%)

The moisture content value in traditional salts can be seen in Table 1, with 9 treatment interactions between 3 temperatures and 3 times. The average water content value is obtained from an average of 3 repetitions. A graph of traditional salt water content can be seen in Figure 1. It shows that there is a real interaction between the temperature and time of making old water with traditional salt water content ($P<0.05$).

Table 1. Results of Analysis of Water Content

Treatment	Water Content (%)		
	Temperature (°C)	Time (Month)	Average ± SD
P1	150	4	$5.14^{f} \pm 0.29$
P2	150	5	$4.29^{e} \pm 0.54$
P3	150	6	$3.93^{ed} \pm 0.55$
P4	200	4	$3.68^{d} \pm 0.40$
P5	200	5	$3.20^{dc} \pm 0.47$
P6	200	6	$2.83^{c} \pm 0.56$
P7	250	4	$1.76^{b} \pm 0.37$
P8	250	5	$1.02^{a} \pm 0.14$
P9	250	6	$0.37^{a} \pm 0.04$

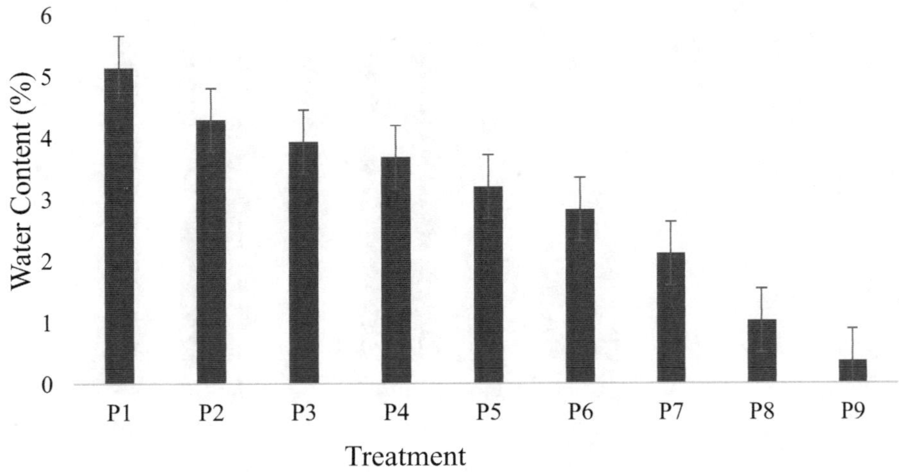

Figure 1. Results of Analysis of Water content

3.2 The results of analysis NaCl content (%)

The value of traditional salt NaCl levels can be seen in Table 2. with 9 interaction treatments between 3 temperatures and 3 times. The average sulfate level is obtained from an average of 3 repetitions. A graph of traditional salt NaCl levels can be seen in Figure 2. It shows that there is a noticeable interaction between temperature and heating time to NaCl ($P<0.05$) levels.

Table 2. Results of Analysis of NaCl

Treatment	NaCl Levels %		
	Temperature (°C)	**Time (Month)**	**Average± SD**
P1	150	4	$89.63^{a} \pm 0.37$
P2	150	5	$89.97^{a} \pm 0.30$
P3	150	6	$90.27^{a} \pm 1.15$
P4	200	4	$97.17^{b} \pm 0.35$
P5	200	5	$97.27^{b} \pm 0.96$
P6	200	6	$97.23^{b} \pm 0.15$
P7	250	4	$97.13^{b} \pm 0.15$
P8	250	5	$97.23^{b} \pm 0.30$
P9	250	6	$97.17^{b} \pm 0.30$

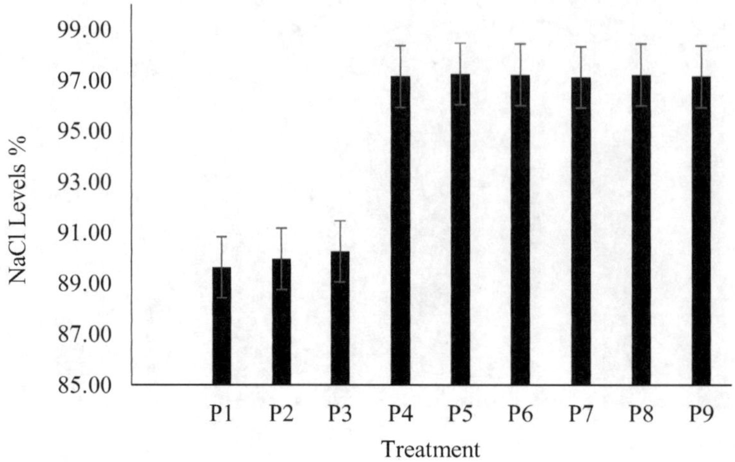

Figure 2. Results of Analysis of NaCl content

3.3 Additional testing of traditional salt sensory test

The results of the traditional salt sensory test can be seen in Table 3 with 9 interaction treatments between 3 temperatures and 3 times. Sensory test results are obtained from the average of each parameter. It shows that there is a noticeable interaction between temperature and heating time to color, taste and texture ($P<0.05$).

Table 3. Results of Analysis of Sensoris

Treatment	Sensory Test Mean Value		
	Color	**Sense**	**Texture**
P1	$8.47^{a} \pm 0.90$	$7.73^{b} \pm 0.98$	$8.67^{a} \pm 0.75$
P2	$8.33^{a} \pm 0.95$	$8.40^{a} \pm 0.93$	$8.27^{a} \pm 0.98$
P3	$8.33^{a} \pm 0.95$	$8.07^{ab} \pm 1.01$	$8.47^{a} \pm 0.90$
P4	$7.80^{ab} \pm 0.99$	$8.33^{a} \pm 0.95$	$7.13^{ab} \pm 1.38$
P5	$7.20^{b} \pm 1.09$	$7.60^{b} \pm 1.30$	$6.87^{ab} \pm 1.38$
P6	$7.33^{ab} \pm 1.29$	$8.20^{ab} \pm 0.99$	$6.80^{b} \pm 0.80$
P7	$6.20^{bc} \pm 0.99$	$8.27^{ab} \pm 0.98$	$6.07^{bc} \pm 1.01$
P8	$6.00^{c} \pm 1.01$	$8.40^{a} \pm 0.93$	$4.87^{c} \pm 0.90$
P9	$6.07^{c} \pm 1.01$	$8.47^{a} \pm 0.90$	$5.07^{c} \pm 1.53$

3.4 Discussion

Water content is one of the chemical properties of materials that indicate the amount of water contained in foodstuffs. According to Hadiwiyoto [15], water is the most abundant component contained in salt. The salt water content obtained from the study has a value of 0.37% to 5.14%. The test results are lower with the condition that the maximum amount of water content for raw materials for the iodine salt industry (SNI 01-3556-2010) and the quality requirement of iodine consumption salt (SNI 01-3556-2010) is a maximum of 7%. Thus, the amount of moisture content in the salt results of this study meet the quality requirements of raw materials for the iodine salt industry and the quality requirements of salt consumption of iodine. This suggests that the lowest moisture content at 250°C, the higher the drying temperature given, exerts a very large influence on the speed of water transfer. According to Yuniarti *et al.*, [19], the higher the drying temperature, the faster evaporation occurs, so that the water content in the material is lower.

The high water content in salt affects the content of NaCl in salt, the higher the water content in salt, the concentration of other compounds will decrease including NaCl compounds [20]. According to Rusiyanto [6] the quality of salt products containing high water content produces low salt quality below SNI quality standards, this can be caused by a salt production process that is still traditional so it is not optimal in the salt crystallization process. Salt yields that contain high water will obtain high value and low quality.

Rachmawan [21], revealed that the higher the temperature and speed of drying air flow the faster the drying process takes place. The higher the dryer air temperature, the greater the heat energy carried by the air so that the more liquid mass is evaporated from the surface of the drained material. If the speed of the dryer airflow is higher, the faster the mass of water vapor transferred from the material to the atmosphere. According to Yuniarti *et al.*, [19], at the beginning of drying the speed of the amount of water lost per unit of time remains, then there will be a decrease in the speed of water removal per unit time. This is related to the type of water that experiences saturation. The results of water quality measurements in the form of water pH show a range between 7.4 - 7.7, this shows that the old water used to be used as salt is good for consumption. According to Prastuti, [22] the pH of the human body is 7.35 so for the pH of water consumed is between 6.5 - 8.5. The highest water temperature is 250°C. While Putri *et al.*, [14] explained that rising temperatures result in an increase in chemical reactions, viscosity, evaporation and voltage.

According to Putri *et al.*, [23] stated based on the chemical theory of salt dissolved in water, it will decompose into electrically charged sodium and chloride particles, because NaCl is a powerful electrolyte that decomposes perfectly. Chloride particles are negative while sodium particles are positive. As a result of this grouping sodium and chloride particles eventually separate and dissolve along with larger amounts of water molecules. The smaller the particles, the more easily dissolved the salt solubility.

NaCl is the main constituent component of salt. Other components are impurities, usually derived from the ions Ca^{2+}, Mg^{2+}, SO_4^{2-}. The impurities are bound to the solvent so that it is suspended and can be separated through filtration. Recrystallization is the technique of purification of a solid substance from a mixture or impurity by crystallizing the substance again after dissolving with solvents. Recrystallization can also be applied in the salt purification process which begins with solubility of salt by using hot water which is then filtered to separate impurities [24].

NaCl levels in the salts tested for the lowest results were P1, P2 and P3 with values of 89.63%, 89.97% and 90.27%. The test results are lower than the minimum sodium clorida (NaCl) requirement for raw materials for the iodine salt industry (SNI 01-3556-2010) and the quality requirement for iodine consumption salt (SNI 01-3556-2010) which is at least 94%. The low percentage of NaCl levels in salt can be caused by the impurities contained in it. According to Martina *et al.*, [3] stated that evaporation of seawater to dry in addition to containing the main mineral NaCl, also contains other minerals or impurities, as well as soil impurities. Six other salt samples, namely P4, P5, P6, P7, P8 and P9, have met nacl standards according to SNI-01-3556-2010. Thus, the amount of NaCl content in salt produced by this refinery already exists that meet the quality requirements of salt raw materials for the iodine salt industry and the quality requirements of salt consumption of iodine. Even the evaporation of clear seawater will only produce NaCl with a purity of 77.74%. Salinity can indicate the level of salt contained in a water. The higher the salinity, the higher the salt content contained and vice versa. Good salinity of seawater for use as raw material for salt making ranges from 35 - 37 ppt [16]. P1, P2 and P3 salts are included in quality 3 salts because NaCl quality ranges from 80-90%. While P9 includes quality salt 2 because the quality of NaCl ranges from 90-95% and P4, P5, P6, P7 and P8 are included in quality salt 1 because the content of NaCl ranges from >95%.

The results of the sensory analysis of salt quality through the Kruskal Wallis test of 30 panelists each can be explained that the color of salt with a drying temperature of 150 °C (P1, P2 and P3) and a temperature of 200 °C (P4, P5 and P6) between each other does not differ markedly at the level of 5%. Judging from the color aspect of the sixth sample does not differ markedly because the number of panelists stated the same is still below the requirement of the smallest amount for a real difference of 5%. While in heating salt with a temperature of 250°C (P7, P8 and P9) has a slightly dull color. This can be seen from the color patterns hanging with heating temperatures and salt washing [1]. According to Martina [3] the color of salt that is dull or not white indicates that there are still insoluble impurities in the form of dust or mud contained in salt. This comes from a heating process that is less than optimal in the crystallization process. So judging from the color quality, P1, P2, P3, P4, P5 and P6 salts are included in the quality of K2 salt, P7, P8 and P9 salts are included in K3 quality.

The taste parameters of all treatments of all salt samples do not differ markedly at the rate of 5% having a very salty value. P1 and P5 have lace niai compared to other samples. According to Abdullah and Susandini [25] the difference in taste in salt is due to the level of impurity content. The results of the panelist's assessment analysis showed that there was no noticeable difference between the salt samples. However, the results of the sensory test are influenced by the level of sensitivity of the taste organs of each panelist so that the results are not completely precise. Sensory test results do not fully guarantee correctness, because these tests tend to be subjective.

Texture parameters of the nine treatments showed different levels of preference. The salt treatment P1, P2 and P3 was preferred by the panelists with the highest scores, while P8 and P8 had the lowest scores. Overall the textures of the nine salts are not much different, the traditional salt, namely P8 and P9, have a coarse texture. This is due to the process of evaporating old water using high temperatures or evaporation is done by boiling old water until crystals form. The basic principle is when water is heated, the temperature will rise and if it reaches the boiling point it will form gas bubbles. The process of heating water to high temperatures will increase the kinetic energy of the particles making up the water [26]. This high kinetic energy makes salt crystals form fine, because there is pressure from the kinetic energy of water particles that break the salt crystals into smaller ones. The influence of temperature in the manufacture of traditional salt produced is rough textured because the temperature in the crystallization process is not maximal [27].

4. Conclusion

The moisture content of traditional salts produced by crystallization method using hotplates with a combination of differences in the time of making old water and different heating temperatures, it was found that all samples met the SNI moisture content standard 01-3556-2010 for the iodine raw material salt industry. NaCl levels obtained from a combination of temperature and time of manufacture of old water, 150°C and time 4 months (P1) with a yield of 89.63%, 150°C and 5 months (P2) with a yield of 89.97%, and 150°C and a time of 6 months (P3) with a yield of 90.97%. The nacl results of the three samples did not meet the standard levels of NaCl SNI 01-3556-2010 for the salt industry of iodine raw materials. Salt sensory test results can be categorized in quality 1 and 2 for raw material salts. This can be proven from the calculation of advanced test variant analysis (ANAVA) which shows the influence of temperature, time and interaction of both on water content and NaCl levels.

5. References

[1] Rositawati, A. L., C. M. Taslim and Soetrisnanto. 2013 *J Teknol Kim dan Ind*. 2(4): 217–225.

[2] Adiraga, Y. and A. H. Setiawan. 2013. *Analysis of the Impact of Changes in Rainfall, Salt Pond Area, and Number of Salt Farmers on the Production of People's Salt Businesses in Juwana*

District, Pati Regency (2003-2012 Period) (Doctoral dissertation, Faculty of Economics and Business). 3(1): 27-29.
[3] Martina, A., J. R. Witono, G. K. Pamungkas and Willy. 2016 *J Tek Kim*. 5(1): 1-6.
[4] Herman, H. and W. Joetra. 2015 *J Momentum*. 17(1): 13-20.
[5] Umam, F. U. 2019 *J Ilm Pangabdhi*. 5(1): 1-4
[6] Rusiyanto, R., E. Soesilowati and J. Jumaeri. 2013 *Sainteknol J Sains dan Teknol*. 11(2): 1-12.
[7] Day, R. A. and A. L. Underwood. 2002. *Quantitative Chemical Analysis Sixth Edition*. Jakarta: Erlangga. pp 95-97.
[8] Kurniawan, T. and A. Azizi. 2012 *J Masy dan Budaya*. 14(3): 499-518.
[9] Utami A.T 2009 Chloride Analysis in Water and Wastewater Using the Argentometric Method. Diploma 3 Study Program in Analytical Chemistry, University of North Sumatra. Scientific Work, pp 13-15
[10] Hopkhar, S. M. 2008. *Basic Concepts of Analytical Chemistry. 4th edition*. University of Indonesia Press. Jakarta. pp 66-70.
[11] Syarif, R. dan Y. Halid, 1993. *Drying Operations In Processing Agricultural Product*s. PT. Mediyatama Sarana Perkasa: Jakarta.
[12] Aisyah, St., Effendi, M., and Triajie, H 2013 *The content of NaCl and H2O of People's Production Salt on Different Storage Media.* Scientific work, Trunojoyo University. Madura
[13] Deglas, W. and F. Yosefa. 2020 *J Agrofood*. 2(1): 16-21.
[14] Kusriningrum. 2008. *Experiment Design*. Airlangga University Press. Surabaya. Hal 15-19
[15] Hadiwiyoto. 2003. *Fishery Product Processing Technology*. Liberty. Yogyakarta. 123-136.
[16] Huljani, M., and Rahma, N 2018 *ALKIMIA: J Ilmu Kim dan Terap*. *2*(2), 5-9.
[17] Daud, A., Suriati, S., & Nuzulyanti, N. 2019 *Lutjanus*, *24*(2), 11-16.
[18] Jolly, W. M., Parsons, R. A., Hadlow, A. M., Cohn, G. M., McAllister, S. S., Popp, J. B., and Negron, J. F. 2012 *For Ecol Manage*. *269*, 52-59.
[19] Yuniarti, D, W., Titik dan Eddy. 2013 *J THPi Stud*. 1(1).
[20] Kurniawan, A., A.A. Abdulah, and N.S. Safitri. 2019 *J Fish Mar Res*. 3(2): 236-244.
[21] Rachmawan, O. 2001. *Drying, Cooling and Packaging of Agricultural Commodities*. Ministry of National Education Bulletin. Jakarta. 1(2): 14-23.
[22] Prastuti, O. P. 2017 *J Tek Kim dan Lingkung*. *1*(1), 35-41.
[23] Putri, L. M. A., P. Trapsilo and B. Susanto. 2017 *J Pembelajaran Fis*. 6(2): 147-153.
[24] Maulana, K. D., M. Jamil., P. E. M. Putra., R. Rahmawati and B. Rohmawati. 2019 *J Creat Student*. 2(1): 42-46.
[25] Abdullah, Z. A., & Susandini, A. 2018 *Eco-Entrepreneur*, *4*(1), 21-36.
[26] Firmansyah, J 2018 *J Filsafat Indones*. *1*(2), 75-79.
[27] Purwati, M. I., A. Gustomi and M. Supratman. 2020 *J Trop Mar Sci*. 3(2): 53-64.

Study of antioxidant activity and microbiological parameters on rosella mangrove syrup preservation

Q S Berliana[1], **D Y Pujiastuti**[2], **S Andriyono**[2,3]

[1]Undergraduated Program of Fisheries Processing Technology, Faculty of Fisheries and Marine, Airlangga University, Campus C Universitas Airlangga Mulyorejo Surabaya, East Java, Indonesia 60115
[2]Department of Marine, Faculty of Fisheries and Marine, Airlangga University, Campus C Universitas Airlangga Mulyorejo Surabaya, East Java, Indonesia 60115

[3]Corresponding author: sapto.andriyono@fpk.unair.ac.id

Abstract. As an archipelagic country, Indonesia has enormous potential for marine and fishery resources. Basalt areas along the coast of Indonesia are habitats for mangrove ecosystems. However, utilising existing mangrove resources still relies on direct benefits such as mangrove wood which causes damage to the existing mangrove ecosystem. Concepts of sustainable use is to process mangrove fruit into functional food, namely mangrove syrup from pedada (*Sonneratia caseolaris*). The syrup from the mangrove fruit is combined with the natural dyes of roselle flowers. In this study, storage temperature testing was carried out by observing the value of antioxidant activity and microbiological contamination. Storage was carried out at temperatures of 10, 25 and 35C, respectively, with a storage time of 28 days. The results showed that as long as storage, the value of antioxidant activity decreased. The lowest value at storage is 10; 25; and 35°C respectively 49.73; 43.72; and 39.92. In addition, storage at 35°C for 28 days showed the highest growth of microorganisms, which was 293.33 CFU. This shows that storing mangrove syrup should be carried out at low temperatures (refrigerator) by reducing direct contact with sunlight to maintain the antioxidant quality of the syrup and other nutritional compositions.

1. Introduction

Indonesia has enormous maritime and maritime potential. As a coastal country, Indonesia has the potential for biological natural resources, non-biological, and artificial resources, and environmental services that are very important for community life [1]. The coastal area consists of several ecosystems, namely coral reefs, beaches, sand, estuaries, and seagrasses. The main ecosystem is mangrove forests which can be called mangrove forests [2].

From an economic perspective, mangrove forests produce several types of wood good quality, and also non-timber products or commonly referred to as non-timber forest products (NTFPs), in the form of wood charcoal, tannins, dyes and cosmetics, as well as food and beverage ingredients. Utilization of mangrove fruit as a food and beverage is a form of diversification of food products [3]. One of the diversifications of processed mangrove fruit products is syrup. Generally, in the manufacture of syrup using the type of mangrove fruit pedada (*Soneratia caseolaris*) [4].

The rosella mangrove syrup product sold to the public should have information on product storage advice [5]. Information advice on safe consumption, especially product storage, needs to be done so that consumers know how a product can be maintained its quality during storage [6], in addition to providing information to distributors or the seller to be able to manage the stock of goods, and can assist in quality control product for the producer [7]. This research will be very useful for consumers to know the safe consumption limit at each storage temperature of rosella mangrove syrup.

2. Material and method

2.1 Place and time of research

This research was conducted from March to May 2022 at the Food Laboratory, Chemistry and Analysis Laboratory, and the Microbiology and Fish Disease Laboratory, Faculty of Fisheries and Marine Affairs, Airlangga University, Surabaya.

2.2 Production of mangrove syrup.

The first step in making mangrove syrup is preparing 1 kg of pedada fruit. The pedada fruit is peeled using a stainless-steel knife until no skin remains. The pedada fruit is then washed thoroughly so that the dirt that sticks out dissolves during washing. The peeled and washed pedada fruit is then put into a pan and added 1 litre of water, then stirred manually until it becomes pedada fruit pulp. Then the first filtering process is carried out using a coarse sieve so that the seeds are set aside. The dirty seeds are added with 1 litre of water and then squeezed using a coarse sieve to produce a second fruit juice. The clean seeds on the results of the second filter are then set aside. The first and second juices were then filtered using golden mella cloth to obtain pure fruit juice. The next step is to dissolve 2 kg of granulated sugar and heat it at 50-60°C. In the dissolution process, stirring is carried out using a stir bar until the sugar is well dissolved.

2.3 Roselle extraction.

Rosella extraction method is based on the maceration method which is a modification from references [8] [9]. Rosella flower extraction is done by drying the rosella flowers first. The dried rosella flowers are then chopped using a chopper machine to become powder. Rosella flower powder is then mixed with water that has been boiled at a temperature of 100°C with a ratio of 1:10 (w/w), and then stirred until homogeneous. Cover the rosella with soaking water for 10 minutes. After 10 minutes, then the rosella soaking water is filtered. The filtered water is 100% rosella flower extract.

2.4 Mixing mangrove syrup with rosella flower.

Extract Mixing is done to combine mangrove syrup with rosella flower extract so that a homogeneous composition is produced from the two ingredients. The formulation for adding rosella flower extract to pedada fruit syrup is 15%. The sterile bottles were added with mangrove syrup and rosella extract according to the formulation and then shaken to make them homogeneous.

2.5 Packaging of rosella mangrove syrup.

The packaging used for rosella mangrove syrup is a brown glass bottle with a size of 300 ml and has a double cap consisting of an inner cap and an outer cap. When packaging rosella mangrove syrup, it must be done aseptically. Rosella mangrove syrup is packaged in sterile bottles. The bottle is first sterilized by boiling the bottle in a saucepan. This is so that the bacteria die and there is no contamination [10]. Then the sterilized bottles are dried. The rosella mangrove syrup is then put into a bottle. After that, the bottle cover is sterilized first by heating on a bunsen, then closing the bottle with a lid. Then the bottle that has been filled with rosella mangrove syrup is then coated with aluminium foil so that the antioxidant content is not reduced due to exposure to sunlight [11].

2.6 Antioxidant activity test.

First, take the supernatant as a sample. This is done to obtain the supernatant as a sample solution. DPPH solution was prepared by weighing 1.97 mg of DPPH and then dissolved in a flask with methanol up to 100 ml until a solution of 50 M was obtained [12]. Furthermore, the sample of mangrove rosella syrup that will be used is 20 mg, then dissolved in 20 ml until the concentration of the solution reaches 1000 ppm. The absorbance of the sample was measured using a UV-VIS spectrophotometer N4S brand with a wavelength of 517 nm.

2.7 Total plate count test.

TPC testing follows the SNI 3544:2013 Syrup, especially in the microbiological contamination section. At the initial stage, the sample was weighed aseptically as much as 50 g, into a diluent bottle containing 450 ml of diluent solution in a ratio of 1:10, then shaken until homogeneous. An amount of 1 ml of the diluted suspension was transferred by pipette into 9 ml of BPW solution. Pipette 1 ml each from each dilution level 10-1, up to 10-4 into petri dishes in duplicate. Then PCA which has been cooled to a temperature of 45°C ± 1°C is poured as much as 15-20 ml into each cup that already contains the suspension. Colony growth (n) was recorded on each petri dish after 48 hours.

2.8 Proximate analysis.

The components tested in the proximate analysis were protein content, carbohydrate content, and fat content based on reference [13]. The first analysis is protein content. A total of 1 g of sample was put into straight digestion tubes with added catalyst (1.5 g K2SO4 and 7.5 mg MgSO4), and 6 ml H2SO4(P) was added. The sample was destroyed at 415°C for 1 hour until the liquid became clear (greenish). Then the sample was cooled, and 30 ml of distilled water was added slowly. The sample was transferred to distillation, and a 125 ml Erlenmeyer was prepared which already contained 25 ml of H3BO3 solution, 7 ml of methylene red and 10 ml of bromine cresol green. The tip of the condenser must be submerged under a 4% H3BO3 solution. Then 30 ml of 40% NaOH solution was added to the Erlenmeyer and distilled for 3-5 minutes. The condenser tube was rinsed with water and accommodated in the same Erlenmeyer. Next, the titration with 0.1 N HCl until a pink colour change occurs.

The second analysis is fat content. The aluminium cup is heated in the oven at 105°C for 1 hour. Cool in a desiccator, then weigh the aluminium cup (Wa). Afterwards, the sample was weighed as much as 2 g (Wb) inserted into the lead and then covered with cotton. The lead containing the sample is put in soxtec, the tool is turned on and heated to a temperature of 135°C and water is drained. Lead is placed in a rinsing position. After the temperature reaches 135°C (the temperature has normalized), put the aluminium cup containing 70 ml of petroleum benzene into the soxtec, then press start to start with the boiling position, carried out for 20 minutes. Then the rinsing position is carried out for 40 minutes, then recovery is 10 minutes with the Soxtec faucet open. The aluminium cup is then placed in the oven at a temperature of 135°C for 2 hours, then cooled in a desiccator and weighed (Wc).

The last analysis is carbohydrate content. Carbohydrate content is usually given as total carbohydrates by difference, meaning that the content is obtained by subtracting the number 100 from the percentage of other components (moisture content, ash content, fat content, and protein content). Carbohydrate content can be calculated by the difference between 100% and the following percentages: % Carbohydrate content = 100% % (protein + fat + water content + ash content)

2.9 Data Analysis.

Data from the study results were obtained using Analysis of Variance (ANOVA) 2 factors to determine the effect of the treatment given. Treatments that show significant results can be continued with Tukey's follow-up test.

3. Results and discussion

3.1 Antioxidant activity test

Testing the antioxidant activity of rosella mangrove syrup was carried out manually every week for 28 days of storage with each temperature treatment is 10°C, 25°C, and 35°C quantitatively using UV-Vis Spectrophotometer. The antioxidant activity of rosella mangrove syrup with storage treatments P1 (10°C), P2 (25°C), and P3 (35°C) expressed in % inhibition Figure 1.

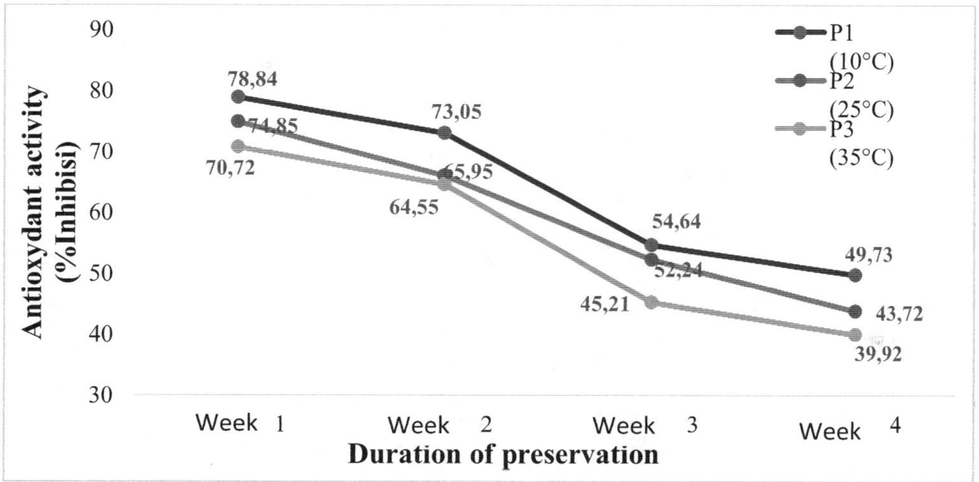

Figure 1. Graph of Antioxidant Activity Value in Rosella Mangrove Syrup during 4-week preservation.

Quantitative antioxidant activity test results from each sample temperature treatment showed that P1 with storage at 10°C has a higher antioxidant value than P2 and P3. On storage temperature of 10°C changes in antioxidant value is insignificant because a storage temperature of 10°C is classified as low. Storage of rosella mangrove syrup at low temperatures can keep the stability of the anthocyanin content in the syrup. Temperature is the one that affects the stability of anthocyanins, other factors that also influence are: consists of changes in pH, sunlight, oxygen, storage time and ions metal [14]. This decrease was also due to the time of preservation. This is because the longer the storage time, the higher the activity of antioxidants decreases due to the oxidation process that occurs during storage. The rate of decline in antioxidant activity is also fast and slowly influenced by temperature and storage time factors [15].

3.2 TPC test

Based on SNI 3544:2013 regarding syrup quality requirements, it is explained that the total plate count for syrup maximum of 5x102. Therefore, it is necessary to do TPC testing on rosella mangrove syrup to determine the development of the number of bacteria during storage at different temperatures Figure 2.

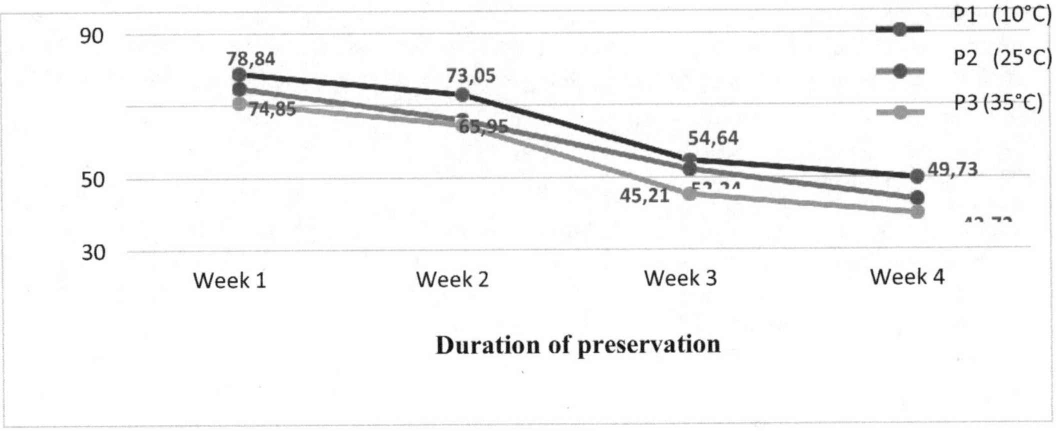

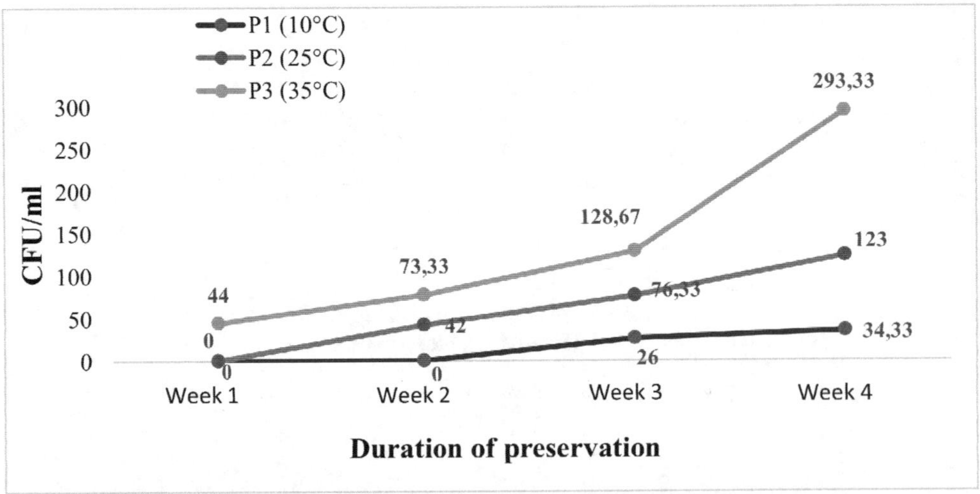

Figure 2. Graph of Test Results Total plate count on Rosella Mangrove Syrup

At 10°C storage, the growth of microorganisms was minimal and very slow. Storage of rosella mangrove syrup at a temperature of 10°C is considered a low temperature so it can withstand the growth rate of microorganisms. Most microorganisms are resistant to low temperatures up to freezing temperatures [16]. However, storage weeks III and IV at a temperature of 10°C still shows the growth of microorganisms although the numbers are very low. The use of low temperature in the preservation of the material cannot cause the death of microorganisms because it continues to grow slow at that temperature and bacterial spores survive so that when food is removed from storage and allowed to thaw again (thawing) the growth of spoilage microbes can run quickly [17].

3.3 Proximate analysis

Proximate analysis was carried out to determine carbohydrate, protein, and fat content before and after different temperature treatments during storage was successfully analyzed. After preservation, carbohydrate and protein were identified to decrease, but for fat, almost all treatments showed no affected Table 2.

Table 2. Proximate test results of rosella mangrove syrup at the beginning and end of preservation

No.	Treatments	Proximate before preservation			Proximate after preservation		
		Carbohydrate	Protein	Fat	Carbohydrate	Protein	Fat
1.	P1 (10°C)				43,32	3,32	2,38
2.	P2 (25°C)	57,41	4,22	2,10	42,58	3,09	2,56
3.	P3 (35°C)				42	2,77	2,97

Decrease in carbohydrate levels can occur drastically and significantly if: storage can reach a high and optimal temperature of 55°C. Decrease Sucrose sugar levels are caused by the occurrence of inversion, namely the presence of the enzyme invertase which breaks down sucrose into inverted sugar [18]. The decrease in protein levels in rosella mangrove syrup was due to the storage process; proteins undergo denaturation and degradation, namely the breakdown of complex molecules into simpler molecules. The protein degradation process is caused by the activity of proteolytic enzymes that play an active role in protein denaturation. This proteolytic enzyme will be activated at low pH so that with the formation of a low pH, the protein is more easily degraded. Change The fat content in rosella mangrove syrup is due to the effect of high fat is the independent variable in this study. In the storage of rosella mangrove syrup, increased fat content was not significantly different, but there was a slight increase from storage temperatures of 10°C, 25°C, and 35°C. The level increase in fat is not too significant because the storage temperature is not too high. Fat will increase significantly when given a temperature treatment of 70°C [19].

4. Conclusion

Based on the results of research on the storage of rosella mangrove syrup in each different temperature treatment, it can be concluded that rosella mangrove syrup with storage treatment at 10°C (P1) has a higher antioxidant value that is 49.73% at week 4. While storage at a temperature of 25°C (P2) and 35°C (P3) has a value of relatively low levels of antioxidants. Storage at low temperatures can maintain the stability of antioxidant activity much better. This study also successfully of microbiology parameters after preservation. The higher the storage temperature of rosella mangrove syrup, the higher the growth of microorganisms will also increase. Storage at 35°C (P3) has significant growth of microorganisms every week. This is because the temperature of 35°C is the optimal temperature for growing bacteria.

5. References

[1] Apriana D S, dan Milla D 2017 *JBT.* 17(1) 15-22.
[2] Dahuri R, Rais J, Ginting S P, dan Sitepu M J 2001 *Integrated Management of Coastal and Marine Resources (Jakarta: Pradnya Paramita) 305 pages.*

[3] Sentoso M S, Ardi A C, Rahmasari N A, Millenia R A, Bangun J A C, Puspasari A D, Budi N P N, Putra R W, Jovenski T W, Dampi A, dan Kristianto I I 2021 *JAI.* 1(1) 20-25.
[4] Rajis, Desmelati, dan Leksono T 2017 *JPK.* 22(1) 51-50.
[5] Sandana F B, Rawung D, Ludong M, dan Mamuaj C 2015 *COCOS.* 5(4) 1-7.
[6] Mulyono A M W, Afriyanti, Basuki J S, dan Sukaryani S 2018 *Pro Food.* 4(1) 277-282.
[7] Wiguna D 2011 Skripsi. Bogor: Institut Pertanian Bogor.
[8] Cisse M, Vaillant F, Acosta O, Mayer C D, and Dornier M. 2009 *J. of Agicultural and Food Chemistry.* 57(14) 6285–6291.
[9] Hastuti N D 2012 *JTP.* 3(1) 29- 63.
[10] Handayani S 2011 Final project. Yogyakarta: Yogyakarta State University. 1-52 p.
[11] Juniarka I G A 2011 *Maj. Obat. Trad.* 16(3): 115–123.
[12] Molyneux P 2004 *JST.* 26(2) 211–219.
[13] AOAC 2005 *Official Methods of Analysis of the Association of Official Analytical Chemists* (Maryland: AOAC International) p 26.
[14] Niendyah H 2004 Skripsi. Malang: Universitas Brawijaya.
[15] Dari D W, Ananda M, dan Junita D 2020 *JTPA.* 24(2) 189-195.
[16] Suwita I K, Kristianto Y, dan Purwaningsih F Y 2012 J. *Ilmiah Fakultas Pertanian Universitas Yudharta Pasuruan.* 3(2) 18- 35.
[17] Santoso S P 2006 *Fresh Ingredients Preservation Technology (Malang: Uwiga Publication Manuscript).*
[18] Winata E D, dan Susanto W H 2014 *JPA.* 3(1) 271-280.
[19] Ahmed R, Moushumi S J, Ahmed, Ali, and Rahmatullah H 2010 *Adv. In Nat. Appl. Sci.* 4(2) 171-173.

6. Acknowledgements

The author gives thanks for this work being supported by the Faculty of Fisheries and Marine, Universitas Airlangga.

Steam Transmission of Bioplastic Straw based Carrageenan and Gelatin

F Amirah[1], E Saputra[2], J Triastuti[2,3]

[1]Fisheries Product Technology Program, Faculty of Fisheries and Marine, Universitas Airlangga, Surabaya, Indonesia
[2]Department of Marine, Faculty of Fisheries and Marine, Universitas Airlangga, Surabaya, Indonesia

[3]Corresponding author: juni.triastuti@fpk.unair.ac.id

Abstract. Straws are included in the microplastic category, the emergence of microplastic waste that is difficult to filter so that it can be toxic and threaten aquatic ecosystems. One of the uses of carrageenan is as a raw material for making bioplastics because it has the best-film-forming ability. The purpose of this study was to determine the best concentration of gelatin in the manufacture of bioplastic straws made from carrageenan that met the standards for water vapor transmission and storability. The research method used was an experiment with a completely randomized design (CRD) with one factor which is gelatin concentration. Parameters observed were moisture content, water vapor transmission, and mass change value in the air resistance test. Data analysis using analysis of variance (ANOVA) to determine the best treatment, a further test was carried out by Duncan's Multiple Range Test with a 95% confidence level. The best result from the four treatments were found in gelatin addition concentration of 0.30 grams. The results showed that the concentration of gelatin affects the yield of bioplastic straws starting from the value of water content, the value of the water vapor transmission test, and the value of the change in mass.

1. Introduction

Plastic waste is currently one of the global problems because some plastics made from petroleum are difficult or even biodegradable in a short period, so they can cause ecosystem damage [1]. Plastics are composed of long polymer chains and are produced on a large scale because of their low production costs and variations in the mechanical characteristics of their application. However, the basic material of plastic consists of polypropylene and polystyrene, making plastic difficult to recycle [2]. In the 1950s plastics were produced on a large scale, which continued to increase to 359 million tons. While 79% of them end up as waste in landfills or nature [3]. Plastic becomes industrial waste in the environment because it is a type of polymer that has resistance to biological decomposition due to its huge molecular weight, this is due to the number of aromatic ring arrangements and complex compound bonds [4]. Straws fall into the macro plastic category, which is a type of plastic waste piece that is more than 5 millimeters long. Macro plastics can decompose into microplastics within a certain period. However, the next problem is the emergence of microplastic waste that is difficult to filter so that it can be toxic and threaten aquatic ecosystems [5].

Although various types of renewable raw materials have been found that have the potential as materials for bioplastic production, there are still other challenges related to the limited availability of raw materials and also the existence of cultivation or synthesis. Seaweed is one of the raw materials that have the potential to be processed into bioplastics. Compared to plant sources on land, seaweed has several advantages such as easy cultivation, low production costs of raw materials, and abundance in nature. In addition, seaweed does not require the provision and clearing of land to grow crops or compete with food production [6]. Red algae (Rhodophyta) is a seaweed that produces polysaccharides through the extraction of carrageenan. Carrageenan are classified into several types such as Kappa (κ), Iota (ι), and Lambda (λ) depending on the amount and arrangement of sulfate esters of 3,6-anhydrous-D or L-galactose. -carrageenan is known to be a type of carrageenan that has the best film-forming ability and is close to commercial plastic products [7]. In the manufacture of bioplastics, materials that are easily degraded are needed, such as gelatin. Easily degradable gelatin is generally used in various applications to manufacturing products that canto decomposes naturally [8]. This study was purposed to determine the effects of the best gelatin concentration addition in straw bioplastic manufacture carrageenan-based which met the standards of water vapor transmission rate and storability.

2. Material and Method

2.1 Materials

Materials used in this study were 90 grams of carrageenan obtained from PT. Kappa Carrageenan Nusantara, 3 liters of distilled water, and 18 grams of pure gelatin obtained from Rousselot Gelatin Co.,Ltd.

2.2 Place and time of research

This reseach was conducted at the Chemistry and Analytics Laboratory of The Faculty of Fisheries and Marine, Universitas Airlangga, Surabaya. This research was carried out in November 2021 until March 2022.

2.3 Reseach design

This study using the experimental method with a completely randomized design (CRD). This study uses five treatments and each treatment was replicated four times. The treatments used is a different addition of gelatin concentration, the treatments were 3 grams of carrageenan without gelatin addition (P0), 0.3 grams gelatin addition (P1), 0.6 grams gelatin addition (P2), 0.9 grams gelatin addition (P3), 1.2 grams gelatin addition (P4).

2.4 Methods

2.4.1 Production of Straw Bioplastic. The production of bioplastic straws refers to [9] where carrageenan is weighed as much as 3 grams for each treatment and given a predetermined concentration of gelatin. Then mixed and dissolved with 100 ml of distilled water under room temperature conditions of 25-30°C. Polyblend was stirred with a glass stirring rod. Homogenization of the polyblend solution was carried out for 15 minutes using a magnetic stirrer and the solution was heated to a temperature of 60°C with a hot plate, the temperature of the solution was controlled every 5 minutes. After reaching a temperature of 60°C, gelatin powder was added to the solution, and homogenization was continued until a gel was formed with a maximum temperature of 80°C. Polyblend was formed using a straw mold with a diameter of 2.5 cm with a length of 20 cm and left to harden for 15 minutes. After hardening, the bioplastic straws were dried using an oven at 45°C for 48 hours [10].

2.4.2 Water Vapor Transmission Rate Test. The water vapor transmission rate test was conducted according to ASTM (American Standard Testing and Material) (1989) by preparing a 10 cm^2 sample 4 times for 5 treatments, then weighing as the initial weight of the sample. Then the bioplastic sheet is placed on the surface of a porcelain cup containing 10 grams of silica gel and put in a desiccator for 24 hours. The calculated value is obtained from the increase in sample mass after testing, time, and area with the calculation formula (1):

$$WVTR = \Delta W / A \times t \quad (1)$$

Where WVTR= Water Vapor Transmission Rate, ΔW= Difference in film weight after being stored in a porcelain cup (grams), A= Surface Area (m^2), and t= testing time (day).

2.4.3 Air Resistance Test. The air resistance test was carried out referring to [11] by placing bioplastic straws for each treatment in an open room. Each treatment used 4 repetitions then weighed to record the initial mass before testing (m_0), then physical observations and data collection on sample weight were carried out for 30 days The mass of the sample will be observed on day 0, 5, 10, 15, 20, 25, until day 30 by testing the moisture content and then calculated as the final mass (m_t).

2.4.4 Moisture Analysis. Moisture analysis of bioplastic straws were measured using moisture analyzer. Bioplastic straw was weighed 0.75 grams then drying is carried out in the tool until the moisture content appears on the moisture analyzer screen so that the lost moisture content is known and the moisture percentage is obtained.

2.5 Data Analysis.

The result data in the form of water vapor transmission test were analyzed using Analysis of Variance (ANOVA) with 95% accuracy level. If the addition of gelatin showed significant results, further tests were carried out using Duncan's Multiple Range Test (DMRT).

3. Results and Discussion

3.1 Bioplastic Straws

The result of bioplastic straws has a different appearance, which is directly proportional to the amount of gelatin concentration added. It affects the color and level of stiffness in each treatment. The higher the concentration of gelatin, the color of the bioplastic straws will be more feculent and turning yellow, also will be harder at the level of stiffness. The results of the bioplastic straws produced can be seen in Figure 1.

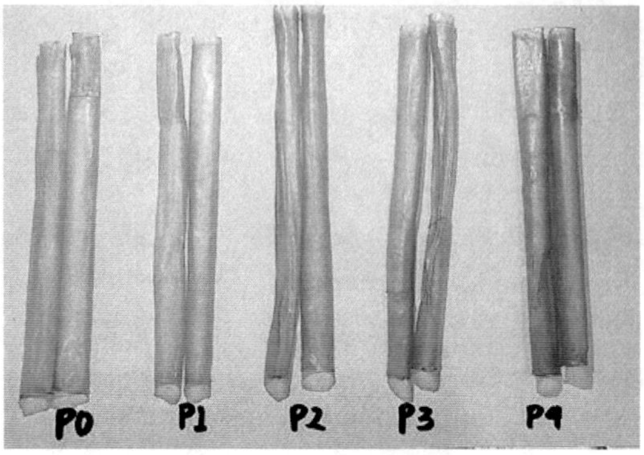

Figure 1. Bioplastic Straw.

3.2 Air Resistance

Air resistance is carried out by testing the air resistance of bioplastic straws to determine the product's resistance to air contained in an open environment. The air resistance test is characterized by a response in the form of changes in shape and color when the sample is treated. It can be seen in Table 1.

Table 1. Mass Change of Bioplastic Straws.

Treatment	Air Resistance
P0	1.703 ± 0.025^{a}
P1	1.855 ± 0.028^{b}
P2	2.126 ± 0.024^{c}
P3	2.154 ± 0.027^{c}
P4	2.261 ± 0.033^{d}

Description: Different superscript means that there is significant difference ($p<0.05$)

The addition of gelatin in the manufacture of bioplastic straws affects the air resistance value of bioplastic straws. Air resistance is one of the important parameters to determine the properties of bioplastics. The lower the degree of impermeability and air resistance of bioplastics, the faster the degradation process. The results of the 30-day test found weight gain in all treatments, which is 0.152 - 0.558 grams from the initial weight. Based on the data obtained, sample P1 has the least weight gain average value of 0.152 grams and sample P4 is 0.558 grams of all gelatin treatments. The addition of weight has a correlation with the level of resistance of the material to open air, the less weight gain reduces the possibility of damage to bioplastics in the storage process. The results of this study are supported by [12], the higher the value of the material's resistance to the environment, the better the bioplastic produced. The humidity factor can also affect the test results because the amount of water content is directly proportional to the growth rate of bacteria causing degradation. Bacteria that cause material damage can grow in environmental conditions with high water content and humidity, resulting in the breaking of protein constituent bonds and accelerating the degradation process [13]. The best air resistance test value was obtained in sample P1 with the addition of 0.3 grams of gelatin and the lowest value was obtained in sample P4 with the addition of 1.2 grams of gelatin. In line with the increase in the concentration of gelatin used.

3.3 Moisture Content

The water content of the bioplastic straws is known by the water content test to determine the weight of the dry sample, analyze the amount of liquid content of a sample, and the level of damage to the sample. The best water content with the lowest average is the addition of 1.2 grams of gelatin, namely sample P4 with an average water content value of 8.806 ± 1.914% which is shown in Table 2.

Table 2. Moisture Content of Bioplastic Straws.

Treatment	Moisture Content
P0	15.775 ± 1.458^{c}
P1	12.830 ± 1.070^{b}
P2	12.090 ± 1.592^{b}
P3	10.224 ± 1.005^{a}
P4	8.806 ± 1.914^{a}

Description: Different superscript means that there is significant difference ($p<0.05$)

The moisture content of protein-based bioplastics can affect the mechanical characteristics of bioplastics. The molecular structure and degree of denaturation of gelatin depend on the interactions between water and protein molecules. Consequently, moisture content has a great influence on the mechanical properties of bioplastics. The results of the test using a moisture analyzer showed that the sample P4 or the addition of 1.2 grams of gelatin had the lowest average value of water content, which was 8.806%. This theory is supported by [14] that the water content in the test sample can be influenced by several factors such as the amount of gelatin concentration added, the thickness of the product, and the drying process.

Gelatin is composed of proteins that have ion groups and certain parts of the collagen sequence that contain hydrophilic and hydrophobic amino acids. Both hydrophilic and hydrophobic parts tend to move to the surface, thereby lowering the surface tension of the solution [15]. The characteristic properties of gelatin are expected to help maintain the stability of the molecular structure of the combination of carrageenan and gelatin and improve the quality of the resulting bioplastic products.

3.4 Water Vapor Transmission Rate

A water vapor transmission rate test was conducted to evaluate the ability of water vapor transfer through semi-permeable and permeable sample surfaces. The water vapor transmission rate is the amount of water vapor that passes through the surface of the material. Factors that can affect water vapor transmission on a surface, namely pore size and hydrophilicity, the polarity of bioplastic raw materials, and cationic properties of bioplastics can also increase the WVTR value [16]. The addition of gelatin can improve the mechanical properties of bioplastic straws so that it affects the ability of the resulting product to absorb moisture so that the shelf life of the product can be determined [17,18].

The lowest decrease in the value of the water vapor transmission rate can be seen in Table 3. Based on the table, shows that sample P1 with a value of 0.000078 g/m2.day has decreased. This decrease in value could be caused by the addition of gelatin concentration in the carrageenan thereby increasing the compactness of the film [19]. The value of the water vapor transmission rate increased to P3 treatment with a value of 0.001400g/m2.day but again there was a decrease in P4 due to the saturation of the gelatin used. This saturation is caused by the capacity to use gelatin and the use of the solution material received during the sample-making process. The addition of additives in the form of gelatin to bioplastics can reduce the value of the water vapor transmission rate with a certain concentration. In accordance with [20] the higher the concentration of gelatin is directly proportional to the amount of water that can be bound to the bioplastic, so that the amount of water vapor that can pass through the bioplastic increases. The water vapor transmission value of carrageenan-based bioplastic from the test results can be seen as the results in Table 3.

Table 3. Water Vapor Transmission Rate of Bioplastic Straws

Treatment	WVTR
P0	$0,000324 \pm 0,00027^{ab}$
P1	$0,000078 \pm 0,00066^{a}$
P2	$0,000730 \pm 0,00016^{c}$
P3	$0,001400 \pm 0,00024^{d}$
P4	$0,000536 \pm 0,00049^{bc}$

Description: Different superscript means that there is significant difference ($p<0.05$)

Water vapor transmission is an important factor in the shelf life of bioplastic-based products. This can be a determinant of the quality of bioplastics because excessive levels of loss or excess moisture can result in a decrease in product quality and reduce shelf life. The water content of protein-based membranes can be influenced by internal and external properties ingredient. Internal factors include amino acid composition (eg, polar and nonpolar amino acid content), heterogeneity, surface charge, pH, molecular size, and volume.

The water vapor barrier properties of protein-based bioplastics still have drawbacks when compared to commercial synthetic polymers such as polyethylene, poly-(vinylidene chloride), and polyester. Based on the results of testing the water vapor transmission rate (Water Vapor Transmission Rate) with the best value of 0.000078 g/m2.day P1, but it does not meet the ISO 1663 standard, namely the standard water vapor transmission rate on commercial plastic (rigid cellular plastic) with a value of 0.0002592 g/m2·day – 0.01728 g/m2·day. This is due to the fact that bioplastics consist of protein hydrophilic functional groups which increase the solubility value of larger water vapor molecules and hence higher permeability [21].

Bioplastics have a low impermeability value when compared to commercial synthetic polymers such as polyethylene, poly-(vinylidene chloride), and polyester, this is due to the water vapor barrier properties of protein-based bioplastics. Based on the results of testing the water vapor transmission rate, the sample that has the best value is P1 with a value of 0.000078 g/m2.day, but the value of P1 does not meet the ISO 1663 standard, namely the standard water vapor transmission rate on commercial plastics. (rigid cellular plastic) which has a value of 0.0002592 g/m2·day – 0.01728 g/m2·day. The WVTR value at P1 does not meet the ISO 1663 standard, this theory is supported by [21] states that bioplastics consist of hydrophilic protein functional groups which increase the solubility value of larger water vapor molecules and cause higher permeability. The rate of water vapor transmission is also closely related to the thickness of the bioplastic, the thicker the bioplastic, the stiffer and tougher the bioplastic will be so that the ability of the bioplastic to hold water vapor is getting better [22].

4. Conclusion

From a series of studies start from production of bioplastics straw, until testing the quality of the straws that has been carried out, There is an effect of adding gelatin to the value of water vapor transmission, moisture content, and air resistance with the best gelatin concentration of 0.30 grams.

5. References

[1] Chae Y, An Y J 2018 *Envi Poll.* 240 387-95.
[2] Polman E M, Gruter G J M, Parsons J R, Tietema A 2020 *Scie of The Tot Envi.* 141953.
[3] Geyer R, Jambeck J R, Law K L 2017 *Scie Adv.* 3
[4] Wahyuningtyas M, Atmaja L 2016 *Ind J of Chem Sci.* 5.
[5] Rice L 2019 *The Plastic Straw Movement How Green Corporate Social Responsibility Impacts Specific Ohio Restaurants* (Kent: Kent State University).
[6] Sedayu B B, Cran M J, Bigger S W 2019 *Carbo Poly* 216 287-302.
[7] Shoojae-Aliabadi S, Hosseini H, Mohammadifar M A, Mohammadi A, Ghasemlou M, Ojagh S M 2013 *Inter J of Bio Macro.* 52 116 – 24.
[8] Sathisaran I, Balasubramanian M 2020 *Heli* 6.
[9] A'yun S N 2020 *Effect of Bioplastic Formulations Based on Carrageenan and Gelatin on the Production of Edible Straws on Water Resistance (Surabaya: Airlangga University).*
[10] Arini D, Ulum M S, Kasman K 2017 *Nat Sci J of Scie and Tech.* 6.
[11] Febriyantoro I, Nuriyah L 2014 *The Effect of Composition of Starch Banana Peel and Cassava Peel as Bioplastic Raw Materials and Measurement of Their Characteristics (Malang: Brawijaya University).*
[12] Syura I 2020 *Manufacture and Characterization of Porang Starch (Amorphophallus sp.) and Chitosan Bioplastic Films with Plasticizer Sorbitol (Medan: University of North Sumatra).*
[13] Santosa B, Saputra D, Pambayun R 2004 *Tek dan Ind Pang.* 15.

[14] Susilawati S, Rostini I, Pratama R I, Rochima E 2019 *World Scientific News* 135 85 – 98.
[15] Remawati, Betha O D 2016 *Extraction and Characterization of Gelatin from Cow Skin Using Acid Hydrolysis Method (Jakarta: Syarif Hidayatullah State Islamic University).*
[16] Suryanegara L, Fatriasari W, Zulfiana D, Anita S H, Masruchin N, Gutari S, Kemala T 2021 *J of Envi Heal Scie and Engine.* 19 415 – 25.
[17] Garcia N L, Ribba L, Dufresne A, Aranguren M, Goyanes S 2011 *Carbo Poly.* 84 203 – 10.
[18] Gonzalez K, Retegi A, Gonzalez A, Eceiza A, Gabilondo N 2014 *Carbo Poly.* 117 83 – 90.
[19] Yi J B, Kim Y T, Bae H J, Whiteside W S, Park H J 2006 *J of Foo Scie.* 71.
[20] Agustin Y E, Padmawijaya K S 2016 *J Tek Kim.* 10 43 – 51.
[21] Bourtoom T 2009 *Inter Foo Res J.* 16 1 – 9.
[22] Jacoeb A M, Nugraha R, Utari S 2014 *J Pengo Has Per.* 17 14 – 21.

6. Acknowledgements

The author gives thanks for this work being supported by Faculty of Fisheries and Marine, Universitas Airlangga. The author extends thanks to Mr. Eka Saputra and Mrs. Juni Triastuti for supporting author to finish the research.

Performance analysis of *Brachionus plicatilis* production with different densities

M B Santanumurti[1,2], D Puspita[3], S Hudaidah[4,5]

[1]Department of Aquaculture, Faculty of Fisheries and Marine, Universitas Airlangga, Surabaya, Indonesia
[2] Department of Marine Biology, Faculty of Marine Sciences, King Abdulaziz University, Jeddah, Kingdom of Saudi Arabia
[3]Aquaculture Program, Faculty of Agriculture, University of Lampung, Lampung, Indonesia
[4]Department of Fisheries and Marine Science, Faculty of Agriculture, University of Lampung, Lampung, Indonesia

[5]Corresponding author: E mail: siti.hudaidah@fp.unila.ac.id

Abstract. Plankton is an aquatic animal that has many benefits such as cosmetics, food, animal feed, and even biodiesel. One of the plankton that has good potential is rotifers (*Brachionus plicatilis*). Rotifers have the advantage that small morphology (no need much space), reproduce quickly, and easy to be cultured. This study aims to determine the effect of differences in density on the performance of rotifers production. The treatments in this study were A (500 kg), B (density 2 tons), and C (density 8 tons). The results showed that the highest density value was found on the 5th day of treatment B (574 ind/ml). This indicates that the culture of rotifers on a mass scale of 2 tons showed the highest value on the 5th day.

1. Introduction

In Indonesia, a lot of fish farming activities are carried out because the market demand for ornamental fish and fish consumption of fresh water and sea water is increasing. At this time, good quality ornamental fish and consumption fish seeds are needed and can grow quickly, so the feed given must also be in accordance with their needs. Aquatic organisms, especially fish/shrimp, need feed that can support their growth and survival. The feed provided can be in the form of artificial feed made of various materials in the form of flour and composed according to the needs of the organism, and there is natural feed in the form of live feed derived from catches in nature or from cultivation.

Natural feed is feed whose availability is produced from nature and can be cultivated by humans. Natural food consists of two types, namely phytoplankton which are aquatic plants and zooplankton which are aquatic animals. Rotifera (*Brachionus plicatilis*) is a zooplankton that is small in size and has the ability to reproduce quickly, and has a density that continues to increase every day, so it can be cultured in large volumes. The size of *Brachionus plicatilis* given to fish/shrimp fry is 40-250 m [1].

The development of *Brachionus plicatilis* culture as a support for aquaculture has been widely carried out in Indonesia, one of which is at the Center for Marine Aquaculture (BBPBL) Lampung. Cultures of Brachionus plicatilis at BBPBL were carried out in various scales, from lab scale to mass scale. The culture of Brachionus plicatilis in general practice is carried out on an intermediate and semi-mass scale, which is expected to increase knowledge and experience that will be useful later.

2. Material and methods

2.1. Sample preparation

This research was conducted at BBPBL Lampung at Jalan Yos Sudarso, Hanura Village, Teluk Pandan District, Pesawaran Regency, Lampung Province. Materials used in the culture of *Brachionus plicatilis* from BBPBL Lampung. The feed given to the culture of *Brachionus plicatilis* was phytoplankton with the species *Nannochloropsis oculata* as much as 1/3 part of the culture container. The treatments in this study were A (500 kg), B (density 2 tons), and C (density 8 tons).

2.2 Research parameter calculation

The parameters in this study were plankton density and water quality. The calculation of the population density of *Brachionus plicatilis* in ind/ml was carried out by taking a sample of ± 20 ml using a 50 ml beaker. Added as much as 6 drops of iodine solution on the surface of the Sedgwick rafter cell. Take and place 1 ml or 20 drops of the sample onto the surface of the Sedgwick rafter cell using a dropper. The sedgwick rafter cell was closed using a glass slide. Place the Sedgwick Rafter cell under a microscope glass. View the density of *Brachionus plicatilis* using a microscope with a magnification of 4 times. Count the number of individual *Brachionus plicatilis* using a hand counter. Parameters observed and measured were temperature, pH, and DO. Water quality measurements (temperature, pH and DO) were carried out once every day for 25 days of maintenance. The method used in the implementation of general practice is the collection of primary data obtained from observations, interviews and active participation as well as secondary data obtained from literature studies ranging from books, journals, reports, theses and so on. Primary data is data that is directly related to field activities, which is obtained from the first source in the form of words, actions, and other data such as documentation, unprocessed information and by making observations [2]. Secondary data is data obtained or collected by researchers from various existing sources [2].

3. Result and discussion

The results of the density of *Brachionus plicatilis* were presented in Table 1. Based on the results, it indicated that on the first day of culture was an adaptation phase or the lag phase. The population density of *Brachionus plicatilis* was still small with a range between 70-110 ind/ml. It was happened because *Brachionus plicatilis* was still in the adjustment stage to the new environment. Previous study stated that *Brachionus plicatilis* experienced low growth and reproduction at the beginning of culture, and lasted for about 2 days due to the adjustment phase to the new culture media environment [3].

On day 3-4, *Brachionus plicatilis* in each culture container, except for a capacity of 2 tons, experienced an exponential phase or rapid growth, ranging from 200-400 ind/ml. An increase in population density was another form of adaptability to feed and a new environment and is a form of energy transfer for the increase in reproductive metabolism. In contrast to other culture containers, the culture container with a capacity of 2 tons of *Brachionus plicatilis* only experienced an exponential phase on day 5 which had the highest total population density [3]. The difference in the location of this exponential phase was thought to be because at the time of culture, *Brachionus plicatilis* with a capacity of 2 tons had the most optimal water temperature. High and stable ambient temperature could increase the density of Brachionus plicatilis.

On day 5 or 6, *Brachionus plicatilis* underwent a stationary phase, i.e. the number of cells tended to be constant and there was no visible increase in population size, resulting in a decrease in nutrients and light intensity of culture [4]. On day 7, *Brachionus plicatilis* experienced a phase of mortality (death). In this phase, no data was collected because of the small population density. The death phase occurred because of the denseness of *Brachionus plicatilis*, so that the space for movement becomes narrow. The lack of space to move caused disruption of the activity of *Brachionus plicatilis* because there can be competence in fighting for food, and changes in water quality in culture media caused by uneaten feed residue. Previous study stated that the decrease in population density in *Brachionus plicatilis* was caused by changes in water quality in the culture media that occurred due to the presence of uneaten feed residues and the absence of water changes [5].

Table 1. *Brachionus plicatilis* density in this study

Day	Total Density in Sample (ind/ml)			Total Density in Culture Container ($\times 10^6$ ind/vol)		
	A	B	C	A	B	C
1	64	71	98	512	142	784
2	127	137	161	1.016	274	1.288
3	279	154	217	2.232	308	1.736
4	201	389	141	1.608	778	1.128
5	113	574	110	904	1.148	880
6	73	351	82	584	702	656

Counting the density of *Brachionus plicatilis* was carried out every day during culture. The density calculation aims to determine the adaptation phase, the exponential phase or the so-called log phase (logarithmic), the stationary phase and the death phase experienced by *Brachionus plicatilis*. The following graph of the density of *Brachionus plicatilis* in culture containers can be seen in Figure 1.

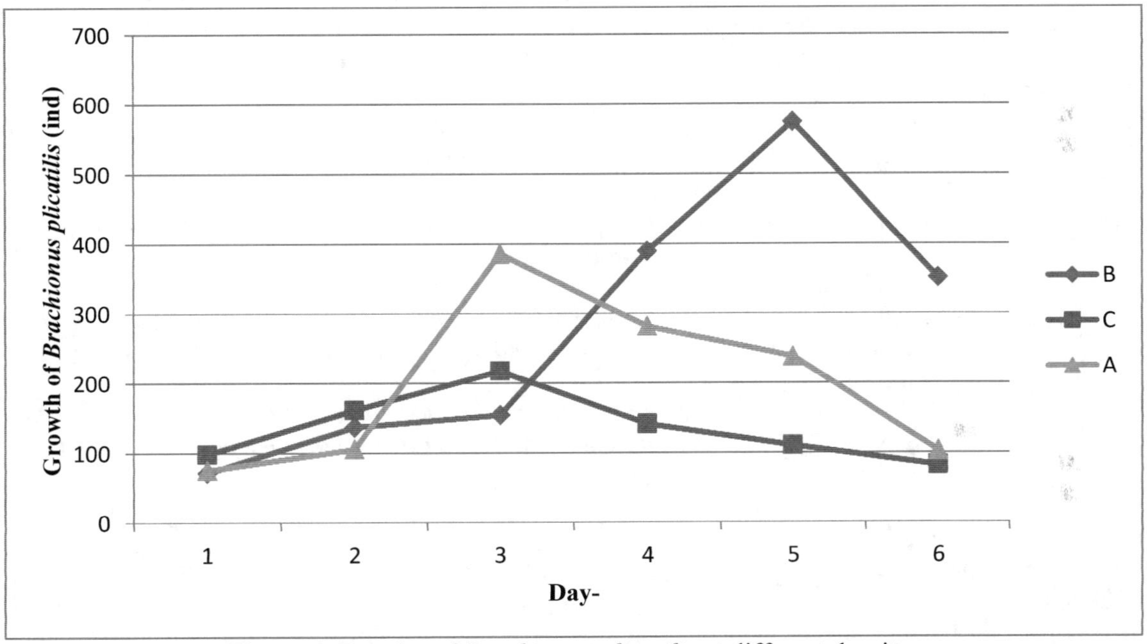

Figure 1. Density of *Brachionus plicatilis* at different density

Measurement of water quality in culture media was carried out at the end of culture before harvesting. The results of water quality measurements can be seen in Table 2. Water quality was important to support the success of *Brachionus plicatilis* culture. Water quality parameters taken in semi-bulk scale of *Brachionus plicatilis* culture consisted of temperature, pH, salinity and DO. Based on secondary data obtained regarding water quality in *Brachionus plicatilis* culture, it could be seen that temperature, pH, salinity and DO were included in the range of good water quality for culture media. If seen in the table, the temperature on the semi-bulk scale *Brachionus plicatilis* culture media ranged from 25-28°C. This value was still in a good range for growth and reproduction of *Brachionus plicatilis*. *Brachionus plicatilis* could grow well in a temperature range of 20-30°C [3].

The pH of the semi-bulk scale *Brachionus plicatilis* culture media ranged from 8-10. This value was a good range for culture. Previous study reported that the pH of seawater in the range of 7-10.5 was a good value for the growth of *Brachionus plicatilis* [6]. The salinity of the semi-bulk scale

Brachionus plicatilis culture media ranged from 31-32 ppt. This value was a good range for the growth and reproduction of *Brachionus plicatilis*. *Brachionus plicatilis* could grow well at a salinity of 20-32 ppt [3]. DO (dissolved oxygen) in semi-bulk scale *Brachionus plicatilis* culture media ranged from 3-5 ppm. This value was a good range for the growth and reproduction of *Brachionus plicatilis*. Previous study reported that *Brachionus plicatilis* can grow well in dissolved oxygen ranging from 2-7 ppm [3].

Table 2. Water quality parameter in this study

Treatment	Parameter			
	Temperature (°C)	pH	Salinity (ppt)	DO (ppm)
A	27-28	9-10	32	4-5
B	26-27	9-10	32	4-5
C	25-26	8-9	31	4-5

4. Conclusion

The results showed that the highest density value was found on the 5th day of treatment B (574 ind/ml). This indicates that the culture of rotifers on a mass scale of 2 tons showed the highest value on the 5th day. It is recommended that using 2 tons of density for rotifers to get optimum value.

5. References

[1] Budi S, dan Aslamsyah S 2011 *JAI.* 10 67-73.
[2] Saputra A R S H, dan Samara S H 2022 *IOP Conf Ser Earth Env Sci*. 1036 012118.
[3] Yoan O 2018 . Malang: Universitas Brawijaya.
[4] Aulia N, Indah R, dan Natsir H 2016 *Quality of Marine Microalgae Chlorella vulgaris as Producer of EPA and DHA Rich Biomass for Sausage Fortification (So-Fit).* Makassar: Hassanudin University.
[5] Yudha A A, Agustriani F, dan Isnaini I 2012 *Provision of microalgae to the population growth of Rotifera (Branchionus plicatilis) on a laboratory scale in BBPBL Lampung.* Palembang: Sriwijaya University.
[6] Genting B P 2012 *Analysis of Water Quality in Mass-Scale Culture of Rotifera Zooplankton (Brachionus plicatilis) at the Lampung Marine Aquaculture Development Center.* Lampung: University of Lampung.

6. Acknowledgement

The authors gratefully acknowledge the collaboration from Universitas Airlangga and the University of Lampung.

IOP Conf. Series: Earth and Environmental Science 1273 (2023) 012032 doi:10.1088/1755-1315/1273/1/012032

Analysis of ram cichlid (*Mikrogeophagus ramirezi*) hatchery performance at Fisheries Aquaculture Laboratory, Faculty of Agriculture, University of Lampung

MB Santanumurti[1,2], Y Sulistyawati[3], S Hudaidah[4,5]

[1]Department of Aquaculture, Faculty of Fisheries and Marine, Universitas Airlangga, Surabaya, Indonesia
[2] Department of Marine Biology, Faculty of Marine Sciences, King Abdulaziz University, Jeddah, Kingdom of Saudi Arabia
[3]Aquaculture Program, Faculty of Agriculture, University of Lampung, Lampung, Indonesia
[4]Department of Fisheries and Marine Science, Faculty of Agriculture, University of Lampung, Lampung, Indonesia

[5]Corresponding author: E mail: siti.hudaidah@fp.unila.ac.id

Abstract. Ram cichlid (*Mikrogeophagus ramirezi*) is one of the ornamental fish that has good development prospects. This is because this freshwater ornamental fish has a very attractive color. This study aims to determine the hatchery performance of ram cichlids at the Fisheries Aquaculture Laboratory, Faculty of Agriculture, University of Lampung, one of the places where this ornamental fish is developed. The results showed that the fertilization rate was 73%, the hatching rate was 62% and the survival rate was 78%. This shows that hatchery activities at the Fisheries Aquaculture Laboratory, Faculty of Agriculture, University of Lampung still need to be developed.

1. Introduction

Indonesia has the potential to become a world exporter of ornamental fish with increasing demand for ornamental fish from importing countries, namely Singapore, Japan, the United States, Malaysia and China. Indonesia has a biodiversity of ornamental fish from both fresh and marine waters [1]. The development of ornamental fish in Indonesia continues to increase, especially freshwater ornamental fish. According to previous study, for the 2014-2019 period, the trend of the export value of Indonesian ornamental fish increased by an average of 10.2% per year, with the dominance of freshwater fish species (73.7%) [2]. The increase in the selling power of ornamental fish becomes an opportunity for aquaculture activities and makes it a very promising business prospect.

One of the ornamental fish that has good prospects for development is ramirezi fish (*Mikrogeophagus ramirezi*). This ornamental fish is a freshwater ornamental fish that is in great demand by ornamental fish lovers in Indonesia. According to previous study, ramirezi fish or Dwarf

Cichlids come from South America [3]. One of the attractions of ramirezi fish is its size and color which is very attractive. Based on KKP data (2018), the production of ramirezi increased by 31.1% on average, namely 81,284 heads in the 2017-1018 period. The increase in the production capacity of ramirezi is projected to continue to increase until it reaches 129,734 tails in 2024 [4]. Efforts that can be made to meet increasing market demand are by conducting ramirezi ornamental fish cultivation activities, the success of a cultivation activity is strongly influenced by several factors including parent handling, spawning, egg management, larval management, seed management and water quality. Therefore, it is very necessary to know about aquaculture, especially in ramirezi fish hatchery techniques.

2. Material and methods

2.1. Sample preparation

This research was conducted in the hatchery of the Aquaculture Laboratory, Faculty of Agriculture, University of Lampung (5°21'54.5"S 105°14'28.7"E). The parent used was ramirezi parent with electric blue ram variety 2.5 months old, weight 0.92±0.16 g and length 3.23±0.25 cm. The number of broodstock used was 30, consisting of 12 males and 18 females. Feeding was carried out with a frequency of 3 times a day, 08.00 AM (silk worm), 01.00 PM (TetraBits), and 04.00 PM (silk worm). The research was carried out for 25 days.

2.2 Research parameter calculation

Observations of fertilization rate, hatching rate, survival rate, growth length, weight were carried out at the beginning and end of the study. Weighing of the test fish was carried out every 7 days using a measuring instrument in the form of a scale. According to previous study, the formula for calculating the weight gain of the test fish was as follows [5]:

$$W = Wt - Wo$$

W : Weight growth (gram)
Wt : Final average weight (gram)
Wo : Initial average weight (gram)

Absolute length measurements were carried out every 7 days. The absolute length growth according to previous study formula was as follows [5]:

$$P = Pt - Po$$

P : Absolute length growth (cm)
Pt : Final length (cm)
Po : Initial lenght (cm)

The number of dead fish was counted every day until the end of the study. According to previous study the survival rate of fish was calculated using the formula [5]:

$$SR = \frac{Nt}{No} \; x \; 100\%$$

SR : Survival rate (%)
Nt : Final number of fish (fishes)
N0 : Initial number of fish (fishes)

The number of fertilization rate was calculated using the formula [6]:

$$Fertilization\ (\%) = \frac{Number\ of\ fertilized\ eggs\ in\ sample}{Number\ of\ total\ eggs\ in\ sample}\ x\ 100$$

The number of hatching rate was calculated using the formula [6]:

$$Hatching\ (\%) \frac{Number\ of\ hatching\ in\ sample}{Total\ number\ of\ fertilized\ eggs\ in\ sample} x100$$

Parameters observed and measured were temperature, pH, and DO. Water quality measurements (temperature, pH and DO) were carried out once every day for 25 days of maintenance.

3. Result and discussion

Based on the observations, the body weight growth value of ramirezi fish was obtained in Figure 1. At week 1, the average body weight of ramirezi broodstock was 3 0.92±0.11 g. In the 2nd week there was a significant increase in body weight of ramirezi broodstock with an average of 3.47±0.20 g. In the 3rd week there was a constant growth of fish brood weight of 3.53±0.20 g.

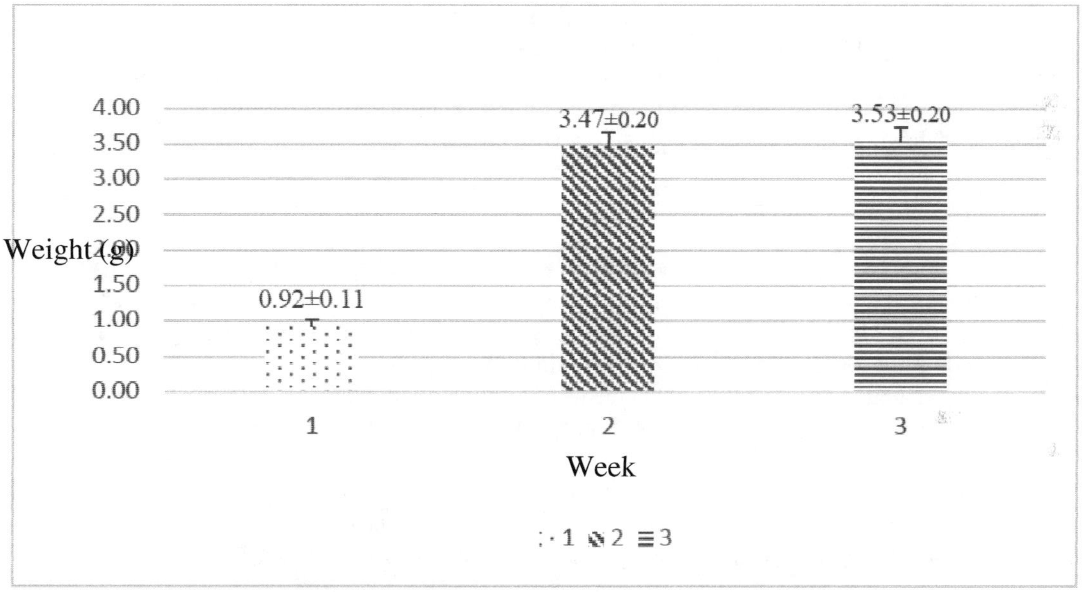

Figure 1. Growth of body weight of ramirezi fish (*M. ramirezi*)

Based on the observations, the growth value of the body length of the ramirezi parent is shown in Figure 2. At week 1, the average body length of the ramirezi parent is 3.42±0.10 cm. In the 2nd week there was an increase in body length of ramirezi broodstock with an average of 3.47±0.32 cm. In the 3rd week there was a constant growth in length of the parent fish of 3.53±0.19 cm.

Fecundity, FR, and HR are very important indicators to know to determine the quantity and quality of larvae that will be produced by broodstock ramirezi fish at spawning time are presented in Table 1. The number of eggs produced was 179 eggs, while 130 eggs were fertilized and 49 eggs were produced. unfertilized eggs. The seeding activity also resulted in the value of the degree of fertilization (FR) which was 73%. The percentage of egg fertilization is influenced by several factors, including egg quality, fish sperm and water quality, especially temperature [7]. 81 eggs of ramirezi fish hatched with a hatching rate of 62%. According to previous study, several factors that can influence the success of hatching eggs are gonad maturity and water quality [8]. During the hatching period, the water temperature was 27.6 C, the pH value was 8.2 and the DO value was 6.1 mg/L.

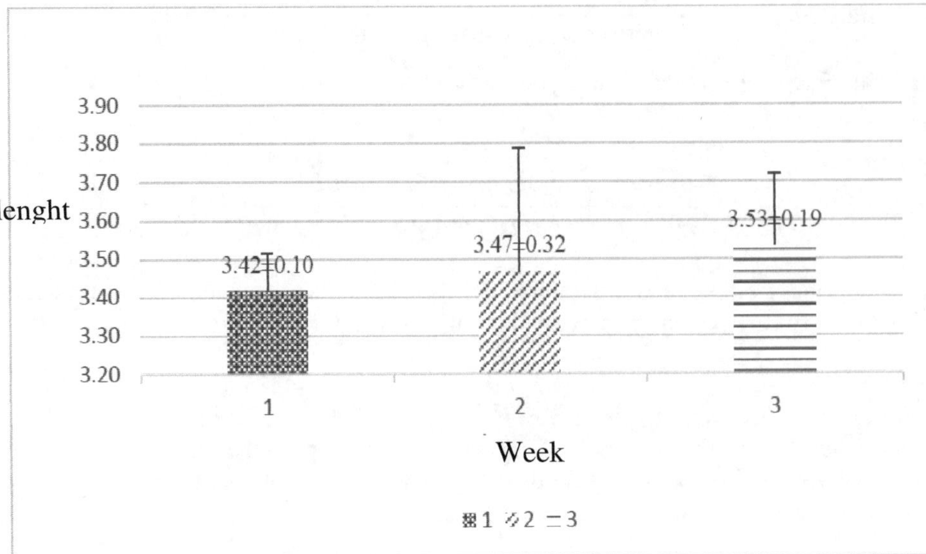

Figure 2. Growth in body length of ramirezi fish (*M. ramirezi*)

Table 1. Hatchery performance in this study

Fecundity	Fertilized Egg	Hatched Egg	Fertilization Rate	Hatching Rate (%)	Survival Rate (%)
179	130	49	73	63	78

The survival rate of ramirezi fish larvae is 78%. The survival rate of larvae is influenced by 2 factors, namely internal and external factors. Internal factors are factors that result from the quality of the larvae itself which is determined by the quality of the parent. External factors are influenced by water and feed quality, good water quality can maintain the survival of larvae.

Water quality management is a way that must be done by ornamental fish maintenance so that the media for ramirezi fish maintenance is always in optimal conditions to support the life and growth of the fish that are kept. To maintain good water quality and support the growth of ramirezi fish, it is necessary to optimize water installations or maintenance media, sedimentation of ornamental fish rearing water is carried out before the media is used in the fish rearing process, regular water quality checks, and regular water changes that can be done. every day as much as 10-25% per day depending on needs [9].

Table 2. Water quality parameter in this study

Parameter	Initial			Final			Optimum
	Water Sources	Hatchery	Broodsctock	Water Sources	Hatchery	Broodsctock	
Temperature (°C)	26.3	26.4	27.8	26	25.2	26.9	25-29
pH	8.4	6.6	6.4	8.4	7.8	6.2	6.5-7
TDS (ppm)	0.81	0.76	0.79	0.81	3.24	5.76	<50
NH_3 (ppm)	0	0	0	0	0	0	
DO (mg/L)	6.8	6.8	6.8	7	7.2	6.9	

4. Conclusion

The results showed that the fertilization rate was 73%, the hatching rate was 62% and the survival rate was 78%. This shows that hatchery activities at the Fisheries Aquaculture Laboratory, Faculty of Agriculture, University of Lampung still need to be developed.

5. References

[1] Prasetyo A B, Sinansari S 2016 *Indonesian Ornamental Fish Research and Development Center for Ornamental Fish Cultivation Depok*

[2] Kelana, Irwan 2020 *Prof Rokhmin Dahuri: Ornamental Fish Contribute IDR 500 Billion in Foreign Exchange* https://www.republika.co.id/berita/qhejy8374/profrokhmin-dahuri-ikan-hias-sumbang-devisa-rp-500-miliar

[3] Padowicz D, Harpaz S 2007 *Israeli J Aquac Bamidgeh* **59** 20536

[4] KKP 2019 *Development of Aquaculture Strategic Leading Commodities and Licensing Management to Stimulate Investment in Sustainable Aquaculture Development Ministry of National Development Planning* /BPS Double Tree by Hilton Jakarta

[5] Effendi H 2003 *Study of Water Quality for Management of Water Resources and Environment* Kanisius Yogyakarta

[6] Biswas A, Choudhury M, Farid M A, Rahman M A, Rahman M A 2021 *Annual Res Rev Biol* **36** 15-25

[7] Keshavanath P, Gangadhara B, Basavaraja N, Nandeesha M 2006 *Asian Fish Sci* **19** 411–422

[8] Saputra S F D 2011 *Application of Controlled Water Recirculation System (SRAT) in Goldfish Cultivation (Cyprinus carpio) Bogor Agricultural University (IPB) Bogor*

[9] Sudrajat M, Setyogati W 2020 *Goldfish Hatchery Chef* Deepublish

6. Acknowledgement

The authors gratefully acknowledge the collaboration from Universitas Airlangga and the University of Lampung.

IOP Conf. Series: Earth and Environmental Science 1273 (2023) 012033 doi:10.1088/1755-1315/1273/1/012033

The Effect of Probiotics to Betta Fish (*Betta splendens* Regan, 1910) Aquaculture Performance

S H Samara[1], F E Saputra[2], H P Fidyandini[3], S Hudaidah[3,4]

[1]Department of Aquaculture, Faculty of Fisheries and Marine, Universitas Airlangga, Surabaya, Indonesia
[2] Aquaculture Program, Faculty of Agriculture, University of Lampung, Lampung, Indonesia
[3]Department of Fisheries and Marine Science, Faculty of Agriculture, University of Lampung, Lampung, Indonesia

[4]Corresponding author: E mail: siti.hudaidah@fp.unila.ac.id

Abstract. Betta fish (*Betta splendens*) is one type of ornamental fish that has high economic value. However, one of the obstacles on its cultivation activity is poor nutrition feed that support the enlargement process. One of the solution is adding probiotics. Probiotics can optimize fish growth, including betta fish. This study aimed to determined the best dose of the probiotics to increased the growth performance of betta fish. This study used a completely randomized design with 4 treatments and 3 replications (C: control, A: 5 ml/kg of feed, B: 7 ml/kg of feed, and C: 9 ml/kg of feed). The data was analyzed by Anova and Duncan's test. Based on the parameters observed, it was shown that the addition of probiotics in commercial feed in treatment C (9 ml/kg feed) had a significant effect on the growth of absolute weight, SGR, and FCR.

1. Introduction

Betta fish (*Betta splendens*) is one type of ornamental fish that has its own uniqueness compared to other ornamental fish. Its uniqueness is its penchant for fighting with the same sex, and also has a variety of unique color patterns and patterns, one of the characteristics of the betta's beauty is when it shows off its tail [1]. Betta fish at the seed stage is the most important stage because betta fish need good feed to support success in the cultivation process. The development of betta fish seeds is currently still categorized as slow, this can be influenced by aspects of the feed used. Therefore, an alternative way is needed to support the growth process of betta fish, namely by giving probiotics into the feed which is known to be able to accelerate fish growth.

Probiotics are micro-organisms that have a beneficial role and are able to survive in the digestive tract. Probiotics are useful for avoiding intestinal pathogenic microorganisms and improving the efficiency of using artificial feed by releasing enzymes that support the digestive process of food in fish. The content of bacteria contained in probiotics creates several important enzymes for feed digestion in the digestive tract of fish such as amylase, protease, lipase and cellulose. These enzymes can help form the process of hydrolysis of nutrients in stored artificial feeds such as carbohydrates, proteins and fats to change into simpler molecules from the start so that it is easy to digest and absorb feed in the digestive tract of fish [2]. This study aims to determine the best dose of probiotics in feed to increase the growth of betta fish (*Betta splendens*).

2. Material and methods

2.1. Sample preparation

The materials used were probiotics (Probio-7), pellet feed (pf 500), and distilled water. The research design used was a completely randomized design. This study used 4 treatments and 3 replications using 20 betta fish for each replication. The concentration used was control treatment = feed without probiotics, A = dose 5 ml/kg feed, B = dose 7 ml/kg feed, C = dose 9 ml /kg feed. Betta fish seeds aged 1.5 months were put into each maintenance container and given feed mixed with probiotics 3 times a day at 08.00, 12.00 and 16.00 WIB.

2.2 Research parameter calculation

Observations specific growth rate, FCR (feed conversion ratio), survival rate, growth length and weight were carried out at the beginning and end of the study. Weighing of the test fish was carried out using a measuring instrument in the form of a scale. According to previous study, the formula for calculating the weight gain of the test fish was as follows [3]:

$$W = Wt - Wo$$

W : Weight growth (gram)
Wt : Final average weight (gram)
Wo : Initial average weight (gram)

Absolute length measurements were carried out every 10 days. The absolute length growth according to previous study formula was as follows [3]:

$$L = Lt - Lo$$

L : Absolute length growth (cm)
Lt : Final length (cm)
Lo : Initial lenght (cm)

The number of dead fish was counted every day until the end of the study. According to previous study the survival rate of fish was calculated using the formula [3]:

$$SR \frac{Nt}{N0} = x\ 100\%$$

SR : Survival rate (%)
Nt : Final number of fish (fishes)
N0 : Initial number of fish (fishes)

The specific growth rate was calculated using the formula [4]:

$$SGR = \frac{(LnWt - LnWo)}{t} x\ 100$$

SGR = *Spesific Growth Rate* (%/hari)
W0 = The average weight of the fish at the start of the study (g)
Wt = The average weight of the fish at the end of the study (g)
t = Maintenance period (days)

The FCR was calculated using the formula [4]:

$$FCR \frac{F}{(Wt + Wd) - Wo}$$

FCR = Feed Conversion Ratio
F = Amount of feed given (g)
Wo = Initial weight (g)
Wt = Final weight (g)
Wd = Weight of death fish (g)

Parameters observed and measured were temperature, pH, and DO. Water quality measurements (temperature, pH and DO). The data obtained from the results of the study were tabulated using the Microsoft Excel program and analyzed using the SPSS program. Data of daily growth rate, absolute weight growth, and feed conversion ratio were tested for homogeneity and normalization. If the data was homogeneous and normal, then a test using variance (ANOVA) was carried out to determine the difference between each treatment, then Duncan's test is carried out at a 95% confidence level.

3. Result and discussion

Provision of probiotics with different doses significantly affected the growth of betta fish seed as shown in Figure 1. Treatment C (0.67 ± 0.04 grams) was significantly different ($p > 0.05$) with control (0.20 ± 0.02 grams), treatment A (0.43 ± 0.04 grams) and treatment B (0.47 ± 0.02 grams). While treatment A was not significantly different from treatment B ($p > 0.05$). This indicates that the administration of the probiotic Probio-7 at a dose of 9 ml/kg of feed was more effective in increasing the absolute weight growth of betta fish. The increase in absolute weight growth in betta fish occurred because the probiotic Probio-7 given contained various bacteria such as *Lactobacillus acidophilus*, *Bacillus subtilis*, *Rhodopseudomonas*, *Aspergillus oryzae*, and *Saccharomyces cerevisiae*. The addition of the probiotic Probio-7 with the greatest concentration (9 ml/kg feed), so that the probiotic bacteria in the fish intestine dominates more than the pathogenic bacteria.

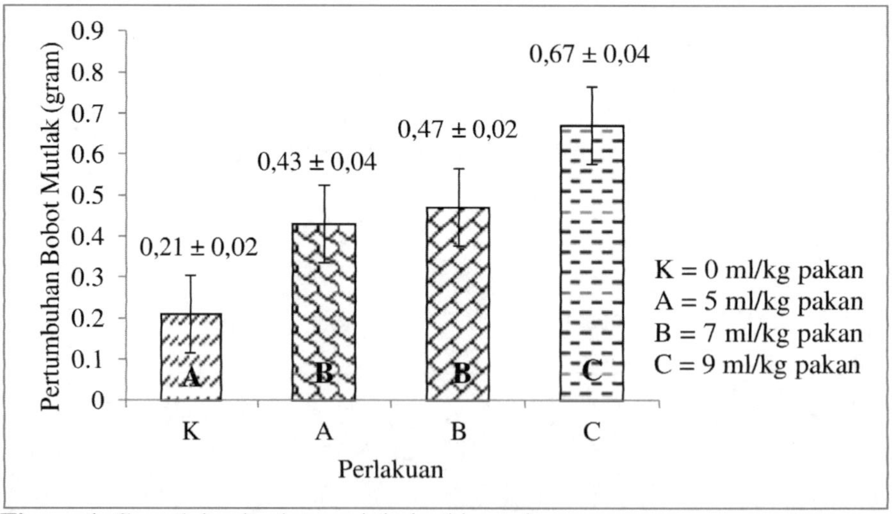

Figure 1. Growth in absolute weight in this study.

Bacillus subtilis is able to improve feed quality by increasing the concentration of proteases in the digestive tract, improving growth and reducing the number of potentially pathogenic bacteria in the digestive tract of fish [5]. The administration of probiotic bacteria containing *Lactobacillus* sp. can provide benefits to the host animal through increasing appetite, increasing gut microbes, synthesizing vitamins and stimulating the immune system [6]. *Aspergillus oryzae* is a type of fungus that is commonly used to increase the nutritional value of feed ingredients, especially to increase protein content. Apart from being non-pathogenic, *Aspergillus oryzae* is also known as a mold that produces the most enzymes, namely -amylase, -galactosidase, glutaminase, protease, and -glucosidase [7].

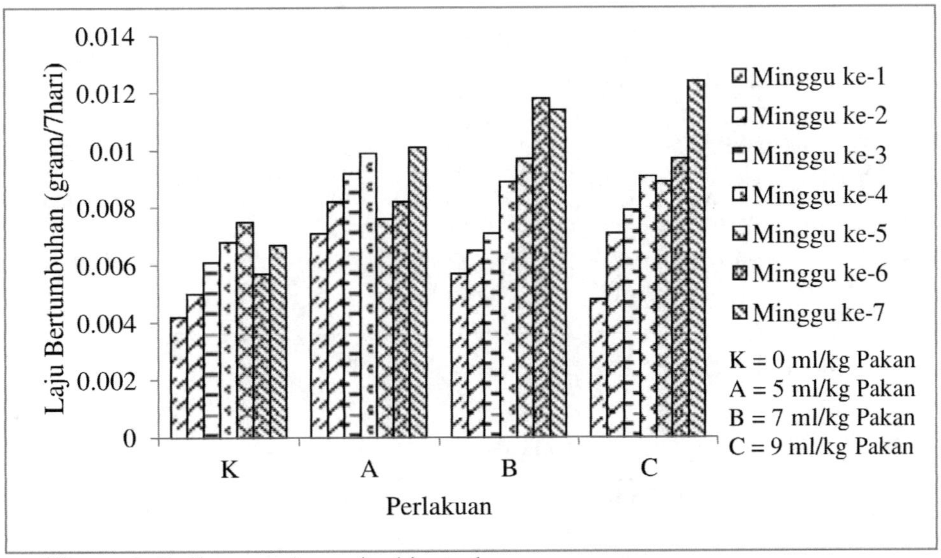

Figure 2. Daily growth rate in this study

The workings of the probiotic bacteria itself is by attaching and colonizing the intestinal wall, probiotic bacteria will compete with pathogenic bacteria, probiotic bacteria will multiply and dominate more than pathogenic bacteria, so that the growth of pathogenic bacteria will be inhibited by probiotic bacteria. These probiotic bacteria are also able to produce antimicrobial substances that are able to kill pathogenic bacteria, thus making the process of digestion and absorption of food essences easier.

The highest daily growth rate was at week 7, which was 0.0124 gram/7days significantly different ($p < 0.05$) treatment B with a value of 0.0114 grams/7 days, treatment A with a value of 0.0101 grams/7 days and control treatment with a value of 0.0067 grams/7 days (Figure 2). The addition of probiotic bacteria to feed will affect the performance of digestive enzymes, so that the digestive process and absorption of feed nutrients by betta fish is better, the use of probiotics in aquaculture can increase fish immune response, resistance to pathogens, survival, growth, digestion and feed efficiency and appetite. [8]. Fish will utilize feed nutrients to be stored in the body and convert into energi [9]. This energy is used by fish for basic metabolism such as movement, production of sexual organs, maintenance of body parts and replacement of damaged cells and more for growth.

The survival rate of betta fish was not significantly different, namely 100 ± 0% ($p>0.05$) (Figure 3). The high level of survival in fish is influenced by environmental conditions and handling at the time of taking water also affects and the maintenance media is still in the optimal category to support the maintenance of betta fish. The addition of probiotic Probio-7 was also able to increase the survival value of fish. This is because the probiotic Probio-7 contains *Nitrobacter* bacteria. *Nitrobacter* bacteria are nitrifying bacteria, because they are bacteria that convert nitrite to nitrate. Nitrification is a biological oxidation process that converts ammonium to nitrate which occurs through a two-step reaction. *Nitrobacter* bacteria play a role in the nitrogen cycle by oxidizing nitrite which is the result of the oxidation of *Nitrosomonas* bacteria to nitrite. *Nitrobacter* uses the oxidation energy of the nitrite ion to become nitrate.

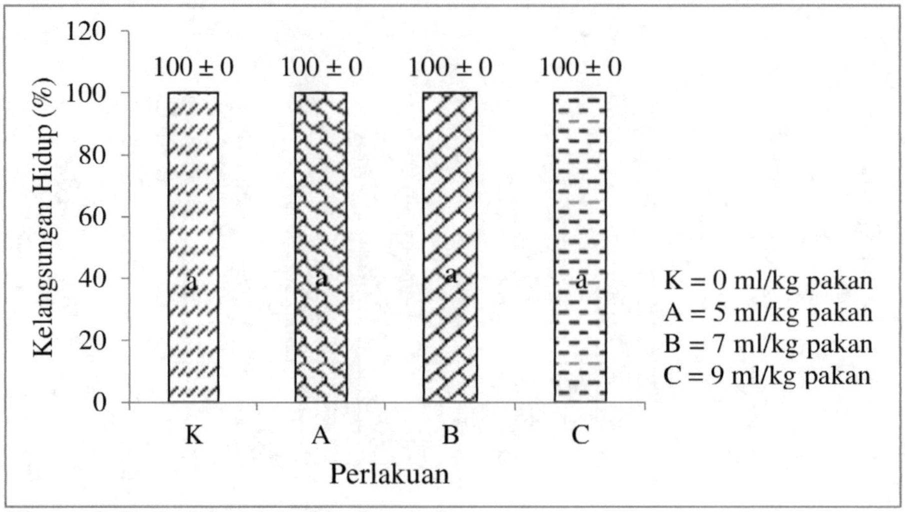

Figure 3. Betta fish survival in this study

The best feed conversion ratio was found in treatment C (9 ml/kg feed) resulting in a feed conversion ratio of (1.24 ± 0.14) significantly different ($p < 0.05$) with the control treatment of (1.89 ± 0.02), treatment A was (1.55 ± 0.15) and treatment B was (1.47 ± 0.11). (Figure 4). Test feed C showed that the addition of probiotics as much as 9 ml/kg of feed had a significant effect on the feed conversion ratio of Betta fish.

Previous study stated that the lower the feed conversion value during the study, the more efficient the feed given to the fish or vice versa [10]. Cultivation should not reduce or excess feed given to fish, if excessive feed is given to fish it will produce leftover feed in the rearing pond so that high ammonia occurs or vice versa if the feed given to fish is reduced there will be cannibalism in the fish. or slow the growth of the fish. The lower the feed conversion value, the better [11]. This is because the amount of feed that must be issued by the farmer to achieve a certain weight on the commodity being cultivated will be reduced, thus affecting the cost of expenditure.

Provision of probiotics in feed aims to use feed more efficiently, which means that less feed is used, because the absorption and fish digestibility of the feed given is optimal so that growth remains good. Factors that affect the level of feed conversion are several factors, including water quality, quantity and quality of feed, type or species of cultivated commodities, age, and quality of cultivated water. The use of probiotics in feed resulted in significantly different feed efficiency values by increasing feed efficiency up to 54.69% compared to the feed efficiency of control fish of 30.27% [12].

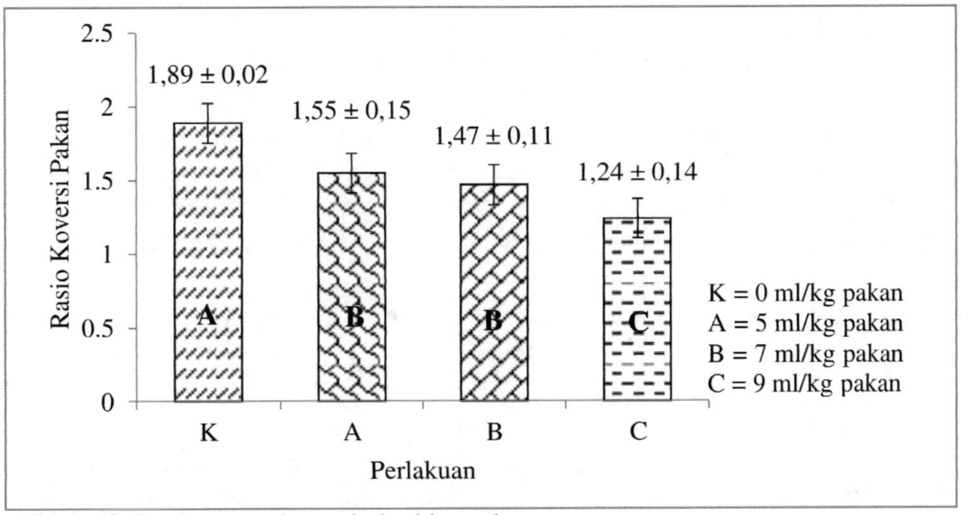

Figure 4. Feed conversion ratio in this study

The results of the measurement of water quality parameters during maintenance can be seen in Table 1. The water quality in the cultivation media during the study was in the optimal category to support fish farming. Water quality is the most important support apart from feed and fish species in supporting the growth and sustainability of cultivated commodities. During the study, the quality of maintenance water during the study was in optimal conditions. This is due to siphon and water changes once a week during the maintenance process. The high and low water temperature will affect the life of the biota that are kept. Water temperature affects the metabolism and development of organisms and affects the amount of feed consumed by fish. In the maintenance of betta fish, the value of oxygen solubility does not have a big influence, because betta fish can take oxygen directly from the air. Betta fish are fish that can live in places with limited oxygen content [13]. Betta fish have a labyrinth in their respiratory system, the labyrinth allows the betta fish to take oxygen directly from the free air.

Table 1. Water quality data in this study

No	Parameter	Sampling		Optimum Value	Souce
		Initial	Final		
1	Temperatur e (°C)	26-31	25-30	25-30 (a)	Atmadjaya dan Sitonggang (2010)
2	DO (ppm)	4-4,5	4,2-5	4-5 (b)	Monalisa (2010)
3	pH	6,9-7	6.9-7,2	6,5-7,5 (c)	Bintang (2007)

4. Conclusion

Based on the parameters observed, it was shown that the addition of probiotics in commercial feed in treatment C (9 ml/kg feed) had a significant effect on the growth of absolute weight, SGR, and FCR.

5. References

[1] Agus M, Yusufi T, Nafi B 2010 *Pen Akua.* 2 21-27.
[2] Sainah, Adelina, Heltonika B 2016 *Ber Peri Teru.* 44 36-50.
[3] Effendi H 2003 *Study of Water Quality for Management of Water Resources and Environment*

Kanisius Yogyakarta. 45-55.
[4] Santanumurti M B, Nugroho A K, Santoso L, Hudaidah S 2022 *IOP Conf Ser: Ear and Envi. Sci.* 1036 012119.
[5] Simanjuntak I C B H, Suminto, Sudaryono A 2016 *Jourl of Aqua Man and Tech.* 5 1-8.
[6] Putri S F, Hasan Z, Haetami K 2012 *Jur Peri dan Kela.* 3 283-291.
[7] Elyana P 2011 *Skrip. 33-43.*
[8] Henky M S, Cindy C, Lantu, Reni L, Kreckhoff, Reiny A, Tumbol 2020 *Budi Pera.* 8 8-18.
[9] Dewi A T, Suminto, Nugroho R A 2019 *Jur Sain Akua Trop.* 3 17-26.
[10] Endraswari L P M D, Cokrowati N, Lumbessy S Y 2021 *Jur Kel: Ind Jou of Mar Sci and Tech.* 14 70-81.
[11] Pascual S 2009 *Nutrition and Feeding of Fish* Van Nostrand Reinhold New York p 34-45.
[12] Arief M, Fitriani N, Subekti S 2014 *Jur Ilm Per dan Kel.* 6 49-53.
[13] Abidin Z, Puspitasari H P 2018 *Mina Betta Fish Business Theory and Technical Application of Cultivation, Disease, Marketing, Financial Feasibility, and Business Development Strategy UB Press Malang page 66-69.*

6. Acknowledgement

The authors gratefully acknowledge the collaboration from Universitas Airlangga and Universitas Lampung.

Antibacterial activity of crude extracts *Ulva lactuca* against *Staphyloccocus aureus* and *Propionibacterium acne*

T I Aprillia[1], M A Adam[2,4], A I. Supii[2], K Mahardika[3], J Kusnadi[1]
[1]Department of Agricultural Product Technology, Faculty of Agricultural Technology, University of Brawijaya
[2]Research Center for Marine and Land Bioindustry, National Research and Innovation Agency
[3]Research Center for Fisheries, National Research and Innovation Agency

[4]Corresponding author: ar.adam87@yahoo.com

Abstract. Sources of antibiotic raw materials as a reliever / acne remover drug is needed. Especially of natural ingredients from the *Ulva lactuca*. This study aims to examine the potential of crude extracts of *U. lactuca* as an antibacterial against *Staphyloccocus aureus* and *Propionibacterium acne*. Extraction of *U. lactuca* used Microwave Assited Extraction (MAE) technique with different types of solvents and extraction time. Analysis of antibacterial activity was carried out by the paper disk diffusion method. The research design used was a Factorial Completely Randomized Design (CRD) with two factors. The first factor of type solvent;ethanol, ethyl acetate and aquades. The second factor of extraction time which consists of 3 different levels;10 minutes, 20 minutes, and 30 minutes. The results of the study The highest yield was obtained in ethanol by 30 minutes extraction time;average of 24.9%. Phytochemical compounds were detected used a color test on the extract of Ulva lactuca namely alkaloids, steroids, flavonoids, and tannins. Whereas the antibacterial test with the highest tasteless zone of *Staphylococcus aureus* bacteria with ethanol solvent extraction time of 30 minutes;average of 17.17 mm. Crude extract of *U. lactuca* has potential as an antibacterial *S. aureus* and *P. acne*.

1. Introduction

Antibacterial is a secondary metabolite product produced by an organism in a certain amount, which is destructive or inhibits other microorganisms [1]. Secondary metabolites generally have antibacterial activity as bioactive compounds [2]. Green algae have the ability to produce secondary metabolites due to the extreme environmental conditions of algae, such as high salinity or defense against predators. Generally, the chemical compounds produced by green algae are terpenoids and aromatic compounds that have antibacterial activity [3,4].

Bioactive compounds resulting from secondary metabolism can be obtained through the extraction process. One method to obtain extracts of bioactive compounds in green algae type *U. lactuca* is to use Microwave Assisted Extraction (MAE). Extraction using MAE is an extraction that utilizes microwaves to speed up the extraction, through a fast and efficient solvent heating process [5]. Extraction is influenced by several factors, including the type of sample preparation, extraction time, solvent quantity, temperature, and solvent type [6].

Crude extract of *U. lactuca* has bacteriostatic properties because it is only able to inhibit the growth of test bacteria. Antibacterial is bacteriostatic properties if the antibacterial compound is only able to inhibit bacterial growth if the compound is continuously administered [7]. Antibacterial is bactericidal if the diameter of the inhibition zone will continue to increase, this is because this compound is able to kill and stop the physiological activity of bacteria, even though the administration of the compound is stopped [8]. *P. acnes* and *S. aureus* are classified as gram-positive bacteria [9,10]. The difference in the inhibition zones produced by *P. acnes* and *S. aureus* is due to the nature and resistance of each bacterium, even though both bacteria are classified as gram-positive [11]. Differences in target bacterial strains cause differences in enzymes or other substances produced as secondary metabolites. The different secondary metabolites produced give different effects on the antibacterial substances contained in the crude extract of *U. lactuca*. Secondary metabolites produced by bacteria can destroy or damage antibacterial substances so that bacteria are resistant to antibacterial substances [12].

The increase in antibacterial resistance and the use of *U. lactuca* is relatively small, so the authors are interested in conducting research on the antibacterial activity test of *U. lactuca* against *Staphyloccocus aureus* and *Propionibacterium acne* bacteria using different types of solvents and also the extraction time. The difference in this type of solvent aims to obtain the nature of the bioactive compounds in *U. lactuca* which have different concentrations, so that it will affect the antibacterial activity. *S. aureus* and *P. acne* are bacteria that can infect the skin, causing acne.

2. Material and methods

U. lactuca was air-dried in an open space at room temperature (±27°C) for 24 hours and followed by a freeze drying process (-40°C) for 2 days. The dried *U. lactuca* was blended until it became a coarse powder. Then the *U. lactuca* coarse powder was sifted with a size of 50 mesh. Next, *U. lactuca* powder was extracted using Microwave Assisted Extraction (MAE). MAE is mixed with 4 g of *U. lactuca* powder, then put into a baking glass. Furthermore, different types of solvents were added according to the treatment, namely 96% ethanol and aquades with 40 ml of each solvent. So that the ratio of *U. lactuca* powder to the solvent is 1:10 (w/v). Extraction was carried out using MAE for 10, 20, and 30 minutes at 50°C with a power of 200 W.

The extraction results were then tested for inhibition using the Kirby-Bauer method, namely the diffusion method with paper discs. The test bacteria cultures of *S. aureus* and *P. acne* were poured aseptically as much as 100 l of NB media into sterile petri dishes containing solid NA medium. The inoculum was then leveled using a spreader. In each petri dish, three quadrants were made, and in each quadrant a paper disc was placed. Previously, disc paper was inserted into *U. lactuca* extract with different solvents and extraction times (e 10 s, e 20 s, e 30 s, a 10 s, a 20 s, a 30 s) and positive control (Medi-Klin). and negative control (aquades and ethanol) for ± 1 minute. The paper discs that had been soaked in the extract were placed aseptically on the surface of the NA medium which had been inoculated with *S. aureus* and *P. acne*. Then incubated at 37°C for 3 hours. After incubation, the clear zone around the paper disc was observed and the diameter was measured. Testing of the inhibitory power of *U. lactuca* extract against *S. aureus* and *P. acne* was carried out three times for each treatment [13].

2.1 Identification of alkaloids
The results of the *U. lactuca* extraction as much as 0.1 g were put into a test tube, then 2 mL of chloroform and 2 mL of ammonia were added, shaken and then 2N HCl was added. The resulting solution is then divided into 3 parts in a test tube where each test tube is added Dragendorf's reagent, Mayer's reagent, and Wagner's reagent. In Dragendroft reagent, if it produces an orange/red precipitate, it indicates a positive result for alkaloid compounds. In Mayer's reagent, if it produces a white precipitate, it indicates a positive result for alkaloid compounds. In Wagner's reagent, if it produces a brown precipitate, it indicates a positive result for alkaloid compounds [14].

2.2 Identification of flavonoids
The results of the *U. lactuca* extraction as much as 0.1 g were put into a test tube, then 0.5 mg of magnesium was added and then 3 drops of concentrated HCl were added. Positive results show yellow, green, black, and orange colors [14].

2.3 Identification of steroids/triterpenoids
The extract of *U. lactuca* as much as 0.1 g was put into a test tube, then added with 2 mL of concentrated H2SO4 and 2 mL of anhydrous acetic acid. The presence of steroids is indicated by a color change from purple to blue or green. The presence of triterpenoids when experiencing a red or brown color change [15].

2.4 Identification of saponins
The extract of *U. lactuca* as much as 0.1 g was put into a test tube, then a few drops of concentrated HCl solution were added. Positive results are indicated by the formation of permanent foam ±15 minutes [16].

2.5 Identification of tannins
The extract of *U. lactuca* as much as 0.1 g was put into a test tube, then added 20 ml of water and brought to a boil. Next, filter and the results of the filter add drops of 1% ferricchloride. A positive result is indicated by a greenish brown or blue-black color indicating the presence of tannins [16].

2.6 Data analysis
The results of the analysis were analyzed descriptively by analysis of variance (ANOVA). Furthermore, the LSD test was carried out to determine the differences between each treatment based on a 95% confidence interval.

3. Results and discussion
The rendemen in extraction process is used to determine the potential of a material when used as a source of antibacterial. While the function of the yield calculation itself aims to measure the effectiveness of the type of solvent in extracting the material. The results of the calculation of the yield of *U. lactuca* extraction using ethanol and aquades solvents with a time of 10, 20, and 30 minutes showed in Table 1.

Table 1. Rendemen extract analysis data

Solvent	Time (minutes)	Average
	10	23,1±1,60
Ethanol	20	24,1±1,01
	30	24,9±0,75
	10	12,3±1,72
Aquades	20	15,8±1,04
	30	17,6±1,44

The data obtained were then analyzed using the ANOVA method, before being tested for normality with the MINITAB software. The results of the Kolmogorov-Smirnov normality test show that the data p value> 0.05 means that the data is normally distributed.

Fresh *U. lactuca* was freeze dried. This drying aims to remove the water content in the sample. The water content can cause enzymatic reactions that result in damage to the sample because the composition of the compounds in the sample has changed [17]. The average moisture content of *U. lactuca* samples using freeze drying was 88.12%. The mechanism for measuring water content using freeze drying is that the material is first frozen using a temperature of -40°C then followed by sublimation using low pressure so that the water content that has become ice will immediately turn into steam [18]. The advantages of drying using freeze drying are that it can retain the compounds contained in the sample, increase product stability in a dry state, handle drying more sterile, and can remove water without excessive heating [19]. To ensure that the treatment groups were significantly different in the diversity of solvent types and extraction time (minutes), it is necessary to carry out further tests using BNJ (5%) showed in Table 2.

Table 2. BNJ Advanced Test Results (5%) on diversity of solvent types

Solvent	Average			BNJ(5%)
Ethanol	24,0667	a		8,7889
Aquades	15,2778		b	

The results of pairwise comparisons using the BNJ further test, the highest average was found in the type of ethanol solvent with an average of 24.0067 and the lowest in the type of aquadest solvent with an average of 15.2778. Judging from this test, it can be seen that the ethanol solvent produces the highest yield, while the aquadest solvent produces the lowest nutrients. For further tests using BNJ (5%) on the diversity of extraction time (minutes) showed in Table 3.

Table 3. BNJ Advanced Test Results (5%) on variation of extraction time (minutes)

Time Extraction (minutes)	Average			BNJ(5%)
30	21,2833	a		
20	19,9833	a		2,2333
10	17,7500		b	

From the data Table 3 explains that the longer the extraction, the greater the rendemen (%) produced. This is due to the length of direct contact between the solvent and the sample. This condition continues until the required contact time with the sample is sufficient and reaches an equilibrium condition [20]. If it has reached equilibrium conditions and the extraction process is not stopped, it will damage the bioactive components contained in the sample. As a result, the sample experienced a decrease in the bioactive components contained. The bioactive components contained in the sample are limited in number and the solvent used has a limited ability to dissolve the existing sample [21].

The amount of extraction yield is influenced by several factors, including material size, solvent type, solvent concentration, extraction method, and extraction time [22]. Selection of the right solvent is one of the most important factors in the extraction process. The factors that must be considered in the selection of the type of solvent are the dissolving power of the compounds contained in the sample, boiling point, toxicity, flammability and the effect on extraction equipment [23]. To obtain an effective extraction of bioactive compounds, an appropriate solvent is needed. This is in accordance with the principle of like dissolves like, which means that a compound will dissolve in a solvent of the same nature, so the use of the type of solvent will have an influence on the yield of the resulting compound [24].

Ethanol solvent has a lower boiling point than aquadest solvent. Ethanol has a boiling point of 78.4°C while distilled water has a higher boiling point of 100°C [25]. Ethanol and aquadest are examples of polar solvents. However, there is a significant difference in yield between ethanol and distilled water. This is because the solvent ethanol is a solvent that has a broad spectrum to dissolve compounds in plants. As a result, polar or non-polar compounds are also extracted [26]. In addition, ethanol solvent can attract flavonoid compounds optimally [27].

Ethanol is usually used as an extractor or solvent of various compounds, the polarity of ethanol is lower than the polarity of distilled water. The nature of ethanol is polar and semipolar, so ethanol is able to extract relatively less polar compounds. Meanwhile, distilled water is a polar compound so it is not able to dissolve less polar compounds properly [28]. Furthermore, the extraction results were analyzed qualitatively by phytochemical screening using a color test. The color test can provide an indication of the types of secondary metabolites, such as alkaloids, flavonoids, tannins, saponins, steroids and triterpenoids. The results of the extract color test showed in Table 4.

Table 4. Phytochemical analysis data

Test		Result (minutes)					
		Ethanol			Aquades		
		10	20	30	10	20	30
Alkaloid	Mayer	+	+	+	+	+	+
	Dagendrof	-	-	-	-	-	-
Flavonoid		+	+	+	-	-	-
Saponin		-	-	-	-	-	-
Steroid		+	+	+	+	+	+
Terpenoid		-	-	-	-	-	-
Tanin		+	+	+	+	+	+

Table 4 shows positive results for alkaloids using Mayer's reagent, flavonoids, steroids, and tannins. In the results of flavonoids, there are differences in results between the use of ethanol and aquadest solvents. This is because each solvent has different abilities in dissolving the content of bioactive compounds in the materials used.

The crude extract of *U. lactuca* is said to be effective if the area is clearer than the surrounding area. This clear zone is formed due to the extract contained in the paper disc diffuses into the agar and prevents the growth of *P. acnes* and *S. aureus* bacteria. This zone is called the inhibition zone [29]. The zone of inhibition can be measured manually using a ruler. The inhibition zone is calculated by subtracting the diameter of the inhibition zone from the diameter of the paper disc, but after that the units obtained must be changed to millimeters (mm) [30,31]. The results of the inhibitory zone of *U. lactuca* crude extract against the growth of *P. acnes* and *S. aureus* bacteria are shown in the Table 5.

Table 5 shows the best results of crude extract of *U. lactuca* using ethanol solvent with extraction time of 30 minutes on *P. acnes* bacteria. Three repetitions resulted in the diameter of the inhibition zone being 9.5, 8, and 8.75, respectively. While for *S. aureus* bacteria with three repetitions, the results of the inhibition zone diameters were 19, 27.5, and 23, respectively.

The mechanism of inhibition of bacterial growth by phytochemical compounds has different antibacterial activity. The antibacterial activity of phytochemical compounds is due to the interaction of phytochemical compounds with bacterial cells. One of the phytochemical compounds in plants is polyphenol compounds and their derivatives. These compounds bind to bacterial cell proteins through non-specific bonds to form protein-polyphenol complexes. At low concentrations of compounds, the bonds formed are weak and undergo decomposition which damages the cytoplasmic membrane so that bacterial growth is inhibited. While at high concentrations of compounds, these compounds coagulate and the cytoplasmic membrane lysis. The mechanism of each of these phytochemical compounds synergizes with each other so as to increase their effectiveness and activity in inhibiting growth or killing bacteria [32,33]. The positive control indicated the formation of an inhibitory zone around the paper disc, so it could be seen that medi-clin had antibacterial activity that could affect the growth of the test bacteria. Medi-Klin is an acne gel that contains clindamycin. Clindamycin is an antibiotic that is generally used as an acne medication because it is able to inhibit inflammation and kill bacteria [11].

Table 5. Antibacterial activity data

Test Bacteria	Solvent	Time	Average
P.acnes	Ethanol	10	1,58±1,13
		20	1,83±1,04
		30	2,75±0,75
	Aquadest	10	1,41±0,95
		20	2,42±0,8
		30	2,75±0,75
	Medi-klin		21±2,00
	Aquadest		0
	Ethanol		1,5±2,12
S.aureus	Ethanol	10	13,5±6,87
		20	15,17±6,00
		30	17,17±4,25
	Aquadest	10	3,34±0,76
		20	3,84±1,04
		30	4,75±1,52
	Medi-klin		24,66±2,51
	Aquades		0
	Etanol		8,95±0,07

Note: Medi-klin = Control positive; Aquades = Control negative; Etanol = Control negative

Based on the results of the analysis, it is known that in all treatments carried out on *P. acnes*, there was no significant difference in the mean at $P<0.05$. so that it is known that in Propionibacterium acnes there is no interaction between the type of solvent and the extraction time. According to Bojar [34], *P. acnes* belongs to the group of anaerobic to aerotolerant bacteria. So that antibacterial testing using the spread plate method in growing Propinibacterium acnes is not an optimum condition for the growth of these bacteria. To find out which treatment groups had significant differences, further tests were carried out using BNJ (5%) which showed in Table 6.

Table 6. Results of Advanced BNJ Test on *S. aureus* bacteria

Solvent	Mean	Grouping		BNJ(5%)
Ethanol	15,2778	a		11,3006
Aquadest	3,9722		b	

Table 6 shows that if the data has a p value <0.05, it means that the data is significantly different between the treatment groups. BNJ further test showed the diameter of the zone of inhibition of *S. aureus* was the highest in crude extract of U. lactucausing ethanol extracted for 30 minutes. *S. aureus* has properties that are very tolerant of antibacterials because this bacterium has a cell wall consisting of 50% peptidoglycan layer and a compact cell wall arrangement that makes a rigid structure [35,36].

4. Conclusions

Based on the results of the study, it can be concluded that the crude extract of *U. lactuca* has antibacterial activity against *Propionibacterium acnes* and *Staphylococcus aureus* bacteria with the highest yield at 30 minutes of ethanol solvent 24.9%. The group of active compounds from crude extract of *U. lactuca* has the potential as an antibacterial from the alkaloids, flavonoids, steroids, and tannins groups.

5. References

[1] Adam M A, Maftuch, Hardoko H 2013 *Natural B.* 2(1) 1-5.
[2] Pangestuti I E, Sumardianto, Amalia U 2017 *Indones J Fish Sci Technol.* 12(2) 98-102.
[3] Liswandari M S, Lantang D, Digantara S 2018 *Pharmavy Med Journal.* 1(1) 9-15.
[4] Wirawan I G P, Dewi N K E S, Sasadara M M V, Sunyamurthi IG N A, Jawi I M, Wijaya I N, Darmawati I A P, Suada I S, and Krisnandika A A K 2022 *Molecules*, *27*(15) 4879.
[5] Widyasanti A, Aryadi H, Rohdiana S 2018 *J Teknol Pertan Andalas*. 22(2) 165-175.
[6] Pendit P C D, Zubaidah E, Sriherfyna F H *J Pangan dan Agroindustri.* 4(1) 400-409.
[7] Fadhilla R R, Diyantoro D W I, and Sundari A S 2020 *Malaysian J Med Heal Sci.* 16, 67-71.
[8] Mycek, M J. 2001. *Farmakologi; Ulasan Bergambar Edisi 2*. Widya Medika., Jakarta
[9] Purwanti T, Rosita N, and Nafisah S 2021 *Trop J Nat Prod Res. 5*(8), 1348-1353.
[10] Khairullah A R, Ramandinianto S C, and Effendi M H 2020 *Syst Rev Pharm. 11*(7) 172-183.
[11] Wahdaningsih S, Untari E K, Fauziah Y 2014 *J Pharm Sci Res.* 1(3) 180-193.
[12] Jawetz M, Adelberg's 2010 *Mikrobiologi Kedokteran. Buku Kedokteran* EGC. Jakarta.
[13] Oroh S B, Kandou F E F, Pelealu J, Pandiangan D 2015 *J Ilm Sains.* 15(1) 52-58.
[14] Kursia S, Lebang J S, Taebe B, Burhan A, Rahim W O R, Nursamsiar 2016 *J Ilmu dan Teknol Farm Indones*. 3(2) 72-77.
[15] Lantah P L, Lita M L A D Y, Reo A R 2017 *J Media Teknol Has Pertan*. 5(3) 167-173.
[16] Siregar A F, Sabdono A, Pringgenies D 2012 *J Mar Res*. 1(2) 152-160.
[17] Ningsih D R, Zusfahair, Kartika D 2016 *J Mol.*11(1) 101-111.
[18] Berk Z 2018 *Food process engineering and technology*, *2*, 567-581.
[19] Nireesha G R, Divya L, Sownya C, Venkateshan N N B M, Babu M N, Lavakumar V 2013 *Int J Nov Trends Pharm Sci*. 3(4) 87-98.
[20] Rezki R S, Dwimas A, Siswarni M Z 2015 *J Tek Kim USU.* 4(3) 29-34.
[21] Yulianti D, Susilo B., Yulianingsih R 2014 *J Biopreses Komod Trop*. 2(1) 35-41.
[22] Saragih J, Assa J, Langi T 2010 *J Media Teknol Has Perikanan*. 6(15) 1-6.
[23] Azis T, Febrizky S, Mario A D 2014 *J Tek Kim.* 20(2):1-6.
[24] Wahyuningtyas S E P, Permana I D G M, Wiadnyani A A I S 2017 *J ITEPA*. 6(2) 61-70.
[25] Hudha M, and Widyaningsih T D 2015 *J Pangan dan Agroindustri.* 3(4) 1412-1422.
[26] Ulaan G A K, Yudistira A, Rotinsulu H 2019 *J Ilm Farm*. 8(3) 119-125.
[27] Arbi B Ma'ruf W F and Romadhon 2016 *Indones J Fish Sci Technol.* 12(1) 12-18.
[28] Fardhyanti DS and Riski R D 2015 *J Bahan Alam Terbarukan*. 4(1) 6-13.
[29] Hakim M L, Susilowati S, Effendi M H, Tyasningsih W, Sugihartuti R, Chusniati S, and Witaningrum A M 2016 *Ecol Environ Conserv*. 26(11), S276-S280.
[30] Mawan A R, Indriwati S E, Suhadi 2018 *Bioeksperimen*. 4(1) 64-69.
[31] Ekawati E R, & Wahyuningsih S P A 2022 *Res J Pharm Technol*. 15(7), 3002-3008.
[32] Marselia S, Wibowo M A, Arreneuz S 2015 *J Kedokt dan Kesehatan*. 4(4) 72-82.
[33] Adam M A, Widiastuti I M, Ernawati E, Yayan A Y, Insivitawati E, Yuliana, Pakaya R F, Soegianto A, Khumaidi A 2022 *J Ilm Perikan dan Kelaut*. 14(1) 160-166.
[34] Bojar R A, and Holland K T 2004 *Clinics in dermatology*, *22*(5), 375-379.

[35] Poeloengan M 2009 *J Biol.* 9(6)715-719.
[36] Tyasningsih W, Ramandinianto S C, Ansharieta R, Witaningrum A M, Permatasari D A, Wardhana D K, Effendi M H, and Ugbo E N 2022 *Vet World.* 15(8). 2021-2028.

6. Acknowledgement
The author would like to thank Faculty of Fisheries and Marine, Airlangga University, for conducting the seminar.

Study on the quality of fesh swanggi (*Priacanthus macracanthus*) at fish auction places (TPI) Brondong, PPN Brondong, Lamongan. East Java.

A H Harahap[1], A M Sahidu[2,3] and W Isroni[2]

[1] Program Fishery Products Technology, Faculty of Fisheries and Marine Universitas Airlangga, Kampus C Unair, Jalan Mulyorejo Surabaya 60115, Indonesia.
[2] Department of Marine, Faculty of Fisheries and Marine Universitas Airlangga, Kampus C Unair, Jalan Mulyorejo Surabaya 60115, Indonesia.
[2] Department of Fish Health Management and Aquaculture, Faculty of Fisheries and Marine Universitas Airlangga, Kampus C Unair, Jalan Mulyorejo Surabaya 60115, Indonesia.

[3] Corresponding author: adriana-m-s@fpk.unair.ac.id

Abstract. This study aims to analyze and describe the quality of swanggi fish and determine the external factors that affect the quality of swanggi fish (*Priacanthus macracanthus*) at the Fish Auction Place (TPI) Brondong, Lamongan, East Java. The type of research used is descriptive survey research. The test parameters in this study were organoleptic test, pH test and formalin test to analyze and describe the quality of swanggi fish. In addition, a causal analysis was also carried out to determine the externalfactors that affect the quality of swanggi fish. This research was carried out at the Fish Auction Place (TPI) Brondong, PPN Brondong, Lamongan, East Java on February 01 to February 24, 2022. The results of organoleptic and pH testing of swanggi fish were declared still fit for consumption but found 68 samples. formalin positive swanggi fishor 24.29% of the total test samples. Meanwhile, the external factor of the decline in the quality of swanggi fish is strongly influenced by the lack of monitoring and evaluationof good fish handling practice (CPIB).

1. Introduction

The freshness or quality of freshly caught fish from the waters cannot be improved and not only that the freshness level will continue to decline. In other words, the decline in fish quality can only be slowed by applying good and correct principles and methods of handling. The decline in fish quality is influenced by various factors, both internal and external. External factors include the process of death, time, method of handling, and fish handling facilities. The decline in quality and post-harvest damage is high, among others, caused by fishing methods, poor handling methods, long supply chains, inadequate handling and storage facilities [1].

The method of catching (type of fishing gear) is directly related to the method of fish death and the method of fish death is related to the physical and chemical processes experienced by the fish's body where these processes directly affect the quality of fish after catching. This is exacerbated by the handling of fish which is classified as poor because it is still carried out improvised without paying attention to the conditions that must be met, both regarding handling facilities and how to handle them, including the use of ice as a fish cooling agent. One of the fishing ports in Lamongan Regency, namely PPN Brondong in 2021 has a total production of 59.024 (Tons) or 73% of the total production of Lamongan Regency with swanggi fish production of 12.000 (Tons) or 20.4% of PPN Brondong production [2], this makes the swanggi fish in PPN Brondong the object of research.

The decline in quality and high damage occurs due to several factors that directly affect, among others: the type of fishing gear used, poorly designed and isolated fishing hatches and crates and poor handling on board, poor loading and unloading and post-unloading. handling system. and processes during the auction. Based on the above background, a study was conducted with the title Study of the Quality of Swanggi Fish (*Priacanthus macranthus*) with organoleptic tests, pH tests and formalin tests. In addition, a Cause and Effect Analysis or Ishikawa Diagram was carried out to analyze external factorsthat affected the decline in the quality of swangi fish (*Priacanthus macranthus*) at brodong fish auctions,the analysis was carried out by examining Good Fish Handling Practices (CPIB).

2. Material and methods

2.1. Material

The equipment used in this study were organoleptic test score sheets, pens, notebooks, cameras, pH meters, formalin test kits, knives, trays, scales, stirring rods, funnels, filter paper, cutting boards, beakers, test tubes, syringes, measuring tube, tissue, mortar and pestle. While the materials used are swanggi fish (*Priacanthus macracanthus*) from the Brondong Fish Auction Place, aquades, carrez 1 & 2, Fo-1 & Fo-2, buffers pH 4 and pH 7.

2.2. Methods

The research method used in this research is descriptive survey research, which is a descriptive survey conducted on a set of objects that aims to see a picture of the phenomena that occur in certain populations [3]. This study aims to describe the condition of the quality or level of freshness of swanggi fish in TPI Brondong by using organoleptic tests, pH and formalin, then describe and identify external factors found to be related and affect the quality of fish which are then presented in the form of a cause and effect analysis diagram. This research was conducted for 7 days by taking 40 samples every day, while the CPIB examination was carried out on 5 ships every day.

3. Result and discussion

3.1 Organoleptic swanggi fish

Table 1. Organoleptic Test Results

	Day							
Parameter	1	2	3	4	5	6	7	Avarage
Eye	7.3	7.3	7.3	7.4	7.3	7.4	7.5	7.3
Gill	6.8	6.8	7.0	7.0	6.7	6.9	7.0	6.8
Meat	6.6	6.6	6.7	6.8	6.6	6.8	6.9	6.7
Odor	6.3	6.3	6.3	6.3	6.2	6.3	6.4	6.3
Texture	6.0	6.1	6.4	6.4	6.2	6.3	6.4	6.2

The quality of swanggi fish marketed at TPI Brondong does not meet the 2013 SNI standard, namelythe organoleptic value of at least 7 with organoleptic values for each indicator as follows: eyes (7.3), gills (6.8), meat (6.7), odor (6.3) texture (6.2) with an average organoleptic value of 6,6. However, according to [4] a value of 7-8 is still fresh, a value of 5-6 is still fresh and 1-4 is not suitable for consumption. Meanwhile, according to Jusrawati. [5] the freshness of fish is assessed with the following criteria: Fresh (7-9), Slightly fresh (4-6) and Not fresh (1-3). So, from the organoleptic assessment, it can be concluded that swanggi fish is still suitable for consumption.

The decrease in eye and gill organoleptic values is influenced by temperature, according to [6]. The higher the temperature, the lower the organoleptic value, this indicates that the fish temperature and organoleptic values have a negative relationship. Changes in body temperature of fish cause the oxidation of hemoglobin to methemoglobin which causes discoloration of the eyes and gills. In addition, it is also caused by fluid dehydration which causes the fish's eyes to become sunken.

The appearance of the meat, especially the color of the incision is influenced by the oxidation reaction between the oxygen and fat components [7]. Furthermore, according to Supartinah. [8] enzymes remodel the structure of muscle protein tissue to become looser so that it is susceptible to bacterial attack which causes fish meat to become somewhat soft. The smell of good quality fish is the one that has a fresh andspecific smell, the longer the odor becomes neutral and the smell is sour or rotten if the quality of the fish is rotten.

Odor is strongly influenced by the degradation products of enzymes and microorganisms. According to Ariyani *et al.* [9] after the fish died, the enzymes found in the fish began to actively degrade the fish flesh into simpler substances and the microorganisms found in the stomach contents, gills and skin multiply rapidly which can cause odor.

The assessment of fish quality from indicators is closely related to its elasticity properties, such as meat in fish that decreases in quality when pressed, the texture of the meat is soft or no longer elastic, as well as Rahmi. [6] Fish that is still fresh when pressed does not leave pressure marks because it has not beencut. fish floss. Fish that begin to rot have a lot of damaged connective tissue and a lot of cell walls are damaged so that the fish meat loses its flexibility [10].

3.2 Swanggi fish pH

The decrease in fish quality is closely related to the pH value, the lower the pH value causes the enzymes in muscle tissue to become active [11]. Furthermore, according to Supartinah. [8] enzymes remodel the structure of muscle protein tissue to become looser so that it is susceptible to bacterial attack which causes fish meat to become somewhat soft. The following table shows the results of testing the pH of swanggi fish at TPIBrondong.

Table 2. pH test results

Day	pH	Average
1	6.6	
2	6.6	
3	6.6	
4	6.5	6.5
5	6.7	
6	6.5	
7	6.5	

The pH value is one of the indicators used to determine the level of freshness of fish and is closely related to the process of quality reduction [12]. Testing the degree of acidity (pH) of swanggi fish using a pHmeter. The pH value illustrates that the susceptible pH value of swanggi fish is between 6.5 to 6.7 with an average value of 6.5, from these data it can be concluded that the freshness level of swanggi fish is still relatively fresh and in the rigor mortis phase. However, if the pH value reaches 7-8 then it is categorized as less good and if more than 8 it is categorized as bad [1]. In general, the acidity value of fish meat is always changing since the fish is caught from the waters to the hands of consumers, this is related to handling, facilities and also the time required during the process flow.

According to Nazrah. [13] the decrease in pH value is related to the length of fish transit time, starting from the length of the transit time of the ship and the process of loading and unloading fish from the ship to the auction place and until the fish are being auctioned. In addition, the decrease in pH was caused by some improper handling methods on board and at TPI and transit times. The decrease in the pH value of fish after death has a certain limit, according to Metusalach *et al.* [1] stated that the pH of fish meat will decrease onlyto a certain extent (around pH 5.5).

The lower the pH value causes the enzymes in muscle tissue to become active. Catechins are proteolytic enzymes that function to break down proteins into simple compounds, remodel the structure of muscle protein tissue to become looser and more susceptible to bacterial attack [14,15]. Similarly, other enzymes present in fish organs, such as the stomach, perform the same activity. This causes the fish meat to become soft. The tissue remodeling phase by enzymes in the fish body is called autolysis [8,16]. When the pH conditions drop to the rigor mortis phase, autolytic enzymes that are active under acidic conditions will degrade protein in fish. Furthermore, alkaline compounds will be produced from the enzyme autolysis process and bacterial activity, the increase in pH continues until decay occurs, this occurs due to the emergence of basic compounds such as ammonia, trimethylamine, and other volatile compounds [4].

3.3 *Swanggi fish formalin*

Freshness or quality of freshly caught fish cannot be increased and not only that the level of freshness will continue to decline, in other words the decline in fish quality can only be slowed down by applying good and correct principles and methods of handling. One of the efforts usually made by fishermen to maintain the freshness of fish is to use the cold storage method. However, because the distance from fishermen to the sea is increasing, the use of ice is increasing which makes some fishermen use fish preservation using formalin. The following table shows the results of qualitative and quantitative formalin testing.

Table 3. Results of formalin qualitative testing

No	Sample	Amount	Percentage
1	Negative	212	75.71%
2	Positive	68	24.29%
3	Total	280	100%

Table 4. Results of formalin quantitative testing.

No	Concentration	Amount	Percentage
1	0.1	9	13.23%
2	0.25	11	16.17%
3	0.4	16	23.52%
4	0.6	16	23.52%
5	0.8	14	20.58%
6	1	2	2.94%
7	Total	68	100%

The results of the formalin qualitative test showed that there were 28 positive swanggi samples (24.29%) and 212 (75.71%) swanggi fish samples that were negative for formalin. While Table 5 quantitative test results show the highest formalin content of 1 ppm in 2 samples and the lowest level of 0.10 ppm in 9 samples, while the highest percentage of samples contains formalin with concentrations of 0.4 and 0.6 with the same percentage, namely 26 samples. (23.52%)

The difference in the concentration of formalin found in swanggi fish samples could be influenced by several factors such as the length of time the fish were immersed in formalin-containing water, the position of the fish in the hold and also the use of uncertain doses of formalin given during storage. . formaldehyde will react between formalin and protein, formalin enters the fish cells binds proteins by denaturing proteins and nucleic acids through an alkaline process between the -NH2 and -OH groups of proteins and nucleic acids, with this formaldehyde hydroxymethyl group what causes fish to stiffen [17]. With the death of the protein after being bound to the chemical element formalin, when pressed, the food that is given formalin feels more chewy, besides that the dead protein will not be attacked by spoilage bacteria that produce acidic compounds, which makes the food more durable.

3.4 Ishikawa CPIB analysis

Fishery products are foods that are easily damaged, damage can be caused by chemical, microbiological and physical activities. In general, this damage is caused by two factors, namely intrinsic factors (especially the fish themselves) and extrinsic factors or external factors that can affect the quality or quality of fish. The extrinsic factors observed in this study are fish loading and unloading, standard handling and storage facilities and standard handling and storage procedures which will then be observed using the examination of Good Fish Handling Practices (CPIB) from the [18]. The following are the results of the CPIB examination made on the Ishikawa diagram.

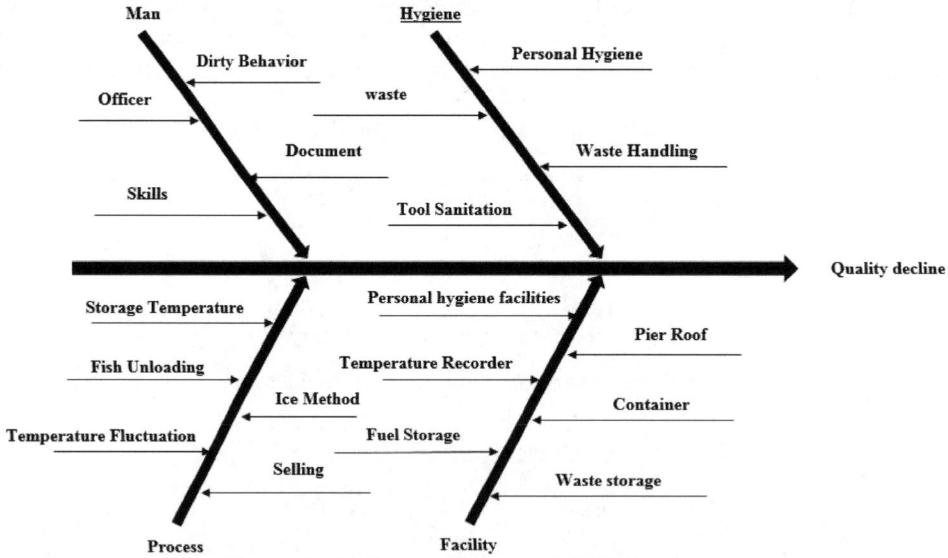

Figure 1. Ishikawa diagram of CPIB inspection

The decline in the quality or quality of swanggi fish (*Priacanthus macracanthus*) at the Brondong Fish Auction Place (TPI) was due to the negligence of crew members in implementing good fish handling procedures. During loading and unloading there was treatment that resulted in physical damage to fish, there was garbage at the loading dock, the temperature during unloading was not controlled, the container was dirty and did not meet the cold chain requirements. Furthermore, in the current standard of facilities there are still several things that have not been met, namely there are no closed bins at least 2 units, personal hygiene facilities are not available, there is no recording of temperature during storagewhich causes the storage temperature to be unattainable and uncontrolled which results in very possiblefluctuations occur often occur.

Meanwhile, from the procedures for handling and storing fish, it was found that several standard aspects were not met, such as the absence of SSOP documents, the absence of recording documents for controlling freezing temperatures and frozen storage, the absence of documents for recording the implementation of standard operating procedures for handling and storing fish for fishing vessels, not testing the organoleptic treatment of the caught fish and temperature requirements that are not guaranteed during storage in the hold, the crew's personal hygiene is not met such as using clean clothes, not smoking, eating and drinking, routine health checks and not having at least 1 crew member who has a certificate fish handling expertise as the person in charge of quality on board. Meanwhile, during sales at TPI, the cold chain does not apply which causes a continuous decrease in quality and personal hygiene such as hand washing facilities equipped with soap, clean water, and disposable drying cloths.

4. Conlusion

The quality of swanggi fish marketed at TPI Brondong does not meet SNI 2013 standards with organoleptic values for each indicator as follows: eyes (7.3), gills (6.8), meat (6.7), odor (6.3) texture (6.2) with an organoleptic mean value of 6.6. The pH value (degree) of swanggi fish with an average value of 6.5 indicates that the freshness level of swanggi fish is still fresh and entering the rigor mortis phase. Meanwhile, from the results of the formalin test, it was found that 68 samples of swanggi fish were positive for formalin or 24.29% of the total test samples. The decline in the quality or quality of swanggi fish (*Priacanthus macracanthus*) at the Brondong Fish Auction Place (TPI) is caused by external factors during the unloading process, standard facilities and handling standards. These external factors are in the form of treatment that results in physical damage, unavailability of garbage collection containers and the amount of garbage scattered, the absence of personal hygiene facilities, how to use ice, fluctuations in loading and unloading temperatures, dirty, damaged storage containers, additional odor and no chains. cold. , wrong way of storing fuel, worker equipment is not in clean condition, no record of resistant temperature, no document (SSOP, temperature control record, SSOP implementation record), crew behavior (eating, drinking, smoking, spitting), none ABK who have certificates of fish handling skills and CPIB inspectors do not control the implementationof Good Fish Handling Practices (CPIB) carried out by ABK so that there are many deviations from quality control requirements.

5. References

[1] Metusalach, Kasmiati, Fahrul and . Jaya 2014 *J IPTEK PSP.* 1 41-50.
[2] Dinas Perikanan Kabupaten Lamongan 2021 *Profil Perikanan*. 14-15.
[3] Notoatmojo S 2010 *Health Research Methodology* Jakarta: Rineka Cipta. 14- 24.
[4] Santhi D G D D 2017 *Organoleptic Examination and pH (Acidity) as Quality Requirements for Safety of Tuna* (*Tunnus* sp). Medical Education Study Program, Faculty of Medicine, Udayana University. Denpasar Bali. p 3-6.
[5] Jusrawati. 2021. *Chemical Quality Characteristics of Fresh Flying Fish (Decapterus macrosoma) using Comparison of Water and Ice Handling Techniques and Storage Time.* Hasanuddin University Makassar. p 12-13.
[6] Rahmi J 2018 *Quality of Fresh Skipjack Fish (Katsuwonus pelamis) After Landing at PPI Lappa Sinjai Until Final Marketing in Sinjai Regency*. Hasanuddin University Makassar. p 10 and 49.
[7] Tamuu H, Harmain R M, and Dali F A 2014 *J Ilm Perikan dan Kelautan.* 2 164-168.
[8] Supartinah 2012 *Descriptive Analysis of Decreased Quality of Offal (Intestine, Liver, Kidney) Milkfish (Chanos chanos) During Chilling Temperature Storage Through Histological Observation*. Bogor Agricultural University, Bogor. p 4-6 and 30-40.
[9] Ariyani F, Murtini J T, Indriati N, Dwiyitno D dan Yenni Y 2007 *Use of Glyroxyl to Inhibit Deterioration of Fresh Goldfish (Cyprinus carpio).* p 125-133.
[10] Prastica E E, Masithah E D, and Pursetyo K T 2021 *World's Vet J.* 11(1) 119-123.

[11] Saputra E, Tjahjaningsih W, and Patmawati 2017 *AIP Conf Proc.* 1813(1) 020016.
[12] Sari M, Triastuti J, and Pramono H 2020 *IOP Conf Ser Earth Environ Sci.* 441(1) 012011.
[13] Nazrah 2013 *Analysis of the relationship between handling methods, handling facilities and transit time with the quality of fish caught with cantrang in the waters of Takalar and Baru districts*. Fishery Resources Utilization Study Program. Faculty of Marine Sciences and Fisheries. Hasanuddin University Makassar. p 23 & 43-45.
[14] Nirmala D, Cahyanto D, Valen F S, and Maulida N 2021 *Ecol Environ Conserv.* 27, S58-S61.
[15] Khiftiyah A M, Hanifah N N, Bachruddin M, Sholichah M A, Istiqomah S, Dewi S R, T Rahayu, I A W Prasetya, L Marjayandari, N Aini, I Tsana, D Triwahyuni and Ni'matuzahroh 2018 *AIP Conf Proc.* 2019(1) 040014.
[16] Azmi K A, Arsad S, and Sari L A 2020 *IOP Conf Ser Earth Environ Sci.* 441(1) 012039.
[17] Wikanta W 2011 *Effect of Addition of Belimbing Wuluh (Averrhoa bilimbi L.) and Boiling on Residual Weight of Formalin, and Protein Profile of Formalin White Shrimp (Letapenaeus vannamei) and Utilization as a Source of Nutrition Education and Food Safety in Society* Malang State University. p 107-108.
[18] Minister of Maritime Affairs and Fisheries NUMBER 7/PERMEN-KP/2019 *Persyaratan Dan TataCara Penerbitan Sertifikat Cara Penanganan Ikan Yang Baik.* p 3-4 & 14.

6. Acknowledgement

The author would like to thank Faculty of Fisheries and Marine, Airlangga University, for conducting the seminar.

Enhancement growth of plantlet *Gracilaria* sp. with additional fermented fertilizer

A L Nurhayati[1,2], D Permana[1], R Aulia[1] and Nurhidayat[1]

[1] Fisheries Research Center, National Research and Innovation Agency, Jl. Raya Jakarta-Bogor No.32, Pakansari, Cibinong, Bogor, West Java 16915 Indonesia

[2]Email : arof003@brin.go.id

Abstract. The purpose of this research is to know the influence of additional fermented fertilizer to the growth of plantlet *Gracilaria* sp. during acclimatization stage. The comparison of composition fermented fertilizer from *Leucaena leucocephala* leaves : *Moringa oleifera* leaves which used were 75%:25%, 50%:50%, 25%:75% and control without additional fermented fertilizer. The best plantlet growth after seven weeks testing, found in comparison 25%:75% with enhancement plantlet growth up till 32,4%, followed with comparison 50%:50% with enhancement plantlet growth up till 27,1%, the last with comparison 75%:25% and control which both show enhancement plantlet growth up till 13,5%. Based on this result, we carried out further testing with comparison 20%:80%, 15%:85%, 10%:90% and 5%:95% but the plantlet growth result only as much 12,67%; 26,67%; 24,67% and -0,09% so we recommend to use combination fermented fertilizer from *Leucaena leucocephala* leaves : *Moringa oleifera* leaves with comparison 25%:75% in acclimatization step since it's not only increase the weight but also the length and number of shoots. However, it's necessary to do further testing using other fertilizer raws to get better plantlet growth result.

1. Introduction

The background of this research is our previous research on 2020, where plantlet *Kappaphycus alvarezii*, showing growth enhancement on the immersion treatment using fermented fertilizer of *Leucaena leucocephala* leaves and *Moringa oleifera* leaves. However the correct combination composition of *Leucaena leucocephala* leaves and *Moringa oleifera* leaves still not found yet. On the other hand, plantlet of *Gracilaria* sp. also need effort to growth enhancement in aclimatozation stage. Therefore, we did research in growth enhancement of plantlet *Gracilaria* sp. using fermented fertilizer from *Leucaena leucocephala* leaves and *Moringa oleifera* leaves with different composition.

The main purpose of research using combination of fermented fertilizer on plantlet *Gracilaria* sp. is enhancing the plantlet growth in acclimatization stage so we can get more seed through tissue culture process in shorter acclimatization stage. Furthermore, we can get the type of fertilizer which can optimize the plantlet enhancement in acclimatization stage with an emphasis on engineering efforts growth enhancement of plantlet using fermented fertilizer. The target of this research is addition of fertilizer in acclimatization stage able to enhancing the growth of thalli up till 33%.

2. Material and methode

2.1 Research material

2.1.1 Cultivant.

Cultivant which is used in this research are plantlets obtained from thalli culture done in seaweed tissue culture laboratory, Main Center of Brackishwater Aquaculture (MCBA) Jepara.

2.1.2 Experiment container.

Fermented fertilizer made in jar volume 10 L while the cultivant maintained in jar volume 3.500 mL which equipped with aeration and in aseptic and controlled room.

2.2 Preparation of tools and materials

2.2.1 Tools.

Tools which is used are glass jar volume 3.500 mL, jar volume 10 L, basin volume 1 L, bowl volume 250 mL, stirring rod, digital scales, aeration system, sampling tools, water quality checking equipment, water, sponge and rag.

2.2.2 Materials.

Materials which is used are plantlet *Gracilaria* sp., fermented fertilizer from *Leucaena leucocephala* leaves, fermented fertilizer from *Moringa oleifera* leaves, fresh water, sea water and dish soap.

2.3 Test object manufacturing activities

2.3.1 Fermentasion fertilizer manufacturing process.

Leucaena leucocephala leaves picked from the branch and collected up till 1 kg. The same also done for *Moringa oleifera* leaves. 1 kg of *Leucaena leucocephala* leaves or *Moringa oleifera* leaves, 333 g of fruit waste, 33,3 g ZA fertilizer, 67 g manure, 13,3 g molasses, 67 mL liquid probiotic, and 1,2 L brackishwater mixed until all component mixed well then put it in to jas. Closed the jar, the part between the lid and the neck of the jar is covered by clear duct tape. The lid of the jar is crushed by concrete brick. The label contains the identity of the fertilizer and the date of manufacture put outside of the jar. The jar kept in dark room in 4 weeks. On the 4th week, we need to check whether all ingredients have been fermented or not. Due to the fermentation process has not optimize, we add 100 mL liquid probiotic and 1.800 mL brackishwater in to the jar. On the 5th week, all ingredients are fermented well so it's ready to be used.

2.3.2 Test treatment.

All of the plantlets for testing 1, 2, 3 and 4 soaked in to fermented fertilizer solution with composition following Table 1.

Table 1. Composition of fermented fertilizer in every treatment.

TESTING	COMPOSITION
1	75% *Leucaena leucocephala* leaves 25% *Moringa oleifera* leaves
2	50% *Leucaena leucocephala* leaves 50% *Moringa oleifera* leaves
3	25% *Leucaena leucocephala* leaves 75% *Moringa oleifera* leaves
4	20% daun lamtoro 80% daun kelor (A), 15% daun lamtoro 85% daun kelor (B), 10% daun lamtoro 90% daun kelor (C) dan 5% daun lamtoro 95% daun kelor (D)

2.3.3 Testing data collection.

Observations were made on the growth of total plantlet weight and quality of maintenance water on a regular basis.

2.3.4 Data analysis.
Collected data, then counted and averaged. For easy viewing analysis result, then a tabulated graph is made to get conclusion.

2.4 Engineering method
2.4.1 Time and place of research. The research done following Table 2.

Table 2. Time and place of research.

TESTING	TIME	PLACE
1, 2 and 3 4	25 of May – 13 of July 2021 14 of September – 26 of October 2021	Tissue culture laboratory, seaweed team work, MCBA Jepara

2.4.2 Testing method.
Fermented fertilizer with composition in accordance to Table 3, dissolved using brackishwater by adding the brackishwater gradually up till the sollution's pH are at normal, i.e between 6-7. The result during testing, we need 2.000 mL brackishwater to neutralize pH 50 g fermented fertilizer from *Leucaena leucocephala* leaves and *Moringa oleifera* leaves.

Table 3. The weight of each composition fermented fertilizer in every treatment.

TESTING	*Leucaena leucocephala* leaves	*Moringa oleifera* leaves
1	37,5 g (75%)	12,5 g (25%)
2	20 g (50%)	20 g (50%)
3	12,5 g (25%)	37,5 g (75%)
	10 g (20%)	40 g (80%) – treatment A
	7,5 g (15%)	42,5 (85%) – treatment B
4	5 g (10%)	45 g (90%) – treatment C
	2,5 g (5%)	47,5 (95%) – treatment D

Plantlet for testing (with weight in accordance to Table 4, put in to fertilizer solution and soaked for 30 minutes while plantlet as control, was not soaed. Plantlet immersion result and control plantlet, put in to different jar volume 3.500 mL, which has been filled with sea water as mush 3 L. Plantlet kept in incubation room of tissue culture laboratory, seaweed team work, MCBA Jepara. Total weight of plantlet and physicochemical quality of water during testing, observed every once in a week.

Table 4. The weight of plantlet in every testing.

TESTING	Weight (g)	Composition of fertilizer from *Leucaena leucocephala* leaves : *Moringa oleifera* leaves
1	10,11	75% : 25%
2	10,18	50% : 50%
3	10,12	25% : 75%
4	15	20% : 80% (A)

15	15% : 85% (B)
15	10% : 90% (C)
15	5% : 95% (D)

2.4.3 Testing schedule.
The schedule of testing operation and data analysis of this research is in accordance to Table 5.

Table 5. Schedule of testing operation and analyse data for every testing.

TESTING	DATE	TIME (WIB)	ACTIVITY
1, 2 and 3	25 of May – 13 of July 2021	07.00 - 09.00	Testing operation
	3rd week of July 2021	13.30 - 15.30	Analyse data of testing result
4	14 of September – 26 of October 2021	07.00 - 09.00	Testing operation
	1st week of November 2021	13.30 - 15.30	Analyse data of testing result

3. Result and discussion

The total weight result of planlet for each treatment during observation are shown in Figure 1. The immersion of plantlet in to fermented fertilizer solution with dosage 25% *Leucaena leucocephala* leaves 75% *Moringa oleifera* leaves, proven most effective to increase the growth of planlet *Gracilaria* sp. compared with other treatments and control. This corresponds to research result of [1] which shows the *Leucaena leucocephala* leaves fertilizer contains 4.6% nitrogen nutrient, so that the apllications of *Leucaena leucocephala* leaves fertilizer can provide nitrogen requirement for plantlet. Addition of enough nitrogen nutrient can improve the growth of plantlet in general because the nitrogen nutrient spur use of photosynthesis for shoot growth in the beginning of plantlet growth become more optimal [2,3]. According to Sutejo and Kartasapoetra [4] nitrogen nutrient directly play a role in formation of protein, clorophyl, fat, enzym and other compound [5,6]. Besides that nitrogen nutrient can increase absorption of potassium and phosphor [7,8].

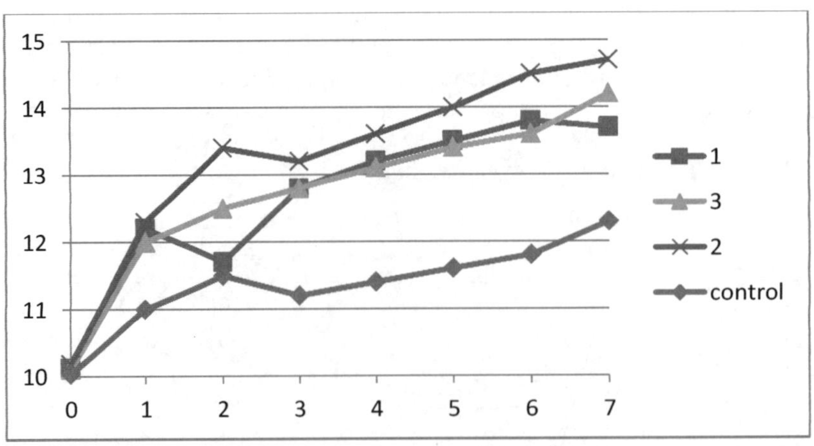

Figure 1. The growth of plantlet for each treatment during observation.

After 7 weeks maintenance, physically seen that plantlet which is maintained in immersion of fermentation fertilizer with dosage 25% *Leucaena leucocephala* leaves 75% *Moringa oleifera* leaves, have more number of shoot and longer plantlet compared with plantlet without immersion of fermented fertilizer (control). The length of plantlet in control ranged between 24 – 26 mm while treatment thallus reach 37 mm. It's caused by fermentation procedure which can decipher nutrient compound become solution form and easily absorbed by plantlet [9,10].

Table 6. Plantlet visualization after maintenance.

Description	Control	1	2	3
Number of shoot	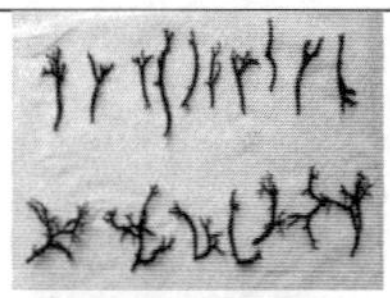		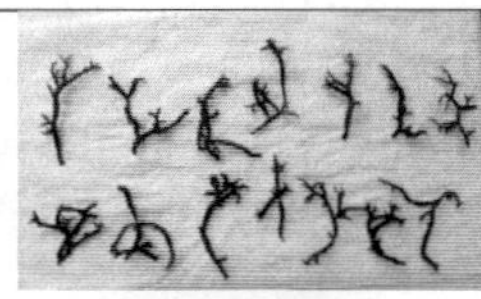	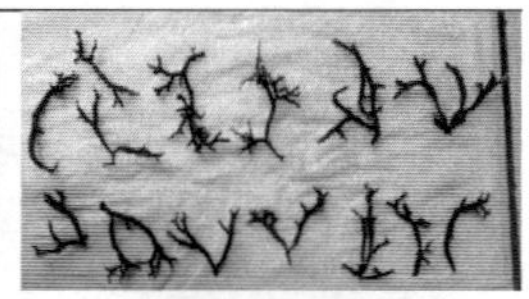
The length of plantlet	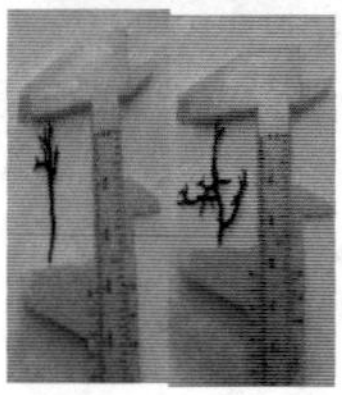	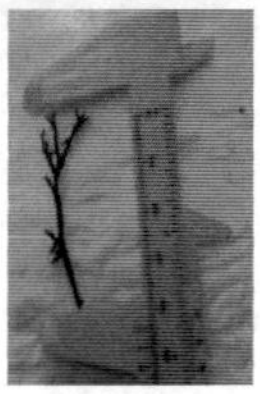		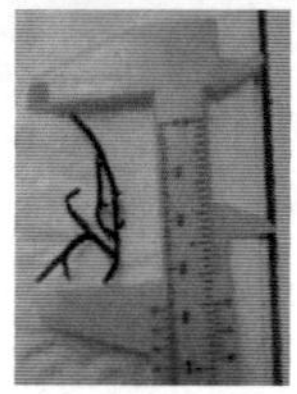

The observation data showing that the daily growth rate of all treatment is more than control (as shown in Table 6. It's in accordance with the statement of Budelman *in* Palimbungan [11] which mention that nutrient compound in *Leucaena leucocephala* leaves consist of 3.84% Nitrogen; 0.2% Phosphate; 2.06% Kalium; 1.31% Calsium; 0.33% Magnesium. Based on the research of Purwani, Achdiat and Dwiwanti (2012), the nutrient compound in *Leucaena leucocephala* leaves fertilizer consist of 4.6% Nitrogen; 0.3% Phosphate; 1.9% Kalium; 2.1% Calsium; 0.4% Magnesium. All these nutrient compound are essential compound which is very needed by plantlet in the beginning of it's growth so that it can grow optimum.

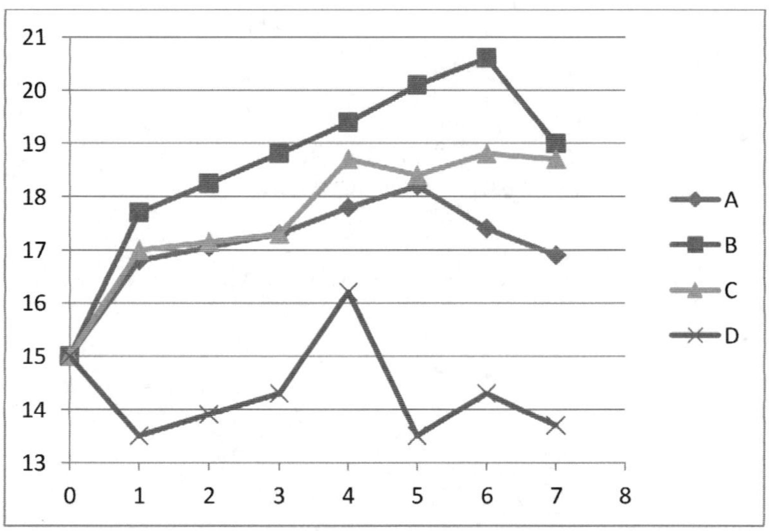

Figure 2. The plantlet growth in the next research.

It's also in accordance to the research result done by Banu *et al.,* [12], which is mention that the making of liquid organic fertilizer by using extract *Leucaena leucocephala* leaves with dosage 75 mL and 150 mL giving best result in dosage 150 mL or in the other word, growth yield will be good along with the increasing of *Leucaena leucocephala* leaves dosage. Therefore, we did the next research by using dosage of fermented fertilizer from *Leucaena leucocephala* leaves : *Moringa oleifera* leaves as much 20% : 80% in treatment A, 15% : 85% in treatment B, 10% : 90% in treatment C and 5% : 95% in treatment D with result growth percentage 12,67% (A); 26,67% (B); 24,67% (C) and -0,09% (D) as shown in Figure 2.

4. Conclusion

The test results resulted in a recommendation, namely plantlet *Gracilaria* sp. can be increased by immersing planlets into a fermented fertilizer solution with a composition of 25% *Leucaena leucocephala* leaves 75% *Moringa oleifera* leaves. Furthermore, further testing is necessary to determine the optimum time for acclimatization plantlet *Gracilaria* sp. using this immersion method, it is necessary to do the same test on *Kappaphycus alvarezii* planlets which have more economic value than *Gracilaria* sp.

5. References

[1] Purwani J, Achdiat and Dwiwanti S 2012 *Composting Time and Application Method of Legume Green Fertilizer on Growth and Yield Caisim (Brassica juncea l)* Vegetable Installation, STPP Bogor.
[2] Utami E S W, and Hariyanto S 2020 *Int J Agron.* 2020 [2795108].
[3] Nisa A K, Lamid M, Lokapirnasari W P, and Amin M 2021 *IOP Conf Ser Earth Environ Sci.* 679(1) 012041.
[4] Sutejo MM dan AG Kartasapoetra 1990 *Fertilizer and Method of Fertilization*. Rineka Cipta Jakarta.
[5] Arofah S, Sari L A, and Kusdarwati R 2021 *IOP Conf Ser Earth Environ Sci.* 718(1) 012018.
[6] Royan M R, Solim M H, and Santanumurti M B 2019 *IOP Conf Ser Earth Environ Sci.* 236(1) 012002.
[7] Bhaskoro P T, Tjahjaningsih W, and Mubarak A S 2020 *IOP Conf Ser Earth Environ Sci.* 441(1) 012144.
[8] Pradipta N, Alamsjah M A, and Masithah E D 2021 *IOP Conf Ser Earth Environ Sci.* 679(1) 012048.
[9] Murbandono L 1996 *Making Compost* Penebar Swadaya Jakarta.
[10] Alamsjah M A, Christiana R F, and Subekti S 2011 *J Ilm Perikanan dan Kelautan.* 3(2) 203-214.
[11] Palimbungan N 2006 *The Effect of Lamtoro Leaf Extract as Liquid Organic Fertilizer on the Growth and Production of Mustard Plants.* Gowa College of Agricultural Extension.
[12] Banu H, Taolin R I, and Lelang M A 2016 *Savana Cendana.* 1(01) 8-12.

6. Aknowledgement

The author would like to thank Faculty of Fisheries and Marine, Airlangga University, for conducting the seminar.

Different omega-3 concentrations in bran suspension toward the firstreproduction time of *Artemia* sp.

R W Pratama[1], L Sulmartiwi[2] and A S Mubarak[2,3]

[1]Marine Department, Faculty of Fisheries and Marine, Airlangga University, Campus C, Surabaya, 60113, East Java, Indonesia
[2]Department of Marine, Faculty of Fisheries and Marine, Airlangga University, Campus C, Surabaya, 60113, East Java, Indonesia

[3]Corresponding author: shofy.ua@gmail.com

Abstract. The need for *Artemia* sp. in Indonesia is very high and still relies on imported products. One of the drawbacks of bran suspension is the absence of Omega-3 fatty acids. Omega-3 is proven to accelerate the maturity and reproduction of *Artemia* sp. The addition of bran suspension with Omega-3 may increase the population of *Artemia* sp. in culture rapidly. This study aimed was to determine the effect of different concentrations of Omega-3 in rice bran suspension on time of the first reproduction *Artemia* sp. and to determine the effect of different concentrations of Omega-3 in rice bran suspension on *Artemia* sp. nauplii production. This study used a Completely Randomized Design which was divided into five treatments and four replications. The concentration of Omega-3 was P1 (3 mL), P2 (4 mL), P3 (5 mL), P4 (6 mL) and using rice bran suspension feed control (P0) 100 gr/ mL. *Artemia* sp. culture was carried out at a densityof 100 ind/L. The results showed that the culture of *Artemia* sp. using bran suspension feed with different concentrations of Omega-3 can affect the first reproduction time of *Artemia* sp. The fastest first reproduction time of *Artemia* sp. was produced from cultured using rice bran suspension feed with a concentration of Omega-3 (4 mL) which was 394 hours.

1. Introduction

Needs of *Artemia* sp. in Indonesia is very high and still relies on imported products. Efforts to overcome this by encouraging the development of naupli production technology in the culture of *Artemia* sp. and cyst production of *Artemia* sp. Constraints in the culture of *Artemia* sp. for the production of cysts is the unavailability of phytoplankton capable of living at high salinity. So, more use of organic material feed such as bran. The results of research conducted by Mubarak *et,al* [17] stated that one of the shortcomings of rice bran suspension is the absence of Omega-3 fatty acids. Omega-3 is proven to accelerate the maturity and reproduction of *Artemia* sp. As well as increasing fecundity and child production [4]. The addition of bran suspension with Omega-3 may increase the population of *Artemia* sp. in culture quickly. Rice bran is a waste from rice milling which is available in large quantities. Bran contains high fat (16-20%) and protein (12-13%), as well as B vitamins and minerals (6-9%) [9,6].Rice bran contains vitamins E and B12 which function to keep the reproductive system running optimally [22]. Rice bran is also used as feed for other cladosera and has been shown to produce high populations [16,6]. The most unsaturated fatty acids found in rice bran are dominated by Omega-6 fattyacids [12], while rice bran does not have Omega-3 fatty acids [18]. In an effort to optimize the culture of*Artemia* sp. that is done by paying attention to the quality and quantity of feed. The potential bran suspension for the culture of *Artemia* sp. quality needs to be improved in order to get better *Artemia*sp.culture results. Increasing the concentration of Omega-3 in bran suspension has been shown to accelerate the time of first reproduction in *D. magna* [23, 16]. Omega-3 fatty acids also have an effect on increasing the reproductive ability of parthenogenesis in cladosera, through increasing fecundity and decreasing reproductive time [1]. Optimization of reproduction of *Artemia* sp. this can be done by improving the quality of the bran suspension feed. Based on the description above, it is necessary to know the optimal concentration of Omega-3 in the bran suspension in order to reduce the length of the first reproduction time and fecundity in the culture of *Artemia* sp. The objectives of this research was to determine the effect of different concentrations of Omega-3 in rice bran suspension on time of the first reproduction *Artemia* sp. and to determine the effect of different concentrations of Omega-3 in rice bran suspension on *Artemia* sp. nauplii production.

2. Materials and Methods

This study was used 5 treatments and 4 replications, namely: concentration of omega-3 o ml, 3 ml. 4 ml, 5 ml and ml in 100 g of rice bran *Artemia sp* culture was carried out with a density of 100 ind/L in 15 liters of saline water (33ppt) with 70 mg/L bran suspension feed [17]. Furthermore, the length of thefirst reproduction time was observed for each treatment.

2.1 Rice bran production

100 g rice bran was in a blender glass, each add with omega-3; 0.0 ml, 3.0ml, 4.0 ml, 5.0 ml and 6.0 ml and filled with 500 mL of water. suspended in a blender at 2000 rpm for 5 minute twice. The second suspended was done after 30 minutes of the first suspension. Furthermore, Furthermore, the suspension was filtered using a 2 mm, 0.1 mm to separate water and rice bran. The suspension was filtered again using ad 40 um filter. to produce a small bran particle size (<40um) The suspension that was passed through the filter then added with water and then the final volume becomes 500 mL [18].

2.2 Analysis of Fatty Acids, Amino Acids, and Proteins

The protein and fat content Rice bran suspension was tested using a proximate based on the AOAC standard method (1995). Amino acids and fatty acids in rice bran were measured using high performance liquid chromatography (HPLC) with the Hewlett Packard Series 1100, based on the method of Fardiaz (1989).

2.3 Data analysis

The data obtained were recorded and then analyzed using *One Way Analysis of Variance (*ANOVA) to determine the effect of the treatment given. Treatment that shows significant results, the calculation is continued with Duncan's Multiple Range Test.

3. Result and Discussion

The rice bran suspension contained 74 mg mL^{-1} organic materials,0.83% protein, and 0.79% fat. The rice bran suspension had a higher content of the amino acids arginine (0.050%) and histidine (0.020%) than the cassava suspension (arginine 0.005% and histidine 0.003%) and low omega-3 (Table 2). In this study, the parameters studied were different concentrations of Omega-3 in the suspension of the first reproduction time of *Artemia* sp. the following results were obtained (Tabel 1):

Table 1. Calculation of the survival rate of *Artemia* sp. using bran suspension feed with different concentrations of Omega-3

Treatment	Omega-3 Concentration in Bran Suspension (mL/100g bran)	SR ± SD (%)
P0	0 mL	85,57 ± 4,41[ab]
P1	3 mL	86,42 ± 6,63[ab]
P2	4 mL	89,90 ± 1,36[a]
P3	5 mL	84,45 ± 2,86[ab]
P4	6 mL	80,40 ± 2,38[b]

The application of concentrations of Omega-3 in the rice bran suspension had a significant effect ($P<0.05$) on the survival rate of *Artemia* sp. so that Duncan's further test can be carried out with a significant level of 5%. Duncan test analisis was resulted that the rice bran suspension without Omega-3 was significantly different from the rice bran suspension with Omega-3 6 ml., but not significantly different Other treatments (P1, P2, and P3). (Table 1.)

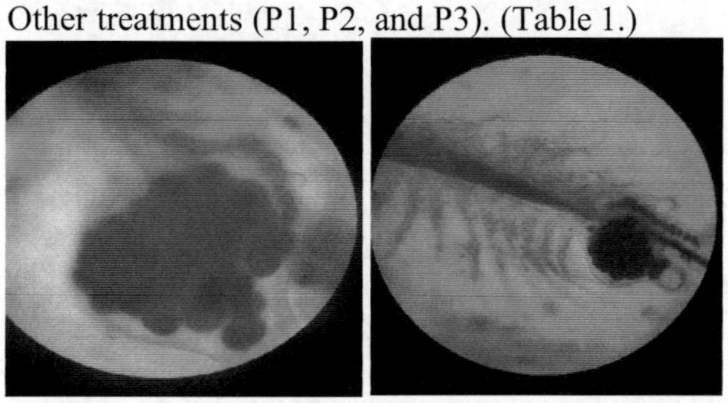

Figure 1. Fecundity process in *Artemia* sp.

Table 2. Amino Acid and Fatty Acid Concentration in Bran Suspension

No	Amino Acid		Fatty Acid	
	Amino Acid	Total	Fatty Acid	Total
1	Protein	12,26	Laurat	0,17±0,01
2	Fat	10.52	Tridecylic	0,04±0,01
3	Carbohidrate	68,72	Miristat	1,25±0,02
4	Leusin	4.27	Pentadecylic	0,30±0,04
5	Arginin	3.85	Palmitat	18,78±0,32
6	Lisin	2.11	Palmitoleic	5,89±0,11
7	Histidin	1.61	Margaric	0,12±0,01
8	Valina	3.02	Stearate	2,37±0,21
9	Fenilalanin	2.90	Vacenat	0,05±0,01
10	Treonin	2.33	Oleat	28,37±1,10

IOP Conf. Series: Earth and Environmental Science 1273 (2023) 012037 doi:10.1088/1755-1315/1273/1/012037

11	Metionin	1.27	Linoleat	23,65±1,00
12	Isoleusin	2.13	Arakidat	0,11±0,01
13	Glutamat	8.60	Gama linoleat	0,2±0,01
14	Aspartat	4.80	Gadoleic	0,25±0,05
15	Glisin	3.18	Alfa linoleat	0,5±0,08
16	Serin	2.69	Eicosadienoic	0,06±0,01
17	Alanin	4.30	Behenat	0,22±0,01
18	Tirosin	3.43	dihomo gama linoleat	0,05±0,02
19			Eicosapentaenoic (EPA)	
20			Docosahexaenoic (DHA)	
			Total Omega-3	0,5
			Total Omega-6	26,90

Table 3. Calculation of the first reproduction time of *Artemia* sp. using bran suspension feed with different concentrations of Omega-3.

Treatment	Omega-3 Concentration in Bran Suspension (mL/100g bran)	The first reproduction time (hours)
P0	0 mL	418.50 ± 1,00^{c}
P1	3 mL	414.75 ± 0,95^{b}
P2	4 mL	394.00 ± 0,81^{a}
P3	5 mL	440.25 ± 1,25^{d}
P4	6 mL	494.50 ± 1,73^{e}

The application of concentrations of Omega-3 in the rice bran suspension had a significant effect ($P<0.05$) with respect to the first reproduction time of *Artemia* sp that the rice bran suspension without Omega-3 was significantly different from Other treatments (P1, P2, and P3). {Table 3.)

Table 4. Calculation of naupli production per parent *Artemia* sp. using bran suspension feed with different concentrations of Omega-3.

Treatment	Omega-3 Concentration in Bran Suspension (mL/100g bran)	Number of naupli ProductionPer Parent ± SD (Ind/Parent)
P0	0 mL	21.00 ± 1,41^{b}
P1	3 mL	22.50 ± 1,29^{ab}
P2	4 mL	24.25 ± 1,25^{a}
P3	5 mL	15.25 ± 0,95^{c}
P4	6 mL	13.25 ± 1,70^{c}

The application of concentrations of Omega-3 in the rice bran suspension had a significant effect ($P<0.05$) on the production of offspring per parent of *Artemia* sp. Duncan test results showed that the rice bran suspension without Omega-3 was significantly different from P2, P3 and P4 but not significantly different from P1 (Table 4). The addition of Omega-3 in rice bran suspension as feed affects the survival rate, time of first reproductionand the number of the offspring produced per brood. Omega-3 are polyunsaturated fatty acids (PUFA) with a triple bond of a terminal methyl group in their chemical structure [20]. Omega-3 fatty acids include *Eicosapentaenoic* fatty acids (20:5 n-3; EPA), and *Docosahexaenoic* fatty acids (22:6 n-3; DHA) [5]. *Eicosapentaenoic* fatty acids (20:5 n-3; EPA) and *Docosahexaenoic* fatty acids (22:6 n-3; DHA) have been shown to be closely related to larval survival [7] and the reproductive success in fish [21]. Omega-3 also functions as a free radical scavenger and anti-inflammatory so that it can minimize the impact of stress on the organism. The availability of Omega-3 can make the availability of fatty acids as a constituent of yolk for the vitellogenesis process. So that the process of egg formation becomes faster. The optimal addition of Omega-3 fatty acids in the *Artemia* sp. feed have a positive impact onthe survival rate of *Artemia* sp. Omega-3 has been shown to increase survival in fish, crustaceans and *Artemia* sp. [13,19]. The addition of Omega-3 at a concentration of 4 mL/99 grams of rice bran (P2) as feed for *Artemia* sp. proven to increase survival by 89.90%. However, the addition of Omega-3 more than 4 mL/99 grams of rice bran caused a decrease in SR. This is because excess Omega-3 fatty acids (EPA and DHA) can increase oxidative stress in sturgeon larvae [11] and decrease metabolism in Atlantic salmon [19]. Oxidative stress due to excess Omega-3 will cause an increase in free radicals anddamage to cells and tissues [14].

The addition of different Omega-3 in bran suspension as feed in the culture of *Artemia* sp. affect the first reproductive time of *Artemia* sp. Culture of *Artemia* sp. which used Omega-3 with a concentration of 4 mL in bran suspension had the fastest first reproductiontime of 394 hours. EPA and DHA in feed ingredients are needed by *Artemia* sp. to speed up reproduction time and maintain the quality of the resulting offspring. The optimal content of Omega-3 in the feed can make the availability of high fatty acids as a constituent of the yolk so as to accelerate the process of vitellogenesis and embryo development. This is in line with researched by Agustin [2] that Omega-3 (EPA and DHA) help accelerate the vitellogenesis process andin research Ma et al [16] DHA/EPA plays an important role in embryo development at the larval stage of Siberian sturgeon. The concentration of feed in the form of EPA and DHA received by female chicksin cladosera shortens the reproductive time of the first parthenogenesis [8] Omega-3 fatty acids also have an effect on increasingthe reproductive ability of parthenogenesis in cladosera, through decreasing reproductive time and increasing fecundity. However, this is different from the statement Balai Besar Perikanan Budidaya Air Payau [3] that *Artemia* sp. will reproduce after 21 days (504 hours).The first reproduction time is causedby differences in the quality and quantity of feed, pH, dissolved oxygen, and temperature [10]. The addition of different Omega-3 in bran suspension as feed in the culture of *Artemia* sp. affected the production of offspring per parent. The highest production of offspring per parent was in the culture of *Artemia* sp. resulting from treatmentusing a concentration of Omega-3 in bran suspension of 4 mL L^{-1}. This is in accordance with research [20], that the production of offspring per parent is highest in the culture of D. Magna (Cladosera) which is in the treatment using an Omega-3 concentration of 4 mL.

The addition of Omega-3 concentrations in the bran suspension gave significant results in increasing the production of offspring per brood produced. This is because the optimal content of Omega-3 in the feed will protect the fatty acids by accelerating the process of embryogenesis. This is in line with research Ma et al [15], that DHA/EPA plays an important role in embryo development at the larval stage of Siberian sturgeon. Theconcentration of fat, protein, and amino acids in the feed can affect the production of offspring per parent and the speed of embryonic development in cladosera [7,12). Production of offspring per parent *Artemia* sp. The lowest occurred in the treatment using Omega-3 concentrations in bran suspension of 6 mL (P4).The addition of a higher concentration of Omega-3 will interfere with the process of embryo development. This is in accordance with research Izquierdo [9]. Excess Omega-3 essential fatty acids infeed will negatively affect reproductive achievement in fish.

4. Conclusion

Culture of *Artemia* sp. using different concentrations of Omega-3 in rice bran suspension feed affected the first reproduction time of *Artemia* sp. culture of *Artemia* sp. using different concentrations of Omega-3 in rice bran suspension feed affected the production nauplii.

5. References

[1] Abrusan, G., P. Fink dan W, Lampert. 2007. *J. Lim and Ocea*, 52(4): 1724–1728.
[2] Agustin L., L Sulmartiwi., and A S Mubarak. 2020.. *IOP Conf. Series: Earth and Environmental Science* 441: 1-4.
[3] Jepara Brackish Water Aquaculture Center. 2017. *Technical Instructions for Artemia Biomass Production Procedures in Bak*. (Jepara:Indonesia) Hal 4.
[4] Beltz BS, Tlusty MF, Benton JL, Sandeman DC. 2007. *Neuroscience Letters*, 415: 154-158.
[5] Damayanti K.Y Mubarak A.S. Sari LA. 2020. *IOP Conference Series Earth and Environmental Science* 441(1):012156.
[6] Faria, S., P. Bassinello, M. Penteado. 2012.. *Brazilian J. of Pharma Scies* 48: 35-45.
[7] Fink, P., C. Pflitsch and K. Marin. 2011. *Ploosone org*, 6(12).
[8] Frost PC, Ebert D, Larson JH, Marcus MA, Wagner MD, Zalewski A. 2010. *Oecologia*, 162:865-887.
[9] Izquierdo, M. 2005. Essential fatty acid requirements in Mediterranean fish species. *CahiersOptions Mediterraneennes,* 63: 91-102.
[10] Jeronimo, F. M. and A. G. Valdivia. 1991. *Journal Hydrobiologia,* 222: 49-55.
[11] Ji, H., Li, J., Liu, P., 2011.. *Comp. Biochem. Physiol. B, Biochem. Mol. Biol.* 159: 49–56.
[12] Jumari, A., A. S. Rahmani and F. R. Riana. 2015. *Equi J. ofChem Engin*, 14(1): 17-22.
[13] Kangpanich C, Pratoomyot J, Siranonthana N, Senanan W. 2016. *PeerJ* 4:e2735.
[14] Kohen, R., Nyska, A., 2002.. *Toxicol. Pathol.* 30: 620–650.
[15] Ma, J., Wang, J., Zhang, D., Hao, T., Sun, J., Sun, Y., & Zhang, L. (2014). *Aquaculture*, 433:
[16] Mehdipour N, Fallahi M, Takami GA, Vossoughi G, Mashinchian A. 2011. *Irania J. of Sci andTech*, 2: 157-163.
[17] Mubarak, A. S., D. Jusadi, M. Z. Junior, and M. A. Suprayudi. 2017. *J. Akuakultur Indonesia* 16(2):234.
[18] Nettleton, J.A. 1995. *Omega-3 Fatty Acids and Health.* (New York: Chapmain and Hall).
[19] Prusinska M, Kushniryk O, Khudyi O, Khuda L, Kolman R. 2015.. *Arch of Pol Fishs* 23: 149-154.
[20] Salvador EM, Steenkamp V, McCrindle CM. 2014. *J.l of Agri and Biotech and Sustai Devt* 6:
29-38.
[21] Tanaka, N., Ishida, T., Nagao, M., Mori, T., Monguchi, T., Sasaki, M., Mori, K., Kondo, K., Nakajima, H., Honjo, T., Irino, Y., Toh, R., Shinohara, M., Hirata, K., 2014. *Atherosclerosis* 237 (2): 577–583.
[22] Wacker A, Creuzburg MD. 2007. *Functional Ecology*, 21: 738-747.
[23] Watanabe, T., and Vassallo-Agius, R., 2003. *Aquaculture,* 227: 35-61.
[24] Watanabe, T., 2007.. *J. of the World Aqua Socie*, 24(2): 152 - 161.

6. Acknowledgments
The authors are grateful to Universitas Airlangga, faculty of fisheries and Marine for all support to thisresearch. Also we thanks to the reviewers for their suggestion, which improved our manuscript

IOP Conf. Series: Earth and Environmental Science 1273 (2023) 012038 doi:10.1088/1755-1315/1273/1/012038

Dynamics of water quality in grouper nurseries with flow through, recirculation and bioremediation systems

S Ismi[1], B Astari[2,3], and I Mastuti[1]

[1] Research Center for Fishery, Research Organization for Earth and Maritime, National Research and Innovation Agency, Cibinong-Bogor, Indonesia

[2] Study Program of Aquaculture, Graduate School, IPB University, Bogor, Indonesia

[3] Corresponding author: belindaastari22@gmail.com

Abstract. The flowthrough system, a daily water exchange, has high risk in fluctuation of water quality. Moreover, the risk could impact the seed productions, especially in grouper nurseries. Recirculation and bioremediation systems are satisfying technique to maintenance the water quality. The aim of this research was to analyse the effectiveness of flowthrough, recirculation and bioremediation systems in grouper nursery for water quality and grouper seed production. The research was conducted with three treatments and four replications as follows: A. flowthrough, B. recirculation and C. recirculation with bioremediation. The research was conducted in a completely randomized design for 40 days. The best water quality in treatment C: Salinity ranged from 30-35 g/L, ammonia reached 0.0726-1.8978 mg/L, and inhibited the growth of *Vibrio* spp. up to 2.580 CFU/mL. The results showed that the best average value was obtained by treatment C with a survival rate of 94.8±0.4%, length growth rate of 0.110±0.003 cm/day, weight growth rate of 0.175±0.013 g/day, specific length growth rate of 2.02±0.04%, specific weight growth rate 5.56±0.28%, total length 8.0 cm, total weight 7.91 g, and feed conversion ratio 0.73±0.07.

1. Introduction

Grouper nursery requires seawater to be filtered first with a filter to clean it from dirt and mud. The filter uses a cement tank with a position higher than the maintenance tank so that seawater from the filter can flow into each maintenance tank following gravity. The water filter consists of stones, coral, and sand. The sea water is then circulated to the maintenance tank with a running water system or routine water changes [1]. Water goes straight out of the system without any recycling and utilization processes [2]. The system requires frequent water changes and is vulnerable to sea conditions, tides, and large waves that can affect water quality. Conditions of tidal sea waters and big waves that cause cloudy and muddy. The resulting sea water can be cloudy due to a lot of mud that escapes and clogs the sand filter. Fish will have a reduced appetite and are susceptible to disease [1]. The use of the Recirculating Aquaculture System (RAS) in each grouper nursery is able to control water quality, salinity, disease, water supply, land availability, and as an alternative to fish maintenance that is more efficient in business [3,4].

Materials that can be used as filters in RAS are dacron, sand, zeolite, activated charcoal and biobole [5]. Dacron and sand act as physical filters, filtering suspended solids in water in the form of leftover feed and feces from fish [6,7]. Zeolite and activated charcoal act as chemical filters that can

IOP Conf. Series: Earth and Environmental Science 1273 (2023) 012038 doi:10.1088/1755-1315/1273/1/012038

absorb toxic substances such as ammonia and nitrite [8]. Zeolite is a hydrated porous alumina silicate mineral formed by tetrahedral alumina (AlO_4^{5-}) and silica (SiO_4^{4-}) which forms a negatively charged and porous structure. Activated charcoal has a chemical content, namely carbon compounds that are useful for the water purification process [9]. Bioball acts as a biological filter, namely a medium for attaching microbes (nitrifying bacteria) that play a role in degrading nitrogen into nitrate which is safer for fish [10]. The use of zeolites, activated charcoal, and bioball can improve water quality through ammonia oxidation and the growth of non-pathogenic nitrifying bacteria in RAS [11]. Nitrogen waste in the form of ammonia is excreted as a result of fish metabolism by 80-90%, while urea is generally 10-15% of dissolved nitrogen waste. The increased accumulation of waste in the maintenance water will reduce the performance of the RAS [12], so there is a need for water remediation. Bioremediation is a water remediation process using microorganisms or their enzymes to degrade contaminants from a cultured water environment [13, 14].

One of the factors that play a role in determining the success of grouper nursery is water quality management, because all life, health and growth depend on water quality as a medium of life. Water quality can be broadly defined as any physical, chemical and biological factor that affects water use [15]. Water quality in general can be defined as any variable (variable) that affects the management and survival, breeding, growth or production [16, 17]. Measurement of water quality during fish rearing is important to know the symptoms that occur as a result of changes in one of the water quality parameters. By knowing these symptoms, an action can be taken to overcome unfavorable changes to the survival and growth of the organisms being reared [18]. The aim of this research was to analyze the effectiveness of flowthrough, recirculation and bioremediation systems in grouper nursery for water quality and grouper seed production.

2. Materials and Methods

2.1. Research Site

This research was conducted in a household scale hatchery CV. Jaya Utama Abadi, Buleleng, Bali from January to February 2021. The tank used is 1 m × 2 m × 0.7 m containing 1 m^3 of seawater. Oxygen supply in each tank is added through aeration using a blower. The test fish used were cantik grouper seeds (*Epinephelus fuscoguttatus* × *Epinephelus microdon*) with an average length and initial weight of 3.6±0.2 cm and 0.92 g/fish with a stocking density of 500 fish/m^3. Before use, the container was disinfected with 100 mg/L chlorine then the bottom and walls of the tub were brushed, rinsed with fresh water and dried.

2.2. Experimental Design

The research was conducted with three treatments and four replications as follows: A. flowthrough, B. recirculation and C. recirculation with bioremediation. During rearing, grouper fry was fed commercial feed containing 48-55% protein with mouth opening adjusted at satiation with a frequency of 3 times a day, in the morning, afternoon and evening. Water management is carried out using seawater that has passed through a water filter consisting of rocks, coral, and sand. The water is then collected into a reservoir and channelled to a maintenance tank. Siphoning is done twice a day about 25% of the initial volume of water. Washing of maintenance tanks and components of the recirculation system is carried out every 10 days. In treatment A, water changes were carried out twice a day, in the morning and evening. In treatments B and C with a recirculation system consisting of dacron, sand, activated charcoal, zeolite, bioball, and P-1800 power-heads water pump with a flow rate of 1,600 L/hour. In treatment C, the addition of 6 mL/m^3 bioremediator containing *Lactobacillus casei* at 2.0×10^6 CFU/mL and *Saccharomyces cerevisiae* at 3.5×10^5 CFU/mL at the beginning of maintenance and after washing the tub and components of the recirculation system. The length of maintenance lasts for 40 days [3].

2.3. Measurement test

Water quality parameters that were monitored, namely temperature, salinity, pH, dissolved oxygen (DO) was carried out every day. Measurement of temperature using a thermometer, salinity using an ATC refractometer, pH using an ATC digital pH meter, and dissolved oxygen using a DO meter TOX-

90i. Measurements of ammonia (NH_3^-) and nitrite (NO_2^-) were carried out before stocking and every 10 days. Total bacteria, and *Vibrio* spp. done before stocking and every 20 days. Ammonia and nitrite measurements were carried out in the water quality laboratory as well as total bacteria and *Vibrio* spp. in the pathology laboratory of the Center for Marine Cultivation Research and Fisheries Extension (BBRBLPP), Gondol, Bali. Growth observations (length and weight) were carried out before stocking and every 10 days as much as 10% of the total grouper seed population in each tank. The dead fish were also weighed. Grouper seed mortality was observed every day during the research. The total length was measured using a ruler with an accuracy of 0.1 cm and the weight was weighed with a digital scale with an accuracy of 0.01 g.

2.4. Data analysis
Data analysis was performed using Microsoft Excel 2013 and SPSS version 23 software. Analysis of variance (ANOVA) at 95% confidence interval was performed to determine whether or not the treatment had a significant effect on the test parameters. If it has a significant effect, then it is continued with Duncan's test to determine the difference between treatments at a 95% confidence interval. The data obtained were analysed descriptively with graphs.

3. Result and Discussion
3.1. Temperature

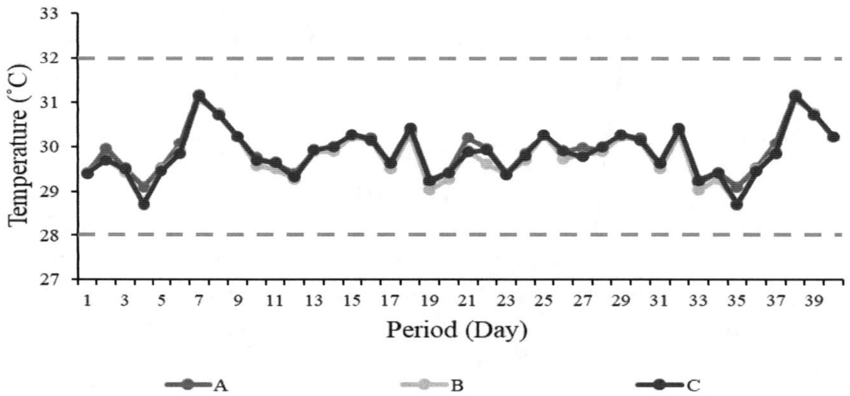

Figure 1. Water temperature dynamics in cantik grouper nursery with A. flowthrough, B. recirculation and C. recirculation with bioremediation system for 40 days of rearing. The green dotted line indicates the optimal value [19].

Fish metabolism, behaviour, growth, morphology, and water temperature can all be impacted. In addition, temperature has an impact on the rate of reaction of elements and compounds in water as well as the solubility of gases. A 40-day observation period with temperature measurements revealed that the water's ideal temperature range was between 28.7 and 31.2°C. While this is happening, 24–32°C is the ideal temperature for the VNN virus to flourish in its host. VNN can replicate at 28°C with a 100% mortality rate in humpback grouper larvae [20]. According to statistics from the Central Agency for Buleleng Regency [21], the maximum rainfall occurs in the beginning of the year (January through June) and towards the end of the year (November-December). The rainy season in that year was essentially identical to the rainy season in 2021, the year the research was conducted. In 2018, a rise in temperature was noted in the waters off North Bali from mid-January (28.20±0.84°C) to April (30.10±1.01°C) [22]. According to Joint & Smalle [23], temperature has an impact on bacterial activity, with high temperatures resulting in the highest levels of microbial activity. High rainfall also causes a decrease in salinity.

3.2. Salinity

Changes in salinity will alter the osmoregulation system of fish, which affects fish growth and metabolism. Salinity is a chemical characteristic of water [24]. Treatment A, which is a flowthrough system with regular water changes, greatly affects the drastic decrease and increase in salinity. This is due to extreme weather changes such as sunny and then heavy rain. The recirculation system used can prevent drastic drops and increases in salinity [25]. The osmoregulation process in fish is also influenced by pH.

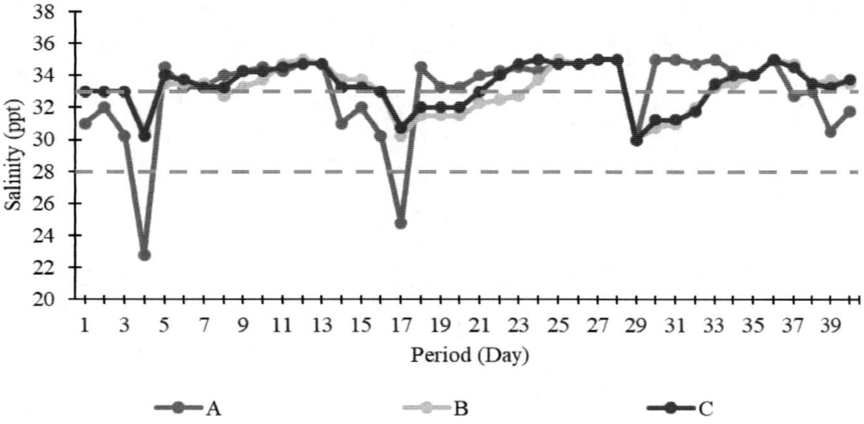

Figure 2. Water salinity dynamics in cantik grouper nursery with A. flowthrough, B. recirculation and C. recirculation with bioremediation system for 40 days of rearing. The green dotted line indicates the optimal value [19].

3.3. pH

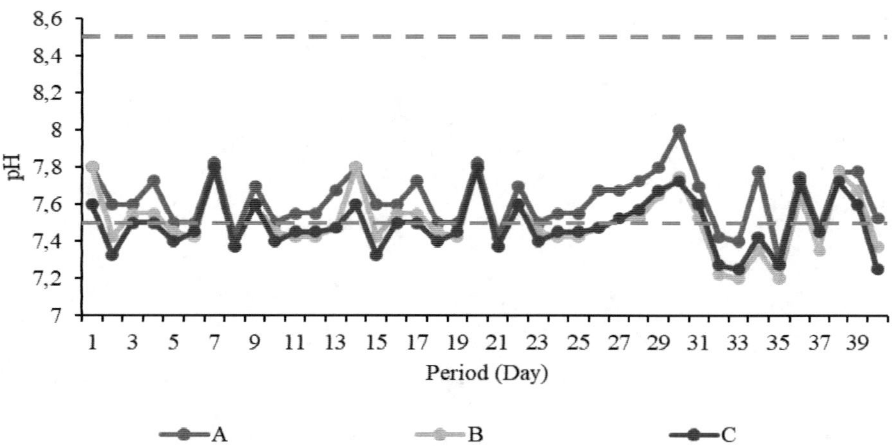

Figure 3. pH dynamics in cantik grouper nursery with A. flowthrough, B. recirculation and C. recirculation with bioremediation system for 40 days of rearing. The green dotted line indicates the optimal value [19].

pH obtained during research tend to be low especially in recirculation system but is still within tolerance for optimal grouper seed growth. Low pH may directly experience alterations in ion regulation and disruption of other metabolic pathways. When exposed to low pH, more energy was needed to accomplish osmorespiration needs. Fish must adapt their physiological requirements to sustain this high energy activity by either increasing feed intake to meet the new high energy demand or limiting energy for essential metabolism demands. However, an increase in feeding activity also caused an increase in endogenous ammonia generation in fish because of the necessity for the body to eliminate the substance as a result of protein catabolism. All fish can develop severe endogenous poisoning from high endogenous ammonia, which can result in gill hypertrophy, slowed growth, seizures, and even death. Ammonia must be effectively eliminated as a waste product of nitrogen metabolism in order to address this issue [26]. The pH value shown in the study is still within tolerance. Besides pH, DO is also an important chemical parameter in water.

3.4. DO (Dissolved oxygen)

The oxidation of waste products and the combustion of food, which generate energy for the life and growth of groupers, are influenced by the DO parameter. Nitrifying bacteria that exist in systems that demand oxygen depend on DO in the maintenance media using a recirculation system [27]. The DO value in the study was at the optimal value because of the use of an aeration blower to supply oxygen.

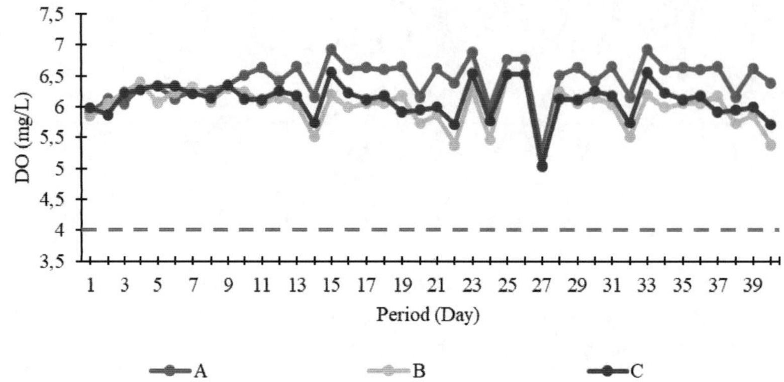

Figure 4. DO (Dissolved oxygen) dynamics in cantik grouper nursery with A. flowthrough, B. recirculation and C. recirculation with bioremediation system for 40 days of rearing. The red dotted line indicates the optimal value [19].

3.5. Ammonia and Nitrite

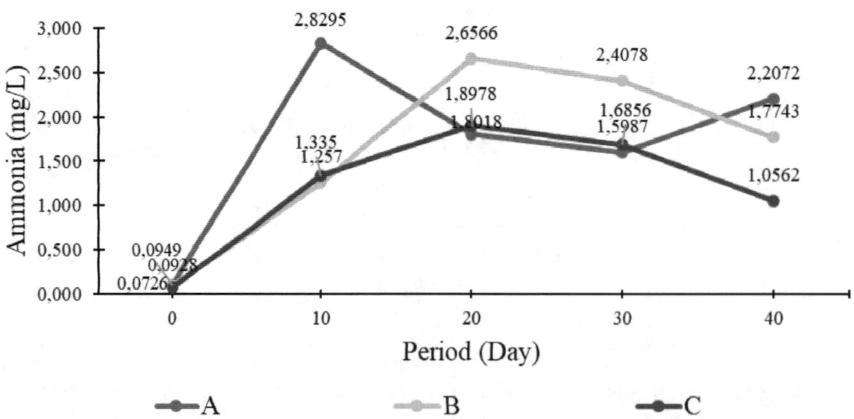

Figure 5. Ammonia dynamics in cantik grouper nursery with A. flowthrough, B. recirculation and C. recirculation with bioremediation system for 40 days of rearing.

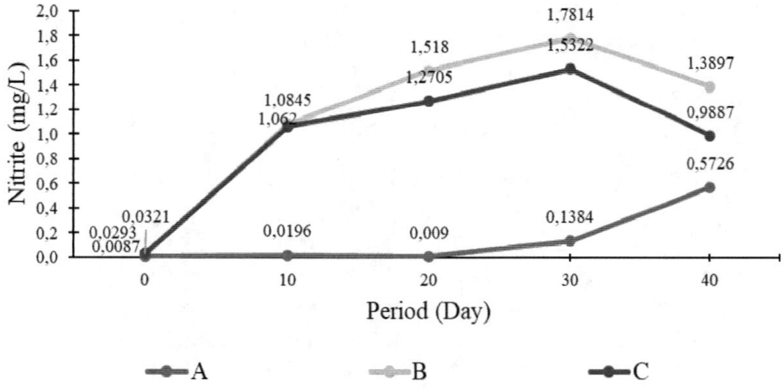

Figure 6. Nitrite dynamics in cantik grouper nursery with A. flowthrough, B. recirculation and C. recirculation with bioremediation system for 40 days of rearing.

Changing the water can help reduce aquatic waste like ammonia by introducing new water. This prevents bacteria from having a chance to perform the nitrification process, preventing the oxidation of ammonia in the water to nitrite. Due to the availability of pores for filtering and as a location for bacteria to carry out the nitrification process, the use of recirculation systems and bioremediators in treatment C can prevent the increase in ammonia in the water [28]. The nitrification process, in which ammonia (NO_3^-) is converted to nitrite (NO_2^-) and then further converted to nitrate, is impacted by the rising water temperature. Ammonia-oxidizing bacteria (AOB) are responsible for the first oxidation phase, while nitrite-oxidizing bacteria are responsible for the second. N_2 gas is produced as a result of the nitrogen cycle's relationship between nitrification and denitrification [29]. In coastal waterways, eutrophication is caused by an increase in organic matter, and nitrogen is the most important factor in this process [30]. According to reports, the bacterial population in coastal waters and the rise in dissolved organic carbon concentration are significantly correlated [31].

3.6. Total bacteria and Vibrio spp.

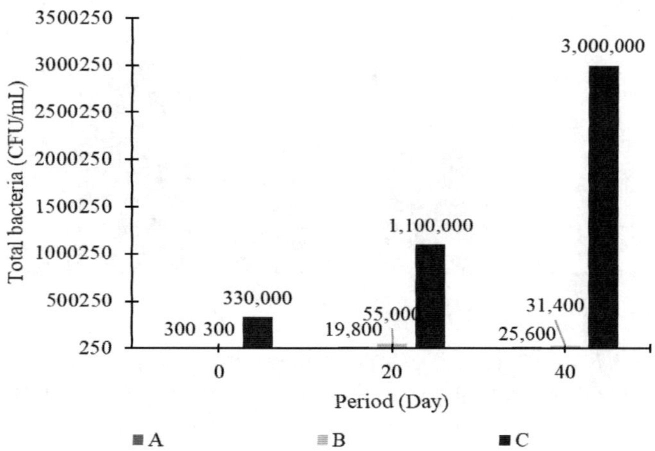

Figure 7. Total bacteria dynamics in cantik grouper nursery with A. flowthrough, B. recirculation and C. recirculation with bioremediation system for 40 days of rearing.

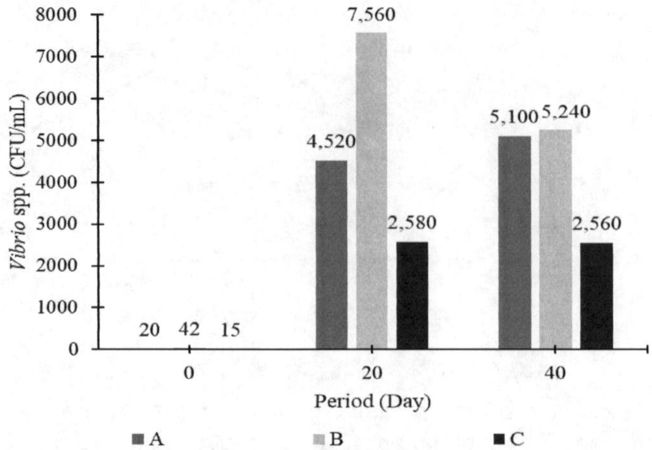

Figure 8. *Vibrio* spp. dynamics in cantik grouper nursery with A. flowthrough, B. recirculation and C. recirculation with bioremediation system for 40 days of rearing.

Deterministic and stochastic causes were considered to be responsible for the large rise in total bacteria. The establishment of microbial communities must take both of these aspects into account. Temperature, salinity, DO, pH, and feed are deterministic elements that work together to create an environment where microorganisms can thrive. The stochastic component provides an opportunity for microorganisms to create a habitat and breed in the ideal circumstances at the ideal time and place [32]. The number of bacteria living on the water's surface can change with the seasons. A rise in the number of *Vibrio* spp. and *Aeromonas* spp. adhered to the copepod's surface is observed during the rainy season [33]. *Vibrio* spp. in water maintenance media can be prevented by the bioremediation in RAS.

3.7. Survival rate

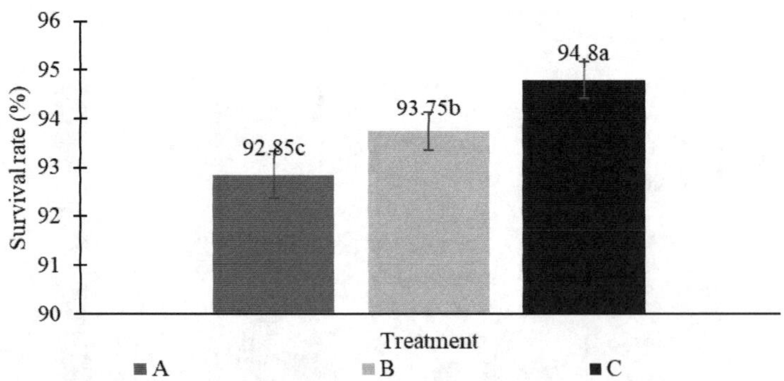

Figure 9. Survival rate in cantik grouper nursery with A. flowthrough, B. recirculation and C. recirculation with bioremediation system for 40 days of rearing.

The highest survival rate was obtained by treatment C. The ideal water conditions employed for grouper life are what account for the high SR value. This is supported by the existence of a recirculation and bioremediation system containing *Lactobacillus casei* and *Saccharomyces cerevisiae*. There are times when strong tides, large waves, clouds, and muddiest conditions present challenges for the grouper nursery industry. As a result, the sand filter becomes clogged with numerous minute particles and loses its effectiveness. The water that will be used gets contaminated, posing a threat to the life and development of the seeds in the nursery tank. At that time, bacterial illnesses typically manifest, causing lesions on the body, redness on the fins and mouth, and numerous deaths [1].

3.8. Length and weight growth

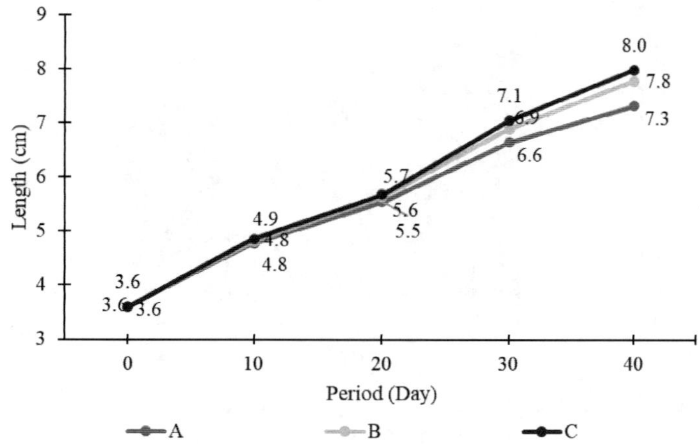

Figure 10. Length growth in cantik grouper nursery with A. flowthrough, B. recirculation and C. recirculation with bioremediation system for 40 days of rearing.

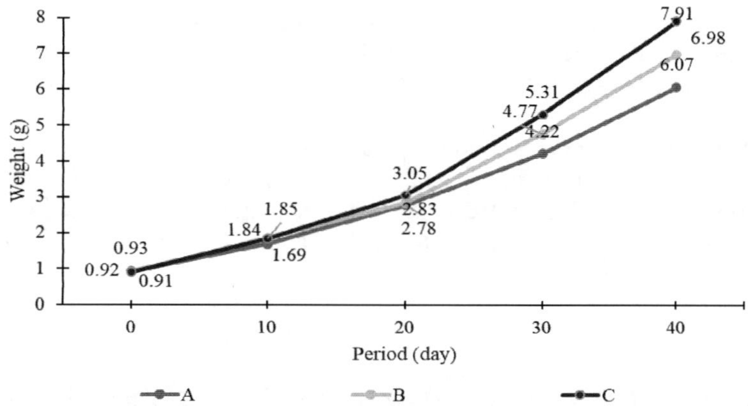

Figure 11. Weight growth in cantik grouper nursery with A. flowthrough, B. recirculation and C. recirculation with bioremediation system for 40 days of rearing.

In recirculation with bioremediation length growth rate of 0.110±0.003 cm/day, weight growth rate of 0.175±0.013 g/day, specific length growth rate of 2.02±0.04%, specific weight growth rate 5.56±0.28%, growth rate was faster than in flow-through systems. This is because the system's ability to maintain continuous water quality, enhance waste management and nutrient recycling, and be hygienic results in more stable water quality. Low energy use and metabolic savings have been linked to this condition. Consumed nutrients can then be used to maximize growth. This study found that fish growth performance, water quality, and feed utilization were all more significantly impacted by RAS with bioremediation. The bioremediator employed in this research comprised *Lactobacillus* sp. as the type of bacteria. These bacteria, according to Akbar *et al.* [34], have lactic acid, which can ferment organic material and hasten the breakdown of organic waste. The pace of fish growth will be accelerated by improved environmental conditions. The results of higher growth of grouper are also shown by research by Samad *et al.* [35] on the recirculation system.

3.9. Feed convention ratio

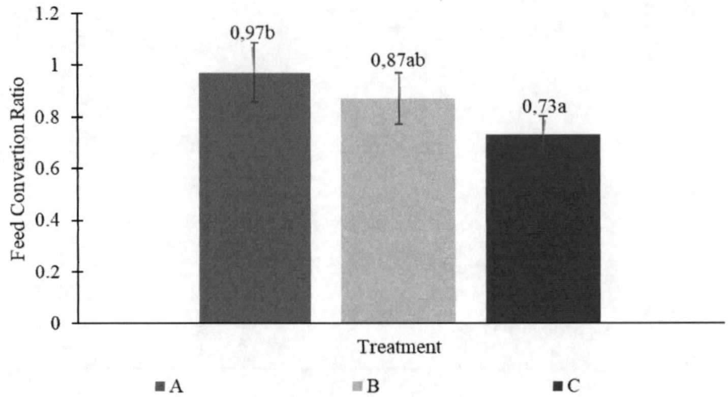

Figure 12. Feed convention ratio in cantik grouper nursery with A. flowthrough, B. recirculation and C. recirculation with bioremediation system for 40 days of rearing.

The FCR value was decreased by Treatment C to 0.73±0.07. *Saccharomyces cerevisiae* was the type of bacterium present in the bioremediator employed in this research. *Saccharomyces cerevisiae* in water can activate protease enzymes that contribute to speeding protein breakdown processes, claim Akbar *et al.* [34]. The remainder of the meal and waste materials in the rearing tank undergo protein breakdown processes, which improve the water quality and increase the fishes' appetite [36].

4. Conclusion

Recirculation with bioremediation can inhibit high risk in fluctuation of water quality. The best average value was obtained with a survival rate of 94.8±0.4%, length growth rate of 0.110±0.003 cm/day, weight growth rate of 0.175±0.013 g/day, specific length growth rate of 2.02±0.04%, specific weight growth rate 5.56±0.28%, total length 8.0 cm, total weight 7.91 g, and feed conversion ratio 0.73±0.07.

5. References

[1] Ismi S 2019 *J Penyul Perikan dan Kelaut.* 13(3) 315–324
[2] Bregnballe J 2015 *A Guide to Recirculation Aquaculture* (Denmark: Food and Agriculture Organization of The United Nations and EUROFISH International Organization) pp 95
[3] Astari B, Budiardi T, Effendi I, and Hadiroseyani Y 2021 *J Ilmu dan Teknol Kelaut Trop.* 13(3) 411–427
[4] Paranita A, Rahardja B S, and Prayogo 2022. *IOP Conf Ser Earth Environ Sci.* 1036(1) 012089.
[5] Scabra A R, and Budiardi T 2019 *J Perikan.* 9(2) 180–187
[6] Alem M D B, Efendi E, Wardiyanto, and Sarwono H A 2018 *J Sains Teknol Akuakultur.* 2(2) 41–47.
[7] Aziz N I A, Othman N, Altowayti W A H, Yunus Z M, Fitriani N, Din M F M, and Fikri F M 2021 *Malaysian J Anal Sci.* 25(4) 605-621.
[8] Supriyono E, Supendi A, and Nirmala K 2007 *J Akuakultur Indones.* 6(2) 135–145
[9] Handoko E S, and Mohammad W R 2017 *J Pendidik Tek Elektro.* 2(1) 26–33
[10] Lukmantoro T A, Prayogo, and Rahardja B S 2020 *IOP Conf Ser Earth Environ Sci.* 441(1) 012121.
[11] Nurhidayat, Nirmala K, and Djokosetyanto D 2012 *J Ris Akuakultur.* 7(2) 279–292
[12] Wahyuningsih S, and Gitarama A M 2020 *Synt Lite J. Ilm Indo.* 5(2) 112–125

[13] Hudaidah S, Putri B, Supono, Larasati E, and Santanumurti M B 2022 *AACL Bioflux*. 15(2) 988-1002
[14] Setyati W A, Habibi A S, Subagiyo, Ridlo A, Nirwani S, and Pramesti R 2016 *J Kelaut Trop*. 19(1) 11–20
[15] Boyd E C 1982 *Water Quality Management for Pond Fish Culture* (Albama: Elsever Scientific Publishing Company) pp 318
[16] Hayati A, Supriyanto A, Suhargo L, Hayaza S, and Ayubu A. 2020 *Pollut Res*. 39(4) 980-984.
[17] Suwoyo H S 2011 *Berk Perikan Terubuk*. 39(2) 25–40
[18] Yulianto B, Sunaryo, Nur T S P J, Djunaedi A, Subagiyo, Santosa A, and Soegianto A 2019 *Ecol Environ Conserv*. 25 S119-S128.
[19] SNI 2014 Hybrid Gouper (*Epinephelus fuscoguttatus*, Forsskal 1775 × *Epinephelus lanceolatus*, Bloch 1970 Section 2: Hybrid Seed Production (SNI 8036.2-2014) (Jakarta: National Standardization Agency) pp 13.
[20] Asri T Y, Harry W P and Thorsten W 2007 *Parasitol Res*. 100 1193–1204
[21] BPS 2017 Average rainfall and rainy days in Buleleng Regency by month 2018
[22] Mahardika K, Mastuti I, Septory R, Roza D, Zafran, Nasukha A 2021 *J. Ris Akua*. 16(1) 49–59
[23] Joint I, and Smale D A 2017 *FEMS Microbiol Ecol*. 93(2) 1–9
[24] Hendriansyah A, Putra W K A, Miranti S 2018 *Intek Akuakultur*. 2(2) 1–12
[25] Noor N M, Das S K, Cob Z C, and Ghaffar M A 2018 *Turkish J Fish Aquat Sci*. 18 1045–1051
[26] Thalib Y A, Razali R S, Mohamad S, Zainuddin R A, Ghaffar M A, Nhan H T, and Liew H J 2021 *Environ Pollut*.271 116375
[27] Pratama W D, and Manan A 2017. *J Aquac Sci*. 1(1) 27-35.
[28] Lu J, Zhang Y, Wu J, and Wang J 2020 *Bioresour Technol*. 305:123037
[29] Isnansetyo A, Getsu S, Seguchi M, and Koriyama M 2014 *Hayati J Biosci*. 21(1) 21–30
[30] Howarth R W, and Marino R 2006 *Limnol Oceanogr*. 51(1 part 2) 164–376
[31] Wang P, Joo J H, Park B S, Kim J H, Kim J H, and Han M S 2017 *Oceanol Hydrobiol Stud*. 46(1) 50–61
[32] Akbar A, Ma'shum M, Dewi N S, and Maha K S 2013 *J Perikan Unram*. 1(2) 60–70
[33] Koji E, Arfao A T, Djimeli C L, Ewoti O V N, Nougang M E, Bricheux G, Nola M, and Simengando T 2015 *Int J Res Stud Biosci*. 3(9) 11–21
[34] Akbar A, Ma'shum M, Dewi N S, and Maha K S 2013 *J Perikan Unram*. 1(2) 60–70
[35] Samad A P A, Hua N F, and Chou L M 2014 *African J Agric Res*. 9(9) 812–822
[36] Arissirajudin R, Hadi S, Safa A, and Purkan P 2019 *IOP Conf Ser Earth Environ Sci*. 217(1) 012032.

6. Acknowledgement
This research is a component of a thesis for the master degree aquaculture study program at IPB University. We appreciate the help from the Institute for Mariculture Research and Fisheries Extension. The research was carried out with help from Jaya Utama Abadi, Bali, Indonesia, and all parties who contributed to its completion.

First photograph record of the segura's throatspine gudgeon *Belobranchus segura* Keith, Hadiaty & Lord 2012 in the south of Java, Indonesia (Teleostei: Gobiiformes: Eleotridae)

V Hasan[1,4], F P Ottoni[2], M S Widodo[3], A R Faqih[3], A L Nafisyah[1]

[1]Department of Aquaculture, Universitas Airlangga, Surabaya, East Java 60115, Indonesia
[2]Laboratory of Systematics and Ecology of Aquatic Organisms, Center of Agrarian and Environmental Sciences, Federal University of Maranhão, Chapadinha, State of Maranhão, Brazil
[3]Department of Aquatic Resources Management, Universitas Brawijaya, Malang, 65145, East Java, Indonesia

[4]Corresponding author: veryl.hasan@fpk.unair.ac.id

Abstract. Segura's Throatspine Gudgeon *Belobranchus segura* Keith, Hadiaty & Lord 2012 is an amphidromous fish from the Indo-Pacific region. In Indonesia, *B. segura* was distributed in Papua, Halmahera, Sulawesi (previously known as Celebes) and Java Island. The finding of this species in the lower stream river of Malang district, south of Java, Indonesia, was firstly recorded here. This study provides the photographs of a single male specimen of *B. segura*, the description of morphological features of a specimen and their habitat conditions.

1. Introduction

Eleotridae (Bully sleepers or Spinycheek sleepers) is the second most diverse Gobiiformes family which is currently comprises of 187 valid species [1]. These species are widely distributed in marine, brackish, and freshwater environments, especially in tropical and subtropical waters worldwide (rarely in temperate waters) [2]. A curious fact is that in the Pacific Ocean eleotrids developed several endemic genera which occur from estuaries to the middle portion of rivers [3].

Belobranchus Bleeker, 1856 is an eleotrid genus, included in the subfamily Eleotridae [1], which was considered to be monospecific until the description of *Belobranchus segura* Keith, Hadiaty & Lord 2012 [3]. Currently it includes only two species, *B. segura* and *B. belobranchus* (Valenciennes, 1837), occurring in Western Pacific freshwater and brackish waters [3]. *Belobranchus segura* is currently reported from estuaries to the middle portion of rivers in Indonesia, Philippines, Papua, Vanuatu and Solomon Islands [4]. In Indonesia, it was previously recorded from Halmahera, Sulawesi, and Papua Barat Province, Lombok, and Java. Especially in Java, the details documentation of specimen's photograph is not available in many major references [5]. In this paper, we reported the first photograph of *B. segura* in Java.

2. Materials and methods

A photograph of single specimen of *Belobranchus segura* (Figure 1) was obtained from two collecting site in the river lower stream, eastern of Java, Indonesia (8°20'22"S 112°34'16"E) (Figure 2). The collection was conducted on September 17, 2020. Specimen, including comparative material, were deposited at the Ichthyological Collection of the Zoology Laboratory Genbinesia Foundation (GBI), East Java, Indonesia. Morphological inspection was done by following Keith et al. [3].

3. Results and discussion

We found one confirmed locality for *Belobranchus segura* in the East Java (Indonesia) in September 2020. The specimen was captured from the eastern Java and was identified as *B. segura*, since it possess the color in life diagnostic characters proposed by Keith et al. [3] that differentiate it from *B. belobranchus*, listed below: i) the specimen of *B. segura* exhibit red brown body coloration, ii) yellowish to orange second dorsal, anal, pectoral and caudal fins, iii) second dorsal and caudal fins never spotted, iv) upper part of first dorsal fin can present orange coloration, and middle and lower parts are mottled greyish to black.

Until this work, *B. segura* was never been photographed from Java (Indonesia), although Dahrudin *et al.* [5] has recorded it in Java but the photograph is not available. Among other biological topics, new records of freshwater gobies are important contributions to improve the knowledge on the species diversity, distribution, and biogeography, and to make decisions related to species conservation [6].

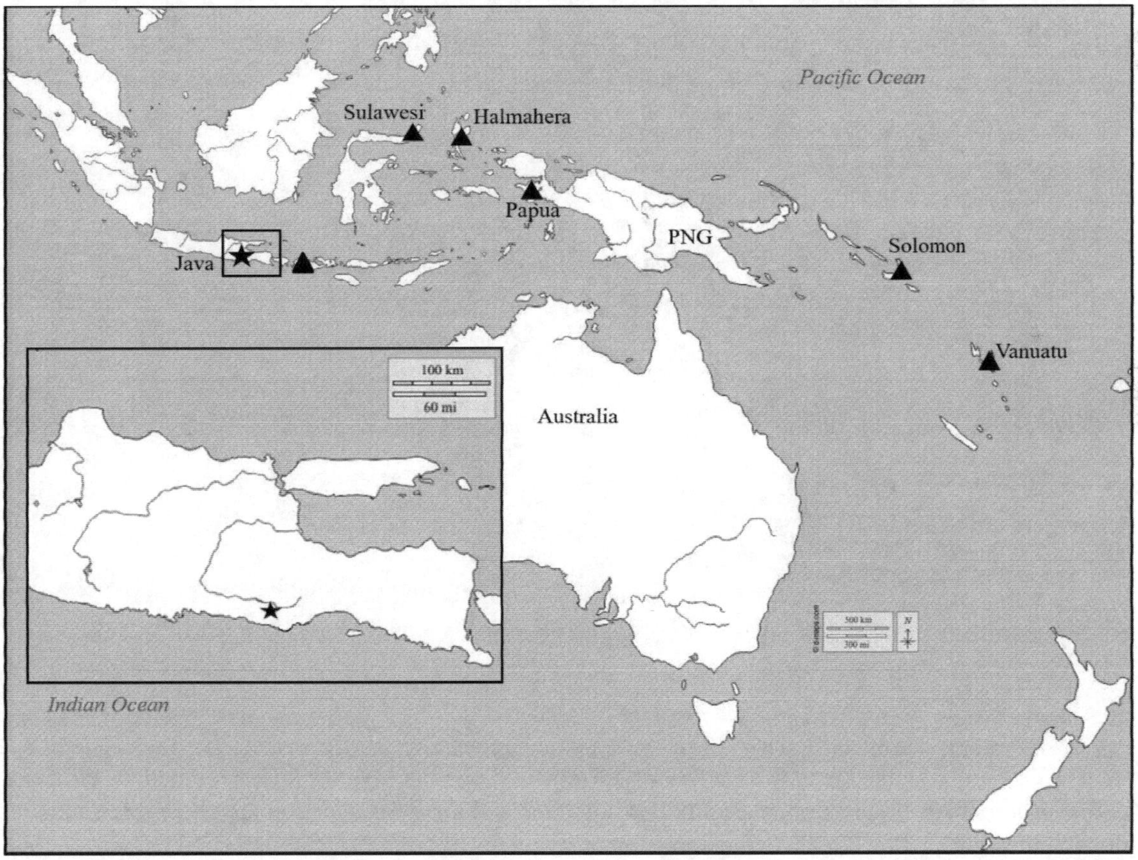

Figure 1. Map with the locality of *Belobranchus segura* in Java, Indonesia (black stars), and the previous localities (black triangles).

Figure 2. Live specimen of *B. segura* collected from a river lower stream, eastern of Java (Photograph by Dicky Fish Spot).

Figure 3. Collecting site river lower stream, eastern of Java, Indonesia (Photograph by Eko Priyo Andoko).

4. Conclusion

Belobranchus segura is an Indonesian native amphidromous fish that is recorded in Java. The habitat of this species is in the river lower stream, eastern of Java. The photograph provided in this paper gives an important description of the morphology and color patterns of this species, which have not documented yet.

5. References

[1] Fricke R, Eschmeyer WN and Fong JD. (Eds) 2021. *Eschmeyer's catalog of fishes: Genera/Species by Family/Subfamily*. Electronic version accessed 24 Jun 2021.

[2] Keith P, Hadiaty RK and Lord C 2012. *Cybium*, 36(3): 479–484.
[3] Nelson JS, Grande TC and Wilson MVH 2016. *Fishes of the World*. John Wiley & Sons, Hoboken, New Jersey.
[4] Larson, H. 2021. *Belobranchus segura. The IUCN Red List of Threatened Species 2021*. Downloaded on 30 August 2021.
[5] Dahruddin H, Hutama A, Busson F, Sauri S, Hanner R, Keith P, Hadiaty R and Hubert N 2017. *Mol Ecol Resour.* 17 (2), 288-299.
[6] Hasan V, Valen F S, Islamy R A, Widodo M S, Saptadjadja A M and Islam I 2021. *Biodiversitas.* 22 (2), 573-581.

6. Acknowledgments

The authors thank Josie South (University of Leeds, UK) for providing suggestions on the manuscript and english language review, Mr. Dicky Eko Febrianto and Eko Priyo Andoko as our guide, and the Universitas Airlangga for funding our research. FPO thanks Coordenação de Aperfeiçoamento de Pessoal de Nível Superior - Brasil (CAPES - Finance Code 001), and Fundação de Amparo à Pesquisa e ao Desenvolvimento Científico e Tecnológico do Maranhão (FAPEMA) for providing the financial and infrastructure support to carry out his scientific research.

IOP Conf. Series: Earth and Environmental Science 1273 (2023) 012040 doi:10.1088/1755-1315/1273/1/012040

Improvement of Red Color Quality in Koi Fish (*Cyprinus carpio*) Through the Addition ff Rosella Extract In Feed (*Hibiscus sabdariffa*)

A A A Amin[1], N N Dewi[2,3], B S Rahardja[2]

[1]Program Study of Aquaculture, Faculty of Fisheries and Marine, Airlangga University, Campus C Mulyorejo, Surabaya 60115 East Java, Indonesia.
[2]Department of Fish Health Management and Aquaculture, Faculty of Fisheries and Marine, Airlangga University, Campus C Mulyorejo, Surabaya 60115, East Java, Indonesia.

[3]Corresponding author: ninanurmaliadewi@fpk.unair.ac.id

Abstract. Koi fish (*Cyprinus carpio*) is one of the ornamental fish that has a beautiful body shape and color, so it has high economic value. To get koi fish production with good color quality, several supporting factors are needed, one of which is feed. In this study, rosella (*Hibiscus sabdariffa*) extract was used as an additive in koi fish feed. The purpose of this study was to determine the effect of rosella extract on increasing the red color of koi fish through the addition of ingredients in the feed. The study was conducted using a completely randomized design with doses of rosella extract in the feed formulation as a source of anthocyanins, namely: 0% (control), 1%, 3%, 5%. Feeding was done three times a day with a feeding rate of 5%. Maintenance was carried out for 30 days and observations of color brightness of koi fish were carried out once a week using a modified TCF (Toca color finder). The results showed that the addition of rosella extract by 5% resulted in the highest value of red brightness every week. For water quality parameters, both temperature, pH, and dissolved oxygen are still within optimal limits for koi fish.

1. Introduction

Koi fish (*Cyprinus carpio*) is one of many ornamental fish known for its beautiful body shape and color so it has high economic value. Koi fish in Latin is Cyprinus carpio L. This fish was first developed in Japan. In the 1820s, Japan succeeded in crossing several sets of carp (carp) into koi fish with interesting color variants. From Japan, the popularity of this freshwater ornamental fish spread to various parts of the world, including Indonesia. Koi fish entered Indonesia in 1962, namely by Prince Akihito with Putri Michoko, Prince Akito visited the Bogor area and saw Indonesian goldfish from the kumpay race. Koi fish have a special attraction for people who love ornamental fish. Indicators of the beauty of ornamental fish can be seen in the brilliant color, physical shape and completeness, behavior, and health conditions or stamina [1]. The increasing market demand for koi carp enourages fish cultivator to improve their cultivation system towards intensive. But on the other hand, this cultivation system creates a new problem : disease [2]

Color is one of the reasons why ornamental fish are in demand by the public so that farmers needto maintain the color of ornamental fish by feeding them with color pigments [2]. Color as an aesthetic value also greatly affects the economic value of ornamental fish, especially koi fish [9].One of the factors that can affect color brightness in koi fish is nutrition in feed [7]. The addition of a color enhancing source in the feed will encourage an increase in color pigment in the body of the fish, or at least the fish is able to maintain the color pigment in its body during maintenance [9].

Color in fish is caused by the presence of pigment cells or chromaphotors in the dermis on the scales, outside or under the scales. Natural dyes (pigments) are dyes that are naturally found in plants and animals. The use of fruits and vegetables as coloring agents is very good and without any adverse side effects [14].

Red dyes that are abundant in nature can be grouped into two groups, namely carotenoids and anthocyanins. Anthocyanins belong to pigments called flavonoids which are usually dissolved in water. The color of the anthocyanin pigment has red, blue, and purple colors which can usually be found in flowers, fruits and vegetables [10]. The important content of anthocyanins that form flavonoids also acts as an antioxidant. Rosella flavonoids consist of flavonols and anthocyanin pigments. Anthocyanins are also useful as antioxidants that can cure various diseases. The extraction process also affects the dye produced [10].

Anthocyanins are safe for consumption by fish, besides that anthocyanins themselves have been shown to produce red, orange, purple, blue colors obtained from various plants or fruits such as grapes, strawberry roses or rosella, The higher number doses of red fruit given to the brightness level of the comet fish will increase [6]. Therefore, it is necessary to feed or supplements that can support and improve the color quality of koi fish. The decrease in color quality in ornamental fish, especially koi fish, greatly affects market interest, the demand for high quality koi fish in the market can be a reference for producing or developing koi fish with high economic value. The ornamental fish business is not enough to just rely on efforts to produce, but also needs to be accompanied by skilled steps in developing the color or appearance of beauty, as well as patterns in ornamental fish [11].

2. Materials and Methods

2.1. Research Methods

The method used in this study was a Completely Randomized Design Environmental Design (CRD) with 4 treatments and 5 replications. In this study, the concentration of rosella extract (*Hibiscus sabdariffa*) used was a concentration of 0% at P0, 1% at P1, 3% at P2, and 5% at P3.

2.2. Preparation of Tools and Materials

Preparation of tools and materials is the first step of a research. The equipment to be used is washed with soap and then rinsed with clean water. After that, all equipment is immersed in a chlorine solution for 24 hours which aims to make the tools to be used free from bacteria. After that, rinse with clean water until the chlorine smell is gone. According to [15] the acclimatization process is carried out with the aim that the koi fish can adapt to their new habitat. After the acclimatization process the fish were not fed for 1-2 days because at this time the fish were still in a state of stress due to not adapting to the new environment [8]. After the fish were acclimatized, the fish were divided into each aquarium with a total of two fish/aquarium for each treatment.

2.3. Making Rosella Extract

Making of roselle extract begins with the selection of quality rosella which has a concentrated color of 3 kg of roselle parts that can be extracted and contains colors including stems, leaves, and flowers [9]. Then washed and drained and continued with cutting. Furthermore, the pieces of rosella are dried in the sun with a black cloth cover. By giving a black cloth aims to speed up drying. Drying is done ± 2 days until the rosella pieces look dry. After drying, the results of dried rosella are obtained whose weight has experienced shrinkage. After that, the maceration process was carried out at room temperature in a dark container for 24 hours using water + 70% ethanol + citric acid as a solvent. After that, filtering is carried out in the form of liquid which is then evaporated [4].

2.4. Feeding Formulations

The feed used is commercial feed, in the form of pellets with a protein content of at least 35%, crude fat at least 5%, crude fiber maximum 3%, and ash content of 12%. Making feed for the treatment was done by mixing rosella extract according to the concentration of each treatment. Then the application of the extract to the feed is done by spraying according to each dose. Feeding three times a day at 08.00; 12.00; 16.00 with a feeding rate of 5% [12].

2.5. Data Analysis

The parameters observed during this study is :

1. *Red color quality level*

   The main observation parameter is the red color change on the body of koi fish during the maintenance of koi fish seeds. Determination of the brightness of the color of the fish was carried out by observing the intensity of the color brightness of the koi using the Toca Color Finder (TCF) which was observed by 5 panelists who had no visual impairment. Observations were made visually by comparing the original color of the fish on TCF paper. Observations were made including observations of color intensity using TCF.
2. *Survival Rate*

   Survival rate (SR) is the ratio between the number of fish that live at the end of the rearing with the number of fish stocked or at the beginning of the rearing. Good aquatic environmental conditions support the survival of fish during the maintenance period.
3. *Water Quality*

   Supporting data that is very necessary is measuring the quality of hatching media water which includes temperature using a thermometer, pH and dissolved oxygen using a DO meter. Water quality tests are carried out on water samples every day with data collection in the morning, afternoon, and evening.

2.6. Statistic Analysis

The data obtained were analyzed using the Anova test (Analysis of Variance) to determine the effect of the treatment given. If there are significant results, the calculation is continued with Duncan's Multiple Range Test with a 5% significance level.

3. Result and Discussion

3.1 Color Brightness of Koi Fish

The results of measuring the brightness of red color in Koi for 28 days can be seen in Table 1. Based on the results of the ANOVA statistical test, the TCF (Toca color finder) value of koi fish at week 1 showed a significant difference ($P < 0.05$) in each treatment with different doses. The highest TCF was at P3 (Treatment 5%) with a value of 8.60±5.225 and the lowest TCF value was at P0 (Treatment 0%) with a value of 24.60±2.074.

Observation of color quality improvement of koi fish (Cyprinus carpio) was observed once every seven days for 30 days using a modified toca color finder which has been given a score from 1 to 30. Observation of color on koi fish was observed starting from the tip of the head to the tip of the tail including the tail. fin. Color changes in koi fish began to be seen on the 7th day of observation. The results of observations of changes in the color of koi fish showed that the addition of 5% rosella extract gave $P<0.05$ or significantly different results compared to the addition of rosella extract below 5%.

Table 1. Average Blood Glucose Levels

Treatments	Day7	Day14	Day21	Day28
P0	$10{,}40^{b} \pm 6{,}107$	$7{,}40^{b} \pm 4{,}879$	$7{,}80^{b} \pm 5{,}119$	$8{,}60^{b} \pm 5{,}225$
P1	$13{,}60^{b} \pm 8{,}264$	$14{,}60^{ab} \pm 8{,}019$	$16{,}20^{a} \pm 7{,}596$	$17{,}40^{a} \pm 7{,}893$
P2	$16{,}60^{ab} \pm 5{,}459$	$17{,}60^{a} \pm 5{,}459$	$19{,}20^{a} \pm 5{,}167$	$20{,}00^{a} \pm 4{,}899$
P3	$20{,}60^{a} \pm 3{,}362$	$21{,}60^{a} \pm 3{,}209$	$22{,}60^{a} \pm 2{,}996$	$24{,}60^{b} \pm 2{,}074$

Note: different superscripts on the same line show significantly different effects ($p<0.05$)

3.2 Survival Rate

The survival rate of koi fish (Cyprinus carpio) is the percentage of the number of fish that survive at the end of the study compared to the number of fish at the beginning of the study. Of the 40 koi fish (Cyprinus carpio) used as research material, 2 fish died during the study so that the presentation of the survival rate (SR) of koi fish in this study reached 95%.

Table 2. *Survival Rate*

Treatment control	*Survival Rate*			
	First Week	Second Week	Third Week	Fourth Week
P0	18	18	18	18
P1	20	20	20	20
P2	20	20	20	20
P3	20	20	20	20

3.3 Water Quality

Water quality is a supporting parameter in the maintenance of catfish for 28 days. The observed water quality parameters include temperature, dissolved oxygen or DO, pH, ammonia which can be seen in Table 3.

Table 3. *Water Quality Parameters*

Treatments	Temperature (°C)	DO (mg/L)	pH
P0	28 – 31,5	5,5-5,7	6,8-7,4
P1	28 – 31,5	5,5-5,7	6,8-7,4
P2	28 – 31,5	5,5-5,7	6,8-7,4
P3	28 – 31,5	5,5-5,7	6,8-7,4
(SNI)	25 – 31,5	5	6,8-7,5

Based on the result observations 28 days on all treatments. During the observation period, the treatment with the addition of rosella extract showed an increase in color. The addition of 1% rosella extract gave an average color increase of 3 scores, as well as the treatment of adding 1% rosella extract to the addition of 3% also gave an impact in the form of an increase in red color by 3 scores. In the treatment of adding rosella extract as much as 5% was obtained. The maximum average color increase is 4 on scores.

In this research, good results were obtained for improving the quality of red color in koi fish because the function of anthocyanin pigments produced from rosella extract can increase the formation of color pigments in koi fish.The mechanism of increasing color intensity towards a brighter direction is basically influenced by chromatophoric cells located in the epidermal layer [5].

The increase in color intensity occurs due to changes in the chromatophores cells, while these changes are divided into two, namely morphological and physiological changes. Morphological changes affect the addition and decrease in the number of chromatophore cells [5] stated that this condition was influenced by the amount and composition of feed containing carotenoid sources in the feed. While physiological changes are changes caused by the movement of the chromatophores pigment cells.

Changes in chromatophoric pigment cells can be physiologically divided into two types, namely chromatophoric pigment cells that gather near the nucleus and chromatophoric pigment cells that are scattered throughout the epidermis. The chromatophoric pigment cells that are scattered in the epidermis layer cause the pigment cell granules to absorb light perfectly, resulting in an increase in the color intensity of the fish's body becoming brighter and clearer. While the pigment cells that gather near the

IOP Conf. Series: Earth and Environmental Science 1273 (2023) 012040 doi:10.1088/1755-1315/1273/1/012040

nucleus cause a decrease in the color of the fish's body so that the fish's body color looks darker and fades.

Water quality is an important role in improving the color quality or color brightness in fish [3] In this research, it has been observed that there are no periodic changes in temperature that can cause stress or even cause fish death and in accordance with the statement according to [13], good water quality for koi maintenance is water temperature 25-29 0C, pH 6, 8 – 7.5 and preferably water in koi maintenance has a minimum DO level of 5 mg/L, the results obtained related to water quality in this study have met the optimum value according to [13]. DO measurements were carried out twice a day in the morning and evening. Each aquarium has been added an aerator that has been adjusted to support the oxygen level and in each aquarium gets the same amount of oxygen. All aspects of water quality in this study have met the same number as [13] statement.

4. Conclusion

The addition of rosella (*Hibiscus sabdariffa*) extract in the feed can improve the red color quality of koi fish (*Cyprinus carpio*) and the optimum concentration for the addition of rosella (*Hibiscus sabdariffa*) extract which can improve the color quality is the highest at the addition of 5%.

5. References

[1] Andriani, Y, A P Wulandari, R I Pratama, dan I Zidni 2019 *Jur Ilm Peng Kep Masy.* 5(1) : 33-38.
[2] Arindita, C, Sarjito, & Prayitno, S B 2014 *Jou of Aqu Mana and Tech.* 3(3):66-75.
[3] Bachtiar, Y dan T Lentera 2004 *Freshwater Ornamental Fish Cultivation for Export* Agro Media Pustaka Tangerang 74-81.
[4] Choiriyah, N A 2017 *Dar Nut Jou.* 1(1): 16-21.
[5] Effendie M I 1979 *Fisheries Biology Method.* Bogor Yayasan Dewi Sri 112.
[6] Rahmadhani L E, L I Widuri and P Dewanti 2020 *J Agro.* 14, 33-43.
[7] Kusrini, E, S Cindelaras dan A B Prasetio 2015 *Med Akua* 10 (2): 71-78.
[8] Lubis, M Z, & Pujiyati, S 2017 *Jur Tek Per dan Kel.* 4(2): 123-129.
[9] Mardiah, L Amalia, dan A Sulaeman 2010 *Jurnal Pertanian* 1(1): 1- 8.
[10] Mastuti E, N P Sari, dan R A Simangunsong 2013 *Jur Uni Seb Mar.* 12(2): 43-47.
[11] Mutiarasari A, 2017 *Thesis*. Universitas Islam negeri Raden Intan Lampung. 146 hal.
[12] Sari N P, L Santoso, dan S Hudaidah 2012. *Jur Rek dan Tek Bud Per.* 1(1): 31-38.
[13] Susanto 2007 *Kiat Budidaya Ikan Mas di Lahan Kritis.* Penebar Swadaya Jakarta 13 – 18.
[14] Winarti, S, dan A Firdaus 2010 *Jur Tek Per.* 11(2): 87-93.
[15] Zidni, I, Ayi Y, Iskandar, dan Yuli A 2017 *Jur Per dan Kel*. 7(2): 125-135

6. Acknowledgement

The authors gratefully acknowledge Faculty of Fisheries and Marine Airlangga University.

Monitoring the distribution of phytoplankton in vannamei shrimp cultivation in intensive system, Banyuwangi, East Java, Indonesia

D A Zulkarnain[1] , L A Sari[2,4], P D W Sari[2], D D Nindarwi[2] and S Arsad[3]

[1]Study Program of Aquaculture, Faculty of Fisheries and Marine, Universitas Airlangga, Campus C Jalan Mulyorejo, Surabaya 60115 East Java, Indonesia
[2]Department of Aquaculture, Faculty of Fisheries and Marine, Universitas Airlangga, Campus C Jalan Mulyorejo,Surabaya 60115 East Java, Indonesia
[3]Department of Aquatic Resources Management, Faculty of Fisheries and Marine Science, Universitas Brawijaya, Jalan Veteran Malang 65145 East Java, Indonesia.

[4]Corresponding author: luthfianaas@fpk.unair.ac.id

Abstract. The increase in Vannamei shrimp farming is closely related to an increase in the demand for shrimp feed. At present there are many technological developments in cultivating natural feed. The natural feed here refers to plankton which can be used as natural food for every organism of larvae in waters. Conditions in water will affect the distribution pattern of phytoplankton either horizontally or vertically. As a result, it affects the abundance of phytoplankton and subsequently also affects the value of the main productivity. The purpose of study is to find out the cultivation process and to find out the types of plankton that exist in Vannamei shrimp farming ponds. Plankton observations were carried out directly with the help of a microscope and haemocytometer, while the measurement of water quality using each instrument had been prepared for each parameter. The results of plankton observations found species of *Chaetoceros sp.*, *Oocystis sp.*, *Oscillatoria sp.* and *Cyclotella sp.* while the water quality measurements obtained temperature (29-31°C); DO during the day (7.6-9.5 ppm) and at night (4.0-4.6 ppm); pH in the morning (7.8-8, 5) and daylight (8.1-9.0); and salinity 26-27 ppt. The application of probiotics used carbon sources derived from molasses and katul.

1. Introduction

Indonesia is known as a maritime country as it is the largest archipelagic country in the world. Indonesia is not an archipelagic country that is surrounded by oceans, but a country with vast oceans surrounded by many islands. Indonesia's ocean area reaches 5.8 million km^2, which is also the largest ocean area in the world. The vast marine area should be a potential resource for the welfare of the community, especially people living in coastal areas [1].

Shrimp farming provides the largest foreign exchange supply in Indonesia in the fisheries sector. Based on the species, shrimp exports consist of Vannamei shrimp, tiger shrimp, and several other types of shrimp. Vannamei shrimp is one of the most popular tyoes of shrimp exports in Indonesia, reaching 85%. Vannamei shrimp in Indonesia are cultivated in 15 provinces. The largest production yield is observed in Lampung Province at 19.43%, West Nusa Tenggara (NTB) Province at 18.89%, and East Java Province at 13.02% [2].

Vannamei shrimp culture using an intensive system has the stocking density of 100-300 individuals/m^2. Intensive ponds are ponds that are facilitated by plastic mulch which covers all parts, water pumps, aerators, waterwheels, dense stocking and feed in the form of pellets. Stocking density plays a major role during cultivation to determine the number of fry used and the area of ponds used [3].

The increase in Vanamei shrimp farming business is closely related to an increase in the demand for shrimp feed. At present there are many technological developments in cultivating natural feed. The natural feed here refers to plankton which can be used as natural feed for every organism of larvae in waters. Phytoplankton are the main producers in waters while the consumer organisms are zooplankton, larvae, fish, shrimp, crabs, and others [4].

In the management of shrimpfarming, plankton is used as a natural food producer, especially at the beginning of fry stocking. In addition, the presence of plankton makes the pond waters shady, and consequently the shrimp can be productive or be active in looking for food during the day. Plant plankton is a producer of oxygen in the water and helps in the absorption of compounds which are harmful to shrimp such as ammonia, nitrite, and nitrate. Plankton can also suppress the growth of klekap and moss at the bottom of the pond [5].

Natural feed is one of the productivity factors of pond waters. Plankton is a natural food in pond waters. Plankton itself is defined as an organism that floats, and its movement depends on water currents. Plankton is divided into 2, namely phytoplankton and zooplankton. Phytoplankton has a role as the main producer, so its existence is very important for the food chain [6].

Conditions in waters will affect the distribution pattern of phytoplankton either horizontally or vertically. Hence, it will affect the abundance of phytoplankton and subsequently also affect the value of the main productivity [7]. Vannamei shrimp are known to be resistant to diseases caused by viruses, but in reality at present many failures in Vannamei cultivation occur due to virus attacks. One of the efforts that need to be made to avoid viruses, is the addition of probiotics. With the addition of probiotics, it is expected that the quality of pond water can be controlled, and probiotic bacteria can suppress the growth of pathogenic bacteria.

The study was conducted in ponds with an intensive system because it has a high density level, so the management in water quality must be optimal. The decrease in water quality values has a negative impact on aquatic organisms. The presence of plankton can have a positive or negative effect on water quality parameters.

2. Material and methods

2.1. Phytoplankton observation

Firt of all, a water sample and haemocytometer were prepared. The water sample was taken and placed into the haemocytometer as much as 1 drop using a clean pipette, and then the haemocytometer was covered with a cover glass. Observations were conducted using a microscope with a magnification of 100x and 400x.

2.2. The making and addition of fermentation

There were 2 types of fermentation used in the study ponds, namely molasses fermentation and bekatul (bran) fermentation.

2.2.1 The making of molasses fermentation

There were some stages in the manufactire of molasses fermentation. First of all, 4 buckets of thick drops were put into a 1200 liter reservoir and added with 4 packs of probiotics. Fresh water was then added to the brim and left for 24 hours. After 24 hours, 2 bottles of EM4 were added and left for another 6-7 days. After being left for 6-7 days, the fermentation was ready to be stocked. Every time fermentation was given, the reservoir was refilled by adding drops.

2.2.2 The making of fermented katul (bran)

The methods used in making katul fermentation included preparing bran (katul), saponins, water, scoop, and a bucket. Next, 5 scoops were put into the bucket, and then 1 scoop of saponins was added in it. Afterwards, 7 scoops of water were added and stirred until evenly distributed. After evenly distributed, the bucket was covered tightly and left for 1 day. Finally, the fermented katul was ready to be stocked into the culture pond.

The addition of saponins was used to eradicate pests in rearing ponds, serve as antimicrobials, and increase immunity in shrimp. As an antimicrobial, saponins are able to denature proteins and cell walls of pathogenic bacteria (Jumah *et al.*, 2020). The growth of bacteria in probiotic fermentation depends on the origin of the material or carbon source used as a substrate for probiotic fermentation. This is because the substrate used can affect the metabolism of bacteria. Different carbon sources can make the number of bacteria that will grow also different [8].

The addition of probiotics through the environment (water and pond bottom) aims to improve and maintain water quality and pond bottom; oxidize organic compounds from leftover feed, shrimp manure, plankton and dead organisms; and reduce toxic metabolites. Meanwhile, the addition of bacteria through feed aims to balance intestinal function so that it can suppress harmful bacteria, form enzymes that help the food digestive system, serve as protein that can be utilized by fish and shrimp that eat it, and increase the immunity of shrimp and fish [9].

3. Results and discusion

a. *Chaetocheros* sp.

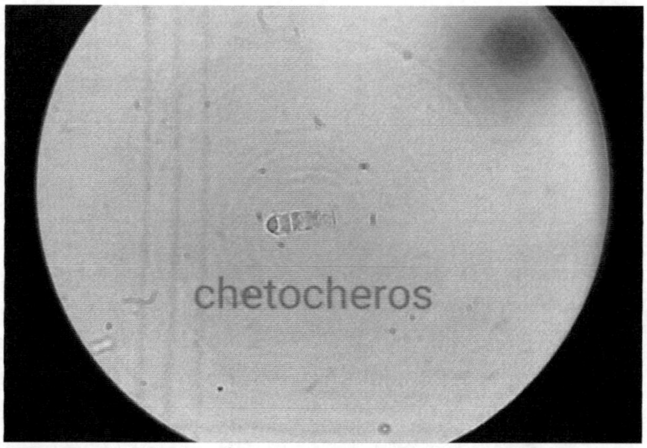

Figure 1. *Chaetocheros* sp magnification 400x

Chaetocheros sp. has a body shape of bilateral symmetry and radial symmetry. On the body of *Chaetocheros sp.*, there is a cell wall known as frustule which is composed of the base of the hypotheca and the covering (epitheca), as well as a belt or cingulum. Frustule is formed by pectin which is covered by silicon [10].

Chaetocheros sp. has characteristics of a round shape measuring between 4-6 microns, but there are also rectangular ones measuring 8-12 x 7-18 microns. *Chaetoceros sp.* belongs to the Diatom species of the order Centrales whose characteristics are cylindrical and have a habitat in sea water [11]. According to Xu *et al.* (2020)[12], *Chaetoceros sp.* has chains that are curved and covered by thin silica with rectangular holes, the cells are rectangular and some are square, and the setae are slightly thin and curved in the same direction. The valves of *Chaetocheros sp.* are elliptical to round.

Chaetocheros sp. is one of the natural feeds commonly used in cultivation. This is because it has a high protein content and is easy to digest [11]. Apart from the high protein content, *Chaetocheros sp.* has other nutritional content, namely 35% protein, 6.9% fat, 6.6% carbohydrate, and 28% ash content [13].

b. *Oocystis* sp.

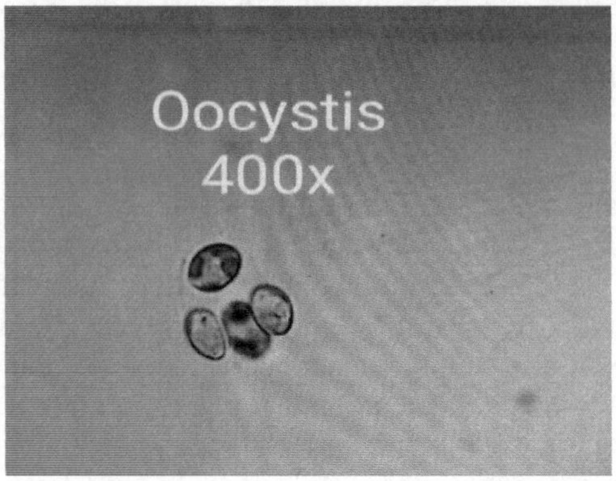

Figure 2. *Oocystis sp.* magnification 400x

Oocystis sp. is one of the species belonging to the *Chlorophyta* family, having 4 to 7 chloroplasts that are shaped like discs in each cell. The cell wall is composed of a layer of cellulose with perpendicular fibrils. There is a discrete barrier sheath formed from a pair of membranes with a width of 11 m [14].

One characteristic of *Oocystis sp.* is the cells are solitary or form colonies into one in the parent cell. The cells are round or ovoid elliptical, and there are some small and slightly large holes [15].In general, the morphology of *Oocystis sp.* shows that it has a rather small elliptical cell shape with a length of 30-40 m and a diameter of 18-30 m. The cell wall is thin measuring less than 1 m. The chloroplast contains essential pyrenoids and granules with a length of < 1-5 m and the inside consists of fibrils while the outside consists of dense starch lamellae. External and internal chloroplasts are tightly packed [14].

O*ocystis sp.* is capable of producing high antioxidant compounds in the form of phenolics and carotenoids. Phenolic compounds are very useful in boosting the immune system and preventing disease transmission. Meanwhile, carotenoids consist of several types of compounds, one of which is β-carotenoids which act as antioxidants[11]. *Oocystis sp.* has an important role in the waters, namely as a producer of oxygen and as natural food [16]. In addition, it is also used as a phycoremediation agent. Phycoremdiation is the process of using algae to treat waste. This is similar to its function to absorb sulfate, nitrate, and phosphate content in waters[11].

c. *Oscillatoria sp.*

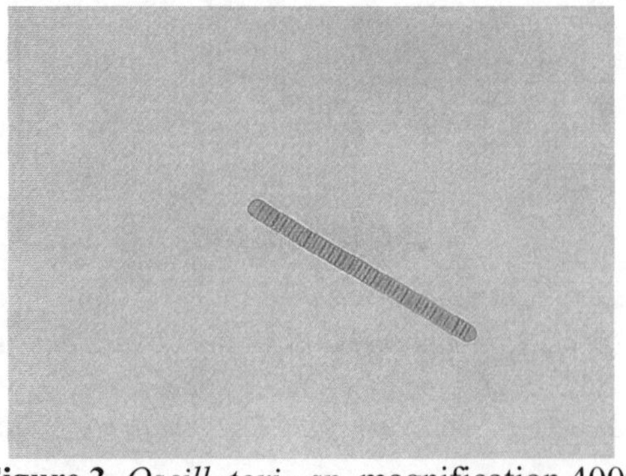

Figure 3. *Oscillatoria sp.* magnification 400x

Oscillatoria sp. has an unbranched cylindrical body shape which is straight and long. There is a wall or partition which separates the cells, and the ends of the cells are round. In the type that floats there are gas bubbles. *Oscillatoria sp.* is green, olive green, reddish or brown in color. The body size is between 8-30 m [15].

Oscillatoria sp. has the ability to glide spontaneously. The gliding movements are very diverse; some slide forward and backward, some slide in a spiral pattern forward or backward, and some move slowly and regularly at the tip of the trichome. Those movements are influenced by external factors, especially temperature and light. The movement will be more active when the light intensity is amplified. Meanwhile, when the temperature increases from 0-30°C the speed of movement doubles for every 10°C increase [17].

Oscillatoria sp. belongs to a group of *Cyanobacteria* that has the ability to adapt well even in polluted waters. *Oscillatoria sp.* is also included as a group that can reduce water quality by changing the biotic components in the waters. *Oscillatoria sp.* can produce secondary metabolites that can be destructive and cause death in aquatic organisms if these secondary metabolites accumulate more in the waters[18].

d. *Cyclotella sp.*

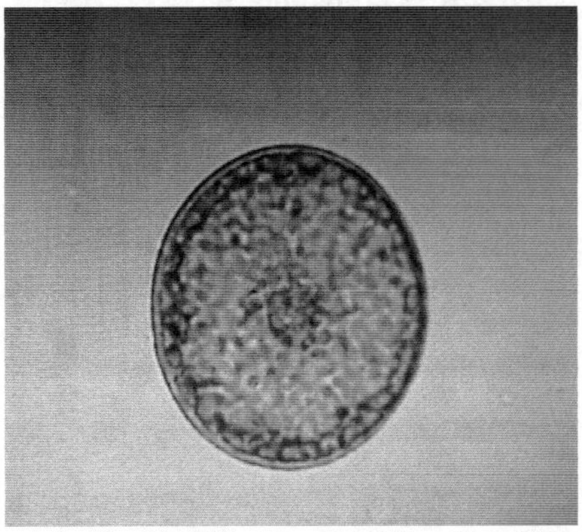

Figure 4. *Cyclotella sp.* magnification 400x

Cyclotella sp. has a diameter of about 3-5 m. Its body size reaches up to 12 m. The body is spherical in shape and has a two-layered cell wall and is composed of silica. The breeding of *Cyclotella sp.* occurs asexually and sexually. It is said to reproduce asexually when the reproduction occurs by means of dividing and forming auxospores, while sexually when the reproduction occurs through oogamy [11].

Cyclotella sp. is a type of diatom in the form of a small disc, the center is flat while the edges are broad bands. Cells have a rectangular shape, where each cell contains chloroplasts in the form of a dicoid [19].

Cyclotella cells are single cells. The cells have a golden-brown color due to the high content of phycoxanthin and photosynthetic accessory pigments. There are also chitin fibers whose length is more than 150 m. The chitin fibers are useful for floating. At the edge of the ring valve there are fultoportula (small tubules) which are sometimes also scattered on the valve surface [20].

Natural feed given during the larval stage is divided into 2 groups of plankton, namely phytoplankton and zooplankton. One of the phytoplankton used as natural feed is *Chaetoceros sp.* because it has a high nutritional content, is easy to digest, and breeds quickly . The nutritional contents of *Chaetoceros sp.* are 35% protein, 6.9% fat, and 28% ash content [13] Meanwhile, the zooplankton given were *Artemia sp.* which also has high nutritional content such as protein 52.7%, carbohydrates 15.4%, fat 4.8%, water 10.3%, and ash content 11.2%. These two types of natural feed have an important role in meeting nutritional needs when the shrimp is still in the larval stage [21].

The type of artificial feed given during rearing was adjusted to the technician's policy, including Lorica, HDV-C, HDV-D, and Witnis brands. Each type of feed used had a different size, and surely the nutritional content was also slightly different. Lorica feed with a size no. 1 was given when the DOC was 1-9 days, and this had a nutritional content of 40% protein, 5-7% crude fat, and 3% crude fiber. The HDV-C feed with sizes no. 1,2, and 3 was given when the DOC was 15-48 days, and this had a nutritional content of 35% protein, 6% fat, 4% crude fiber, 15% ash content, and 12% water content. The HDV-D feed with a size no. 3 was given when the DOC was 49-64 days , and this had a nutritional content of 32% protein, 6% fat, 5% crude fiber, 15% ash content, and 12% water content. The HDV-D feed with a size no 4 was given when the DOC was 71 until harvest. Meanwhile, the Witnis feed was given when the DOC was 65-75 days with a nutritional content of 34% protein, 6% fat, 4% crude fiber, 15% ash, and 12% water content. During feeding there was a reduction in the proportion of protein aimed at reducing production costs as the most expensive feed composition is in its protein content. The reduction in the proportion of protein has no impact on the growth rate and synthesis of Vannamei shrimp production [22].

In shrimp feed, chitosan was also added with a composition of 25 grams of chitosan mixed with 25 kg of feed. According to Ekaputri *et al.* (2018)[23] the addition of chitosan in feed aims to stimulate immune function, improve digestion and absorption of nutrients. Thus, it can increase the growth rate. The more protein that can be stored in the body and the less protein that is converted into energy, the higher the growth rate.

4. Conclusion

The increase in Vanamei shrimp farming business is closely related to an increase in the demand for shrimp feed. At present there are many technological developments in cultivating natural feed. The natural feed here refers to plankton which can be used as natural food for every organism of larvae in waters, such as *Chaetoceros sp.*, *Oocystis sp.*, *Oscillatoria sp.*, *Cyclotella sp.*, and *Artemia sp.* Apart from being used as natural food, plankton can also affect the quality of aquaculture water. Water quality parameters for Vannamei shrimp culture in the ponds in this research were still within normal limits and could still support the growth and survival of shrimp.

5. References

[1] Amrial Y, dan Rahayu E 2021 *JPM.* 2(2) 4.
[2] Aji J M M, dan Supriono A 2018 *JEPA.* 2(4) 255-266.
[3] Purnamasari I, Purnama D, dan Utami M A F 2017 *J. Enggano*. 2(1) 58-67.
[4] Madinawati M 2012 *Media Litbang Sulteng*, 3(2).
[5] Samadan G M, Supyan S, Andriani R, dan Juharni J 2020 *JIKK.* 3(2).
[6] Pagoray H, dan Udayana D 2018 *JPT.* 6(1) 30-38.
[7] Rokhim K, Arisandi A, dan Abida I W 2009. *J. Kel.* 2(2) 7-16.
[8] Citria I 2018. Doctoral dissertation. Mataram: Universitas Mataram.
[9] Burhanuddin B, Wahyu F, dan Suratman S 2016 *Octopus: JIP.* 5(1) 462-465.
[10] Pratama A F 2015. Doctoral dissertation. Surabaya: Universitas Airlangga.
[11] Kusriawan H 2016. Skripsi.Makassar: Universitas Muhammadiyah Makassar.
[12] Xu X, Lundholm N, and Li Y 2020 *J. of phycology*. 56(3) 784-797.
[13] Sopian T, Junaidi M, dan Azhar F 2019 *J. Kel.: Ind. J. of Mar. Sci. and Tech.* 12(1) 36-44.
[14] Arsad S, Aprilianita L, Herawati E Y, Musa M, Hertika A M S, Putra R B D S, dan Siswanto D P 2021 *Distribusi Mikroalga di Perairan Indonesia*. (Malang: Universitas Brawijaya Press).
[15] Sulastri 2018 *Fitoplankton Danau-danau di Pulau Jawa: Keanekaragaman dan Perannya sebagai Bioindikator Perairan* (Jakarta: LIPI Press).
[16] Silaban W, dan Silalahi M V 2021 *JST.* 10(2) 299-307.

[17] Saptasari Murni 2007 *Buku Ajar Botani Tumbuhan Bertalus Alga* (Malang: Fakultas Matematika dan Ilmu Pengetahuan Alam Universitas Negeri Malang).
[18] Rahma Y A, Wihelmina G, Sugireng S, dan Ardiyati T 2020 *Biotropika: J.of Trop. Bio.*8(3) 135-143.
[19] Mardana M A 2019 Skripsi. Banda Aceh: Universitas Islam Negeri Ar-Raniry.
[20] Graham L E, dan Wilcox L W 2000 *Algae* (USA: Prentice Hall Publisher).
[21] Putri T, Supono S, dan Putri B 2020 *JARI.* 8(2) 176-192.
[22] Mansyur A, Markus M, Hidayat S S, Brata P, dan Rachman S 2014 *Petunjuk Teknis: Strategi Pengelolaan Pakan Pada Budidaya Udang Vaname Litopenaeus vannamei* (Balai Penelitian dan Pengembangan Budidaya Air Payau).
[23] Ekaputri R A, Arief M, dan Rahardja B S 2018 *JMCS.* 7(2).

6. Acknowledgements
The author is very grateful to Universitas Airlangga for organizing InCoFims 2021.

Application of combination feed of fish viscera silage and fermented rice bran on biomass and population of silk worms *(Tubifex tubifex)*

L Febrianti[1], L Sulmartiwi[2,4] and N N Dewi[3]

[1]Undergraduated Program of Fisheries Processing Technology, Faculty of Fisheries and Marine, Airlangga University, Campus C Universitas Airlangga Mulyorejo Surabaya, East Java, Indonesia 60115
[2]Department of Marine, Faculty of Fisheries and Marine, Airlangga University, Campus C Universitas Airlangga Mulyorejo Surabaya, East Java, Indonesia 60115
[3]Department of Aquaculture, Faculty of Fisheries and Marine, Airlangga University, Campus C Universitas Airlangga Mulyorejo Surabaya, East Java, Indonesia 60115

[4]Corresponding Author : laksmi-s@fpk.unair.ac.id

Abstract. Silkworms are an alternative feed to grow and develop for fish seeds. Silkworms have the advantage of being high in nutrients. The purpose of this study was to determine the effect of using fish offal silage combination feed on silkworm biomass and population as well as the optimal concentration of fish offal silage and fermented rice bran to increase silkworm biomass and population. The method used in this study was experimental using a completely randomized design (CRD), namely P0: paddy mud, P1: paddy mud + combined feed (75% fish offal silage and 25% fermented rice bran), P2: paddy mud + combined feed (50% fish offal silage and 50% fermented rice bran), and P3: paddy mud + combined feed (25% fish offal silage and 75% fermented rice bran). Data analysis used Analysis of Variance (ANOVA) and continued with Duncan's test. The results showed that silkworm cultivation using a combination feed (25% fish offal silage and 75% fermented rice bran) produced the highest absolute biomass and absolute population of 22 grams and 7,147 individuals, significantly different from absolute and absolute biomass. produced from cultivation in paddy field mud with control treatment.

1. Introduction

Aquaculture is one of the activities that utilize the results of aquatic resources. Feed is one of the determining factors for the success of aquaculture, both in terms of quality and quantity. There are two types of feed, namely natural feed and artificial feed. The availability of natural feed plays an important role in cultivation activities, especially in the hatchery phase. One of the natural feeds used in fish farming is silk worms. Silkworms are an alternative food to grow and develop for fish fry [1].

Silkworms are widely used as feed for ornamental fish and fish for consumption. The advantages of silk worms are that they have high nutritional content, slow movement, small size according to the mouth opening of fish seeds and easy to digest [2]. Silkworms contain 48.63% protein and 19.45% carbohydrates [3]. However, the current production of silk worms still relies on catches from nature, so the quality and quantity cannot be ascertained. [4], stated that silk worms from nature can potentially carry harmful pollutants that can accumulate in fish. In addition, silk worms in nature are not available throughout the year, during the rainy season silk worms can be carried away by high water currents. The solution to overcome the dependence of silk worms on catches from nature is by cultivating silk worms.

Research on the cultivation of silk worms using waste has been widely carried out such as fish waste [5], paper and compost factory solid waste [6], rejected milk powder [3], manure chicken [7], bagasse [8], fermented rice bran [9] and palm kernel cake [10]. Another waste that can be used for the cultivation of silk worms is silage of fish offal mixed with fermented rice bran. Catfish offal is one of the fishery wastes that can be processed into silage. [11], stated that the silage of catfish offal contains 54.17% protein. High protein can be used as a source of N-organic. High N-organic content will increase the bacterial population in the rearing media so that the availability of silk worm food will increase. In addition, the cultivation of silk worms also requires organic C which is a carbohydrate-forming element [6]Alternative materials that can be used as a source of organic C can be obtained from fermented rice bran. Fermented rice bran contains carbohydrates 35.04% [12]. Therefore, this study intends to examine the combination of silage feed of fish offal and fermented rice bran in silkworm cultivation. silkworm biomass and population.

2. Material and method

The method used in this study is an experimental method using Completely Randomized Design (CRD). This study used four treatment concentrations of combined feed with five replicates, consisting of one control treatment using rice field mud and three treatments using a combination feed of fish offal silage and fermented rice bran. The treatments were as follows: P0: paddy field mud (control), P1: paddy mud + combination feed (75% fish offal silage and 25% fermented rice bran), P2: paddy mud + combination feed (50% fish offal silage and 50% rice bran). fermented rice bran), and P3: paddy mud + combination feed (25% fish offal silage and 75% fermented rice bran).

The main parameters in this study were the silkworm biomass obtained by weighing the silkworms using an analytical balance with an accuracy of 0.01 grams and the silkworm population was determined by direct counting from one gram of sampling and converted to the amount of biomass obtained from each maintenance container. While the supporting parameters in this study are water quality data consisting of temperature, pH, and DO which are measured every seven days.

The data obtained from the research results will be analyzed using ANOVA (Analysis of Variance) variance to determine the effect of treatment. If there is an effect of treatment, it is continued with the Duncan Multiple Distance test with a 95% confidence level and an error interval of 5% which aims to determine the differences between treatments. The water quality data obtained are presented in the form of a graphic table and analyzed descriptively.

3. Result and Discussion

3.1. Silkworm (Tubifex tubifex) Biomass and Population

Silkworm cultivation using a combination of silage feed of fish offal and fermented rice bran with several different concentrations produced a final biomass of 17-32 grams and a final population of 7,013-11,397 individuals. The biomass and population of silkworms can be seen in Figures 1 and 2.

The growth of silk worms can be influenced by the availability of organic matter in the waters, because organic matter is a nutrient needed by silk worms to grow and develop [13]. Catfish innards which were processed into biological silage (fermentation) and rice bran which were also fermented using probiotic EM4 were able to increase the nutrition of the silkworm culture media. [14], stated that fermentation is a way to increase the nutrition of feed ingredients through the help of microorganisms. The bacteria found in probiotics (EM4), namely Lactobacillus and Streptomyces bacteria will produce protease enzymes that are able to break down peptide bonds in proteins into oligopeptides and amino acids [15]. Then the amino acids will be utilized by bacteria to grow and develop so that bacteria on cultivation media will increase. These bacteria will decompose organic matter, so that the results of the reshuffle can become nutrients for worms to grow and develop [16].

Silkworm biomass cultivated with a combination feed of 25% fish offal silage and 75% fermented rice bran (P3) produced the highest absolute biomass of 22 grams. This could be due to the fact that in P3 the percentage of rice bran used was more than the other treatments, so it had sufficient nutrient content for silkworms to grow and develop. Rice bran is an alternative food source for silk worms to increase growth [9]. According to [12], fermented rice bran has a fairly high carbohydrate content of 35.04%. Carbohydrates are used as a carbon source that will be used by microorganisms as an energy source to grow and develop [17]. The faster the bacteria in the cultivation media grow, the faster the organic matter will decompose. So that the availability of food for silk worms will be formed more quickly which will affect the high biomass.

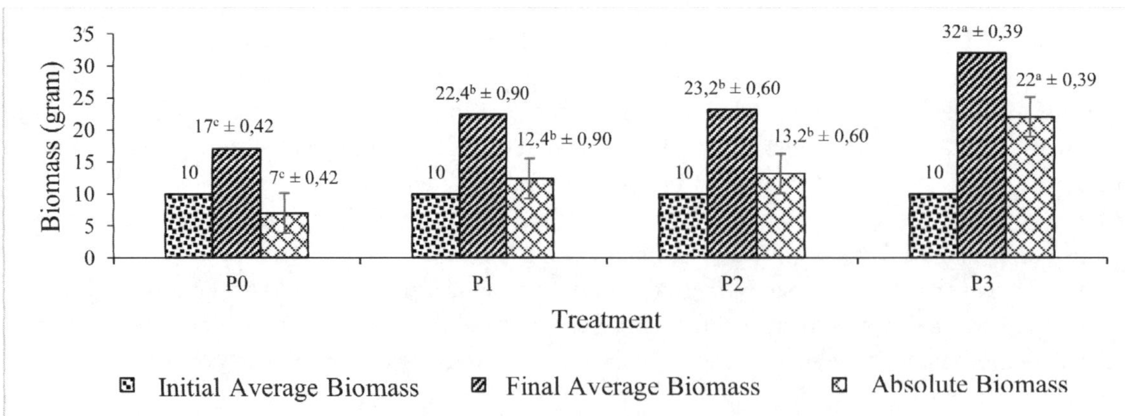

Figure 1. Biomass of Silkworm

The population of silk worms cultivated with a combination feed of 25% fish offal silage and 75% fermented rice bran (P3) resulted in the highest absolute population of 22 grams as many as 7,147 individuals. This could be due to the fact that P3 was able to meet nutrient needs and affect the growth of silk worms. The number of populations is closely related to the reproductive process, in addition to the quantity of food available, the quality of food must also be considered, so that it can meet the needs for both growth and reproduction [18]. The protein contained in the silage of fish offal can be used as a nitrogen source that can be utilized by microorganisms, then these microorganisms become a food source for silk worms. Microorganisms utilize nitrogen as a source of protein that is used to grow and develop [17]. Good protein content for worms is in the range of 9-15%. Protein content that is too high can cause poisoning in worms. Protein is used as a source of energy in the body needed in the process of forming proteins in the body which will become energy reserves for the animal. Feed ingredients for reproduction need to contain protein, because the amino acid content in protein is needed in the formation of male gametes and female gametes of worms [19].

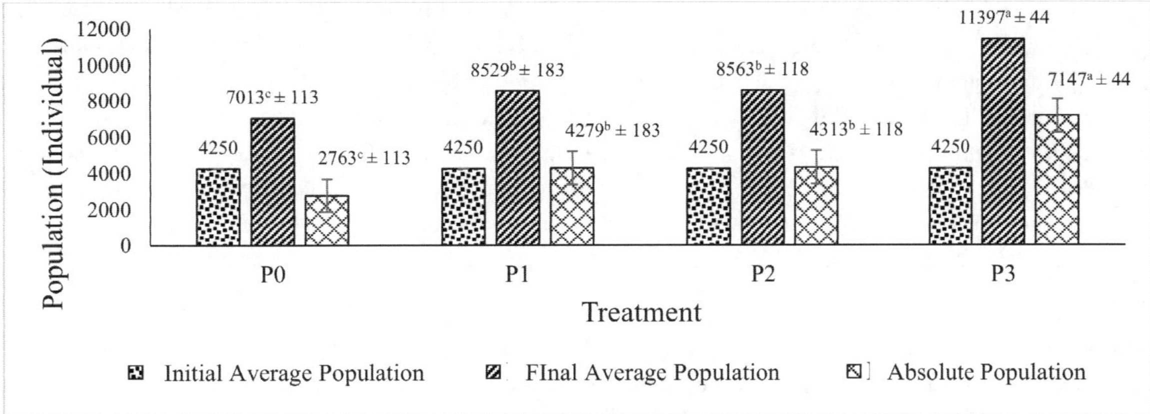

Figure 2. Population of Silkworm

The silk worms used in this study were adult silk worms. The use of adult silk worms can speed up the breeding process to form new cocoons and individuals. The adult silk worms are ready to reproduce and produce cocoons which then hatch into young worms to become adult worms. Adult silk worms can sever their body segments and reproduce [20]. The difference in population produced by silk worms in each treatment can be caused by differences in biological abilities, such as the number of cocoons of each individual, as well as hatching rates and growth rates [13].

Absolute biomass and absolute population of silkworms in P1 and P2 using fish offal silage with a higher percentage resulted in lower absolute biomass and absolute population. This can be caused by the use of acidic silage, causing the pH of the culture medium to be low. According to [21], silage is said to be good if it has a pH ranging from 3-4. The pH obtained in P1 cultivation media was in the range of 4.2 4.9 and P2 was in the range of 5.1 5.6. The degree of acidity or pH of the medium affects the growth of microorganisms. Each microorganism has its own minimum, maximum and optimal pH for growth [22]. The pH that is generally favored by microbes is a neutral pH, namely pH 7. Some bacteria grow at pH 6, but there are also found bacteria growing at pH 4 or pH 5. Very few bacteria grow well at pH 4, except for certain autotrophic bacteria because bacteria produce products. metabolic acid or base [23].

The effect of pH on bacterial growth is related to enzyme activity. Enzymes are needed by bacteria to catalyze reactions associated with bacterial growth. When the pH decreases, the physiological state of the cell will change, resulting in cytoplasmic acidification which causes inhibition of enzyme activity, as a result, the catabolic flux through glycolysis decreases and the rate of biochemical energy synthesis decreases. Decreased energy production is followed by increased energy use to overcome cytoplasmic acidification, causing limited energy for biomass synthesis. Under these conditions, the specific growth rate decreases progressively, and growth eventually stops [24]. Bacterial growth that is not optimal causes nutrients in the silkworm culture medium to be low. This is because the process of decomposition of organic matter by bacteria is slow, so the availability of silk worm food in the culture media is little or reduced. The availability of food in cultivation media is very necessary to meet the needs of silk worms for both growth and reproduction [18].

3.2. Water Quality

The water quality measured during this study was temperature, pH and DO. The water temperature during the study ranged from 27 30 , in accordance with the statement of [25], the optimal temperature for silkworm life ranges from 25–30°C. The pH value of the water obtained ranged from 4.2 to 6.2, but the optimal pH for silkworm cultivation was 5.5 to 8.0 [25]. At neutral pH bacteria can break down organic matter normally into simpler ones that can be used by silk worms as food. The lower the pH value, the lower the organic matter content. [26], stated that at low pH will decrease the process of mineralization of organic matter. Decrease or increase in pH can be influenced by the composition of organic matter because the decomposition process of organic matter can produce acid. Dissolved oxygen content is in the range of 3.51 3.77 mg/L, which is still optimal for the life of silk worms. Silkworms grow well at DO in the range of 2.5 5.5 mg/L [27]. Silkworms are able to survive in low oxygen content because of their ability to carry out respiration.

4. Conclusion

Silkworm cultivation using a combination feed of fish offal silage and fermented rice bran with different concentrations affects the biomass and population of silkworms. The use of rice field mud media with a combination of 25% fish offal silage and 75% fermented rice bran (P3) resulted in the highest absolute biomass and absolute population, which were significantly different from the other treatments.

5. References

[1] Hamron, N., Y. Johan dan B. Brata 2018 *J Penelit Pengelolaan Sumberd Alam dan Lingk*, 7 (2): 79-89
[2] Suharyadi 2012 Study on the Growth and Production of Silkworms (Tubifex sp.) with Different Fertilizers in Recirculating Systems. Thesis. Open University. Jakarta. 116 pp.
[3] Mi'raizki, F., Suminto dan D. Chilmawati 2015 *J Aquac Manag Technol*, 4 (2): 82-91.
[4] Supriyono, E., D.Pardiansyah., D. S. Putri dan D. Djokosetianto.2015. *J Ilmu-Ilmu Perairan, Pesisir dan Perikan*, 4 (1): 8-14.
[5] Ardana, P. P. N., Z. Abidin dan N. Diniarti 2018 *J Perikan*, 8 (1): 55-64.
[6] Bintaryanto, B. W dan T. Taufikurohmah 2013 *J Chem*, 2(1): 1-7.
[7] Putri,D. S., E. Supriyono dan D. Djokosetiyanto 2014 *J Akuakultur Indones*, 13 (2): 132- 139.
[8] Syaputra, S.E., H. W. Maharani dan B. Putri 2016 *J Rekayasa dan Teknol Budid Perair*, 5 (1): 597-603.
[9] Laheng, S., I. W. Putri dan Riskasari 2019 *J Agroqua*, 17 (2): 126-133.
[10] Putri, B., S. Hudaidah dan W. I. Kesuma 2018 *J Rekayasa dan Teknol Budid Perair*, 6(2): 729-738.
[11] Probosasongko, D. A. M 2003 Effect of Different Levels of Catfish Offal Silage in Feed on the Growth of Catfish (Pangasius hypopthalmus) Sejari Size. Thesis. Bogor Agricultural Institute. Bogor
[12] Umidayati 2021 *J Sains Akuakultur Trop*, 5(2): 179-189.

[13] Poluruy, S., M. Idris dan A. Rahman 2019 *Media Akuatika*, 4 (3): 103-109.
[14] Dawood, M. A. and S. Koshio. 2010 *Rev Aquac*, 12 (2): 987-1002.
[15] Puspitasari, F. D., M. Shovitri dan N. D. Kuswytasari. 2012 *J Nutr Ternak Trop*, 2 (1): 42-52.
[16] Raharjo, E. I., Z. Islami dan Farida. 2018 *J Ruaya*, 6 (2): 56-62.
[17] Syam, F. S., G. M. Novita dan S. N. Kusumastuti. 2011. The Effectiveness of Fertilization with Chicken Manure in Efforts to Increase Population Growth and Biomass Silkworm Limnodrilus sp. through Daily Fertilization and Fermentation Results. Faculty of Fisheries. Bogor Agricultural University, Bogor, 8 pp.
[18] Cahyono, E. W., J. Hutabarat dan V. E. Herawati 2015 *J Aquac Manag Technol*, 4 (4): 127-135.
[19] Pangestika, D. K., Nurwidodo, dan L. Chamisijatin. 2016 *J Pendidik Biol Indones*, 2 (2): 168-179.
[20] Armin., M. Idris dan M. Hamzah. 2019 *Media Akuatika*, 4 (4): 133-141.
[21] Sulistyoningsih, M 2015 *Pros Semin Nas Masy Biodiversitas Indones*, 1 (2): 378-382.
[22] Yuda, I. G. Y. W., I. M. Wijaya dan N. P. Suwariani 2018 *J Rekayasa dan Manaj Agroindustri*, 6 (2): 115-124.
[23] Oktavia, E. 2017. Effect of Temperature and pH on Antibiotic Production from Mutant Beam-19. Diploma Thesis. Faculty of Mathematics and Natural Sciences. Andalas University, Padang.
[24] Subagiyo., S. Margino., Triyanto dan W. A. Setyati 2015 *Ilmu Kelaut*, 20 (4): 187-194.
[25] Sriwahyuni, E., Mahendra dan S. Diansyah 2019 *J Akuakultura*, 3 (2): 35-39.
[26] Yuningsih, H. D., P. Soedarsono dan S. Anggoro 2014 *Diponegoro J Maquares*., 3 (1): 37-43.
[27] Fadhlullah, Muhammadar, S. A. Rahimi 2017 *J Ilm Mhs Kelaut dan Perikan Unsyiah*, 2(1): 41-49.

6. Acknowledgements
The author gives thanks for this work being supported by Faculty of Fisheries and Marine, Universitas Airlangga. The author extends thanks to Mrs.Laksmi Sulmartiwi and Mrs. Nina Nurmalia Dewi for supporting author to finish the research.

Water quality and fish health in the upper asahan river in supporting the sustainability of fish resources

D P Anggraeni[1,2], Y C Ditya[1], S Makmur[1] , Makri[1], K Fatah[1], and M Ali[1]

[1]Research Center for Conservation of Marine and Inland Water Resources, National Research and Innovation Agency, Cibinong 16911, Indonesia

[2]Corresponding author : nebula_dpa@yahoo.com

Abstract. This research aims to analyze condition of the aquatic environment from the physical and chemical aspects of the waters and analyze the health of fish in supporting the sustainability of fish resources in the upper part of the Asahan River. Sampling was carried out by purposive sampling method at nine stations obtained in situ and secondary data. The results showed that in general the quality of the waters upstream of the Asahan River was below the second class of river water quality standard. However, it is necessary to monitor regularly the parameters of ammonia and oil content in the waters of the Asahan River so as not to exceed the second class of river water quality standard, because this had implications for the development of the parasite population of *Alitropus sp.* In addition, the emergence of a population of fish lice *(Alitropus sp.*) has also caused many fish that used to inhabit the upper reaches of the Asahan River, there is only one species, namely Asa Fish (*Osteochillus vittatus*) and freshwater lobster (*Cherax quadricarinatus*). Several physical and chemical parameters of the waters have a positive effect on the development of the *Alitropus typus* parasite, including temperature, brightness, ammonia and pH.

1. Introduction

The Asahan Watershed is a priority watershed under the supervision of the Ministry of Environment and Forestry with a total area of 353,770.34 Ha and includes a regional watershed that passes through two regencies (Asahan and Toba Samosir regencies) and one municipality, namely Tanjung Balai City with an area of dominant is in Asahan Regency with an area of 311,689.88 (88.11%). The Asahan Watershed contains a Hydroelectric Power Plant (PLTA) which has been operating since 2011 with a capacity of 180 MW with the aim of meeting electricity needs in North Sumatra Province. In addition, the existence of the industrial sector is a distinctive feature of the existence of the Asahan watershed as part of the hydrological system so that multi-sectoral management becomes an important thing to maintain its sustainability[1,2,3].

The Asahan watershed includes Lake Toba which is its headwaters, along with the Asahan River as its main river outlet, and flows into the Malacca Strait. The length of the Asahan river is 147 km with six main tributaries, and the area of the Asahan watershed reaches 3,741 km^2. The average rainfall in the Asahan watershed is around 2,112 mm per year. The main cities covered by the Asahan watershed include Parapat, Porsea, Balige, Kisaran, and Tanjung Balai. In the lower reaches of the river, flooding often occurs due to the overflow of the Asahan River.

The Asahan River is one of the largest rivers in North Sumatra, Indonesia. This river flows from the mouth of Lake Toba, across the city of Tanjung Balai and ends at Teluk Nibung, the Malacca

Strait. This area is bounded by a high elevation contour that surrounds the lake and crosses Porsea Village where the 150 km long Asahan river drains water out of Lake Toba.

The Asahan River is a lotic ecosystem where as an open ecosystem, the lotic ecosystem receives shipments of organic material carried by water flows from upstream or mainland areas, for example, in the form of carcasses, garbage, or fallen leaves into the river[4,5,6]. Based on the condition of the aquatic environment with vegetation in the form of shrubs, the flow of water is quite heavy, and the level of clarity is high, it is possible for the waters of the upstream of Asahan River to have a moderate to high diversity of fish species[7]. Research from [8] suggested that 12 species of fishes were found in the waters upstream of the Asahan River with a low diversity index (H') of 1,968 – 2,084. These 12 fish species are *Cyprinus carpio, Mystacoleucus padangensis, Osteochilus* sp., *Puntius* sp., *Aplocheilus panchax, Gambusia* sp., *Trichogaster* sp., *Trichopsis* sp., *Tilapia Mossambica, Channa* sp., *Oxyeleotris marmorata,* and *Clarias batrachus.* Furthermore, it was said that there was a fish farming business in Porsea Village.

The local government's policy to increase fishery production through floating net cage farming is a challenge and opportunity for fishermen to fulfill market needs and increase income, this can be seen from the existence of fish farming in Porsea Village. However, the farming of floating net cages is highly dependent on the conditions of the aquatic environment and the maximum production number of floating net cages that can be accommodated. Unfavorable aquatic environmental conditions will affect the health condition of the fish, while the floating net cage quota that exceeds the carrying capacity has the potential to cause mass fish mortality.

2. Materials and methods

This research was conducted in 2018 in the upper reaches of the Asahan River with a length of 11.5 km, starting from the outlet of Lake Toba to approaching the Siruar Dam which includes North Siantar Village and Offering Village, Porsea District. Sampling at four stations was determined using a purposive sampling method, which included Station 1 Outlet TPL (Toba Pulp Lestari), Station 2 North Siantar Village, Station 3 in Gala-gala Pakailan Village, and Station 4 Outlet Lake Toba. Observations of water quality parameters were carried out *in situ* which included parameters of temperature, brightness, pH, conductivity, alkalinity, dissolved oxygen, and carbon dioxide levels. In addition, several samples were analyzed in the laboratory, namely total suspended solids (TSS), ammonia, and COD. Meanwhile, fish and lobster samples were taken directly from the Asahan River. Likewise for parasitic samples taken from the body of the fish when the parasite is attached to the body of the fish. This sample is then analyzed in the laboratory to determine the type of parasite attached to it.

Water quality data was also taken through secondary data from Inalum Company, especially for locations that are not allowed for sampling. There are 5 stations designated for research sampling, namely (1) Porsea Bridge, (2) Before TPL, (3) After TPL, (4) Sigura-gura DAM, and (5) Tangga DAM. Inalum Company carried out water quality measurements because it was worried about paper mill waste (TPL) which was not far from the Siruar Dam which was the first dam to be encountered from the Lake Toba outlet (Upper Asahan River). The reason this company conducted water quality measurements twice a month was because the paper mill effluent creates an unpleasant odor in the Asahan River water along the dam (DAM) which is close to the factory effluent discharge.

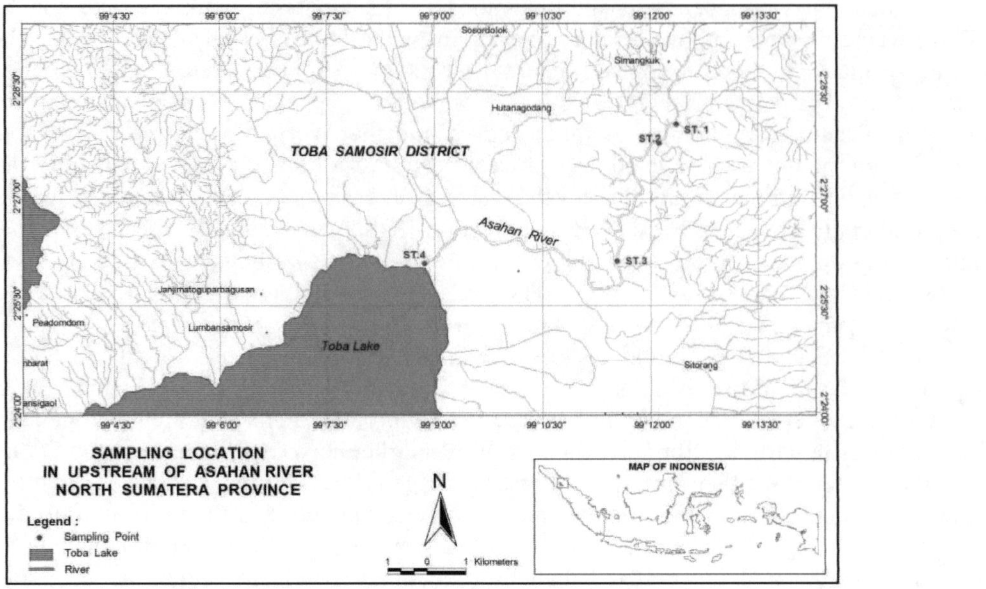

Figure 1. Sampling location in upper Asahan River

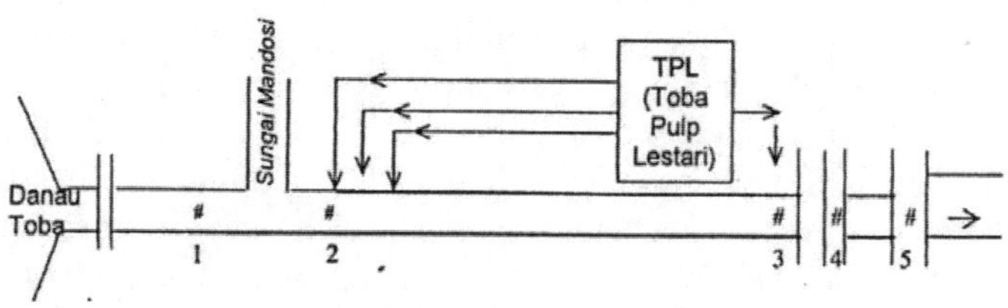

Figure 2. The sampling location scheme from Inalum Company

3. Results and discussion

The Upper Asahan River is an outlet of Lake Toba which is located in Porsea District, Tobasa Regency. About ± 11.5 km downstream, the first dam from the Sigura-gura project was found, namely the Siruar Dam, about ± 5 km, the second dam was found, namely the Sigura-gura Dam. The waters in the Sigura-gura Dam area are still beautiful, both for their aquatic biota and the vegetation that protects them, and are untouched by fishermen because fishing was not allowed in the Inalum Company area. And for security reasons too, it was not allowed to take samples in the waters that were included in the location of Inalum Company. Therefore, Inalum Company provided some water quality data from the Siruar DAM – Tangga DAM which could be taken into consideration in the analysis.

The availability of the elements needed by river water is influenced by the amount of pH in the river water. The degree of acidity (pH) of water needs to be measured so that we are able to determine the quality of the water in the river. If the amount of pH in the river is still normal, it means that the life in it is still well maintained. The degree of acidity (pH) of the water sample of the upper Asahan River was in the neutral range, with a range between 7.5 – 8.1. In general, according to second class of river water quality standards, the pH value was around 6 - 9. This indicated that the acidity of the Asahan River water sample was still good.

IOP Conf. Series: Earth and Environmental Science 1273 (2023) 012043 doi:10.1088/1755-1315/1273/1/012043

Dissolved oxygen value was in the range 6 - 7.2 mg/L, where the minimum value for second class of river water quality standards is 4 mg/L. Carbondioxide levels in the water samples of the upstream of Asahan River ranged from 1.76 – 3.52 mg/L, and the carbondioxide levels that were still tolerable for river water quality were around 9 – 100 mg/L. While the alkalinity values ranged from 75 – 125 mg/L. The alkalinity level required for river water is around 80 - 250 mg/L.

Table 1. *In situ* water quality data for the upstream of Asahan River

Parameter	Sampling Location			
	Station 1	Station 2	Station 3	Station 4
Water Temperature (°C)	25,4 - 25,9	25,3 - 25,9	25,1 - 27,6	25,1 - 26,5
Brightness (cm)	90 - 110	180 - 300	110 - 240	50 - 370
pH	7,5 - 7,8	7,5 - 7,8	7,5 - 7,7	7,5 - 8,1
Alkalinity (mg/L)	97 - 110	75 - 123	77 - 125	79 - 111
DO (mg/L)	6 - 6,66	6 - 6,47	6 - 6,09	6,98 - 7,2
CO_2 content (mg/L)	1,76 - 3,52	1,76 - 3,168	1,76 - 3,168	0 - 1,76

Table 2. Laboratory analysis results from upstream of Asahan River

Parameter	Sampling Location			
	Station 1	Station 2	Station 3	Station 4
Total Phosphat (mg/L)	0,046 - 0,207	0,055	0,107 - 0,177	0,203 - 0,220
TSS (mg/L)	0,33 - 1,27	0,33 - 0,57	0,23 - 1,67	0,33
COD (mg/L)	2,66 - 3,33	0,33 - 2,83	0,83 - 3,49	1,50 - 3,33
NH_3 (mg/L)	0,011 - 0,879	0,071 - 0,118	0,052 - 0,130	0,055 - 0,101
Turbidity (NTU)	7,84 - 9,7	3,26 - 8,4	1,63 - 10,6	1,54 - 47

When referring to Government Regulation Number 22 of 2021 concerning the Implementation of Environmental Protection and Management, the maximum Total Phosphate value is 0.2 mg/L, maximum TSS is 50 mg/L, COD is 25 mg/L, and ammonia is 0.2 mg/L. L. In data collection at the four stations, the values of the four parameters were below or smaller than the second class of river water quality standard. The category of water used was second class where the designation can be used as infrastructure/facilities, water recreation, freshwater fish farming, animal husbandry, and to irrigate crops.

Based on Government Regulation Number 22 of 2021 too, the TDS value at the five stations ranges from 0 - 15 mg/L, and the maximum quality standard value is 1000 mg/L. For other parameters, namely pH, COD, chloride, and sulfate, the values were far below the class 2 river water quality standard. As for the ammonia and oil content parameters, at some stations the values were more than the second class of river water quality standard. And all the locations were after TPL Company.

At the time of the research, there were Asa Fish (*Osteochilus vittatus*) and freshwater lobster (*Cherax quadricarinatus*). Asa fish are mostly caught in the waters near the upstream (outlet) of Lake Toba with a consumption size of 200 - 250 grams/head (Figure 1) and 33-40 grams of lobster/head

(Figure 2). Asa fish migrated from Lake Toba following the current into the Asahan River to the Siruar Dam. However, Asa Fish chicks or juveniles are already very difficult to find, this was also reinforced by information from local fishermen who suspected that Ikan Asa migrated to the Asahan River in search of food, and when ripe the gonads returned to the lake to spawn. Asa fish when experiencing gonad maturity, the weight ranges from 200-250 grams / head. Fish that live around the dam are not touched by fishing gear, so they can grow optimally. Likewise, freshwater crayfish, migrate into the Asahan River in search of food. The size of the lobster in the Asahan River is larger than the lobster that lives in Lake Toba.

Table 3. Water physical-chemical parameters from Porsea Bridge to Tangga DAM

No.	Parameter	1 Porsea Bridge	2 Before TPL	3 After TPL/Reg DAM	4 Sigura-gura DAM	5 Tangga DAM
	PHYSICAL					
1.	Total Suspended Solid, (mg/L)	1 - 7	0 - 11	0 - 12	4 - 15	5 - 15
	CHEMICAL					
1.	pH	8,1 - 8,6	8,1 - 8,4	8,1 - 8,3	8,2 - 8,3	8,1 - 8,2
2.	Amonia (NH_3), (mg/L)	0,0 - 0,4	0,0 - 0,3	0,2 - 0,3	0,1 - 0,7	0,1 - 0,4
3.	COD, (mg/L)	1,9 - 4,7	2,2 - 5,0	2,6 - 6,6	3,6 - 4,7	3,2 - 4,3
4.	Chloride (Cl), (mg/L)	3,2 - 4,1	2,8 - 3,6	5,1 - 5,6	4,1 - 5,6	5,6 - 6,2
5.	Oil content, (mg/L)	0,0 - 0,8	0,1 - 0,7	0,4 - 1,3	0,1 - 1,1	0,1 - 1,1
6.	Sulfate (SO_4), (mg/L)	1,4 - 4,5	0,5 - 4,8	1,8 - 6,6	3,4 - 8,8	2,9 - 8,1

Source : Inalum Company

The catch of fish in the upstream of Asahan River were only two types, namely Asa Fish (*Osteochilus vittatus*) and freshwater lobster (*Cherax quadricarinatus*) which can be caught throughout the year. Even so, it can provide added value because Asa Fish and freshwater lobster are very popular with local consumers. The catch can reach 1 - 5 kg for Asa Fish, and 1 - 7 kg for freshwater lobster. For waters near TPL (paper mills), catches range from 1 kg for Asa Fish, and 1 - 2 kg for lobster. This decrease in catch was due to the unpleasant smell of water and an outbreak of fish lice. From the 12 species of fish found in the upper part of the Asahan River based on research by [8], only Asa Fish and the freshwater lobster that live there. Even the fish farming businesses that used to be widely available in the upstream of Asahan River are no longer there.

Figure 3. Asa Fish (*Osteochillus vittatus*) caught in the upstream of Asahan River (North Siantar Village and Persembahan Village)

Figure 4. Lobster (*Cherax quadricarinatus*) on the Asahan River

This unpleasant smell was thought to have come from the exhaust of TPL Company. In the chemical pulping process, there was the use of sulfur-based chemicals to extract resin and lignin. In this process, odorous gas is produced in the form of hydrogen sulfide (H_2S), methyl mercaptan, dimethyl disulfide, and sulfur gas components which have a very negative impact on human health. Industrial liquid waste was sourced from production processes, utilities, and domestic activities. In addition, the source of the liquid waste production process was also found in the pulp bleaching process, mixing chemicals, and washing raw materials. Sources of utilities, for example, include steam boilers (boilers), demin plants (BFW), laboratory waste water, sealing water, condensate water, cooling water and so on. The last source was domestic activities, namely industrial canteens, floor cleaning, and household waste disposal. While the unpleasant smell of water can come from the ammonia contained in river water. At some stations the value reached 0.4 - 0.7 mg/L, where the maximum value is 0.2 mg/L.

TPL Company claimed to have treated the waste before it was dumped into the Asahan River. They carried out various processes to treat waste before it was disposed of into the Asahan River which is located around the community area, and is located right next to the location of TPL Company. After carrying out various processes, first tested in a small pond filled with goldfish and no dead fish. Around the fish pond, various kinds of vegetables are planted which indicated that there was no air pollution or acid rain around the area, which can damage or die plants like before the waste recycling process. Therefore, the testing process proved that factory waste from TPL Company was an environmentally friendly waste [9]

From table 3, 5 more parameters of laboratory analysis results from Inalum Company showed a value that was below the environmental quality standard, except for ammonia and the oil content which reached 0.4 - 0.7 mg/L and 1 mg/L. These results were obtained at stations 3 to 5, which are located after TPL. According to [10] in TPL Company, hazardous and toxic class waste can come from pulp production and factory maintenance in the form of iron, pipes, iron plates, wires, used oil drums, used mercury lamps, and others. The oil is thought to have come from hazardous and toxic class waste in the form of oil.

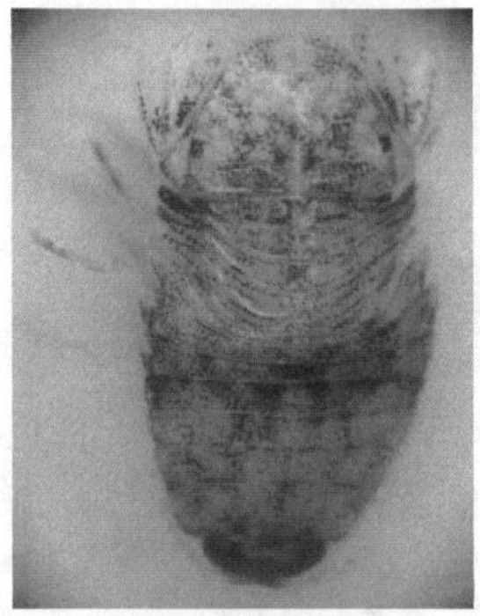

Figure 5. Fish lice (*Alitropus* sp.) that live in the waters of the upstream Asahan River starting from North Siantar Village – Siruar Dam

The results of laboratory analysis showed that the type of parasite that attacked fish in the upstream of Asahan River is *Alitropus typus*. *Alitropus typus*, an isopod parasite belonging to the order Isopoda and family Aegidae, is a common crustacean ectoparasite distributed in freshwater and brackish water ecosystems, affecting many freshwater fish and causing significant economic losses to the aquaculture industry [11]. These fish lice have destroyed many species of fish in the upstream of Asahan River, where now only a few types of fish live. Decreased water quality results in a high stress response in fish making them susceptible to parasitic infections.

[12] explained that rapidly rising temperatures and increased organic loads in inland waters led to a significant increase in parasite populations. The isopod infestation of *Alitropus typus* in fish in these waters is becoming more and more frequent, and causes the death of both wild and fish farming. If you look at the water quality of *in situ* data, the water temperature values ranged from 25.1 - 27.6 ^{o}C. There is no significant increase in temperature which indicates such a rapid temperature increase. As for the indicator of the increase in organic load, the research has not conducted an analysis of the waste discharged from the waste processing of TPL Company. Research from [13] confirmed that several physical and chemical parameters of the waters have a positive effect on the development of the *Alitropus typus* parasite, including temperature, brightness, ammonia and pH. The pH element which tends to be alkaline is a good condition for the life of fish lice.

4. Conclusion

Asahan River water quality when viewed from several parameters, both carried out *in situ*, laboratory analysis, and also secondary data from Inalum Company showed still good. The value of each parameter is also still below the second class of river water quality standard based on Government Regulation Number 22 of 2021 concerning the Implementation of Environmental Protection and Management. The results of the analysis showed a supportive value for fish life. However, it is necessary to monitor regularly the parameters of ammonia and oil content in the waters of the Asahan River so that they do not exceed the class 2 river water quality standards.

5. References

[1] Triatmaja R A, Pursetyo K T, and Triatuti J 2019 *AACL Bioflux*. 12(4) 1025-1030.

[2] Hasan V, Widodo M S, Faqih A R, Mahasri G, Arief M, Valen F S, Tamam M B, Yonarta D, Pratama F S, and Fitriadi R 2020 *Eco. Env. And Conservation*. 26 S73-S75.

[3] Cahyonugrono O H, Hariyanto S, dan Supriyanto G 2022 *Global J. Of Env Sci and Management*. 8(1) 59-74.

[4] Siswanto, Taftazani, and Prasetyo D 2020 *Asian J. Of Water, Env and Pollutan*. 17(3) 19-23.

[5] Hasan V, and Widodo M S 2020 *J. Biodiversitas*. 21(9) 4433-4439.

[6] Chalabia C D, Zohra K T F, Saida B, Fayrouz S, and Boucena M 2022 *JAFH*. 11(2) 201-209.

[7] Nindarwi D D, Samara S H, and Santanumurti M B 2021 IOP Conf. Series: EES. 679(1) 012063.
[8] Ambarita R 2009 Skripsi. Medan: Universitas Sumatera Utara.
[9] Siagian A R 2018 Skripsi. Medan: Universitas Sumatera Utara.
[10] Aritonang A G, Hardjanto U S, A Soemarmi 2016 *Dip. Law J.* 5(3) 1-19.
[11] BunkleyWilliams L, Williams E H, Jr Bashirullah A K 2006 *Rev de Biol Trop* . 54:175–188
[12] Suresh K, Gopil S, Rakesh C G, Ittoop G, Pillai D 2022 *India. J Parasit Dis.* 46(3) 695–703.
[13] Ahmad, Afshan I, Ramzan K, Hayat M, Rizvi S, and Qayyum 2016 *Pakistan J. Zool.* 48(3) 769-779.

6. Acknowledgement

The author is very grateful to Universitas Airlangga for organizing InCoFims 2021.

The Effectiveness of Regulation of the Minister of Marine and Fisheries Number 2 of 2015 Concerning the Utilization of Demersal Fish Landed at PPN Brondong, Lamongan

DN Jannah[1,2], **RA Perdanawati**[1], **MY Fahmi**[1]
[1]Marine Science Program, Faculty of Science and Technology, State Islamic University of Sunan Ampel, Surabaya, Indonesia

[2]Corresponding author : imatin2001@gmail.com

Abstract. The Nusantara Fishing Port of Brondong in Lamongan, is one of the largest fishing bases in East Java. Types of fish caught by fishermen are dominated by *Nemipterus japonicus*, *Upeneus moluccensis*, and *Gerres oyena*. The purpose of this study was to determine the utilization rate of demersal fish landed at PPN Brondong, Lamongan. This research was conducted using quantitative and descriptive statistical methods. Maximum sustainable yield analysis uses the production surplus model from Schaefer (1954). Based on the model of Schaefer (1954) demersal fish are stated in a fully exploited to over exploited condition. Analysis of the Schaefer model (1954) found the F_{msy} value of 12,457 trips, the C_{msy} value of 25,216 tons and the allowable catch of 20,173 tons. PERMEN KP Number 2 of 2015 in terms of response effectiveness according to respondents obtained a value of 52%. Based on research, the issuance of PERMEN KP Number 2 of 2015 was not effective on the level of utilization of demersal fish landed at Nusantara Fishing Port of Brondong, Lamongan.

1. Introduction

The northern coastal water of Java it is included in WPPNRI 712. Many demersal fish are found which will become the main target of fishermen in their fishing business. Based on Kepmen-KP No. 47 of 2016 the potential for demersal fish in WPPNRI 712 was recorded at 320,432 tons [1]. The Nusantara Fishing Port (PPN) in Brondong District, Lamongan Regency is one of the largest fish landing bases in East Java. In 2010 this Nusantara Fishing Port was designated as a Minapolitan area based on the Decree of the Minister of Marine and Fisheries No. 32/MEN/KP [2].

Cantrang is one of the most widely used fishing gear to catch demersal fish on the North Coast of Java. Based on data from the Ministry of Marine and Fisheries in 2015, there were 13,300 cantrang operations in the North Sea of Java. In 2008 fishermen in Brondong District used 1,441 units of cantrang out of a total of 1,528 units of fishing gear [3,4]. The government has issued the Minister of Marine and Fisheries Regulation Number 02 of 2015 concerning the Prohibition of the Use of Trawls and Seine Nets in the Indonesian Fisheries Management Area .

If fishing is carried out beyond the limit, there can be an exploitation level that exceeds the maximum point or is called the Maximum Sustainable Yield (MSY) [5]. The level of utilization of fishery resources can be seen from the catch of production (catch) and the effort made to catch fish (effort). If the production of fishery products has reached the Maximum Sustainable Yield value, the level of production per unit will decrease [6]. If over exploitation has been carried out, fish production will decrease and result in a decrease in fishermen's income [7,8]. Based on data from the Ecosystem Approach to Fisheries Management (EAFM) listed in KEPMEN 50/2017, the stock status of demersal fish resources in WPP-NRI 712 is moderate. The purpose of this study was to determine the level of utilization of demersal fish landed at PPN Brondong, Lamongan and to determine the implementation of the Minister of Marine and Fisheries Regulation Number 2 of 2015 in terms of effectiveness according to respondents' responses.

2. Methods

The research method used is quantitative and descriptive statistics. The data used includes primary and secondary data. Primary data collection through interviews with fishermen who operate cantrang fishing gear to look for demersal fish. Information from interviews is used to determine PERMEN KP Number 2 of 2015 concerning the utilization of demersal fish. The secondary data used include data on the production of demersal fish catches and fishing efforts in 2011-2020. Secondary data was obtained from the statistical data of the Nusantara Fishing Port of Brondong, Lamongan.

2.1. Utilization Analysis Method

Data analysis is sourced from statistical data of the Nusantara Fishing Port of Brondong, Lamongan in 2011-2020. Data in the form of production and effort to catch demersal fish. According to (Sparre & Venema, 1999; Nugraha et al., 2012) the formula for calculating the standardization of effort is:

$$CPUEs = Cs/Fs$$

Information :
CPUEs = Catch per unit of effort to catch standard fishing gear
Cs = Catch of standard fishing gear
Fs = Number of attempts to catch standard fishing gear

2.1.1. Catch Per Unit Effort (CPUE)

According to Nugraha *et al.* [9] the calculation of CPUE for each period of the year can be calculated by using the following formula:

$$CPUEi = Ci/Fi \quad (1)$$

Information :
CPUEi = Catch per unit effort in period-i
Ci = Catch of period-i
Fi = Catch effort of period-i

2.1.2. Maximum sustainable yield (MSY)

Maximum Sustainable Yield obtained from the calculation of fish production and efforts to catch demersal fish landed at the Nusantara Fishing Port (PPN) Brondong in Lamongan was carried out using the production surplus model from Schaefer (1954). It is necessary to calculate slope and intercept with the following formula:

$$b = \frac{n\sum xy - \sum x \sum y}{n\sum x^2 - (\sum x)^2} \quad (2)$$

$$a = \frac{\sum y - b\sum x}{n} \quad (3)$$

$$r = \frac{n\sum XiYi-(\sum Xi\sum Yi)}{\sqrt{n\sum Xi^2-(\sum Xi)^2}\sqrt{n\sum Yi^2-(\sum Yi)^2}} \quad (4)$$

Information :
b = Slope of the regression line
a = Intercept of the regression line and the y axis
r = Correlation coefficient
n = Time period (years)
x = Catch effort (trip)
y = Catch (tons/trip)

Schaefer's Model (1954)

The relationship between fishing effort and catch per trip using the Schaefer's model (1954) can be calculated by using the following formula:

$$CPUE = a + bf \quad (5)$$

Information :
CPUE= Catch per unit effort (Kg/trip)
a = Intercept of the regression line and the y axis
b = Slope of the regression line
f = Effort to catch (trip)

The relationship between catch and effort is:

$$C = af + bf^2 \quad (6)$$

Information :
C = Catch (Kg)
a = Intercept (regression line intersection point) Schaefer's model
b = Slope of the Schaefer's model regression line
f = Effort to catch (trip)

The optimum fishing effort (fopt) can be calculated through the equation of the first derivative of the catch with catch effort = 0 is:

$$C = af + bf^2$$
$$C^1 = a + 2bf = 0$$
$$f_{opt} = -(\frac{a}{2b}) \quad (7)$$

Information :
C = Catch (Kg)
a = Intercept (regression line intersection point) Schaefer's model
b = Slope of the Schaefer's model regression line
f = Effort to catch (trip)

The calculation of the Maximum Sustainable Yield value using the Schaefer's model (1954) obtained by substituting the optimum catch effort value into equation (2) is:

$$C_{max} = a\left(-\frac{a}{2b}\right) + b\left(-\frac{a}{2b}\right)^2$$
$$MSY = C_{max} = -\frac{a^2}{4b} \quad (8)$$

2.1.3. Total Allowed Catch (TAC)

Based on the Decree of the Minister of Agriculture No. 473a/1985 the value of the number of fish that can be caught in a fishery area is 80% of the Maximum Sustainable Yield (MSY).

$$TAC = MSY \times 80\% \quad (9)$$

TAC = Total Allowed Catch (ton)
MSY = Maximum Sustainable Yield

2.1.4. Utilization Rate (E)

According to (Sparre & Venema, 1999; Nugraha et al., 2012) the formula for calculating the utilization rate can be expressed using percent is:

$$TPi = (Ci/MSY) \times 100\% \quad (10)$$

Information :
TPi = Period-i utilization rate (tons)
Ci = Period-i catch (tons)
MSY = Maximum sustainable yield

Based on PERMEN KP Number 50 of 2017 the level of fish utilization is grouped into 3 is :

a. Moderate ($E < 0.5$), is the status of fishery resource utilization with catch effort still can be increased
b. Fully exploited ($0.5 \leq E < 1$), is the status of the utilization of fishery resources with fishing effort maintained with strict monitoring, because if the fishing effort is increased it will threaten the sustainable potential of the fish.
c. Over exploited ($E \geq 1$), is the status of the utilization of fishery resources with the fishing effort that must be reduced, because if it continues to be caught it can cause extinction.

2.2. Implementation of the Minister of Marine and Fisheries Regulation Number 2 of 2015

2.2.1. Respondent Sample

The sample of respondents used is part of the population. The population of cantrang boat fishermen in Nusantara Fishing Port of Brondong is 6,689 people, so the error tolerance limit is 10%. The number of samples used in this study was obtained through calculations from the Slovin formula is :

$$n = \frac{N}{1 + Ne^2}$$

$$n = \frac{6.689}{1 + (6.689 \times 0{,}1^2)}$$

$$= \frac{6.689}{67{,}89}$$

$$= 99$$

Information :
n = Number of samples
N = Total population
e = Allowance rate or percent error (10%)

2.2.2. Hypothesis Testing

The hypothesis test used is descriptive hypothesis testing of one or more variables with data in the form of intervals or ratios, so that one sample t-test is used. Test one sample t-test using the right side test with an Alpha of 5%. This hypothesis test was conducted to determine PERMEN KP Number 2 of 2015 in terms of effectiveness according to respondents to the use of demersal fish landed at Nusantara Fishing Port of Brondong, Lamongan.

The hypothesis proposed by the researchers in this study refers to the effectiveness scale set by the Research and Development Ministry of Home Affairs, Minister of Home Affairs Regulation Number 20 of 2011 is :

a. H0; $\mu \leq 60\%$ then PERMEN KP Number 2 of 2015 in terms of effectiveness according to respondents on the level of utilization of demersal fish landed at Nusantara Fishing Port of Brondong is declared ineffective.
b. Ha ; $\mu > 60\%$, then PERMEN KP Number 2 of 2015 in terms of effectiveness according to respondents to the level of utilization of demersal fish landed at Nusantara Fishing Port of Brondong is declared effective.

3. Results and Discussion

3.1. Production and effort of demersal fish in Nusantara Fishing Port of Brondong

Based on statistical data from the annual report at the Nusantara Fishing Port of Brondong, Lamongan demersal fish production from 2011 to 2020 [10] has fluctuated. Production and effort to catch demersal fish on land at Nusantara Fishing Port of Brondong can be seen in Table 1.

Table 1. Production and effort of demersal fish in Nusantara Fishing Port of Brondong

Years	Catch (tons)	Effort (trip)	CPUE (tons/trip)
2011	24.187	12.949	1,8679
2012	25.339	10.525	2,4075
2013	22.837	8.309	2,7484
2014	26.730	8.474	3,1543
2015	24.408	9.195	2,6545
2016	23.547	9.643	2,4418
2017	23.328	9.758	2,3906
2018	20.517	8.436	2,4321
2019	19.651	7.566	2,5973
2020	20.205	7.090	2,8498

Based on the Table 1 , the highest CPUE value of demersal fish landed at Nusantara Fishing Port of Brondong from 2011-2020 was in 2014 which was 3.1543 and the lowest CPUE value was in 2011 which was 1.8679. The increase or decrease in the CPUE value is strongly influenced by the fishing effort in each year [11]. In addition, according to Budiman *et al.* [12] explained that the decrease in CPUE value was influenced by the increasing number of fishing gear. In this research conducted at Nusantara Fishing Port of Brondong, the CPUE value from 2011-2020 is relatively increasing, due to the decreasing number of fishing gear from year to year. The CPUE value will be inversely proportional to the effort value, because any increase in effort will affect the decrease in catch per unit of effort (CPUE) [13]. This can reduce the stock of fish resources if fishing efforts are carried out continuously.

Factors that can affect fluctuations in the production of fish catches include:

a. Fishing gear,
Catching demersal fish that landed at Nusantara Fishing Port of Brondong is done using cantrang fishing gear. In accordance with the Regulation of the Minister of Marine and Fisheries number 59 of 2020, the mesh size of the cantrang is 2 inches, but the cantrang used at Nusantara Fishing Port of Brondong has a mesh size of 1 inch. The difference in size makes it easier for fishermen to find fish.

b. Fishing boat,
The majority of the fishing boat in Nusantara Fishing Port of Brondong are boats that use outboard motors. There are two categories of cantrang fishermen at Nusantara Fishing Port of Brondong, including fishermen who use boats with a size of 10 GT and 10-30 GT. Fishermen who use boats 10 GT go to sea in an average of one day, while those who use boats with a size of 10-30 GT go to sea weekly.

c. Fishing effort,
The occurrence of fluctuations every year is because the number of fishing gear used in Nusantara Fishing Port of Brondong has decreased, thus affecting the number of fishing efforts made by fishermen to find fish. According to (Sparre and Venema, 1999; Nugraha et al., 2012) an increase in the number of fishing efforts can reduce catches in the following years. This is because stock biomass is a limited resource pursued by ships, so the catch is affected by the number of vessels operating.

d. Hydro oceanographic parameters
According to [14] explained that currents and wind direction can affect catch production. Fish and fishing gear can move due to the influence of ocean currents. The fish will move against the current, so the mouth of the net must block the movement of the fish. The effect of the wind on the production of the catch is causing the movement of the ship when it is operated.

3.2. Estimation of Maximum Sustainable Yield (MSY)

Analysis of MSY using the Schaefer's model (1954) requires two types of data, including catch data and fishing effort data. The data is then carried out by linear regression, variable x is the data of fishing effort, and variable y is the data of CPUE. The relationship between CPUE and demersal fishing effort can be seen in Figure 1.

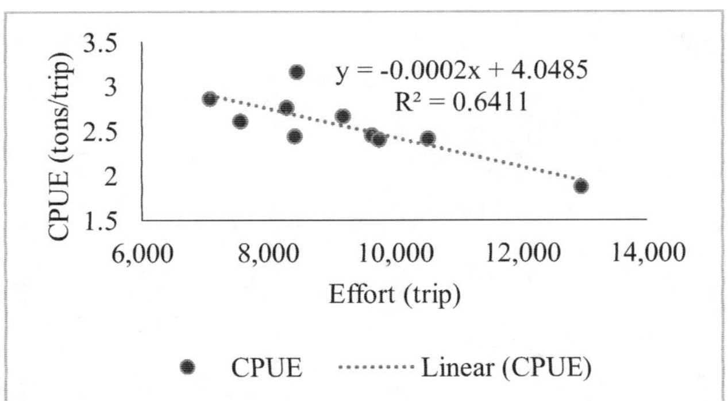

Figure 1. CPUE relationship and capture effort demersal fish

Based on the Figure 1, the regression coefficient value is 0.0002, which means that every reduction (marked by minus) 1 trip will increase the CPUE value by 0.0002 tons/trip. If there is an increase of 1 trip, the CPUE value can have an impact on decreasing catch production by 0.0002 tons/trip.

Table 2. Results of Regression Analysis

No.	Parameters	Demersal fish
1	F_{msy} (trip/year)	12.457
2	C_{msy} (tons/year)	25.216,6
3	TAC (tons)	20.173

The Maximum Sustainable Yield potential is 25,216.6 tons/year with an optimum effort of 12,457 trips/year. The allowable catch is 80% of the potential value, so that a value of 20,137 tons/year is obtained. Analysis of sustainable potential using the Schaefer's model (1954) is presented in the form of a curve that shows the relationship between effort (trip) as the x-axis and catch (tons) as the y-variable which can be seen in Figure 2.

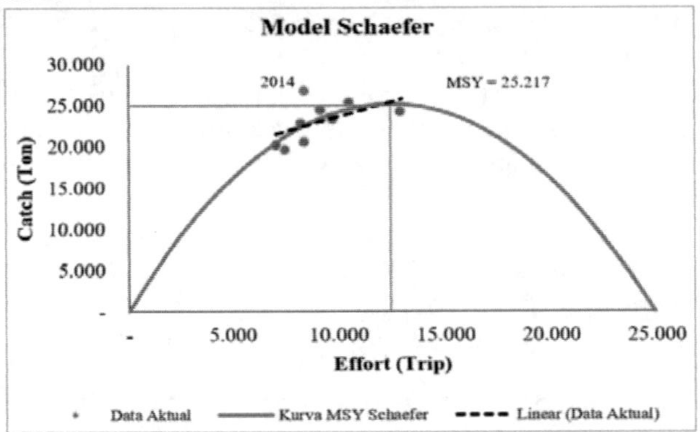

Figure 2. Effort and Catch Demersal Curves

Table 3. Percentage of utilization rate and effort to catch demersal fish

Year	Catch (ton)	Utilization rate (%)	E	Effort (trip)	Effort rate (%)
2011	24.187	95.92	0.96	12.949	103.95
2012	25.339	100.49	1.00	10.525	84.49
2013	22.837	90.57	0.91	8.309	66.70
2014	26.730	106	1.06	8.474	68.03
2015	24.408	96.80	0.97	9.195	73.81
2016	23.547	93.38	0.93	9.643	77.41
2017	23.328	92.51	0.93	9.758	78.33
2018	20.517	81.37	0.81	8.436	67.72
2019	19.651	77.93	0.78	7.566	60.74
2020	20.205	80.13	0.80	7.090	56.92

The highest utilization rate occurred in 2014 and 2012 on a scale of 1.06 and 1. The status of utilization of demersal fish resources in these conditions based on PERMEN KP Number 50 of 2017 is over exploited. The catch of demersal fish landed at Nusantara Fishing Port of Brondong in 2011 to 2020 tends to be stable, but in 2014 the catch has exceeded the Maximum Sustainable Yield value of 26,730 tons. If the level of resource utilization exceeds the Maximum Sustainable Yield value, it can have an impact on the sustainability of these resources, so that better fisheries management is needed so that they can be used sustainably. According to According to Yuliono. [11] the ecosystem of fishery resources will be disrupted due to overfishing, which will result in a decrease in fish resource stocks [15,16,17].

3.3. Implementation of the Minister of Marine Affairs and Fisheries Regulation Number 2 of 2015
PERMEN KP Number 2 of 2015 is known through the results of interviews with 99 cantrang fishermen at Nusantara Fishing Port of Brondong. The theory of effectiveness used is Duncan's theory cited by Steers (1985). There are three aspects of measuring effectiveness, including the achievement of goals, aspects of integration and adaptation.

The results of hypothesis testing using the *One Sample t-test* method through the SPSS version 16. Application can be seen in Figure 3.

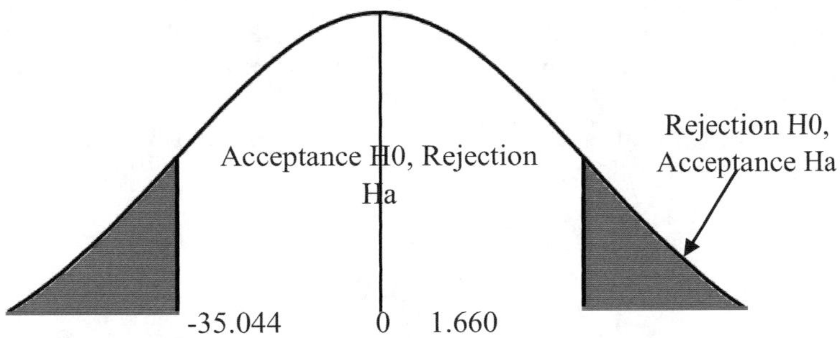

Figure 3. The curve of the acceptance area of the research hypothesis

Based on Figure 3, the value of –t count ≤ t table (-35,044 ≤ 1.660) so that H0 is accepted and Ha is rejected. This means that based on the research hypothesis, the effectiveness of PERMEN KP Number 2 of 2015 on the utilization rate of demersal fish landed at Nusantara Fishing Port of Brondong is declared ineffective because the assessment is ≤ 60%. To find out the percentage value of PERMEN KP Number 2 of 2015 in terms of effectiveness according to respondents on the level of utilization of demersal fish landed at Nusantara Fishing Port of Brondong, the following calculations can be made:

$$\text{Percentage} = \frac{\text{Actual score}}{\text{Ideal score}} \times 100\%$$

$$= \frac{3.143}{5.940} \times 100\%$$

$$= 52\%$$

According to Rafida [18], it is explained that continuous measurement is related to a variable. In this study, the variable measured is PERMEN KP Number 2 of 2015 in terms of effectiveness, which can be seen in Figure 4.

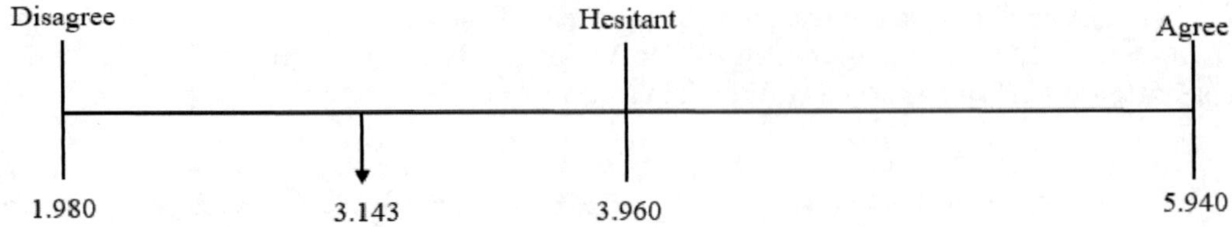

Figure 4. Measurement of effectiveness on a continuum

Based on the picture 4, the actual total score is 3,143 which shows that the assessment of respondents who work as cantrang fishermen at Nusantara Fishing Port of Brondong is in the category of assessment that tends to be hesitant. The results of the assessment of PERMEN KP Number 2 of 2015 on a continuum and the results of hypothesis testing are in harmony. This is evidenced by hypothesis testing which is declared ineffective because the percentage value is ≤ 60% and according to the responses of respondents from cantrang fishermen at Nusantara Fishing Port of Brondong after evaluating the research questionnaire, it shows a category that tends to be hesitant.

The results of responses from respondents who were willing to fill out research questionnaires were then analyzed which averaged each question indicator used to determine PERMEN KP Number 2 of 2015.

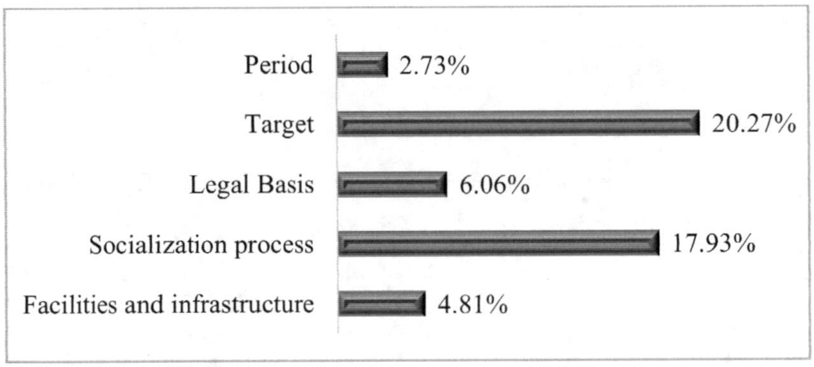

Figure 5. Percentage of measurement of sub-indicators of PERMEN KP Number 2 of 2015

Based on the picture above, the time period, goals and legal basis are sub-indicators of the aspect of achieving the objectives of the issuance of PERMEN KP Number 2 of 2015. According to data from the Research and Development Ministry of Home Affairs, Minister of Home Affairs Regulation Number 20 of 2011 if the effectiveness indicator has a value (21 < N < 40%) then it is included in the ineffective criteria, whereas if the value (N ≤ 20) is categorized as very ineffective.

The issuance of PERMEN KP Number 2 of 2015 related to the utilization rate of demersal fish landed at Nusantara Fishing Port of Brondong was declared ineffective. The number of utilization of demersal fish in 2011-2020 has reached the status of fully exploited to over exploitation. The ban on the operation of cantrang in WPPNRI has also not been implemented at Nusantara Fishing Port of Brondong, as evidenced by the use of cantrang to this day. These two aspects support the ineffectiveness of PERMEN KP Number 2 of 2015.

4. Conclusion

Based on the results of this study, it can be concluded as follows:

a. The result of estimating the Maximum Sustainable Yield (MSY) value using the Schaefer model is 25,216.6 tons/year with an optimum catch effort of 12,457 trips/year.
b. In 2012 and 2014 the utilization rate of demersal fish has reached over exploited status, while in other years the utilization of demersal fish is in fully exploited status.
c. Based on the determination of the Total Allowed Catch (TAC) of 80% of the MSY value, the TAC value of demersal fish is 20,173 tons/year.
d. PERMEN KP Number 2 of 2015 in terms of effectiveness according to respondents on the use of demersal fish landed at Nusantara Fishing Port of Brondong which was measured using Duncan's theory obtained a percentage value of 52%, and was declared ineffective.

5. References

[1] Prayitno, M. A. 2019. *Biological Aspects and Utilization Status of Kuniran Fish (Mullidae) Landed at PPN Brondong, Lamongan*. Brawijaya University.
[2] Giamurti, anis syahfitri rilia, Bambang, aziz nur, & Fitri, aristi dian purnama. 2018. *Marketing Analysis of Catches of Red Snapper (Lutjanus sp.) at Nusantara Fishing Port of Brondong, Lamongan, East Java*. *7*, 19–28.
[3] Riyanto M, Purbayanto A, Mawardi W, and Suheri N 2011 *Bul PSP*. XIX(1) 97–104.
[4] Aulia A, and Sahidu A M 2019 *IOP Conf Ser Earth Environ Sci*. 236(1) 012131.
[5] Fitriani R, Hartati R, Sunaryo S, Irwani I, Ario R, and Soegianto A. 2020 *Ann Biol*. 36(2) 304-310.
[6] Damora A, Fadli N, Andriyono S, Suman A. 2021 *IOP Conf Ser Earth Environ Sci*. 869(1) 012049.
[7] Gunawan, Y. N. 2013. *Estimation Study of Maximum Sustainable Yield (MSY) and Gonad Maturity Level of Red Snapper (Lutjanus sp) at the Nusantara Fishing Port of Brondong, Lamongan, East Java*. Brawijaya University.
[8] Dewi N N, Pursetyo K T, Darmono O P, Fachr F R, Puspitasari S, and Damora, A 2019 *IOP Conf Ser Earth Environ Sci*. 236(1)1 012039.
[9] Nugraha E, Koswari B, Yuniarti 2012 *J Fish Mar*. 3(1) 91–98.
[10] Nusantara Fishing Port of Brondong, 2011-2020. Nusantara Fishing Port of Brondong Statistics Data. Lamongan
[11] Yuliono G M A 2021 *Utilization Rate of Swanggi Fish (Priacanthus Spp.) in North Waters of East Java*. Brawijaya University.
[12] Budiman, Supriharyono, & Asriyanto. 2006. *Analysis of the Distribution of Demersal Fish as a Basis for Management of Coastal Resources in Kendal Regency*. *2 (1)*, 52–63.
[13] Rahmawati M, Fitri A D P, and Wijayanto D 2013 *J Fish Resour Util Manag Technol*. *2*(3) 213–222.
[14] Hakim, L., & Nurhasanah. 2016. *Cantrang : Problem and solution.*
[15] Lelono T D, Rahman M A, Bintoro G, Setyowati N H, and Wulandari N N 2021 *J Aquac Sci*. 6(1IS) 61-76.
[16] Isroni W, Samara S H, and Santanumurti M B 2019 *Biodiversitas*. *20*8 2273-2278.
[17] Pamungkas G Y, and Sari L A 2021 *IOP Conf Ser Earth Environ Sci*. *679*(1). 012030.
[18] Rafida, H. 2019. *Effectiveness of Online Service Management Information System Application (Simponie) in Online Business Licensing Services at the Serang Regency Investment and One Stop Integrated Service Office in 2018*. Sultan Ageng Tirtayasa University.

Study of Bawal Bintang fish (*Trachinotus blochii*) hatchery at the Lombok marine aquaculture, West Nusa Tenggara, Indonesia

A A Ramadhani [1], L A Sari[2,4], P D W Sari[2], D D Nindarwi[2] and S Arsad[3]

[1] Study Program of Aquaculture, Faculty of Fisheries and Marine, Universitas Airlangga, Campus C Jalan Mulyorejo, Surabaya 60115 East Java, Indonesia
[2] Department of Aquaculture, Faculty of Fisheries and Marine, Universitas Airlangga, Campus C Jalan Mulyorejo,Surabaya 60115 East Java, Indonesia
[3] Department of Aquatic Resources Management, Faculty of Fisheries and Marine Science, Universitas Brawijaya, Jalan Veteran Malang 65145 East Java, Indonesia.

[4]Corresponding author: luthfianaas@fpk.unair.ac.id

Abstract. Bawal Bintang fish (*Tracinotus blochii*) has high economic value, to meet both domestic and foreign consumption needs. Bawal Bintang fish cultivation has become a commercial business to be developed because of its relatively fast growth, relatively easy maintainance, and public interest in it. The objective of this study is to find out the hatchery techniques of Bawal Bintang Fish and the obstacles that arise in their hatchery at the Lombok Marine Cultivation Fisheries, West Nusa Tenggara. The method used was descriptive with data collection including primary data and secondary data. Data collection was done by means of active participation, observation, interviews, and literature study. Bawal Bintang fish hatchery technique began with tank preparation, spawning, egg distribution, and maintenance for 20-25 days. Bawal Bintang fish hatchery included feed management, water quality monitoring, and pest and disease control. After rearing the larvae for 20-25 days or reaching seedling size, then harvesting was carried out to transfer them to the nursery tank using the RAS system. The problem that arose in the hatchery of Bawal Bintang fish was the frequent occurrence of Bawal Bintang larvae that fell from a pond.

1. Introduction

The potential of fishery resources in Indonesia is quite large, both capture fisheries and aquaculture resources. Those fishery resources are one of the national assets which must be managed properly. Indonesia consists of islands that can support aquaculture activities. The potential for developing aquaculture include: aquaculture potential for commercial purposes, aquaculture potential for conservation purposes, potential for the application of germplasm storage, potential natural feed sources, and potential use of GIS and 3D technology to support aquaculture. The amount of potential possessed is expected to provide maximum cultivation results [1].

Marine aquaculture activities are an alternative which be a solution to deal with fishermen's dependence on fishing business. One of the profitable marine cultured fish that has good marketing prospects is the Bawal Bintang (*Tracinotus blochii*). The Bawal Bintang cultivation business has developed well in Kepulauan Seribu, Batam, and Bali which are managed by private parties [2].

Bawal Bintang or golden pompano (*Tracinotus blochii*) is one of the marine commodities which are cultivated in Indonesia. This fish belongs to the *Carangidae* family that lives in river mouths, corals, and the high seas. Bawal Bintang is one of the consumption fish that has economic value. This fish also has good marketing prospects in the Asia Pacific region as it has a high price and does not require a long time to be cultivated until it reaches consumption size. The Bawal Bintang is a fish produced from Taiwan and has good prospects. Bawal Bintang is a very active pelagic fish because it always moves (rotates) on the surface, so it requires an adequate location/place[3].

Bawal Bintang is a type of high economic marine fish which has very good market prospects in Indonesia (Jakarta and other big cities) and the Asia Pacific zone. The soaring market demand for consumption (live) start pomfret, especially in Jakarta at a relatively high price, opens up great opportunities in an effort to increase their production, especially for cultivators in the capital's buffer zones such as in Kepulauan Seribu, Lampung and Karawang. Problems that will be faced next year related to the regional free market (AFTA) require us to make efficiency, as an effort to anticipate AFTA (the entry of start pomfret from neighboring countries) [4].

The Bawal Bintang (*T. blochii*), is a leading commodity in marine aquaculture. Bawal Bintang has a big market potential. Market demand for Bawal Bintang is quite high, from local to international levels such as Taiwan, Hong Kong, and Singapore. In addition to its high economic value, Bawal Bintang is also disease resistant and easy to maintain [5].

Fish larvae are the stages or phases of fish after the eggs hatch. The beginning of the fish life cycle includes the egg stage and its development, namely the larval and juvenile stages [6]. The egg and larval stages of fish can be classified as plankton, where part of their life cycle is temporary plankton or meroplankton [7]. Newly hatched Bawal Bintang larvae are usually characterized by the presence of egg yolk and oil globules used as endogenous feeding, which will shrink for 3 days. The Bawal Bintang larvae will eat natural food after the egg yolk runs out and will eat artificial food when it reaches the size of the seed.

Bawal Bintang cultivation activities include spawning, egg hatching, larval stocking, feed management, water quality management, health and environmental control, grading and sampling, and harvesting. Fish will experience a phase called larval stage in their life cycle. One of the aspects that must be considered is the maintenance technique because this phase is a very critical phase so it must be handled with care in order to increase production yields, which will eventually be able to support the sustainability of cultivation.

2. Material and methods

2.1. Preparation of fish cultivation tanks

The initial activity carried out in broodstock maintenance was tank preparation. The tank used for broodstock maintenance was a fiber tank with a diameter of 3 m and a height of 2 m, equipped with 7 aeration points and 2 outlet channels and 2 inlet channels. The 2 outlet channels in the tank were the water drainage and egg outlet channels that were connected to the egg collection tank. Meanwhile, the 2 inlet channels were filtered water flow channels and sea water channels.

The preparation of the tank at the study site was carried out by chlorinating the tank using a diluted dose of 100 ppm. The diluted chlorine was then splashed all over the tank. After that, it was let stand for 2-3 hours. Chlorine $Ca(ClO)2$ serves to reduce organic substances, oxidize metals, and disinfect microorganisms [8]. After settling, then the tank was scrubbed using a brush and then rinsed with running water. After the chlorination process was complete, the tank was filled with water again through the inlet channel as much as 50-80%.

2.2. Broodstock stocking
White snapper broodstock at the study site were intensively reared in fiber tanks. The number of broodstock in the study area was 27 with 13 males and 14 females. The broodstock were stocked at a density of 1 individual/m^3. The purpose of rearing broodstock was to get mature broodstock that were ready for spawning in order to produce high-quality eggs.

The prospective broodstock in the research site was reared semi-intensively in the KJA (Floating Net Cage). The number of prospective broodstock in the KJA of the research site was 240 individuals. Prospective broodstock were stocked with a density of 6 individuals/m^3.

2.3. Feed management
The success of a cultivation is strongly influenced by feeding. White snapper brooders were fed 2 times a day with a minimum feed containing 42-45% protein to support gonadal maturation. The dose of feed given was 3-7% of the weight of the biomass [9]. Meanwhile, in the research site itself the dose of feed given was 5.22%. The feed provided was in the form of fresh lemuru fish and squid. Fresh lemuru fish contains 51-70% protein [10] and squid contains 68% protein[11]. Feeding was done at satiation in the morning and afternoon. Every week, an immunostimulant was added in the form of multivitamins and vitamin E by inserting a vitamin capsule in the mouth of fresh lemuru fish.

2.4. Water quality management
Broodstock rearing activities in the research site have direct water sources from the Lombok Strait and require water quality management. Water quality management in broodstock rearing activities was carried out by means of filtration, water changes and water quality measurements. Filtration was carried out physically using a sand filter machine which aimed to filter the remaining sand and dirt.

The water system used in the main rearing tank was flow through system. Water would be flown continuously through the inlet channel so that the water in the tank kept changing. The daily water change carried out in tank of the Bawal Bintang broodstock at the study site was 245% every day. Cultivation with flow through system is able to improve water quality and suppress metabolic waste produced [12]. In addition, water quality measurements in the main tank were also carried out every two weeks, to be precise on Mondays. This water quality measurement was carried out in a water quality laboratory.

3. Result and discussion
3.1. Spawning
Spawning is the initial stage in the hatchery of Bawal Bintang fish. Spawning was carried out in a fiber tank with a volume of 10m^3. The first stage in spawning was parent selection. Parent selection began with anesthesia in an acclimatization tank measuring 1.5 x 1 x 0.4 m. The acclimatization tank was filled with fresh water serving to neutralize parasites attached to the fish skin. The anesthesia was only done using fresh water, in contrast to the one of white snapper broodstock that used additional alcohol and clove oil. This was done because white snapper b broodstock have a high tolerance; in addition, factors that cause start pomfret broodstock did not use additional alcohol and clove oil was due to prevention of blindness to the eyes of the Bawal Bintang which are quite sensitive. Moreover, there was additional treatment for Bawal Bintang brooders to make them immediately get passed out or limp, namely by tilting the Bawal Bintang's body. After the fish were affected by anesthesia (passed out), then the selection was carried out. The selection was done by cannulation technique using a 1.2mm diameter catheter which was inserted into the male and female parent genital openings as deep as 5- 10cm. This cannulation technique was carried out to determine the level of gonad maturity. Male broodstock that were ready for spawning had milky white sperm with a thick texture. Meanwhile, the female broodstock had eggs that were uniform in size, yellow in color, and separate. The ratio of male and female broodstock was 3:1, with 9 males and 3 females.

The spawning of Bawal Bintang in the research site was carried out semi-naturally with the stimulation of HCG (Chorionic Gonadotropin Hormone). HCG contains the active ingredients of LH and FSH which function to improve the performance of growth and development of gonads in the Bawal Bintang broodstock. Before being injected, the HCG hormone would be mixed with 5% DEXTROSE IV fluid which was then homogenized. The dose of HCG that was given to the broodstock was 250IU/kg. Injections using the hormone hCG at a dose of 250 IU/kg and Ovaprim 250 IU/kg were done on the right back, and a second injection was carried out to Bawal Bintang in the following morning (23-24 hours) on the left back or vice versa alternately with the first injection with the same dose. Injections were carried out 2 times with a 24-hour range using a syringe, and the injection of the Bawal Bintang broodstock was carried out intramuscularly 3-4cm from the dorsal fin.

During spawning, water was allowed to flow and the environment was controlled. Fertilization in the spawning stage usually occurs at 18.00-24.00 [13]. When fertilization had been carried out, the eggs would come out to the tank that had been installed with an egg collector. The number of eggs obtained during the spawning process was approximately 601,315 fertilized eggs.

3.2. Egg Development

The development of Bawal Bintang eggs passes through several stages. At the age of 1 hour, the egg is in the fertilized phase, where the new egg is fertilized. Then, at the age of 2 hours the egg is in the multi- cell division phase, where the egg has undergone many divisions until it has many cells. Furthermore, at the age of 3 hours the egg enters the morula stage, where the preimplantation embryo develops between the division of the blastomere and blastocyst. At the age of 4 hours the egg is in the blastula stage. At the age of 5 hours the egg enters the gastrula stage, where there is a blastula cell remodeling into 3 germ layers. At the age of 6 hours entering the neoul stage, the organs beginto form until the embryo has eye organs. Lastly, at the age of 7 hours the embryo has formed an organ.

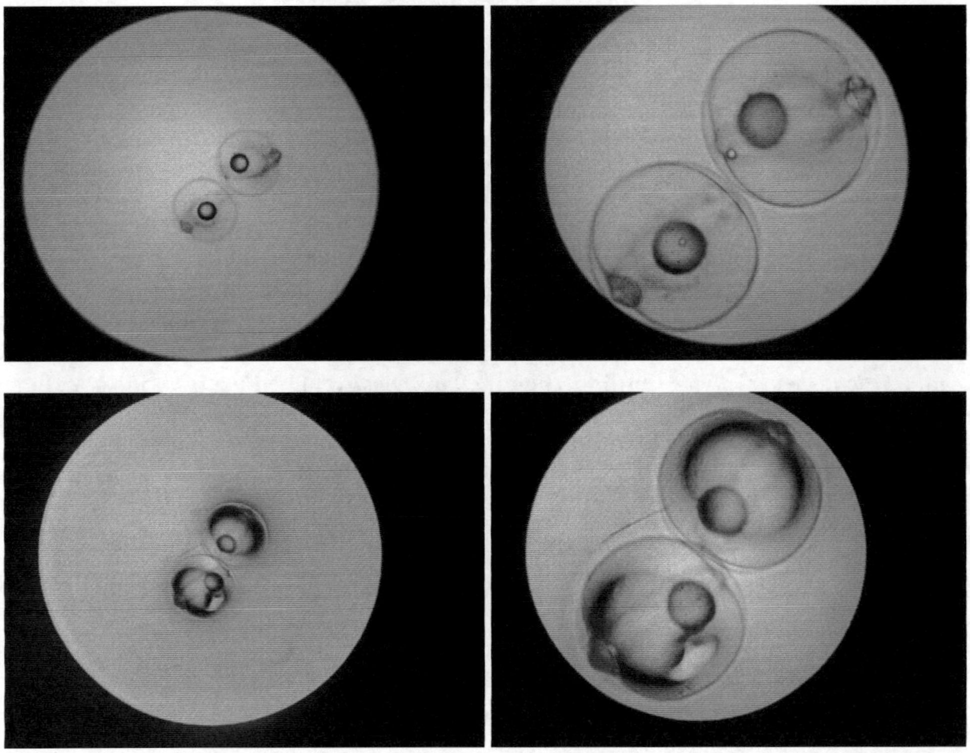

IOP Conf. Series: Earth and Environmental Science 1273 (2023) 012045 doi:10.1088/1755-1315/1273/1/012045

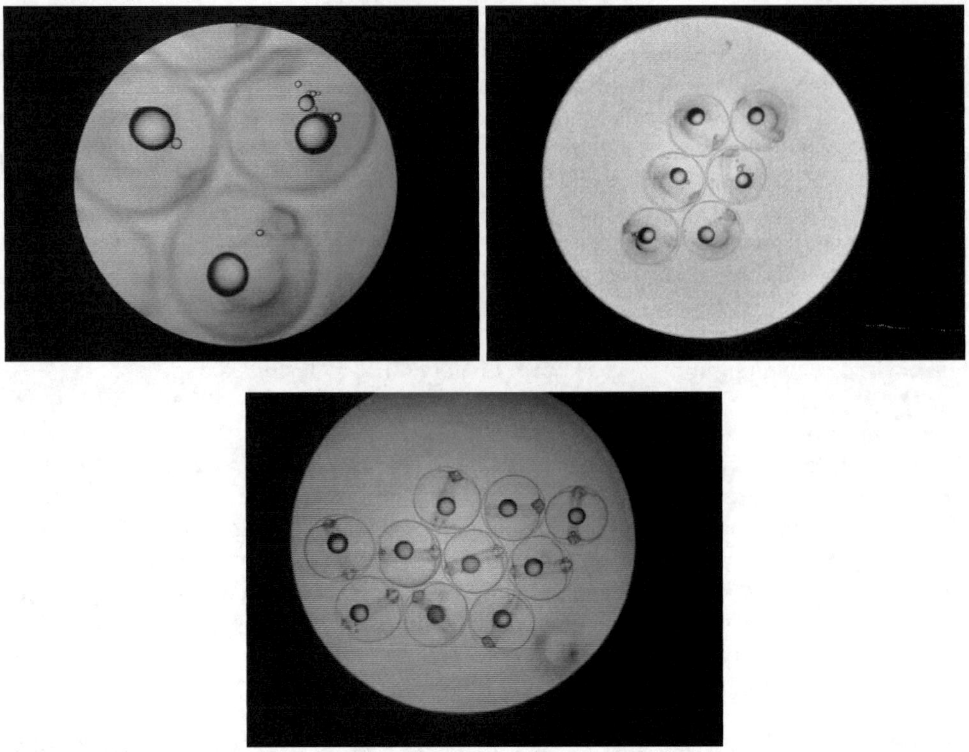

Figure 1. The Development of Bawal Bintang Fish Eggs

3.3. Larvae management

3.3.1. Tank preparation and egg stocking

The initial activity carried out in larval rearing was tank preparation. The tank used in rearing larvae was a round concrete tank with a diameter of 2.5 m with a height of 1.5 m equipped with 7 aeration points and 1 outlet and 2 inlet channels. 1 outlet channel on the tank was a water drainage. Meanwhile, the 2 inlet channels were filtered water flow channels and sea water channels.

The tank preparation activity at the research site was carried out by chlorinating the tank using a diluted dose of 100ppm. The diluted chlorine was then splashed all over the tank. After that, it was let stand for 2-3 hours. Chlorine Ca(ClO)2 functions to reduce organic substances, oxidize metals, and disinfect microorganisms [8]. After settling, then the tank was scrubbed using a brush and then rinsed with running water. After the chlorination process was complete, the tank was filled with filtered water again through the inlet channel.

Egg harvesting was done after spawning by taking eggs that were accommodated in the egg collector, and then the eggs were temporarily accommodated in an aquarium measuring 60 cm x 30 cm x 30 cm, done to identify and distinguish fertilized and unfertilized eggs. Eggs stocked were about 124,410 eggs/tank. The eggs would hatch into larvae within 18-20 hours after the spawning process.

3.3.2. Feed management

Feeding is important in larval rearing. The feed given to larval rearing was natural feed (*Rotifera sp.*) and artificial feed (pellet). As an immunostimulant, the larvae were given vitamin E with the trademark “Nature E” to support their growth, with a dose of one capsule for one tank. Feeding the larvae would be different at each age. Feeding started from the age of 2 days of larvae because the larvae of the first day of age still had food reserves (yolk eggs). Feed in the form of *Rotifera sp.* as much as 4-12 scoops was given according to the age of the larvae. Before giving Rotifera, the tank would be filled with 200-300 liters of *Nannocloropsis*. Natural feed would stop being given when H-3 harvest. In addition to natural feed, Bawal Bintang larvae were also given artificial feed in the form of "love larvae" powder pellets from the age of 11 days, with a number of 1 or 2 microns. The size of the pellet was adjusted to the larva's mouth opening, and the artificial feeding was carried out every 2 hours from 6 am to 6 pm.

The specifications for the quality content or quality of feed pellets with 1 micron in size were 48% protein and 10% fat. Feed pellets with 2 micron in size also had the same protein and fat content as feed pellets with 1 micron in size. According to the 2013 SNI, the protein content of the feed pellets of Bawal Bintang larvae must have a protein content of at least 37%.

3.4. Natural feed management

3.4.1. Mass scale culture of Nannochloropsis sp.

Phytoplankton production in the research site focused on the type of algae (*Nannochloropsis sp.*). Algae culture was carried out in 2 stages, namely the intermediate scale and the mass scale. Intermediate scale algae culture was carried out in containers with a volume of 95 L, while mass scale algae culture was carried out in concrete tanks with a volume of 22.500 L. Culture on a mass scale used two methods, namely the direct (full) method and the gradual method. The gradual culture method was carried out as a solution to the obstacles faced during the rainy season. The algae *Nannochloropsis sp.* was used as feed for *Rotifera sp.* and as natural feed for *barramundi* larvae.

The intermediate culture activities of *Nannochloropsis sp.* used containers of 20 units. Before used, the container would be disinfected first using chlorine at a dose of 1 ppm and then rinsed and dried for 1 day. The container was later filled with 100 L of water, and then 100 ml of chlorine was added and allowed to stand for 12-24 hours. Afterwards, thiosulfate as much as 10 g was added and allowed to stand for 2 hours, which served to neutralize the algae culture water. Next, fertilizer in the form of urea and comey with a dose of 80 mL each was added. Following this, inoculants of *Nannochloropsis sp.* were put into a 2.5 L container. The algae culture of *Nannochloropsis sp.* was left for 6 days until the density reached 20 million cells/tank, and afterwards it would be harvested. After this, they would be transferred to a concrete tank of mass-scale algae culture.

Mass algae culture activities included tank cleaning, water filling, inoculant distribution, and fertilization. The cleaning was carried out starting from the culture tank, equipment, aeration installations to other work equipment, by washing and then drying them. The cleaning of the tank was carried out using a 50 ppm dose of chlorine. Then, the tank was rinsed using running water. The filtered seawater would be put into the culture media and filtered again using a sand filter or filterbag, and then 5,000 L of culture media water was added. Inoculant distribution was carried out by transfering algae using a pump and a hose connected from the container. Inoculants were added as much as 2,500 L. Fertilization was carried out only once during algae culture. The dose of fertilizer used was ZA 250 grams, 135 grams, and Urea 350 grams. TSP was soaked first before being mixed with other ingredients, and if the TSP had dissolved then other fertilizers could be mixed and they were stirred until smooth. Fertilizer formulations might change depending on weather conditions and phytoplankton needs. The addition of fertilizers was done using a basin and a scoop. Fertilizers were distributed evenly in the culture tank. In one pond there were 20 aeration points that were turned on and adjusted to the age of the culture. After 5-6 days of mass culture, *Nannochloropsis sp.* could be harvested, and the maximum age of mass culture was 7 days. The color of each pond could vary depending on the conditions of *Nannochloropsis sp.* If the pool water is green and does not look thick, then the condition of *Nannochloropsis sp.* can be said to be good; if the color is dark green, then the condition is not good; and if the color is yellow to clear, it shows that a lot of *Nannochloropsis sp.* die and the culture is said to have failed. The density of *Nannochloropsis sp.* after 5 days was as much as 3.125 x 107 cells/mL.

3.4.2. Mass culture of Rotifera sp.

Culture of *Rotifera sp.* began with the preparation of the tank. The preparation of the tank was done by cleaning the tank using chlorine and then rinsed it clean. Next, the tank was filled with the filtered seawater as much as 45% of the volume of the tank. Then, *Nannochloropsis* was added as much as

50% of the volume of the tank. Then, the stocking of *Rotifer sp.* was done as much as 3000ml with a density of 5350 individuals/ml, with the density in a tank with a diameter of 1.6m and a water height of 1m being 39 individuals/ml. Density calculations were carried out using a sedgewick rafter with a volume of 1ml. *Rotifera sp.* would be harvested every day, and the harvesting was done by draining as much as 50% of the contents of the tank, which was then filtered. The harvest of *Rotifera sp.* was ready to be used as natural feed for larvae. After that, *Nannochloropsis* was replenished as feed for *Rotifera sp.* Then, *Rotifera sp.* would continue to multiply every second, so the density would also change every day. The density of *Rotifera sp.* each day is presented in the following table:

Table 1. *Calculation of density and growth phase of Rotifera* sp.

Age (Day-)	Density	Phase
1	39 individuals/ml	Lag phase
2	786 individuals/ml	Exponential phase
3	2889 individuals/ml	
4	4456 individuals/ml	
5	6897 individuals/ml	
6	5670 individuals/ml	Stationary phase
7	3897 individuals/ml	
8	2097 individuals/ml	
9	1234 individuals/ml	

4. Conclusion

During 3 cycles of rearing Bawal Bintang larvae, it was found that the seeds produced were about 24% in each tank, but in each cycle, it was not uncommon for 1 tank to experience a decrease in the number of seeds produced. This could be caused by water quality, feed quality or pathogen attack. Therefore, the next solution is to conduct more frequent control to the quality of water, feed, and also pathogens if they are present.

In the Standard Operational Procedure (SOP) for rearing Bawal Bintang larvae, there is additional natural feed in the form of Artemia, but it is not added because the Bawal Bintang larvae that are fed with Artemia will grow faster but have a weaker survival compared to those that are not.

5. Reference

[1] Sahfitri I A H, dan Haji A 2018 *FIKP UMRAH J.* 1-12.
[2] Junianto N M S, Akbar, dan Zakimin 2008 *Aqua. Asia Mag.* 46-49.
[3] Sudiyatno, Nainggolan A 2021 *JISM.* 06(02) 106-11.
[4] Wahyudi D, Kurnia B, Triana A, Sutisna T, dan Aris M N 2016. *Business Potential of Star Pomfret Fish Cultivation.* 1(1).
[5] Retnasari H T, dan Abdulgani N 2013 *J. Sains dan Seni Pomits.* 2(2) 2337-3520.
[6] Effendie M I 1997 *Fisheries Biology (Yogyakarta: Nusatama Library Foundation).*
[7] Odum E P 1993 *Fundamentals of Ecology (Yogyakarta : Gajah Mada University Press)..*
[8] Herawati D, dan Yuntarso A 2017 *JSH.* 1(2) 66-74.

[9] SNI 2014 Barramundi (Lates calcarifer, Bloch 1790)-Part 3: Broodstock Production (Jakarta: National Standardization Agency).
[10] Fauzi I A, Mokoginta I, Yaniharto D 2008 *JAI.* 7(1) 65-70.
[11] Jumiati, Ratnasari D, dan Sudianto A 2019 *JIPK.* 11(1) 55-6.
[12] Jumaidi A, Yulianto H, dan Efendi E 2017 *e-Jurnal Rek. dan Tek.Bud. Perairan.* 5(2) 587-596.
[13] Mulah A, Raza'i T S, dan Putra W K A 2017 *Intek Akuakultur.* 1 1-6.

6. Acknowledgement

The author is very grateful to Universitas Airlangga for organizing InCoFims 2021.

IOP Conf. Series: Earth and Environmental Science 1273 (2023) 012046 doi:10.1088/1755-1315/1273/1/012046

Feeding management of vannamei shrimp (*Litopenaeus Vannamei*) with immunostimulant addition

K Daniel[1] and P D W Sari[2,3]

[1]Study Program of Aquaculture, Faculty of Fisheries and Marine, Universitas Airlangga, Surabaya 60115, Indonesia
[2]Department of Aquaculture, Faculty of Fisheries and Marine, Universitas Airlangga, Surabaya 60115, Indonesia

[3]Corresponding author : putri.dw@fpk.unair.ac.id

Abstract. Giving immunostimulants is one of the preventive measures against infectious diseases. This activity aims to know the feed management of vannamei shrimp with adding immunostimulants on the culture of vannamei shrimp. The parameter observed were Total Haemocyte Count (THC) to know the effect of immunostimulant. The value of total haemocytes in the DOC 31-35 was 15.87×10^6 cells/ml - 25.56×10^6 cells/ml. The second measurement of total haemocyte count was done for the vannamei shrimp in the DOC 59-63 and showed 14.58×10^6 cells/ml - 27.45×10^6 cells/ml.

1. Introduction

The potential of shrimp commodity resources is quite large with a high selling value and very good market opportunities, because the demand for shrimp is high both for domestic and international market [1]. Vannamei shrimp is a high-value shrimp commodity compared to other species, regarding their fast growth, tolerance to fluctuations in water conditions, resistance to disease, high productivity levels, availability of brood stock or disease-free seed production, and relatively low protein requirements in feed [2].

In carrying out Vannamei shrimp cultivation, the emergence of disease attacks is one of the limiting factors that can cause failure in culture production [3]. White Faeces Disease (WFD) caused by *Vibrio* sp., for example, which causes clinical symptoms of decreased shrimp appetite, the colour of the intestine becomes white and even seems empty due to lack of dietary intake. The presence of white faeces floating on the surface of the waters and abnormal growth [4]. Meanwhile, diseases caused by an *Infectious Myo Necrosis Virus* (IMNV) can cause mortality in prawns from 40% to 70% before harvest [5].

One of the preventive measures against infectious diseases carried out is the provision of immunostimulants. Immunostimulants can be drugs, chemical compounds, or materials with the function to increase response mechanisms, either specific or non-specific [6]. The immunostimulant given is a commercial immunostimulant consisting of herbal ingredients containing Anti-Quorum Sensing substances and as a source of energy and selenium (Se).

The purpose of giving immunostimulants is to activate the non-specific immune system such as haemolytic cells in invertebrates. In giving immunostimulants, it is necessary to provide the optimal dose, where the dose can affect the immune response in shrimp.

To observe the effect of giving immunostimulants, the value of Total Haemocyte Count (THC) presents the total number of haemocyte cells consisting of hyaline, semi granular, and granular cells. THC is one of the parameters used to determine shrimp immunity. The increase of THC can be caused by the function of shrimp immune which initiated by lectin molecules to do adaptation with foreign body (non-self-recognition) that enters the body of the shrimp [7]. According to Lopez et al. [8], haemocytes will increase in shrimp that have immunostimulants-fed contains.

The purpose of this study is to observe the value of Total Haemocyte Count (THC) of vannamei shrimp that were given immunostimulant in order to be used as reference in monitoring vannamei shrimp health.

2. Material and methods

a. Research methods

This research was conducted at the Vannamei Shrimp Pond (TWB) in Merak Belantung, Kalianda, South Lampung in September – December 2022. The study used observation and survey method.

b. Immunostimulant and dietary blend

Commercial immunostimulant and selenium were used as immunostimulant. with 5 g/kg of food and selenium at 0.5 g/kg of feed given daily. The amount of food given per day is calculated from the biomass multiplied by FR%.

c. Sample collection

Vannamei shrimp were taken from the pond, each of five samples, and then a random vannamei shrimp was taken from the sample to collect the haemocytes.

d. Haemocyte examination

Vannamei shrimp haemocytes were collected from the base of the pleopod in the abdominal segment near the genital opening by means of a syringe filled with anti-coagulant solution (EDTA 2%). The haemocytes that have been taken are drained on a haemocyte meter and counted with a hand counter and observed under a microscope with a magnification of 400 x. Then, shrimp blood samples that had been collected in the microtube were taken to observe the shape of the haemocyte cells (granular, semi-granular, and hyaline) and counted the number of haemocyte cells under a microscope with the help of a haemocytometer and a hand counter.

e. Data analysis

Data on the results of the haemocyte examination in vannamei shrimp are presented in the form of tables and are described. To determine the correlation between effect of giving immunostimulants and haemocyte levels using simple correlation analysis.

3. Result and discussion

The highest total haemocyte in the first examination of DOC 31-35 was found with a total haemocyte of 25.56 x 10^6 cells/ml, while the lowest total haemocytes were found with a total of 16.29 x 10^6 cells/ml. Then, in the second examination (DOC 59-63), the highest total haemocytes were found with a total haemocyte of 27.45 x 10^6 cells/ml. Meanwhile, the lowest total haemocytes were found with a total of 15.93 x 10^6 cells/ml. The dose of immunostimulants in each pond was identical: commercial immunostimulants of 5 g/kg of food and selenium of 0.5 g/kg of food. For more details the examination of Total Haemocyte Count (THC) value of vannamei shrimp samples are shown in Table 1.

Table 1. The value of Total Haemocyte Count (THC) Value of Vannamei Shrimp

Step of observation	*Total Haemocyte Count* (THC) (cell/ml)				
	Pond 3	Pond 5	Pond 8	Pond 10	Pond 11
1	16,29 x10^6	18,33x10^6	15,87x10^6	23,16x10^6	25,56x10^6
2	17,40x10^6	23,10x10^6	14,58x10^6	16,59x10^6	27,45x10^6

Calculation of the total haemocytes of white shrimp was carried out in two examinations observed from five different ponds, namely ponds 3, 5, 8, 10, and 11 as presented in Table 1. Observations of the total value of vannamei shrimp haemocytes showed a variety of patterns. In the first observation, the pond 11 had the highest total haemocyte count of 25.56 x 10^6 cells/ml and pond 8 was the plot with the lowest haemocyte value of 14.58 x 10^6 cells/ml.

The second observation showed a different pattern of total haemocytes. There are three ponds that have a trend of increasing total haemocyte (pond 3,5, and 11). While, two ponds had a decreasing trend in the total number of haemocyte (pond 8 and 10).

According to Van De Braak [10], changes in THC values could be caused by infection, then increasing the the granular cells in the infected shrimp body parts. The decrease in THC in shrimp blood circulation occurs because the cellular defence system against infection by migrating to the infection areas [11]. In addition, according to Sakai [12] the ability of immunostimulants to increase immune responses and develop protection against pathogenic infections is affected by the dose of application. Application of immunostimulants at concentrations below the minimum value for the occurrence of an immune response will not influence increasing the number of haemocytes.

Haemocytes is commonly used as quantitative parameters in measuring the stress condition in shrimp [13]. An increase in total haemocytes, the health of vannamei shrimp also improve because of the form of phagocytic cells which play a very important role in defending themselves from microorganism attacks [14]. Total Haemocyte Count (THC) data collection in this study was carried out in DOC 31-35 shrimp, also for vannamei shrimp in DOC 59-63. Total Haemocyte Count (THC) is the total number of haemocyte cells consisting of hyaline, semi granular, and granular cells. THC is one of the parameters used to determine shrimp immunity. The amount of THC of normal white vannamei shrimp ranges from 20x10^6 – 40x10^6 cells/ml [15].

Immunostimulants are chemical compounds, drugs, or other materials that can increase specific and non-specific response mechanisms. Shrimp immune system is not the same as fish, shrimp only have a non-specific immune system. The non-specific immune system found in shrimp is the first line of defence against pathogens. When pathogens attack, the shrimp immune system will be stimulated to produce responses including agglutination, phagocytosis, production of antimicrobial peptides or antimicrobial peptides (AMPs) and the humoral immune system [16].

The immunostimulants given to the shrimp feed mixture in TWB ponds were commercial immunostimulants and selenium (Se). The commercial immunostimulants provided are made from herbal ingredients containing Anti Quorum Sensing (AQS) substances and as an energy source. AQS is a compound or molecule that degrades autoinducer in bacteria, thereby preventing quorum sensing in bacteria [17]. According to Hentzer & Givskov [18], the application of AQS is very beneficial because it can reduce the emergence of resistant properties in bacteria without affecting the survival of individual cells, but rather on the virulent nature of the bacterial population as a whole.

4. Conclusion
The addition of immunostimulants can give affect the THC value, but other factors, such as the environment, also need to be considered in determining the health status of shrimp.

5. References
[1] Putri D S, Affandi M I, dan Sayekti W D 2020 *JIIA* 8 (4) 625-632.
[2] Sudrajat, A., dan Wedjatmiko. 2010. Budidaya Udang di Sawah dan Tambak, Penebar Swadaya. Jakarta.
[3] Ismawati, Destryana R A, Huzaimah N 2019 *Jur Kelau* 12 (2) 201-206.
[4] Jayadi M, Prajitno A, dan Maftuch 2016 *Intern Jour of ChemTe Resea* 9 448-452.
[5] Loy J D 2011 Infection and Immunity in the Pasific white shrimp, *Litopenaeus vannamei*. Dissertation. Iowa State University. Ames, Iowa.
[6] Putri F M, Sarjito dan Suminto 2013 *Jour of Aquac Manag and Techno* 2 (1) 102-112.
[7] Rodriguez J and Le Moullac G 2000 State of The Art of Immunological Tools and Health Control of Penaeid (*Penaeus monodon*). PhD Thesis, Wageningen University. Netherland.
[8] Lopez N, Cuzon G, Gaxiola G, Taboada G, Valenzuela M, Pascual C, Sánchez A and Rosas C 2003 *Aquac* 224 223-243.
[9] Ismawati R A, Destryana, Huzaimah N 2019 *Jurn Kelau* 12 (2) 201-206.
[10] Van De Braak K 2002 Haemolytic defence in black tiger shrimp (*Penaeus monodon*), Dissertation, Van Wereningen Universiteit, Germany.
[11] Darwantin K, Sidik R, dan Mahasri G 2016 *Jur Bios Pasca* 18 (2) 1-17.
[12] Sakai M 1999 *Aqua* 172 63-92.
[13] Hartinah, Sennung L P, dan Hamal R 2014 *Jurn Bion*15 (2) 104-110.
[14] Tenriulo A, Parenrengi A, and Tampangallo R 2014 Respons Imun Udang Windu, *Penaeus monodon* yang Membawa Marker DNA Tahan Penyakit Setelah Dipapar Bakteri Patogen *Vibrio harveyi*. In Forum Inovasi Teknologi Akualtur 991–999.
[15] Putri F M, Sarjito, dan Suminto 2013 Jour of Aquac Manag and Techno 2 (1) 102-112.
[16] Pang H, Wang G and Zhou, S 2019 *Jour of Fis and Shellf Immu* 92 712-718.
[17] Novita H, Rusmana I dan Yuhana M 2016 *Jurn Ris Akua* 11 (3) 291-296.
[18] Hentzer M, and Givskov M 2003 *J Clin Invest* 112 1300-1307.

6. Acknowledgements
The author appreciates the cooperation of all parties involved in this research.

The effect of formulated diets with different protein sources on feed consumption, feed conversion ratio, and nutrient retention of scalloped spiny lobster (*Panulirus homarus*)

R Agustin[1], M Amin[2,4] and M Lamid[3]

[1]Study Program of Aquaculture, Faculty of Fisheries and Marine, Airlangga University, Surabaya 60115 Indonesia
[2]Department of Aquaculture, Faculty of Fisheries and Marine, Airlangga University, Surabaya 60115 Indonesia
[3]Department of Animal Husbandry, Faculty of Veterinary Medicine, Airlangga University, Surabaya 60115 Indonesia

[4]Corresponding author: muhamad.amin@fpk.unair.ac.id

Abstract. Scalloped spiny lobster (*Panulirus homarus*) is a type of marine biota that has high economic value. Lobster cultivation must to be developed, which is in terms of feed. Currently, the majority of cultivators still depend on trash fish as the main food for lobsters. This study was conducted to find the best feed formulation as an alternative to trash fish. The method used is an experimental method with five treatments and four replications where trash fish as control (T0), fish meal formula (T1), rebon shrimp flour formula (T2), maggot flour formula (T3), and snail flour formula (T4). Lobster rearing was carried out for 30 days with the observed parameters including the feed consumption, feed conversion ratio, and nutrient retention. The data obtained were analyzed using ANOVA and continued with Duncan's Multiple Range Test with a level of 5% to determine the best treatment. The results showed that formulated diets with different protein sources was able to affect the feed consumption and nutrient retention of scalloped spiny lobster, but was not affect on the feed conversion ratio. These results indicate that the best formula feed is rebon shrimp flour formula which shows the best results from the four artificial feeds.

1. Introduction

Lobster (Palinuridae) is one type of marine biota that has high economic value [1]. One type of lobster that is in great demand by the public is scalloped spiny lobster (*Panulirus homarus*) because it has a characteristic delicious meat taste. This is what causes the export value of lobster to increase every year. Based on data from the Ministry of Maritime and Fisheries (MMF), in 2019 the export volume of lobster was recorded at 1.633.220 kg or 33.189.390 US$. Meanwhile, in 2020 the export volume of lobster increased from the previous year to 2.150.420 kg or 76.106.250 US$ [2].

The highest cost incurred in lobster cultivation is the cost of feed because the feeding takes place continuously every day [3]. Currently, cultivators still use trash fish as the main feed for lobsters [1]. In

IOP Conf. Series: Earth and Environmental Science 1273 (2023) 012047 doi:10.1088/1755-1315/1273/1/012047

addition to its cheap price, trash fish contains 57% protein, 13,65% fat, 0,17% crude fiber and 7,69% water [4]. However, the problem is that trash fish is not always available and only exists in certain seasons. In addition, the high water content in fresh fish causes the rest of the feed to decompose more easily which can lead to a decrease in water quality and susceptibility to disease [1]. Based on description, the development of artificial feed for scalloped spiny lobster needs to be done to reduce the use of trash fish. There is a need for research on the use of alternative feeds to replace trash fish using different protein sources, which are expected to meet the nutritional needs of lobsters and to determine their effect on feed consumption, feed conversion ratio, and nutrient retention of scalloped spiny lobster.

2. Materials and methods

2.1. Time and place

This research was carried out in the Anatomy and Aquaculture Laboratory, Faculty of Fisheries and Marine, Airlangga University from June to July 2022 (East Java, Indonesia).

2.2. Tools and materials

The tools used include 20 aquariums measuring 40 x 30 x 30 cm, shelter pipes, aerator, tray, filter, DO meter, thermometer, pH meter, refractometer, test kits (nitrate, nitrite, ammonia), and digital scale. While the materials used were scalloped spiny lobster (puerulus phase) with an average weight of 0,19 ± 0,01 gr, sea water, trash fish, fish meal, rebon shrimp flour, maggot flour, snail flour, corn flour, agar powder, fish oil, and premix.

2.3. Work procedures

Feed formulations using protein ranged from 28-30% with different protein sources. First, analyze proximate the ingredients used to determine the nutritional content. Nutrient content is used in the preparation of feed formulations using the *Solver* program in *Microsoft Excel.*

Table 1. Feed formulation calculated using *Solver* program.

No	Materials	Proportion (%)				
		T0	T1	T2	T3	T4
1.	Trash fish	100	-	-	-	-
2.	Fish meal	-	59,14	-	-	-
3.	Rebon shrimp flour	-	-	55,21	-	-
4.	Maggot flour	-	-	-	92,50	-
5.	Snail flour	-	-	-	-	37,51
6.	Corn flour	-	32,86	36,79	3,00	48,49
7.	Agar powder	-	5,00	5,00	3,00	8,00
8.	Fish oil	-	2,00	2,00	0,50	5,00
9.	Vitamin Mix	-	0,50	0,50	0,50	0,50
10.	Mineral Mix	-	0,50	0,50	0,50	0,50
TOTAL			100	100	100	100
Protein content (%)		62,06	30	30	28,04	30
Fat content (%)		11,98	6,16	8,46	16,73	5,98
Carbohydrate content (%)		9,41	38,66	28,63	43,19	41,59
ME (Kcal/kg)		3193,12	2959,14	2768,67	2708,11	3124,97

Before use, seawater and aquariums are sterilized using chlorine. Next, put shelters that aim to minimize contact between lobsters, reduce stress and reduce the level of cannibalism during moulting and maximize growth [5]. Lobster rearing was carried out for 30 days by feeding 15% of the biomass per day. Feeding was carried out 2 times a day in the morning and evening.

2.4. Experimental design

The research design used was a completely randomized design (CRD). The method used is an experimental method with five treatments and four replications where trash fish as control (T0), fish meal formula (T1), rebon shrimp flour formula (T2), maggot flour formula (T3), and snail flour formula (T4).

2.5. Observed parameters and data analysis

The observed parameters were feed consumption, feed conversion ratio, protein retention, fat retention and energy retention of scalloped spiny lobster. Observed parameters were assessed by using the following formula :

Feed consumption = Total initial feed (gr) – Total leftover feed (gr) [6]

$$FCR = \frac{\text{Total feed consumed (gr)}}{\text{(Final weight + Dead weight) - Initial weight (gr)}} \; [3]$$

$$PR = \frac{\text{(Final body protein weight - initial body protein weight) x 100\%}}{\text{Total protein feed consumed}} \; [7]$$

$$FR = \frac{\text{(Final body fat weight - Initial body fat weight) x 100\%}}{\text{Total fat feed consumed}} \; [7]$$

$$ER = \frac{\text{(Final body energy weight - Initial body energy weight) x 100\%}}{\text{Total energy feed consumed}} \; [7]$$

Where FCR, PR, FR, ER are feed conversion ratio, protein retention, fat retention, and energy retention. The data obtained were analyzed using Analysis of Variance (ANOVA) with a 95% confidence level. This aims to determine the effect of the treatment given. If the results of the analysis show a significantly different, continued with Duncan's Multiple Range Test (DMRT).

3. Results and discussion

The results showed that formulated diets with different protein sources was able to affect ($p < 0,05$) the feed consumption and nutrient retention of scalloped spiny lobster, but was not affect ($p > 0,05$) on the feed conversion ratio.

Table 2. The calculation result of Feed Consumption (FC), Feed Conversion Ratio (FCR), Protein Retention (PR), Fat Retention (FR), and Energy Retention (ER) of scalloped spiny lobster reared for 30 days.

Parameter	Treatments				
	T0	T1	T2	T3	T4
FC (gr)	$3,16 \pm 0,33^{b}$	$2,75 \pm 0,18^{a}$	$2,80 \pm 0,21^{ab}$	$2,44 \pm 0,29^{a}$	$2,59 \pm 0,19^{a}$
FCR	$2,56 \pm 0,40^{a}$	$2,82 \pm 0,46^{a}$	$2,61 \pm 0,70^{a}$	$3,25 \pm 0,32^{a}$	$3,26 \pm 0,88^{a}$
PR (%)	$38,36 \pm 6,49^{b}$	$20,63 \pm 3,24^{a}$	$23,56 \pm 6,35^{a}$	$20,11 \pm 1,78^{a}$	$19,18 \pm 5,02^{a}$
FR (%)	$77,03 \pm 11,70^{d}$	$38,70 \pm 5,55^{c}$	$31,07 \pm 7,89^{bc}$	$9,45 \pm 0,88^{a}$	$24,49 \pm 7,34^{b}$
ER (%)	$51,34 \pm 8,37^{c}$	$14,09 \pm 2,16^{a}$	$23,99 \pm 5,60^{b}$	$15,79 \pm 1,33^{a}$	$9,55 \pm 2,82^{a}$

Values (mean ± SD) with different superscripts in the same row are significantly different ($p < 0,05$).

Feed consumption is the total feed consumed in a certain time and used by fish to fulfill the needs of life [8]. Feed consumption in this study ranged from 2,44 – 3,16 gr, while the research of Kristiana *et al.* resulted in feed consumption between 1,45 – 2,00 gr [6]. The high feed consumption in the control treatment (T0) was because trash fish had an odor that was favored by lobsters, thereby increasing the lobster's response to feed. This is in accordance with the statement of Anggraini *et al.* which states that if the taste of the feed is in accordance with the lobster's desire, the feed will be consumed, on the contrary if the taste and aroma of the feed are not suitable, the feed will be left or not consumed [3].

Feed conversion ratio is the amount of feed given during maintenance to bolster lobster weight gain [3]. The result showed that formulated diets with different protein sources did not significantly affect the feed conversion ratio. This indicates that the scalloped spiny lobster in each treatment has almost the same ability to utilize feed optimally. The results of the feed conversion ratio in this study ranged from 2,56 – 3,26, while the result of research conducted by Nugraha *et al.* shows the value of the feed conversion ratio ranging from 1,73 – 2,22 [9], and the research of Prariska *et al.* shows the value ranging from 3 – 9 [10]. This indicates that the value of feed conversion ratio in this study is considered optimal. The value of the feed conversion ratio is influenced by several factors, namely the quality and quantity of feed, species, lobster age, body size, and water quality [11].

Retention is the amount of nutrients stored in the form of body tissues divided by the amount of feed nutrients consumed [12]. The protein retention value in this study ranged from 19,18 – 8,36%. While the result of Minarni's research stated that protein retention of lobster ranged from 15,25 – 26,69% [13]. The high feed consumption at T0 is related to the amount of feed consumed and the absorption of protein in the body. In addition, the high retention value was also due to the amino acid profile in trash fish according to the lobster's body, and containing higher levels of arginine, histidine, and valine compared to rebon shrimp, maggot, and snail. Arginine, histidine, and valine are amino acids that play an important role in growth [14]. The more suitable the profile of essential amino acids in the feed to the fish body, the more parts of the amino acids synthesized into protein [13].

The value of fat retention in this study ranged from 9,45 – 77,03%. Fat retention at T0 showed significantly different result with all treatments, as well as feed with maggot flour protein source (T3). Meanwhile, the fat retention of T2 was not significantly different with T1 and T4. The results were not significantly different because the fat content contained in the T1, T2, and T4 was not much different, which was in the range of 5.98 – 8.46%. The low value of fat retention in T3 is due to the high amount of fiber crude in the feed which results in decreased feed consumption. Based on the proximate, crude fiber contained in maggot flour was 32.41%. Yudiarto *et al.* stated that the more fiber consumption, the higher proportion of fat that is wasted and causes less fat to be absorbed [15].

The value of energy retention in this study ranged from 9,55 – 51,34%. Meanwhile, the results of Aji's research stated that energy retention ranged from 51,95 – 64,75% [16]. The high energy retention at T0 was due to trash fish containing higher isoleucine and leucine amino acids than rebon shrimp, maggot, and snail. Isoleucine is an amino acid that plays a role in metabolic processes, while leucine is an amino acid that plays a role in the body's energy production and protein synthesis [14].

4. Conclusion

The results showed that formulated diets with different protein sources was able to affect the feed consumption and nutrient retention of scalloped spiny lobster, but was not affect on the feed conversion ratio. These results indicate that the best formula feed is rebon shrimp flour formula which shows the best results from the four artificial feeds.

5. References

[1] Ridwanudin A, Fahmi V, and Pratama S 2018 *Oseanologi dan Limnologi* 3 95–103.
[2] Ministry of Maritime Affairs and Fisheries 2020 Export-import data from the Ministry of Maritime Affairs and Fisheries Jakarta.
[3] Anggraini W, Abidin Z, and Waspodo S 2018 *J Perikanan* 8 20–29.
[4] Ningtias A S, Junaidi M, and Rahman I 2019) *J Perikanan* 9 144–152.
[5] Adiyana K E, Supriyono, Junior M Z and Thesiana L 2014 *JKN* 9 1–9
[6] Kristiana R, Arini E, and Hastuti S *J Aqua Manag and Technol.* 3 95–104
[7] Lupatsch I, Santos G A, Schrama J W and Verreth J A J *Aquaculture.* 298 245–250
[8] Amalia R, Hastuti S, and Sudaryono A 2019 *Ind J Trop Aqua.* 3.
[9] Nugraha M D, Setyowati D N A, and Waspodo S 2019 *Jurnal Perikanan* 9 153–159.

[10] Prariska D, Supriyono E, Soelistyowati D T, Puteri R E, Sari S R, Sa'adah R, and Guttifera 2020 *JIPAT (clarias).* 1 1–7.
[11] Zainudin M and Pujiyati S 2013 *JTPK.* 4 123–129.
[12] Subekti S, Prawesti M, and Arief M 2011 *J Kel: Ind J. of Mar Sci and Technol.* 4 90–95.
[13] Minarni E, Yudha I G, and Suparmono S 2017 *Proc. Semnas Perikanan dan Ilmu Kelautan* 45–58.
[14] Subagio A, Windrati W S, Fauzi M, and Witono Y 2004 *JTIP.* 15 70–78.
[15] Yudiarto S, Arief M, and Agustono 2012 *JIPK.* 4 135–140
[16] Aji F S, Rahardja B S, and Agustono 2019 *J. of Mar Coast Sci.* 8 119–129

6. Acknowledgements

The authors express their gratitude to all parties involved in this research and all colleagues on the Faculty of Fisheries and Marine Airlangga University for their cooperation and technical support.

The effectiveness of additional silage of fish ofal and bagasse fermentation on biomass and population of silk worm (*Tubifex tubifex*)

H Ismawati[1], L Sulmartiwi[2,3], N N Dewi[2],

[1]Undergraduated Program of Fisheries Processing Technology, Faculty of Fisheries and Marine, Airlangga University, Campus C Universitas Airlangga Mulyorejo Surabaya, East Java, Indonesia 60115
[2]Department of Marine, Faculty of Fisheries and Marine, Airlangga University, Campus C Universitas Airlangga Mulyorejo Surabaya, East Java, Indonesia 60115

[3]Corresponding author: laksmi-s@fpk.unair.ac.id

Abstract. The purpose of this study was to determine the effect of giving catfish offal silage and sugarcane bagasse fermentation to the biomass The method used in this research is experimental using Completely Randomized Design (CRD). The treatments were 413 gram (100%) rice field mud, 206.5 gram rice field mud + 75% fish offal silage + 25% bagasse fermentation, 206.5 gram rice field mud + 50% fish offal silage + 50% bagasse fermentation, rice field mud 206.5 grams + 25% silage of fish offal + 75% bagasse fermentation. Each treatment was repeated five times. The results showed that the highest peak of silk worm biomass with 206.5 grams of rice field mud culture + 25% fish offal silage + 75% bagasse fermentation with an average of 22.25 grams and the lowest biomass was produced on 406 grams (100%) mud media. Rice fields with an average of 7.15 grams. The highest population of silkworms was found in 206.5 gram rice field mud media + 25% fish offal silage + 75% bagasse fermentation obtained 6651 ind, the lowest population in 406 gram (100%) rice field mud media with an average of 2.263 ind.

1. Introduction

Natural food that can be given to larvae include phytoplankton, zooplankton and benthos [1]. One of the natural food that can be given to freshwater fish and shrimp is silk worms (Tubifex tubifex). Giving in the larval stage can support growth and has a size that is in accorandce with the mouth opening [2]

Silkworm (Tubifex tubifex) is a type of natural food that is widely used as feed for fish larvae, both ornamental fish and consumption [3]. The content of silk worms is 57% protein, 13.3% fat, 2.04% crude fiber, 3.6% ash content and 87.7% water [4]. Silkworms live at the bottom of the water with calm water flow [5]. live in groups, easy to breed and fast growth [6].

The availability of silk worms is currently still lacking. The large demand for silk worms has not been matched by the availability from cultivators who are only able to meet 35-37% of market demand [7]. However, silk worms are not available throughout the year, especially during the rainy season because they will be carried away by currents due to high rainfall [8].

Utilization of waste is a solution to increase nutrients such as protein, carbohydrates, nitrogen, carbon and overcome environmental pollution. Waste that can be used to increase the nutrition of silk worms includes tofu dregs, chicken manure and rice bran [9]. In this study, the waste that will be used to increase the nutrition of silk worms is catfish innards and bagasse. The use of catfish innards and bagasse because these two materials have low economic value, contain C and N required by silk worms and reduce environmental pollution. However, in its utilization it needs to be processed to increase the digestibility value so that silk worms can make good use of it.

Silage of catfish offal is one way that can be done. Silage of catfish offal can increase protein content as the main constituent of organic N by 54.17% [10]. Bagasse fermentation has an organic C content of 30.79% [11]. Silage of catfish offal and fermented bagasse were combined to determine the carbohydrate and protein content to support the growth of silk worms.

2. Material and Method

2.1 Materials

The tools used in this study consisted of a fine sieve, plastic container measuring 7 cm x 13.5 cm x 13.5 cm, stacking rack, water pump, plastic bucket, spatula, paralon pipe, PVC glue, ruler, blender, measuring cup, digital scale with 0.01 gram accuracy, pH paper, thermometer, ammonia test kit, tissue, hand counter, magnifying glass, DO meter, label paper, and stationery. The materials used are bagasse from sugarcane sellers in the Sidoarjo area, catfish innards obtained from the PT. Dimas Reiza Perwira Surabaya, silk worm (Tubifex tubifex) from natural feed cultivators in Surabaya, Effective Microorganisms (EM4), molasses, mud from rice fields in Sidoarjo area.

2.2 Place and time of research

This research was conducted at the Anatomy and Aquaculture Laboratory, Faculty of Fisheries and Marine Affairs, Airlangga University, Surabaya from March to April 2022.

2.3 Methods

This Research Uses Experimental Method. The experimental method is a quantitative method used to determine the effect of treatment on the results under controlled conditions. This study used Completely Randomized Design Model Analysis (CRD) with four treatments. Each treatment was repeated five times so that 20 research units were obtained. The treatments used in this study were as follows: P0 = Rice Mud 413 grams (100%), P1 = Rice Mud 206.5 grams + 75% silage of fish offal + 25% sugarcane bagasse fermentation, P2 = Rice Mud 206.5 grams + 50 % silage of fish innards + 50% fermented bagasse, P3 = Rice Mud 206.5 grams + 25% silage of fish innards + 75% fermented bagasse.

3. Results and Discussion

The highest peak of biomass and silkworm population with 206.5 grams of paddy mud culture media + 25% fish offal silage + 75% sugarcane bagasse fermentation with an average of 22.25 grams, then cultivation media with 206.5 grams of paddy mud media + 50% fish offal silage + 50% sugarcane bagasse fermentation with an average of 13.29 grams, 206.5 grams of paddy mud + 75% fish offal silage + 25% sugarcane bagasse fermentation produces an average of 11.95 grams The lowest biomass was in the media 406 grams (100%) of paddy mud with an average of 7.15 grams

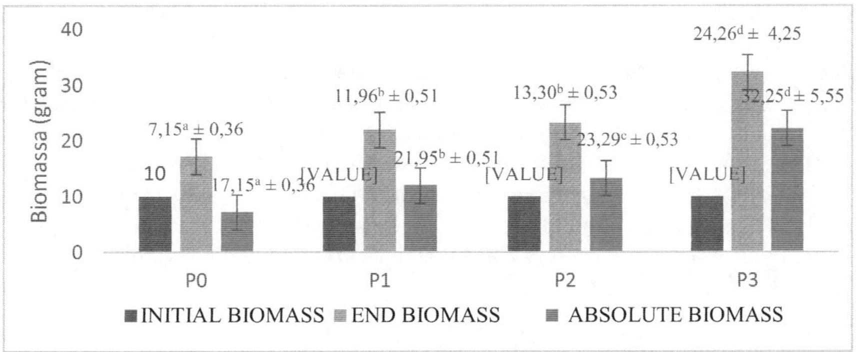

Figure 1. Silkworm biomass

Silkworm population increased during maintenance. The highest mean population of silk worms was in 206.5 gram rice field mud media + 25% fish offal silage + 75% sugarcane bagasse fermentation obtained 6651 ind, then 206.5 gram rice field mud media + 50% fish offal silage + 50% sugarcane bagasse fermentation with an average of 3,864 ind, 206.5 grams of paddy mud + 75% fish offal silage + 25% sugarcane bagasse fermentation produced an average of 3,356 ind, the lowest population was in the media 406 grams (100%) rice field mud with an average of 2,263 ind.

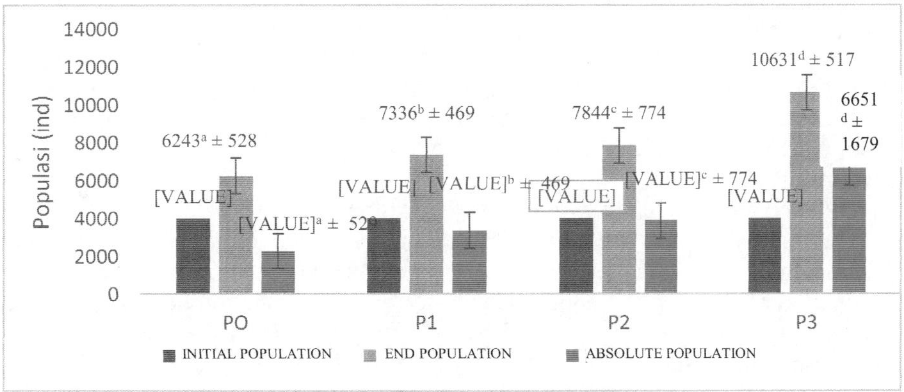

Figure 2. Silkworm population

The rearing medium affects the success of silk worm cultivation. The nutrition of silk worms from culture is determined by the media used as food for silk worms to survive during rearing. After 21 days of rearing silk worms, the biomass increased. This is obtained from the calculation of silk worms at harvest minus at the time of initial stocking. Silkworms utilize organic materials that have been decomposed by bacteria as a source of nutrients for survival. High protein content as a nitrogen source that is utilized by microorganisms to grow and develop. The high availability of organic N will increase the population in the media used for cultivation so that the availability of silkworm food is abunandt. In addition to protein, carbohydrates are also needed to increase biomass. Carbohydrates are composed of C-organic which is needed by microorganisms in producing energy and metabolic processes, so that bacterial growth is faster and organic matter is decomposed faster [12].

The process of decomposition of organic matter by bacteria utilizes the carbon content in bagasse and fish innards so that the carbon content of organic matter decreases. The length of the fermentation time is influenced by the C/N content in the basic ingredients [13] The content of bagasse C in bagasse is 30.17 and N in silage of fish offal is 54.17%. The P3 treatment with rice mud cultivation media + 25% fish offal silage and 75% sugarcane bagasse fermentation got the highest score because of the high organic matter content this is due to the high nutrient content in P3, from the proximate test of fish innards silage and bagasse fermentation that has been carried out showed that both materials have high nutritional value so that silk worms can utilize these nutrients to develop.

The P0 treatment using only paddy field mud got the lowest biomass value because of the low organic matter content without additional admixtures. The organic matter content of paddy fields is 2-5% [14]. The organic matter content of paddy fields varies depending on weather, time and environmental conditions, irrigation management is carried out. Rice fields that are flooded with irrigation water cause the transfer of organic matter carried by the flow of water and settles through the sediment process [15]. Treatment P1 is rice field mud cultivation media + 75% fish offal silage and 25% sugarcane bagasse fermentation and P2 rice field mud cultivation media + 50% fish innards silage and 50% sugarcane bagasse fermentation of nutrients in the media until death due to too acidic pH an increase in biomass but not much different from the P3 treatment. Fish offal silage has a pH of 3-4 [15].

This is not in accorandce with the living environment of silk worms which are in the pH range of 5.5-8.0 [16]. If the environmental conditions are too acidic then silk worms have difficulty in absorption.

The increase in population is suspected when the silkworm research used for 21 days of maintenance is in the adult phase and ready to reproduce to produce new individuals so that the population increases. Silkworms become new individuals after the eggs hatch and emerge from the cocoon on the 10th to 20th day [17]. The increase in population is due to the reproductive process of adult worms. The female silkworm releases mature eggs and is fertilized by another worm. Adult silkworms produce eggs which are stored in the cocoon and then hatch into young worms. Young worms need 21 days for embryos to develop. At that time the young worms become adults and are ready to produce cocoons and then hatch to produce young worms again.

The highest population increase of silk worms was in treatment P3 with the number of silkworms 6651 ind, while the lowest population in treatment P0 resulted in 2.263 ind. The factor that affects the population is the amount of organic matter used by bacteria in the decomposition process so as to produce detritus for silkworm nutrition [18] The appropriate dose for the biomass and population of silk worms with the best results was P3 treatment with 206.5 grams of paddy mud media + 25% fish offal silage + 75% sugarcane bagasse fermentation.

Table 1. Average measurement of water quality in each maintenance medium.

Treatment	Temperature	pH	DO (mg/L)
P0	27 – 31	6,0 – 6,2	3,61 – 3,68
P1	27 – 32	5,2 – 5,4	3,65 – 3,86
P2	28 – 30	5,3 – 5,4	3,65 – 3,88
P3	28 – 31	6,0 – 6,3	3,63 – 3,88

Temperature is very influential on the growth and survival of silk worms because it is related to metabolism. In this study, the temperature fluctuated with the lowest temperature being 27 and the highest being 32. The temperature was too high because at the time of measurement the sun was very hot and the weather was hot. An increase in temperature results in the rate of metabolism and respiration of water

IOP Conf. Series: Earth and Environmental Science 1273 (2023) 012048 doi:10.1088/1755-1315/1273/1/012048

metabolism which results in increased oxygen consumption. While the low temperature is obtained when measuring cloudy weather so that the temperature drops. Optimal temperature for silkworm growth 25 – 30 °C.

The pH range during the study was still on the verge of normal. The optimal pH for silkworm cultivation is 5.5 – 8.0. The pH value affects the biochemical processes of the waters, for example the nitrification process will end if the pH conditions are low. Changes in pH values during maintenance indicate the activity of microorganisms in decomposing organic matter and an increase in population. In this study, each treatment was still normal according to the optimal pH. Dissolved oxygen is an important parameter in water quality analysis. DO value indicates the amount of oxygen available in a waters. During maintenance, dissolved oxygen is obtained from water droplets flowing through the rearing medium. The DO range in this study is the lowest 3.61 and the highest 3.88. The optimum DO requirement for the maintenance of silkworms is 2.4-7 mg/L. This shows that the water conditions in the rearing media can be used for silk worms to develop. The low DO value is caused by the increasing number of silkworm populations and the respiration process. Silkworms take oxygen from the surface of flowing water by shaking their bodies. The lower part of the body is buried in the substrate to take in food.

4. Conclusion

Utilization of silage of catfish offal and fermentation of bagasse has an effect on biomass and silkworm population. The doses that gave the best results were rice field mud cultivation media + 25% fish offal silage and 75% sugarcane bagasse fermentation.

5. References

[1] Nuraini, S. Nasution, A. tanjung and H. Syawal 2019 *J Rural Urban Community Empower.* 1(1), 9-14.

[2] Muria, E. S., E. D. Masithah and S. Mubarak 2012 The Effect of Using Media with Different C:N Ratios on the Growth of Tubifex. Thesis. Faculty of Fisheries and Maritime Affairs. Airlangga University, 2 pp.

[3] Suharyadi. 2012. Study of Development and Production of Silk Worms (Tubifex sp.) With Different Fertilizers in Recirculation Systems. Open University Masters Program Final Project. Jakarta. 116 p.Mandila, S. P. and N. Hidajati 2013 *J Chem.* 2 (1) : 103- 108.

[4] Suharyadi. 2012. Study of Development and Production of Silk Worms (Tubifex sp.) With Different Fertilizers in Recirculation Systems. Open University Masters Program Final Project. Jakarta. 116 hal.

[5] Agusnaidi, A 2020 *Arwana J Ilm Progr Stud Perair.* 2(1): 53-62.

[6] Setyawati, R. 2014. Complete Guide to Silk Worm Cultivation and Business.Jakarta: Flash Book.

[7] Cahyono, E. W., Hutabarat, J., Herawati, V. E 2015 *J Aquac Manag Technol.* 4(4), 127-135.

[8] Afifi, R and B. Setia 2017 *J Hexagro*, 1(2): 7-13.

[9] Prabusasongko, D. A. M. 2003. The Effect of Different Levels of Catfish Offal in Feed on the Growth of Catfish (Pangasius hypothalmus). Thesis. Faculty of Fisheries And Marine Sciences. Bogor Agricultural Institute. Bogor.

[10] Syaputra, S. E., H. W. Maharani and B. putri 2017 *e-Jurnal Rekayasa Teknol Budid Perairan.* 5(2): 597-604.

[11] Syam, F. S., G. M. Novia and S. N. Kusumastuti 2011 Effectiveness of Fertilization with Chicken Manure in Efforts to Increase Population and Biomass Growth of Silkworms Limnodrilus sp. Through Daily Fertilization and Fermentation Results. Thesis. Department of Aquaculture. Faculty of Fisheries and Maritime Affairs. Bogor Agricultural Institute. Bogor. 8 p.Tangketasik, A., N. N. Wikarniti., N. N. Soniari and I. W. Narka 2012 *J Agrotop.* 2(2): 101-107.

[12] Sukristiyunobowo 2008 *J Tanah Iklim*. 28(1): 39-54.
[13] Sulistioningsih, M 2015 *Pros Sem Nas Masyarakat Biodiv Indonesia*. 1(2): 378-382.
[14] Effendi, M. 2013. Modern way of raising silkworms. Jakarta: penebar swadaya. Jakarta.
[15] Nascimento, H. L. S and R. G. Alves 2008 Rev. Bras. Zool 25(1), 16-19.
[16] Safrina, B. Putri and H. Wijayanti 2015 *Pros Semin Nas Pengemb Teknol Pertan*. 2011;4(1). 520 – 525.
[17] Marian, M. P and Pandian, T. J 1984 *Aquaculture*, 42 (3-4) : 303-315.

6. Acknowledgements

The author gives thanks for this work being supported by Faculty of Fisheries and Marine, Universitas Airlangga. The author extends thanks to Mrs. Laksmi Sulmartiwi and Mrs. Nina Nurmalia for supporting author to finish the research.

The addition of *Mucuna pruriens* seed extract with different dosage on feed conversion ratio and feed efficiency of *Rasbora argyrotaenia*

Y Husnacahya[1], T A Chandrawati[1], S H Samara[2], M Amin[2], A S Mubarak[3,4]
[1]Aquaculture Program, Faculty of Fisheries and Marine, Universitas Airlangga, Surabaya, Indonesia
[2] Department of Aquaculture, Faculty of Fisheries and Marine, Universitas Airlangga, Surabaya, Indonesia
[3]Department of Marine, Faculty of Fisheries and Marine, Universitas Airlangga, Surabaya, Indonesia

[4] Corresponding author: shofy.ua@gmail.com

Abstract. The objectives of this research were: 1) determine the effect of feed supplementation using velvet beans (*Mucuna pruriens*) seed extract with different doses on the feed conversion ratio and feed efficiency of silver barb (*R. argyrotaenia*), 2) determine the optimal dose of feed supplementation using velvet beans seed extract in silver barb on the value of feed conversion ratio and feed efficiency. This research used an experimental method with a completely randomized design consisting of five treatments of commercial feed without any addition (P0), 0.1 ml/kg velvet beans seed extract addition (P1), 1 ml/kg velvet beans seed extract addition (P2), 10 ml/kg velvet beans seed extract addition (P3), 35 ml/kg velvet seed beans extract addition (P4) in four replication for each treatments. The results of the supplementation with a different concentration of velvet beans seed extract addition showed significant difference ($P < 0.05$) on the specific growth rate, feed conversion ratio, and feed efficiency of *R. argyrotaenia*. The highest efficiency and feed conversion ratio of the supplementation of *R. argyrotaenia* feed obtained from addition of 0.1 ml/kg velvet beans seed extract.

1. Introduction

Silver barb (*R. argyrotaenia*) is a local fish that lives in domesticated river waters [1]. Silver barb production in Indonesia comes from nature [2]. The demand for silver barbs is increasing, but growth is slow and the ideal feed supplementation for silver barbs has not been found [3]. The solution of this problem is enrichment in feed through nutritional engineering with the addition of velvet beans extract (*Mucuna pruriens*) [4].

Velvet beans (*Mucuna pruriens*) contain bioactive compounds including phenols, tannins, phytic acid, and L-Dopa [5]. The L-Dopa compounds in velvet beans are 0.58-6.42 % [6]. L-Dopa is a non-protein isomer of natural aromatic amino acids that functions for the synthesis of dopamine in dopaminergic neutrons which stimulates dopamine to increase the production of Growth Hormone so that it can increase growth[5].

Fish growth is directly measured through the increase in body weight and length. Meanwhile, feed conversion ratio (FCR) and feed efficiency can be measured indirectly. FCR is the amount of feed required with an increased body mass of one kilogram [7]. Feed efficiency is the utilized feed in the body [8]. FCR and feed efficiency can be affected by giving Growth Hormone which can reduce FCR and increase feed efficiency [9], [10].

The feed conversion ratio of tilapia (*O. niloticus*) with the addition of *M. pruriens* fermentation showed an FCR value of 1.19 and feed efficiency of 84% [11]. Manwhile, the addition of velvet bean flour to *Pangasius* sp. seed feed showed a higher feed efficiency value of 67.25% compared to the control treatment, which was 66.10% [12]. Low feed conversion value and high feed efficiency are associated with a large fish weight growth rate so that the quality of the feed consumed is getting better [13].

Research on addition of velvet beans extract to the silver barb feed on feed conversion rate (FCR) and feed efficiency has not been carried out out so that the addition of velvet beans extract in the diet of silver barb is expected to determine the ideal feed supplementation that can improve the cultivation of silver barbs.

2. Material and method

2.1 Materials

Materials used in this study are *R. argyrotaenia* size 3-5 cm, *Mucuna pruriens* extract, ethanol, commercial feed PF 500 and distilled water.

2.2 Place and time of research

This reseach was conducted at the Chemistry and Analytics Laboratory of The Faculty of Fisheries and Marine, Universitas Airlangga, Surabaya also at UPT Laboratory of Fish Health and Environment, Pasuruan, East Java. This research was carried out in January until March 2022.

2.3 Reseach design

This study using the experimental method with a completely randomized design (CRD). This study uses five treatments and each treatments was replicated four times. The treatments used is a different doses of feed supplementation without any addition (P0), 0.1 ml/kg velvet beans extract addition (P1), 1 ml/kg velvet beans extract addition (P2), 10 ml/kg velvet beans extract addition (P3), 35 ml/kg velvet beans extract addition (P4).

2.4 Methods

2.4.1 Production of velvet beans (Mucuna pruriens) extract

Velvet beans are cleaned of dirt that sticks to them using running water, then dried in the sun for 3-4 days. After drying, the velvet beans were chopped into flour and mashed using an 80 mesh sieve until the size was homogeneous [14]. The mashed velvet beans were extracted using the maceration method, 750 grams of velvet beans flour was soaked in 1875 ml of 96% ethanol and 1875 ml of distilled water in a ratio of 1:1 [15]. Immersion was repeated 3 times. The resulting marinade was combined and then evaporated at a temperature of 40°C at a speed of 50 rpm to become a thick velvet bean extract [16].

2.4.2 Feed supplement production

The commercial feed used is PF 500 in the form of pellets. The spray method was used for the manufacture of feed by mixing velvet bean extract at doses of 0, 0.1, 1, 10, and 35 ml/kg on feed with 0.5% CMC and aquades to reach a volume of 70 ml. The solution was homogenized and sprayed on commercial feed evenly and air-dried. Then the feed can be applied to the silver barb [17].

IOP Conf. Series: Earth and Environmental Science 1273 (2023) 012049 doi:10.1088/1755-1315/1273/1/012049

2.4.3 Silver barb cultivation

The silver barb was reared for 45 days in a 16-liter jar and the silver barb density was about 3 fish/liter [3]. A commercial feed with the addition of velvet bean extract by spraying was previously given twice a day. Prior to cultivation, it is necessary to calculate silver barb biomass. The amount of feed given was 4% of the weight of the biomass of the test fish on an ad-libitum basis. Cultivation of the test media was done by siphoning every 2-3 days to remove the remaining feed and dirt and change the water as much as 50%. Sampling was done by measuring the weight and length of silver barbs every 10 days. Water quality parameters such as temperature, DO, and pH was carried out every day during the study, while ammonia levels were measured once a week.

2.4.4 Parameter

Feed conversion ratio and feed efficiency of silver barbs were the main parameter. Feed conversion ratio, feed efficiency, dan survival rate was calculated by equation (1), (2), and (3) respectively.

$$FCR = F/(Wt - Wo) \quad (1)$$

Where FCR is Feed Conversion Ratio, F is amount of feed given to silver barb in gram, Wo is the weight of silver barb before cultivation in gram, Wt is the weight of silver barb after cultivation in gram.

$$FE = (Wt - Wo)/F \times 100\% \quad (2)$$

Where FE= Feed Efficiency.

$$SR = (Nt/No) \times 100\% \quad (3)$$

Where SR= Survival Rate, No is the number of samples at the beginning of study, Nt is the number of samples survived until the end of study.

2.4.5 Data analysis

The observation data were analyzed using Analysis of Variance (ANOVA). If the results from the analysis of variance showed that the treatment had significantly different results, then it was followed by Duncan's Multiple Range Test with a 5% error rate to determine the best treatment.

3. Results and discussion

3.1. Survival rate

Supplementation of velvet bean extract in feed did not affect the survival rate (SR) of silver barb. While feeding using velvet bean extract supplementation affects the specific growth rate, feed conversion ratio, and feed efficiency of silver barb. Survival rate of silver barbs after the supplementation of velvet beans extract are illustrated in Table 1 respectively.

Table 1. Average result of Silver Barbs Survival Rate

Supplementation (ml/kg feed)	SR (%) ± SD
P0 (0)	95 ± 10.00
P1 (0.1)	99.5 ± 1.00
P2 (1)	99 ± 2.00
P3 (10)	96.75 ± 6.50
P4 (35)	100 ± 0.00

IOP Conf. Series: Earth and Environmental Science 1273 (2023) 012049 doi:10.1088/1755-1315/1273/1/012049

Description : P0 commercial feed (control); P1 commercial feed+0.1 ml velvet bean extract; P2 commercial feed+1 ml velvet bean extract; P3 commercial feed+10 ml velvet bean extract; P4 commercial feed+35 ml velvet bean extract.

Velvet beans contain 0.58-4.32% L-Dopa [6]. L-Dopa is a precursor of dopamine to carry out the synthesis of dopamine which is assisted by the enzyme dopa decarboxylase. Dopamine will stimulate the hypothalamus to increase GHRH which is passed on to the pituitary gland for growth hormone production so that it can increase growth [18]. However, dopamine will inhibit the production of FSH-RH/LH-RH found in the hypothalamus. Decreased production of FSH-RH/LH-RH will inhibit the secretion of gonadotropin hormones so that FSH and LH decrease. Growth hormone will go into the blood circulation of fish and go to target organs such as the liver, muscles, and bones to support fish growth [19]. Growth hormone also affects the process of osmoregulation, nutrient metabolism, and formation in bone and soft tissue [20].

Based on the analysis of variance in Table 1, shows that velvet bean extract supplementation was not significantly different from the survival rate of silver barbs. The survival rate of silver barbs fed without velvet bean extract supplementation was 95%. Meanwhile, the average survival rate of silver barbs given commercial feed supplementation with velvet bean extract during rearing ranged from 96.75 - 100%. Extraction using 96% ethanol is not toxic to the fish body and the compounds contained in velvet beans extract are not harmful so they can support the survival of fish [21]. Factors that affect the survival rate in fish include factors originating in the fish body and factors originating from outside such as the environment, feed, water quality, pests, and diseases [22].

3.2 Silver barb biomass

The average data of silver barbs biomass during 40 days of rearing with velvet bean extract supplementation are shown in Table 2. The results of statistical calculations using Analysis of Variance (ANOVA) showed that supplementation of velvet bean extract with different doses had no significant effect on 1st-week biomass but significantly affected the biomass of silver barbs at 2nd, 3rd, and 4th weeks ($p<0.05$).

The results of Duncan's multiple range test on the biomass of silver barb showed that the highest biomass in the second week was found in wader fish with 0.1 mL/kg of feed supplementation with velvet bean extract (P1) of 14.32 grams, which was not significantly different from that of silver barb with supplementation. Velvet bean extract 1 ml/kg feed (P2) of 14.13 grams. The lowest biomass of silver barbs in the second week was found in silver barb with feeding without supplementation of velvet bean extract (P0) of 12.05 grams, which was not significantly different from supplementation with 10 ml/kg of feed (P3) and 35 ml/kg feed (P4) of 12.28 grams and 12.24 grams.

Table 2. Data of Silver Barb (*R. argyrotaenia*) Biomass Average

Velvet bean extract dose (ml/kg feed)	Biomass Weekly in gram				
	Initial	1	2	3	4 (final)
P0 (0)	8.92±1.03	10.56±1.85	$12.05±2.30^a$	$13.24±2.10^a$	$13.87±1.59^a$
P1 (0.1)	8.79±0.67	11.81±0.71	$14.32±0.51^b$	$15.57±0.42^b$	$18.49±0.57^c$
P2 (1)	8.91±0.43	11.77±0.30	$14.13±0.29^b$	$15.16±0.81^b$	$17.17±0.30^b$
P3 (10)	8.97±0.64	11.24±1.16	$12.28±1.03^a$	$13.48±0.51^a$	$14.26±0.43^a$
P4 (35)	8.96±0.24	10.56±0.35	$12.24±0.18^a$	$13.45±0.37^a$	$14.50±0.68^a$

Description : Different superscripts (a, b, c) showed significant differences ($p<0.05$) in each treatment. P0 commercial feed (control); P1 commercial feed+0.1 ml velvet bean extract; P2 commercial feed+1ml velvet bean extract; P3 commercial feed+10 ml velvet bean extract; P4 commercial feed+35 ml velvet bean extract.

The results of Duncan's multiple range test on the biomass of silver barb showed that the highest biomass in the third week was found in silver barb with 0.1 ml/kg feed (P1) of velvet bean extract supplementation (P1) of 15.57 grams, which was not significantly different from that ofsilver barb with supplementation. Velvet bean extract 1 ml/kg feed (P2) of 15.16 grams. The lowest biomass of silver barbs in the third week was found in silver barb with feeding without supplementation of velvet bean extract (P0) of 13.24 grams, which was not significantly different from supplementation with 10 ml/kg of feed (P3) and 35 ml/kg feed (P4) of 13.48 grams and 13.45 grams.

Duncan's multiple range test results on the biomass of silver barbs showed that the highest biomass in the fourth week was found in silver barb with supplementation of velvet bean extract 0.1 ml/kg of feed (P1) of 18.49 grams, which was significantly different from the supplementation of velvet bean extract 1 ml/kg feed (P2) of 17.17 grams. The lowest biomass of silver barbs in the fourth week was found in silver barb with feeding without supplementation of velvet bean extract (P0) of 13.87 grams which was not significantly different from supplementation of velvet bean extract with a dose of 10 ml/kg of feed (P3). and 35 ml/kg of feed (P4) of 14.26 grams and 14.50 grams.

3.3 Silver barb (R. argyrotaenia) growth performance

Supplementation of velvet bean extract which contains L-Dopa affects the growth and development of fish. The advantage of L-Dopa is that it is able to become a dopamine precursor which is able to stimulate growth hormone so that it can increase the rate of fish growth. The specific growth rate (SGR), feed conversion ratio, and feed efficiency of silver barb during rearing with feed supplementation using velvet bean extract can be seen in Table 3.

However, there is a lack of L-Dopa, at a concentration of 4-7% it is toxic so it can be harmful to fish life [23]. High concentrations of L-Dopa can increase the production of testosterone, FSH, LH, and the hormone prolactin so that it can inhibit fish growth [24]. Silver barbs fed with velvet bean extract supplementation at a dose of 1 ml/kg of feed had lower growth rates. However, in feeding using velvet bean extract 1 ml/kg feed had the highest growth. This is presumably because the stingray wader has a limited number of enzymes in the body so that it cannot be maximized in fish growth [25].

Table 3. Specific growth rate (SGR), initial biomass, final biomass, feed conversion ratio, total feed consumption (F), feed conversion ratio (FCR), and feed efficiency (FE) of silver barbs during each treatment on rearing.

Parameter	Velvet bean extract dose (ml/kg feed)				
	P0 (0)	P1 (0.1)	P2 (1)	P3 (10)	P4 (35)
SGR (%/day)	0.82±0.56^a	1.87±0.4^b	1.75±0.30^b	1.46±0.09^b	1.44±0.22^b
Initial Biomass (grams)	8.92±1.03	8.79±0.67	8.91±0.43	8.97±0.64	8.96±0.24
Final Biomass (grams)	13.87±1.59^a	18.49±0.57^c	17.17±0.30^b	14.26±0.43^a	14.50±0.68^a
Total feed consumption (grams)	15.99±2.44	18.00±0.16	17.77±0.243	16.33±1.13	16.11±0.28
FCR	3.34±0.73^b	1.87±0.21^a	2.15±0.14^a	3.12±0.48^b	2.92±0.21^b
Feed Efficiency (%)	30.90±5.92^a	53.81±6.35^b	46.51±3.29^b	32.64±5.93^a	34.36±2.70^a

Description : Different superscripts (a, b, c) showed that there was a significant difference ($p<0.05$) in each treatment.

Silver barbs growth can be analyzed using a specific growth rate (SGR). The specific growth rate is a measurement of the daily growth rate during the cultivation period [26]. SGR of silver barbs fed with velvet bean extract supplementation swelling 0.1-35 ml/kg of feed by 1.44-1.87%/day. While the SGR of silver barb with feeding without supplementation of velvet bean extract (P0) was 0.82%/day and the lowest among several treatments. The SGR of silver barbs is reported to be 0.51-1.98% [3] . Factors that affect SGR are feeding, water quality, and stocking density in cultivation [27]. Feeding is related to the size of the feed, the amount of feed, and the nutritional content that can affect growth [28]. Supplementation of velvet bean extract does not improve the nutritional quality of feed but aims to increase growth hormone in fish. Growth hormone will enter the blood circulation of fish and go to target organs such as the liver to stimulate the production of IGF-I. IGF-I plasma will stimulate muscles and bones in the growth process to increase fish body length and weight so that it can affect fish SGR [29].

3.3.1 Feed conversion ratio
The lowest FCR of silver barbs on feeding using velvet bean extract supplementation was 0.1 ml/kg of feed (P1) and 1 ml/kg of feed (P2) of 1.87 (P1) and 2.15 (P2). The FCR value is still in the range of the optimal fish FCR value according to Fahrizal and Nasir (2017) ranging from 1.5 to 2.5. While the value of FCR with feeding without supplementation (P0) had the highest value of 3.34 and feeding with supplementation of velvet bean extract of 10 ml/kg of feed (P3) and 35 ml/kg of feed (P4) of 3.12 and 2.92. Factors that affect FCR are environmental conditions, quality, and quantity of feed [30].

3.3.2 Feed efficiency
One indicator that can determine good feed quality for fish is the value of feed efficiency [31]. The value of feed efficiency is a comparison of the increase in the amount of fish weight with the amount of feed consumed [32]. The supplementation of velvet bean extract with different doses can affect the feed efficiency of silver barb. Where the highest feed efficiency was found in feeding using supplementation of velvet bean extract 0.1 ml/kg (P1) of 53.81%, while the lowest feed efficiency was found in the provision without supplementation of velvet bean extract (P0) of 30, 90%. The value of feed efficiency is related to the growth rate, where the higher the growth rate, the greater the body weight gain in fish, this will cause the feed efficiency value to be greater. In addition, the value of feed efficiency is also related to FCR because it is inversely proportional. The lower the FCR, the less feed is needed to produce 1 kg of fish meat, and the higher the feed efficiency value so that the quality of fish feed is good [33].

Good feed quality in feed efficiency means that the feed can be easily digested by fish. There are several factors that affect the digestibility of fish including the type of feed, environmental conditions, age of fish, nutrient content, and frequency of feeding the fish [34]. GH in the body of fish in addition to playing a role in the growth process but will also act in the process of absorption of nutrients. Feed that can be digested well in the fish body with the help of GH-stimulated enzymes from complex compounds to simple compounds can affect FCR and feed efficiency [35]. The GH is able to increase protein synthesis, stimulate metabolism and stimulate insulin synthesis and release [36]. The difference in feed efficiency values is thought to be related to differences in feeding doses, so fish are only able to take advantage of the availability of nutrients according to the treatment given. In addition, negative feedback occurs because there is an excess of GH which provides feedback impulses to the pituitary gland so that it cannot secrete GH, resulting in stunted growth [35].

3.4 Silver barb (R. argyrotaenia) meat proximate analysis
The results of the proximate analysis of silver barb (*R. argyrotaenia*) meat during 40 days of rearing with supplementation of velvet bean extract with different doses can be seen in Table 4.

Table 4. Silver barb meat proximate analysis (% gross weight)

Parameter	Velvet bean extract dose (ml/kg feed)				
	P0 (0)	P1 (0.1)	P2 (1)	P3 (10)	P4 (35)
Carbohydrates (%)	0.29	0.4	0.42	0.43	0.31
Protein (%)	28.72	29.02	28.89	29.34	28.97
Fat (%)	9.82	10.07	9.92	9.89	9.88
Water (%)	57.75	57.15	57.49	56.99	57.53
Ash (%)	3.42	3.36	3.29	3.35	3.31

Description : P0 commercial feed (control); P1 commercial feed+0.1 ml velvet bean extract; P2 commercial feed+1 ml velvet bean extract; P3 commercial feed+10 ml velvet bean extract; P4 commercial feed+35 ml velvet bean extract.

Based on Table 4, the meat content of silver barb with feeding using supplementation of velvet bean extract with different doses showed that the carbohydrate content was 0.31% - 0.43%, protein 28.89% - 29.34%, fat 9.88 % -10.07%, water content 56.99% - 57.53% and ash content 3.29% - 3.36%. Meanwhile, the silver barbs which were cultured by feeding without supplementation had a carbohydrate content of 0.29%, protein 28.72%, fat 9.82%, water content of 57.75% and ash content 3.42%.

The results of the proximate analysis of silver barbs meat with feed without supplementation had a carbohydrate content of 0.29%, protein of 28.72%, fat of 9.82%, and water content of 57.75%, and ash content of 3.42%. While the feeding using supplementation of velvet bean extract with different doses showed that the carbohydrate content was 0.31%-0.43%, protein 28.89%-29.34%, fat 9.88%-10.07%, and protein content was low. Water content was 56.99%-57.53% and ash content of 3.29%-3.36%. This study showed that feeding using velvet bean extract had an increase in carbohydrates, protein, and fat compared to feeding without supplementation with velvet bean extract. The nutritional requirements of cyprinidae fish range from 5-15% [37] while protein requirements range from 25-35% [38].

3.5 Water quality

Water quality in this study as a supporting parameter by measuring four parameters, temperature 23.1-25.6°C (morning) and 26.2-30.6°C (afternoon), DO (dissolved oxygen) 3.2-4 0.7 mg/l (morning) and 2.8-4.6 mg/l (afternoon), pH 7-8.3 and ammonia 0-0.5 mg/l. Measurements of temperature, DO (dissolved oxygen), and pH was carried out every day in the morning and evening, while ammonia measurements were measured once a week. Data on the range of water quality during maintenance can be seen in Table 5.

Table 5. Data of Water Quality Parameter

Treatment	Parameter					
	Temperature (°C)		DO (ppm)		pH	NH_3 (mg/L)
	Morning	Afternoon	Morning	Afternoon		
P0	23.1-25.1	26.7-30.6	3.2-4.7	3.1-4.3	7-7.9	0-0.05
P1	23.2-25.6	26.2-30.6	3.4-4.7	3.0-4.38	7-8.2	0-0.25
P2	23.3-24.5	26.6-30.6	3.4-4.6	3.3-4.6	7-8.1	0-0.25
P3	23.3-24.5	26.2-30.5	3.6-4.7	3.02-4.6	7-7.8	0-0.5
P4	23.1-24.6	26.6-30.4	3.5-4.7	3.1-4.5	7-8.3	0-0.5

Description : P0 commercial feed (control); P1 commercial feed+0.1 ml velvet bean extract; P2 commercial feed+1 ml velvet bean extract; P3 commercial feed+10 ml velvet bean extract; P4 commercial feed+35 ml velvet bean extract.

Water quality parameters measured in the study were temperature obtained from 23.1-30.6ºC, dissolved oxygen (DO) in the morning ranging from 3.2-4.7 mg/l while in the afternoon between 3.0-4.6 mg /l, pH is around 7-8.3 and ammonia is around 0-0.5 mg/l. Temperature, DO and pH values have been in accordance with the optimal range of growth silver barb (*R. argyrotaenia*), the temperature is around 20-30ºC, DO >3 mg/l [39], and the optimal pH is between 6.5-8.5 [40]. While the measurement results of ammonia during the cultivation of silver barb obtained 0-0.5 mg/l. The ammonia level is not suitable because it exceeds the threshold of eligibility according to SNI 7550:2009 for fish farming is < 0.02 mg/l. Factors that affect the high levels of ammonia are thought to come from the metabolism of fish and the rest of the feed that is not consumed. High ammonia levels tend to cause death in fish because it can inhibit the absorption of blood hemoglobin to oxygen so that it cannot be maximized in the body's metabolism and affects the growth rate [41].

4. Conclusion

From a series of studies started from extraction of velvet beans until final cultivation that has been carried out, supplementation of velvet bean extract in fish with a dose of 0.1-35 ml/kg feed had an effect on the conversion ratio and feed efficiency of *R. argyrotaenia*. Also, the optimal dose of velvet bean extract that produced the highest FCR and feed efficiency was 0.1 ml/kg feed (P1).

5. References

[1] Retnoaji B, Wijayanti N, Zulfadhli 2016 *Pendekatan Histol.* 3 32-39
[2] Ulum M, Mun'im A, Juliyani E, Nisa K 2019 *DIMAS.* 19 1-14
[3] Dewi S A, Mubarak A S, Mukti A T 2020 *IOP Conf Ser Earth Environ Sci.* 441 1-6
[4] Musthafa M S, Asgari S M, Kurian A, Elumalai P, Ali A R J, Paray B A, Al-Sadoon M K 2018 *J Fish Shelfish Immun.* 75 374-380
[5] Ojha M L, Chadha N K, Saini V P, Damroy S, Chandraprakash, Sawant P B 2013 *J Ecol Environ Conserv.* 19 287-293
[6] Puli kkalpura H R, Kurup R, Mathew P J, Baby S 2015 *Sci Rep.* 5 1-9
[7] Fry J P, Mailloux N A, Love D C, Milli M C, Cao L 2018 *Environ Res Lett.* 13 1-8
[8] Sunarto, Sabariah 2009 *J Akuakultur Indones.* 8 67-76
[9] Apriliani R, Fajar B, Ristiawan A 2017 *J Sains Akuakultur Trop.* 2 49-58
[10] Kling P, Jonsson E, nilsen O T, Einarsdottir I E , Ronnestad I, Stefansson S O, Bjornsson B T 2012 *Gen Comp Endocrinol.* 175 153-162
[11] Woke G N, Aleleye L P, Komi G W, Bekibele D O 2013 *Glob J Pure Appl Sci.* 19 9-15
[12] Veronika S, Santoso L 2011 *Berk Perikan Terubuk.* 39 9-16
[13] Ihsanudin I, Rejeki S, Yuniarti T 2014 *J Aquac Manag Technol.* 3 94-102
[14] Mardina P, Sari L A, Noerhayati E 2020 *Konversi.* 9 19-27
[15] Sardjono R E, Khoerunnisa F, Musthopa I, Akasum N S M, Rachmawati R 2018 *4th Int Semin Math Sci Comput Sci Educ.*
[16] Agbafor K N, Nwachukwu N 2011 *Biochem Res Int*
[17] Islamiyah D, Rachmawati D, Susilowati T 2017 *J Aquac Manag Technol.* 6 67-76
[18] Pangestiningsih T W, Susmiati T, Wijayanto H 2016 *J Veteriner.* 18 116-120
[19] Bjornsson B T, Ingibjorg E, Marcus J, Ningping G 2018 *Front Endocrinol.* 9 1-10
[20] Reineche M 2005 *Gen Comp Endocrinol.* 142 20-24

[21] Hikmawanti N P E, Fatmawati S, Asri A W 2021 *IOP Conf Ser Earth Environ Sci*
[22] Arzad M, Ratna R, Fahrizal A 2019 *Median.* 11 39-47
[23] Lampariello L R, Cortelazzo A, Guerranti R, Sticozzi C, Valachi G 2012 *J Tradit Complement Med.* 2 331-339
[24] Das A 2018 *Prep Biochem Biotechno.* 48 671-682
[25] Huisden C M, Szabo N J, Arriola K G, Adesogan A T 2019 *Trop. Subtrop. Agroecosystems.* 22 45-53
[26] Mulqan M, Rahimi S S, Dewiyanti I 2017 *J Ilm Mhs Kelaut dan Perikan Unsyiah.* 2 183-193
[27] Karimah U, Samidjan I, Pinandoyo 2018 *J Aquac Manag Technol.* 7 128-135
[28] Hariawan h, Tatisina C M 2020 *J Pengabdi Masy Sasambo.* 1
[29] Simbolon S, Iraan H, Putra W K A 2019 *Intek Akuakultur.* 3 53-68
[30] Arifin Z, Rumondang R 2017 *Fisherina* 1
[31] Islama D, Diana F, Yunanda S, Saputra F, Zulfaldhi, Febrina C D 2021 *J Akuakultura.* 4 46-53
[32] Saputra F, Achmadi J, Pangestu E 2013 *Anim Agric J.* 2 137-147
[33] Shofura H, Suminto S, Chilmawati D 2018 *Sains Akuakultur Trop.* 1
[34] Melati R R 2021 *Universitas Muhammadiyah Malang*
[35] Pratama A E, Lumbessy S Y, Azhar F 2021 *J Kelaut.* 14 164-174
[36] Fissabela F A, Nugroho R A 2016 *J Aquac Manag Technol.* 5 1-9
[37] Webster C D, Lim C 2002 *Nutrient Requirements and Feeding of Finfish for Aquaculture* (New York: CABI Publishing)
[38] Rohma A, Agustono, Arief M 2012 *J Aquac Fish Heal.* 2 22-29
[39] Suryani F Y, Setyawanti T R, Yanti A H 2019 *J Protobiont.* 8 74-81
[40] Andria A F, Rahmaningsih S 2018 *J Ilm Perikan Kelaut.* 10 95-105
[41] Firmansyah W, Cokrowati N, Scabra A R 2021 *J Ilm Perikan Kelaut.* 26 85-93

6. Acknowledgements

The author gives thanks for this work being supported by Faculty of Fisheries and Marine, Universitas Airlangga. The author extends thanks to Mr. Ahmad Shofy Mubarak and Mrs. Syifania Hanifah Samara for supporting author to finish the research.

Water Quality Analysis for Vannamei Shrimp Culture in Coastal of Banyuwangi Regency

E W Setyaningrum[1,2], M Yuniartik[1], S H Yuniari[1]

[1]Faculty of Agriculture and Fisheries, Universitas 17 Agustus 1945 Banyuwangi, Jl. Adi Sucipto 26 Banyuwangi East Java, Indonesia

[2]Corresponding author : ervinawahyu@untag-banyuwangi.ac.id

Abstract. The coastal of Banyuwangi Regency is the longest coast and has a high potential for vannamei shrimp culture with the largest production in East Java Province, Indonesia. The sustainability of vannamei shrimp culture depends on seawater input, where water quality is the main thing that must always be considered. The purpose of this study was to analyze water quality for the benefit of vannamei shrimp culture. The method used is descriptive. While data collection using surveys and direct observations, with water quality parameters taken there are 14 types, namely temperature, salinity, pH, DO, alkalinity, ammonia, nitrate, nitrite, phosphate, TOM, TSS, vibrio, redox and plankton. Data analysis used factor analysis based on the Principal Component Analysis (PCA). The results from the calculation of PCA analysis, it can be seen that there are 5 components that are able to describe the overall water quality on the coast of Banyuwangi Regency for the benefit of vannamei shrimp cultivation with a total variance of 80,522%. Component 1 filled with parameters Salinity, Nitrite, Nitrate and Vibrio. Component 2 are pH, TOM and Redox. Component 3 are DO, TOM, TSS and Plankton. Component 4 are Alkalinity, Ammonia, Nitrite and Phosphate. Component 5 are Temperature, Salinity and Plankton.

1. Introduction

Coastal areas have abundant biological natural resources such as fish, coral reefs, seagrass beds, mangroves, etc., as well as non-biological natural resources such as sand, seawater, seabed minerals, etc. Which then causes damage to the coastal environment, including biological factors such as tsunamis, alteration, and earthquakes, use of explosives and poisons in fishing, chemical pollution, brackish water aquaculture, floating net cages, exploitation of coral reefs, and other coastal activities [1].

The coast of Banyuwangi Regency is the longest beach in East Java, which is 175.8 km, including facing the Indonesian Ocean, some facing the Bali Strait and even the Gulf. The coastal area here is a priority in the development of industry, agribusiness, agro-industry, housing, transportation, ports, and tourism. Increasing the use of water as a means of various community activities can cause changes to these factors [2,3].

Damage and pollution of the coastal environment can make a direct impact on decreasing the quality of aquatic habitats, reducing fish stocks, and disrupting the function of the coastal environment [4,5]. Optimal water quality is clearly important for the environment [6]. Water pollution is not only detrimental to aquaculture but also affects human health [7]. The physical and chemical parameters of water that can determine whether they are contaminated include dissolved oxygen, pH, total nitrogen, phosphorus, nitrite, natural, ammonia, etc. And poor management practices can lead to various physiological changes in the growth and even survival of aquatic animals [8].

Brackish water aquaculture is one of the fastest-growing food sectors in the world. Among various fields of cultivation, shrimp farming is developing rapidly due to the fast growth rate of shrimp, short cultivation period, high export value, and high market demand all over the world. Shrimp farming has become an important industry in developing countries in East and Southeast Asia, which make up the majority of exports in both volume and value. Currently, more than 70% of the world's shrimp production is produced in Asian countries, and most of the shrimp produced is exported for foreign currency acquisition and economic development of these countries. In relation to production, water quality plays an important role in increasing pond productivity. In this case, cultivation requires a balanced and healthy environment [9].

Surface water quality provides important information about the availability of natural resources to support ecosystem living. Characterization of site variability and water quality parameters can improve understanding of environmental conditions and policy decisions based on site management development priorities. Knowing the characteristics and variability of seawater quality parameters is one of the most important management issues in aquaculture operations. Based on data on water quality conditions, input was made to select a suitable location [10].

Water quality analysis helps monitor changes in water quality values, and because global changes are elusive, multivariable water quality assessments provide a complete understanding of water quality and help manage the coastal environment. On the other hand, water quality in cultivated land is usually degraded by human activities such as sanitation, disposal of crop residues, and ecosystem degradation. Therefore, the development of an efficient method for comprehensively assessing the quality of coastal water is very important for industrial and aquaculture management in Banyuwangi Province. Water quality monitoring typically results in large and complex databases of biological, physical, and chemical variables that are difficult to manage [11].

As an anthropogenic activity, shrimp farming can contribute to the input of nutrients into the ecosystem through the disposal of wastewater containing nitrogen and phosphorus. These nutrients are obtained from the feces used, shrimp droppings, and food scraps. The accumulation of organic matter and nutrients in shrimp ponds causes extensive phytoplankton growth, increasing the potential for wastewater contamination and anoxic conditions in the influent [12].

Sustainable development requires internalizing externalities in certain production decisions. Regulation is often seen as a necessary tool for regulating aquaculture production. Knowing how farmers internalize environmental impacts related to land use change, water, and soil pollution, and mangrove destruction is very important, especially when production takes place in developing countries without strict environmental regulations [13]. The purpose of this study was to analyze the water quality along the coast of Banyuwangi Regency for the benefit of vannamei shrimp culture.

2. Material and method

The study was conducted along the coast of Banyuwangi Province which is spread across eight coastal districts with 24 sampling locations. The survey period is from May to August 2022. The method used to determine the state of coastal water quality in Banyuwangi is descriptive multivariate analysis. The data collection method was carried out by direct observation in the field, as shown in Table 1 below. Table 1 lists the materials and tools used in the study.

Table 1. The materials and tools used in the study.

Parameter	Tools/Method	Information
Temperature (^{0}C)	Termometer	Insitu
Salinity (PPT)	Refractometer	Insitu
pH	pH meter	Insitu
DO (mg/l)	Winkler	Laboratory Analysis
Alkalinity (mg/l)	Titration	Laboratory Analysis
Amonia (ppm)	Nessler	Laboratory Analysis
Nitrite (ppm)	Titration	Laboratory Analysis
Nitrate (ppm)	Titration	Laboratory Analysis
Phosphate (ppm)	Titration	Laboratory Analysis
TOM	Titration	Laboratory Analysis
TSS (mg/l)	Titration	Laboratory Analysis
Vibrio (CFU/ml)	Total Plate Count	Laboratory Analysis
Potencial Redox	ORP meter	Laboratory Analysis
Plankton	Sedgewick Rafter Cell	Laboratory Analysis

The data analysis used is principal component analysis (PCA). PCA is the best multivariate statistical analysis technique for extracting linear relationships between different groups of variables. The results of the analysis come in the form of numerical data and images, which are interpreted based on the physical and chemical parameters of the existing water bodies. PCA analysis provides important parametric information with little loss of original information. The principal components can be expressed by the formula:

$$Zij = ai1x1j + ai2x2j + \cdots + aimxmj$$

Where:

Z = score of component
a = component load
x = Variable measurement result
i = component number
j = sample number
m = total number of variables

3. Result and Discussion

3.1. Banyuwangi Regency Coastal Water Quality

Table 1 shows the results of direct observations of water quality in 21 locations along the coast of Banyuwangi Province and measurements of 14 parameters.

Table 2. Banyuwangi Regency Coastal Water Quality Data.

Location	Temp (° C)	Salinity (PPT)	pH	DO (mg/L)	Alkalinity (mg/L)	Amonia (ppm)	Nitrite (ppm)	Nitrate (ppm)	Phosphate (ppm)	TOM	TSS (mg/L)	Vibrio (CFU/ml)	Potencial Redox	Plankton
1	28	33	7,9	9,57	184	0,069	0,141	5	0,078	7,584	0,69	420	108	523.412
2	28	32	7,8	6,097	108	0,04	0,069	5	0,026	6,32	0,37	220	103	169.556
3	29	35	7,8	6,6	196	0,051	0,11	10	0,076	10,112	0,29	100	102	563.958
4	29	33	7,8	8,04	60	0,023	0,068	10	0,024	20,224	0,51	200	102	455.000
5	29	35	7,7	8,32	124	0,036	0,072	7	0,026	36,656	0,42	320	107	265.000
6	29	33	7,7	8,13	124	0,026	0,078	9	0,022	37,92	0,47	400	106	570.000
7	30	33	7,7	9,34	12	0,037	0,086	10	0,027	11,376	0,75	50	106	240.000

Location	Temp (° C)	Salinity (PPT)	pH	DO (mg/L)	Alkalinity (mg/L)	Amonia (ppm)	Nitrite (ppm)	Nitrate (ppm)	Phosphate (ppm)	TOM	TSS (mg/L)	Vibrio (CFU/ml)	Potencial Redox	Plankton
8	29	32	7,7	9,75	164	0,065	0,123	5	0,056	25,28	1,12	260	123	695.000
9	29	29	7,7	9,34	116	0,049	0,083	5	0,057	37,92	0,89	170	114	400.000
10	28	18	7,8	9,75	104	0,027	0,107	22	0,078	13,904	0,05	150	95	760.000
11	28	28	7,8	8,94	116	0,017	0,082	15	0,041	18,96	0,55	380	110	875.000
12	28	28	7,8	9,07	96	0,012	0,078	10	0,037	25,28	0,56	2360	101	540.000
13	29	27,5	7,7	9,34	96	0,028	0,155	25	0,063	46,768	0,65	3930	99	635.000
14	28	29	7,8	6,5	124	0,027	0,081	15	0,059	25,544	0,78	430	121	405.000
15	28	28	7,7	9,29	108	0,019	0,136	15	0,062	37,92	0,66	210	116	735.000
16	29	30	7,7	8,94	164	0,031	0,079	10	0,037	18,96	0,19	70	98	405.000
17	28	29	7,7	8,13	140	0,015	0,125	15	0,064	1,264	0,05	290	110	260.000
18	28	31	7,6	7,72	140	0,031	0,081	10	0,045	3,792	0,52	50	109	245.000
19	29	32	8	9,1	148	0,028	0,087	10	0,038	68,256	0,68	320	93	350.000
20	28	33	8,1	9,83	132	0,022	0,058	5	0,024	61,936	0,65	30	85	245.000
21	28	32	8,1	9,75	132	0,016	0,053	5	0,025	64,464	0,81	150	91	1.350.000

3.2. Principal Component Analysy (PCA)

Based on water quality parameters, principal component analysis (PCA) was conducted to identify water quality components that affect the coast of Banyuwangi Regency. As a first step, the Bartlett test was carried out to test whether the variables measured were correlated with each other in calculating the quality of coastal water in Banyuwangi. The process of calculating the Bartlett test is carried out using SPSS computer software, resulting in Table 2 below.

Table 3. Barlett Test Result.

Bartlett's Test of Sphericity	Approx. Chi-Square	171.402
	df	91
	Sig.	.000

The results of the Barlett test analysis showed that the correlation matrix data met the Bartlett test, with a significance less than 0.05 (zero value). From this, we can conclude that the water quality data meet the requirements for principal component analysis (PCA).

It also takes the number of components with eigenvalues greater than 1, or 5 components as shown in the scree plot image. The scree plot itself is a plot of the eigenvalues against the number of components extracted. The resulting diagram form is used to determine the number of components. The problem graph is shown in Figure 2.

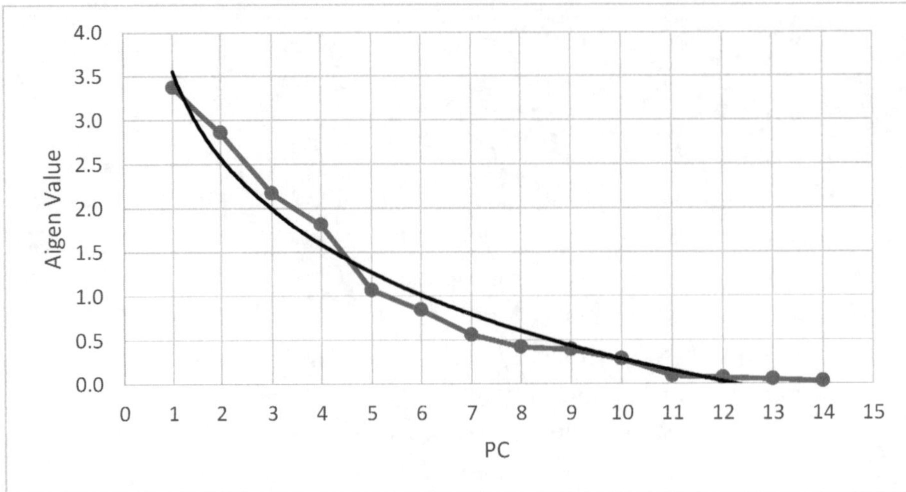

Figure 1. Grafik Scree Plot.

The figure above shows the line breaks starting to fall out in components 4 and 5. Therefore, the premise of this analysis is that the five components obtained are components.

Based on the scree plot diagram, Banyuwangi coastal water quality parameters are divided into 5 components with an overall variance of 80,522%. See Table 2 for details on eigenvalues and variances.

Table 4. Tabel Scree Plot.

Component	Initial Eigen values		
	Total	% of Variance	Cumulative %
1	3,373	24,092	24,092
2	2,856	20,403	44,495
3	2,169	15,493	59,988
4	1,809	12,922	72,911
5	1,066	7,611	80,522

The table above shows that the variance of the first principal component is 3,37 or equal to the proportion of 24,09% of the total variance of all data. The variance of the second component is 2,86 and accounts for 20,.40 of the total variance of the data. The third component with a variance of 2,17 and a share of 15,49%. The four components have a variance of 1.81 and a share of 12,92%. And the fifth component has a variance of 1,07 and a share of 7,61%. In this case, the sum of the five principal components accounts for 80,52% of the overall power, indicating that the five principal components are sufficient to explain most of the overall variance in the data.

The process of separating the parameters into each component is carried out by component loading. Component loading is a value that shows the correlation between the values of the variables formed by the preliminary data. The sum of the squares of each component is the previously obtained eigenvalues. This is what determines the parameters that will include to the five components. The results of the component loading are shown in Table 5.

Table 5. Component Loading of Each component.

	Component				
	1	2	3	4	5
Temperature	-0.136	-0.339	0.035	0.708	0.388
Salinity	-0.592	-0.619	0.113	0.071	0.217
PH	-0.632	0.420	0.296	-0.405	0.126
DO	-0.100	0.501	0.535	0.261	0.008
Alkalinity	0.082	-0.248	0.465	-0.689	0.250
Amonia	0.154	-0.687	0.563	-0.004	0.203
Nitrite	0.795	0.019	0.435	0.100	0.238
Nitrate	0.700	0.559	-0.236	0.166	0.096
Phosphate	0.814	-0.005	0.399	-0.287	0.095
TOM	-0.581	0.508	0.377	0.180	0.113
TSS	-0.310	-0.115	0.670	0.447	-0.358
Vibrio	0.335	0.457	0.083	0.431	0.306
Redox	0.485	-0.551	0.165	0.210	-0.560
Plankton	0.001	0.548	0.468	-0.123	-0.354
Eigen Value	3.373	2.856	2.169	1.809	1.066
Extraction Method: Principal Component Analysis.					

The results of component loading above show that not all parameters can be determined to be in the correct component. This is because there is only a slight difference in value between one component and another. Only a few parameters show a significant difference in values between components. Therefore, the rotation process must be able to display the parameters that enter each component with the highest component loading value. The rotation results are shown in Table 5.

Table 6. Component Loading Post Rotation.

	Component				
	1	2	3	4	5
Temperature	0.014	0.083	0.083	-0.099	0.873
Salinity	-0.713	-0.102	-0.018	0.081	0.523
PH	-0.309	-0.736	0.356	0.043	-0.279
DO	0.264	-0.217	0.705	0.034	0.010
Alkalinity	-0.272	-0.189	-0.079	0.809	-0.224
Amonia	-0.293	0.314	0.085	0.709	0.400
Nitrite	0.623	0.297	0.097	0.628	0.097
Nitrate	0.913	0.066	-0.133	-0.087	-0.177
Phosphate	0.472	0.285	-0.041	0.739	-0.246
TOM	-0.053	-0.563	0.652	-0.181	0.077
TSS	-0.294	0.272	0.821	0.005	0.224
Vibrio	0.705	-0.110	0.203	-0.035	0.235
Redox	-0.078	0.943	0.047	0.159	-0.008
Plankton	0.159	-0.086	0.613	0.052	-0.498
Eigen Value	2.894	2.214	2.191	2.188	1.785
% Variance	20.673	15.816	15.653	15.628	12.752
Extraction Method: Principal Component Analysis.					

The table above explains that the figures displayed tend to be towards the component with a more significant component load value, so it can be concluded that the parameters that go into each component are as follows:

- The first component with a variance of 20.673% includes the parameters of Salinity, Nitrite, Nitrate, Phosphate, and Vibrio.
- The second component with a variance of 15.816% includes the parameters of pH, TOM, and ORP.
- The third component with a variance of 15.653% includes DO, TOM, TSS, and Plankton parameters.
- The fourth component with a variance of 15.628% includes the parameters of Alkalinity, Ammonia, Nitrite, and Phosphate.
- The last component with a variance of 12.752% includes temperature, salinity, and plankton.

Based on the results of the analysis, it can be concluded that in measuring the quality of coastal water in Banyuwangi Province for shrimp farming, the indicators are salinity, nitrite, nitrate, phosphate, and vibrio. Basically, the characteristics of coastal waters in Banyuwangi Province are most influenced by these parameters compared to others. The smaller the ratio of N and P, the higher the the vibrio abundance. Conversely, the higher the ratio of N and P, the lower the frequency of the vibrio. The higher the salinity, the higher the vibrio pathogenicity. And the lower the salinity, the lower the pathogenicity of Vibrio. Vibrio is a salt-tolerant bacteria.

Furthermore, the determination of the linear combination of the principal components can be determined based on Table 6. The linear combination of the principal components in question is as follows:

a. PC1 = -0,713 Salinity 0,623 NO_2 + 0,913 NO_3 + 0,427 PO_4 + 0,705 Vibrio

Components Salinity, Nitrite, Nitrate, Phosphate, and Vibrio occupy the first component because in shrimp culture, the N and P ratio affect the abundance of vibrio, and the salinity affects the pathogenic ability (lethality) of vibrio bacteria.

Understanding the characteristics of different *V. parahaemolyticus* strains under different salinities can better prevent and control the disease caused by this bacterial pathogen [14]. To produce healthy shrimp as a safe food for human consumption, maintaining high water salinities seems to be recommendable in Recirculating Aquaculture System [15].

b. PC2 = PC2 = -0,736 pH – 0,563 TOM + 0,943 Redox

Absorbance of organic matter can increase pH value [16]. The pH, TOM, and Redox values occupy the second component because the pH affects the redox potential in terms of electron conductivity that occurs in water, and an increase in the PH value also increases the absorbance of organic matter in the water.

c. PC3 = 0,705 DO + 0,652 + 0,821 TSS + 0,613 Plankton

DO, TOM, TSS, and Plankton parameters occupies the third component because the increase in organic matter affects the abundance of plankton and has an impact on changes in the DO value in the waters. In addition, phytoplankton serves as a barrier to fluctuations in of organic matter and DO.

Phytoplankton inhabiting coastal areas experience frequent fluctuations in many parameters (i.e. temperature, dissolved nutrient concentrations, and salinity) [17].

d. PC4 = 0,809 Alkalinity + 0,709 NH_3 + 0,628 NO_2 + 0,739 PO_4

Alkalinity, Ammonia, Nitrite, and Phosphate occupy the fourth position because alkalinity is directly related to nitrogen removal. And in the nitrification reaction the amount of ammonia that affects the results of the nitrite parameters. And the increase in alkalinity value will also affect the removal of phosphate in water.

Nitrification is a two-step reaction: ammonium (NH^{4+}) is first oxidized to nitrite (NO^{2-}) by autotrophic ammonia oxidizers, nitrite is then oxidized to nitrate (NO^{3-}) by autotrophic nitrite-oxidizers [18].

e. PC5 = 0,873 Temperature + 0,523 Salinity – 0,498 Plankton

Temperature, Salinity and Plankton occupies the fifth component because phytoplankton functions as a barrier to fluctuations in temperature. And fluctuations in salinity, can have a negative impact on the growth of phytoplankton, depending on the origin of the community.

And fluctuations in salinity, can have a negative impact on the growth of phytoplankton, depending on the origin of the community [19]. These changes in atmospheric CO2 will also impact the world's oceans with anticipated increases in SST of 1 to 4°C, decreases in ocean pH of 0.3 units, and major impacts on upper-ocean stratification [20].

4. Conclusion

The conclusion of the research is there are 5 components that are able to describe the overall water quality on the coast of Banyuwangi Regency for the benefit of vannamei shrimp cultivation with a total variance of 80,522%. Component 1 filled with parameters Salinity, Nitrite, Nitrate and Vibrio. Component 2 are pH, TOM and Redox. Component 3 are DO, TOM, TSS and Plankton. Component 4 are Alkalinity, Ammonia, Nitrite and Phosphate. Component 5 are Temperature, Salinity and Plankton. Sedangkan untuk kombinasi linier maka dihasilkan model komponen untuk PC1 = -0,713 Salinity 0,623 NO_2 + 0,913 NO_3 + 0,427 PO_4 + 0,705 Vibrio; PC2 = PC2 = -0,736 pH – 0,563 TOM + 0,943 Redox; PC3 = 0,705 DO + 0,652 + 0,821 TSS + 0,613 Plankton; PC4 = 0,809 Alkalinity + 0,709 NH_3 + 0,628 NO_2 + 0,739 PO_4; dan PC5 = 0,873 Temperature + 0,523 Salinity – 0,498 Plankton.

5. References

[1] Saraswati N L G R A, Rustam A, Salim H L, Heriati A, and Mustikasari E 2017 *J. Segara.* 13 (1) 37–47.
[2] Setyaningrum E W, Masithah E D, Yuniartik M, Nugrahani M P, and Dewi A T K 2020 *IOP Conf Ser Earth Environ Sci. 441*(1) 012129.
[3] Setyaningrum E W, Maghdalena, Dewi A T K, Yuniartik M, and Masithah E D 2019 *IOP Conf Ser Earth Environ Sci. 236*(1) 012045.
[4] Vernanda O, Sari L A, Cahyoko Y, Arsad S, Pursetyo K T, Dewi N N, and Mustafa S 2022 *Int J Adv Sci Eng Inf Technol. 12*(2) 826-833.
[5] Melati V H, Sari L A, Cahyoko Y, Arsad S, Pursetyo K T, Dewi N N, and Idris M H 2021 *Ecol Eng Environ Technol. 22*(3) 82-90.
[6] Hutami W W, Sari L A, Masithah E D, Sahidu A M, and Pursetyo K T 2020 *Asian J Water, Environ Pollution.* 17(3) 81-86.
[7] Kurniawan S B, Said N S M, Imron M F, and Abdullah S R S 2021 *Environ Technol Innov. 23* 101790.
[8] Ni M, Yuan J L, Liu M, and Gu Z M 2018 *Aquac. Reports.* 11 53–58,
[9] Venkateswarlu V, Seshaiah P V, Arun P, and Behra P C 2019 *Int. J. Fish. Aquat. Stud.*, 7(4) 394–399.
[10] Wiyoto W, and Effendi I 2020 *J. Aquac. Fish Heal. 9*(2) 143-154.
[11] Ma Z, Li H, Ye Z, Wen J, Hu Y, and Liu Y 2020 *Mar. Pollut. Bull.* 157 111285.
[12] Bauer W, Abreu P C, and Poersch L H 2017 *Brazilian J. Oceanogr.* 65(3) 495–508.
[13] Hukom V, Nielsen R, Asmild M, and Nielsen M 2020 *Ecol. Econ.* 176 106717.
[14] Ma R, Wang Y, Huang L, Zhao S, Li L, Yin M, and Fang W 2021 *J. Fish Dis.* 44(12) 2055–2066.
[15] Bauer J, Teitge F, Neffe L, Adamek M, Jung A, Peppler C, Steinhagen D, and Jung-Schroers V 2021 *J. Fish Dis.* 44 (1) 89–105.
[16] Timko S A, Gonsior M, and Cooper W J 2015 *Water Res.* 85 266–274.

[17] Groß E, Di Pane J, Boersma M, and Meunier C L 2022 *J. Plankton Res.*

[18] Akcil A and Koldas S 2006 *J. Clean. Prod.* 14(12-13) 1139–1145.

[19] Lionard M, Muylaert K, Gansbeke D V, and Vyverman W 2005 *Hydrobiologia* 540 (1–3) 105–115.

[20] Sarmiento J L, Hughes T, Stouffer R J, and Manabe S 1998 *Nature. 393*(6682) 245-249.

6. Acknowledgments

The author would like to thank the Ministry of Education, Culture, Research and Technology for the research grant on the national competitive basic research scheme.

IOP Conf. Series: Earth and Environmental Science 1273 (2023) 012051 doi:10.1088/1755-1315/1273/1/012051

Different stocking density on growth and survival rate of tilapia seeds (*Oreochromis niloticus*)

S B Maharani[1] **and A H Fasya**[1,2]

[1] Aquaculture Departmen, School of Health and Natural Sciences. Airlangga University, Banyuwangi, Jl. Wijaya Kusuma No. 113, District. Giri, Banyuwangi Regency, Tel: 0333 - 417788, Fax: 0333 – 428890

[2]Corresponding author: arifhabibfasya@fpk.unair.ac.id

Abstract. Stocking density is one aspect of cultivation that needs to be known because it affects the growth rate and survival rate that determines the level of production. High stocking density will result in competition in getting feed and space, resulting in growth variations. The purpose of this study was to determine the effect of stocking density on growth and survival of tilapia (*Oreochromis niloticus)* seeds. The observation method was carried out experimentally with the treatments given in this observation: treatment P1 = 100 head stocking density; P2= 200 head stocking density; p3 = 300 head stocking density. The results showed that different densities resulted in different growth and survival rates, treatment 1 had an SGR value of 5.58% and an SR of 94%; Treatment 2 obtained an SGR value of 4.96% with an SR value of 88% and the lowest results were obtained in treatment 3 with an SGR value of only 2.02%/day and an SR of 83%. Water quality at the time of maintenance showed optimum results with a temperature range between 27-29°C, pH around 6.9-7.7 and DO around 3-5 Mg/l

1. Introduction

Tilapia (*Oreochromis niloticus*) is one of the leading freshwater fish commodities because tilapia is favored by various groups of local and international communities [4]. Tilapia production increased from 2010 to 2013 with an average value of 34.85%. The comparison of the total national tilapia production to the total world tilapia production shows that in 2011 Indonesia was the third largest producer of tilapia products with a percentage of around 20.3% of the total production of tilapia in the world [7].

The demand for tilapia continues to increase, which encourages intensive cultivation. The consistency of increasing tilapia production can be done through intensive cultivation by paying attention to various aspects that support the survival of the fish such as water availability, cultivation area, and good environmental quality [9]. Stocking density is one aspect of cultivation that needs to be known because it affects growth rates, feed conversion ratios and survival rates that determine production levels [6].

According to [11], added that the stocking density is influenced by the size of the seed, the type of fish and the cultivation system carried out.

The increase in stocking density is directly proportional to the increase in the amount of feed, the rest of the body's metabolites and the consumption of dissolved oxygen, but inversely proportional to the water quality [3].High stocking density will result in competition in getting feed and space, resulting in growth variations. In addition, high density will affect water quality due to the frequent accumulation of organic matter originating from the waste of fish metabolism and uneaten food residue [5]. Based on the description above, the purpose of this study was to determine the effect of different stocking densities on the growth rate and survival rate of tilapia seeds raised in aquariums.

2. Material and methods

The method of observation was carried out experimentally. The tools used are an aquarium with a size of 80×60×60 cm, analytical balance, ruler, siphon, aerator, pH meter, thermometer and DO test kit. The materials used are tilapia seeds measuring 1-1.5 cm, reservoir water and commercial feed. The treatments given in this observation were: treatment T1 = stocking density of 100 fish (0.96 liters/head); T2= stocking density of 200 fish (0.48 liters/head); T3 = stocking density of 300 fish (0.32 liters/head).

Maintenance is carried out for 30 days. Feeding was carried out three times a day in the morning, afternoon and evening with the amount of feeding as much as 3% of body weight. Water quality measurements were carried out 2 times a week in the morning and evening to determine the DO content, pH and temperature. Sampling of fish seed length and weight growth was carried out once a week.

The variables observed were the specific growth rate (SGR) and survival rate (SR) of tilapia seeds raised at different densities. Specific Growth Rate (SGR) is the percentage of the difference between the final weight and the initial weight divided by the length of maintenance time. SGR can be calculated using the formula [8].

$$\mathbf{SGR = In(Wt) - In(W0)T \ x\ 100\%}$$

Information:
SGR = specific growth rate
Wt = final average weight
W0 = initial average weight
T = maintenance time

Survival rate (SR) is the percentage of the ratio between the final population and the initial population. SR can be calculated using the formula [2].

$$\mathbf{SR = NtN0\ x\ 100\%}$$

Information:
SR= rate of survival
Nt = number of final rearing fish
N0 = number of initial rearing fish

3. Result and discussion

3.1 Growth

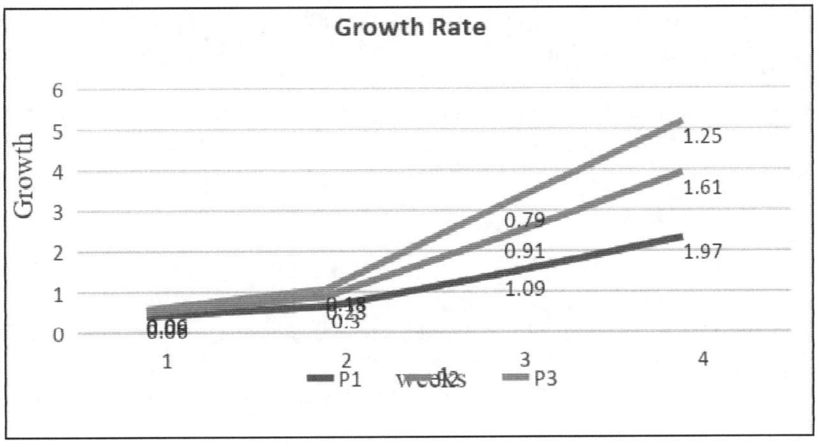

Figure 1. Tilapia seed growth

Based on the observations made, it can be seen that treatment 1 gave a significant increase in weight and length. From the results of the SGR calculation, the SGR in treatment 1 was 5.58%/day, treatment 2 was 4.96%/day and treatment 3 was 2.02% day. This shows that the treatment with a low density level gets enough space compared to the treatment with a high density level, so that individuals in the low density treatment can grow more optimally than individuals reared with a higher density. According [11], revealed that a high stocking density will lead to the competitiveness of fish in utilizing food and space which can affect their growth. according to [1] added that when there is competition for space, individuals who are not able to compete will have their growth disturbed, besides that the narrow space of motion will reduce fish appetite and slow down the growth rate.

3.2 Survival Rate

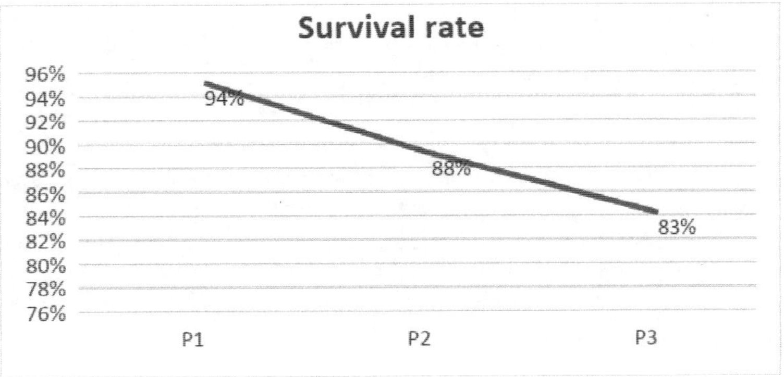

Figure 2. Survival rate of tilapia seed

Based on the observations made, different survival results were obtained in each treatment. Treatment 1 has a high SR value of 94%, treatment 2 has an SR value of 88% and treatment 3 has the lowest SR value of 83%. The SR value in the observations made is still quite good, because according to [2], an SR value of more than 80% is a good SR value, while those less than 80% can be said that the survival rate is low. This difference in SR values is thought to be due to differences in stocking density. According to [1], high density will result in an increase in ammonia concentration and a decrease in water quality. Decreased water quality can cause stress to fish, even if the decrease in water quality has exceeded the tolerance limit it will result in death. In addition, a decrease in water quality can also affect the appetite of fish. When the appetite decreases, the intake of feed into the fish's body also decreases. If this goes on for a long time, it will cause death.

3.3 Water Quality

Water quality is a very important factor in fish farming because it is needed as a living medium. Measurement of water quality during observations showed that the water quality conditions in the aquarium were in normal conditions, meaning that the water quality conditions were appropriate and within the normal range for tilapia seed. The results of water quality measurements can be seen in table 1.

Table 1. Parameters of maintenance water quality

Parameters	Unit	Value
Temperature	27-29	°C
pH	6,9-7,7	
DO	3-5	Mg/l

The temperature in the aquarium shows a temperature range between 27-28°C. The temperature value in each treatment showed good results, because based on SNI 6141-2009 the optimum temperature value for rearing tilapia fry ranges from 25 - 30° C. The optimal temperature will support the life and development of organisms well [10]. The pH range is between 6.9-7.7 and is still within the normal range that can be tolerated for rearing tilapia fry according to [8]'s statement that the optimal pH range for rearing tilapia is 6-8 ,5.

Based on the results of observations, obtained DO values ranging from 3-5 ppm. Tilapia can survive at dissolved oxygen (DO) content of more than 0.3 mg/l, well below the tolerance limit for most cultured fish. Although tilapia can survive at low oxygen levels for several hours, tilapia ponds must be adjusted to maintain dissolved oxygen content above 1 mg/l [1].

4. Conclusion

Breeding at different densities results in different growth and survival rates. Based on the observations that have been made, the best tilapia seed growth is in treatment 1 with a stocking density of 100 fish (0.96 liters/head) getting an SGR value of 5.58% and an SR of 94%; Treatment 2 with a stocking density of 200 fish (0.48 liters/head) obtained an SGR value of 4.96% with an SR value of 88% and the lowest yield was obtained in treatment 3 with the highest density level of 300 fish (0.32 liters/head) with an SGR value of only 2.02%/day and an SR of 83%.

5. References

[1] Alfia A R, Arini E and Elfitasari T 2013 *J of Aquaculture Mng and Tech*. 2(3): 86-93.
[2] Darwis, Joppy D M and Sammy N J L 2019 *Aquaculture* 7(2): 15 -21.
[3] Diansari V R, Endang A and Tita E 2013 *J of Cult Mng*. 2 (3): 37-45.
[4] Fadri S, Muchlisin Z A and Sugito 2016 *Sci J of Marine and Fisheries Students Unsyiah*. 1(2): 210-221.
[5] Hakim A R 2019 *Effect of Stocking Density on Growth and Survival of Tilapia Seeds* (Medan: University of North Sumatra Press).
[6] Ismael F 2018 *Effect of Stocking Density on Growth and Survival of Baung Fish (Hemibagrus mumurus) Seeds Using Biofloc Tech in Peat Swamp Water* (Pekanbaru: University of Riau Press).
[7] Marie R, Syukron M A and Rahardjo S S P 2018 *J of Natural Resources and Env*: 1-6.
[8] Mulqan M, El-Rahimi S A and Dewiyanti I 2017 *Sci J of Marine and Fisheries Unsiyah*. 2(1): 183-293.
[9] Putra I, Setiyanto D D and Wahyuningrum D 2011 *J of Fisheries and Marine Affairs*. 16(1):56-63.
[10] Sutiana S, Erlangga E and Zulfikar Z 2017 *Acta Aquatica: Aquatic Sci J*. 4(2) pp 76-82.
[11] Yaningsih N 2018 *Effect of Stocking Density on Growth and Survival of Red Tilapia (Oreochromis niloticus) Seeds with Biofloc Technology in Peat Swamp Water*. (Pekanbaru: Riau University Press).

6. Acknowledgement

This research was conducted jointly with the team at the SIKIA (Sekolah Ilmu Kesehatan dan Ilmu Alam), Airlangga University.

Effect of Different Natural Feeding (Blood Worm, Silk Worm, and Water Fleas) on The Growth of Betta Fish (*Betta* sp.)

M S M Iqbal[1], Prayogo[2], S Han Samara[3,4]

[1]Aquaculture Program, Faculty of Fisheries and Marine, Universitas Airlangga, Surabaya, Indonesia
[2]School of Health and Natural Science (SIKIA), Universitas Airlangga, Banyuwangi, Indonesia
[3]Department of Aquaculture, Faculty of Fisheries and Marine, Universitas Airlangga, Surabaya, Indonesia

[4]Corresponding author: E mail: syifania.hanifah@fpk.unair.ac.id

Abstract. One type of freshwater ornamental fish that is in great demand by the public is *Betta* fish (*Betta* sp.). The problem faced by *Betta* fish cultivation activity is inappropriate feed, for example feed with low protein content. Low protein causes slow growth in fish. This study aims to determine the effect of different natural feeds (silk worms, blood worms, water fleas, and controls) on the growth of *Betta* fish. The results showed that there was a significant effect ($p<0.05$) on the growth and specific growth rate of betta fish. However, there was no significant effect ($p>0.05$) of different natural feeding on the survival rate parameter. The highest result showed in treatment B (silk worm) with a weight value of 1.09 g, a length of 4.48 cm, and specific growth rate value of 1.77 ± 1.768. It can be concluded that silk worm is the best natural feeding for betta fish.

1. Introduction

One type of freshwater ornamental fish that is in great demand by the public is betta fish (*Betta* sp.). This is because betta fish have beautiful body colors and fins [1]. In addition, betta fish have a high price, ranging from IDR 5,000-1,000,000.00 [2]. Betta fish are quite easy to maintain because these fish have an additional organ in the form of a labyrinth so they do not need an aerator [3]. Therefore, many people cultivate betta fish.

The problem faced by betta fish cultivators is inappropriate feed, for example feed with low protein content. Low protein causes growth in fish to be slow [4]. Protein plays an important role in fish growth as a substance to build muscles, cells and body tissues [5]. This causes losses because cultivators need longer time and higher maintenance costs. One way to overcome this is by providing appropriate and quality natural feed [6]

The natural food used in this study were blood worms, silk worms and water fleas. Natural feed contains nutrients, especially high protein. Blood worms contain up to 56.60% protein and 2.80% fat. The nutritional content of silk worms is 57% protein, 13.30% fat, 2.04% carbohydrate [7]. While water fleas have a nutritional content of 50% protein from dry weight, 20-27% fat [8]. Based on the description above, research is needed to determine the best type of natural food for the growth of betta fish.

2. Material and methods

2.1. Sample preparation

This research will be carried out for 2 months from March 15, 2022 to May 15, 2022 at Bluerim Surabaya. The material used in this study was betta fish with the same multicolor and genetic type (homogeneous) obtained from the Bluerim farm, Surabaya (2 months old) with an initial length of ± 2.5 cm and an initial weight of ± 0.23 g as many as 60 fish. (3 tails/jar). The research design used was a completely randomized design (CRD) consisting of 4 treatments (A: control, B: silk worms, C: blood worms, D: water fleas) and 5 replications. Fish were treated for 2 months by feeding twice a day, at 10.00 WIB and at 16.00 WIB [5]. The dose of feed given for each treatment was 4% of body weight per day [9].

2.2 Research parameter calculation

Observations specific growth rate, survival rate, growth length and weight were carried out at the beginning and end of the study. Weighing of the test fish was carried out using a measuring instrument in the form of a scale. According to previous study, the formula for calculating the weight gain of the test fish was as follows [10]:

$$W = Wt - Wo$$

W : Weight growth (gram)
Wt : Final average weight (gram)
Wo : Initial average weight (gram)

Absolute length measurements were carried out every 10 days. The absolute length growth according to previous study formula was as follows [10]:

$$L = Lt - Lo$$

L : Absolute length growth (cm)
Lt : Final length (cm)
Lo : Initial lenght (cm)

The number of dead fish was counted every day until the end of the study. According to previous study the survival rate of fish was calculated using the formula [10]:

$$SR = \frac{Nt}{No} x\ 100\%$$

SR : Survival rate (%)
Nt : Final number of fish (fishes)
No : Initial number of fish (fishes)

The specific growth rate was calculated using the formula [11]:

$$SGR = \frac{(LnWt - LnWo)}{t} x\ 100$$

SGR = *Spesific Growth Rate* (%/hari)
W0 = The average weight of the fish at the start of the study (g)
Wt = The average weight of the fish at the end of the study (g)
t = Maintenance period (days)

Parameters observed and measured were temperature, pH, and DO. Water quality measurements (temperature, pH and DO). The analysis used in this study is the ANOVA test (Analysis of

IOP Conf. Series: Earth and Environmental Science 1273 (2023) 012052 doi:10.1088/1755-1315/1273/1/012052

Variance).Analysis of variance was carried out to determine whether there was an effect of treatment on the growth of betta fish [1]. The data obtained during the study were first tested for normality and homogeneity. The results of the ANOVA test showed that there was an effect between treatments. To find out more about the difference in effect between treatments, a DMRT test (Duncan's Multiple Range Test) was performed [12].

3. Result and discussion

The results of observations for 7 weeks obtained data on the weight gain of betta fish which can be seen in Figure 1. The highest average weight growth of betta fish was achieved in treatment B (silk worms) of 1.09 g, then followed by treatment C (blood worms) and treatment D (water fleas) which were 1.02 g and 0.93 g, respectively. The lowest growth was found in treatment A (pellets) of 0.71 g. The results of the normality test on the weight growth of betta fish showed that the data obtained were normally distributed. According to previous study, the data can be tested with ANOVA test and continued with DMRT test (Duncan's Multiple Range Test) [12]. Duncan test results can be seen in Table 1.

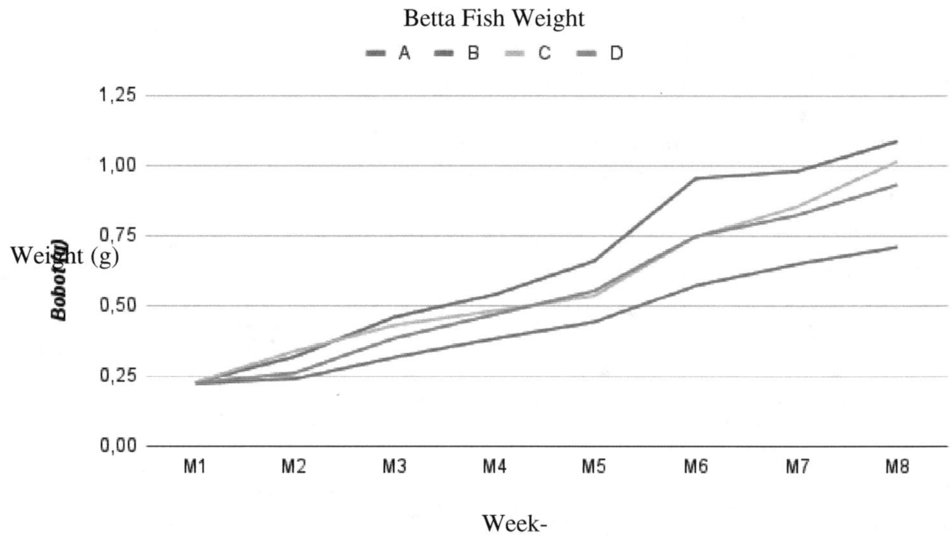

Figure 1. Betta fish weight gain in this study

The results of Duncan's analysis showed that treatment A was significantly different from treatment B, C, and D. Furthermore, treatment B was significantly different from treatment A, C, and D. The weight calculation was carried out by preparing a plastic cup filled with water on the scales, then setting the scales digital pocket to zero. Next, the betta fish is put into a plastic cup that has been prepared for counting. The results obtained in treatment B gave the greatest influence on absolute weight growth in betta fish (*Betta* sp.) aged 2-4 months when compared to treatments A, C, and D. This was because treatment B used natural feed in the form of silk worms. . Silkworms have a higher nutritional content than other betta fish feeds such as blood worms, water fleas and pellets. The nutritional content of silk worms consists of 57.00% protein and 13.30% fat [13]. This value is higher than the nutritional content of blood worms (47.8% protein and 9.7% fat), water fleas (37.38% protein and 13.29% fat), and pellets (39% protein, 5% fat) [14]. The nutritional content of silk worms has met the nutritional needs of betta fish. Feeds with high protein content can stimulate growth [4].

Table 1. Betta fish absolute weight gain (*Betta* sp.)

Treatment	Initial Weight (g)	Final Weight (g)	SGR (g) ± SD	DMRT 5% test
A (Pellet)	0,22	0,71	0,48 ± 0,08	0,4420[a]
B (Silk worm)	0,22	1,09	0,87 ± 0,10	0,6500[c]
C (Blood worm)	0,23	1,02	0,79 ± 0,21	0,5780[b]
D (Water fleas)	0,22	0,93	0,71 ± 0,08	0,5500[b]

During the study, betta fish experienced an increase in body length. Figure 2 shows a graph of the difference in the average body length of betta fish (*Betta* sp.). The results showed that treatment A (control) had a body length of 3.89 cm, treatment B (silk worms) was 4.48 cm, treatment C (blood worms) was 4.37 cm and treatment D (water fleas) was 4.21 cm. The results could be seen in Table 2.

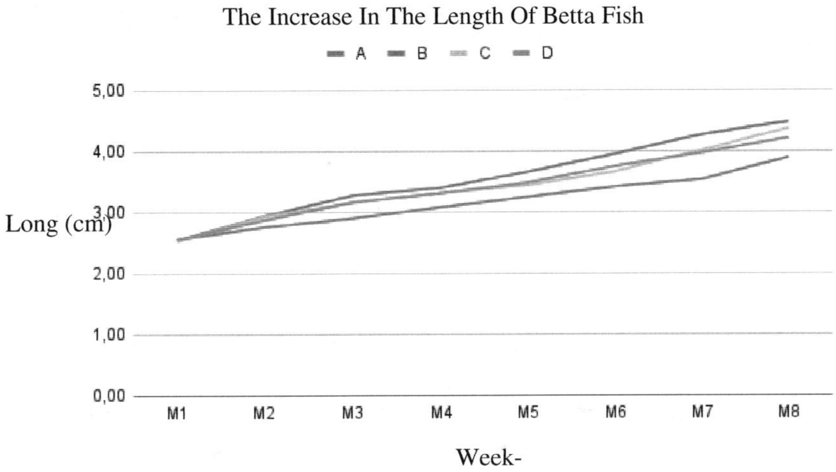

Figure 2. Betta fish body length growth

Based on Anova analysis, different natural feeds for betta fish had a significant effect (F hit>F0.05) on absolute length growth. Duncan's analysis showed that treatment A was significantly different from treatment B, C and D. Furthermore, treatment B was significantly different from treatment A, C, and D. The increase in body length in treatment B was caused by an adequate amount of feed nutrients. In treatment B the amount of carbohydrates and fats can meet the body's caloric needs, so the protein is only slightly oxidized to add calories but is used as a building block for fish growth (Buwono, 2000). Feeding silk worms in live conditions causes the enzymes contained in the feed to work optimally and fish can take advantage of the nutrients contained in the feed [16]. In addition, silk worms are relatively more digestible in the intestines of fish, which is about 2 hours, while water fleas and blood worms can be digested for more than 2 hours [17].

Table 2. Betta fish length increase

Treatment	Initial Length (cm)	Final Length (cm)	Length increase (cm) ± SD	DMRT test 5%
A (Pellet)	2,56	3,89	1,34 ± 0,20	3,1740[a]
B (Silk worm)	2,55	4,48	1,93 ± 0,17	3,5640[c]
C (Blood worm)	2,53	4,37	1,84 ± 0,10	3,4260[b]
D (Water fleas)	2,55	4,21	1,66 ± 0,15	3,4160[b]

Specific growth rate is the daily growth rate or the percentage of fish weight gain per day. The increase in growth can be identified by increasing the growth rate and specific growth rate. Based on statistical analysis using Anova, it was shown that there was an effect of feeding natural silk worms (*Tubifex* sp.), blood worms (*Chironomus* sp.), water fleas, and pellets on the growth of betta fish. The results can be seen in Table 3.

Table 3. Betta fish daily growth rate

Treatment	Initial Weight (g)	Final Weight (g)	SGR (%/day)
A (Pellet)	0,22	0,71	0,99 ± 0,9860[a]
B (Silk worm)	0,22	1,09	1,77 ± 1,768[b]
C (Blood worm)	0,23	1,02	1,61 ± 1,6112[b]
D (Water fleas)	0,22	0,93	1,44 ± 1,444[b]

Silkworms contain high enough protein, namely 57.00% protein and 13.30% fat [13]. Protein and fat are good nutritional content for betta fish during their growth period. Therefore, natural feed in the treatment of silk worms, perfectly meets the needs of betta fish, so that the treatment of silk worms has better growth than the treatment of blood worms, water fleas and pellets. This is in accordance with the research conducted by previous study, *Tubifex* sp. worm feed gave a significantly different specific weight growth (%) in guppies [18].

The results of the Anova analysis showed that the survival rate of Betta fish on different natural feeds had no significant effect (F count < 0.05) (Figure 3). The best survival rate of betta fish was found in treatments B, C, and D with a percentage of ±93%.

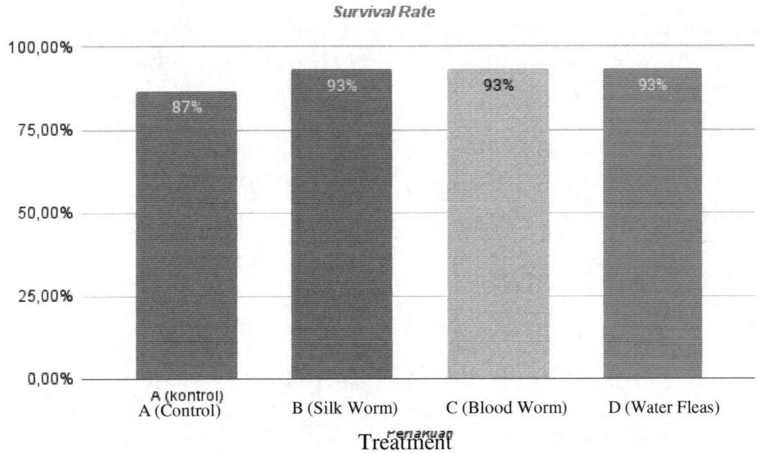

Figure 3. Survival rate result in this study

Based on Figure 3, the survival rate during the study was to obtain nutrients that meet their needs and a suitable aquatic environment during the study. The results of the SR values obtained in this study ranged from 87 to 93%. This is in accordance with the statement of previous study, in betta fish cultivation, the SR value which ranges from 87-100% is still quite good [19]. Feeding pellets, silk worms, blood worms, and water fleas did not have a significant effect on the Survival Rate value. This is because the size of the feed used is in accordance with the mouth opening of the betta fish so that it can be used by the fish to grow and live. The ability of fish to receive feed depends on the type of feed and the size of the feed [20].

Water quality is a limiting factor in the growth of cultured fish, including betta fish so that the quantity and quality of water used in fish farming activities must be optimal to meet the living needs of fish. The water used in this study came from water that has been deposited in the aquarium for 2-3 days and then filled into treatment jars. Measurement of water quality in this study was carried out once a week using water quality measuring instruments, namely, pH meter, DO meter, and thermometer. Water quality measurements were carried out in the morning and evening. The condition of the water in this research media looks brownish yellow due to the effect of giving ketapang leaves. This is because the water soaked in ketapang leaves contains tannins [21]. In addition to containing tannins, ketapang leaves also contain flavonoid compounds that function as antibacterial, antimicrobial, and antioxidant [22]. This can increase fish immunity because flavonoid compounds can help increase the number of leukocytes [23]. Leukocytes are one of the blood components that function as non-specific defenses that will localize pathogens by phagocytosis [24]. The results of water quality measurements during the study can be seen in Table 4.

Table 4. Results of observation of water quality

Parameter	A	B	C	D	Standard
	Mean ± SD	Mean ± SD	Mean ± SD	Mean rata ± SD	
pH	7.5 ± 0.2	7.4 ± 0.2	8 ± 0.6	7.4 ± 0.6	6,5 – 9,0*
Temperature (°C)	28.3 ± 0,4	28.1 ± 0.3	28.3 ± 0.4	28.4 ± 0.5	24 – 30**
DO (ppm)	7.4 ± 0.1	7.5 ± 0.1	7.5 ± 0.2	7.5 ± 0.1	> 4***

Noted : (*) Agus (2010); (**) Sunari (2008); (***) Arman (2001)

4. Conclusion
The highest result showed in treatment B (silk worm) with a weight value of 1.09 g, a length of 4.48 cm, and specific growth rate value of 1.77 ± 1.768. It can be concluded that silk worm is the best natural feeding for betta fish.

5. References
[1] Tarigan M R M, Masnadi M 2018 *Jur Edu Dan Sain Bio.* I 1 – 7.
[2] Diani S, Mustahal, Pramu S 2005 *Jur Pengk dan Penge Tek Pert.* 8 292-299.
[3] Atmadjaja, Sitanggang M 2008 Complete Guide to Ornamental Hickey Cultivation and Care Agromedia Jakarta 67-78.
[4] To'bungan N 2016 *Jur Bio.* 1 111−116.
[5] Sutresna Y 2005 *Skrip*. 50-76
[6] Simbolon S M, Mulyani C, Febri S P 2021 *Jur Kel Dan Peri Indo.* 1 1-9.
[7] Weismann G F, Lucas, Ockstan Kalesaran J, Lumenta C 2015 *Jur Bud Per* 3 19-28.
[8] Pangkey H 2009 *Jur Per dan Kel* 5 33-36.
[9] Permatasari U 2016 *Skrip.* 55-60.
[10] Effendi H 2003 Study of Water Quality for Management of Water Resources and Environment Kanisius Yogyakarta 23-30.
[11] Effendie M I 1997 *Fisheries Biology Method Pustaka. Nusantara Foundation* Jakarta 45-54.
[12] Agus M, Mardiana T Y 2015 *Pen Aku: Jur Ilmi Per dan Kel.* 2 21-29
[13] Mandila S P, Hidajati N 2013 *Jour of Chem.* 2 103-108.
[14] McWilliams D A 2008 *Nutrition Recommendations for Some Captive Amphibians Spesies (Anura and Caudata)* Canadian Association of Zoos and Aquariums Canada p 76-83.
[15] Buwono I D 2000 *Kebutuhan Asam Amino Esensial dalam Ransum Ikan* Kanisius Yogyakarta 34-40.
[16] Nurhayati N, Utomo N B P, Setiawati M 2014 *Jur Ikti Indo.* 14 167-178.
[17] Hadiroseyani Y, Dana D 1994 Provision of disease-free silkworms as food healthy fish through improved cultivation systems, Bogor Agricultural Institute, Bogor 33-40.
[18] Görelşahin S, Yanar M, Kumlu M 2018 *Aqu.* 493 153-157.
[19] Nengtias R A, Watiniasih N L, Ayu P W K D 2021 *Adv in Trop Bio and Envi Sci.* 5 78-82.
[20] Situmorang T S, Barus T A, Wahyuningsih H 2013 *Jur Peri dan Kel.* 18 48-58.
[21] Rochmat A, Liantony G, Septiananda Y D 2019 *Jur Inte Pro.* 8 45-50.
[22] Manzur A, Raju A, Rahman S 2011 *Pharm and Phar.* 2 299-305.
[23] Wahjuningrum D, Ashry N, Nuryati S 2008 *Jur Aku Indo.* **7** 79–94.
[24] Sukenda L, Jamal D, Wahjuningrum, Hasan A 2008 *Jur Aku Ind.* **7** 159– 169.

6. Acknowledgement
The authors gratefully acknowledge the collaboration from Universitas Airlangga and Bluerim Surabaya.

IOP Conf. Series: Earth and Environmental Science 1273 (2023) 012053 doi:10.1088/1755-1315/1273/1/012053

Morphological characteristics of *Aeromonas, Plesiomonas* and *Pseudomonas* bacteria in tilapia (*Oreochromis niloticus*) cultivation in Bangli Regency

S A Saraswati[1,4], Y Jasmanindar[2] E W Suryanigtyas[3] , D A Pebriani[3], N P Wijayanti[3], A Djonu[2] dan O Arimardewi[3]

[1] Water Resources Management Study Program, Faculty of Animal Husbandry, Marine and Fisheries, University of Nusa Cendana, Kupang-NTT
[2]Aquaculture Study Program, Faculty of Animal Husbandry, Marine and Fisheries, University of Nusa Cendana, Kupang-NTT
[3]Water Resources Management Study Program, Faculty of Marine Affairs and Fisheries, Udayana University, Bukit Jimbaran, Bali 80361, Indonesia

[4]Correspondence author: suprabadevi@staf.undana.ac.id

Abstract. Aquaculture is a popular activity in Indonesia, with one of the leading commodities, namely tilapia. After shrimp and salmon, tilapia as a successful example of world aquaculture set by the Department of Fisheries and Aquaculture FAO (Food and Agriculture Organization). In line with the development of the cultivation business, there are also several disturbing problems such as pests and diseases that hinder the development of the cultivation business. Monitoring activities were carried out in one of the districts, namely Bangli district. Monitoring activities are carried out by identifying pathogenic bacteria found in tilapia. The steps in identification are bacterial isolation, purification, presumptive test, biochemical test and reading of the results. The bacteria identified were *Aeromonas hydrophila, Plesiomonas shigelloides, Pseudomonas putida, Aeromonas sobria* and *Pseudomonas alkaligenes*.

1. Introduction

Tilapia as a type of freshwater fish that has high economic value, so it is necessary to strive for its utilization and management, one of which is in aquaculture activities [1]. After shrimp and salmon, tilapia as a successful example of world aquaculture set by the Department of Fisheries and Aquaculture FAO (Food and Agriculture Organization)

One type of disease that attacks cultured fish is a disease caused by bacteria [2,3]. Symptoms caused by diseases caused by bacteria include loss of appetite [4], wounds on the surface of the body, bleeding from the gills, the condition of the stomach is enlarged and filled with fluid [5], loose scales, loose tail fin, there is swelling and damage to the liver, kidneys and spleen when surgery is performed [6]. This bacterial disease occurs in a relatively short time and can cause death above 80% [7]. Fish diseases arise due to an imbalance between fish as hosts, pathogens (disease-causing microorganisms)

IOP Conf. Series: Earth and Environmental Science 1273 (2023) 012053 doi:10.1088/1755-1315/1273/1/012053

and the environmen [8,9,17]. The defense system of the fish body can be disrupted due to environmental changes to the development of pathogens in a container.

2. Material and methods

2.1. Sampling Time and Location

Tilapia fish samples were taken as many as 4 samples from aquaculture ponds in Bangli Regency. The sample consisted of 4 fish taken at the location of aquaculture ponds.

2.2. Work Procedure

Fish samples are parts (gills, kidneys, liver, spleen, brain, eyes, scales) that will be analyzed. (each sample is 25 g) taken from the sampling location. The samples were stored in a cooling box that had been given ice water to be brought to the Denpasar Class I KIPM Laboratory. Furthermore, samples of tilapia will be isolated and bacterial identification.

2.3. Research Materials

Bacterial identification includes: observation of bacterial cell morphology by simple staining method, and Gram stain reaction test, observation of cell shape using a microscope at 1000 times magnification, biochemical reaction test using IMVIC method (indole, methyl red, voges proskauer-VP, and citrate), catalase, oxidase, urease, motility and gelatin. Identification of bacteria based on; Bergey's Manual of Systematic Bacteriology and Bacteria from Fish and Other Aquatic Animals: A Practical Identification Manual [10]. The vibrio genus was identified and continued based on the modified method of West and Colwell (1984).

3. Result and discussion

3.1. Bacteria Isolation

Isolation was carried out to pick up bacteria on target organs and grow them on TSA media. Bacterial purification was carried out to take the dominant bacterial colonies from the results of bacterial growth through the isolation process and then incubated at 28 0C-30 0C for 24 hours. The growing bacterial colonies were selected based on the shape and size of the color, inoculated on TSA for further identification of the genus. The results of bacterial isolation and purification can be seen in Tables 2 and 3.

Table 1. Isolation in sample 1.

No.	Sample Code	Sample Type	Targets Organ	Incubation
1.	PM 314	Nila	Kidney	✔
2.	PM 315	Nila	Kidney	✔
3.	PM 316	Nila	Kidney	✔
4.	PM 317	Nila	Heart	✔
5.	PM 318	Nila	Kidney	✔
6.	PM 319	Nila	Heart	✔

Table 2. Results of isolation and purification of samples 1.

No.	Isolation Result		Purification	Incubation
	Grow	Not Grow		
1.	✔		✔	
2.	✔		✔	
3.		✔		✔
4.		✔		✔
5.		✔		✔
6.	✔		✔	

Table 3. Isolation of bacteria sample 2.

No.	Sample Code	Sample Type	Targets Organ	Incubation
1.	PM 320	Nila	Heart	✔
2.	PM 321	Nila	Kidney	✔
3.	PM 322	Nila	Heart	✔
4.	PM 323	Nila	Heart	✔
5.	PM 324	Nila	Kidney	✔
6.	PM 325	Nila	Kidney	✔

Table 4. Results of isolation and purification of samples 2.

No.	Isolation Result		Purification	Incubation
	Grow	Not Grow		
1.	✔		✔	✔
2.	✔		✔	✔
3.		✔		
4.		✔		
5.		✔		
6.		✔		

Table 5. Isolation of sample bacteria 3.

No.	Sample Code	Sample Type	Targets Organ	Incubation
1.	PM 326	Nila	Kidney	✔
2.	PM 327	Nila	Kidney	✔
3.	PM 328	Nila	Kidney	✔
4.	PM 329	Nila	Heart	✔
5.	PM 330	Nila	Heart	✔
6.	PM 331	Nila	Kidney	✔

Table 6. Results of isolation and purification of samples 3.

No.	Sample Code	Hasil Isolasi		Purification	Incubation
		Grow	Not Grow		
1.	PM 326		✔		
2.	PM 327		✔		
3.	PM 328	✔		✔	✔
4.	PM 329		✔		
5.	PM 330	✔		✔	✔
6.	PM 331		✔		

Table 7. Isolation of sample bacteria 4.

No.	Sample Code	Isolation Result	Targets Organ	Incubation
1.	PM 332	Nila	Kidney	✔
2.	PM 333	Nila	Heart	✔
3.	PM 334	Nila	Heart	✔
4.	PM 335	Nila	Kidney	✔
5.	PM 336	Nila	Kidney	✔
6.	PM 337	Nila	Kidney	✔

Table 8. Results of isolation and purification of samples 4.

No.	Sample Code	Isolation Result		Purification	Incubation
		Grow	Not Grow		
1.	PM 332	✔		✔	✔
2.	PM 333		✔		
3.	PM 334	✔		✔	✔
4.	PM 335	✔		✔	✔
5.	PM 336		✔		
6.	PM 337	✔		✔	✔

3.2. Bacteria Morphological

Characteristics Identification of Aeromonas, Plesimonas, Pseudomonas bacteria in tilapia was preceded by isolation of organs (gills, kidneys, liver, spleen, brain, eyes, scales) on TSIA media as selective media. The distribution and growth of colonies can be seen in Figures 1 and 2. The results of tilapia in sample 1, on TSIA media, the results of colonies of Aeromonas hydrophila and Plesimonas shigelloides bacteria were marked by a change in the red color of the media. In sample 2 obtained the results of bacterial colonies of Plesimonas sp and Pseudomonas putida which were marked by changes in the red and yellow media.

The results of tilapia in sample 3 obtained the results of bacterial colonies of Aeromonas hydrophila and Plesimonas shigelloides characterized by a change in the red media, while in sample 4 the results of colonies of Aeromonas sobria and Pseudomonas alkaligenes were obtained by changing the media to red.

The results of the characteristics of Aeromonas and Plesimonas bacteria have been subjected to further biochemical tests. In the indole test, Aeromonas hydrophila and Plesimonas shigelloides showed positive results and were motile. Meanwhile, in the sucrose and mannitol fermentation test, Aeromonas bacteria also gave positive results, namely being able to ferment sucrose and mannitol, but in the lactose test, negative results were obtained, namely unable to ferment lactose using TSIA media. , and the bottom is yellow which means it is acidic, and no H2S is formed. The lysine decarboxylation test against Aeromonas also showed a positive result in the form of a purple color, the 4% NaCl test gave a positive result in the form of high turbidity. Aeromonas has the highest activity in producing the enzymes lecithinase and hemolysin.

Chitinase, lecithinase, and hemolysin enzymes are an association that work synergistically, can be produced at the same time or separately. Del Coral el al., [11] stated that the main factors that determine the high pathogenicity of a disease include the type of disease-causing microorganism, weak fish immunity, and poor water environmental conditions compared to Pseudomonas sp. the ability of A. hydrophila and A. sobria to infect the host's body. This is through the mechanism of action of toxins that are released when bacteria attach to the surface of fish skin.

Characteristic results of Pseudomonas bacteria In the indole test Pseudomonas putida and Pseudomonas alkaligenes showed positive results and were motile. Meanwhile, in the sucrose and mannitol fermentation test, Pseudomonas bacteria also gave positive results, namely being able to ferment sucrose and mannitol, but in the lactose test, negative results were obtained, namely unable to ferment lactose using TSIA media. , and the bottom is red which means it is acidic, and no H2S is formed. Pseudomonas sp. including Gram negative bacteria. These bacteria are aerobic, short rod-shaped, catalase positive, oxidase positive, unable to ferment but can oxidize glucose/other carbohydrates, do not have spores, have no sheath (sheat) and have a monotric flagellum (single flagellum at the poles) so that they are always moving.

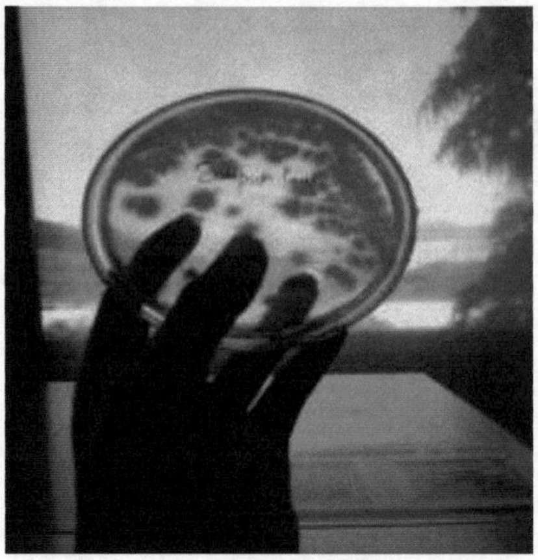

Figure 1. Colonies spread due to the presence of water.

Figure 2. Growth of bacterial colonies.

3.3. Clinical Symptoms of Bacterial Infected Tilapia

Clinical symptoms in fish are a sign of disease. This is in accordance with what was reported that Aeromonas sp. which attack freshwater fish (C. carpio) showed clinical symptoms of fin flakes, hemorrhagic, and swelling of internal organs [12,13].

Clinical symptoms in freshwater fish indicated by Pseudomonas sp. namely ulcers/wounds on the fish's body, bloating, protruding eyes (exopthalmia), dark body color, bleeding, sluggish movement, fins flaky, pale body color, gills and body surface wounds, hemorrhagic, excess mucus production, and loose scales and rough and followed by hemorrhagic which forms a white spot surrounded by a red zone, and bleeding in internal organs [14,15,16]. Several research results showed that bacteria of the genus Aeromonas and Pseudomonas attack freshwater fish including P. putida bacteria that attack tilapia with clinical symptoms of pale liver with red dots and hemorrhagic by Badjoeri [14] and Aydin *et al.*, [16].

4. Conclusion

Some of the bacteria obtained were *Aeromonas hydrophila, Plesiomonas shigelloides, Pseudomonas putida, Plesiomonas shigelloides, Aeromonas sobria* and *Pseudomonas alkaligenes*. In bacterial testing, it must be carried out aseptically to avoid contamination. Constraints faced when carrying out bacterial analysis are media contamination, which can be overcome by increasing accuracy in the manufacture of media.

5. References

[1] Agustono, A, Wicaksono, R, & Lokapirnasari, W P 2013 *J Ilmi Peri Dan Kel.* 5 2 201–204.
[2] Hayati, A, Nurbani, F A, Amira, M, Seftiarini, W, Wanguyun, A P, & Muchtaromah, B 2020 *Asi J of Wat Envi and Poll. 17* 3 61-64.
[3] Farizky, H S, & Satyantini, W H 2019 *IOP Conf Ser: Ear and Envi Sci. 236* 1 012118.
[4] Putra, T F, Suprapto, H, Tjahjaningsih, W, & Pramono, H 2018 *IOP Conf Ser: Ear and Envi Sci. 137* 1 012060.
[5] Rahmadini, S, Sudarno, & Subekti, S 2019 *IOP Conf Ser: Ear and Envi Scie. 236* 1 012078.
[6] Austin, B and Austin, D A 2007 *Bacterial Fish Pathogens. Disease in Farmed and Wild Fish Fourth edition* Ellis Horword Limited Chichester England 552 p.
[7] Kamiso HM, Triyanto 1993 *Direct of Res and Com Ser.* 23-30.
[8] Elisafitri, M, Satyantini, W H, Arief, M, & Sulmartiwi, L 2021 *IOP Con Ser: Ear and Envi Scie. 718* 1 012050.
[9] Rukmana, N R, Mahasri, G, Hidayah, S. N, Ulkhaq, M. F, & Kenconojati, H 2019 *IOP Conf Seri: Ear and Envi Scie. 236* 1 012107.
[10] Buller, B, N 2004 *Bacteria from fish and other aquatic animals: a practical* identification *manual.* CABI Publishing Massachusetts Avenue Cambridge 358 pp.
[11] Del Coral, F Shotts, E.B, & Brown, J 1990 *Jou of Fis Dise.* 13 255- 268.
[12] Camus A, Durborow R, Hemstreet W, Thune R, Hawke J 1998. *Jour of fis Disea.* 13-15.
[13] Hambali S, Widiyati, Sunarto, Heru 2005 *Indo Jour of Fish Res.* 11 34-37.
[14] Badjoeri, M 2008 *Jour Peri.*34 2 169-184.
[15] Anugarah, 1990 *Potential and Development of Aquaculture in Indonesia* Indonesian Research Institute Jakarta.
[16] Aydin, S, Ciltas, A, and Erman, Z 1998 *Symp Proc.* 8 p.
[17] Arimardewi O, Restu W,and Saraswati S A. 2018 A Preliminary Study Of Lead And Cadmium Content In Water And Tilapia (Oreochromis Niloticus, Linn.) For Water Quality Assessment Of Telaga Tunjung Dam Bali. Metamorphosis Journal, 85-93

6. Acknowledgement

The authors are grateful to University of Nusa Cendana Kupang and Udayana University for all support to this research. We thanks to the reviewers for their suggestion, which improved our manuscript.

IOP Conf. Series: Earth and Environmental Science 1273 (2023) 012054 doi:10.1088/1755-1315/1273/1/012054

The Effectiveness of selective breeding on the productivity of giant freshwater prawns (*Macrobrachium rosenbergii*) population

L E Hadie[1,2], E K M Kontara[1], Kusdiarti[1], R Zulkarnain[1]

[1]Research Centre for Fishery, Agency for National Research and Innovation, Cibinong- Bogor, Indonesia

[2]E-mail: emmalitbang@gmail.com

Abstract. Selection programs for giant freshwater prawns stocks from various locations have been implemented by several government institutions. Some of the results of selective breeding of giant freshwater prawns have been released by the Indonesian Ministry of Marine Affairs and Fisheries. Several giant prawn strains have been released, including GIMacro, GIMacro II, and Siratu. The development of giant prawn strains after release has not been significantly observed among shrimp farmers. Therefore, it is necessary to evaluate its productivity in order to observe its impact on the development of giant prawn aquaculture. The study was conducted using a meta-analysis method with data sources derived from searching the database on Google.com. The result s of this study indicate that selective breeding of giant prawn can successfully increase the productivity of their farming ($P < 0.001$). It has also been demonstrated that the effectiveness of family selection significantly accelerates the growth rate of giant prawns. It is anticipated that the findings of this study will help formulate policy solutions for reversing the current drop in giant prawn production.

1. Introduction

Giant freshwater prawns (*Macrobrahium rosenbergii*) are one of the leading freshwater fishery species that can be relied upon as a source of income for the farming community. Technology for growing giant prawns has advanced pretty quickly in the area. This is evidenced by the existence of centres for the cultivation of giant freshwater prawns, among others, in the district of Sleman, the Special Region of Yogyakarta, the provinces of West of Java, East of Java, Centre Java, South of Kalimantan, West Kalimantan, Banten, Aceh, island of Bali [1]. Giant freshwater prawns have a comparative advantage as a centre for geographic distribution and potential ecosystem diversity in cultivation [2,3]. Potential lands for the cultivation of giant freshwater prawns include ponds, rice fields, ponds field, and freshwater ponds.

Giant freshwater prawns can be cultivated in monoculture, polyculture with rice known as the giant freshwater prawn and rice polyculture system, or polyculture with other freshwater fish. The FAO has suggested that other agrarian nations establish the rice – freshwater prawn culture, a growing method for giant prawns that is based on the blue economy. For the rice – freshwater prawn culture, which the agricultural community has successfully adopted, Indonesia has emerged as a role model. Several regions, including Sleman, Yogyakarta, Central, and East Lombok, have seen the development of the rice – freshwater prawn culture [4,5].

The development of giant prawn cultivation will be able to develop optimally if it is supported by hatchery technology that is able to produce shrimp seeds continuously. Shrimp seed production requires a large number of quality brooders [6]. Shrimp seed production requires a large number of quality brooders [7,8].

However, the need for parent prawn for seed production cannot rely on nature alone. The lack of appropriate quality and quantity of giant prawns is now impeding the development of giant prawn production. This is a significant issue because, without good brooders, it will be impossible to produce the volume of superior seeds required for giant prawn aquaculture on a national scale. This issue demands immediate response. The amount of giant prawns produced will decrease if it is not handled. A series of prawn breeding programs have been initiated by several government institutions and universities. The results of breeding giant freshwater prawns have also produced superior strains that have been released a department of the Ministry of Marine Affairs and Fisheries. Several superior strains that have been released include giant freshwater prawns GIMacro, GIMacro II, and Siratu [9,10,11]. Some of these superior strains of giant freshwater prawns are genetic resource assets that can be utilized to optimize comparative advantage into competitive advantage [12]. However, there are obstacles to exploiting this comparative advantage of giant freshwater prawns.

It has not been possible to successfully transfer to the aquaculture community the mass production of superior giant prawns that have been launched. A comprehensive investigation is necessary to pinpoint the areas where these better breeds are not being used to their full potential.

The goal of this study was to assess how well various superior strains produced and how well selective breeding strategies supported the production of giant prawns. The consistency of seed production will assure the growth of giant prawn aquaculture thanks to the outstanding broodstocks productivity performance, it has both quality and quantity assurance.

2. Materials and Methods

2.1. Method for selecting and using the literature

A review of the literature was done. through a literature database on Google.com. The search was conducted using the following search terms : selective breeding, giant freshwater prawn, strains, growth rate, phenotypic plasticity, culture. As many as 102 references were disqualified based on the initial title screening since the subject matter wasn't pertinent to the study. Twenty five of the 38 were chosen after the abstracts were reviewed. After screening, 12 publications were found that may be used for later data coding and statistical data analysis.

2.2. Coding studies

For the study's initial data and coding, some moderator variables were used. These factors were the research location, institution type, and strain type. A thorough analysis of the 13 chosen publications produced a total of 25 studies. For subsequent data analysis, the units of measurements were homogenized together with the mean value and standard deviation from the growth rate parameters.

2.3 Analytical statistics

The parameter distance between the two was quantified using the effect size as' hedges' d', selected and farmers' or synthetic stock of giant freshwater prawn. The farmers'/synthetic stock group was combined to form a control group (C), whereas the chosen group was combined to form an experimental group (E). Statistical analysis using open-source MEE [13].

Table 1. A list of the giant freshwater prawn comparison studies that were used in the meta-analysis.

Code study	Location	Kind of institution	Kind of strains	Xc	Xe	References
1	Sukamandi-Subang, West Java	Research	GIMacro	8.59	11.96	14
2	Sleman, Yogyakarta	University	Farmers stocks	11.1	14.15	15
3	Sleman, Yogyakarta	University	Farmers stocks	11.7	14.05	16
4	Sukabumi, West Java	University	Siratu	6.73	13.32	17
5	Sukabumi, West Java	Research	GIMacro II	10.61	12.15	17
6	Sukamandi-Subang, West Java	Research	Synthetic stock	8.99	11.93	18
7	Aceh, Aceh	University	Farmers stocks	11.06	14.0	18
8	Ciamis, Sukamandi, West Java	Research	Synthetic stock	12.75	13.5	19
9	Soppeng, South Sulawesi	Research	GIMacro	5.18	18.24	20
10	Soppeng, South Sulawesi	Research	GIMacro	8.69	17.6	20
11	Soppeng, South Sulawesi	Research	GIMacro	9.48	16.8	20
12	Soppeng, South Sulawesi	Research	GIMacro	7.66	14.02	20
13	Sukamandi-Subang, West Java	Research	GIMacro	5.06	11.88	21
14	Sukamandi-Subang, West Java	Research	GIMacro	4.46	8.35	22
15	Sukamandi-Subang, West Java	Research	GIMacro II	11.49	14.37	23

Xc: giant freshwater prawn growth rate for farmers/synthetic stocks
Xe: growth rate of giant freshwater prawn for strains based on selection

3. Results and discussion

3.1. Result

The results of the meta-analysis show the effect of the selection program on the growth rate of giant prawn characters as shown in Table 2 and Table 3 below. The forest plot of giant freshwater prawn (*Macrobrachium rosenbergii*) growth rate performance for the selection program can be shown in Figure 1.

Table 2. Effect size on the growth rate character of giant prawns (*Macrobrachium rosenbergii*) as a result of the selection program compared to farmers/synthetic stock.

Estimate	Lower bound	Upper bound	Standard error	p-value
1.887	1.350	2.425	0.274	< 0.001

Table 3. Population heterogeneity of giant prawns (*Macrobrachium rosenbergii*) as a result of the selection program on growth rate characters.

τ^2	Q (df=14)	Het. p-value	I^2
0.967	126.746	< 0.001	88.954

The forest plot of giant freshwater prawn (*Macrobrachium rosenbergii*) growth rate performance from the selection program can be seen in Figure 1.

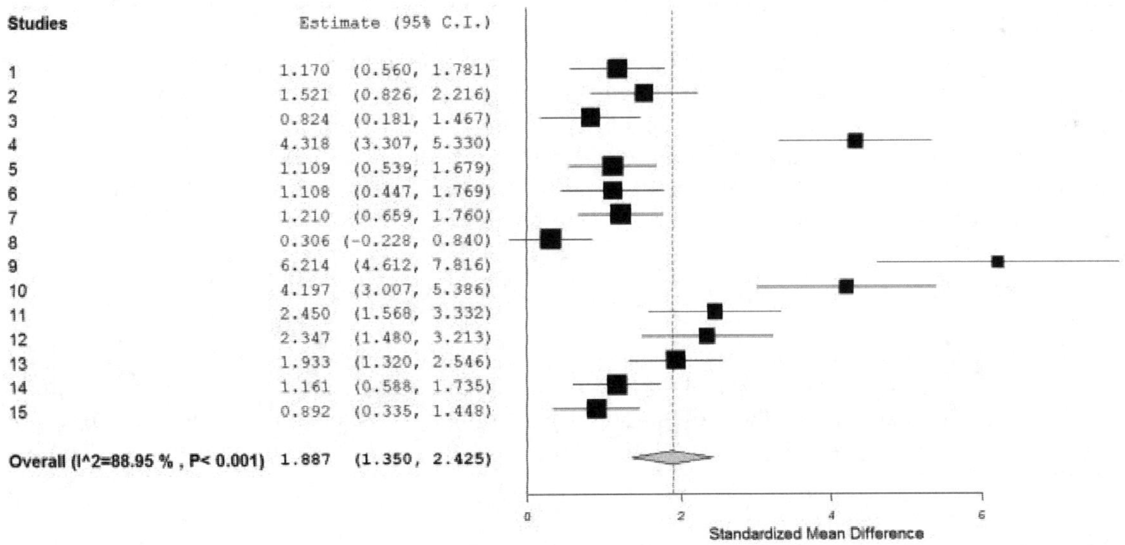

Figure 1. Comparing specific strains and farmer's stocks, a forest plot of the cumulative effect size (d++) and 95% confidence interval (CI) of the growth rate parameter was created (synthetic).

IOP Conf. Series: Earth and Environmental Science 1273 (2023) 012054 doi:10.1088/1755-1315/1273/1/012054

3.2. Discussion

The selection program for giant freshwater prawns has been applied with various selection techniques. Some breeding research uses individual selection, family selection, or a combination of several selection techniques [24]. Likewise, the formation of the basic population in the selection program also shows variations in several government and university institutions. The basic population formed will determine the response that can be achieved through the selection program. The analysis's findings are shown in Table 2 and Table 3 showed that selection program applied to the growth rate character of giant freshwater prawns had a significant effect with confidence level 95% ($P< 0.001$).

Some of the locations used in this giant prawn selection program are quite varied in conditions. However, the selection applied to the growth rate character of giant freshwater prawns still showed effective results. This condition can occur because the selection on the growth rate character affects the frequency of gene to giant prawn population. The results obtained are in accordance with the statement of [25,26], which states that the selection function on certain characters will change how frequently genes occur that regulate several qualitative and quantitative traits. Environmental differences in each population will also affect the quantitative traits of the production character. For example, in a bad environment, a character that has a plus gene frequency cannot be expressed optimally. This condition can be seen in the results of the forest plot in the selection of giant freshwater prawns (Fig. 1).

The GIMacro strain showed significant advantages at locations in Sukamandi (West Java) and Soppeng (South Sulawesi). The Siratu strain proved to be superior in Sukabumi (West Java). This can happen because it is related to the adaptability of the giant prawn population, which has undergone changes in gene frequency. This adaptability is formed because a population will maintain its life by flexing other characteristics. This is known as phenotypic plasticity [27,28,29].. Phenotypic plasticity is a phenomenon of adaptation mechanism that involves a number of genes that control the average character in a certain environment [30].

This is in accordance with the research that has been carried out on giant freshwater prawns, which shows the presence of phenotypic plasticity on growth characters in a saline environment. Giant freshwater prawns grew well in media with salinities ranging from 0 to 10‰ and declined at 15‰ salinity. This shows the potential for salinity-tolerant genes in giant freshwater prawns [21,31]. Potential genes that control the growth rate of giant freshwater prawns in different environments can be exploited in developing giant prawn culture in land ponds and ponds with salinities of 0–10‰.

4. Conclusion

The study's findings suggest that selective breeding of giant prawns can successfully increase the productivity of their farming ($P< 0.001$). It has also been demonstrated that the effectiveness of family selection significantly accelerates the growth rate of giant prawns. It is anticipated that the findings of this study will help formulate policy solutions for reversing the current drop in giant prawn production.

5. References

[1] Directorate General of Aquaculture 2013 *Book of Aquaculture Statistics Ministry of Marine Affairs and Fisheries Republic of Indonesia*

[2] Hadie L E, Hadie W, Praseno O 2001 The geographic distribution and ecological characteristics of giant freshwater prawns *Proceedings of the Workshop on Research Results of Giant Prawn Cultivation*

[3] Indarjo A, Salim G, Nugraeni C D, Prakoso L Y, Soejarwo P A, Rukisah R, Pham Y T H, Daengs G S, A Hariyadi and Jabarsyah A 2020 *J Ilm Perikan Dan Kelaut*., 12(2), 236–249.
[4] Tahir A G and Pasaribu A M 2003 *J Agric Technoml Assess Dev.*
[5] Sarifin S, Sutomo H, Wibowo KT, Rosellia S 2013 *The cultivation of giant prawns, rice and fish Proceedings of the aquaculture innovation and technology forum*
[6] Wiguno T A A, and Dewi N N 2022 *IOP Conf Ser Earth Environ Sci.* 1036(1) 012101.
[7] Wardani A I, Sari L A, Sari P D W, Nindarwi D D, and Arsad S 2021 *IOP Conf Ser Earth Environ Sci. 718*(1). 012016.
[8] Nashrullah S, Sari L A, and Arshad S 2022 *IOP Conf Ser Earth Environ Sci. 1036*(1). 012064.
[9] Ministry of Marine Affairs and Fisheries Republic of Indonesia (MMAF-RI) 2001 Decree of the. Minister of Marine Affairs and Fisheries Number Decree 42/MEN/2001 regarding the release of giant freshwater prawns as superior varieties (GIMacro).
[10] Ministry of Marine Affairs and Fisheries Republic of Indonesia 2014 Decree of the Minister of Marine Affairs and Fisheries Number Decree 23/MEN-KP/2014 regarding the release of giant freshwater prawns as superior varieties (GIMacro II).
[11] Ministry of Marine Affairs and Fisheries Republik Indonesia 2015 Decree of the Minister of Marine Affairs and Fisheries.Number Decree.25/KEPMEN-KP/2015 regarding the release of giant freshwater prawns as superior varieties (Siratu).
[12] Samara S H, Fathurrozi A W, and Sutarno S 2020 *IOP Conf Ser Earth Environ Sci.* 441(1), 012006.
[13] Wallace, Byron C, Marc JL, George D, Issa J D, Thomas AT, Christopher HS, and Jessica G 2016 Open MEE *Intuitive, open-source software for meta analysis in ecology and evolutionary biology Methods in Ecology and Evolution.*
[14] Kasani I, Imron , Suprapto R, Himawan Y 2010 *Proc Aquac Technol Innov.* 581-590.
[15] Priyono S B, Sukardi S, Harianja BSM 2011 *J Fish Sci.* 13 2 78-85.
[16] Herawati M 2018 *Natl Semin Proc.* 8 184-194.
[17] Manurung A P, Indah A Y I A, Haris RBK 2018 *J Fish Aquac Sci.* 13 1.
[18] Putri F P, and Dewi N N 2019 *IOP Conf Ser Earth Environ Sci. 236*(1) 012016.
[19] Hadie L E and Hadie W 2012 *Biol News.* 11(2) 211-219.
[20] Zamri 2015 *Postgraduate Thesis* Open University Jakarta.
[21] Hadie L E, Hadie W, Kusmini I I, Jaelani J, Sularto S, and Hikmayani Y 2000 *Indones Fish Res J.* 6 3-4.
[22] Hadie W,Sumantadinata K, Noor R R, Subandriyo S, Carman O, and Hadie L E 2017 *Indones Fish Res J.* **5** 33 – 46.
[23] Hadie L E, Hadie W, Praseno O 2001 *The geographic distribution and ecological characteristics of giant freshwater prawns Proceedings of the Workshop on Research Results of Giant Phenotypic plasticity of giant freshwater prawns (Macrobrachium rosenbergii) strains of Musi, Prawn Cultivation.*
[24] Sopian A, Krettiawan H, Anggraeni F 2014 *Proc Aquac Technol Innov forum.*
[25] Warwick E J, Astuti J M, Hardjosubroto W 1995 *Livestock Breeding* Gadjah Mada University Press Yogyakarta Indonesia 152-178.
[26] Falconer D S and Mackay T F C 1996 *Introduction to Quantitative Genetics* Longman Malaysia 184 – 201.
[27] Swaina D P and Foote C J 1999 *Fish Res* 43 113-128.
[28] Noor RR 2004 *Livestock Genetics* Third printing Penebar Swadaya Jakarta.
[29] Hadie W and Lystianto N 2006 *Aquac Media.* 1 2 49 – 54.
[30] Agil M, Kusumawati I, and Purwitasari N 2017 *J Bot.* 2017 7232171.
[31] Putranto T W C, Andriani R, Munawwaroh A, Irawan B, and Soegianto A 2014 *Mar Freshw Behav Physiol.* 47(5) 349-360.

Evaluation of body performance of tilapia (*Oreochromis niloticus*)

Wahidah[1,3], A Rusli[2], Hartinah[1], Amrullah[1], I Dwinhoven[1]

[1]Department of Aquaculture, Pangkep State Polytechnic of Agriculture, Pangkep, South Sulawesi, Indonesia
[2]Department of Processing Technology, Pangkep State Polytechnic of Agriculture, Pangkep, South Sulawesi, Indonesia

[3]Corresponding author: ida_wahidah@yahoo.co.id

Abstract. Efforts to improve the quality of fish seeds can be done by using quality broodstock which includes having the right body shape. This study aimed to evaluate the body performance of male and female Sultana strain tilapia (*Oreochromis niloticus*) which were to be used in hatchery activities. A total of 35 male and 35 female broodstock were evaluated for body size using a ruler. Measurements of 12 morphological characters: total length (TL), standard length (SL), head length (HL), pre-dorsal length (PrDL), pre-pectoral length (PrPL), pre-pelvic length (PrPeL), pre-pelvic length (PrPeL), anal length (PrAL), anal fin length (AFL), dorsal fin base length (DFL), body height (BH), caudal peduncle height (CPH), and dorsal anal length distance (DAL). The ratio between the character values of SL, HL, PrDL, PrPL, PrPeL, PrAL, AFL, DFL, BH, CPH, DAL, and TL were analyzed descriptively. The results of the analysis showed that the average value of the ratio of length characters (SL/TL, HL/TL, DF, and AFL/TL), height character (BH/TL), and diagonal character (DAL/TL) were higher in the male broodstock than in the female broodstock. On the other hand, other length characters (PrDL/TL, PrPL/TL, PrPeL/TL, PrAL/TL) and height characters (CPH/TL) had the same values in the male and female broodstock. In general, the size of the male broodstock used tended to be larger than the female broodstock.

1. Introduction

Efforts to improve fish seed quality can be conducted through the use of high-quality broodfish such as using broodfish with ideal/appropriate and superior body size. Broodstock improvements have undergone rapid advancements. Several studies that have produced various strains include BEST (Bogor Enhanced Strain Tilapia) tilapia, Nirwana II tilapia (Tilapia Ras Wanayasa), Jati Umbulan tilapia (Umbulan East Java), and Sultana tilapia (Salabintana Superior Selection), each strain having its advantages. The Sultana strain has advantages such as a fairly high growth rate. [1,2,3] stated that tilapia generally reach a size of 250 g in 5 - 6 months, but Sultana tilapia only requires 3-4 months; its growth is 40% faster. In addition, its survival can reach 70-90% [4,5].

The government has released several tilapia strains superior in growth as a result of breeding activities. The superior tilapia produced was used as broodstock in hatchery activities. The fish

IOP Conf. Series: Earth and Environmental Science 1273 (2023) 012055 doi:10.1088/1755-1315/1273/1/012055

hatchery program is conducted by taking the performance of the fish to be used into account; therefore, it is necessary to evaluate the phenotypic performance. Measurements of the fish phenotype can be done through morphometric measurements. Morphometrics is a field that studies variations and changes in shape, namely the size and shape of an organism [6,7,8]. Morphometric evaluation of fish reveals relationships between body parts, such as the length of the head, snout, eyes, body, fins, and tail. Morphometrics also describes the distance between different points or landmarks [9,10,11]. These landmarks are closely related to the fish's body structure and are easily observed by sight [5].

The problem encountered in tilapia is a decrease in the quality of superior strains due to mismanagement or inappropriate broodstock selection (related to morphology). To overcome this, it is necessary to evaluate the performance of the prospective broodstock that will be used to produce seeds. This study aimed to evaluate the body performance of Sultana strain male and female tilapia (*Oreochromis niloticus*) broodstock to be used in hatchery activities.

2. Material and methods

The research was conducted in June 2022. The broodstock of strain Sultana tilapia used came from the Sukabumi Center for Freshwater Aquaculture (Balai Besar Perikanan Budidaya Air Tawar/BBPBAT), which had been prepared to be a prospective broodstock.

Measurements of the morphometric characters using a ruler (cm) were taken on 35 male and 35 female broodstock. Measurements of 12 morphological characters included total length (TL), standard length (SL), head length (HL), pre-dorsal length (PrDL), pre-pectoral length (PrPL), pre-pelvic length (PrPeL), pre-pelvic length (PrPeL), anal length (PrAL), anal fin length (AFL), dorsal fin base length (DFL), body height (BH), caudal peduncle height (CPH) and dorsal anal length distance (DAL) (Figure 1).

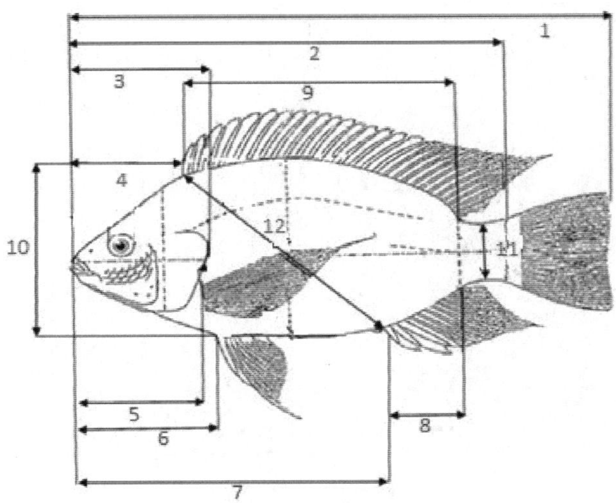

1 : total length
2 : standard length (SL)
3 : head length (HL)
4 : pre-dorsal length (PrDL)
5 : pre-pectoral length (PrPL)
6 : pre-pelvic length (PrPeL)
7 : pre-anal length (PrAL)
8 : anal fin length (AFL)
9 : dorsal-fin base length (DFL)
10 : body height (BH)
11 : caudal peduncle height (CPH)
12 : dorsal-anal length distance (DAL)

Figure 1. Morphological Characters of Tilapia

The morphometric character data (SL), (HL), (PrDL), (PrPL), (PrPeL), (PrAL), anal fin length (AFL), (DFL), (BH), (CPH), and (DAL) were divided by TL, resulting in the tilapia character ratio. Morphometric character ratio data were analyzed descriptively.

3. Results and Discussion

The results of the analysis showed that the average value of the ratio of length characters (SL/TL, HL/TL and DFL/TL, and AFL/TL) (Figure 2), the height character (BH/TL) (Figure 3), and the diagonal character (DAL/TL) (Figure 4) were higher in the male broodstock than in the females. Meanwhile, other length characters (PrDL/TL, PrPL/TL, PrPeL/TL, and PrAL/TL) (Figure 2) and height characters (CPH/TL) (Figure 3) had similar values in the male and female broodstock.

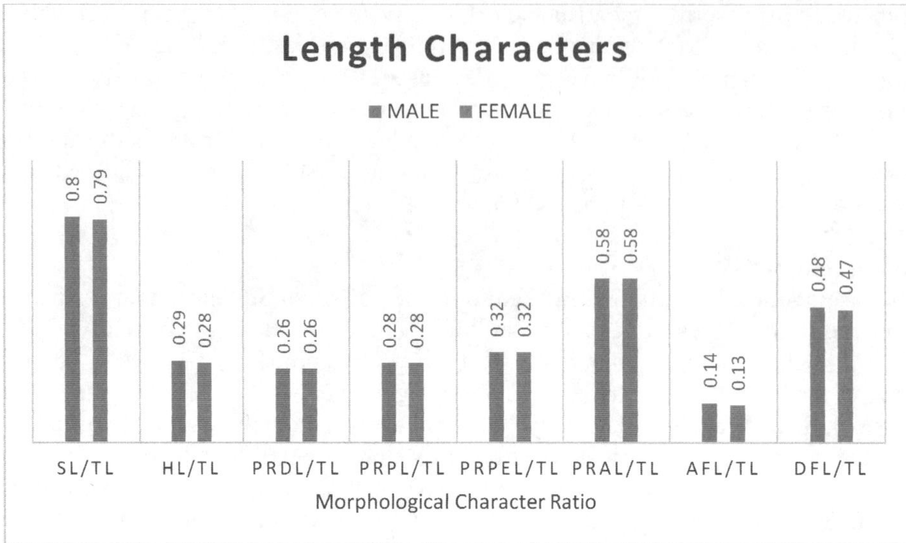

Figure 2. Average Value of the Ratio of Length Characters of Tilapia

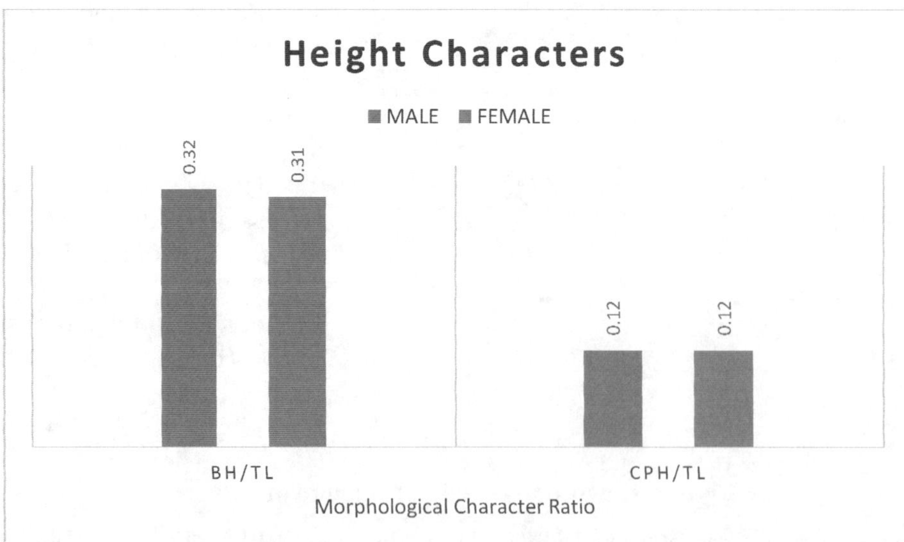

Figure 3. Average Value of the Ratio of Height Characters of Tilapia

IOP Conf. Series: Earth and Environmental Science 1273 (2023) 012055 doi:10.1088/1755-1315/1273/1/012055

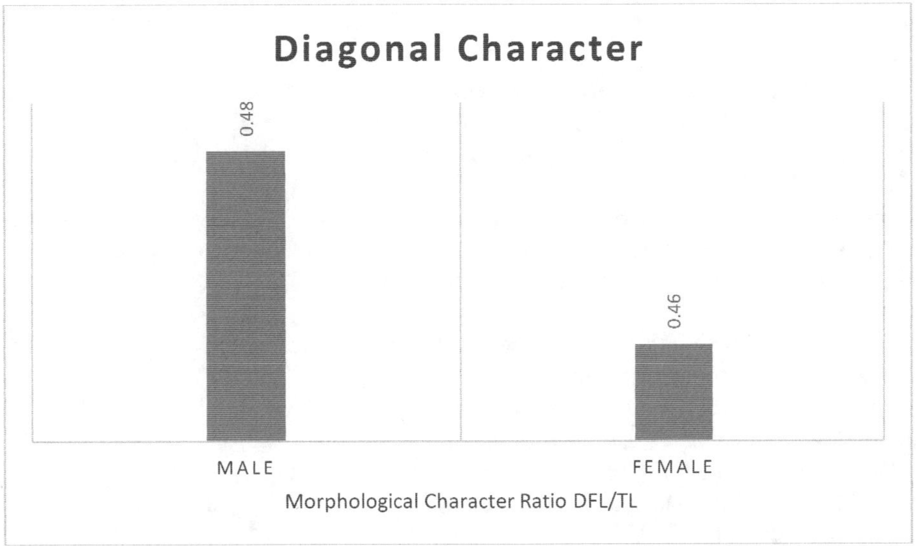

Figure 4. Average Value of the Ratio of Diagonal Character of Tilapia

Evaluation of the morphometric character ratio in the strain Sultana tilapia to be used showed that it was following the Decree of the Minister of Maritime Affairs and Fisheries of the Republic of Indonesia Number Kep.28/Men/2012 for Sultana tilapia, with a morphometric description of total length (cm) 20.48 ± 0.80, standard length (cm) 17.08 ± 0.80, head length (cm) 5.57 ± 0.25, the distance between anal fin to the tip of mouth (cm) 11.58 ± 0.53, caudal peduncle height (cm) 2.38 ± 0.11; body height (cm) 6.80 ± 0.33; standard length/height ratio 2.51 ± 0.11. These values showed that the parent morphometric ratio obtained from the analysis approaches the ratio of the morphometric descriptive characters to the total length. The ratio values between descriptive morphometric characters and the analysis results in the present study were: standard length (0.83 and 0.80); head length (0.27 and 0.28); caudal peduncle height (0.11 and 0.12); body height (0.33 and 0.32). Strain Sultana tilapia has been under development by the Sukabumi Center for Freshwater Aquaculture (BBPBAT) since 2001. This tilapia variety is recognized by the MMAF through the issuance of the Decree of the Minister of Marine Affairs and Fisheries Number KEP.28/MEN/2012 concerning the Release of Strain Sultana Tilapia on 7 June 2012 [12].

This study showed that, in general, the ratio value of several morphometric characters in males was higher than in females, especially in the body, while the length character in the head showed the same value between males and females. This shows that the ratio of the length of the head of the male tilapia tended to be larger than that of the female.

The broodstock weight measured in males ranged from 330-566 g/individual and 246-387 g/individual for females. According to [13], tilapia can spawn at a weight of 250 g and the age of 4 - 6 months. This indicates that the broodstock used complied with the requirements for broodstock. The broodstock reared was healthy with no morphological defects [14].

4. Conclusion

The morphological ratio values of Sultana tilapia which will be used in hatchery activities are almost the same as the morphological ratio value of the description of Sultana tilapia released by the MMAF. In general, the size of the male broodstock used tended to be larger than the female broodstock.

5. References

[1] Kurniawan A, Riana 2015 These are the advantages of the sultana fish compared to other fish.

[2] Anggraeni I, and Sari P D W 2020 *Eart and Enviro Scie* 441 (1) 12133.

[3] Triastuti J, Kintani D, Luqman E M, and Pujiastuti D Y 2018 *Ear and Enviro Scie* 137 (1) 12023.

[4] Sukabumi Center for Freshwater Aquaculture (BBPBAT), West Java 2015 Introduction to the Salabintana Superior Selection Sultana strain tilapia. Sukabumi(ID): BBPBAT.

[5] Cahyani R, Satyantini W H, Nindarwi D D, and Cahyoko Y 2021 *Ear and Enviro Scie* 679 (1) 12042.

[6] Webster M 2006 Introduction to geometric morphometrics. Department of the Geophysical Sciences, University of Chicago.

[7] Kurniawan D, and Fasya A H 2021 *Ear and Enviro Scie* 679 (1) 12039.

[8] Asmamaw B, and Tessema M 2021 *J Aquac and Fis Heal* 10(3) 341-355.

[9] Simon K D, Bakar Y, Temple S E, Mazlan A G 2010 *J Zheji Univ Sci B* 11 871-879.

[10] Hayati A, Pramudya M, Supriyanto A, Nurhariyati T, Zahra P F, and Hayaza S 2021 *Ear and Enviro Scie* 718 (1) 12005.

[11] Mukti A T, Carman O, Alimuddin A, and Suprayudi M A 2020 *Turk Jour of Veter & Anim Scie* 44 (2) 290-298.

[12] Kosai P, Sathavorasmith P, Jiraungkoorskul K, Jiraungkoorskul W 2014 *Walailak J Sci & Tech* 11 (10) 857-863.

[12] Ministry of Marine Affairs and Fisheries (MMAF) 2012 Decree of the Minister of Maritime Affairs and Fisheries of the Republic of Indonesia Number Kep.28/Men/2012 regarding the Release of Sultana Tilapia.

[13] Polonia H 2015 Red tilapia (*Oreochromis niloticus*) hatchery techniques [Final assignment]. Sorong (ID): Sorong Marine and Fisheries Polytechnique.

[14] National Standardization Agency (BSN). Indonesian National Standard (SNI) 6141:2009. Production of stocking class black tilapia (*Oreochromis niloticus* Bleeker) seed. http://sisni.bsn.go.id.

6. Acknowledgments

We thank the Directorate of Resources, Directorate General of Higher Education, Ministry of Education and Culture, Research and Technology for the financial support for this research through Penelitian Produk Vokasi Unggulan Perguruan Tinggi (PPVUPT) scheme, based on the Contract Number 095/PL.22.7.1/SP-PG/2022.

IOP Conf. Series: Earth and Environmental Science 1273 (2023) 012056 doi:10.1088/1755-1315/1273/1/012056

Intensive enlargement management of vannameii shrimp (*Litopenaeus vannamei*) in Banyuwangi

A R Agustin[1] **P D W Sari**[2,3]
[1]Study Program of Aquaculture, Faculty of Fisheries and Marine, Universitas Airlangga, Surabaya 60115, Indonesia
[2]Department of Aquaculture, Faculty of Fisheries and Marine, Universitas Airlangga, Surabaya 60115, Indonesia

[3]Corresponding author: putri.dw@fpk.unair.ac.id

Abstract. Vannamei shrimp (*Litopenaeus vannamei*) becomes main commodities aquaculture productions in Indonesia which has high economic value. The production strategy and management of vannamei shrimp culture must be considered and developed mainly at the enlargement stage to produce a consumption measure and export size. Feeding management was done by blind feeding program and demand feeding with ancho as monitoring tool, also biomass sampling control. The frequency of feeding is 2-5 times a day with a distance of 3-4 hours. Water quality monitoring is carried out in situ and ex situ including 7.6-8.1 in pH, temperature was 26-28°C, brightness was 28-40 cm, dissolved oxygen was 3.76-4.66 ppm, salinity was 27-29 ppt, and water height was 110-140 cm.

1. Introduction

Vannamei shrimp as the main commodities aquaculture productions in Indonesia which has high economic value. Based on the Directorate General of Strengthening the Competitiveness of Marine and Fishery Products (2021), the export value of Indonesian fishery products in June 2021 has reached 2.6 billion USD with shrimp as the main export product worth 1 billion USD or 40.1% of the total export value. Shrimp fishery products, especially vannamei shrimp, have become a priority program that can drive the Indonesian economy. Therefore, the production strategy and management of vannamei shrimp culture must be considered and developed, especially at the rearing stage to produce the expected consumption and export sizes.

Vannamei shrimp production strategy must be considered both in brood stock rearing, hatchery, and rearing activities. Enlargement is one of the stages of aquaculture production to produce the consumption size and export size of vannamei shrimp. The productivity of the super-intensive vannamei shrimp (*Litopenaeus vannamei*) aquaculture system can reach more than 50 tons per hectare of land [1]. One of the pond systems for vannamei shrimp (*Litopenaeus vannamei*) cultivation is a super intensive aquaculture system.

The super-intensive vannamei shrimp aquaculture system puts forward the concept of low volume high density, which produces high productivity without a large area of land with the application of a stocking density range of 400-600 fish/m [2]. At the grow-out stage, farmers strive to produce

vannamei shrimp with maximum growth in a short maintenance time and optimize operating costs [3]. Factors that must be considered at the rearing stage of vannamei shrimp include the application of aquaculture systems, selection of aquaculture sites, cultivation land management, water quality management, feed management, WWTP management (Wastewater Treatment Installation), plankton diversity, bacterial abundance, and disease control management cause of production failure.
Standard operating procedures for vannamei shrimp rearing management that have been implemented by several aquaculture companies cannot be a sustainable reference for a strategy to increase production but can be a foundation that needs to be developed. This is because there are still many problems that arise without or with a prediction that can be identified from the water quality data on vannamei shrimp farming environments.

2. Material and methods

This reserch was conducted at vannamei shrimp in Bomo, Banyuwangi, East Java, Indonesia Java in Augustus – December 2021. This research used descriptive methods which primary and secondary data were collected. Data on the results of water quality measurement is carried out in situ and ex situ tables and are described.

3. Result and discussion

3.1. Pond preparation

The vannamei pond preparation includes drying the pond bottom, and cleaning the remaining mud and barnacles that have died at the pond bottom. Preparation of vannamei pond lasts for 3-4 weeks. The pond bottom drying is accompanied by the application of chlorine in the amount or dose adjusted to the pond bottom area to kill the remaining germs from the previous cycle. Mud cleaning is carried out by direct extraction and followed by spraying using a diesel engine and spiral hose with reservoir water to remove the remaining mud that is not collected.

The water source used comes from Bomo Beach seawater which is taken through suction with pipes and underground pumps. The reservoir filtration system used is a physical filtration system, a chemical filtration system and a biological filtration system. A physical filtration system is applied by placing seawater on a plot and allowing it to undergo sedimentation of organic matter which will be decomposed by decomposing bacteria. Biological filtration system was applied using green mussels, gastropods, barnacles, milkfish, snapper, and grouper. Chemical filtration system with the addition of H_2O_2 and molluscicide chemicals. H_2O_2 is applied to the reservoir for water sterilization at a dose that is safe for organisms. Meanwhile, molluscicide was applied to the reservoir as a molluscicide to kill mollusc organisms that were not expected to be present in the vannamei shrimp rearing plot system because there would be competition for feed and habitat with white shrimp.

3.2. Installation of the aerator

A total of 4 aerators with a power of 1 HP can move 4 propellers and are placed at each corner of the pond plot. At more than 30 days of DOC until harvest time, the aerators applied are 10 aerators and 4 aerators single chain (ankle).

3.3. Plankton growth

Fertilizers used in vannamei shrimp cultivation are inorganic fertilizers including KCL, NPK and urea which contain compounds of carbon (C), nitrogen (N), phosphorus (P) and potassium. Fertilizers are used to grow beneficial phytoplankton and zooplankton species in the waters. One of the factors that influence the abundance and dominance of plankton is the nutrients including Carbon (C), Nitrogen (N), Phosphorus (P), Potassium (K) and Silicate (S).

3.4. Fry Stocking

The stocking of the fry was carried out three days after the fertilization process where plankton had grown to dominate the waters as one of the natural food sources for the vannamei shrimp fry. Prior to stocking, the fry was acclimatized for 30 minutes on the outskirts of the pond bordered by a bamboo. The acclimatization process was carried out by observing the swimming activity of the fry after being stocked.

3.5. Feed Management and Growth Monitoring

Feed management for vannamei shrimp rearing consists of two stages, namely blind feeding, and demand feeding. Blind feeding is the initial feeding based on the estimated ABW (Average Body Weight) or the average body weight of shrimp and biomass. Demand feeding is a follow-up feed based on ancho adjustment and biomass sampling. For DOC 1-6 the feed used was Tomboy O with a flour-like form, DOC 7-23 the feed used was Lorica 1 with a small crumble form, DOC 23-42 the feed used was Gamma 2 with a medium crumble form, DOC 42-68 the feed used was Gamma 3 in the form of pellets measuring 1.2 mm and DOC 69 until harvest (DOC 100-120), the feed used was Gamma 4 with pellets measuring 1.4 mm. The proximate protein content in Lorica and Gamma feeds had a significant difference where Lorica feed for early DOC had a higher protein content. Feeding during the blind feeding period was carried out by spreading evenly over the plot area using a pond boat.
Artificial feed is a feed that is intentionally made by humans with nutritional formulations that are suitable for the needs of aquaculture commodities [4]. Crude fibres and fat are the main energy sources for vannamei shrimp with optimal fat content of 4.5% and the optimal content of crude fibre is not more than 6% [5]. The feeding program based on biomass sampling is used to determine the next feed requirement by sampling, which is presented in the table of biomass sampling results in the Table 1.

Table 1. ABW and ADG measurement result

DOC	ABW (g)	ADG (g)
20	1,11	
30	2,68	0,16
35	3,78	0,22
40	5,23	0,29
45	6,58	0,27
50	7,71	0,23
55	8,86	0,23

Note: DOC (*Day of Culture)*; ABW (*Average Body Weight)*; ADG (*Average Daily Gain)*

Demand feeding is a follow-up feed program that is given based on calculation of the remaining feed on ancho and biomass sampling. The feed program based on ancho control is the determination of the amount of feed needed from the rest of the feed in the ancho to control the level of feed consumption in vannamei shrimp. Ancho control is carried out 2-3 hours after feeding, it aims to determine the amount of reduction and addition of feed for the next feeding.

3.6. Auto feeder used

The used of auto feeder has the advantage of evening the amount of feed given and the time interval of feeding can be controlled. In addition, in shrimp growth, auto feeder can reduce FCR (*Food Conversion Ratio*) or feed efficiency level, increase ADG (Average Daily Gain) or shrimp body weight gain per period, SR (*Survival Rate*) and biomass. After using the auto feeder, the ABW (*Average Body Weight*) of DOC shrimp was 3.78 so that the ADG (*Average Daily Gain*) of shrimp per sampling period increased to 0.22 g. This number continues to increase every sampling so that it has the potential to increase biomass and reduce FCR (*Food Conversion Ratio*) in shrimp. Feed with high protein content can be applied less to produce 1 kg of fish meat so that it is considered more efficient in terms of production operational costs [6].

IOP Conf. Series: Earth and Environmental Science 1273 (2023) 012056 doi:10.1088/1755-1315/1273/1/012056

3.7. Water quality measurement

Water quality management was measured to maintain optimal water quality for the culture of vannamei shrimp. Water quality parameters that must be considered consist of physical parameters, chemical parameters, and biological parameters (Table 2).

Table 2. Results of water quality measurement

Parameters	Measurement method	Unit	Result	
			Morning	Afternoon
pH	Ex situ: PH paper		7,6-8,1	7,9-8,7
Salinity	Ex situ: Refractometer	ppt	27-29	
Temperature	In situ: Termometer	◦C	26-28	29-32
Brightness	In situ: Secchi disk	cm	28-40	
DO	Ex situ: DO meter	mg/l	3,76-4,66	6,37-7,72
TOM	Ex situ: Titrasi	mg/l	56,25-65,10	
Ammonia	Ex situ: Test Kit	mg/l	0,2-0,8	
Nitrite	Ex situ: Test Kit	mg/l	0,005-1	
Nitrate	Ex situ: Test Kit	mg/l	1-2	
Phosphate	Ex situ: Test Kit	mg/l	0,2-0,8	
Weather	In situ		Sunny, cloudy, rainy	
Water colour	In situ		Green, yellowish green, brownish green	
Water level	In situ	cm	110-140	

Based on the results of the table, the pH of these waters in the morning is 7.6-8.1 and in the afternoon is 7.9-8.7. Fluctuations in increasing pH are influenced by liming while fluctuations. A low pH value indicates a decrease in water quality which has an impact on the life of the biota in it [7]. The daily salinity value from the measurement using a refractometer is 27-29 ppm. Fluctuations in salinity values are influenced by environmental factors such as rain and metabolic activity of plankton, bacteria and shrimp. High salinity fluctuations cause shrimp growth due to disturbed osmoregulation process and increased density of *Vibrio* sp.

The temperature in the morning measurement is 26-28 ◦C and during the day it is 29-32 ◦C. Temperature fluctuations can affect the appetite and metabolic processes of the shrimp body when the temperature is high. The daily brightness value of the waters shows 28-40 cm. Fluctuations in brightness can be caused by accumulation of organic matter from feed input or water treatment. The daily DO value in the morning is 3.76-4.66 mg/l and in the afternoon is 6.37-7.72 mg/l. Fluctuations in the increase in DO value during the day are caused by the process of photosynthesis of plankton and the decrease at night is caused by the respiration process of all organisms and microorganisms that require oxygen. The daily TOM value is 56.25-65.10. Fluctuations in the TOM value can be caused by the residual shrimp feed shrimp mortality, plankton death and shrimp excretion results. Bacteria play a very important role in fluctuations the TOM (Total Organic Matter) or aquatic organic matter. According to Arifin [8], one of the signs of decreasing water quality can be seen from the total bacterial population, which is more than 106 CFU/ml.

Fluctuations in the value of nitrogen and phosphorus in the waters cause differences in the abundance of plankton that dominates so that the colour of the waters varies including green, yellow, brown, yellow green, yellow brown, and green brown. Nitrogen is a plankton nutrient that is utilized in the form of free ammonia (NH_3) and nitrate (NO_{3-}) while phosphorus is in the form of orthophosphate (HPO_{-4}) [9]. Plankton is one of the biological parameters that have a very important organisms in the life cycle of vannamei shrimp.

Sunny, cloudy, and cloudy weather will affect the intensity of light penetration that enters the vannamei pond. Meanwhile, rainy weather will affect the decrease in temperature, salinity, alkalinity, pH, and dissolved oxygen or DO (Dissolved Oxygen). Efforts that can be made after the rain is to dispose of rainwater through runoff or top disposal because it is known that rainwater has a lower density than seawater so that rainwater tends to be on the surface of the water. In this situation, feeding reduced by 10-20%, operating all aerators and lime treatment of 10 kg-20 kg/ha with the used of aerators can prevent the accumulation of mud at the bottom of the pond.

4. Conclusion

Feed processors with blind feeding program and demand feeding with auto feeder feeding technique resulted in DOC shrimp ABW (*Average Body Weight*) of 3.78 so that shrimp ADG (*Average Daily Gain*) per sampling period increased to 0.22 g. This number continues to increase every sampling so that it has the potential to increase biomass and reduce FCR (Food Conversion Ratio) in shrimp. Water quality monitoring is carried out in situ and ex situ including 7.6-8.1 in pH, temperature was 26-28°C, brightness was 28-40 cm, dissolved oxygen was 3.76-4.66 ppm, salinity was 27-29 ppt, and water height was 110-140 cm.

5. References

[1] Tahe, S, Suwoyo 2011 *J Ris Akua*. 6 1 31-40.
[2] Rahim, Rukmana, M R A, Landu, A, Asni 2021 *J of Fish and Mar Res*. 5 3 596-602.
[3] Rejeki, S, Aryati, R W, Widowati, L L 2019 *Introduction to Aquaculture Undip Press Semarang.*
[4] Isnawati, N, R Sidik, G Mahasri 2015 *J Ilmi Per dan Kel*. 7 2 121-124.
[5] Rahman, R, Lahming, Fadillah, R 2018 *J Pend Tek Pert.* 4 101-11.
[6] Hasan, M R, Soto, S 2017 *FAO Non-Seri Publi Rom FAO*. 33 pp.
[7] Susana, T 2009 *J Tekno Ling*. 5 2 33-39.
[8] Arifin, Z, Andrat, K, Subiyanto 2005 *Simple Vannameii Shrimp (Litopenaeus vannameii) Production Techniques BBPBAP Jepara Ministry of Maritime Affairs and Fisheries Jepara 56.*
[9] Edhy, W A 2002 *Dolomite in Our Cultivation Collection of Mitra Bahari Magazine 1996 – 2002 Edition Year VII No 3.*

6. Acknowledgement

We thank the editor and reviewers for their insightful comments on the manuscript.

IOP Conf. Series: Earth and Environmental Science 1273 (2023) 012057 doi:10.1088/1755-1315/1273/1/012057

Microalgal in waters and sediments of Surabaya mangrove areas

A L Nafisyah[1,8], A Anggraeni[2], E D Masithah[3], S Sugumar[4], A K Jayanthy[4], K D Arunachalam[4], W A A Q I W Mohtar[5], K Viswanathan[6,7]

[1] Department of Aquaculture, Faculty of Fisheries and Marine Science, Universitas Airlangga, Surabaya, Indonesia
[2] Study Program of Aquaculture, Faculty of Fisheries and Marine Science, Universitas Airlangga, Surabaya, Indonesia
[3] Department of Marine, Faculty of Fisheries and Marine Science, Universitas Airlangga, Surabaya, Indonesia
[4] SRM Institute of Science and Technology, India
[5] Institute of Biological Sciences, Faculty of Science, University of Malaya, Malaysia
[6] Department of Chemical Engineering, National Cheng Kung University, Tainan City, Taiwan
[7] Department of Mechanical Engineering, Sri Krishna College of Technology, Coimbatore, India

[8] Corresponding author: ayulana@fpk.unair.ac.id

Abstract. Mangrove areas stored the highest carbon content in the tropics which is three times higher than the average carbon in mainland tropical forest areas. The nutrients surround may support the growth of primary producers such microalgae. The community of microalgae was observed during the high and low tide in wet season (December 2021). Five water samples were collected during high tide, while during low tide three stations was visited to obtain the sediment samples. Supporting parameters include chlorophyll *a*, nutrient content, ORP, DO, and pH was measured in situ. The results showed that the microalgae communities both in waters and sediments of Surabaya mangrove areas was always consisted of Chaetoceros sp., *Pleurosigma* sp. and *Skeletonema* sp. in each sample. The abundance of microalgae in waters and sediments was ranged from 32 – 7,898 and 15,200 – 17,200 cells mL^{-1}, respectively. Based on the Simpson's dominance index (D), waters and sediments samples were resulting a range from 0.15 – 0.19 and 0.17 – 0.23, respectively. Whereas the Shannon's diversity index (H') ranged from 1.94 – 2.04 for water samples and 1.99 – 2.17 for sediment samples. The equitability index (E) of both waters and sediments was range from 0.13 - 0.63 and 0.42 - 0.86, respectively. These results show that Surabaya mangrove areas is classified as an oligotrophic area.

1. Introduction

Mangrove ecosystem is known as the ecosystems containing highest amount of carbon in the tropics, at where those carbon can be utilized by microalgae living in the sediment particles [1,2]. The total mangrove areas worldwide are around 15.9 million ha, while the mangrove area in Asia reaches about half of mangrove worldwide (i.e., 7.4 million ha). From total, most of mangrove ecosystems are scattered in Indonesia [3]. In East Java, particularly in the city of Surabaya, mangrove ecosystems are divided into several locations, including the East and North Coasts of Surabaya and along the Suramadu Bridge with different mangrove areas and densities. Mangrove ecosystems in Wonorejo – Surabaya are mangrove areas with total areas of 73 ha that is also utilized for ecotourism [4].

Microalgae are microscopic unicellular organisms that is classified as eukaryotic organisms or prokaryotic cyanobacteria [5]. They are divided into four groups, i.e., Chromophyta, Chkorophyta, Rhodophyta, and Cyanophyta which exists in freshwater to marine ecosystems and have the potential to be developed [6]. Microalgae found in waters are also called phytoplankton, which have a role as organisms that convert inorganic into organic materials through the photosynthesis process and support water productivity [7]. Microalgae found among sedimentary particles are also called microphytobenthos (MPB) and are known to have a major contribution to the ecosystem [8].

The role of microalgae in the waters is being the primary producer and the main oxygen producer [9]. In addition, MPB is also being the main carbon producer for higher trophic organisms and macrofauna which plays a role for the welfare of mangrove ecosystems [10]. Therefore, the mangrove ecosystem, especially in Indonesia is a fertile area that is often used as a location for fish and shrimp farming [11]. It was reported in Indonesia mangrove ecosystem for the occurrence of harmful algal species, named *Chattonella marina* var. *marina* [12]. The use of mangrove ecosystem commercially has been widely used, but the fertility monitoring of the area has not been carried out well. Nevertheless, it is necessary to check the health of the mangrove ecosystem regularly so that its cultivation productivity [4]. This study was aiming to analyze the density and diversity of microalgae in the waters and sediment of mangrove ecosystems.

2. Materials and Methods

2.1. Study areas

This study was done by field survey in December 2021 in Surabaya mangrove ecosystems (Wonorejo) during high tide (P1-P5) and low tide (S1-S3) as shown in Figure 1. Sediment sampling was done during low tide, while the water samples was taken during high tide. Observation and identification of microalgal communities were done under the microscope in the Biotechnology Laboratory, Faculty of Fisheries and Marine, Universitas Airlangga. Water quality was observed in situ, such as pH, DO, salinity, chlorophyll *a*, temperature, ORP, and light intensity during high (water column) and low tide (overlying water).

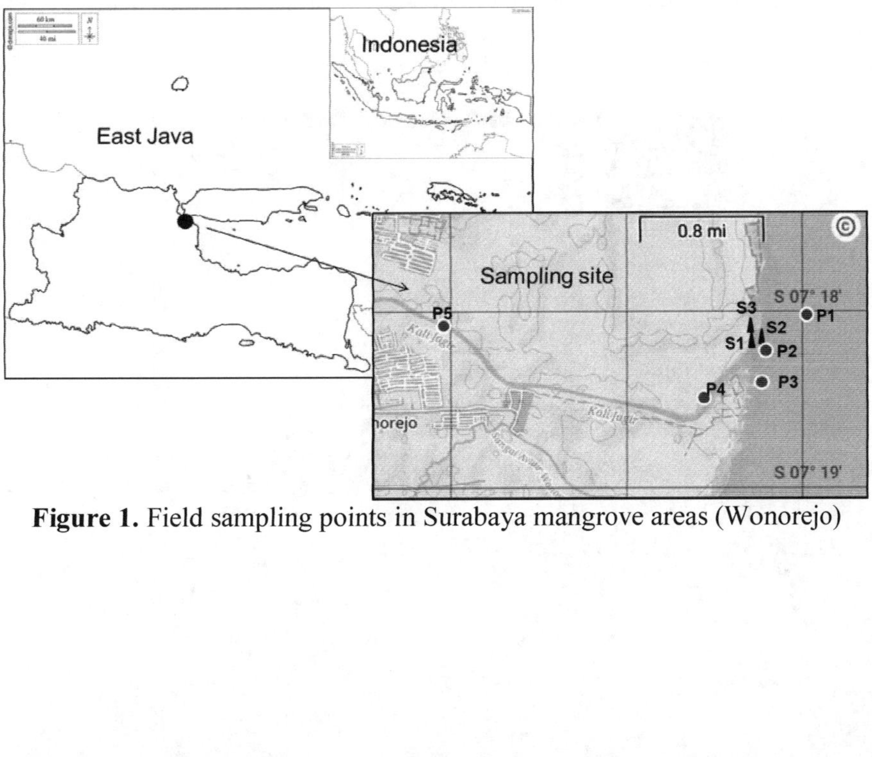

Figure 1. Field sampling points in Surabaya mangrove areas (Wonorejo)

2.2. Sampling methods and samples preparation

The water samples were taken from two different depth, i.e. 0 and 1 m by using a Van Dorn water sampler. While the sediment samples were taken by using an azunol Petri dish (10 cm^3). Soon after arriving in the laboratory, sediment and water samples was fixed by glutaraldehyde 1%.

3. Results and Discussion

3.1. Microalgal abundance in the waters

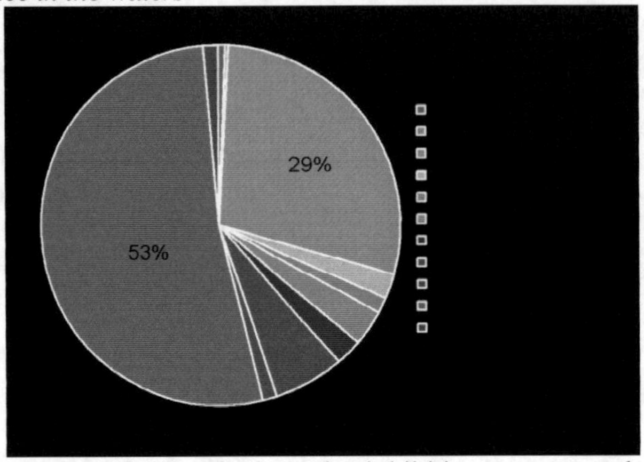

Figure 2. Percentage of microalgal species inhibiting water samples in Surabaya mangrove areas (Wonorejo)

Based on the enumeration on water samples, the microalgal communities are consisted of total 11 species (Figure 2) w ith range of abundance 4,400 – 115,280 cells mL^{-1} (Figure 3). The highest abundance of microalgal species in Surabaya mangrove ecosystems from water samples are *Skeletonema* sp. and *Chaetoceros* sp. as the second abundant species (29%). The enumeration results showed that the lowest abundance was existed in P5 at 1 m depth (4,400 cells mL^{-1}). Whereas the highest abundance is found in P1 at 0 m depth with the values of 115,280 cells mL^{-1}. All samples showed highest abundance of microalgal in the 0 m depth of water column, except P3 (Figure 3). The Shannon's diversity index (H') was in the range of 1.06 – 1.35, the similarity index (E) is in the range of 0.53 – 0.83. The dominancy index (D) is in the range of 0.31 – 0.44.

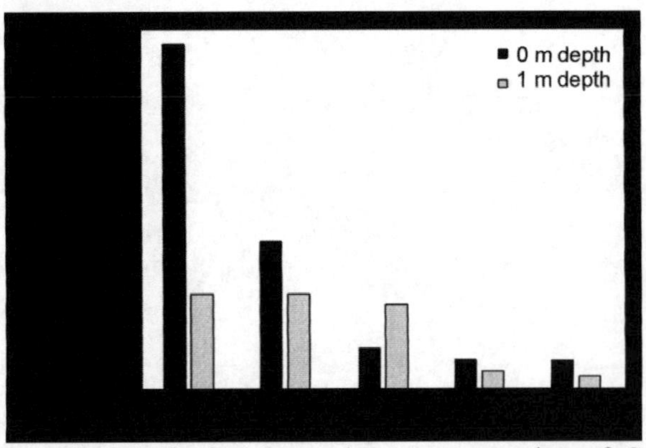

Figure 3. Total microalgal abundance in water samples of Surabaya mangrove areas (Wonorejo) from 0 and 1 m depth

3.2. Microalgal abundance in the sediments

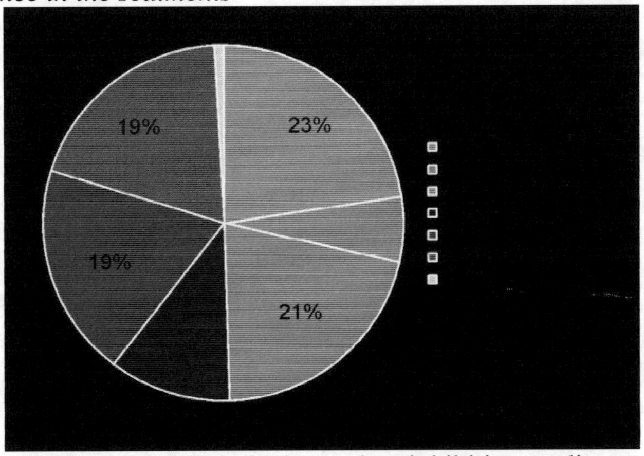

Figure 4. Percentage of microalgal species inhibiting sediment samples in Surabaya mangrove areas (Wonorejo)

Based on the enumeration on sediment samples, the microalgal communities are consisted of total 7 species (Figure 4) with range of abundance 123 x 10^4 – 554 x 10^4 cells mL^{-1} (Figure 5). The results showed that the lowest abundance was existed in S1 with the values of 123 x 10^4 cells mL^{-1} while the highest abundance is found in S2 with the values of 554 x 10^4 cells mL^{-1}. The Shannon's diversity index (H') was in the range of 1.16 – 1.55, the similarity index (E) is in the range of 0.72 – 0.88. The dominancy index (D) is in the range of 0.23 – 0.25.

Figure 5. Total microalgal abundance in sediment samples of Surabaya mangrove areas (Wonorejo)

3.3. Water quality surround the field sampling points

The water quality was measured in situ during high tide (Figure 6) and low tide (Figure 7). The light intensity during high tide was low in a range of 11 - 46 ∫mol photons m^{-2} s^{-1} (Figure 6B). This low light intensity due to the sampling points that was underneath the mangrove trees, except for P1 and P2. The temperature both in low and high tide showed in a range of 27 - 32°C (Figure 6A and 7A). The salinity was varied particularly during high tide sampling (Figure 6D), this is because the sampling points P1 - P2 was located in the offshore (ocean waters) while the sampling points P3 - P5 was located along the riverine (freshwater systems). The oxygen reduction-potential (ORP) showed in a high value both in high and low tide sampling (Figure 6F and 7E) with the values ranged for 169 - 257 mV.

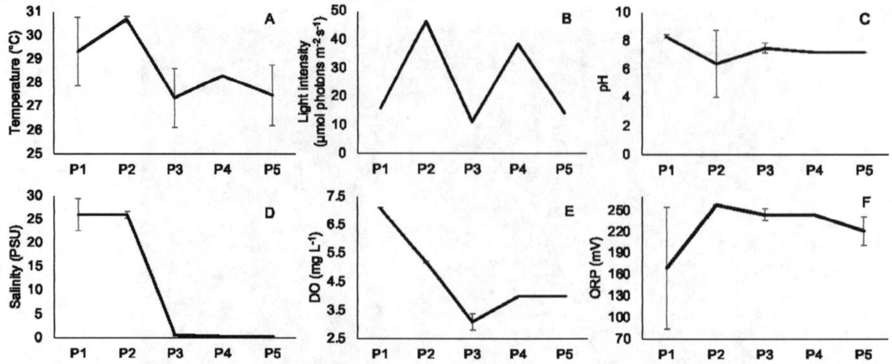

Figure 6. Water quality parameters of the water column during high tide; A)temperature, B)light intensity, C)pH, D)salinity, E)DO, F)ORP

In the similar manner on describing the water quality, the light intensity during low tide sampling was not showed in the graph due to the result of 0 μmol photons m^{-2} s^{-1} in all three sampling points (S1 - S3). The results of pH were showed in a range of 7.2 - 8.3, while the dissolved oxygen values showed in a range of 2.3 - 7.1 mg L^{-1} (Figure 6E and 7D). The lowest DO values was found during low tide sampling (S2 and S3) with the values of 2.3 and 2.7 mg L^{-1}, respectively.

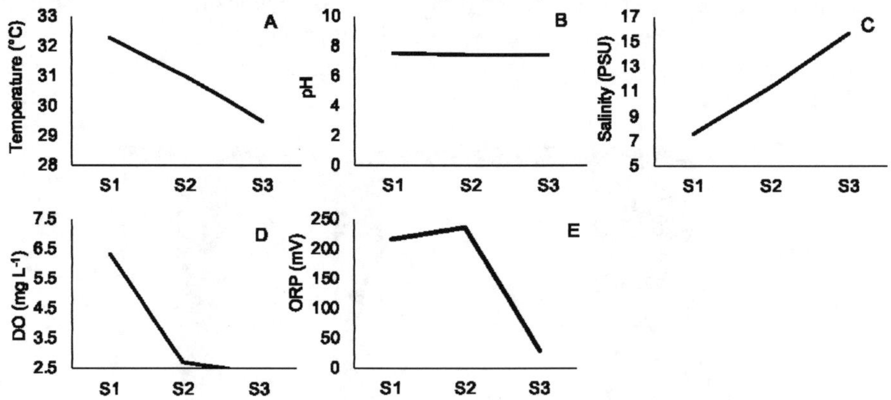

Figure 7. Water quality parameters of the sediments during low tide; A)temperature, B)pH, C)salinity, D)DO, E)ORP

3.4. Chlorophyll a biomass during high and low tide sampling

The chlorophyll *a* value in Surabaya mangrove areas (Wonorejo) are shown in Figure 8. In the water column, the chl *a* was ranged from 1.3 to 7.3 mg m^{-3} (Figure 8A). Higher chl *a* was located in P1 and P2, both are stations towards the ocean. While in the sediment samples, chl *a* was much higher than the water column samples with the values of 95-344.5 mg m^{-3} (Figure 8B).

Microalgal composition in Surabaya mangrove areas during wet season was consisted of 13 species in total from water and sediment samples. *Skeletonema* sp. and *Chaetoceros* sp. was the dominant species with presence occurrence in water column as much as 53 and 29%, respectively. High abundance of these species is describing that the Surabaya mangrove areas has suitable conditions for beneficial microalgal species. Both species, *Skeletonema* sp. and *Chaetoceros* sp., were known to be the natural feed for shrimp culture (*Litopenaeus vannamei*) [13,14]. Besides, *Scenedesmus* sp. was also found in the sediment samples wherein this species is having high protein (47%) and fat content (4%) which support the growth of snapper (*Sparus aurata*) [15]. The abundant of microalgal in this study was mostly found at the depth of 0 m in water samples. This is plausible since the light penetration is higher in the water surface which is used by microalgae for photosynthesis [16]. Chlorophyll *a* value in P2 showed in a high amount (6.9 mg m) at where the light intensity was also high (46.4 umol photons m-2 s-1). Bernard *et al.* [17] reported that the light intensity value is correlated with the amount of chlorophyll *a* as the basis of photosynthesis. The microalgal in sediments (microphytobenthos/MPB) of Surabaya mangrove areas showed in a high abundance compared to thos in Pemalang, Indonesia, with the value of 90,499 - 153,132 cells L^{-1} [18].

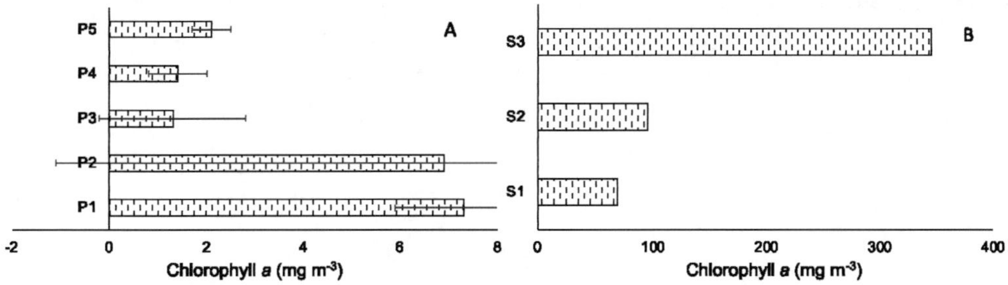

Figure 8. Chlorophyll *a* analysis; A)water samples, B)sediment samples

4. Conclusion

Microalgal communities during wet season in Surabaya mangrove areas (Wonorejo) are comprises by 13 species in total. The sediment samples are showing higher dominancy of microalgal communities compared to the water samples. The species found is potentially explore for the natural feed in aquaculture system, especially those culture maintaining nearby the mangrove areas.

5. References

[1] Donato D C, Kauffman J B, Murdiyarso D, Kurnianto S, Stidham M, Kanninen M 2011 *Natur Geosc* 5(4) 293–297.
[2] Jeslin J, Benny N, Thomas L C, Padmakumar K B 2021 *Wetla Ecol and Manag,* 29(5) 689-702.
[3] Giri C, Ochieng E, Tieszen L L, Shu Z, Singh A, Loveland T, Duke N 2011 *Glob Ecol and Biog* 20 154-159
[4] Ula M 2016 *Jurn Kebij dan Manaj Publ* 4(3) 1 - 8
[5] Day J G, Benson E E, Fleck R A 1999 *Biol Plant* 35: 127-136
[6] Morancais M, Mouget J L, Dumay J 2018 *Protei and Pigments*. Chapter 7: Microalgae in Health and Disease Prevention. Universite de Nantes, France, pp 31.
[7] Ridho MR, Patriono E, Mulyani YS 2020 *Jurnal Ilmu dan Teknologi Kelautan Tropis*, 12(1): 1-8
[8] Curry T, Krause J, Baker R 2020 G*ul and Carib Resea* 31 1
[9] Chindanonda C, Saramul S, Bhattarakosol P, Puimsomboon A 2019 *Trop Natu Hist* 19(2) 51-69
[10] Stocks K and Grassle J 2001 *Journal Mar Ecol Prog Ser* 221 93-104
[11] Pramudji 2000 *Osea* 25 (2) 13-20
[12] Nafisyah A L, Masithah M E, Matsuoka K, Lamid M, Alamsjah M A, Hara O S, Koike K 2018 *Fishe Scie* 84 877-887
[13] Arifin B N, Fakhri M, Yuniarti A, Hariati A 2018 J*urn Ilm Perika dan Kelau* 10(1) 46-53
[14] Ratnawati N 2018 Natural Feed Management Techniques *Chaetoceros* sp., *Thallasiosira* sp. and *Artemia* sp. in the Maintenance of White Shrimp Larvae (*Litopenaeus vannamei*) at PT Central Pertiwi Bahari Takalar South Sulawesi. Thesis. Pangkajene Pangkep Archipelago State Agricultural Polytechnic. pp. 44.
[15] Vizcaino A J, Lopez G, Zaez M I, Jimenez J A, Barros A, Hidalgo L, Camacho J, Martinez T F, Ceron M C, Alarcon F J 2014 *Aquac* 431 34-43
[16] Asriyana, Yuliana 2012 Waters Productivity PT Bumi Aksara, Jakarta Timur. 320 pp.

[17] Bernard O, Mairet F, Mart C 2018 IFAC-Papersonline 51(2) 735-740.

6. Acknowledgments

This study was done under the project Number 1003/UN3/2022 funded by LPPM Universitas Airlangga Surabaya, Indonesia.

Monitoring of microbial, water quality, and production performance of *Litopenaeus vannamei* on intensive pond in Bulukumba, South Sulawesi, Indonesia

P A Wiradana[1] , A Y Maharani[2] , M D Sani[2] , R J K Susilo[3] , M I Riandi[2] , I G Widhiantara[1] , P E Sudaryatma[4] , T Okabayashi[5,6] , and A T Mukti[7,8]

[1]Study Program of Biology, Faculty of Health, Science, and Technology, Universitas Dhyana Pura, Jalan Raya Padangluwih, North Kuta, Badung Regency, Bali, Indonesia
[2]Study Program of Biology, Faculty of Science and Technology, Universitas Airlangga, Kampus C Universitas Airlangga, Surabaya, East Java, Indonesia
[3]Department of Nanotechnology, Faculty of Advanced Technology and Multidiscipline, Universitas Airlangga, Surabaya, Indonesia
[4]Fish Quarantine and Inspection Agency Denpasar Office, Ministry of Marine Affairs and Fisheries, Indonesia
[5]Department of Veterinary Science, Faculty of Agriculture, University of Miyazaki, Miyazaki, Japan
[6]Center for Animal Diseases Control, University of Miyazaki, Miyazaki, Japan
[7]Department of Aquaculture, Faculty of Fisheries and Marine, Universitas Airlangga, Kampus C Unair Jalan Mulyorejo, Surabaya (60115), East Java, Indonesia

[8]Corresponding author: akhmad-t-m@fpk.unair.ac.id

Abstract. The aim of the research is to monitor the dynamics of microbial number, water quality, and production performance in Pacific white shrimp (*Litopenaeus vannamei*) cultivation in Bulukumba, South Sulawesi Province, Indonesia, utilizing an intensive system. This research was carried out on seven Pacific white shrimp ponds over the duration of one maintenance cycle (84 days), from November 2019 to February 2020. Interestingly, the results showed that the highest total bacterial counts (TBC) were obtained in pond P3, namely in the range of $0.12\text{-}98.0\times10^5$ CFU/mL, the highest total *Vibrio* counts (TVC) were obtained in pond P5, namely in the range of $0.09\text{-}2.8\times10^5$ CFU/mL, luminous bacterial counts (LBC) were only found in ponds P1, P4, and P5 with the same range values, namely $0\text{-}100\times10^3$ CFU. The best performance of Pacific white shrimp production was found in pond P2, with a percentage of survival rate (SR) and productivity of 65 percent and 4074.5kg/ha, respectively, and the worst was found in pond P5, with a percentage of survival rate (SR) and productivity of 35 percent and 2583.66 kg/ha, respectively. Finally, the intense culture system used for Pacific white shrimp accumulates dynamics that may alter microbiology, water quality, and production performance.

1. Introduction

Pacific white shrimp *(Litopenaeus vannamei)* is one of the seafood commodities with the highest percentage of productivity in the world, with a total production of 3,879,786 metric tons (MT) and nearly half of the total harvest of other crustacean commodities (7,351,350 metric tons) [1]. This demonstrates that Pacific white shrimp have the highest economic value in the aquaculture industry [2]. One of the major components for productivity in various Asian nations, including Indonesia, is the use of an intensive method in Pacific white shrimp aquaculture [3–5].

This intensive system's use is produced with high stocking density, optimal utilization of supplementary feed, and low water change activity [6]. Bulukumba, South Sulawesi, is one of the Indonesian locations that has implemented a comprehensive system for promoting Pacific white shrimp cultivation. According to data from the Office of Marine Affairs and Fisheries, Bulukumba Regency has 6,030 Ha of land used for aquaculture development, and the production of Pacific white shrimp in Bulukumba Regency reached 2,241.4 tons in 2015 and 2,591.8 tons in 2016 [7].

However, increasing the quantity of artificial feed input with a high protein content into intensive culture ponds might have an impact on water quality in the form of suspended particles as well as an increase in nitrogen waste from metabolic products and feed residue that is not used by shrimp [8,9]. Chronic ammonia and nitrite contaminating may induce disease outbreaks and economic losses in the aquaculture of Pacific white shrimp [10,11]. As a result, research on the formation of microbial communities in aquaculture conditions and their interaction to the environment, such as water quality, is crucial [10,12,13].

Changes in the culture environment may cause fast changes in microbial dynamics [2,14]. Despite the fact that various research have been published on disease agents such as bacteria in shrimp, attempts to reduce disease outbreaks in shrimp farming [15–18] remain inefficient. Similarly, the aquaculture sector's sustainability efforts are dependent on the stability of bacteria, which may be important in delivering effective organic food sources and ensuring ideal water quality. As a result, an evaluation of the major parameters, such as water quality, is required in order to manage the dynamics of bacteria in the intense system of Pacific White Shrimp culture ponds and achieve optimum production performance. The objective of this research is to investigate at changes in the quantity of microbials, water quality, and production performance in farmed Pacific white shrimp in Bulukumba, South Sulawesi, Indonesia.

2. Materials and methods

2.1. Study area

The period 84 days, or one maintenance period, in an intensive Pacific white shrimp pond in Bulukumba Regency, South Sulawesi, Indonesia, from November 2019 - February 2020. The ponds employed as study subjects were 7 units of pond plots ranging in size from 2,900 - 4,000 m^2 and with an average stocking density of 208 shrimp fry / m. All pond plots contain high-density polyethylene (HDPE) walls and plastic mulch at the bottom of the pond, as well as 26 turbines as an aeration system. Every week, about 8–10% of the total water volume is added in an attempt to restore water lost due to sludge disposal, evaporation, and seepage [19,20].

2.2. Procedure

The pond is prepared in the case mentioned by Shaari et al. [21]. Liming the pond bottom with activated lime CaO at an average dosage of 39.717.58 kg, drying for two weeks, and putting mulch plastic for the pond bottom are the first steps in pond preparation. The water utilized originates from a reservoir that has been cleaned with $CuSO_4$ at up to 1.5 mg/L and left for 24 hours. Following that, 1.5 mg/L of dichlorvos was added and aerated for 4 hours with 4 windmills before being dispersed across the ponds. When the water level was about 80 cm, disinfection was performed using chlorine at a concentration of 30 mg/L and homogenized with a pinwheel for 4 hours before adding up to 1.5 mg/L of $CuSO_4$ and 250 Ec of Pondfos and remaining for 3 days.

The chlorine was added again at a concentration of 30 mg/L, followed by the addition of 40 L of saponins the next day. The addition of fermented products is again adjusted to plankton development in the pond plots and may be repeated when the quality of water increases and plankton density starts to decline. Herobacillus probiotic bacteria (commercial items) were supplied to the ponds 24 hours after the fermentation product was applied. After the plankton has grown to a water brightness of 50-60 cm, the fry is supplied. The fry is dispersed using post larvae (PL) stadia. Feed is supplied up to 3-5% of the biomass at a frequency of 3-5 times per day, which is modified according to the age of the increase. Table 1 contains information on the processes for maintaining Pacific white shrimp.

Table 1. Details of Pacific white shrimp farming on the intensive culture ponds systems of Bulukumba, South Sulawesi, Indonesia

Particulars	Pond 1	Pond 2	Pond 3	Pond 4	Pond 5	Pond 6	Pond 7
Pond area (m^2)	4,700	4,600	3,400	4,400	4,300	3,800	2,600
Number of fry (pcs)	889,536	1,084,800	803,520	863,784	798,000	763,800	760,000
Density (pcs/m^2)	189	236	236	196	200	200	200
Herobacillus stocking culture	50 L	50 L	50 L	50 L	50 L	50 L	50 L
Molase for C/N ratio	10.4	12	10.1	8.6	9.6	7.9	8.4
Herococcus	3.5	3.2	2.5	3.3	2.7	2.6	1.8
Aeration	+	+	+	+	+	+	+

2.3. Water sampling

Water samples were collected from seven pond plots, and parameter testing was performed on a daily, weekly, and monthly, either directly (in situ) or by laboratory testing. Water samples were collected at a depth of 20 cm from the pond's surface. Water samples were collected in a 500 mL sterile plastic container using a test-specific device. Dissolved oxygen (DO), pH, salinity, temperature, transparency, and total organic matter are all water quality indicators that are assessed (TOM) [3].

2.4. Biological parameter analysis

Total bacterial counts (TBC), total *Vibrio* counts (TVC), luminous bacteria count (LBC), and *Vibrio* ratio were monitored every week until shrimp harvest time. Bacteria were tested using a spread plating approach, which included diluting water samples in % saline and spreading them on nutrient agar (NA), seawater complex (SWC) agar, and thiosulfate citrate bile salt sucrose (TCBS) agar to evaluate the development of LBC, TBC, and TVC. The sample-containing medium was then incubated at 30±2°C for 24-48 hours to determine the colony forming unit (CFU) [22].

2.5. Production performance

The performance of Pacific white shrimp production is measured to evaluate the performance of aquaculture production at the end of the maintenance period by measuring total harvest (kg), feed conversion ratio (FCR), survival rate (%), MBW (g), ADG (%/day), and productivity (kg/ha) [10,23].

2.6. Data analysis

The obtained data on water quality parameters and biological parameters were collected and processed in Microsoft Excel 2016 (Microsoft, USA). In addition, the data were entered into SPSS 23.0 (IBM,USA) and the ANOVA test was used to determine the difference in each parameter in each pond, followed by the Duncan's test with a confidence interval of 0.05.

3. Results and discussion

A heatmap of bacteriological information obtained from each Pacific white shrimp rearing pond is shown in Figure 1. Total bacterial counts (TBC) were found to be significantly higher ($p<0.05$) in ponds P2 and P3. For this parameter, the greatest total *Vibrio* counts (TVC) were recorded in pond P5, which varied relatively from other rearing ponds. In this investigative process, there was no difference in luminous bacterial counts (LBC) across the ponds. The proportion of *Vibrio* count (%) was found to be greatest in ponds P6 and P2. According to Alfiansah et al. [14] organic matter breakdown, feed leaching, and the release of ammonium from shrimp faeces may all promote inorganic nutrients in the water column. These nutrients promote bacterial growth, which causes changes in physical parameters such as pH, DO, suspended particulate matter (SPM), and turbidity to act synergistically [24–26]. Based on the statistics, it is clear that the P3 pond has a high overall abundance of bacteria. It is also predicted that there is a considerable quantity of faeces generated by shrimp, or feed that is not digested by the shrimp, which accumulates in the pond substrate and is subsequently utilized as nutrients by bacteria to develop [14].

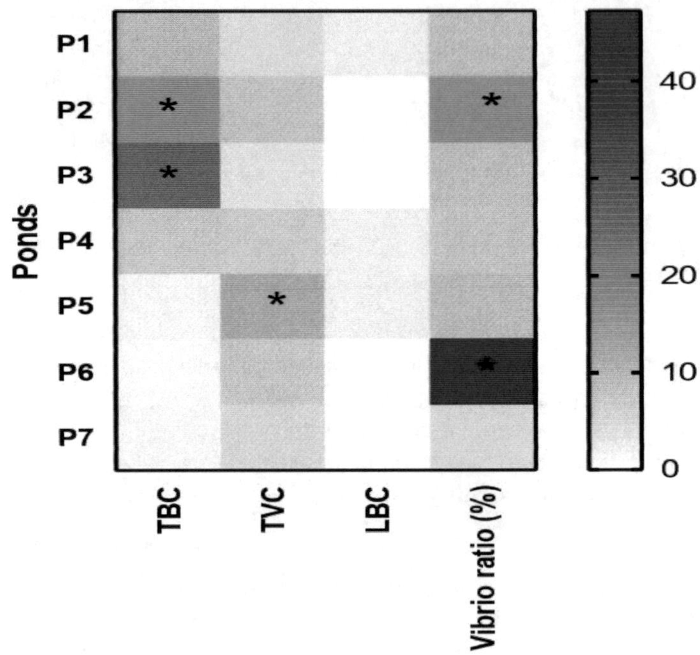

Figure 1. Heatmap visualisazation value of total bacteria counts (TBC), total *Vibrio* counts (TVC), luminous bacteria counts (LBC) and *Vibrio* ratio (%) in the maintenance medium of Pacific white shrimp in intensive ponds in Bulukumba, South Sulawesi, Indonesia. The black asterisk indicates a significant difference between each rearing pond with bacteriological parameters ($p<0.05$).

Moreover, according to Alfiansah et al. [14] the heterotrophic halophilic bacterial genera *Alteromonas*, *Erythrobacter*, *Halomonas*, *Pseudoalteromonas*, *Salegentibacter*, *Sulfitobacter*, and *Psychrobacter* represent a majority of bacterial communities in intensive and semi-intensive pond systems. *Halomonas* and *Psychrobacter* are heterotrophic nitrifying bacteria that are salt tolerant. Under saline circumstances, these bacteria may oxidize significant amounts of particles. *Sulfitobacter* has the ability to oxidize sulfur as well as decompose dissolved organic materials with a high molecular weight. According to Marwiyah et al. [22] multiple occurrences of vibriosis in shrimp cultivation in ponds occurred when the total vibrio level in pond water exceeded 1.0×10^3 CFU/mL. The presence of *Vibrio* spp. in water may be induced by a variety of reasons, including an increase in total organic matter in water from excrement and waste transported by rainfall currents, as well as abrupt fluctuations in water temperature [27–29].

Based on sources from the Indonesian National Standard (SNI) no 7310 of 2009 the optimal temperature of the water quality for rearing pacific white shrimp is 28-32°C, where if the temperature is too low than the standard it will easily cause abundance of *Vibrio* sp. water. The optimal pH is between 7.5-8.5, according to the average pH of water, which is 7.58, but there is a pH that fluctuates up to 7.02. Based on chemical parameters, the maximum amount of ammonia is <0.01 mg/L, the maximum nitrite is <0.01 mg/L, and the maximum nitrate is <0.5 mg/L, and if the level exceeds the standard limit, it can stimulate the growth of *Vibrio* bacteria. Luminous bacteria are those that have the lux gene, which contains protein involved in photophysical [30]. The majority of luminous bacteria can release light that may be seen in laboratory cultures and in environment [31]. The presence of LB in pond water and sediment samples may indicate nutrient enrichment and the absence of relatively low levels of pollutants [31,32].

The presence of luminous bacteria in ponds implies that the environment is healthy and optimal. These microorganisms have been suggested as helpful indicators for measuring the health of ponds [33]. Season, temperature, nutrient content, depth, geographic location, kind of food ingested by shrimp, and shrimp farming practices all have an impact on LB distributions [34]. The majority of Luminous bacteria are in the Vibrio group, which is typical microflora in the shrimp gut [35]. However, if the number of bacteria is too great, it will induce an imbalance in the autochthonous microflora, resulting in diseases development.

According to the findings of the *Vibrio* bacterial ratio assay (see Figure 1), the P6 had the greatest proportion of *Vibrio* bacteria, followed by the P2. *Vibrio* bacteria are essentially indigenous bacteria and microflora that are constant in pond waters as a consequence of fecal discharge and pellets that shrimp do not absorb. *Vibrio* bacteria are not harmful to ponds in concentrations less than 10%, but if the concentration reaches 10%, it stimulates the pathogenicity of *Vibrio* bacteria, resulting in significant shrimp mortality rates [36]. According to the results in Figure 1, ponds P6 and P2 have high *Vibrio* ratios, which may be related to temperature changes, weak shrimp feed consumption activities, resulting in feed buildup in the sediment, and a lack of filtration of shrimp waste or environmental waste surrounding the pond.

Table 2 demonstrates this. According to the findings of this investigation, the P4 group had the highest levels of DO and TOM, namely 5.35±0.65 ppm and 69±13.1 mg/L, respectively. Meanwhile, the P3 group had the highest pH and temperature values of 7.3±1.30 and 29.1±1.24, respectively. The P1 group had the greatest salinity level at 37.8±1.11 ppt. The maximum transparency level, however, was 48.1±133.8 cm in the P2 group. The data is shown in Table 2 below. On P4, which had a high area, stocking density, and probiotic supply, had the highest DO and TOM levels. This state does not degrade the quality of oxygen contained, indicating that there is enough oxygen in the shrimp [37] and the importance of probiotics in preserving water quality [38].

Furthermore, a major area might decrease the degree of competition for shrimp when it comes to accessing oxygen supply. However, due to the high shrimp stocking density and the leftover feed supply squandered in the culture pond, the TOM content created as a consequence of the shrimp secretion results may rise [36]. High density, on the other hand, may actually enhance a high pH value (alkaline conditions). The presence of this molasses serves to activate the probiotics that are added to the pond, with one of the objectives being to promote shrimp fry development [39]. The optimization of this growth results in extensive secretion from shrimp, which has an impact on the increased pH level [11,40]. Additionally, the limited pond compared to other groups caused the temperature of the maintenance pond to rise [41].

Table 2. Water quality parameters (mean ± SD, and range) in the Pacific white shrimp rearing ponds for 84 days of maintenance (n = 8).

Parameters	P1	P2	P3	P4	P5	P6	P7
DO (ppm)	5.10±0.72 (4.1-6.3)	4.72±0.75 (4.1-6.1)	5.03±0.55 (4.3-5.9)	5.35±0.65 (4.2-6.3)	4.90±1.00 (3.0-6.3)	4.87±0.70 (4.0-5.9)	4.72±0.70 (4.1-6.0)
pH	7.11±1.46 (4.4-8.4)	6.95±1.67 (4.1-8.1)	7.30-1.30 (5.1-8.2)	7.25±1.24 (5.0-8.0)	7.11-1.49 (4.2-8.1)	7.03±1.78 (4.1-8.1)	6.86±1.64 (4.1-7.9)
Salinity (ppt)	37.8±1.11 (36-39)	37.7±0.70 (37-39)	35.7±0.70 (35-37)	36.7±0.70 (36-38)	38.2±0.70 (37-39)	35.5±0.53 (35-36)	36.5±0.92 (35-38)
Temperature (°C)	29.0±1.19 (27-31)	29.1±1.24 (27-31)	29.1±1.24 (27-31)	28.8±1.12 (27-30)	28.6±0.91 (27-30)	28.5±0.92 (27-30)	28.2±1.03 (27-30)
Transparency (cm)	33.74±6.4 (25-45)	48.1±13.8 (30-75)	36.2±10.2 (30-60)	34.3±10.8 (25-60)	33.7±9.1 (25-55)	38.1±11.9 (30-60)	33.1±15.3 (25-70)
TOM (mg/L)	61.4±14.6 (37.5-79.6)	62.2±14.6 (38.6-79.6)	64.2±11.9 (43.2-78.3)	69.0±13.1 (38.6-78.3)	68.6±11.5 (43.2-78.3)	63.3±13.6 (42.1-77.1)	68.4±17.8 (44.3-93.5)

Note: Dissolved Oxygen (DO); Total Organic Matter (TOM)

Based on the data shown in Table 3, it can be seen that the highest stocking density is in P7 as much as 297/m^2 and the lowest stocking density is in ponds P1 and P4 as much as 189/m^2. The stocking density in the cultivation pond will affect the survival rate. In pond P7 it has a survival rate of 43% and in pond P1 it has a survival rate of 62%. This is in accordance with the results of research by Tantu et al.[42], who reported that the stocking density of 150 individuals/m^2 resulted in a survival rate of 92.0%, followed by the stocking density of 300 and 450 individuals/m^2, respectively 81.2% and 75.0%. However, there are conditions that are inconsistent with the P2 pond with a stocking density of 235/m^2 and P4 with a stocking density of 189/m^2 which has a survival rate of 65% and 45%, this can be influenced by various factors, namely environmental conditions in cultivation, level of cannibalism, competition for feed and shrimp physiological [43].

Table 3. Production performance parameters of Pacific white shrimp in each rearing pond after 84 days of maintenance in Bulukumba, South Sulawesi, Indonesia

Parameters	P1	P2	P3	P4	P5	P6	P7
Stocking density (ind/m^2)	889.54 (189/m^2)	1084.80 (235/m^2)	894.228 (263/m^2)	831.84 (189/m^2)	883.20 (205/m^2)	728.64 (191/m^2)	772.80 (297/m^2)
Ponds area (m^2)	4,700	4,600	3,400	4,400	4,300	3,800	2,600
Total harvest (kg)	5336.32	6156.66	4123.64	3720.34	3075.96	4433.70	2856.09
FCR	2.50	2.60	2.90	2.90	3.70	2.20	3.50
Survival rate (%)	62.00	65.00	45.00	45.00	35.00	60.00	43.00
MBW (g)*	6.43	6.44	6.99	6.44	6.28	7.17	6.47
ADG (%/day)*	0.17	0.20	0.18	0.19	0.22	0.24	0.19
Productivity (kg/ha)	3581.80	4074.50	2530.64	1954.04	2583.66	2732.00	1467.27

Note: FCR = feed conversion ratio, MBW = mass body weight, ADG = average daily growth

4. Conclusions

In conclusion, in each intense Pacific white shrimp pond in Bulukumba Regency, there are dynamics in total bacteria count, total *Vibrio* count, luminous bacteria count, and *Vibrio* ratio. Water quality has a significant function in regulating bacterial dynamics and optimizing the production performance of Pacific white shrimp produced in an intensive system. Further study on the molecular identification of the dominant bacteria and the amount of plankton in the intense system of grown Pacific white shrimp ponds is needed. We advise shrimp producers to give careful consideration to bacteria and water quality in order to reduce the incidence of harmful bacterial outbreaks in their ponds. It is also proposed that probiotics and immunostimulants be used to give protection and increase shrimp immunity.

5. References

[1] Muhammad, Situmorang and Djohan 2016 .*J. Fish. Livest. Prod.* 04
[2] Yang W, Zhu J, Zheng C, Lukwambe B, Nicholaus R, Lu K and Zheng Z 2020. *Aquaculture* 520 734733
[3] Riandi M I, Susilo R J K, Sani M D, Maharani A Y, Soegianto A, Putranto T W C and Wiradana P A 2021. *Poll Res* 40 611–6
[4] Wiradana P A, Sani M D, Mawli R E, Ashshoffa F N D, Widhiantara I G and Mukti A T 2022 . *IOP Conf. Ser. Earth Environ. Sci.* 1036 012003
[5] Wiradana P A, Mahasri G, Sari R E R, Marwiyah U C and Prihadhana R 2019. *IOP Conf. Ser. Earth Environ. Sci.* 236
[6] Esparza-Leal H M, Ponce-Palafox J T, Álvarez-Ruiz P, López-Álvarez E S, Vázquez-Montoya N, López-Espinoza M, Mejia M M, Gómez-Peraza R L and Nava-Perez E 2020. *Aquac. Int.* 28 1473–83
[7] Sipahutar Y H, Ramli H K, Kristiani M G E, Guntur Prabowo D H, Suryanto M R and Pratama R B 2019. *IOP Conf. Ser. Earth Environ. Sci.* 399 012125
[8] Dauda A B, Ajadi A, Tola-Fabunmi A S and Akinwole A O 2019. *Aquac. Fish.* 4 81–8
[9] Fira D, Wiradana P, Ansori A, Susilo R J K and Sabdoningrum E 2021. *Iraqi J. Vet. Sci.* 35 605–9
[10] Kamaruddin A, Nurhudah M, Rukmono D and Wiradana A 2021. *Iraqi J. Vet. Sci.* 35 757–63
[11] Alves Neto I, Brandão H, Furtado P S and Wasielesky Jr W 2019. *Ciência Rural* 49
[12] Bentzon-Tilia M, Sonnenschein E C and Gram L 2016. *Microb. Biotechnol.* 9 576–84
[13] Luu Q H, Nguyen T B T, Nguyen T L A, Do T T T, Dao T H T and Padungtod P 2021. *Aquac. Reports* 20 100711
[14] Alfiansah Y R, Hassenrück C, Kunzmann A, Taslihan A, Harder J and Gärdes A 2018. *Front. Microbiol.* 9
[15] Sahandi, Sorgeloos, Xiao, Wang, Qi, Zheng and Tang 2019. *Antibiotics* 8 95
[16] Soto-Rodriguez S A, Gomez-Gil B, Lozano-Olvera R, Betancourt-Lozano M and Morales-Covarrubias M S 2015. *Appl. Environ. Microbiol.* 81 1689–99
[17] Xiao J, Liu L, Ke Y, Li X, Liu Y, Pan Y, Yan S and Wang Y 2017. *Sci. Rep.* 7 42177
[18] Lee C-T, Chen I-T, Yang Y-T, Ko T-P, Huang Y-T, Huang J-Y, Huang M-F, Lin S-J, Chen C-Y, Lin S-S, Lightner D V., Wang H-C, Wang A H-J, Wang H-C, Hor L-I and Lo C-F 2015. *Proc. Natl. Acad. Sci.* 112 10798–803
[19] Sani M D, Wiradana P A, Maharani A Y, Mawli R E and Mukti A T 2022. *IOP Conf. Ser. Earth Environ. Sci.* 1036 012057
[20] Sani M D, Maharani A Y, Riandy M I, Joko R, Susilo K, Wiradana P A and Soegianto A 2020. *Ecol. Environ. Conserv.* 26 1271–5
[21] Shaari A L, Surif M, Latiff F A, Omar W M W and Ahmad M N 2011. *Trop. Life Sci. Res.* 22 51–69
[22] Marwiyah U C, Mahasri G, Ratnasari R E and Wiradana P A 2019. *IOP Conf. Ser. Earth Environ. Sci.* 236
[23] Rahardjo S, Vauza M, Rukmono D and Wiradana P 2022. *J. Adv. Vet. Anim. Res.* 9 78
[24] Fleckenstein L J, Kring N A, Tierney T W, Fisk J C, Lawson B C and Ray A J 2020. *Aquac. Eng.* 90 102093
[25] Lumbantoruan G S G and Azizah FFN M H 2015. *J. Aquac. Res. Dev.* 06
[26] Suantika G, Situmorang M L, Kurniawan J B, Pratiwi S A, Aditiawati P, Astuti D I, Azizah F F N, Djohan Y A, Zuhri U and Simatupang T M 2018. *Aquac. Eng.* 82 12–24
[27] Cabral J P S 2010. *Int. J. Environ. Res. Public Health* 7 3657–703
[28] Kopprio G A, Neogi S B, Rashid H, Alonso C, Yamasaki S, Koch B P, Gärdes A and Lara R J

2020. *Front. Microbiol.* 11
[29] Thickman J D and Gobler C J 2017. *PLoS One* 12 e0185994
[30] Brodl E, Winkler A and Macheroux P 2018 Molecular Mechanisms of Bacterial Bioluminescence *Comput. Struct. Biotechnol. J.* 16 551–64
[31] Dunlap P V. and Urbanczyk H 2013. (Berlin, Heidelberg: Springer Berlin Heidelberg) pp 495–528
[32] Zulkifli S N, Rahim H A and Lau W-J 2018. *Sensors Actuators B Chem.* 255 2657–89
[33] Abraham T J, Shanmugam S A, Palaniappan R and Dhevendaran K 2003. *Indian J. Mar. Sci.* 32 208–13
[34] Ramesh C and Mohanraju R 2019. *J. Terr. Mar. Res.* 3 1–32
[35] Soonthornchai W, Chaiyapechara S, Jarayabhand P, Söderhäll K and Jiravanichpaisal P 2015. ed G Dimopoulos *PLoS One* 10 e0135783
[36] Wang R, Guo Z, Tang Y, Kuang J, Duan Y, Lin H, Jiang S, Shu H and Huang J 2020. *AMB Express* 10 109
[37] Nimrat S, Tanutpongpalin P, Sritunyalucksana K, Boonthai T and Vuthiphandchai V 2013. *Aquac. Int.* 21 655–66
[38] Ringø E 2020. *Aquac. Fish.* 5 1–27
[39] Anand P S S, Kumar S, Panigrahi A, Ghoshal T K, Syama Dayal J, Biswas G, Sundaray J K, De D, Ananda Raja R, Deo A D, Pillai S M and Ravichandran P 2013. *Aquac. Int.* 21 511–24
[40] Furtado P S, Fugimura M M S, Monserrat J M, Souza D M, Garcia L de O and Wasielesky W 2015. *Mar. Freshw. Behav. Physiol.* 48 417–29
[41] Brito L O, Chagas A M, da Silva E P, Soares R B, Severi W and Gálvez A O 2016. *Aquac. Res.* 47 940–50
[42] Tantu A G, Salam S and Ishak M 2020. *J. Aquac Res Dev.* 11 1–6
[43] Xie F, Zeng W, Zhou Q, Wang H, Wang T, Zheng C and Wang Y 2012. *Aquaculture* 358–359 116–21

6. Acknowledgments

This research was supported by Universitas Airlangga, Universitas Dhyana Pura (UNDHIRA) Bali, and Miyazaki University, Japan. Researchers also thank Pacific white shrimp farmers who have contributed and granted permission for this research.

IOP Conf. Series: Earth and Environmental Science 1273 (2023) 012059 doi:10.1088/1755-1315/1273/1/012059

Microplastic abundance in Surabaya mangrove areas during the wet season

A L Nafisyah[1,6], H S Iswandi[2], A D Karisma[3], D D Nindarwi[1], S M Praveena[4], F Schneider[5]

[1] Department of Aquaculture, Faculty of Fisheries and Marine Science, Airlangga University, Surabaya, East Java 60115, Indonesia
[2] Undergraduate School of Aquaculture, Faculty of Fisheries and Marine Science, Airlangga University, Surabaya, East Java 60115, Indonesia
[3] Department of Industrial Chemical Engineering, Faculty of Vocational Studies, Institut Teknologi Sepuluh Nopember, Surabaya, Indonesia
[4] Department of Environmental and Occupational Health, Faculty of Medicine and Health Sciences, Universiti Putra Malaysia 43400 Serdang, Selangor, Malaysia
[5] National Cheng Kung University, Department of Environmental Engineering, No.1 University Road, Tainan City 701, Taiwan

[6] Corresponding author: ayulana@fpk.unair.ac.id

Abstract. Plastic waste pollution is the most dominant problem in various parts of the world, and it has increased since 2020. Degraded plastic waste can accumulate in the environment, especially the mangrove ecosystem, which is a sediment trapper, allowing for the accumulation of various materials. This study was done during the wet season in December 2021 at Wonorejo mangrove areas, Surabaya, Indonesia. The aims of study were to observe the abundance of microplastic that accumulates not only in waters during high tide but also in sediments during low tide. The results will provide information about environmental pollution of the Surabaya mangrove ecosystem. Four types of microplastics were observed in water and sediment samples under the stereo microscope, i.e. fragments, fibers, films, and granules. The average abundance of microplastics in waters from five stations was 8×10^6 particles km^{-2}, while in sediments (three stations) was counted as 1.5×10^9 particles km^{-2} on average. The Fourier Transform Infra-Red (FTIR) analysis was done resulting in the characteristic of microplastics. From both waters and sediments, contamination of microplastics consisted of Polyvinyl alcohol (PVA), Polyvinyl chloride (PVC), Polypropylene (PP), Polystyrene (PS), and Polyethylene (PE).

1. Introduction

Plastic waste accounts for 14% of the total municipal waste management in Surabaya, which is around 111,300 tons of plastic waste per year [1]. As much as 27% of plastic waste is coming from the single-use plastic bags, 18% is the single-use food service plastics, 17% is the plastic sanitation products, and 14% is the PET bottles. Uncollected waste is the largest source of plastic emissions each year releasing 8,775 tons of plastic waste into the environment [1]. Plastic waste in the environment will quickly accumulate in the water and accumulate in the estuary. Mass transportation along the river during the rainy season is higher than the dry season, so that Visible Plastic Debris (VPD) in the mangrove environment increases [2].

The mangrove environment is an ecosystem that has high potential because of the variety of living organisms in that environment [3]. The presence of microplastics in sediments will have an impact on increasing ammonium concentrations in the air column, thus affecting organisms in the waters [4]. The presence of microplastics in the aquatic environment can cause several disturbances to fish and other biota. Carbery [5] reported that some plastic particles are ingested because organisms cannot distinguish between food and plastic, so microplastics accumulate in the body of organisms. This study aimed to identify the potential for microplastic contamination in the Surabaya mangrove ecosystem at Wonorejo as a sediment trap that allows the accumulation of various materials. The observation results are expected to be a benchmark in predicting plastic waste, especially during the rainy season.

2. Materials and methods

2.1 Field sampling

Field sampling was carried out along the riverine towards the oceans at high tide (P1-P5) and low tide (S1-S3) in Wonorejo mangrove areas, Surabaya, Indonesia as shown in Figure 1. The detail location of each sampling station is shown in Table 1.

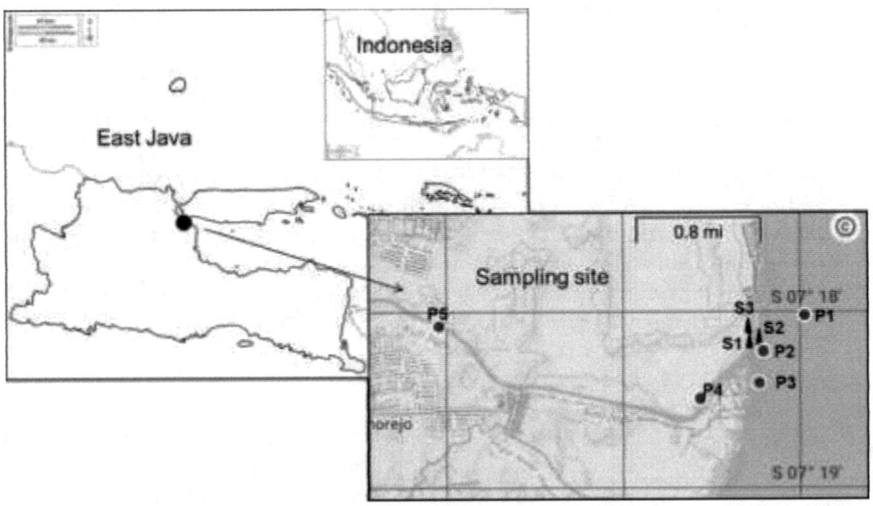

Figure 1. Field sampling points (P1-P5 and S1-S3) in Surabaya mangrove areas (Wonorejo) during the wet season

Water samples were collected at high tide (P1-P5) by using a manta net. The manta net was placed beside the boat on the water surface and pulled by the boat as far as 30 m (with a boat speed of ±1 knots). The collected waters were then poured into a dark bottle (600 ml). Sediment samples were collected at low tide (S1-S3) by the small Azunol Petri. Plastic waste found along the sampling site was also collected and stored in a plastic container. Water quality data was taken in situ, consisting of DO, temperature, pH, salinity using multiparameters and flow velocity using the floating method.

Table 1. The detail sampling station in Surabaya mangrove areas (Wonorejo)

Stations	Latitude (N)	Longitude (E)	Notes
P1	S 07°18.112'	E 112°51.046'	Ocean
P2	S 07°18.225'	E 112°50.744'	Close to the ocean
P3	S 07°18.577'	E 112°50.414'	Estuary
P4	S 07°18.462'	E 112°49.729'	Riverine
P5	S 07°18.051'	E 112°48.906'	Riverine
S1	S 07°18.223'	E 112°50.670'	Mangrove site around estuary
S2	S 07°18.216'	E 112°50.662'	Mangrove site around estuary
S3	S 07°18.210'	E 112°50.568'	Mangrove site around estuary

2.2 Microplastic Sample Preparation and Identification
Directly after the field sampling, all samples were processed in the laboratory. Water samples (P1-P5) were poured into a 50 ml beaker glass and added by 30% H_2O_2 before stand for 24 hours [6]. Water samples were then filtered by using a filter paper (Whatmann 47 mm). The sediment samples were oven-dried at 90°C until dry (± 15 hours). Density separation was carried out by mixing 50 g of dry sediment with 150 ml of 30% NaCl solution. Then stirred evenly and allowed to stand for 24 hours until it settles and the small microplastic will float. The supernatant portion was taken and filtered by using filter paper. The identification process of microplastics uses a stereo microscope to determine the characteristics and abundance of microplastics.

3. Results and discussion
3.1 Types of microplasctics found in Surabaya mangrove areas
The microplastics found in the waters and sediments of the Wonorejo Mangrove during the rainy season are fragments, granules, films, and fibers (Figure 2.)

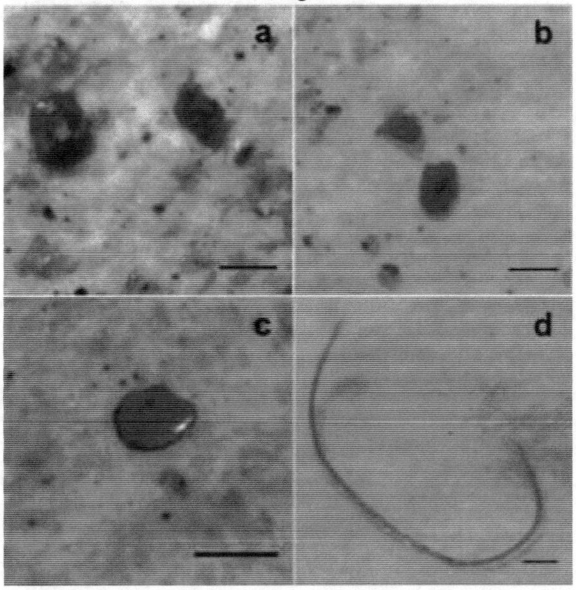

Figure 2. Types of microplastics **a)** fragment, **b)** granule, **c)** film, **d)** fiber. Scale bar 0.40 mm

The fragments (Figure 2a) are thick, irregular in shape, and predominantly dark in color. Granules (Figure 2b) are shaped like seeds/small granules that have various colors. The film (Figure 2c) has similar characteristics to fragments, irregular, but its color is transparent. The type of fiber (Figure 2d) is a type of microplastic such as thin fibers that easily accumulate in sediments [7].

3.2 Microplastics abundance in waters and sediments of Surabaya *mangrove areas*

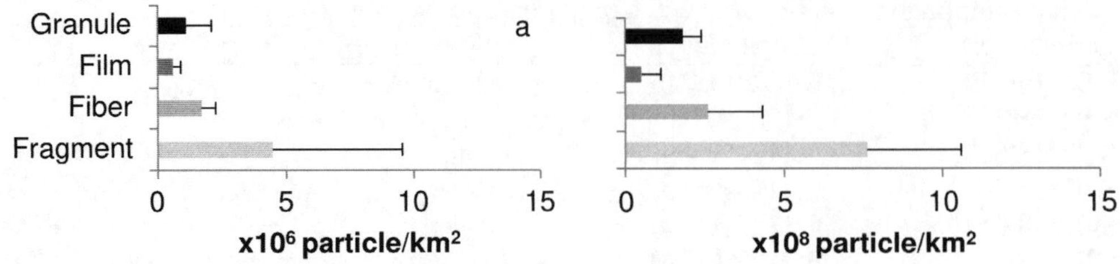

Figure 3. The average abundance of microplastics found in Surabaya mangrove areas from **a)**waters samples, **b)**sediment samples

Based on the Figure 3, the fragment type is known to have the highest average abundance both in waters (4,4 × 10^6 particles/km^2) and sediments (7,6 × 10^8 partikel/km^2). The second abundant type in both waters and sediments samples are type of fiber with average of 1.7 × 10^6 particles/km^2 in water samples and 2.6 × 10^8 particles/km^2 in sediment samples. The lowest abundance type that is found in Surabaya mangrove areas during wet season are film (0.6 × 10^6 particles/km^2 in waters and 0.5 × 10^8 particles/km^2 in sediments)

3.3 Types of microplastic polymers found in waters and sediments in Surabaya mangrove areas

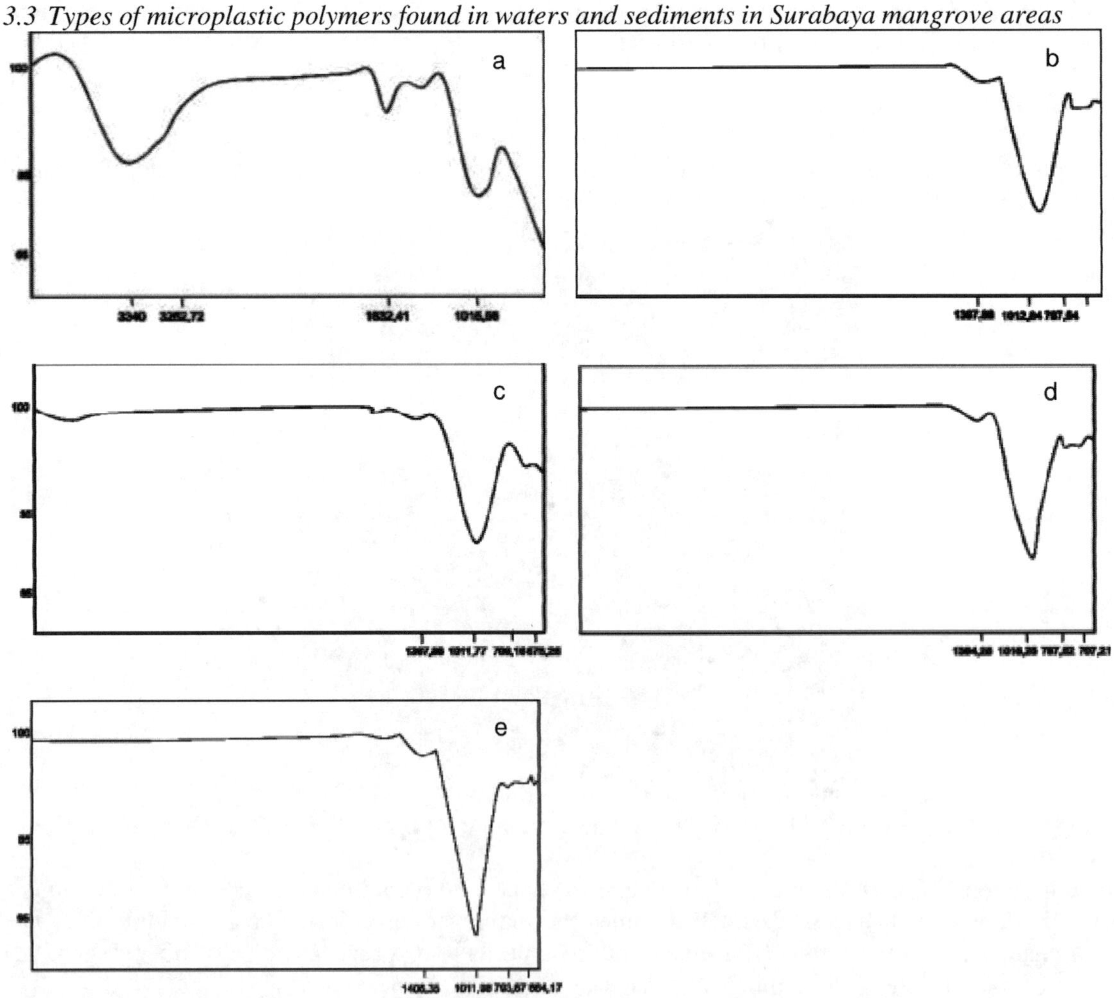

Figure 4. Fourier Transform Infra Red (FTIR) results analyzing the types of microplastic polymers in water samples a)P1, b)P2, c)P3, d)P4, e)P5

IOP Conf. Series: Earth and Environmental Science 1273 (2023) 012059 doi:10.1088/1755-1315/1273/1/012059

The Fourier Transform Infra Red (FTIR) analysis was resulting as shown in Figure 4 for water samples and Figure 5 for sediment samples. The values of the P1 waves (Figure 4a) are 1015,56 cm^{-1} (CH_2), 1632,41 cm^{-1} (C=C), 3352,72 cm^{-1} (OH *stretch*) and 3334 cm^{-1} (OH *stretching*). It is analyzed as the characteristic of polyvinyl alcohol (PVA) polymers. The wave values of the P2 (Figure 4b) are 684,37 cm^{-1} (*Alkyne* C-H *bend*), 767,54 cm^{-1} (*skeletal* C-C), 1012,84 cm^{-1} (CH_2), and 1398,19 cm^{-1} (CH_3) which is a characteristic polymer of polypropylene (PP). The wave values of P3 (Figure 4c) are 709,16 cm^{-1} (C-H *bend*), 676,26 cm^{-1} (*aromatic* C-H *out-of plane bend*), 1011,77 cm^{-1} (CH_2 *methylene*) and 1397,88 cm^{-1} (-CH_3). This is a characteristic of polystyrene (PS) polymers. The wave values of the P4 (Figure 4d) are 707,21 cm^{-1} (CH_2), 787,52 cm^{-1} (C-Cl), 1016,25 cm^{-1} (C-C), and 1394,26 cm^{-1} (-CH_3 atau OH *str*) which is a characteristic of polyvinyl chloride (PVC). The wave values of P5 (Figure 4e) are 664,17 cm^{-1} (*Alkyne* C-H *bend*), 793,67 cm^{-1} (*skeletal* C-C), 1011,98 cm^{-1} (CH_2) and 1406,35 cm^{-1} (OH *bend*) that is a characteristic of polyvinyl alcohol (PVA) polymers.

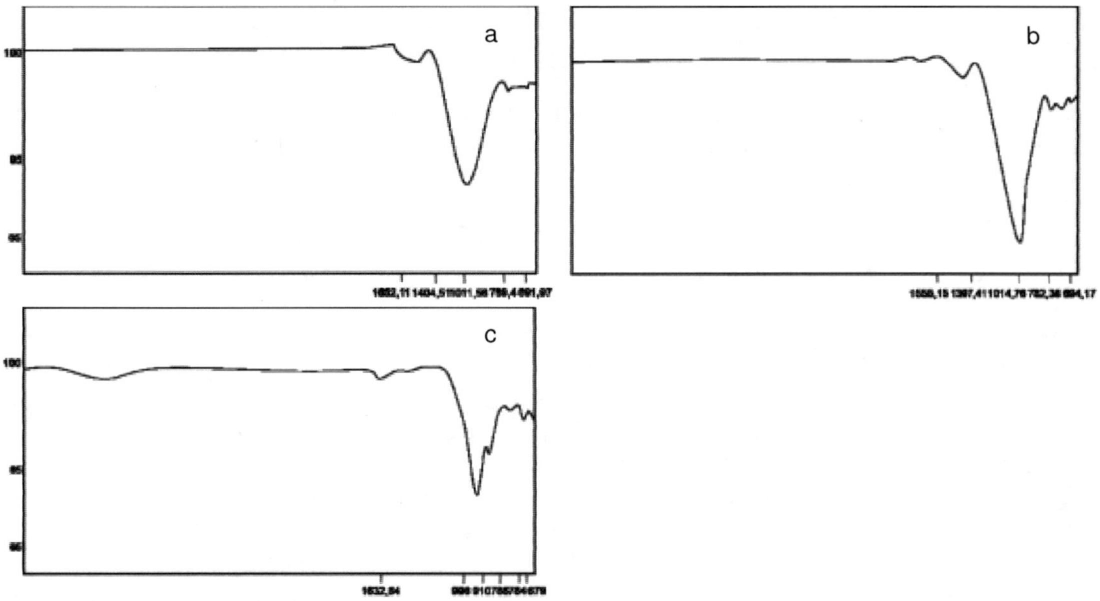

Figure 5. Fourier Transform Infra Red (FTIR) results analyzing the types of microplastic polymers in sediment samples a)S1, b)S2, c)S3

From the sediment samples, the FTIR results shown in Figure 5. S1 wave values (Figure 5a) are 691,97 cm^{-1} (*alkyne* C-H bend), 780,40 cm^{-1} (C-Cl), 1011,58 cm^{-1} (CH_2), 1404,51 cm^{-1}(OH), 1562,11 (N-H *bend*) and 1647,12 cm^{-1} (C=C *stretching*) which is the characteristics of polyvinyl alcohol (PVA) and polyvinyl chloride (PVC). While S2 wave values (Figure 5b) are 694,17 cm^{-1} (C-H *bend*), 782,38 cm^{-1} (*aromatic* C-H), 1014,76 cm^{-1} (CH_2), and 1397,41 cm^{-1}(OH) is a polymer characteristic of polystyrene (PS). The wave values of S3 (Figure 5c) are 679,39 cm^{-1} (C-H *bend*), 784,77 cm^{-1} (*methylene* CH_2 *bend*), 788,77 cm^{-1} (C-Cl), 910,14 cm^{-1} (CH_2), 996,59 (CH_2) and 1632,84 cm^{-1} (C=C), which are polymer characteristics of polypropylene (PP) and polyvinyl chloride (PVC).

Fragments are the type of microplastic with the highest abundance found in water and sediment samples. Fragments are the result of the fragmentation of plastic products and the breakdown of rigid plastics with very strong synthetic polymers. The source of the fragments found in the sediment can come from plastic waste generated by local human activities [8]. Most of the types of fragments were found at P1 in the ocean. The microplastic polymer at P1 is polyvinyl alcohol (PVA) which is the polymer materials of fishing nets. This is plausible since the station P1 is located in the ocean where many fishing activities take place. The high number of fragments-type in marine waters is caused by fishing activities and the waste produced by local residents in the form of food packaging waste carried by currents at low tide [9].

Film is the least type of microplastic found in the water and sediment samples of the Surabaya mangrove areas (Wonorejo) with most films were found at P1. The high current velocity at P1 significantly resulted in an increase in microplastic particles in these water locations ($p < 0.05$) from low to high density. Microplastic film is a constituent of polyethylene (PE) and polypropylene (PP) that is mostly found in food packaging and plastic bags which have low density [10]. The lowest abundance of microplastics was found at P4 which is a riverine close to the tourist pier. Station P4 is known to have the lowest current velocity of the 5 sampling locations. Since the current velocity significantly affects the abundance of microplastics ($p < 0.05$), the higher current velocity affecting the higher abundance of microplastics, and vice versa [11]. The abundance of microplastics in sediments is 98% higher than the abundance of microplastics in waters. Kilponen [12] was reported that the degradation process on land occurs 4 to 32 times higher than the degradation process in the waters. Microplastic particles that have formed on land areas will be carried into water bodies due to rain and wind. In the wet season, contamination of estuary plastic waste increases [2], and results in more degradation processes occurring in estuary and sea flows.

4. Conclusion

The types of microplastics found in water and sediment samples in Surabaya mangrove areas (Wonorejo) during wet season are fragments, fibers, films, and granules. The average abundance of microplastics from all stations in water and sediment samples was 8×10^6 particles/km^2 and 1.5×10^9 particles/km^2, respectively. The polymers resulting from the interpretation of the FTIR test on water and sediment samples are polyvinyl alcohol (PVA), polyvinyl chloride (PVC), polypropylene (PP), polystyrene (PS), and polyethylene (PE).

5. References

[1] UN-ESCAP 2020 *Handling plastic pollution in Surabaya, Indonesia (Menangani polusi plastik di Surabaya, Indonesia)* Baseline report (Thailand: United Nations Economic and Social Commission for Asia and the Pacific). https://www.unescap.org/sites/default/d8files/event-documents/SB%20Baseline%20Report_Bahasa.pdf (In Bahasa Indonesia).

[2] SB Kurniawan and MF Imron 2019 *Environ. Technol. Innov.* 16.

[3] Kathiresan K and Bingham BL 2001 *Advances in Marine Biology*. 40: 81-251.

[4] Song YK, Hong SH, Jang M, Han GM, Rani M, Lee J, Shim WJ 2015 *Marine Pollution Bulletin*: 202-209.

[5] Carbery M, O'Connor W, Palanisami T 2018 *Environmental Int.* 115: 400-409.

[6] Wong G, Lowemark L, Kunz A 2020 *Environmental Pollution* 260.

[7] Warsidah W 2019 *Jurnal laut khatulistiwa* 2(3):96-101.

[8] Eriksen M, Mason S, Wilson S, Box C, Zellers A, Edwards W 2013 *J. Mar. Pollut. Bull.* 77:177–182.

[9] DI Sari, AA Budiarsa, IR Ritonga 2015 *Jurnal Ilmu Perairan, Pesisir, dan Perikanan.*

[10] Kingfisher J 2011 *Microplastic debris accumulation on puget sound beaches*. Port Townsend Marine Science Center.

[11] Harpah, Novrida, Ageng P, Addauwiyah R, Rizki A, Perdana Z, Suryati I, Leonardo R, Husin A, Faisal M 2019 *Microplastic pollution in Deli River Medan* International Conference on Research Collaboration of Environmental Science.

[12] Kilponen J 2016 *Microplastics and Harmful Substances in Urban Runoffs and Landfill Leachates, Possible Emission Sources to Marine Environment*. Bachelor's Thesis in Environmental Engineering, Lahti University of Applied Sciences 8.

6. Acknowledgments

This study was conducted under project Number 525/UN3/2021 funded by LPPM Universitas Airlangga Surabaya, Indonesia.

The effect of using different types of fertilizer on plankton abundance and growth rate of backyard shrimp farm

L B R Ritonga[1], Nasuki[1], K Primasari[1], P N Rizky[1,2], M H Edy[1], T Harijono[1]

[1]Fisheries Cultivation Techniques, Marine and Fisheries Polytechnic Sidoarjo, Jl. Raya Buncitan, Gedangan, Dusun Kp. Baru, Buncitan, Sidoarjo, East Java

[2]Corresponding author : putrimarine92@gmail.com

Abstract. The application of fertilizers in aquaculture ponds is one of the ways to increase water fertility which is characterized by an abundance of phytoplankton. However, further research on the effect of adding organic fertilizer to the abundance of plankton and the productivity of cultivation is necessary. This study aims to determine the effect of the use of liquid organic fertilizer and ceraclean on the abundance of plankton in the vannamei shrimp rearing media, and the survival and growth of vannamei shrimp. There were two treatments in this study, namely the application of ceraclean fertilizer 20 ppm/week in treatment 1 and the application of liquid organic fertilizer from chicken manure 20 ppm/week in treatment 2. Treatment 1 with addition of ceraclean fertilizer showed the highest abundance of plankton (262 Ind/l) compared with treatment 2 (225 Ind/l) ($P < 0.05$). However, the number of beneficial plankton identified in ponds with organic fertilizer application (treatment 1) was greater (80.0%) than ceraclean fertilizer application (treatment 2) (67.5%). This explained that organic fertilizer could significantly increase the growth and survival rate of shrimp by increasing the production of beneficially plankton which lead the lower of feed intake (FCR value of 1.1).

1. Introduction

The volume of vannamei (*Litopenaeus vannamei*) production in 2019 was 517,397 tons and experience significantly increase of 120% in 2024 to 1.290.000 tons [1]. The increasing productivity of shrimp culture sector is line with the high intensification of shrimp farming systems which causes the environmental degradation [2,3,4]. Various efforts and innovation have been made, especially through the development of aquaculture technology based on the land characterstics and infrastructure availability and feed efficiency [5,6]. The existences of phytoplankton in shrimp culture have several functions as natural feed with highest of protein and amino acid content (20-35%) [7,8,9] and maintaining the stability of the pond ecosystem [10,11,12]. Phytoplankton as a primary producer and water quality act as oxygen sources in the waters, reduced the klekap and moss production at the bottom of pond, and stabilize the ammonia, nitrite, and nitrate content in the water [13]. However, the growth performance of in shrimp ponds are influenced by several factors such as temperature, light intensity, nutrients, salinity, pH, and pollutants [14].

Fertilizers is commonly used in shrimp pond culture to stimulate phytoplankton productivity and enhance the availability of natural food. Organic fertilzer (livestock manure and other agricultural wastes) can be applied to shrimp pond culture as biofeltilizer that improve chemical, physical and biological soil properties and stimulating algae, diatoms, bacteria and micro-organisms growth [15]. Moreover, organic fertilizers have intended to reduce the long-term effects of the overuse chemical fertilizers on the soil organisms that are similar to human overuse of antibiotics. In addition to organic fertilizers, silicate is used for performed the cell walls and shells in diatoms. Small particle sizes and easily soluble in water, making it easier for the absorption of nutrients by plankton cells [16]. The nutrient content contained in liquid organic fertilizer and silicates is expected to be able to meet the nutrient needs of plankton so as to reduce feed costs and increase the productivity of vannamei shrimp culture. However, further research needs to be done on the effect of adding organic and silicate fertilizers to the plankton abundance. The aims of this study are determining the effect of liquid organic fertilizer and ceraclean application on the abundance of plankton in the vannamei shrimp rearing media, and the survival and growth of vannamei shrimp which equipped with an innovative system of intensive technology in household scale in vannamei shrimp culture.

2. Materials and Methods

2.1 Research design

This research was conducted in December 2021, at Teaching Factory of Marine and Fisheries Polytechnic, Sidoarjo. The shrimp ponds were used in this research is household scale vaname shrimp culture pond types with tarpaulin pond. The pound is round with 4 m of diameter and 1,2 m height. The ponds equipped with a microbubble aeration system, high blower aerator, and oxygen diffusion. Descriptive statistics was used to describe general description of abundance and number genrea, biological index of plankton. In addition, to analyzes the difference of number between treatment groups T-test statistic were used. There were two treatments in this study, namely the application of ceraclean fertilizer 20 ppm/week in treatment 1 and the application of liquid organic fertilizer from chicken manure 20 ppm/week in treatment 2.

2.2 Research methods

2.2.1 Backyard Shrimp Farm Design

The pond used in this study is a circular pond with the concept of household scale culture. A total of 2 tarpaulin ponds with a diameter of 4 m was equipped with supercharger aerator technology with a power of 1 HP, an AMR 500 compressor with 150 aeration points output and a microbubble aerator with a power of 0.5 HP which is capable of producing dissolved oxygen 7 – 8 ppm and a current velocity of 0.3 m/s to increase shrimp productivity. The outlet system uses a central drain. The design and technology applied to the backyard shrimp farm in this study showed in Figure 1.

Figure 1. Backyard Shrimp Farm Design. (a) Microbubble technology application and (b) additional aerator and central drain

2.2.2 Fertilizer application

Organic fertilizer used in this research are came from chicken manure. About 5 kg of dry chicken manure were fermented anaerobically for 7 – 10 days into a plastic drum with a capacity of 150 L which contained 100 L of fresh water; 5 L of Molasses; 5 L of Decomposing bacteria (*Bacillus sp*); 15 grams of tape yeast (*Saccharomyces seriviceae*); 15 grams of fish meal, and 5 L of coconut. Meanwhile, silicate fertilizers were obtained from the market with the trademark Ceraclean. Plankton growth was carried out in two treatments. Ceraclean fertilizer with 40% silicate and 25% Ca were added in treatment 1 for 20 ppm/week of doses. In addition, organic fertilizer from fermented chicken manure were added in treatment 2 for 20 ppm/week of doses.

2.2.3 Vaname Shrimp rearing

The post larvae 18 of vanamei shrimp were tested as specific pathogen-free (SPF) were used in this study and carried out in the trampaulind pond for 75 days. This study using a super intensive system with a stocking density of 700 shrimp/m^3. The shrimp were fed using crumble feed with 40% of protein content in the initial phase, then changed to a pellet with 36% of protein content after 20 days with blind feeding method. Feeding frequency was carried out five times a day. During the rearing period, the ponds were applied with 5 ppm of probiotics for 3 days respectively and 20 ppm fertilizer every week. Several parameters such as average body weight (ABW), survival, feed conversion ratio (FCR), and productivity were measured regularly by collecting 30 shrimp in each pond every two weeks. the water quality such as temperature, salinity, pH, TOM, nitrite, alkali and phytoplankton were measured every week. The following formulas are used to calculate ABW, Survival, FCR and Productivity (Maica *et al.*, 2014).

$$ABW = \frac{\Sigma \text{ Shrimp weight (g)}}{\Sigma \text{ Shrimp population (g)}} \qquad (1)$$

$$SR\ (\%) = \frac{\text{Nt}}{\text{No}} x\ 100 \qquad (2)$$

2.2.4 Phytoplankton analyses

Samples were collected for phytoplankton analyses by filtering 500 liters of media's water using a plankton net (250µm) weekly. The identification of plankton is carried out at the genre level. Plankton abundance was observed using a SRC (Sedgwick rafter counter cell) under microscope and expressed in individual per ml following the formula;

N= T/L x P/p x V/v x 1/Wn (3)

Where :
N = abundance of phytoplankton (ind/l)
T = Number of squares in SRC (1000)
P = Number of observed phytoplankton
p = Number of observed SRC boxes
V = Amount of water in the sample bottle
v = Volume of water in box SRC
W = Volume of filtered water pond

2.3 Statistic analysis

Plankton abundances and shrimp growth rate were analyzed statically using repeated measures T-test through graphpad prism 9 to determine the significant difference among the treatment in 95% of confidence level.

3. Results and Discussion

Treatment 1 presented the highest of plankton abundance with 262 Ind/l plankton consists of 67.5% of beneficial plankton and 21.3% of detrimental plankton. Meanwhile, about 225 Ind/l plankton were identified in treatment 2, consist of 80.0% of beneficial plankton and 20.0% of detrimental plankton (Table 1).

Table 1. The amount of plankton found in vaname shrimp culture ponds with the addition of ceraclean and liquid organic fertilizer

Variable	Ceraclean	Poultry fertilizer	p-value*
Plankton density (Ind/l)	262	225	0.000
Benefical plankton (%)	67,5	80	0.000
Chlorophyta (%)	57,2	73,4	
Harmful plankton (%)	32,5	20	0.000
Blue Green Algae (Cyanophyta) (%)	37	21,3	
Crysophyta (%)	5	4,4	
Dinoflagellata (%)	0,8	0	

*independence T-test with level significant of $p < 0.05$

Organic fertilizers is enriched with organic matter and nutrients which greatly stimulating the growth of bacteria and fungi in pond that used to decompose toxic substances or waste that accumulates in waters than inorganic fertilizers [17]. Organic fertilizers contain animal-based materials that are end product of naturally occurring processes such as animal manure (goats, sheep, cattle, and chickens) contained high protein, carbohydrates and fat. The application of liquid organic fertilizers has the advantage of being able to increase nutrient uptake and quickly overcome nutrient deficiencies because the nutrients in it have broken down so that they are more easily to absorb efficiency [18].

Livestock waste, especially chicken manure, has a high nutrient content compared to cow/goat manure. Macro and micro elements in chicken manure consist of: N(1.72%), P(1.82%), K(2.18%), Ca(9.23%), Mg(0.86%), Mn(610%), Fe(3475%), Cu(160%), and Zn(150%) [12]. Iyiola and Ojo (2015) explains that poultry manure is able to stimulate the phytoplankton production and maintained the food chain in the waters. The results showed that poultry manure from chicken content high nutrient (2.34%), Phosphorus (0.24%), Potassium (0.03%), Calcium (0.10%), Magnesium (0.04%), Sodium (64.91 mg/kg), and Iron (15.63 mg/kg) compared to cow dung (nutrient (0.57%), Phosphorus (0.04%), Potassium (0.03%), Calcium (0 .10%), Magnesium (0.03%), Sodium (57.20 mg/kg), and Iron (13.91 mg/kg)). In addition, poultry manure fertilizer produce the highest plankton with 5141 Ind/mL of plankton compared to fertilizer from cow manure (3674 Ind/mL).

Silicate fertilizer contain a significant amount of nutrients (Nitrogen, Phosphorus, Iron, Sulfur, Magnesium, Potassium, and Calcium) with small particle sizes and easily soluble in water so as to accelerate the process of nutrient absorption by plankton cells. The dominant type of phytoplankton found during the research in treatment 2 consists of high benefical plankton (80%) compare to treatment 1. Moreover, treatment consist of high detrimental plankton such as Blue Green Algae (Cyanophyta*)* (37%), Crysophyta (5%), and Dinoflagellates (0.8%) compare to treatment 2 with Cyanophyta (21.3%), Crysophyta (4.4%), and Dinoflagellates (0%) (Figure 2).

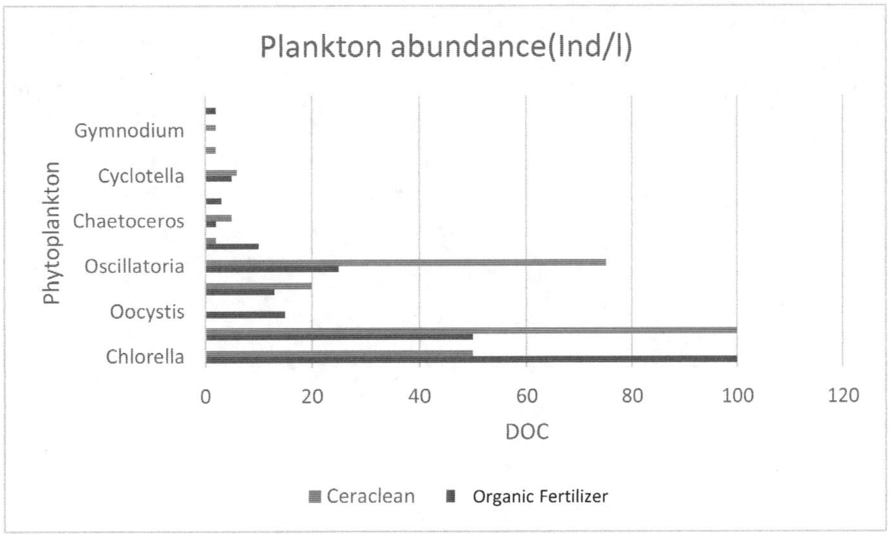

Figure 2. Plankton Abundance

High levels of nutrients in the aquatic environment have a positive impact in the form of increasing phytoplankton productivity and total shrimp/fish production. However, this also has a negative impact in the form of decreasing oxygen content, biodiversity and increasing the potential for the emergence and development of harmful phytoplankton species or Harmful Algal Blooms (HABs) [19]. Increasing levels of silica in pond water could affected the growth performance of diatoms. Therefore, the higher silica content in the pond, could decrease the concentration of nitrogen and phosphorus of water since the diatoms required more silica for cell division than nitrogen and phosphorus [20]. This results in changes the structure and diversity of plankton which lead to HAB [21]. Moreover, the high number of cyanophyta significantly increase the possibility of algal blooms [22].

Treatment 2 showed high performance of shrimp body weight compared to treatment 1 (Figure 3) with the daily growth rate was 3.74 grams and 9.8 grams of ABW in treatment 1. Meanwhile, the daily growth rate in treatment 2 was 4.00 grams and 20 grams of ABW. The total cumulative feed used in these two treatments was 299.64 kg with an FC Ratio of 1.1. The standard FCR for shrimp farming is 1.2 – 1.5.

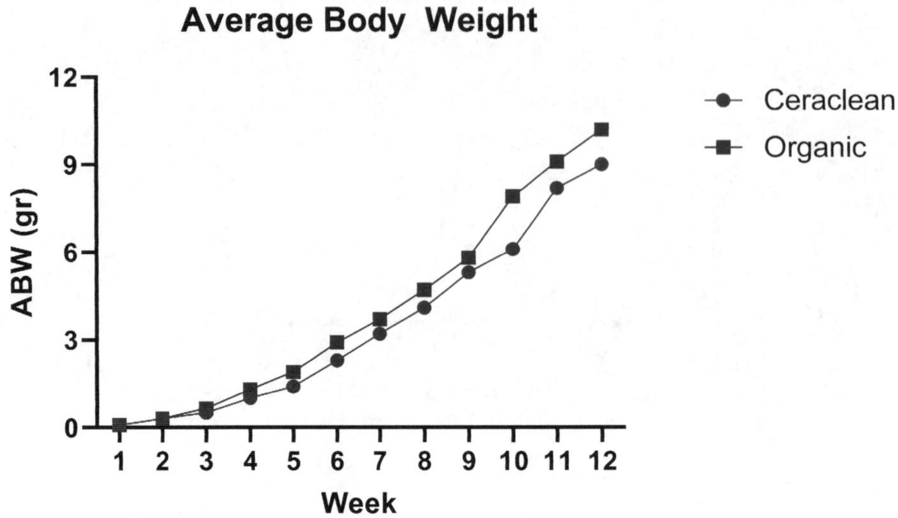

Figure 3. Vaname shrimp growth rate in treatment 1 (Ceraclean) and treatment 2 (organic fertilizer)

The growth performance of vanamei shrimp was strongly influenced by water quality, dissolved oxygen, and the availability of appropriate feed in terms of quality and quantity (Huang et al., 2021). Excess amount of feed has increased the content of ammonia and nitrite results in decreasing the oxygen levels due to the decomposition of organic matter [23]. The availability of dissolved oxygen in aquaculture pond water bodies is an important factor to support the growth, development, and life of biota in rearing ponds [24]. Especially in the household-scale vaname shrimp culture system that uses small pond sizes and high stocking density, the consistency of the presence of dissolved oxygen in the waters is very necessary to increase the productivity of vaname shrimp. The applied microbubble aerator is capable of producing small bubble sizes, tends to be stable, and does not break easily, so that it can expand oxygen transfer in the water column [25].

In general, the mean value of water quality observations in treatments 1 and 2 has not significantly differences. The microbubble aeration technology applied to household-scale cultivation is capable of producing relatively high current velocity (0.299 m/sec) and DO (5.05 ppm) compared to conventional aeration using a blower. The average temperature observation in treatment 1 was 28.9°C and 28.00°C for treatment 2. The nitrite value was 0.6 ppm (treatment 1) and 19.58 ppm (treatment 2); TAN 4.58 (treatment 1) and 2.7 (treatment 2) and average salinity of 21 ppt in both treatments.

4. Conclusion

The application of a microbubble aerator is effective for maintaining the stability of dissolved oxygen in the maintenance pond. The use of organic fertilizer (Ceraclean and chicken manure) significantly increasing the productivity of phytoplankton. Application of chicken manure showed the high production of beneficially plankton (80%) compare to ceraclean application (67,5%). In addition, the detrimental plankton were found in this research was dominated by blue green algae, crysophyta, and dinoflagellate.

5. References

[1] KKP 2020 *Household Scale Shrimp Cultivation*. (Jakarta: KKP Press).
[2] Flegel T W 2012 J. *Invertebr Pathol* 110 166–173.
[3] Nindarwi D D, Masithah E D, Zulian D, dan Suyoso A L 2019 *Eart and Environ Scie* 236 (1) 12072.
[4] Jefri M, Satyantini W H, Sahidu A M, Nindarwi D D dan Rozi R 2020 *Jurn Ilmi Perik dan Kel* 12 (1) 97-104.
[5] Joffre O M, Klerkx L, Khoa T N D 2018 *Agron Sustain Dev* 34 – 38.
[6] Larasati L G, Nimitkul S, dan Dewi N N 2021 *Eart and Environ Scie* 718 (1) 12097.
[7] Abo-Taleb H 2019 Importance of plankton to fish community Biological Research in Aquatic Science IntechOpen p. 83.
[8] Daruti D N dan Rahayu K 2018 *Eart and Environ Scie* 137 (1) 12028.
[9] Azmi K A, Arsad S dan Sari L A 2020 *Eart and Environ Scie* 441 (1) 12039.
[10] Brigs M, Smith S F, Subasinghe R, Phillips M 2004 Introduction and movement of penaeus vannamei and penaeus stylirostris in asia and the pacific RAP Publication 2004/10.
[11] Masithah E D, Nindarwi D D dan Rahma T 2019 *Eart and Environ Scie* 236 (1) 12017.
[12] Ulumiah M, Lamid M, Soepranianondo K, Al-arif M A, Alamsjah M A dan Soeharsono S 2020 *Jour of Aquac and Fis Heal* 9(2) 95-103.
[6] Utojo 2015 The diversity of plankton and water conditions in intensive and traditional ponds of the Biosphere. South Sulawesi Brackish Water Cultivation Research and Development Center.
[7] Chilmawati D 2016 *Aquat Proce* 7 39-45.
[8[Tackaert W and Sorgeloos P 2018 Semi-intensive culturing in fertilized ponds (In Artemia biology (pp. 287-316) CRC Press).
[9] Taucher J Bach L T Prowe A E Boxhammer T Kvale K and Riebesell U 2022 *Decli Natu* 605 (7911) 696-700.
[10] Iyiola A O and Ojo-Awo A P 2015 *J Agricul and Ecol Resea Intern* 3(4) 141 – 146.
[11] Yusnaini Y, Nur I and Baheri B 2022 Intern Confer on Improv Tropic Anim Produc for Fo Sec 461-466.
[12] Herawati V E dan Hutabarat J 2015 *Pena Akuatik* 12(1) 1 – 12
[13] Pal M, Yesankar P J, Dwivedi A and Qureshi A 2020 *J environ manag* 268 110687.
[14] Carey J C, Kathijo J, Pail J I, Lienne R S, Patrick K T and Jason R 2019 *Front in Ecol and Evolu* 7
[15] Weirich C A and Miller T R 2014 *Heal Car* 44 (1) 2 – 24.
[16] Huang H H, Luo T, Lei Y J, Kuang W Q, Zou W S and Yang P H 2021 *Aqua Rep* 21 100894
[17] WWF 2014 Vannamei shrimp cultivation, semi-intensive ponds with wastewater treatment plants (WWF-Indonesia Jakarta).
[18] Alday-Sanz V, Brock J, Flegel T W, McIntosh R, Bondad-Reantaso M G, Salazar M, Subasinghe R *Rev Aquac* 12 76–84.
[19] Basso A Hamad F and Ganesan P 2018 *J Chemi Engine* 13 (6) 2259.

6. Acknowledgements

The author expresses many thanks to the supervisors who have given a lot of knowledge and energy. The author also thanks the editors and reviewers for their suggestions and input on our manuscript.

IOP Conf. Series: Earth and Environmental Science 1273 (2023) 012061 doi:10.1088/1755-1315/1273/1/012061

Water profile of white shrimp (*Litopenaeus vannamei*) infected with white feces disease (WFD)

K K E Rahardjo[1], W H Satyantini[1,2], M Amin[1] and A T Mukti[1]

[1]Faculty of Fisheries and Marine, Universitas Airlangga, Surabaya, Indonesia

[2]Corresponding author: woro_hs@fpk.unair.ac.id

Abstract. In mid-2021, shrimp farmer reported case of mass deaths in white shrimp (*Litopenaeus vannamei*) in Pasuruan Regency, reaching 2.6-3 kg/day in Kalianyar Village, Bangil District. To determine the main pathogenic bacteria that cause mass mortality and the composition of microbes on rearing media infected with certain diseases and healthy ones, this study observed the water profile of ponds which include microbial composition in it using next-generation sequencing (NGS) technology. The research was conducted by taking healthy shrimp pond water samples and infected shrimp pond water samples. These results generally indicate that the disease is similar to white feces syndrome (WFS). The results of the NGS test showed that the main causative pathogen is *Vibrio vulnificus* because its presence in infected pond water is quite dominant.

1. Introduction

In the last year, Indonesia is the fifth-largest shrimp exporter in the world [1] where white shrimp (*L. vannamei*) is a commodity that has high economic value in Indonesia. In the last few years, white shrimp production has continued to increase by 15.7 %per year (2013–2017). It was recorded that in 2020, Indonesia's total shrimp exports reached 881,559.16 tons [2]. Seeing the high market demand for white shrimp and the many high potency areas that have not been utilized optimally, Indonesia has developed and implemented various shrimp farming systems, one of which is the intensive shrimp farming system. However, imprecise and prudent implementation of intensive cultivation can lead to various disease attacks due to bacterial and viral [3].

Shrimp farmers in Kalianyar, Bangil, Pasuruan reported mass deaths in mid-August 2021. In Kalianyar, clinical symptoms were detected when the shrimp entered 51 days of culture. Symptoms that appear include empty shrimp intestines, pale hepatopancreas, a continuous shrimp death of approximately 2-3 kg per day, decreased food appetite and growth retardation. The survival rate value obtained at the end of the shrimp farming cycles is 30%.

Several studies have stated that the source of pathogens is generally rearing media [4]. Therefore, this study aimed to identify the type of disease. The main pathogen causing mass death of white shrimp in shrimp ponds located in Kalianyar, Bangil, Pasuruan, East Java Province. The methods used in this study is next-generation sequencing (NGS) by comparing the composition and types of microbes in healthy and infected shrimp rearing media. This method allows researchers to know the predominance of bacteria and compare the composition of microbes in rearing media that are infected with the disease and the healthy ones. The results of this comparison will be able to detect pathogens that cause disease [5].

2. Material and Methods

2.1. Sampling and research location

The research was carried out for 6 (six) months starting from August 2021 - February 2022. Water samples were obtained from intensive shrimp ponds belonging to the shrimp farmer group located in Kalianyar, Bangil, Pasuruan, East Java. The DNA extraction process from water samples was carried out aseptically at the Microbiology Laboratory of the Department of Aquaculture, Faculty of Fisheries and Marine, Universitas Airlangga Surabaya. Meanwhile, the DNA concentration test was carried out at the Biomolecular Laboratory of the Faculty of Veterinary Medicine, Universitas Airlangga Surabaya. The DNA sequencing process was carried out at Novogene AIT Genomics Singapore PTE LTD Singapore. In general, this research will be carried out by comparing the microbial composition in WFD-infected shrimp and normal pond water (not infected with the disease).

2.2. Water sampling

The water samples were collected from the two intensive shrimp ponds, which consist of WFD-infected water sample and healthy water sample. For each pond, 180 ml of water was taken and put into three different sterile bottles with 60 ml for each. After that, 90 ml of ethanol was added to each bottle and then homogenized. The samples were stored at -80°C or -20°C until DNA extraction was carried out [6].

2.3 DNA extraction

The extraction bacterial DNA in the rearing water used the QIAamp DNA Mini Kit (Qiagen, Germany), as previously described by [7]. Then, the concentration of DNA was measured using a DNA-RNA protein Quantification Spectrophotometer (NanoDrop spectrophotometer). The microtubes were sterile at -20°C. The microtubes used DNA extraction.

2.4 Amplification of 16S rRNA gene

Gene amplification and microbial community analysis were carried out at the Novogene Biological Information Technology Co. 50 L PCR amplification reaction containing Phusion® High-Fidelity PCR Master Mix and 10 ng DNA template. PCR amplification was carried out under initial denaturing conditions at 98°C for 2 minutes. Next, followed by 35 cycles at 94°C during 30 seconds, 55°C during 30 seconds, 72°C during 90 seconds and final used condition 72°C during 10 minutes. The quantification and quality mixing used equal volumes of 1x loading buffer containing SYB green followed by electrophoretic mixing on a 2% agarose gel. Samples between 400 bp–450 bp for purification by Qiagen Gel Extraction Kit (Qiagen, Germany). The PCR product with same concentration and then the analysis of microbial in the Genome Sequencing Company (Novogene, China). Paired libraries were generated using NEBNext® UltraTM DNA Library Prep Kit, then quantified by Qubit and Q-PCR.

2.5 Data treatment

The final pairwise readings obtained to sample group on unique barcode used by barcode truncation and primary sequence. Paired final reads were combined using FLASH V1.2.7 to generate raw tags. The raw tags were filtered in several conditions to produce high quality clean tags according to Qiime V1.7.0. The clean tag was compared SILVA database with a confidence interval of 80% and resulted in an Effective Tag. All effective tag sequences were then analyzed using UPARSE software [8]. Sequences data with 97% similarity were assigned to OTU for annotation [10].

3. Results and discussion

3.1 Results

3.1.1 Clinical signs

Before the mass deaths, there were accompanying clinical signs such as empty shrimp intestines, pale hepatopancreas, decreased food appetite, growth retardation and shrimp swimming on the surface of the pond. In addition, white shrimp feaces appeared on the surface of the rearing media water.

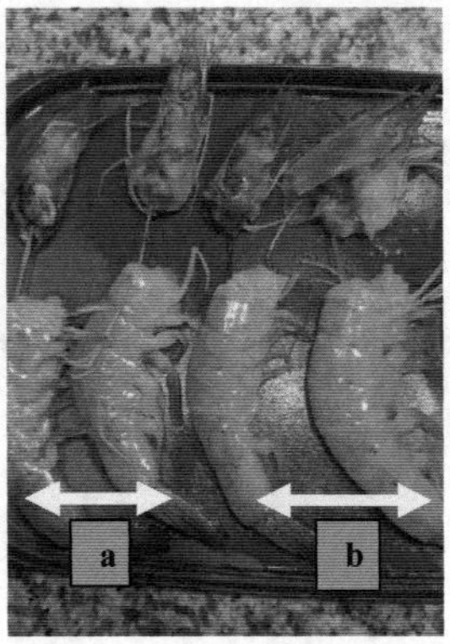

Figure 1. (a)Intestines and hepatopancreas of WFD- infected shrimp (b) Normal shrimp intestine and hepatopancreas

Figure 2. White feaces from white shrimp infected with WFD

In Figure 1, it can be seen the difference in the intestine and hepatopancreas between infected shrimp and normal shrimp. The intestines of infected shrimp look empty and white; this indicates that the shrimp have not eaten for several days so that no more metabolic waste is excreted as feaces. While in the normal shrimp's intestines look full and brown in colour which indicates that the shrimp are absorbing nutrients in the feed well. There are also differences in shrimp's hepatopancreas, where infected shrimps have pale hepatopancreas and have a smaller size than healthy shrimps. On the rearing media, white shrimp faecal were floated on the surface (Figure 2). In addition to the clinical signs of white shrimp, indications of disease attack can be seen from a continuous shrimp death, with the number of deaths as much as 2.6-3 kg/day at the farming location in Kalianyar village.

3.1.2 Bacteria profile of rearing water

Based on biometric analysis results of relative abundance, it was found that the main phyla that dominate in normal shrimp rearing media include Proteobacteria (54.16%), Actinobacteriota (37.70%), Patescibacteria (2.96%) and Bacteroidota (2.19%). Meanwhile, the main phyla found in WFD-infected rearing media were Proteobacteria (65.77%), Bacteroidota (16.11%) and Actinobacteriota (9.86%). In both groups of samples, there is one phylum that is quite dominating, namely the Proteobacteria phylum (Figure 3).

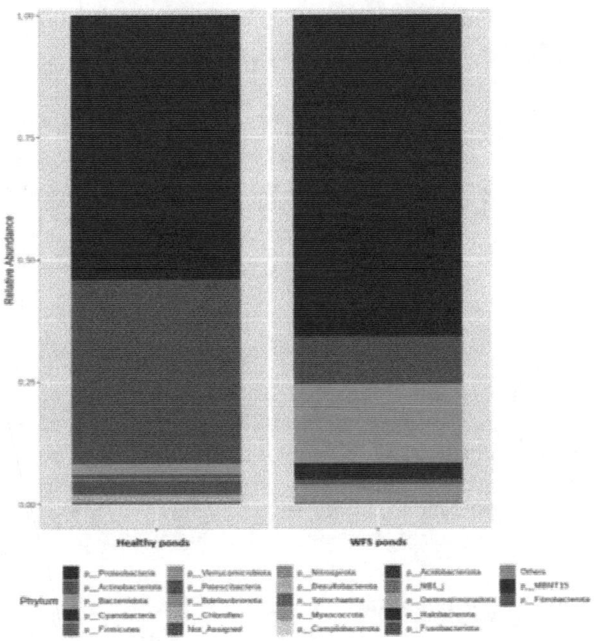

Figure 3. Bacterial composition at the phylum level in normal (healthy) and WFD-infected vaname shrimp rearing water.

From biometric analysis at the genus level in healthy pond, there were found that Paracoccus become the most dominant genus with relative abundance 27.39% and followed by PeM15 14.67% (Figure 4). Meanwhile in WFD-infected pond, found that the number of relative abundance of Vibrio is quite dominant that is up to 6.68% (Figure 5).

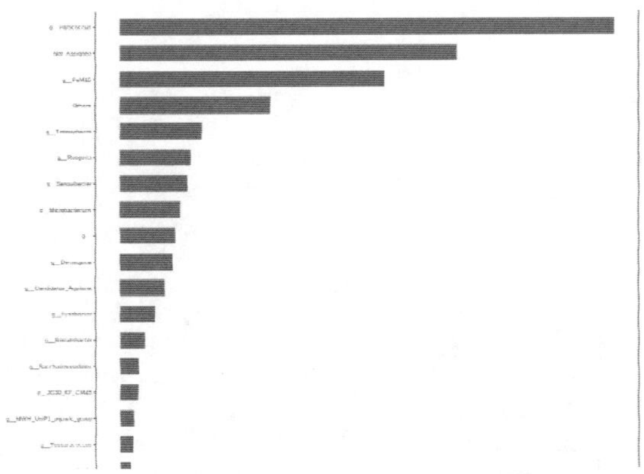

Figure 4. Bacterial composition at the genus level in healthy rearing media

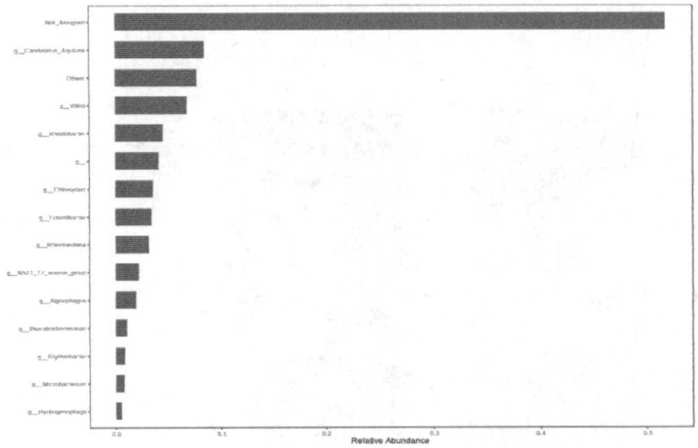

Figure 5. Bacterial composition at the genus level in WFD-infected rearing media

At the species level, from the biometric analysis results of healthy rearing media, there were found two types of bacteria that predominate in it, namely *Ruegeria lacuscaerulensis* and followed by *Sanguibacter* sp. with relative abundance respectively 3.90% and 3.70% (Figure 6). Meanwhile, in WFD-infected rearing media, it was found that the bacteria from the Vibrio group, namely *Vibrio vulnificus* which dominated with its relative abundance 5.14% (Figure 7).

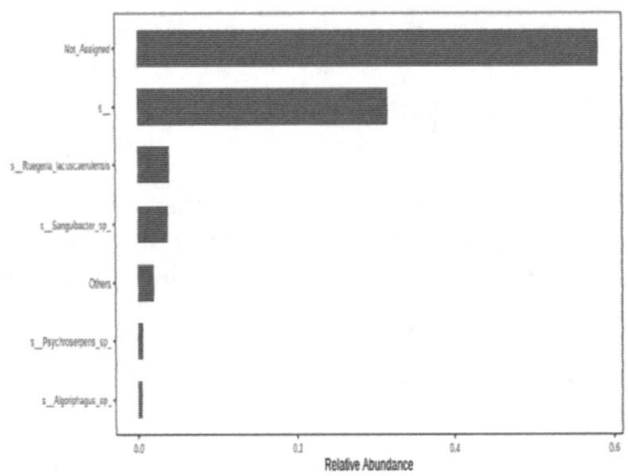

Figure 6. Bacterial composition at the species level in healthy rearing media

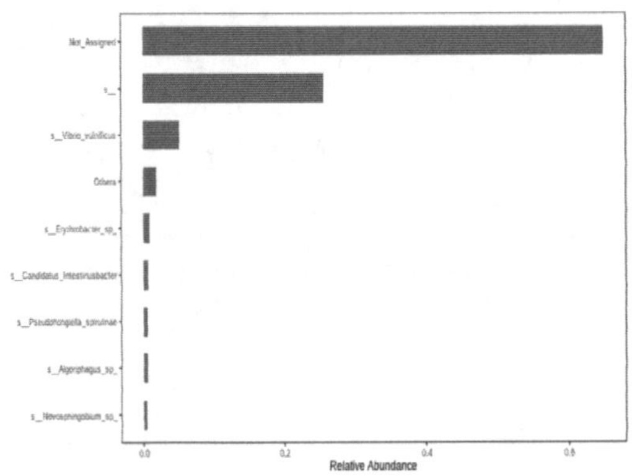

Figure 7. Bacterial composition at the species level in WFD-infected rearing media

3.2 Discussion

The aquatic environment is an integral component in aquaculture activities. Rearing media also contains several microorganisms that can cause to certain diseases. Based on the clinical signs that appear, the incidence of shrimp death is associated with the presence of white feaces disease (WFD) infection, which is related to the condition of rearing water quality. WFD spreads rapidly through the water, especially in ponds with high stocking densities, so rapid treatment is necessary [9]. Optimal pond water quality will provide living space for shrimp so that they can maximum growth. If the quality of the rearing water is appropriate to the needs of the shrimp, the survival rate will be high and the shrimp growth will be optimal.

Proteobacteria have been identified as the most dominant phyla in both sample groups. These results are in accordance with the research conducted by [10], that the Proteobacteria is a phylum that has a high levels of abundance in the white shrimp's intestine with an average relative abundance 40.83% and more than 80% in the *Penaeus monodon*'s intestine [11,12]. Based on these studies, *Proteobacteria* became the dominant phylum in several species of shrimp and became the only bacteria that commonly found in the digestive tract of shrimp from all farming locations. The phylum

Proteobacteria consisting of aerobic and facultative bacteria. Due to their diversity, proteobacteria have been shown to make up more than 40% of all prokaryotic genera of gram-negative bacteria.

Paracoccus become the most dominant genus at healthy pond. *Paracoccus* is a genus of bacteria in the family Rhodobacteraceae. One of species from *Paracoccus* that is *Paracoccus denitrificans* has an ability to reduce nitrate concentration in the water [13]. So, the nitrate content can be controlled. *P. denitrificans* can survive in ecosystems with fluctuating aerobic or anaerobic conditions.

In this study, Vibrio bacteria can be found at WFD-infected rearing media and it is in accordance with the statement of [12], that Vibrio is a potential opportunistic pathogenic bacteria for white shrimp (*L. vannamei*) Vibrio bacteria can convert from non-virulent to viral because it is influenced by two factors, such as significant environmental changes and being at a certain threshold of cell density [14]. The abundance of vibrio bacteria in an environment can cause vibriosis.

In the healthy rearing media, found that *Ruegeria lacuscaerulensis* become the most dominant species. *Ruegeria* constitutes one of the oldest genera of the so-called *Roseobacter* group, a collection of mostly marine members of the family *Rhodobacteraceae* in the class [15]. *Ruegeria lacuscaerulensis* possessed genes encoding the Dimethylsulfoniopropionate and cleavage pathways. *R. lacuscaerulensis* consumed only 28% of the DMSP [16]. Not many halophilic and mesophilic strains have optimum growth at 26 °C, pH 7.0, 3% salinity, require sodium and magnesium ions for growth so as to reduce nitrate to dinitrogen [17]. Based on previous studies, WFD is caused by different pathogens, such as *Vibrio parahaemolyticus* [18], *Vibrio anguillarum*, *Vibrio fluvialis*, *Vibrio alginolyticus* and *Vibrio mimicus* [19,20]. However, there are also who state that WFD is caused by the *Enterocytozoan hepatopenei* [21]. This study found that the dominant *Vibrio* bacteria in infected rearing media was *Vibrio vulnificus* and presumed to be the causative agent of the disease in this case. The results obtained are in accordance with the statement of [22] that *Alteromonas, Photobacterium, Pseudoalteromonas and Vibrio* are the dominant bacteria that are found when shrimp disease occurs, both in infected and healthy pond water. The emergence of WFD cases in Kalianyar village is in August, when in that month has entered the summer season. *Vibrio* bacteria will have an increase in growth when entering the summer. This is to be expected as a cause of WFD attacks that occur in two regions at the same time. The presence and dominance of Vibrio bacteria cause these bacteria are easy to penetrate the shrimp immune system and reach the hepatopancreas tubular cells and then disrupt the epithelial cells.

V. vulnificus belongs to halophilic and mesophilic Gram-negative bacteria that are naturally found in estuary waters and marine environments [23,24]. These opportunistic pathogenic bacteria have been isolated from various seafood such as shrimp, fish, oysters and shellfish. The level of salinity and the temperature of the water media greatly affect the amount of *V. vulnificus* in its natural habitat. The suitable salinity and temperature for the growth of these bacteria respectively are 5-25 ppt and >20°C [25]. In the study conducted by Li *et al.* (2019), showed that the GXFL1-3 strain was pathogenic to zoea and post-larvae of *Macrobrachium rosenbergii* with an LD_{50} dose respectively 1.16×10^6 CFU/ml and 1.45×10^6 CFU/ml. [26] in their study stated that *Litopenaeus vannamei* juvenile that infected with *V. vulnificus* MTCC 1145 with a lethal dose of 1×10^7 CFU/shrimp, will died after 7 days of injection. *V. vulnificus* produces protease, amylase, urease, lecithin and hemolysin activities which can support the strong virulence of this bacterium against *Macrobrachium rosenbergii*. In *V. vulnificus* also found the TCP gene associated with virulent regulation [27]. The growth of *Vibrio* bacteria is also determined by the water quality of the rearing media. Poor water quality will support the quick proliferation of bacteria.

4. Conclusion

Clinical symptoms of infected shrimp in Kalianyar Village showed that the shrimp were infected with white faeces disease. The bacterial profile of pond water shows that the bacterial phylum that dominates the WFD-infected rearing media is Proteobacteria. At the lower taxa level, *Vibrio vulnificus* was found. However, to confirm the level of bacterial pathogenicity, further tests are needed.

5. References

[1] Annisa K, Sutinah S and Jusni J 2022 *Intern J Soc Sci Res and Rev.* 5(1) 134-143
[2] KKP 2022 *Ministry of Fisheries and Maritime Affairs of the Republic of Indonesia.*
[3] L Arafani, M Ghazali, dan A Muhamad 2016 *J Vet.* 17(1) 88-95.
[4] Hamzah, Srinawati, Jumriadi, A A Afandi, M Asa'at, Herawaty dan N Hartanto 2019 *Pros. Simp. Nasional Kelautan dan Perikanan* VI. pp 5-9. (Makassar: Universitas Hasanuddin).
[5] W Gu, S Miller and CY Chiu 2019 *Annual Rev of Path: Mec Dis.*14(1) pp 319-38
[6] IR Pedrosa-Gerasmio, R Nozaki, H Kondo and I Hirono 2020 *Aqua Res.* 51(10) pp 4075-86
[7] M Amin, R R C Kumala, A T Mukti, M Lamid, D D Nindarwi 2022 *J Aqua.* 550 737849
[8] S Gao, L Pan, F Huang, M Song, C Tian and M Zhang 2019 *Aquaculture* 499 109-118.
[9] D Rusadi, Wardiyanto and R Diantari *E-Jurnal Rekayasa dan Teknologi Budidaya Perairan* 7(2) 909
[10] L Fan, Z Wang, M Chen, Y Qu, J Li, AP Zhou and JL Zou 2019 *Sc. Total Environ.* 657 1194–1204
[11] W Rungrassamee, A Klanchui, S Maibunkaew, S Chaiyapechara, P Jiravanichpaisal and N Karoonuthaisiri 2014. *PLoS ONE* 9(3) 1-11
[12] W Rungrassamee, A Klanchui, S Maibunkaew and N Karoonuthaisiri 2016 *J Invertebrate Pathol.*133 12–19
[13] B Bauman, M Snozzi, A J B Zehnder and J R van deer Meer 1996 *J Bacteriol.*178(15) 4367-74.
[14] J Zhou, W Fang, X Yang, S Zhou, L Hu, X Li, X Qi, H Su and L Xie 2012 A *PLoS One* 7(2) e29961
[15] M J Pujalte, T Lucena, M A Ruvira, D R Arahal and M C Macián 2014 The Family *Rhodobacteraceae.* In : Rosenberg E, DeLong EF, Stackebrandt E, Lory S, Thompson F (eds). The *Prokaryotes-Alphaproteobacteria and Betaproteobacteria* 4th ed (Berlin : Springer Verlag) pp. 439-512
[16] J S Wirth, T Wang, Q Huang, R H White and W B Whitman *mBio Res Article App and Environ Sci.* 11(2) e00329-20
[17] D R Arahal, T Lucena, L Rodrigo-Torres and M J Pujalte 2018 *Inter J Systematic and Evol Microbiol.* 68 2515-22
[18] X Zhang , J Sun, Z Han, F Chen, A Lv, X Hu, X Sun, H Qi and Y Guo 2021 *Aquaculture* 544 737061
[19] S Supono, W Wardiyanto and E Harpeni 2019. *AACL Bioflux* 12(2) 417-25
[20] M Jayadi, A Prajitno and M Maftuch 2016. *InterJ of Chem Tech Res. 9*(7) 448-52
[21] W M H W Sajiri, M H Borkhanuddin and B C Kua 2021 *J Dis Aquatic Organism.* 144 1-7
[22] Maftuch, I Kurniawati, A Adam and I Zamzami *Egyp J of Aquatic Res.* 42(4) 405–10
[23] S L Drake, A DePaola and LA Jaykus 2007 *Comp Rev in Food Sci and Food Saf.* 6(4) 120-44
[24] D A Linkous and J D Oliver *FEMS Microbiol Let.* 174(2) 207-14
[25] M L Motes, A DePaola, D W Cook, J E Veazey, J C Hunsucker and W E Garthright 1998 *App and Environ Microbiol.* 64 4 1459-65
[26] Patel M P, Vadher K H and Tank P R 2018 *J Entomol and Zoo Stud.* 6(6) 1258-1262
[27] X Li, Y Zhou, Q Jiang, H Yang, D Pi, X Liu, X Gao, N Chen and X Zhang 2019 *Macrobrachium rosenbergii Microbial Path.* 127 166-71

6. Acknowledgements

I'm grateful for the patience of my lecturers at the Faculty of Fisheries and Marine, and their invaluable feedback. I do not forget to thank my parents because their belief in me has kept my spirit and motivation throughout this process.

Assessment of coral reef condition using upt (underwater photo transect) method on gili labak island madura indonesia

D Irawan[1], A T Mukti[2,5], S Andriyono[3], F F Muhsoni[4]

[1]Study Program of Master Program of Fisheries Sciences, Department of Aquaculture, Faculty of Fisheries and Marine, Universitas Airlangga, Surabaya, East Java, Indonesia
[2]Department Aquaculture, Faculty of Fisheries and Marine, Universitas Airlangga, Surabaya, East Java, Indonesia
[3]Department of Marine, Faculty of Fisheries and Marine, Universitas Airlangga, Surabaya, East Java, Indonesia
[4]Study Program of Water Resource Managenement Faculty of Agriculture, Universitas Trunojoyo Madura, Bangkalan, East Java

[5]Corresponding author: akhmad-t-m@fpk.unair.ac.id

Abstract. Coral reefs are unique, complex and high-productivity marine ecosystems that make coral reefs a place to grow and breed most fish and biota in the waters, and physically the function of coral reefs is to withstand coastal abrasion, break waves and so on. To be able to manage coral reefs properly, valid data support is needed as a basis for formulating a policy. Gili Labak Island is one of the islands that is currently a tourist attraction because it has a very wide expanse of coral reefs. The UPT method is a method that utilizes technological developments, because in this method digital camera technology and computer software are used. Data retrieval in the form of underwater photos is done by shooting using an underwater digital camera. The photographs were then analyzed using computer software to obtain quantitative data. This study aims to calculate the biota in the transect. The results obtained Coral 72.92%, Dead coral (DC) 2.03, Dead Coral Algae (DCA) 9.09%, Soft Coral (SC) 0.33%, Rubble (R) 6.15%, Sand (S) 8.98%, Rock (RK) 0.52%.

1. Introduction

Coral reefs are ecosystems that have important and complex economic values and have high diversity, but are very vulnerable to natural factors and human activities [1], Coral reef tourism also has economic value that can create new jobs for the community [2]. As tourism activities are carried out by utilizing coral reef areas for tourism activities, will always experience pressure and lead to damage [3]. The development of methods used to assess the condition of coral reefs in the world in recent decades continues to grow. The various methods used are inseparable from weaknesses, both in terms of technical field implementation, human resource capabilities, and the amount of budget required for the implementation of each method. Using the manta tow method English *et al.* [4]. can reach a wider study area in a shorter time, but it will be difficult and dangerous if done in an area full of massive hard corals with large sizes. Using the Line Intercept Transect (LIT) method. It takes longer time than the manta tow method, and requires one's ability to dive and the ability to identify coral to the species level. The UPT (Underwater Photo Transect) method was developed to speed up and simplify coral reef analysis using software.

Sumenep Regency is one of the regencies in East Java which has the most islands (approximately 115 named islands). This condition makes this area has great potential for coral reefs [5]. These islands have great potential to be developed as tourist areas, including Gili Labak [6,7,8]. Gili Labak Island has great potential for the development of diving ecotourism, snorkeling ecotourism, and beach ecotourism. The results of the 2016 survey, the area of coral reefs on Gili Labak Island reached 66 ha. The condition of live coral reached 48.7% and dead coral reached 51.3%. The dominant type of coral is Stylophora (Branching) reaching 29.27% [6].

Gili Labak Island tourism has been developed by the Sumenep Regency government since 2014, causing many visitors to come to see coral reef tourism so that the risk of damage to coral reefs is very high [9] . Gili Labak Island is a favorite tourist destination on the island of Madura and has a coral reef cover of 73% and is in the good category [10]. The percentage of coral lifeform cover on Gili Labak Island, especially in the snorkelling area, is dominated by live coral as much as 74% and abiotic elements by 22%. The coral species that dominate Gili Labak Island are Acropora Branching by 19.88% and Coral Foliose by 10.25%. During 6 weeks of observation, there was a decrease in total coral by 0.64% which was categorized as lightly damaged, where most of the damage occurred to corals with branching growth such as Acropora Submassive and Coral Submassive. The decline in the percentage of high coral cover occurred after snorkeling activities (after) carried out by tourists. Analysis of the potential impact of marine tourism (DWB) snorkeling on coral reefs in Gili Labak waters for 6 weeks of observation is in the low category, ranging from 0.052% to 0.085% [9].

2. Materials and methods

The study was conducted in July 2022 in the waters of Gili Labak Island (Figure 1). The percentage of coral cover data was collected using the UPT (Underwater Photo Transect) method taken at the place where tourists go snorkeling with a 50 meter transect length [11]. The measurement of supporting parameters includes: temperature, sanitation, pH, DO, and brightness at the point of observation.

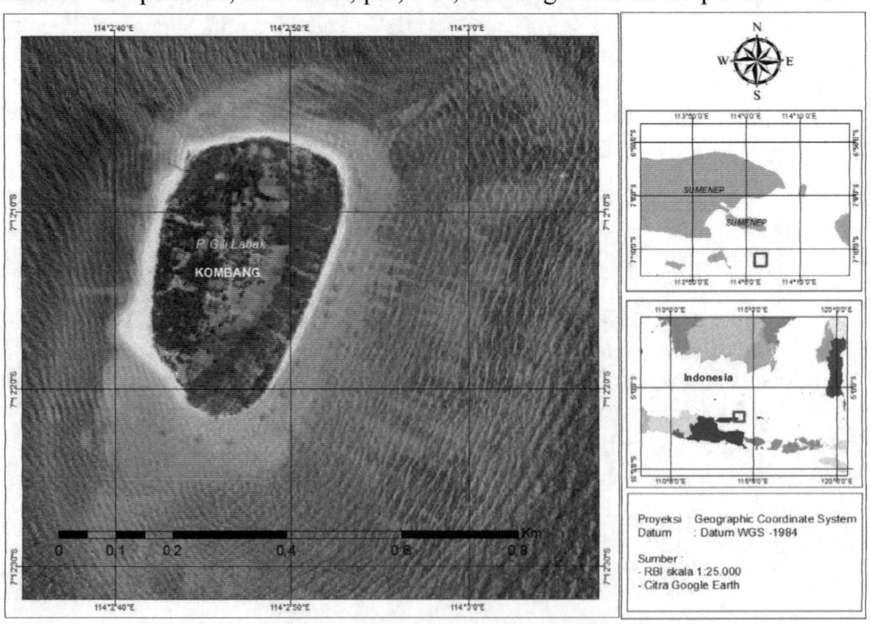

Figure 1. Map of research locations on Gili Labak Island, Madura, Indonesia

Coral reef data analysis was analyzed using CPCe software computer software. Calculation of the condition of coral reefs using the formula:

Percentage of coral reef cover = $\frac{\text{Number of Category i}}{\text{Number of random points}}$ x100% [11].

Table 1. Coral cover categories follow [12].

No.	Percentage (%)	Category
1	0-24,9	Bad
2	25-49,9	Currently
3	50-74,9	Well
4	75-100	Very well

3. Results and discussion

Table 2. Physical and chemical parameters of the waters

No.	Time	Temperatur (°C)	pH	Brightness (%)	Sanility (‰)	DO (mg/l)	TSS	Current Speed (m/s)	CD
1	Morning	28,8	8,29	100	31	11,2	0,0422	21	51,4
2	Afternoon	31,9	8,47	100%	30	14,4	0,0504	17,22	52,2
3	Afternoon	31	8,48	100	34	13,9	0,047	11,49	52

Measurement of the physical and chemical parameters of the waters in this study aims to determine the suitability of the physical and chemical parameters of the water with the needs of corals. The results of the measurement of the physical-chemical conditions of the Gili Labak island waters are in good condition. This is in accordance with the Regulation of the Minister of the Environment No. 51 of 2004.

The water parameter of acidity (pH) obtained from the results of the average time measurement is 8.41. This value is said to be good for coral reef growth. According to Madduppa and Zamani. [13] The range of pH values suitable for coral reefs is 7 to 8. Apart from measuring chemical parameters, physical parameters of the waters such as brightness and water temperature are also carried out. According to Nyabakken. [14] The optimal temperature for the growth of coral biota is between 25-30°C. However, the temperature of 30-35°C can be tolerated by coral reefs. Based on observations, the average water temperature is 30°C. High brightness levels can affect the growth of coral reefs, brightness is directly related to the level of light intensity that enters the water column. High brightness has a positive impact on zooxanthellae because it can facilitate photosynthesis and can help growth. The average brightness level of observations has a value of 100% at a depth of 8 meters. The condition of the waters at the study site has currents that are influenced by tides. However, corals still survive to live, because currents are needed by corals to obtain food in the form of zooplankton and oxygen and can clean corals from sediment.

The results showed that the highest coral cover was in the category of hard coral (HC) 72.92%, Dead Coral Algae (DCA) 9.09%, Sand (S) 8.92%, Rubble (R) 6.15%, Dead Coral (DC) 2.03%, Rock (RK) 0.52%, and Soft Coral (SC) 0.33%. The results of the analysis show that the percentage of live coral on Gili Labak Island is in the good category. Benthic cover on Gili Labak Island can be seen in Figure 2.

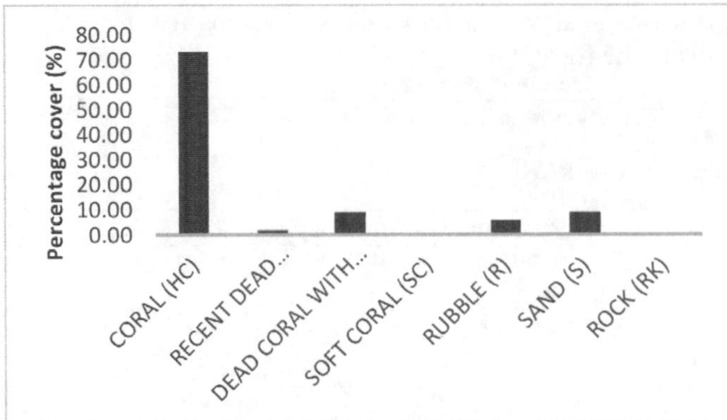

Figure 2. Percent closure on Gili Labak Island

The highest category of coral coral cover (HC) at the study site was due to the observation location being in waters that were directly opposite a closed area with weak waves. The coral was found growing in deeper places. It is natural selection for these animals to protect their colonies from damage.

The results of live coral data on Gili Labak Island reached 72.92% consisting of Acropora Branching (ACB) 19.88 %, Acropora Digitate (ACD) 23.87%, Acropora Encrusting (ACE) 1.44%, Acropora Submassive (ACS) 2.09 % , Acropora Tabulate (ACT) 10.27 %, Coral Branching (CB) 0.07%, Coral Encrusting (CE) 0.20%, Coral Foliose (CF) 6.80%, Coral Massive (CM) 3.40 % , Coral Mushroom (CMR) 1.90%, and Coral Submassive (CS) 3.01%. The results of live coral on the island of Gili Labak can be seen in Figure 3.

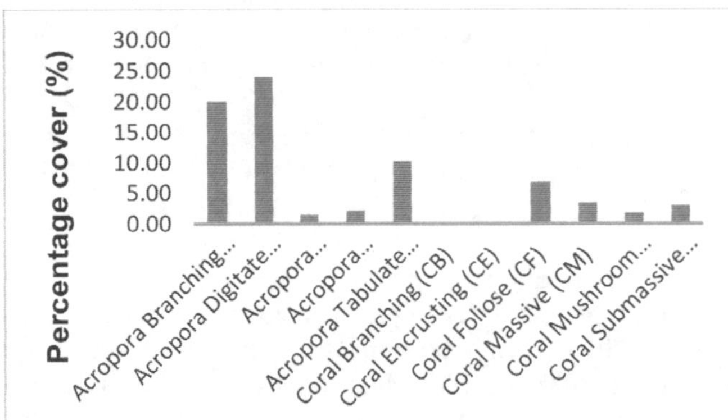

Figure 3. Results of live coral analysis on Gili Labak Island

Types of coral reefs in the waters around Gili Labak Island are fringing reefs. The basic forms of complete coral colonies are massive, branching, encrusting, and foliaceous corals. The type and number of life forms, especially those with branched and massive colonies, are high enough to support the life of marine organisms. The high percentage of CB is supported by clear water conditions and far from sedimentation so that sunlight penetration can penetrate the waters up to 100% at a depth of 3-10 m. According to , in clear waters or low sedimentation, corals will be found in branched and tabulated forms.

4. Conclusion

Based on the results of monitoring the condition of coral reefs on Gili Labak Island, it can be concluded that: The percentage of live coral cover using the UPT method is 72.92% so it is categorized as good. This shows that Gili Labak Island has quite good coral reefs but is under pressure from destructive human activities or due to pressure from natural factors.

5. References

[1] Nurrahman Y A, Faizal I 2020 *Akuatika Indones.* *5*(1) 27–32.
[2] Akhmad D S, Supriharyono and Purnomo P W 2018 *J Ilmu dan Teknol Kelaut Trop.* 10(2) 419–430.
[3] Roche R C, Harvey C. V, Harvey J. J, Kavanagh A P, McDonald M, Stein-Rostaing V R, and Turner J R 2016 *Environ Manage.* 58(1) 107–116.
[4] English, S., Wilkinson, C., & Baker, V. 1998. Survey manual for tropical marine resources. Second edition. *Survey Manual for Tropical Marine Resources. Second Edition.*
[5] Muhsoni F F, Syarief M, and Effendi M 2011 *J Ilmu Kelaut.* *4*(1), 96–101.
[6] Muhsoni F F 2016 *Semin Nas Perikan dan Kelaut VI, Fak Perikan dan Ilmu Kelautan, Univ Brawijaya Malang*. 1–6.
[7] Muhsoni, F. F. 2017. *Potential and Management of Gili Gili Labak Island*. UTM PRESS.
[8] Muhsoni F F, and Nuraini C 2006 *J Protein*., 13(1) 87–94.
[9] Insafitri I, Asih E N N, and Nugraha W A 2021 *Bul Oseanografi Mar.* 10(2), 151–161.
[10] Insafitri I, Asih E N N, and Nugraha W A 2020 *AACL Bioflux*, 13(6), 3789–3797.
[11] Giyanto. 2013. *Underwater photo transect method for assessing the condition of coral reefs. Oseana, Volume XXX.*
[12] Gomez Yap, H. T., E. D. 1984. Monitoring reef conditions. In *Coral reef management handbook.*
[13] Madduppa H, and Zamani N P 2011 *J Indones Coral Reefs*. 1(2). 137-146
[14] Nyabakken, J. W. 1992. Marine Biology an Ecological Approach Translated by H.M. Eidman, Koesbiono *et al. Jakarta : Gramedia.*

6. Acknowledgement

This study was partially supported by the Directorate of Research, Technology, and Community Services (DRTPM), General Directorate of Higher Education, Research, and Technology, Ministry of Education, Culture, Research, and Technology, Republic of Indonesia through the Grants of Penelitian Tesis Mahasiswa (PTM) with Decree Number: 1004/UN3/2022 and Contract Number: 085/E5/PG.02.00.PT/2022 and 978/UN3.15//PT/2022.

Physicochemical characteristics and organoleptic catfish sausage (*Clarias* sp.) with the additionof gelatin as binding agent

R F Yuzi[1], E Saputra[2], and L Sulmartiwi[2,3]

[1]Fishery Product Technology Study Program, Faculty of Fisheries and Marine, Airlangga University
[2]Department of Marine, Faculty of Fisheries and Marine, Airlangga University, Surabaya, East Java, Indonesia.

[3]Corresponding author: laksmi-s@fpk.unair.ac.id

Abstract. This study aims to determine the effect of adding gelatin as a binding agent on the physicochemical and organoleptic characteristics of catfish sausage (*Clarias* sp.). This study used a completely randomized design (CRD) consisting of 5 treatments with commercial fish sausage and a gelatin concentration of 0.25%, 0.5%, 1%, and 1.5% with 4 replications. The method used in this research includes proximate analysis, texture analysis, and organoleptic testing. The results of this study indicated that the addition of gelatin had a significant effect ($P < 0.05$) on the value of gel strength, product texture and consumer acceptance of catfish sausage (*Clarias* sp.), As well as the resulting moisture and protein content. The best gelatin concentration was produced by the addition of gelatin of 1.5%. Gelatin with a concentration of 1.5% produces the best gel strength with the characteristics of a compact and not too hard gel, has a water content of 68.82% and has a protein content of 16.95%.

1. Introduction

Catfish (*Clarias* sp.) is a freshwater fish that is widely cultivated in almost all of Indonesia. This is because catfish is a leading commodity, and has good market prospects [1]. Some of the advantages of catfish are that it has a fairly high nutritional content, economic value, fast growth and easy maintenance [2]. Catfish (*Clarias* sp.) contains 17.7% protein, 4.8% fat content, 1.2% minerals, 76% water, and 0.3% carbohydrate content [3]. Because of its advantages and nutritional content, increasing catfish consumption can be developed by diversifying product processing

One of the efforts to diversify the processing of catfish (*Clarias* sp.) is fish sausage. Sausage is a mixture of mashed meat and mixed with salt, spices and gelatin then put into a sausage mold and processed through heating by boiling. The process of making sausages is strongly influenced by the ability of the meat with the addition of salt, water and fillers such as tapioca flour and other ingredients that function to form an emulsion with stable fat. The stability of this emulsion is indicated by the non-separation of fat from the sausage [4].

It is necessary to have a binding agent that has high protein, can form and stabilize emulsions, increase water binding capacity, and reduce cooking losses [5]. Another use of the addition of a binder is to increase the flavor and improve the characteristics of a food product [6]. One of the materials that can be used as a binder is gelatin [6]. Gelatin functions as a thickener, stabilizer, emulsifier, gelling agent, adhesive agent and foam-forming agent [7]. Gelatin as a binder (binder) can work at low temperatures to glue mashed meat. Gelatin also improves the physicochemical and organoleptic qualities so that it can be accepted by consumers.

2. Material and methods

2.1. Material

The material used in this study refers to a similar study by [8] which is catfish meat (*Clarias* sp.) 1000 gr, 15% tapioca flour, 20% ice water, 5% vegetable oil, 4% garlic, ground pepper. 0.5%, 1% sugar, 2% salt, commercial gelatin derived from animal by-products 0% 0.25% 0.5% 1% and 1.5%. The chemicals used in this study consisted of NaOH, 95% - 97% H_2SO_4, 30% H_2O_2, methyl red indicator, protein base, 4% H_3BO_3, aquades, chloroform, K_2SO_4, and $Na_2S_2O_3$ [8].

2.2. Making catfish sausage

Grind the fish meat then add salt, sugar, on, tapioca flour, and gelatin (0.25%; 0.5%; 1%; 1.5%). Then stir until homogeneous. During kneading, ice water is added to keep the temperature cool, which is below 20°C this is to prevent denaturation of proteins which are very important as emulsifiers. The homogeneous dough is put into a piping bag or triangular plastic and then put into a sausage mold accompanied by binding, the ends of the mold are tied with thread, the length of the sausage is made about 10 centimeters. After that, it was boiled at 90°C for 20 minutes [8].

2.3. Water content analysis

Moisture content was analyzed using the gravimetric method according to the Association of Official Analytical Chemists (1995) in [9]. The porcelain dish was cleaned and dried in an oven at a temperature of 105°C - 110°C for one hour. Cooling in a desiccator was carried out for 30 minutes and the weight of the sample was weighed (A gram). The sample is weighed as much as two grams and placed in a porcelain dish with a known weight (B grams). Drying was carried out in an oven at a temperature of 105°C - 110°C for 24 hours. Cooling was carried out in a desiccator for 30 minutes and the weight of the sample was weighed (C grams).

$$Water\ content = \frac{(\mathrm{B}-\mathrm{C})}{(\mathrm{B}-\mathrm{A})} \times 100\%$$

Information:
A = dry weight of the cup (grams)
B = dry weight of the cup and initial sample (grams)
C = dry weight of the cup and sample after drying (grams).

2.4. Ash content analysis

The sample was weighed 5 g in a porcelain dish whose weight was known. The sample is illuminated. Ashing is carried out in a kiln at a temperature of 100°C for 1 hour, then the temperature is raised to 300°C for 2 hours and the final step is to increase the temperature to 500°C for 2 hours. The ash obtained was cooled in a desiccator for 15 minutes and weighed.

$$Ash\ content = \frac{\text{ash weight}}{\text{sample weight}} \times 100\%$$

2.5. Protein content analysis
Protein content testing was carried out using the AOAC (2005) method. The sample was weighed as much as 0.2 g which had been mashed into a Kjeldhal flask, added 2 g of catalyst K2SO4: CuSO4 (1:1) then added with 3 ml of concentrated H2SO4. The sample is boiled for 1- 1.5 hours or until the liquid is clear. The flask and its contents were cooled and the contents were transferred to a 500 ml Erlenmeyer. The flask containing the sample from the destruction was mounted on a distillation apparatus and 15 ml of 40% NaOH solution was injected until a black precipitate was formed. Prepare a container containing 25 ml of 0.02N H2SO4 in an Erlenmeyer and drop it with the Mengsel indicator (a mixture of 0.02% methyl red in alcohol and 0.02% methyl blue in alcohol in a ratio of 2:1) as much as 2-4 drops and then placed under condenser. Then removed when the volume has reached 125 ml. The results of the distillation were then titrated using 0.02N NaOH until the color changed to bluish green. The determination of the blank is carried out in the same way.

$$Protein\ content = \frac{(\mathrm{A} - \mathrm{B}) \times \mathrm{N} \times 0{,}014}{\text{sample weight}} \times \text{conversion factor} \times 100\%$$

Information :
A = ml NaOH for blank titration
B = ml NaOH for sample titration
N = Normality of NaOH
Conversion factor = 6.25

2.6. Fat content analysis
The fat content test was carried out using the AOAC method (2005). The sample was weighed as much as 5 g wrapped in filter paper, then placed in a Soxhlet extraction apparatus. The condenser is installed above it and the fat flask is below it. Hexane fat solvent was put into the fat flask as much as part of the flask, then reflux was carried out for ± 6 hours until the solvent dropped back into the fat flask and became clear in color. Then the sample in the shell after extraction was dried in an oven at 70 oC until it reached a constant weight, then cooled in a desiccator.

$$Fat\ content = \frac{\text{weight before extraction} - \text{weight after extraction}}{\text{sample weight}} \times 100\%$$

2.7. Carbohydrate content analysis
Carbohydrate content of the product is calculated based on the formula:

$$Carbohydrate\ content = 100\% - (\%\text{water} + \%\text{protein} + \%\text{fat} + \%\text{ash})$$

2.8. Texture analysis
Measurement of the physical properties of the texture of catfish sausage (*Clarias* sp.) with the addition of gelatin using a tensile strength tool. The tensile strength device was turned on and waited five minutes. The measured material is placed directly under the tool needle. The load is released and the indicator scale is read after the tool stops. The value listed on the monitor is the texture value expressed in Newton (N) units [10].

2.9. Organoleptic analysis
According to [11] organoleptic analysis of catfish sausage (*Clarias* sp.) with the addition of gelatin was carried out by a hedonic test involving 30 panelists because this catfish sausage was intended for consumers. The test was carried out by giving 20 samples at random, each sample had been assigned a different code. Panelists are required to give an assessment of the sample according to the existing hedonic scale.

3. Result and discussion

3.1. Proximate analysis of fish sausage

Proximate analysis is a chemical test to determine the nutrient content of a feed or feed raw material [12]. Proximate analysis has advantages such as cheaper analysis costs, and can produce broad analysis. The proximate analysis of catfish sausage using commercial fish sausage and the addition of gelatin with concentrations of 0.25%, 0.5%, 1%, and 1.5% ranged from 66.66 to 68.82% (water content), 1.58 to 4.24% (ash content), 13.95-16.95% (protein content), 3.41 to 6.38% (carbohydrate content), and 10.94-16.37% (fat content) (Table 1).

Table 1. Proximate analysis of fish sausage

	Proximate Analysis				
Treatment	Water Content (%)	Ash Content (%)	Protein Content (%)	Carbohydrate Content (%)	Fat Content (%)
Commercial fish sausage	66,66^{d} ± 0,51	4,24^{a} ± 0,60	13,95^{d} ± 0,33	6,38^{a} ± 0,55	16,37^{a} ± 0,56
P1 (0,25%)	67,23^{cd} ± 0,68	3,07^{b} ± 0,71	14,71^{c} ± 0,47	5,38^{ab} ± 1,26	15,36^{b} ± 0,62
P2 (0,5%)	67,64^{bc} ± 0,48	2,56^{bc} ± 0,78	15,89^{b} ± 0,54	4,36^{bc} ± 1,27	13,78^{c} ± 0,70
P3 (1%)	68,34^{ab} ± 0,56	2,20^{bc} ± 0,92	16,64^{a} ± 0,35	3,97^{bc} ± 1,04	11,88^{d} ± 0,80
P4 (1,5%)	68,82^{a} ± 0,73	1,58^{c} ± 0,60	16,95^{a} ± 0,42	3,41^{c} ± 0,61	10,94^{d} ± 0,47
SNI 7755:2013	Max 70	Max 5	Min. 9	Max 8	Max 25

Moisture content is a factor that can affect the boiling process that takes place in a material. The water content in foodstuffs is influenced by raw materials and processing processes, the steaming process with hot steam tends to increase the water content [13]. Water content has a role in food quality. The water content in food determines the acceptability, freshness and durability of food [14]. The function of water is as a medium for transporting nutrients, regulating body temperature, maintaining a balance of blood volume [15].

The water content in cooked sausages is approximately 45.55% of the total weight of sausages. Water affects the elasticity and tenderness of the sausage at the final stage. Water and fat are the determinants of quality, along with the increase in water and fat sausage, the elasticity and tenderness will also increase. The addition of water is also important to give the characteristics of the dough, in this case to the emulsion [16].

The water content of sausages is closely related to the binding of water by protein, namely high water binding will reduce the release of water during boiling, thus the water content of sausages will be high. Vice versa, low water binding ability will cause high levels of water loss during boiling, so that the sausage water content will be low [17].

The ash content aims to determine whether or not a food ingredient is good, distinguish between native and synthetic foods as well as a parameter of a material [18]. The higher the ash content, the worse the quality of the food. Ash consists of mineral components. The mineral content in the ash content cannot explain the amount and type of minerals contained in it [12].

The ash content of meat is closely related to the water content and protein content of a fat-free tissue [19]. Insoluble minerals are associated with protein, because minerals are mainly associated with non-fat parts, lean meats usually have a high mineral or ash content. In a previous study conducted by [8], the ash content of fish sausages ranged from 5.08%-5.44%. According to meat sausage quality standards, the ash content should not be more than 5% [20]. This indicates that the ash content of catfish sausage in this study met the specified quality standards. The high ash content in the treatment was due to the addition of spices such as salt, garlic, and pepper in the catfish sausage batter. These materials contain inorganic compounds that do not burn when ignited.

Several important functional properties of protein in food relate to water and protein. These functional properties include solubility, water absorption and binding, viscosity, and gelation [21]. During the preparation of sausage dough or emulsion, meat protein has two functions, namely it can coat or emulsify fat and can bind water. If both are not met the sausage will not be stable and break during boiling. The muscle fraction of myofibrils, which are salt-soluble proteins, is more important than the sarcoplasm, which contains water-soluble proteins [21].

Denaturation is one of the causes of decreased protein levels (changes in protein molecules). Sometimes this change is desirable in food processing, but it is often considered detrimental so that it needs to be prevented [14]. Denaturation occurs when the bonds that make up the molecular configuration are broken, causing the molecule to expand. Denaturation itself is related to gel formation and water and protein interactions, which can be interpreted as the process of breaking hydrogen bonds, hydrophobic interactions, salt bonds.

The development of denatured protein molecules will open the reactive groups present in the polypeptide chain. Furthermore, there will be binding back to the same or adjacent groups. When a sufficient number of units are formed and the protein is no longer dispersed as a colloid, the protein retains all of the liquid and forms a gel. Meanwhile, if the liquid is separated from the coagulated protein, the protein will precipitate [14]. Denaturation can occur in the treatment during heating, which is local heating of the meat. In addition, the treatment during frozen storage and the dehydration process or freeze drying can cause denaturation.

Carbohydrates with multiple functions play an important role in various food processing. Carbohydrates are ingredients whose function is naturally to provide a good texture. In various new products, carbohydrates are still designed as components that strengthen the structure of food products [17]. Carbohydrates, apart from being the main source of calories, also have an important role in determining the characteristics of foodstuffs, such as taste, color, and texture [14]. The results showed that the more use of tapioca flour in making catfish sausage, the higher the carbohydrate content. This is due to the carbohydrate content in the tapioca flour.

Fat is an important food substance for the body and is a more effective source of energy compared to carbohydrates and protein. Fats provide flavor and improve texture in foodstuffs, as well as a source of energy and solvent for vitamins A, D, E and K. Fat is a certain organic compound and is insoluble in water [14].

The addition of fat affects the texture and taste of sausages, but can also be a problem in processing, so that the sausage processing must be maintained so that there is no fat separation. Softness and hardness are also influenced by fat content, such as in cooked sausages which the US government has limited fat content of 30% [16]. The results showed that the higher the amount of tapioca flour used in the processing of catfish sausage, the higher the fat content of catfish sausage. This is because tapioca flour has a maximum vegetable fat content of 25%. When compared with [22], the maximum fat content for fish sausage is 25%, meaning that the research conducted is accepted because it has met the SNI setting.

According to [2] gelatin is a colloidal solution, when mixed with water it forms a hydrocolloid. The mechanism of gelatin gel formation which is hydrocolloid occurs because of the ability of proteins to form a three-dimensional network as a result of interactions between proteins and protein-water formed by trapping a certain amount of water in it. Water serves to prevent the disintegration of the three-dimensional matrix into a compact mass and the formation of a strong gel.

3.2. Texture analysis

The texture value of catfish sausage using commercial fish sausage and the addition of gelatin with concentrations of 0.25%, 0.5%, 1%, and 1.5% ranged from 2.11-10.00%. The highest value is owned by catfish sausage with the addition of 1.5% gelatin, which is 10% and the lowest texture value is owned by commercial fish sausage, which is 2.11% (Tabel 2).

Table 2. Proximate analysis of fish sausage.

Treatment	Mean ± SD
Commercial Fish Sausage	$2{,}11^{e} \pm 0{,}76$
P1 (0,25%)	$3{,}19^{d} \pm 0{,}71$
P2 (0,5%)	$5{,}03^{c} \pm 0{,}69$
P3 (1%)	$7{,}14^{b} \pm 0{,}53$
P4 (1,5%)	$10{,}00^{a} \pm 0{,}36$

Texture analysis was conducted to determine the level of hardness and compactness of commercial fish sausage products and catfish sausage (*Clarias* sp.) with different additions of gelatin. The texture of the product is formed during the boiling and cooling process. Texture analysis was carried out a week after the boiling and cooling process took place [17].

3.3. Organoleptic analysis

The organoleptic analysis of catfish sausage using commercial fish sausage and the addition of gelatin with concentrations of 0.25%, 0.5%, 1%, and 1.5% ranged from 6,43-7,88% (appearance), 6,44-7,81% (odor), 6,02-7,84% (flavor), and 6,06-7,86%. (texture) (Table 3).

Table 3. Organoleptic analysis of fish sausage

Treatment	Organoleptic analysis			
	Appearance	Odor	Flavor	Texture
Commercial fish sausage	$7{,}16 \pm 1{,}20^{a}$	$6{,}44 \pm 0{,}96^{a}$	$6{,}02 \pm 0{,}95^{a}$	$6{,}94 \pm 0{,}95^{a}$
P1 (0,25%)	$7{,}88 \pm 0{,}56^{b}$	$7{,}81 \pm 0{,}53^{b}$	$7{,}84 \pm 0{,}52^{b}$	$7{,}86 \pm 0{,}56^{b}$
P2 (0,5%)	$6{,}69 \pm 0{,}88^{c}$	$7{,}22 \pm 0{,}90^{c}$	$7{,}51 \pm 0{,}70^{b}$	$6{,}54 \pm 1{,}12^{c}$
P3 (1%)	$6{,}43 \pm 0{,}93^{c}$	$6{,}91 \pm 1{,}03^{c}$	$7{,}29 \pm 0{,}88^{b}$	$6{,}06 \pm 1{,}06^{d}$
P4 (1,5%)	$7{,}33 \pm 0{,}75^{d}$	$6{,}50 \pm 0{,}97^{c}$	$6{,}48 \pm 1{,}10^{c}$	$6{,}32 \pm 1{,}16^{d}$

Organoleptic testing of catfish sausage (*Clarias* sp.) on four aspects, namely the appearance, odor, taste, and texture of catfish sausage (*Clarias* sp.) with different concentrations of gelatin (0.25%, 0.5%, 1%, 1.5%) and commercial fish sausage, showed a significant difference in the appearance aspect of catfish sausage (*Clarias* sp.). It is presumed that the addition of oil to catfish sausage was not too significant to the color changes that occurred when the catfish sausage was boiled and the addition of gelatin was not too significant.

The results of the analysis of the smell of catfish sausage (*Clarias* sp.) which significantly affected the preference of the panelists, which was possible because the researchers used or added seasoning to the catfish sausage (*Clarias* sp.) used for research. The taste aspect also experienced a significant difference due to the concentration of gelatin used. Panelists were influenced by the addition of spices by researchers. The taste you get comes from the garlic used. Salt is one of the spices used and can affect the taste of catfish sausage (*Clarias* sp.). Salt increases the volatility of the most hydrophobic compounds by decreasing the availability of water molecules. Meat protein binds volatile compounds

IOP Conf. Series: Earth and Environmental Science 1273 (2023) 012063 doi:10.1088/1755-1315/1273/1/012063

and salts to reduce the ability to modify the surface protein polarity and prevent protein denaturation which can change the taste [23].

The organopeltic test on the texture aspect showed a significant difference with the value which indicated that the panelists somewhat liked the texture of catfish sausage (*Clarias* sp.). The addition of gelatin into catfish sausage (*Clarias* sp.) causes changes in hardness and compactness. The higher the texture and juiciness value of fish meat, the higher the water content of fish meat [24]. This can be influenced by the addition of gelatin which helps maintain the product structure which acts as a binding agent so that the final product is firm, consistent, easy to cut, and does not release water [7]. This proves that the addition of gelatin affects the preference or level of consumer acceptance of the texture of the catfish (*Clarias* sp.) sausage product.

4. Conclusion

The conclusion obtained from this study is that the concentration of addition of gelatin used does not have a significant effect on the physicochemical characteristics and texture analysis of catfish sausage (*Clarias* sp.) give a real influence on consumer acceptance (appearance, smell, taste, and texture). The addition of gelatin to catfish sausage (*Clarias* sp.) has an effect on the value of preference or the level of consumer acceptance. From the whole catfish sausage (*Clarias* sp.) the use of gelatin with a level of 1.5% can increase the proximate analysis maximally in the product so that it can increase the texture value of catfish sausage (*Clarias* sp.).

5. References

[1] Yunus T, Hasim dan Rully 2014 *Faculty of Agricultural Sciences, University of Gorontalo p 8*

[2] Prayogo B S, Rahardja dan Manan A 2012 *Faculty of Fisheries and Maritime Affairs, Airlangga University, Surabaya p 5.*

[3] Hadiwiyoto S 1993 *Fishery Product Processing Technology* (Yogyakarta: Liberty).

[4] Rahayu 2008 *JPG.* Vol. 35 No. 15. Surabaya.

[5] Syafiqoh F 2014 *Faculty of Pharmacy Syarif Hidayatullah State Islamic University p 107.*

[6] Dewita, Syahrul dan Loekman S 2012 *Universitas Riau Riau* p 55.

[7] Schrieber R and Herbert G 2007 *Gelatine Handbook Theory and Industrial Practice* (Wiley VCH: Germany) p 63 and 213.

[8] Shierly 2002 *Pelita Harapan University, Faculty of Agricultural Technology, Department of Food Technology.*

[9] Bawinto A S, Eunike M dan Kaseger B E 2015 *J MedTekno Hasil Perikanan* 3 (2) 55-65.

[10] Midayanto D N dan Sudarminto S Y 2014 *Jurn Panan dan Agro.* 2 (4) 259-267.

[11] Widyaningtyas M dan Wahono H S 2015 *Jurn Panan dan Agro.* 3 (2) 417-423.

[12] Suparjo 2010 *Chemical Analysis of Feed Materials Animal Feed Laboratory, Faculty of Animal Husbandry, Jambi University, Jambi.*

[13] Nugroho A, Fronthea S dan Apri D A 2014 *J Pengolahan dan Bioteknologi Hasil Perikanan.* 3 (4) 140-149.

[14] Winarno F G 2008 *Food Chemistry and Nutrition* (Jakarta: PT Gramedia)

[15] Irianto K. 2006 *Microbiology Reveals the World of Microorganisms volume 1* Yrama Widya Bandung.

[16] Price J F dan Bernard S S 1987 *The Science of Meat and Meat Product Third Edition.*

[17] Rompis J E G 2018 *Bogor Bogor Agricultural University Postgraduate Program.*

[18] Irawati 2008 *Modul Pengujian Mutu 1* Diploma IV PDPPTK Vedca Cianjur

[19] Forrest J C, Aberle E B, Hedrick H B, Judge M D and Merkel R A 1975 *Principles of Meat Science* W H Freeman and Co San Fransisco

[20] Indonesian National Standard [SNI] 2013 Requirements for Quality and Safety of Fish Sausage 7755 2013. Jakarta: National Standardization Agency.H
[21] ardman T M 2008 *Water and Food Quality* (London: Elsevier Applied Science).
[22] Indonesian National Standard [SNI] 2006 Instructions for Organoleptic and or Sensory Testing. 01-2346-2006. Jakarta: National Standardization Body.
[23] Yogesh K., Ahmad T, Manpreet G, Mangesh K And Das P 2013 *J Fo Sci Techno.* 50 (1) 191-196.
[24] Kang H Y and Hui-Huang C 2014 *Fo Sci and Techno* 30 1-6.

6. Acknowledgement
This research was supported by Faculty Fisheries and Marine Airlangga University.

IOP Conf. Series: Earth and Environmental Science 1273 (2023) 012064 doi:10.1088/1755-1315/1273/1/012064

Microbial Diversity of Biological Filters in Recirculating Aquaculture Systems for Sand Lobster (*Panulirus homarus*) Rearing

B W Prastowo[1,2], **T Penataseputro**[1]

[1] National Research and Innovation Agency, Research Centre for Fishery, Cibinong, Indonesia[2] National Research and Innovation Agency, Research Centre for Veterinary, Cibinong, Indonesia

[2]E-mail: bambang.widyo.prastowo@brin.go.id

Abstract. The recirculating aquaculture system (RAS) is a technology that reuses water (often less than 5%) in land-based production for farming aquatic organisms' water treatment. This technology uses filters, either mechanical or biological, for any aquaculture species rearing. In lobster rearing using RAS, one of the most important factors is the quality of rearing water, which depends primarily on the biochemical activities of the microbial communities in the biofilter. There is a significant correlation between the abundance of nitrifying microbial communities and high total ammonia nitrogen (TAN) removal rate. These organisms can convert ammonia accumulates in the system into nitrate. The application of molecular tools (Metagenome 16S Full-Length Barcoding using Oxford Nanopore Platform) to characterize microbial community diversity has been conducted for this study. The diversity of microbes in RAS biofilter for sand lobster rearing is summarized as follows: Nitrification Bacteria group (β- Proteobacteria, g-Proteobacteria, Nitrospirae); Denitrification/Nitrate Reducing Bacteria (α- Proteobacteria, δ-Proteobacteria, Bacteroidetes/Chlorobi, Firmicutes), Sulfide Oxidation Bacteria (ε-Proteobacteria), Probiotic Bacteria (*Bacillus subtilis*, *Pseudomonas* sp., *Roseobacter* sp.). This study's results indicate a potential for RAS in sand lobster-intensive aquaculture applications and also discuss future studies' directions.

1. Introduction

The development of an efficient, productive, biologically safe, and disease-free Recirculation Aquaculture System (RAS) requires a thorough understanding of all processes in RAS system consisting of mechanical filtration, biological filtration, oxygenation, UV/ozone and denitrification systems. Mechanical filtration process can be monitored and controlled. On the contrary, the process of biological filtration depends on the interaction between microbial communities and the environment as a result of nutrient input (feed residues, fish waste) are not easy to control. Microbial composition of biological filters which often form a highly organized biofilm structure only known about ten years ago. New technologies development based on the activities of its inhabitants is not possible, this is because the initial studies that focused on microbial cultivation and enrichment techniques, turned out to be biased in the results where <1% of microbial species were able to grow under standard laboratory conditions. Recently, molecular research has not only made it possible to evaluate the diversity of microbes but has

IOP Conf. Series: Earth and Environmental Science 1273 (2023) 012064 doi:10.1088/1755-1315/1273/1/012064

also given some insights into their activities that should lead to an understanding of the interactions of the community. This approach has undoubtedly provided new settings for RAS biofilters and insights into new processes and equipment for systems upgrading and monitoring.

This study focuses only on the diversity of bacterial organisms in biofilters RAS, which discusses and highlights several microbial species that could serve as critical microbial candidates for their future use. Even though eukaryotic microorganisms (algae, fungi, and rotifers) also have complex filtration functions and as a feed source (as components of bioflocs or green water technologies we will not discuss it. A molecular method was used in this study to identify microbial species diversity using the Next Generation Sequencing (NGS) method with 16S rRNA gene analysis. This independent bacterial culture method has successfully identified the microbial phylogenetic diversity presence in functionally distinct regions of RAS system. This community of bacteria hoped has an essential role in improving the performance of RAS biofilter system for lobster rearing.

2. Materials and Methods

This research was carried out at the Field Test Facility of Fish Health and Environmental Assessment Center (BPKIL) Serang in June-November 2021.

2.1. System Construction

The components of RAS used for rearing sand lobster use a series of containers, compartments, tanks and equipment for water treatment with the following purification stages: (1) sand lobster rearing tanks, (2) sedimentation tank, (3) physical filter, (4) biological filter, (5) UV irradiation for water disinfection, (6) CO_2 removal; (7) clean water reservoir and (8) nano-bubble air generator. The tanks/compartments, functions and specifications in this RAS can be seen in Figure 1.

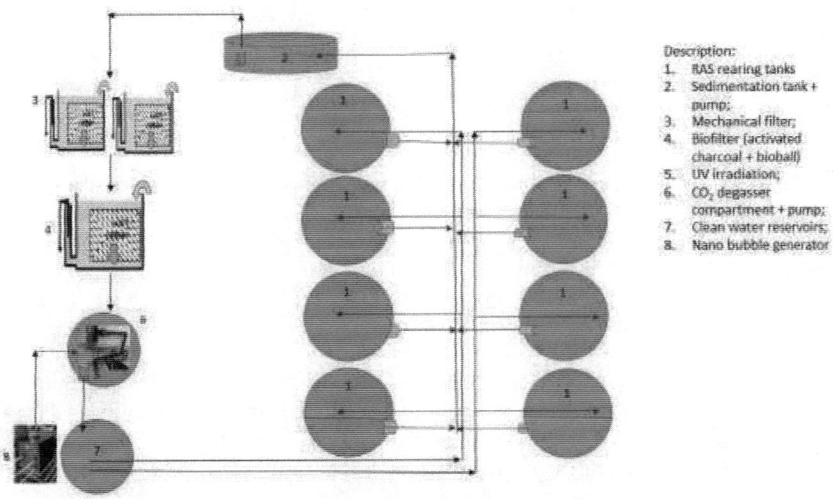

Figure 1. Components and layout of RAS for sand lobster rearing.

In this research, the tank for sand lobster rearing is a fiber glass tank with a volume of 1000-L with an initial size of 50-100 grams. The stocking density was 25 sand lobster/1000-L. Feed in the form of trash fish with a rate of ± 5-12%/day/sand lobster biomass, feeding is done twice a day (morning and evening). Siphoning of feed wastes in the sand lobster rearing tanks were conducted every day, and new water add (± 5%/day) to replace the wasted water.

2.2. Water sample analysis and RAS optimization

The concentration of total ammonia nitrogen (TAN/NH_4^+), nitrite (NO_2^-), nitrate (NO_3^-), active phosphate (PO_4^{3-}), alkalinity, and total organic matter were analyzed every week. The concentration of pH, temperature, and dissolved oxygen (DO) were examined daily. The availability of suitable organic matter in the RAS system was carried out by adding carbon source (molasses) at a dose of 64 g/gr TAN once a week to promote the stable growth of bacteria.

2.3. Sample collection, DNA extraction and sequencing

Samples collected from water contained in the biofilter compartment and from the sand lobster rearing tank with the RAS system. The sample was filtered first using a 0.22 m filter. ZymoBIOMICS DNA Miniprep Kit from Zymo Research used for isolating Genomic DNA from each sample. The concentration of DNA was determined by NanoDrop spectrophotometers and a Qubit fluorometer and using Oxford Nanopore Technology's kit for the preparations of the library. The workflow is as follows:

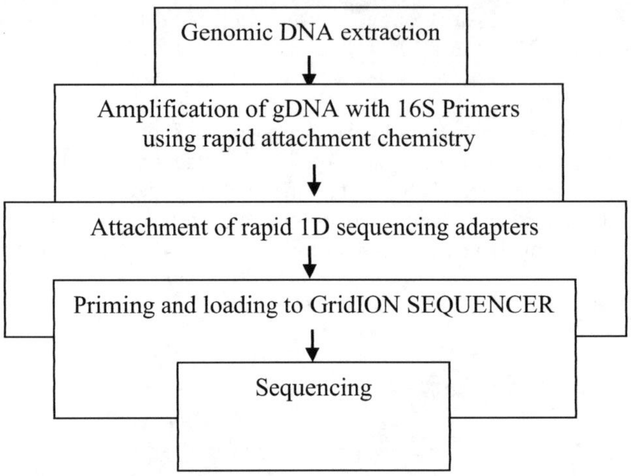

Figure 2. The workflow for sample collection, DNA extraction, and sequencing.

Every step from raw DNA samples to final data, including testing of samples, PCR, preparing the library, and sequencing, affects the quality of data that directly impacts the results of analysis. Quality control is carried out at every step of the procedure to maintain data reliability.

2.4. Processing of sequencing data

GridION sequencing was operated by MinKNOW software version 20.06.9. Base calling was performed by Guppy version 4.0.11 with highly accurate mode. Visualization of the quality of FASTQ files by NanoPlot [3]. Classification of filtered reads by the Centrifuge classifier. The Centrifuge website (https://ccb.jhu.edu/software/centrifuge) to download the index of bacteria and Archaea. Pavian (https://github.com/fbreitwieser/pavian) and Krona Tools (https://github.com/marbl/Krona) for downstream analysis and visualizations. For the correctness of the sequencing data analysis, the raw data will be combined and filtered to get clean data. The effective data used to perform OTU clusters and species annotations for each OTU sequence. Thus, relative species, evenness, and abundance distribution analyzed with Alpha diversity. Graphs of alpha, beta or interest diversity and others. Next, the OTUs were picked, and the taxonomic tree construction phylogenetic task was carried out by downstream statistical analysis to describe differences in community construction between samples or groups through PCoA, PCA, and NMDS. Statistical methods for example T-test, MetaStat, LEfSe, Anosim, and MRPP can analyse the community composition and structural differences between groups' significance. Meanwhile, the analysis results incorporate CCA/RDA to study the key environmental factors. The analysis of data workflow is shown as follows:

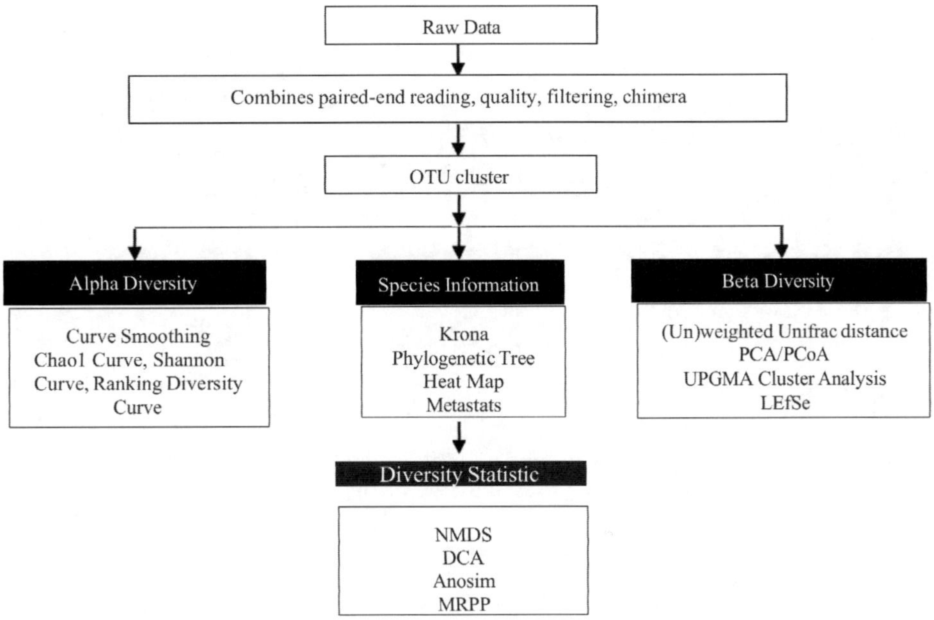

Figure 3. The workflow for processing sequencing data.

Data analysis
All data obtained from this study were analyzed statistically and descriptively.

3. Results and Discussion

3.1. Optimization of Nitrifying Bacteria Performance in RAS System
Nitrifying bacteria in RAS biofilter compartment were grown by adding molasses with a dose of 64 g molasses/gr TAN every two weeks. The indication of nitrifying bacteria to have grown and reproduced well in RAS biofilter compartment by the increasing concentration of nitrate in the water. The molasses treatment was stopped when the nitrate concentration had increased and TAN and nitrite concentration in RAS system below the optimal value and was monitored constant during the test. The results of monitoring the chemical parameters in RAS system involved in the conversion of ammonia into non-toxic materials, showed a pattern of nitrate accumulation in the biofiltration system starting on day 7 to day 22. This phase we call the start-up (or preparation) phase, where nitrifying bacteria first start to grow due to increased excretion of metabolites that can be assimilated by the biofilter system. In this phase, the nitrite concentration is still very high in RAS system. Subsequent increase in nitrate levels began to be seen starting at the age of 23-71 days with nitrate concentrations ranging from 8-17 mg/L. This phase is called the "Transition Phase", where at this phase the levels of ammonia and nitrite have decreased significantly as nitrifying bacteria begin to develop, in particular, in RAS biofilter compartment. The last phase of the growth of nitrifying bacteria in RAS system started to occur on day 72, which is called the ―Stabilization Phase‖. This phase is characterized by an increase in concentration of nitrate until it reaches a peak (15-23 mg/L). This indicates that there has been stability in the growth of nitrifying bacteria in the entire RAS system. As an important evaluation of the performance of RAS biofilter in this research activity, RAS biofilter becomes effective after 10 weeks since the operation of RAS system.

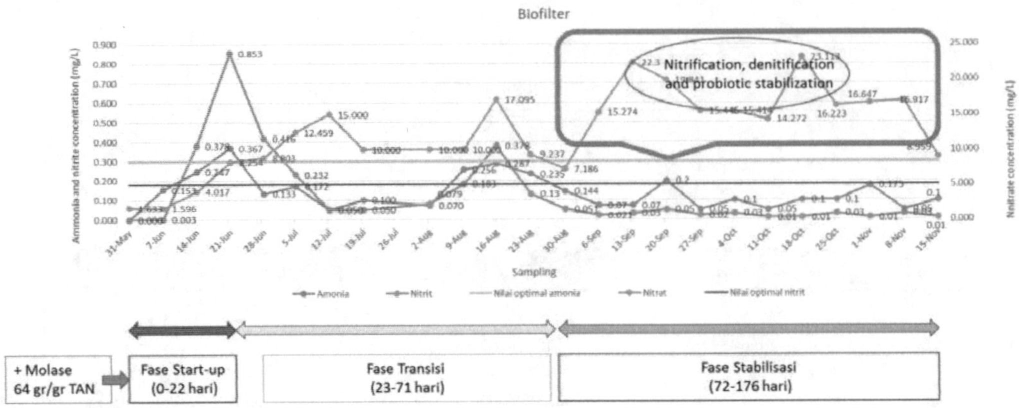

Figure 4. Stages of the maturation process of nitrifying bacteria in a RAS biofilter compartment for sand lobster rearing.

3.2. *Isolation and Identification of Nitrifying Bacteria in RAS System*

Bacterial diversity assessment, produced in RAS system for sand lobster rearing, was carried out using the sequencing method with 16S ribosomal RNA (16S rRNA) gene. This method has become a widely used approach in environmental microbiology, especially with the development of complete genome-reading sequencing technology for living things. Oxford Nanopore Technology (ONT) provides long read sequencing covering the 16S rRNA gene full length sequence (region V1-9) through an express, inexpensive and high-yield read procedure. As all the 16S rRNA gene informative sites were considered, the sequencing method with full-length 16S rRNA offers an advanced taxonomic and phylogenetic resolution level for bacteria identification.

To assess bacteria that grow in RAS system, 2 samples were prepared and coded according to the source of collection. Samples taken from seawater in RAS biofilter compartment were coded with barcode23 and samples taken from one of the sand lobster rearing tank was coded with barcode24. Prior to testing using the Next Generation Sequencing (NGS) method, both samples were tested using the polymerase chain reaction (PCR) method with 16S 27F – 1429R primers for tested bacteria size determination. The samples were also tested using the NanoPlot method to test the quality of the samples. From the test results using PCR (Figure 3), one band from each sample was obtained with a size of about 1000 bp.

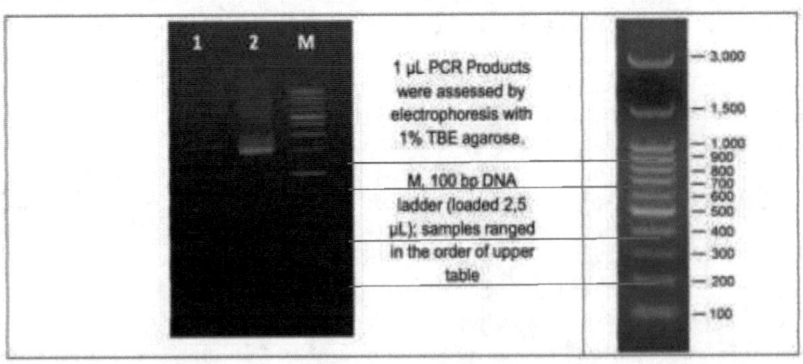

Figure 5. Photo of Agarose Gel Electrophoresis from PCR products (gDNA amplification with 16S 27F – 1429R Primer).

The Nanoplot is using for further test (Table 1). From the table, it can be seen that the total base length of the samples taken from RAS biofilter compartment is actually much longer than the base length of the samples taken from the sand lobster rearing tank. However, from the purity of the sample (mean read quality) there is almost no difference between the two samples, so that the two samples can be continued with the process of identifying bacteria using the NGS method.

Table 1. Quality of FASTQ files calculated by NanoPlot.

Quality Element	Biofilter	RAS rearing tank
Mean read length	1,540.7	1,443.9
Mean read quality	11,1	11,8
Number of reads	123,637.0	94,243.0
Read length N50	1,571.0	1,544.0
Total bases	190,488,225.0	136,074,789.0

Sequencing data analysis is carried out after the sequencing process is complete. This data will be used as bioinformatics material. The first bioinformatic is information on alpha diversity which describes the average species diversity at a location or habitat diversity at a local scale or species number in a given community and can be applied to make comparisons between species number in different biological groups or geographic regions. From the various indexes shown in Table 2, it is known that the microbial species in RAS biofilter compartment have a higher diversity of microbial species when compared to the microbial species in the sand lobster rearing tank. The high diversity of microbial community in RAS biofilter compartment can be understood because these bacteria initially grow in RAS biofilter compartment and then after growing more in number then spread throughout the sand lobster rearing tank.

Table 2. Alpha diversity of microbial species from water samples in RAS biofilter compartment and sand lobster rearing tanks.

Index	RAS Biofilter	Rearing tank
Berger_parker	0.111	0.098
Buzas_gibson	0.00189	0.00156
Chao1	2124.2	1949.8
Dominance	0.974	0.969
Equitability	0.659	0.622
Jost	60.3	44.8
Jost1	130.3	88.1
Reads	68836	56637
Richness	1616	1334
Richness2	1154	849
Robbins	0.286	0.363
Simpson	0.0259	0.0315
Shannon_e	4.87	4.48
Shannon_2	7.03	6.46
Shannon_10	2.11	1.95

The logical relationship between the samples in the RAS biofilter compartment and sand lobster rearing tank can be determined using the Venn Diagram (Figure 4). It can be seen that the microbial species originally came from RAS biofilter compartment (651) has been mixed with other microbial species (965) that exist in the sand lobster rearing tanks (369). These results indicate that not all microbial species from the RAS biofilter compartment can migrate or even existent in the sand lobster rearing tanks.

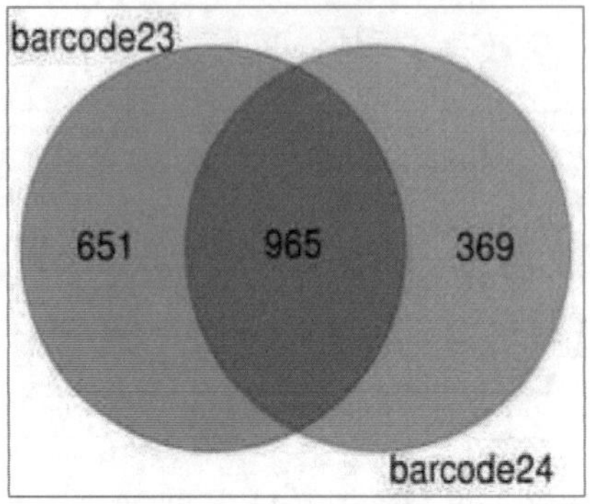

Figure 6. Venn diagram illustrating the logical relationship between samples in the RAS biofilter tank and sand lobster rearing tank.

Krona Visualizations allows us to intuitively explore the relative abundances in the complex hierarchies of metagenomic classifications. In this research, Krona Visualization is used to see the relative abundance of microbial species that develop in the RAS biofilter compartment and sand lobster rearing tank. (Figure 5a and b).

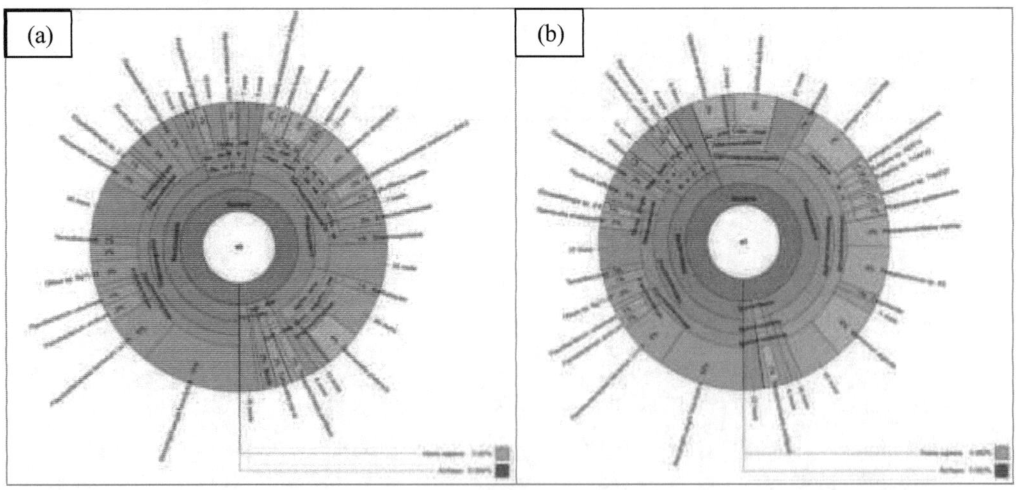

Figure 7. Corona visualization of the relative abundance of microbial species from (a) RAS biofilter compartment, (b) sand lobster rearing tank.

The visualization of Sankey Diagram is to find out the classification description that shows the kinship relationship of a species with its ancestors and the evolutionary relationship between organisms (Figure 5). In this diagram, the width of the arrow is proportional to the quantity to illustrate the change in the time hierarchy between clusters.

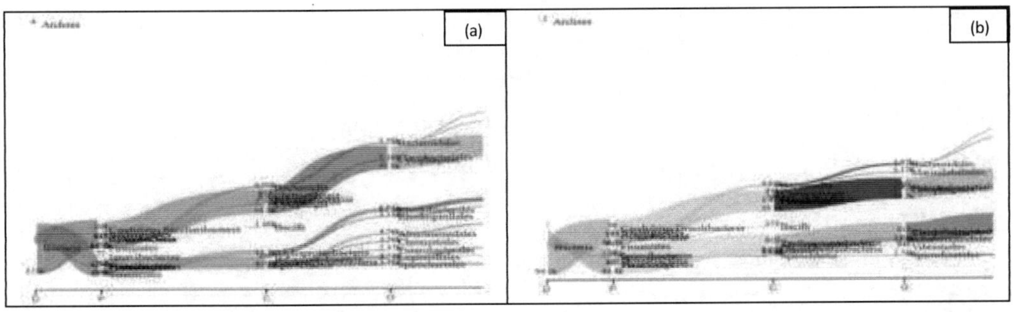

Figure 8. Sankey diagram of microbial species relative abundance from (a) RAS biofilter compartment, (b) sand lobster rearing tank.

Substantial differences in microbial species representing unique and complex environments have been found in RAS biofilter and sand lobster rearing tanks. However, in this study only the taxonomic affiliation of microbial species identified from RAS biofilter compartment was shown (Table 3).

Table 3. Taxonomic affiliation of microbial species identified in RAS biofilter compartment.

No	Group	Microorganism	Reference
1	α-Proteobacteria	*Aminobacter* sp.	[2]
		Hyphomicrobium sp.	[25}
		Paracoccus sp.	[4]
		Rhodobacter sp.	[15, 4]
		Rhodovulum sulfidophilum	[4, 25]
		Rhizobium sp.	[18]
		Roseobacter sp.	[2, 4, 25]
		Ruegeria sp.	[15]
		Sphingomonas sp.	[15]
		Sulfitobacter sp.	[15, 24]
2.	β-Proteobacteria	*Nitrosomonas* sp.	[6, 8, 17, 24]
		Thauera sp.	[4]
3	γ-Proteobacteria	*Acinetobacter* sp.	[18, 9]
		Aeromonas hydrophila	[18]
		Marinobacter sp.	[3, 12]
		Microbulbifer sp.	[25]
		Nitrosococcus sp.	[15]
		Pseudoalteromonas sp.	[15]
		Pseudomonas sp.	[18, 1, 12, 25]
		Pseudomonas stutzeri	[15]
		Vibrio sp.	[15, 12]
4	δ-Proteobacteria	*Desulfovibrio* sp.	[2]
5	ε-Proteobacteria	*Thiomicrospira* sp.	[4]
6	Actinobacteria	*Streptomyces* sp.	[7]

		Arcobacter sp.	[18]
7	Bacteroidetes/Chlorobi	*Bacteroides* sp.	[15, 4]
		Chryseobacterium sp.	[1]
8	Firmicutes	*Bacillus subtilis*	[15]
		Sarcina sp.	[18]
9	Planctomycetales	*Pirellula* sp.	[15]

The oxidation of TAN is correlated with 16S rRNA cultivars of *Nitrosomonas* sp. (β- Proteobacteria) and *Nitrosococcus* sp. (γ-Proteobacteria) in biofilter system tested (Table 2), both microorganisms dominating or marine nitrification systems [6, 8, 17, 24, 15]. Biofilter in all RAS systems is critical for retaining TAN (NH_3-N) sub-lethal levels (0.1 -<3 mg/L) depending on parameters including pH, salinity, and species (23). Nitrite (NO_2^--N) is the first product of ammonia oxidation, although aquatic species are adversely affected by long-term exposure to low levels (<0.2 mg/L) but may be tolerated at high levels. [11]. TAN and nitrite oxidation occur simultaneously in RAS biofilters with *Nitrosomas* sp. (β-Proteobacteria) as a dominant nitrite oxidizer. However, due to its low affinity of substrate, *Nitrobacter* sp. (α-Proteobacteria) has not been detected in RAS. [5].

Several phylotypes belong to marine RAS nitrification communities (*Pseudomonas* sp. and various α-Proteobacteria) which related to denitrification (nitrate-reducing activities) found in the anaerobic stimulated condition [7,13]. A high levels of nitrate/nitrite combination and sources of organic carbon (uneaten feed and fish excess) is provided as a supplement to promote the heterotrophic denitrification process in RAS biofilter compartment or deep within the biofilm layer under anaerobic conditions. [14, 4]. *Pseudomonas* sp. (with a strain related to *Pseudomonas stutzeri* identified in 16S rRNA), capable of denitrification, appears to be the most abundant γ-Proteobacteria in RAS biofilter

RAS biofilters, in addition to heterotrophic denitrifying bacteria, include bacteria from sulfate-reducing fermentative (Bacteroidetes and Desulfovibrio (δ-Proteobacteria)) and sulfide-dependent autotrophic denitrification (Firmicutes) (Table 2), intensified by the high level of sulfate and is a cycle part of an oxidation/reduction of sulfur-based formed in RAS biofilter compartments drives both organic waste digestions and nitrate reduction [2,15,3,4,25]. Bacteria, in this cycle, produce N_2 gas by oxidizing organic waste to change sulfate to sulfide, used for sulfide-dependent denitrification [2,23]. This process for removing toxic sulfide gas - and, consequently, re-forming sulfate gas is carried out by sulfide oxidizing bacteria (*Thiomicrospira* sp.) aerobically [6]. Oxidation of sulphide, under anoxic conditions, is probably assisted by chemolithotrophic sulfide-oxidizing bacteria reducing nitrate, including *Thiomicrospora* sp. (ε-proteobacteria) and *Rhodobacter* sp. (α-proteobacteria) with high sequence similarity to *Paracoccus* sp., *Rhodovulum sulfidophilum*, *Roseobacter* sp., and *Ruegeria* sp. [3,4].

The evidence for anammox microorganisms (*Pirellula* sp.), in addition to denitrifiers, has been provided by 16S rRNA sequences in marine biofilters RAS for sand lobster rearing [15]. Anammox can partially bypass the reaction of denitrification with the conversion of ammonia and nitrite to N2 directly, this can result in operating costs savings and reduce the emissions of N_2O significantly [16]. In RAS performance, anammox pathway is crucial for TAN removal, however its organic acids sensitivity and very time-consuming of doubling time (>11 days) makes it unable to completely replace the denitrification process. Therefore, in the management of nitrogen waste in the RAS biofilter, the contribution of the anammox and denitrification combination (what is called 'denammox' or 'deamox') is indispensable [10,12, 25] because of its similar function in the environment or in the treatment plant of industrial wastewater [17,22].

In Table 2, probiotic bacteria in the RAS biofilter have been identified (*Bacillus subtilis*, *Pseudomonas* sp., and *Roseobacter* sp.), however its role remains to be studied. RAS biofilter, however, also allows for the accumulation of pathogens within its protective biofilm, cultivars and 16S rRNA sequences have identified several pathogenic bacteria (Vibrio and *Aeromonas hydrophila*) among others in RAS biofilter. This condition can result in disease contamination in fish and its consumer. Pathogenic bacteria are usually treated by antibiotic therapy, increasing antibiotics resistance, and their consequences for farmed fish necessitate the use of additional equipment, such as UV for protection against infection in RAS. Recent methods for pathogen colonization inhibition have focused on the application of probiotic bacteria for nutrients competition, growth inhibitors production, and squelch cell-to-cell communication that allows them to settle within the biofilm [19,20,21].

Although the design of the RAS system has been developed in a generic manner, especially for the assembly of biofilters, the function of the system as a whole has been adapted to remove harmful metabolic products or wastes in aquaculture of fish species. However, the discovery of microbial communities in RAS biofilters, regardless of their application in aquaculture production of small or large seawater or freshwater systems, has led to the need for a metagenomic approach that will provide detailed information in designing efficient RAS systems and their bioaugmentation strategies. This metagenomic approach has also identified the presence of important bacteria (ex. anammox and probiotics) for effective, ecologically aware, and economically RAS technology for sustainable aquaculture

4. Conclusion

In this study, the full maturation activity of nitrifying bacteria in RAS biofilter was about ten weeks (3 weeks of start-up phase and 7 weeks of transition phase). The microbial community of this RAS biofilter for sand lobster rearing is consist of nitrification bacteria group (β-Proteobacteria, γ- and Proteobacteria); denitrification/nitrate reducing bacteria (α-Proteobacteria, δ-Proteobacteria, Bacteroidetes/Chlorobi, Firmicutes), sulfide oxidation bacteria (ε-Proteobacteria) and probiotic bacteria (*Bacillus subtilis*, *Pseudomonas* sp., *Roseobacter* sp.).

5. References

[1] Bowman J P and Nichols D S 2005 *Int J Syst Evol Microbiol* **55** 1471–1486
[2] Cytryn E, Gelfand I, Barak Y, van Rijn J and Minz D 2003 *Environ Microbiol* **5** 55-63
[3] Cytryn E, van Rijn J, Schramm A, Gieseke A, de Beer D and Minz D 2005 *Appl Environ Microbiol* **71** 6134-6141
[4] Cytryn E, Minz D, Gelfand I, Neori A, Gieseke A, De Beer D and Van Rijn J 2005 *Environ Sci Technol* **39**1802-1810
[5] Ebeling J M, Timmons M B and Bisogni J J 2009 *International Journal of Recirculating Aquaculture* **10** 63-89
[6] Foesel B U, Gieseke A, Schwermer C, Stief P, Koch L, Cytryn E, de la Torre JR, van Rijn J, Minz D and Drake H L 2008 *FEMS Microbiol Ecol* **63** 192-204
[7] Guttman L and van Rijn J 2008 *Aquaculture* **279** 5-91.
[8] Hovanec T A and DeLong E F 1996 *Appl Environ Microbiol* **62** 2888-2896
[9] Itoi S, Ebihara N, Washio S and Sugita H 2007 *Aquaculture* **264** 297-308
[10] Kalyuzhnyi S, Gladchenko M, Mulder A, Versprille B: DEAMOX — new biological nitrogen removal process based on anaerobic ammonia oxidation coupled to sulphide- driven conversion of nitrate into nitrite. Water Res 2006, 40:3637-3645.
[11] Kirchman D L 2002 *FEMS Microbiol Ecol* **39** 91–100
[12] Lahav O, Tackoubov D, Zelikson R, Mozes N, Tal Y and Tarre S 2009 *Aquaculture* **288** 76-82.
[13] Leonard N, Blancheton J P, Guiraud J P 2000 *Aquacult Eng* **22** 109-120
[14] Liu Y, Liu, L-Z, Liu, H-C, Zhou Y-G, Q, F-J and Liu Z-P 2014 *International Journal of Systematic and Evolutionary Microbiology* **64** 469–474
[15] Mancuso M, Zaccone R, Carella F, Maiolino P and De Vico G 2013 *J Aquaculture Res Dev* **4**(5)
[16] Panjaitan P 2010 *Journal of Coastal Development Vol.* **14** (1) 35 – 44
[17] Paredes D, Kuschk P, Mbwette T S A, Stange F, Muller R A and Koser H 2007 *Eng Life Sci* **7** 13-25.
[18] Paungfoo C, Prasertsan P, Burrell P C, Intrasungkha N and Blackall L L 2007 *Biotechnol Bioeng* **97** 985-990
[19] Ray A J, Drury T H, and Cecil A 2017 *Aquacultural Engineering* **77** 9–14.
[20] Sartika D, Harpeni E dan Diantari R 2012. *e-Jurnal Rekayasa Dan Teknologi Budidaya Perairan Volume* **1** (1) 57-64
[21] Schneider O, Chabrillo-Popelka M, Smidt H, Haenen O, Sereti V, Eding E H, Verreth J A J 2007 *FEMS Microbiol Ecol* **60** 207-219
[22] Schmidt I, Sliekers O, Schmid M, Bock E, Fuerst J, Kuenen J G, Jetten M S and Strous M 2003 *FEMS Microbiol Rev* **27** 481-492.
[23] Sher Y, Schneider K, Schwermer C and van Rijn J 2008 *Water Res* **42** 4386-4392
[24] de Souza Valente, C and Wan A H L 2021 *Journal of Invertebrate Pathology* **181**
[25] Suantika G., Situmorang M L, Kurniawan J B, Pratiwi S A, Aditiawati P, Astuti D I, Azizah F F N, Djohana Y A, Zuhrid U and Simatupang T M 2018 *Aquacultural Engineering* **82** 12–24 DOI:10.1016/j.aquaeng.2018.04.002

6. Acknowledgments
Authors wishing to acknowledge BPKIL Serang for their facilities and financial support for this research. Grateful thanks extended to colleagues and technical staff for the assistance and encouragement in conducting the research and in preparing this article.

The effect of squid ink (*Loligo* sp.) utilization in feed on improvement black colour and growth of koi fish (*Cyprinus carpio*)

K T Laksono[1], N N Dewi[2,3], B S Rahardja[2]

[1]Program Study of Aquaculture, Faculty of Fisheries and Marine, Universitas Airlangga, Surabaya 60115 Indonesia
[2]Department of Aquaculture, Faculty of Fisheries and Marine, Universitas Airlangga, Surabaya 60115 Indonesia

[3]Corresponding author: ninanurmaliadewi@fpk.unair.ac.id

Abstract. Koi fish is one of the ornamental fish that is in demand by ornamental fish lovers. Color quality is a determining factor for the economic value of koi fish production. The purpose of this study was to determine the effect of adding squid ink supplements to feed on increasing black color and growth in koi fish. This study used a completely randomized design with 4 different doses of treatment with 3 replications. Treatment P0 without the addition of squid ink (control), P1 was 1% squid ink, P2 was 3% squid ink and P3 was 5% squid ink in the feed. Koi fish are kept in an aquarium with feeding rate of 5%. The main parameters observed were increasing black color and absolute growth of koi fish. The black color assessment was observed using a modified Toca Color Finder. Supporting parameters were water quality. The results showed that 5% squid ink treatment resulted in the highest black color score. The absolute weight growth between treatments showed a significant difference ($p<0.05$) and the absolute length growth did not show a significant difference ($p>0.05$). It can be concluded that the addition of squid ink with a concentration of 5% is the best treatment.

1. Introduction

Ornamental fish is one of the favorite fishery commodities of the Indonesian people. Ornamental fish are not only in demand in the local market, but also in the export market. Therefore, the opportunity for ornamental fish as a source of foreign exchange for a country is getting wider [1]. The beauty of ornamental fish can also increase the commercial value of ornamental fish exports [2]. The export value of Indonesian ornamental fish reached 13.26 million US dollars in 2011 and 5.24 million US dollars in 2012 [3]. According to Ministry of Marine and Fisheries [4], during the period 2012-2018 ornamental fish production increased annually by around 5.05%, in 2012 production reached 938.47 million fish and in 2018 increased to 1.19 billion fish. One of the popular ornamental fish is the koi fish (*Cyprinus carpio*). Koi fish production has increased every year. The production of koi fish itself in 2019 was 523,775 individuals, exceeding the target of 350,000 individuals.

Koi fish is an ornamental fish known for its beautiful shape and color. In general, indicators of the beauty of koi fish and other ornamental fish can be seen from the brilliant color, shape and completeness of the physical, behavior, and health conditions [5]. The color and size of the koi fish is the main attraction. The color of koi fish is a reference in the high price of koi fish which can be caused by the consumption of feed containing color pigments. Color cells (pigment substances) consist of five different types of color cells, namely melanophore (black), xanthophore (yellow), erythrophore (red), leukophore or guanuphore (white) and iridophore (reflecting color/sheen) [6].

Koi fish cannot synthesize pigments in their bodies, so additional supplements are needed to improve color quality [7]. Pigments that play a role in fish coloring are divided into two types, namely carotene and melanin. Carotene will cause orange, yellow and red colors on fish scales, while melanin will produce brown to black colors on fish skin surface scales [8]. Currently, natural dyes, especially the pigment ®-carotene, are widely used to increase the yellow to red color of various types of koi. While the increase in black color has not been done much. One source of natural black dye is squid ink. Squid ink is one of the ingredients that is rarely used in people's daily lives, even though the melanin content in squid ink reaches 90%. Melanin is a black dye that is used by the epidermal tissue to dye the body of organisms [9].

The ink sac is a place for squid that has a high melanin content, in addition to melanin there is also a protein content of 10-15%, which consists of essential and non-essential amino acids and sulfated polysaccharides [10]. The complete and balanced protein and amino acid composition have better quality for growth in fish [11]. Therefore, through this squid ink supplementation, it is hoped that it can improve the color quality and growth of koi fish. The purpose of this study was to determine the effect of squid ink in feed on the growth and increase in the black color of koi fish also knowing the optimal concentration of adding squid ink in feed to increase the black color of koi fish.

2. Materials and Methods

2.1. *Place and time of research*

The research activity was carried out from March 2021 to May 2021, at the Faculty of Fisheries and Marine, Universitas Airlangga.

2.2. *Tools and Materials*

The tools used in the study were an aquarium measuring 50x30x30 cm, aerator, water pump, thermometer, pH pen, DO meter and ruler, TCF. Materials in the research will include koi fish with a size of 5-10 cm, water, commercial feed and squid ink supplements.

2.3. *Experimental Design*

This study used a completely randomized design (CRD) with 4 different dose treatments with 3 repetitions. Repetition is done with the aim of increasing the accuracy of the research results. The research treatment:

P0 = Control, without the addition of squid ink (feed pellets only)
P1 = Addition of feed and squid ink with a concentration of 1%
P2 = Addition of feed and squid ink with a concentration of 3%
P3 = Addition of feed and squid ink with a concentration of 5%

2.4. *Work Procedure*

The initial preparation of the research was carried out by preparing tools as maintenance media. The aquarium used is 45 liters in size, each aquarium is filled with 22.5 liters of water and is given an aerator that is turned on for the first 24 hours to supply dissolved oxygen and eliminate chemical odors remaining in the aquarium. Before being stocked, the koi fish are acclimatized for 15 minutes, according to Zidni et al. [12] the acclimatization process is carried out with the aim that the koi fish can adapt to their new habitat. After the acclimatization process, the fish are not fed for 1-2 days because at this time the fish are still in a state of stress due to not adapting to the new environment [13].

After the acclimatization process, the fish were put into an aquarium, each with two fish per aquarium with a predetermined treatment group. The placement of two fish per 22.5 liters is intended for maximum feed absorption and there is no competition in the feeding process [14]. The feed used is commercial feed, in the form of pellets with 31% protein content, 4% fat, 5% fiber and 13% ash content. The daily protein requirement of young koi fish is 30% [15]. Making feed for treatment is done by mixing squid ink according to the concentration of each treatment. If the squid ink liquid is too thick, it diluted with distilled water. Then the application of squid ink supplements to the feed is done by spraying according to each dose. Feed is given three times a day at 08.00, 12.00 and 16.00 with a feeding rate of 5% [16].

2.5. *Growth Measurement*

There are two kinds of absolute growth, namely absolute weight growth and absolute length growth. The absolute length calculation can be calculated by the formula [17] :

$$Lm = Lt - L0$$

Information :
Lm = Absolute length growth (cm)
Lt = total length of the fish at the end of the treatment (cm)
L0 = total length of the initial test fish rearing (cm)

The absolute weight calculation can be calculated by the formula [17]:

$$Wm = Wt - W0$$

Information :
Wm = Absolute weight growth (grams)
Wt = Total weight of test fish at the end of treatment (grams)
W0 = Total weight of test fish at the beginning of rearing (grams)

2.6. *Colour Quality*

Determination of the brightness of the color of the fish was carried out by observing the intensity of the color brightness of the koi using the Toca Color Finder (TCF) which was observed by 5 panelists who had no visual impairment. Observations were made visually by comparing the original color of the fish on TCF paper. Observations were made including observations of color intensity using TCF.

2.7. *Water Quality*

Supporting parameters in this research are water quality including DO (Dissolved Oxygen), temperature, pH. Measurement of dissolved oxygen and temperature using a DO meter, pH was measured using a pH meter. DO, temperature and pH measurements are carried out every day, namely in the morning and evening

2.8. *Statistic Analysis*

Data analysis consisted of fish growth and black color quality. The variables obtained were then statistically analyzed using the Analysis of Variance (ANOVA) test with a 95% confidence level. If there are significant results, the calculation is continued with Duncan's Multiple Range Test.

3. Result and Discussion

3.1. *Colour quality*

The results of measuring the black color of koi fish using the Toca Color Finder can be seen in Table 1, Table 2, Table 3, and Table 4. Based on the results of the ANOVA statistical test, the TCF (Toca color finder) value of koi fish at week 1 showed no significant difference ($p>0.05$) in each treatment with different doses. The highest TCF was at P3 (Addition of 5% squid ink) with a value of 7.80±1.304 and the lowest TCF value was at P0 (Control) with a value of 5.60±1.817.

Tabel 1. Average of black color value in week 1

Treatments	TCF value
P0	$5,60^{b}\pm1,817$
P1	$6,00^{b}\pm2,236$
P2	$6,60^{b}\pm0,548$
P3	$7,80^{b}\pm1,304$

Description: The same superscript in the column and row shows the effect that is not significantly different ($p>0.05$)

Based on the results of the ANOVA statistical test, the TCF (Toca color finder) value of koi fish at week 2 showed a significant difference ($p<0.05$) in each treatment with different doses. The highest TCF was at P3 (Addition of 5% squid ink) with a value of 8.80±1.304 and the lowest TCF value was at P0 (Control) with a value of 6.00±1.581.

Tabel 2. Average of black color value in week 2

Treatments	TCF value
P0	$6,00^{b}\pm1,581$
P1	$6,40^{b}\pm1,949$
P2	$7,00^{ab}\pm0,707$
P3	$8,80^{a}\pm1,304$

Description: The same superscript in the column and row shows the effect that is not significantly different ($p>0.05$)

Based on the results of the ANOVA statistical test, the TCF (Toca color finder) value of koi fish at week 3 showed a significant difference ($p<0.05$) in each treatment with different doses. The highest TCF was at P3 (Addition of 5% squid ink) with a value of 9.20±1.095 and the lowest TCF value was at P0 (Control) with a value of 7.00±1,000.

Tabel 3. Average of black color value in week 3

Treatments	TCF value
P0	$7,00^{b}\pm1,000$
P1	$7,00^{b}\pm1,581$
P2	$7,80^{ab}\pm0,447$
P3	$9,20^{a}\pm1,095$

Description: The same superscript in the column and row shows the effect that is not significantly different ($p>0.05$)

Based on the results of the ANOVA statistical test, the value of TCF (Toca color finder) of koi fish at week 4 showed a significant difference ($p<0.05$) in each treatment with different doses. The highest TCF was at P3 (Addition of 5% squid ink) with a value of 9.40±0.894 and the lowest TCF value was at P0 (Control) with a value of 7.40±1.140.

Tabel 4. Average of black color value in week 4

Treatments	TCF value
P0	$7{,}40^{b}\pm1{,}140$
P1	$7{,}40^{b}\pm2{,}074$
P2	$7{,}80^{ab}\pm0{,}447$
P3	$9{,}40^{a}\pm0{,}894$

Description: The same superscript in the column and row shows the effect that is not significantly different ($p>0.05$)

3.2. *Absolute Length Growth*

The average absolute length growth data are shown in Table 5. The results of statistical tests using Analysis of Variance (ANOVA) showed that the addition of Squid Ink to the feed showed no significant effect on the absolute length growth of koi fish ($p>0.05$). Based on the results, it can be seen that the Absolute Length Growth for 28 days in treatment P1, P2 and P3 increased and was greater than in treatment P0.

Tabel 5. Average of absolute length growth of Koi fish

Treatments	Absolute Length Growth (cm)
P0	$0{,}596 \pm 0{,}1734^{a}$
P1	$0{,}880 \pm 0{,}2636^{a}$
P2	$0{,}810 \pm 0{,}1193^{a}$
P3	$0{,}730 \pm 0{,}2196^{a}$

Description: The same superscript in the column and row shows the effect that is not significantly different ($p>0.05$)

3.3. *Absolute Weight Growth*

The average absolute weight growth data are shown in Table 6. The results of statistical tests using Analysis of Variance (ANOVA) showed that the addition of Squid Ink to the feed showed a significant effect on the absolute weight growth of koi fish ($p<0.05$). Based on the results, it can be seen that the growth of absolute weight for 28 days in treatment P1, P2 and P3 increased and was greater than in treatment P0.

Table 6. Average of absolute weight growth of Koi fish

Treatments	**Absolute Weight gowth (gram)**
P0	$0.698 \pm 0{,}1721^{b}$
P1	$1{,}382 \pm 0{,}5639^{ab}$
P2	$1{,}230 \pm 0{,}3113^{a}$
P3	$1{,}093 \pm 0{,}3455^{a}$

Description: The same superscript in the column and row shows the effect that is not significantly different ($p>0.05$)

3.4. Water Quality
Water quality parameters observed in this study include pH, DO (dissolved oxygen) and temperature. Measurements of pH and DO were carried out twice a day, in the morning and evening, while the temperature was measured in the morning and evening. The results of water quality measurements which include pH, dissolved oxygen and temperature range from pH values of 6.8-7.4, DO 5-5.7mg/l and temperatures of 28-31.5 °C (Table 7).

Tabel 7. Water quality measurement

Treatments	Parameters			
	temperature(°C)		pH	*Dissolved Oxygen* (mg/l)
	morning	evening		
P0	28-29	28-31,5	6,8-7,4	5,5-5,7
P1	28-29	28-31,5	6,8-7,4	5,5-5,7
P2	28-29	28-31,5	6,8-7,4	5,5-5,7
P3	28-29	28-31,5	6,8-7,4	5,5-5,7
Optimal range [18]	25-29	28-31,5	6,8-7,4	5

3.5. Discussion
This research used Showa Koi fish. Showa koi fish is koi fish with 3 color patterns, with a black base color combined with red and white patterns. This type of koi fish has a characteristic red stripe on the head and black stripes [19]. The reason for using this type of fish is because Showa koi fish are basic fish in the koi world and are still very common in some farmers. Showa koi fish are also fish that are often found in various ornamental fish contests. Showa koi fish are the result of the second marriage after Kohaku koi fish [20]. The brightness of the color of the koi fish is influenced by several factors, including breeders (genetic), foodd and the environment [14]. In this study, feed was used as a factor for increasing the color of koi fish. The feed was added with squid ink with several concentrations.

The maintenance of the research fish was carried out for 30 days, during the rearing period it showed an increase in black color. At the addition of 1%, 3% and 5% ink there was an increase in color quality with an average of 1 score per week. At the end of the study, P3 gave the highest black color value for koi fish. The highest and most effective color enhancement to increase the color pigment of Koi fish was the addition of 5% squid ink which had a darker level of color and a faster increase in color brightness. This can be due to the absorption of melanin from squid ink, such as the statement from Putriana et al., [14] that the formation of color in fish is caused by the presence of pigment cells or chromatophores contained in the dermis on the scales, outside and under the scales.

This is supported by the statement of Effendie [21] which states that the appearance of color in fish is an expression of the movement or distribution of pigment grains (chromatophores) contained in the dermis layer of the skin or under and outside the scales. These chromatophores can spread throughout the cell or accumulate at one point. If the chromatophores are spread throughout the cell, the color display will be clearly visible depending on the pigment grains, otherwise if the chromatophores accumulate at one point, the resulting color display will look faded [22].

This squid ink liquid is dark in color which contains grains of melanin or black pigment. This natural melanin is a melanoprotein that contains 10-15 percent protein. This melanin binds to proteins through sulfur-containing amino acids, such as cysteine which can bind to white blood cells and is useful for preventing anticancer [23]. The study used commercial feed and the addition of squid ink with several doses and based on the results it was found that the addition of squid ink to the feed did not significantly affect growth. Because based on Indonesia National Standard [24] protein requirements for *Cyprinus carpio* at least 30%.

Water quality management factors also play an important role in improving color quality or color brightness in fish [1]. In this research, it has been observed that there are no periodic changes in temperature that can cause stress or even cause fish death and in accordance with the statement according to Susanto [18], good water quality for koi maintenance is water temperature 25–29°C, pH 6 ,8 – 7.5 and preferably water in koi maintenance has a minimum DO level of 5 mg/L. The results obtained related to water quality in this study still in optimum value. DO measurements were carried out twice a day in the morning and evening. Each aquarium has been added an aerator that has been adjusted to support the oxygen level and in each aquarium gets the same amount of oxygen.

4. Conclusion

The addition of squid ink (*Loligo sp.)* in the feed did not significantly affect the growth of koi fish (*Cyprinus carpio*). The addition of squid ink in the feed can improve the black color quality of koi fish (*Cyprinus carpio*) and the optimum concentration for increasing color is the highest concentration of 5%.

5. References

[1] Bachtiar Y and Lentera 2004 *Freshwater Ornamental Fish Cultivation For Export (*Depok : Agromedia Pustaka) p 108.
[2] Arulvasu C, Meena S R, Chandhirasekar D, and Sivaganam S 2013 *Euro Jour of Biolog Sci* 5 (2) 44-49.
[3] Utami S W 2013 *Ornamental Fish Export Opportunities* (Jakarta : Warta Ekspor) p 19.
[4] Ministry of Maritime Affairs and Fisheries 2018 Indonesian Fisheries Productivity (Jakarta: KKP)
[5] Pradana H, Prajitno A, and Maftuch 2022 *Jur Ilmi Peri dan Kela* 14 (2) 246-259.
[6] Lesmana D S and Dermawan I 2001 *Popular Freshwater Ornamental Fish Cultivation* (Jakarta: Penebar Swadaya) p 160.
[7] Amin M I, Rosidah, and Lili W 2012 *J Perik dan Kela* 3 (4) 243-252.
[8] Malide S M, Hendri A, and Budiman 2018 *J Akuak* 2 (2) 64-71.
[9] Mimura T, Maeda K, Aunoma S, Sataka M, and Fujita T 1982 *Chem Pharm Bull* 30 (4) 1381-1386.
[10] Luo P and Liu H 2013 *Afri Jour of Pharm and Pharma* 7 (21) 138-1388.
[11] Raharjo E I and Tampubolon T P 2016 *Jurn Rua: Jurn Pene dan Kaj Ilm Perik dan Kel* 4 (2) 28-33.
[12] Zidni I, Ayi Y, Iskandar and Yuli A 2017 *Jur Perik dan Kela* 7 (2) 125-135.
[13] Lubis M Z and Pujiyati S 2017 *Jur Tekno Perik dan Kela* 4 (2) 123-129.
[14] Putriana N, Tjahjaningsih W, and Alamsjah M A 2015 *Jur Ilmi Perik dan Kela* 7 (2) 189-194.
[15] Agus G T K, Agus K A, Dianawati A, Dipo U T, Irawan E S, Miharja K, Gusyadi L, Luluk A M, Maman N, Karno P S, Dachlan P, Udin S, Ujang J M, Yana T, and Sastro Y 2002 *Koi* (Tangerang: PT AgroMedia Pustaka) p 23 – 46.
[16] Sari N P, Santoso L, and Hudaidah S 2012 *e-Jur Reka dan Tekno Budi Per* 1 (1) 31-38.
[17] Effendie M 1997 *Fisheries Biology Method (Bogor: Dewi Sri Bogor Foundation*) p 162.
[18] Susanto 2007 *Tips for Cultivating Goldfish on Critical Land* (Jakarta: Penebar Swadaya) p 13 – 18.
[19] De Kock S and Watt R 2006 *Koi: A Handbook on Keeping Nishikigoi* (New York : Firefly Books Inc.) p 115 – 130.
[20] Kusrini E, Cindelaras S, and Prasetio AB 2015 *Med Akua* 10 (2): 71-78.
[21] Effendi 2003 *Study of Water Quality for Management of Water Resources* (Yogyakarta: Kanisius) p 257.

[22] McCoy M 1999 *Astaxanthin Market A Hard One To Crack* (USA: Chem and Eng News).
[23] Hajime M, Yoshiaki T, Hidenitsu U, Tetsushi N, Bun-ichi O, Fumiaki N, Kunio L, Jin-ichi S 1997 *Antitumor Peptidoglycan with New Carbohydrate Structure from Squid Ink*. In Ohigashi H, Osawa T, Terao J, Watanabe 5, Yoshikawa T (Eds.). Food Factors for Cancer Prevention. Tokyo: SpringerVerlag. p 331-336.
[24] Indonesia National Standard 2006 Artificial Feed for Goldfish (Cyprinus carpio L.) in Intensive Aquaculture SNI 01-4266-2006 (Jakarta: SNI) p 20.

6. Acknowledgement

The author expresses many thanks to the supervisors who have given a lot of knowledge and energy. The author also thanks the editors and reviewers for their suggestions and input on our manuscript.

IOP Conf. Series: Earth and Environmental Science 1273 (2023) 012066 doi:10.1088/1755-1315/1273/1/012066

Analysis on intensive program of vaname shrimp *(Litopenaeus vannamei)* enlargement in the technical implementation unit of brackish and sea water cultivation (UPT BAPL) Bangil, Pasuruan Regency, East Java

M Marista[1], R M Khatami[1], J C Santoso[1], M H Fauzi[2], S H Samara[3,4]

[1]Aquaculture Program, Faculty of Fisheries and Marine, Universitas Airlangga, Surabaya, Indonesia
[2]Unit Pelaksanaan Teknis Budidaya Air Payau dan Laut (UPT BAPL) Bangil, Pasuruan Regency, East Java
[3]Department of Aquaculture, Faculty of Fisheries and Marine, Universitas Airlangga, Surabaya, Indonesia

[4]Corresponding author: syifania.hanifah@fpk.unair.ac.id

Abstract. Vannamei shrimp (*L. vannamei*) is one of the popular aquaculture commodities in Indonesia. This is because vaname shrimp has various advantages compared to other species, such as fast growth rate, can be cultivated with high stocking density, tolerant of high salinity ranges, relatively uniform harvest size, and easy to cultivate. Vannamei shrimp also has many benefits. For example, its meat can be processed as nuggets or food and the shell can be used for medical, cosmetic and even biotechnology. This study aims to determine the production of vaname shrimp at UPT Bangil, one of the largest shrimp productions in East Java. The results of this study showed that the average daily gain of vaname shrimp at UPT Bangil had a yield of 0.33 g/day, and the weight of each shrimp increased to 15.7 g/head. Vannamei shrimp could be harvested at DOC 83 and yields 193.5 kg, with a size of 58. This shows good performance of vaname shrimp enlargement in UPT Bangil.

1. Introduction

Vannamei shrimp (*L. vannamei*) is one of the cultivation commodities in Indonesia that is growing rapidly recently. Vannamei shrimp production in Indonesia according to the Ministry of Maritime Affairs and Fisheries (2018) in 2017 reached 400,073 tons with a total national scale value of IDR 22,009,626,415 [1]. Vannamei shrimp has various advantages compared to other species, such as fast growth, can be cultivated with high stocking density, tolerant of high salinity ranges, relatively same harvest size, and easy to cultivate [2].

Intensive vaname shrimp cultivation is inseparable from aspects of water quality and disease potential. According to previous study, water quality in vaname shrimp culture includes DO, organic matter (nitrite, nitrate, ammonia), salinity, ph, brightness, and temperature [3]. If the water quality in vaname shrimp culture is not controlled properly, it can reduce the shrimp's immune system and cause death. The steps that must be taken to deal with shrimp infected with parasites are controlling water quality in intensive aquaculture ponds and observing vaname shrimp commodities.

UPT BAPL Bangil is one of the centers that developed the technique of rearing vaname shrimp. Based on this, it is necessary to carry out field work practices at UPT BAPL Bangil. Mastery of

IOP Conf. Series: Earth and Environmental Science 1273 (2023) 012066 doi:10.1088/1755-1315/1273/1/012066

vaname shrimp rearing techniques is obtained through practical field work activities. With the practice of field work, it is expected to increase knowledge and skills about rearing techniques for white shrimp (L. vannamei) in intensive cultivation systems and the factors that influence it. The results of a good and sustainable system can produce vaname shrimp that are not attacked by disease and have high economic value.

2. Material and methods

2.1. Research method

This research was conducted at UPT BAPL Bangil, Pasuruan Regency, East Java. This study uses primary and secondary data collection as described in the previous study [4]. According to Sangadji and Sopiah (2010) data were collected either through collection and observation instruments, as well as through documentation data. According to the source, the data can be classified as primary data and secondary data. There are three methods that can be used in primary data collection, namely survey methods, observation methods, and participation methods.

2.2 Research parameter calculation

The parameters in this study were survival rate, average daily gain, and water quality parameters (temperature, salinity, nitrate, nitrite, ammonia). According to previous study, the average daily gain was calculated using the formula [5]:

$$W = \frac{\mathrm{W_t} - \mathrm{W_o}}{\mathrm{day}}$$

W : Weight growth (gram)
Wt : Final average weight (gram)
Wo : Initial average weight (gram)
Day : Time of cultivation (day)

According to previous study, the survival rate of shrimp was calculated using the formula [6]:

$$SR = \frac{\mathrm{N_t}}{\mathrm{N_o}} \times 100\%$$

SR : Survival rate (%)
Nt : Final number of shrimp
N0 : Initial number of shrimp

Water quality parameter was conducted as previous study [7].

3. Result and discussion

The results of the shrimp sampling were counted and the weight of the shrimp was weighed. From the results of the sampling calculation, it can be seen that Average Body Weight (ABW), growth or Avarage Daily Growth (ADG), Specific Growth Rate (SGR), Food Conversion Rate (FCR), Survival Rate (SR) and the total shrimp biomass contained in the aquaculture pond [8]. The percentage of shrimp survival rate (SR) in this cycle is 70%. Previous study stated that SR is categorized as good if the SR value is > 70%, for the medium category SR is 50-60%, and in the low category the SR value is <50% [9]. The high SR of vaname shrimp is thought to be caused by the availability of nutrients, which include protein, carbohydrates and fat, which are in accordance with the needs of white vaname shrimp. In addition, optimal environmental factors are important factors because they affect the growth and physiological processes of white vaname shrimp [10].

Average Body Weight (ABW) increases every week, with a difference of 1-2 g/week. Average Daily Gain (ADG) in round pools fluctuates up and down slightly. At the 61st DOC there was a spike in weight gain. This was caused by the increase in the dose of feed given every day, in the range of DOC 51 to DOC 61. In addition, vaname shrimp had a good appetite, seen from checking on anco.

The feed consumed during one cycle was 425 kg, with a total biomass of 323.5 kg. Feed Consumption Rate (FCR) in this cycle has a result of 1.3 so it can be said that the efficiency of feed in ponds is good. The FCR value is said to be good if it is in accordance with the standard FCR or the targeted FCR. Based on data collection on intensive ponds under normal harvesting conditions, the FCR achieved was between 1.5–2.0 (depending on the media/pond land condition used) [11]. The graph of periodic weight gain and ADG can be seen in Figure 1.

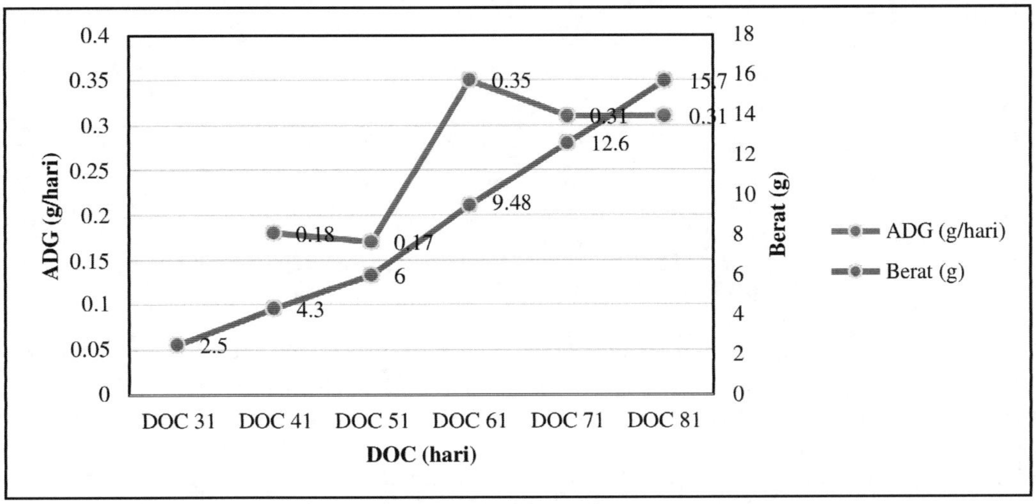

Figure 1. Weight gain and ADG in this study

The pH measurement can be seen in Figure 2. The degree of acidity (pH) is one of the important parameters in a water that describes how much acidity or wetness a water is. Water conditions are said to be neutral if the pH value is equal to 7 and acidic water conditions if the pH is less than 7 while the water conditions are alkaline if the pH is more than 7 [12]. The high pH value is influenced by the oxygen content that comes from the photosynthesis process so that it affects the pH value in the waters. The pH value is also influenced by the decomposition process of organic matter such as phytoplankton which produces a lot of $CO2$, resulting in a decrease in pH [13].

The results of pH measurements in vaname shrimp ponds obtained the highest value of 8.98 and the lowest value of 7.23. Based on the measurement data obtained an average of 7.66 in the morning, 7.83 in the afternoon and 7.8 in the afternoon. The increase in pH occurs due to sunny weather so that the sunlight needed by plants and phytoplankton is sufficient to carry out photosynthesis. The process of photosynthesis that occurs produces oxygen in the waters so that the pH concentration will also increase. According to previous study, the pH in ponds can change throughout the day due to photosynthesis [14]. Phytoplankton and aquatic plants utilize CO2 during the day so that the pH increases while at night CO2 is produced by organisms through respiration so that the pH drops and is acidic. The higher the CO2, the lower the pH. The optimum pH for shrimp ranges from 7.5 to 8.5 [15].

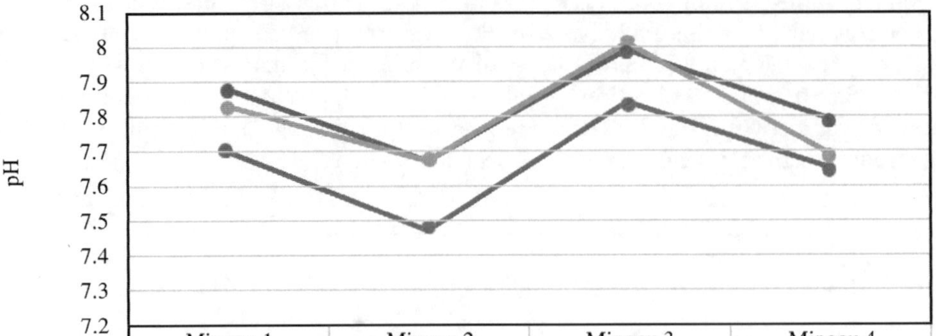

	Minggu 1	Minggu 2	Minggu 3	Minggu 4
Pagi	7.7	7.47	7.84	7.65
Siang	7.88	7.67	7.99	7.79
Sore	7.83	7.67	8.01	7.69

Figure 2. pH measurement in this study

Oxygen is one of the important factors in aquaculture activities so that if the availability in water is not sufficient for the needs of aquaculture biota, then all biota activities will be hampered [16]. Oxygen is needed by fish for respiration and metabolism in the fish body for swimming, plants, reproduction and others. In addition, oxygen is also needed for the oxidation of organic and inorganic materials. The value of oxygen in activities is very important because conditions that are less than optimal for growth and development can cause stress [17]. The main source of oxygen in waters comes from the process of diffusion of free air and the results of photosynthesis of organisms that live in these waters [18]. Dissolved oxygen content in waters is influenced by the presence of organisms, photosynthesis, temperature, light penetration rate, the flow rate of the amount of organic matter described in the water [19]. Dissolved oxygen levels in waters fluctuate daily and seasonally, depending on the level of pollution and mass movement of water, photosynthetic activity, respiration and wastewater entering water bodies [6]. The graph of Dissolved Oxygen (DO) can be seen in Figure 3.

Based on the results of the dissolved oxygen measurement above, the average measurement in the morning is 4.45 mg/l, 5.76 mg/l in the afternoon and 3.82 mg/l in the afternoon. The highest dissolved oxygen value was 7.28 mg/l in the morning and the lowest dissolved oxygen value was 3.04 mg/l in the afternoon. The measurement of dissolved oxygen values in the morning is higher than in the afternoon due to the reduction of phytoplankton during the process of replacing pond water with water entering from bore wells, so that in the afternoon the photosynthesis process by phytoplankton is not as much as in the morning and afternoon.

The increase in dissolved oxygen in the waters is caused by the weather at the time of measurement so that the intensity of sunlight entering the waters is also higher so that the photosynthesis process by phytoplankton increases, besides that the wind is blowing hard so that it can be seen visually on the surface of the water there are very small waves.

The increase in dissolved oxygen is also influenced by the fertilization process, which is done so that phytoplankton photosynthesize and produce O2. During the day oxygen is produced through the process of photosynthesis, while at night the oxygen formed will be reused by algae for metabolic processes when there is no light [20]. The optimal dissolved oxygen concentration in the maintenance of vannamei shrimp ranged from 4-8 ppm [21]. This value indicates that the oxygen content contained in the rearing media is still optimal and good enough to support the growth of cultured vannamei shrimp.

IOP Conf. Series: Earth and Environmental Science 1273 (2023) 012066 doi:10.1088/1755-1315/1273/1/012066

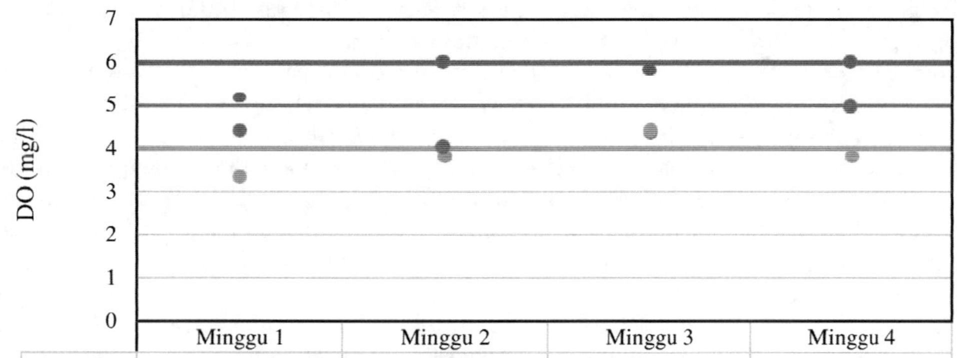

	Minggu 1	Minggu 2	Minggu 3	Minggu 4
pagi	4.37	4.08	4.36	5.01
siang	5.1	6.06	5.82	6.06
sore	3.3	3.8	4.41	3.8

Figure 3. DO measurement in this study

Water temperature is a physical parameter that has a major influence on the growth and life patterns of shrimp. The effect of temperature on the growth of vannamei shrimp is on the specificity of the stage and size. Young shrimp can grow well in water with warm temperatures, but the larger the shrimp, the lower the optimum water temperature [22]. Temperature also greatly affects the activities of organisms such as metabolism and the spread of organisms in the sea and organisms in fresh water. Temperature measurement is very necessary to determine the characteristics of waters, water temperature is an abiotic factor that plays an important role for the life of organisms in the waters [14]. The temperature graph can be seen in Figure 4.

Based on the results of temperature measurements in vaname shrimp ponds, the average measurement for 2 months is 28.38°C in the morning, 30.85°C in the afternoon and 29.72°C in the afternoon. The highest temperature reached 31.7 °C during the day, and the lowest temperature occurred in the morning, namely 25.9 C. This value indicates that the temperature contained in the rearing media is in the optimal range to support the growth of cultured vannamei shrimp. The pond water temperature requirements for optimal vannamei shrimp culture are between 26°C - 32°C [23]. Fish farming requires the availability of sufficient water with good quality. Sudden changes in temperature adversely affect the life of the cultivated biota. At temperatures below 15.5°C – 12°C generally fish cannot live well [24].

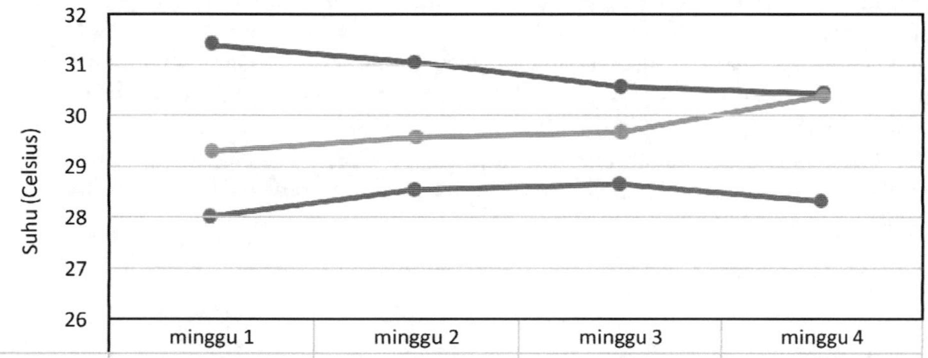

	minggu 1	minggu 2	minggu 3	minggu 4
pagi	28.01	28.54	28.65	28.32
siang	31.38	31.05	30.57	30.43
sore	29.29	29.57	29.67	30.37

Figure 4. Temperature measurement in this study

Salinity is the total concentration of all dissolved ions such as sodium, potassium, calcium, magnesium, chloride, sulfate, and bicarbonate contained in a water which is expressed in units of milligrams per liter (mg/l). Salinity or salt content is an important parameter in shrimp culture because it is closely related to the osmotic and ionic pressure of water, as well as the level of shrimp osmoregulation. The results of salinity measurements in vaname shrimp ponds can be seen in Figure 5. Based on the results of salinity measurements in vaname shrimp ponds, the average measurement for 2 months is 10.68 ppt in the morning, 10.66 ppt in the afternoon and 10.59 ppt in the afternoon. The highest salinity occurred in the morning at 12 ppt, and the lowest salinity was 10 ppt. This value indicates that the salinity contained in the rearing media is in the optimal range to support the growth of cultured vannamei shrimp.

The increase and decrease in salinity can be influenced by temperature values that affect the dissolved oxygen concentration in pond waters. According to (Badruddin, 2014), generally the salinity range for shrimp culture is between 0-35 ppt and optimally 10-30 ppt. Meanwhile, according to Hendrawati and Rohmah (2008), adult and brood shrimp require higher salinity > 28 ppt, while the larval to post larval stages require salinity < 28 ppt.

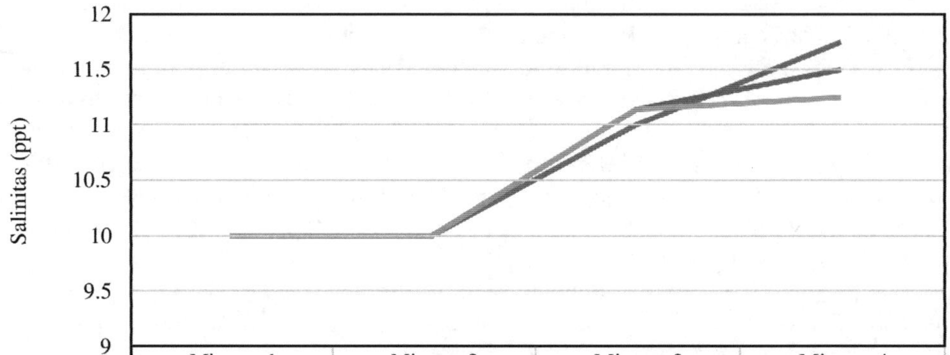

	Minggu 1	Minggu 2	Minggu 3	Minggu 4
pagi	10	10	11	11.75
siang	10	10	11.14	11.5
sore	10	10	11.14	11.25

Figure 5. Salinity measurement in this study

Nitrate (NO_3) is the product of nitrification which is the most oxidized form of nitrogen in water. The end result of the nitrification process is the formation of nitrate. Good nitrite and nitrate content in shrimp ponds, namely nitrite <0.1 mg/l and nitrate <1 mg/l, if not appropriate, it will cause toxins that have an impact on fish health [26]. Based on the test results of nitrate content in vaname shrimp ponds, an average measurement of 21.47 mg/l was obtained with the highest value of 25.5 mg/l and the lowest value of 12.9 mg/l, although the concentration of NO_3 was in the high range. On observation, nitrate is a form that is not dangerous, so its presence does not disturb the biota that are kept. N-inorganic compounds are relatively non-toxic to shrimp life compared to nitrite and ammonia [27]. Nitrate data can be seen in Figure 6.

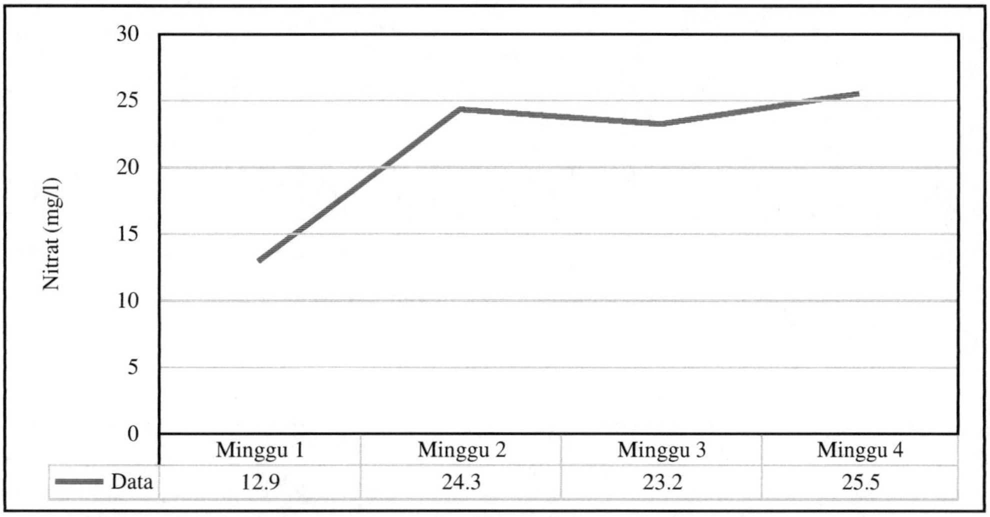

Figure 6. Nitrate measurement in this study

Nitrite is one of the chemical contaminants in water. Besides being caused by human activities, the increase in nitrite in water can also be caused by the activity of bacteria that can reduce nitrate to nitrite and oxidize ammonia to nitrite by Nitrosomonas bacteria [28] Nitrite is an intermediate form between ammonia and nitrate (Nitrification), it is also an inorganic nitrogen compound that can endanger shrimp life when present in high amounts [27].

Nitrite is toxic because of its ability to bind to hemoglobin, thereby interfering with the absorption of oxygen in the blood. The nitrite content in the waters is influenced by the DO content in the waters. The high DO concentration causes the formation of nitrite to take place more quickly. High nitrite concentrations are not good for shrimp culture because nitrite can cause methemoglobin (the oxidation process of Fe2+ in hemoglobin or Cu in hemocyanin) in fish so that the O2 binding process is inhibited. Although shrimp do not have red blood cells, there is an effect on shrimp when nitrite is high [29].

Based on the results of the nitrite content test in vaname shrimp ponds, an average measurement of 1.97 mg/l was obtained with the highest value of 3.16 mg/l and the lowest value of 0.71 mg/l, this value is classified as dangerous to the waters. Good nitrite content for shrimp ponds is nitrite <0.1 mg/l. Vannamei shrimp feces are suspected to be a substrate for nitrifying bacteria, so that nitrifying bacteria grow a lot which causes the nitrification process to run well and the accumulation of NO2 becomes less. Nitrite data can be seen in Figure 7.

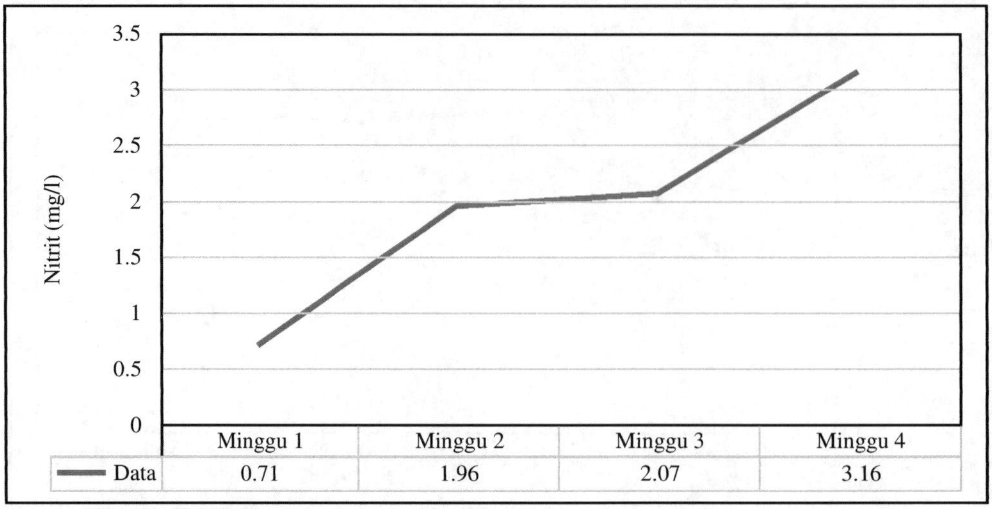

Figure 7. Nitrite measurement in this study

Ammonia (NH_3) and other salts are easily soluble in water. Ammonia in waters comes from the breakdown of organic nitrogen (protein and urea) and inorganic nitrogen found in soil and water from the decomposition of organic matter (dead plants and aquatic biota) by microbes and fungi (Effendi, 2003). Ammonia research data fluctuations in NH_3 concentrations are different. Based on research data, the lowest NH_3 value is 0.021 mg/l and the highest value is 3.3 mg/l, this value is classified as dangerous to the waters. Good NH_3 content for shrimp ponds is <0.05 mg/l.

The high content of NH_3 is influenced by the rest of the metabolism and the rest of the feed in the water. Most of the feed eaten by fish and shrimp will be converted into meat or body tissue, while the rest is disposed of in the form of solid waste (faeces) and dissolved (ammonia) [17]. The high content of NH_3 comes from the accumulation of feed residues in the maintenance of vannamei shrimp which may not have undergone the ammonification process completely, the increase in ammonia concentration can occur because the decomposition of organic N-compounds continues to increase due to the degradation of accumulated feed residues. Meanwhile, some things that are thought to have an effect on the accumulation of ammonia are the low dissolved oxygen content.

If the dissolved oxygen available in the waters is only sufficient for the formation of ammonia, the oxidation process of ammonia, nitrite and nitrate (nitrification) will not run perfectly [27]. In addition, the density of microbes using ammonia and the presence of compounds inhibiting the nitrification process caused high levels of ammonia.

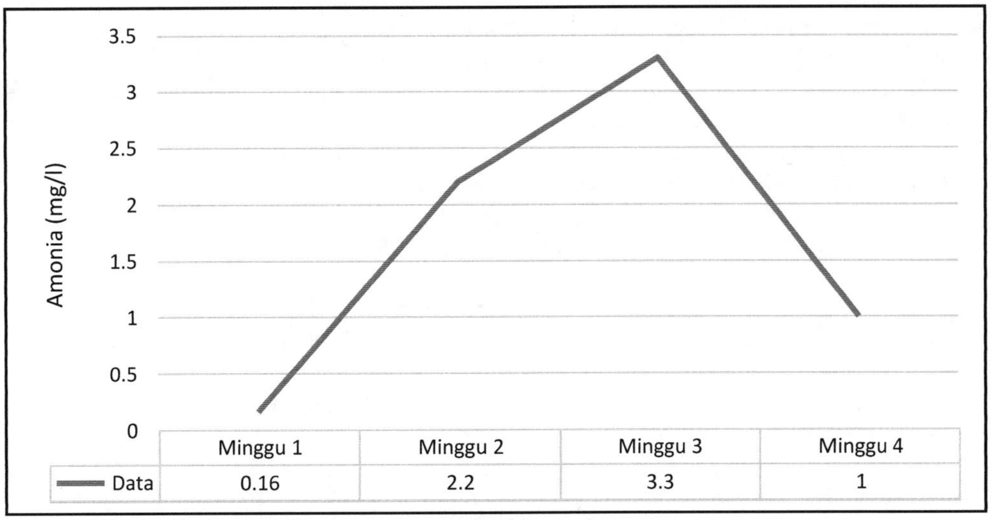

Figure 8. Ammonia measurement in this study

4. Conclusion

The results of this study showed that the average daily gain of vaname shrimp at UPT Bangil had a yield of 0.33 g/day, and the weight of each shrimp increased to 15.7 g/head. Vannamei shrimp could be harvested at DOC 83 and yields 193.5 kg, with a size of 58. This shows good performance of vaname shrimp enlargement in UPT Bangil.

5. References

[1] Ruliaty L, Sumarwan J, Handayani R, dan Susanto A 2014 *Scoring Method: A Measurable Way to Get Quality Shrimp Seeds Ministry of Maritime Affairs and Fisheries Directorate General of Aquaculture Fisheries Center for Brackish Water Aquaculture Jepara.*

[2] Suriawan A, Efendi S, Asmoro S, dan Wiyana J 2019 *J. Perekayasaan Budidaya Air Payau dan Laut.* 14 6-14.

[3] Bhaskoro R G E, Ramadhan T E 2018 *J. Presipitasi: Media Komunikasi dan Pengembangan Teknik Lingkungan* 15 62-68.

[4] Laksono M G, Santanumurti M B 2021 *IOP Conf Ser Earth Env Sci***.** 679 012018

[5] Dedi D 2018 *Intek Akuakultur.* 2 33-48.

[6] Effendi H 2003 *Study of Water Quality for Management of Water Resources and Environment (Kanisius: Yogyakarta).*

[7] Kewcharoen W, Srisapoome P 2019 *FSI.* 94 175-189.

[8] Suciyono S A, Mahmudi M 2013 *Online J Anim Feed Res.* 3 222-229.

[9] Arsad S, Afandy A, Purwadhi A P, Saputra D K, dan Buwono N R 2017 *JIPK* . 9 1-14.

[10] Zainuddin, Haryati, Aslamyah S, dan Surianti 2014 *J. Fish. S.ci.* 16(1) 29-34.

[11] Marini A 2019 Skripsi. Surabaya: Universitas Airlangga Surabaya.

[12] Irianto, Triweko 2011 *Eutrophication of Reservoirs and Lakes, Problems and Control Measures (Bandung: ITB Press) 1-150 pages.*

[13] Parawita D, Insanfitri W A, dan Nugraha 2009 *J. Kelautan.* 2 34-42.

[14] Suherman R 2011 Skripsi. Jakarta: Universitas Islam Negeri Syarif Hidayatullah. 1-68 hal.

[15] Erlangga E 2012 *Intensive Vannamei Shrimp Cultivation* (Tangerang Selatan: Pustaka Agro Mandiri).

[16] Kordi K M, Ghuffran H, dan Andi B T 2007 *Water Quality Management in Aquaculture* (Jakarta: Rineka Cipta) 1-208 hal.

[17] Jayadi M, Prajitno A, dan Maftuch 2016 *Int. J. of ChemTech Res.* 9 448-452.

[18] Salmin 2005 *Oseana.* 30(3) 21-26.

[19] Nabilah 2012 Skripsi. Depok: Universitas Indonesia. 1-99 hal.
[20] Tatangdatu F, Kalesaran O, dan Rompas R 2013 *J. Budidaya Perairan.* 1 8-19.
[21] Fegen D F 2003 *Cultivation of Vanname Shrimp (Litopenaeus vannamei) at Asia Gold Coin Indonesia* (Jakarta: Specialities).
[22] Wyban J, Sweeney J N 1991 *Intensive shrimp production technology: the Oceanic Institute shrimp manual*(United States: The Institute). p 158.
[23] Utojo U, Tangko A M 2008 *J. Media Akuakultur.* 3 118-125.
[24] Cahyono B 2000 *Freshwater Fish Cultivation* (Yogyakarta: Kanisius).
[25] Hendrawati, Prihadi T H, dan Rohmah N N 2007 *J. Valensi.* 1 106716.
[26] Kuhn D D, Smith S A, Boardman G D, Angier M W, Marsh L, dan Flick Jr G J 2010 *Aquaculture* . 309 109-114.
[27] Komarawidjaja W 2006 *J. Hidrosfir Ind.* 1(1) 32-37.
[28] Amri C, Siswanta D, dan Mudasir 2009 *Indo. J. Chem.* 9(2)254-260.
[29] Budi N A R S 2019 Skripsi. Yogyakarta: STIE YKPN. 1-25 hal.

6. Acknowledgement
The authors gratefully acknowledge the collaboration from Universitas Airlangga and the University of Lampung.

IOP Conf. Series: Earth and Environmental Science 1273 (2023) 012067 doi:10.1088/1755-1315/1273/1/012067

Simple additive weighting – a recommended seeds selections system for nile tilapia fish farmer

H R Hatta[1,4], A Wijayanti[2,3], A Septiarini[1], N Puspitasari[1], and M Wati[1]

[1]Department of Informatics, Mulawarman University, Jl. Kuaro Gn. Kelua,Samarinda, 75119 Indonesia
[2]Department of Ecology and Evolutionary Biology, University of California Santa Cruz, Santa Cruz, CA 95060, USA
[3]Generasi Biologi Indonesia, Zoology Division, Gresik 61171, East Java, Indonesia

[4]Corresponding author : heliza.rahmania@gmail.com

Abstract. Nile tilapias are known to be an economically important farmed fish in Indonesia. Due to its adaptability, this tilapia fish can thrive in various aquatic conditions, making them an important commercial fish to produce locally or export internationally. Because of this potential, this study aims to provide recommendations using the decision support system to facilitate fish farmer decision-making with the Simple Additive Weighting model. This approach is expected to provide recommendations to fish farmers, or the general public interested in having their own Nile tilapia brood stock. The fish in this research was selected randomly following six criteria, namely fish weight, length, movement, fin or body deformation, fish colour and fish stomach condition. The result-based decision making model in this study provides an accuracy of 86.6%.

1. Introduction

Nile tilapia (*Oreochromis niloticus*) is a widely introduced and cultured fish worldwide. It has become one of the most important fish, with a global harvest of over 6 million tons (MT) [1,2]. Although this fish originated from Africa, the number of countries that farm and culture this fish has increased. In 2009 reported that 80 countries had farmed tilapia, and it has expanded to over 140 countries ever since, including Indonesia [1,3]. In Indonesia, the freshwater and marine aquaculture industry is crucial to the country's economy. Indonesia's most common freshwater commodities are catfish, common carp, and Nile tilapia [4,5].

Aquaculture has continued to expand more rapidly than other significant food-producing industries. It exists on a variety of scales and takes on several different shapes. Broadly it can be classified as small, medium, and large scale [1]. The Nile tilapia has overtaken carp as the second-most significant aquaculture product [2, 6]. Although it has importance as a commodity, it is a known invasive species [7,8,9] and might threaten native species if it escapes from aquaculture facilities or is accidentally released into the wild [10,11,12]. Responsible aquaculture practice is important in managing Nile Tilapia even in the small-scale aquaculture level.

However, because of its importance as a food supply [13,14,15], Nile tilapia fish production was predicted to increase significantly with an expected amount of 1.15 million tons [16,17]. A careful selection of fish seeds is critical for stocking material production to meet increasing demand. A system that can aid fish farmers in making better decisions is needed. This study aims to provide recommendations on Nile tilapia seeds selection using the decision support system to facilitate fish farmer decision-making with the Simple Additive Weighting model.

2. Materials and methods

Simple additive weighting is one of the multi-criteria decision-making models whose focal point lies in the performance rating summation for each alternative on all the attributes. This technique involves a number of steps. However, normalizing the choice matrix (X) to a scale that is proportionate to all current alternative ratings using Equation 1 is the main stage of the SAW approach. Equation 1 will get different treatments if j/criteria are of benefit or cost type. Therefore, before performing calculations, determine the type of each criterion to determine which Equation 1 will be used. After getting the normalized data, do a ranking of the results of the preference values for each alternative with Equation 2. The greater the preference value, the higher the ranking of the alternative, and the more recommended the alternative. In carrying out the process, the SAW method requires several alternatives (A), criteria (C) and weights (W) [11,12].

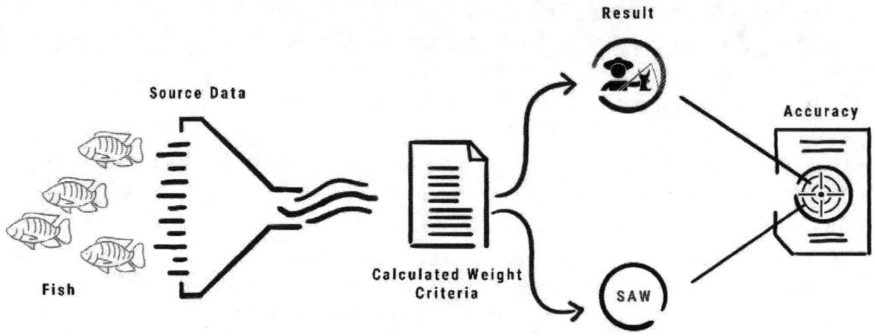

Figure 1. Research framework.

Following are the steps in the SAW method:

1. Decision-making gives preference weights to be processed in ranking with a normalized matrix

$$X = \begin{bmatrix} X_{11} & X_{12} & X_{1n} \\ X_{21} & X_{22} & X_{2n} \\ \vdots & \vdots & \\ X_{m1} & X_{m2} & X_{mn} \end{bmatrix}$$

W = {W1, W2,.... ,Wn}

2. The matrix is formed from the Alternate (A) and Criterion (C) match table
3. The matrix normalization is performed based on the following equation after obtaining the value of the X matrix.

$$_{ij} = \begin{cases} \frac{_{ij}}{_{i} \; _{ij}} & \text{If j is a benefit attribute} \\ \frac{_{i} \; _{ij}}{_{ij}} & \text{If j is a cost attribute} \end{cases} \quad (1)$$

4. Following the creation of a normalized matrix (R), a ranking process is conducted using the preference values (Vi) for each choice.

$$V_i = \sum_{j=1} \; _j \; _{ij} \quad (2)$$

Information:
Vi = the ranking of every alternative
Wj = weight value of each criteria

rij = value of a normalized performance rating

The alternative (Ai) with the highest Vi value is picked because it has the highest probability of being selected as the best option.

3. Results and discussion

Information on Nile tilapia was gathered at a fish farm in Samarinda City, East Kalimantan, Indonesia. The data was obtained in the form of criteria, criteria weights and sub-criteria values. Table 1 contains the data on the weight of the criteria, and Table 2 has the data on the sub-criteria.

Table 1. Criteria weight

Code	Criteria	Weight	Description
C1	Fish weight	4	Benefits
C2	Fish length	5	Benefits
C3	Fish move	3	Benefits
C4	Physical disability	3	Benefits
C5	Fish colour	3	Benefits
C6	Stomach condition	4	Benefits

Table 2. Subcriteria value

Criteria	Subcriteria	Value
Fish weight (C1)	250 – 500 g	1
	501 – 750 g	2
	751 – 1000 g	3
Fish length (C2)	10 – 15 cm	1
	16 – 20 cm	2
	21 – 25 cm	3
Fish move (C3)	Not agile	1
	Agile	2
	Very agile	3
Physical disability (C4)	Yes	1
	No	2
Fish color (C5)	Dull	1
	Bright	2
Stomach condition (C6)	Fluid	1
	No fluid	2

Table 3. Nile tilapia data

No	Code	Fish weight (gr) (C1)	Fish length (cm) (C2)	Fish movement (C3)	Physical disability (C4)	Fish color (C5)	Stomach condition (C6)
1	NT01	605	20	Agile	No	Bright	No fluid
2	NT02	650	23	Very agile	No	Bright	No fluid
3	NT03	740	23	Very agile	No	Bright	No fluid
4	NT04	655	22	Very agile	No	Bright	No fluid
5	NT05	600	21	Agile	No	Bright	No fluid
6	NT06	650	22	Agile	No	Bright	No fluid
7	NT07	635	21	Very agile	No	Bright	No fluid
8	NT08	700	23	Agile	No	Bright	No fluid
9	NT09	715	23	Not agile	No	Bright	No fluid
10	NT10	580	19	Not agile	No	Bright	No fluid
11	NT11	750	23	Very agile	No	Bright	No fluid
12	NT12	725	23	Agile	No	Bright	No fluid

13	NT13	680	22	Very agile	No	Bright	No fluid
14	NT14	630	23	Very agile	No	Bright	No fluid
15	NT15	805	24	Agile	No	Bright	No fluid

Fish data obtained from Nile tilapia cultivation in Samarinda City, East Kalimantan, Indonesia based on criteria, namely fish weight, fish length, fish movement, physical disability, fish color, stomach condition as shown in table 1. Nile tilapia data was obtained as in table 3, then changed according to table 2 so that it can be input into the system. Table 4 contains the converted data.

Table 4. Nile nile tilapia fish data conversion results

No	Code	Fish weight (gr) (C1)	Fish length (cm) (C2)	Fish movement (C3)	Physical disability (C4)	Fish color (C5)	Stomach condition (C6)
1	NT01	2	2	2	2	2	2
2	NT02	2	3	3	2	2	2
3	NT03	2	3	3	2	2	2
4	NT04	2	3	3	2	2	2
5	NT05	2	3	2	2	2	2
6	NT06	2	3	2	2	2	2
7	NT07	2	3	3	2	2	2
8	NT08	2	3	2	2	2	2
9	NT09	2	3	1	2	2	2
10	NT10	2	2	1	2	2	2
11	NT11	2	3	3	2	2	2
12	NT12	2	3	2	2	2	2
13	NT13	2	3	3	2	2	2
14	NT14	2	3	3	2	2	2
15	NT15	3	3	2	2	2	2

The first step making the decision matrix in the SAW method based on table 4 as follows:

$$X = \begin{bmatrix} 2 & 2 & 2 & 2 & 2 & 2 \\ 2 & 3 & 3 & 2 & 2 & 2 \\ 2 & 3 & 3 & 2 & 2 & 2 \\ 2 & 3 & 3 & 2 & 2 & 2 \\ 2 & 3 & 2 & 2 & 2 & 2 \\ 2 & 3 & 2 & 2 & 2 & 2 \\ 2 & 3 & 3 & 2 & 2 & 2 \\ 2 & 3 & 2 & 2 & 2 & 2 \\ 2 & 3 & 1 & 2 & 2 & 2 \\ 2 & 2 & 1 & 2 & 2 & 2 \\ 2 & 3 & 3 & 2 & 2 & 2 \\ 2 & 3 & 2 & 2 & 2 & 2 \\ 2 & 3 & 3 & 2 & 2 & 2 \\ 2 & 3 & 3 & 2 & 2 & 2 \\ 3 & 3 & 2 & 2 & 2 & 2 \end{bmatrix}$$

With W = {4, 5, 3, 3, 3, 4} based Table 1.

X matrix normalization using Equation 1 denotes the cost values
$|C_1|$ max = 3 from {2; 2; 2; 2; 2; 2; 2; 2; 2; 2; 2; 2; 2; 2; 3}

$R_{11} = 2/3 = 0{,}67$
$R_{21} = 2/3 = 0{,}67$
$R_{31} = 2/3 = 0{,}67$
$R_{41} = 2/3 = 0{,}67$
$R_{51} = 2/3 = 0{,}67$
$R_{61} = 2/3 = 0{,}67$
$R_{71} = 2/3 = 0{,}67$
$R_{81} = 2/3 = 0{,}67$
$R_{91} = 2/3 = 0{,}67$
$R_{101} = 2/3 = 0{,}67$
$R_{111} = 2/3 = 0{,}67$
$R_{121} = 2/3 = 0{,}67$
$R_{131} = 2/3 = 0{,}67$
$R_{141} = 2/3 = 0{,}67$
$R_{151} = 3/3 = 1$

$|C_6|$ max = 2 from {2; 2; 2; 2; 2; 2; 2; 2; 2; 2; 2; 2; 2; 2; 2}
$R_{16} = 2/2 = 1$
$R_{26} = 2/2 = 1$
$R_{36} = 2/2 = 1$
$R_{46} = 2/2 = 1$
$R_{56} = 2/2 = 1$
$R_{66} = 2/2 = 1$
$R_{76} = 2/2 = 1$
$R_{86} = 2/2 = 1$
$R_{96} = 2/2 = 1$
$R_{106} = 2/2 = 1$
$R_{116} = 2/2 = 1$
$R_{126} = 2/2 = 1$
$R_{136} = 2/2 = 1$
$R_{146} = 2/2 = 1$
$R_{156} = 2/2 = 1$

From the normalization calculation, the normalized matrix R is obtained as follows:

$$= \begin{bmatrix} 0{,}67 & 0{,}67 & 0{,}67 & 1 & 1 & 1 \\ 0{,}67 & 1 & 1 & 1 & 1 & 1 \\ 0{,}67 & 1 & 1 & 1 & 1 & 1 \\ 0{,}67 & 1 & 1 & 1 & 1 & 1 \\ 0{,}67 & 1 & 0{,}67 & 1 & 1 & 1 \\ 0{,}67 & 1 & 0{,}67 & 1 & 1 & 1 \\ 0{,}67 & 1 & 1 & 1 & 1 & 1 \\ 0{,}67 & 1 & 0{,}67 & 1 & 1 & 1 \\ 0{,}67 & 1 & 0{,}33 & 1 & 1 & 1 \\ 0{,}67 & 0{,}67 & 0{,}33 & 1 & 1 & 1 \\ 0{,}67 & 1 & 1 & 1 & 1 & 1 \\ 0{,}67 & 1 & 0{,}67 & 1 & 1 & 1 \\ 0{,}67 & 1 & 1 & 1 & 1 & 1 \\ 0{,}67 & 1 & 1 & 1 & 1 & 1 \\ 1 & 1 & 0{,}67 & 1 & 1 & 1 \end{bmatrix}$$

Looking for the best alternative using Equation 2 based on the values of W and R. The results of Vi for all alternatives can be seen in Table 5.

$V_1 = (0,67 \times 4) + (0,67 \times 5) + (0,67 \times 3) + (1 \times 3) + (1 \times 3) + (1 \times 4) = 18,04$

$\vdots$

$V_{15} = (1 \times 4) + (1 \times 5) + (0,67 \times 3) + (1 \times 3) + (1 \times 3) + (1 \times 4) = 21,01$

Based on the Vi value generated by equation 2, it was found that the superior fish seeds with the highest relative preference value were Nile tilapia with code NT15 in V15 with a value of 21.01. They were followed by Nile tilapia with codes NT02, NT03, NT04, NT07, NT11, NT13, and NT14, with a relative preference value of 20.68. The results Vi value will be sorted from the highest value to the lowest to get a recommendation for superior Nile tilapia seeds. First, the lower threshold value of preference is established as 19.69 for the superior fish seed. This value was obtained after consulting with fish farmers in Samarinda City, East Kalimantan, Indonesia.

Table 5. Result comparison

No	Alternative	Vi	Code	SAW method result	Farm result	Description
1	V_{15}	21,01	NT15	Chosen	Chosen	Match
2	V_2	20,68	NT02	Chosen	Chosen	Match
3	V_3	20,68	NT03	Chosen	Chosen	Match
4	V_4	20,68	NT04	Chosen	Chosen	Match
5	V_7	20,68	NT07	Chosen	Not elected	Mismatch
6	V_{11}	20,68	NT11	Chosen	Chosen	Match
7	V_{13}	20,68	NT13	Chosen	Chosen	Match
8	V_{14}	20,68	NT14	Chosen	Not elected	Mismatch
9	V_5	19,69	NT05	Not elected	Not elected	Match
10	V_6	19,69	NT06	Not elected	Not elected	Match
11	V_8	19,69	NT08	Not elected	Not elected	Match
12	V_{12}	19,69	NT12	Not elected	Not elected	Match
13	V_9	18,67	NT09	Not elected	Not elected	Match
14	V_1	18,04	NT01	Not elected	Not elected	Match
15	V_{10}	17,02	NT10	Not elected	Not elected	Match

$$= \frac{\mathit{same\ amount\ of\ data}}{\mathit{amount\ of\ data}}\ 100\% = \frac{13}{15}\ 100\% = 86.6\%$$

The results of superior fish seeds and the value obtained from the SAW will be compared with the results in fish farms. Out of 15 Nile tilapia SAW candidates, 13 matched result criteria as the fish farms, and produced an accuracy value of 86.6%, which can be seen in table 7. The two sets not mirrored in result selection were NT07 and NT14.

4. Conclusion

The accuracy of the results from the multi-criteria decision-making model based on the Simple Additive Weighting (SAW) method of 15 Nile Tilapia was 86.6%. This research provides careful recommendations for preference seed in Nile Tilapia brood stock for selective breeding to fish farmers or the general public.

5. References

[1] Munguti J M, Nairuti R, Iteba J O, Obiero K O, Kyule D, Opiyo M A, Abwao J, Kirimi J G, Outa
[2] N, Muthoka M, Githukia C M and Ogello E O 2022 *Aqua Fis and Fish* 2 265–76.
[3] Mengistu S B, Mulder H A, Benzie J A H and Komen H 2020 *Rev Aquacult* 12 524–41.
[4] Henriksson P J G, Banks L K, Suri S K, Pratiwi T Y, Fatan N A and Troell M 2019 *Environ. Res. Lett.* 14 124062.

[5] Fisheries Division, Food and Agriculture Organization of the United Nations (FAO), Rome, Italy and Miao W 2020 Trends of Aquaculture Production and Trade: Carp, Tilapia, and Shrimp AFS 33S.
[6] Zaman Y R, Wiguno T A A, Dewi N N, Wulansari P D, and Fauzan A L 2021 *Eco Envir and Conser* 27 (2) 752-758.
[7] Indriawati V, and Rahardja B S 2021 *IOP Con Seri Eart and Env Sci* 679 (1) 012054
[8] Hasan V, Mukti A T, and Putranto T W C 2019 *IOP Con Seri Eart and Env Sci* 25 187 - 189.
[9] Sunarto A, Grimm J, McColl K A, Ariel E, Krishnankutty Nair K, Corbeil S, Hardaker T, Tizard M, Strive T and Holmes B 2022 *Biol Con* 174 105020
[10] Insani L, Hasan V, Valen F S, Pratama F S, Widodo M S, Faqih A R, Islamy R A, Mukti A T, and Isroni W 2020 *Ecol Environ Conserv* 26 (3) 1115-1118
[11] Hasan V, and Tamam M B, 2019 *Che Lis* 15 (1) 225-227.
[12] Wijayanti A, Hasan V, and Tamam M B, 2021 *IOP Con Seri Eart and Env Sci* 718 (1) 012096.
[13] Hasan V, and Tamam M B 2019 *Che Lis* 15 (1) 225-227.
[14] Firdaus R M, Mulyono M, and Farchan M 2020 *J Aqua Sci* 5(2) 88-98.
[15] Setiawan D, and Rahardja B S, 2021 *IOP Con Seri Eart and Env Sci* 718 (1) 012098.
[16] Saha P, Hossain Md E, Prodhan Md M H, Rahman Md T, Nielsen M and Khan Md A 2022 *Profit and loss dynamics of aquaculture farming Aquaculture* 561 738619
[17] Hayati R L, Hakim L, Prihanto A A and Nurjannah N 2022 Forecasting of Indonesian Annual Protein Demand for Nile Tilapia (Oreochromis niloticus) Using Exponential Smoothing ujar 10 363–70
[18] Amin, M., Musdalifah, L., & Ali, M. 2020 *IOP Con Seri Eart and Env Sci* 441(1), p. 012135).
[19] Donovan J J 1976 Databsse system approach the management decision support ACM Trans. Database Syst. 1 344–69
[20] Tzeng G and Huang J-J 2011 Multiple attribute decision making: methods and appliations (Boca Raton, Fla. London: Chapman & Hall/CRC)

6. Acknowledgments

We would like to thank the Nile Tilapia farmers in Samarinda City for providing information for the completion of this research.

Growth rate and survival rate of montipora coral transplant on faba material in Baluran National Park

E Y Herawati[1], S Arsad[1], R I Khasanah[2,6], Yudatomo[3], E F Juniarsa[4], M C Ma'arif [2], D R W Nugraha[2], A D Aisyah[2], D Sofarini[5]

[1]Faculty of Fisheries and Marine Sciences, Universitas Brawijaya. Jl. Veteran, Malang 65145, East Java, Indonesia.
[2]Marine Science Program, Faculty of Science and Technology, State Islamic University of Sunan Ampel, Surabaya, East Java, 60237. Indonesia
[3]PT.YTL Jawa Timur –PT. Jawa Power.Paiton Coal Fired Power Station Unit 5&6, Probolinggo 67291, East Java, Indonesia
[4]Baluran National Park, Ministry of Environment and Forestry of the Republic of Indonesia, Situbondo 68374, East Java, Indonesia
[5]Department of Aquatic Resource Management, Faculty of Fisheries and Marine, University of Lambung Mangkurat, Indonesia

[6]Corresponding author: ulick.isfatul@gmail.com

Abstract. Coral transplantation is one of the rehabilitation efforts that can be applied to accelerate the coral reef recovery process. FABA (*fly ash bottom ash*) is an environmentally friendly material and is declared suitable for use as a coral transplantation medium. This study aims to determine the growth rate of transplanted corals as an effort to rehabilitate damaged coral reef ecosystems and restore coral reef habitat and biota associated with coral reef communities in Bama waters, Baluran National Park, East Java. The corals studied were Montipora sp. The research location has a salinity range of around 31-34 psu, temperature 28-30 °C, 100% brightness, turbidity (turbidity) 1.2-1.6 NTU, current velocity between 0.07 m/s to 0.31 m/s , and sedimentation ranged from 2.81 mg/cm^2/day to 5.92 mg/cm^2/day, nitrate 0.004-0.026 mg/l, ammonia 0.048-0.189 mg/l, and orthophosphate 0.008-0.028 mg/l. In general, the water conditions in Bama are still in quite good condition for coral growth. During the 4 months of observation, the achievement of growth in length and height of coral species of Montipora sp. reach 60.56 mm and 21.31 mm, respectively. The growth rate of the average length and height of Montipora sp. respectively reached 15.14 mm per month and 5.33 mm per month.

1. Introduction

One of the tropical aquatic ecosystems that has the highest productivity is coral reefs. One of the most important components in this ecosystem is a group of coral animals belonging to the order Scleractinia. This coral animal has a skeleton made of calcium carbonate (CaCO3) [1]. Coral reef ecosystems have an ecological role as a carbon store as well as a provider of food, care, growth and protection for various associated biota [2]. Coral reefs have a unique symbiotic relationship between coral animals and zooxanthella which are able to absorb carbon for photosynthesis and produce oxygen [3,4]. In addition, coral reefs are also very likely to store carbon in their skeletons [5].

The fatal damage that occurred to the coral reef ecosystem was partly caused by human activities, namely fishing using dangerous fishing gear, namely bombs, the use of ship anchors so that corals in Baluran National Park were mostly estimated to reach 65% in damaged condition [6,7].

Based on this situation, it is very necessary to restore and enrich marine ecosystems, namely by one way of accelerating the recovery of coral reef ecosystems. For this process, it is necessary to develop coral transplantation technology [8]. Efforts to rehabilitate coral reefs, especially to restore function and coral reefs need to be carried out. One of the efforts to overcome the problem of damage to coral reef ecosystems and find alternatives to reduce pressure on coral reef destruction can be done by means of coral transplantation [9].

This research activity carried out transplantation of coral reefs in coastal waters in the Baluran National Park area using kanstin faba as a medium for growing coral transplants. Kanstin faba is a product innovation similar to paving blocks/brick made from fly ash and bottom ash (FABA) which is the dominant waste from PLTU which is used as raw material for making Kanstin faba. Faba's ingredients are declared non-toxic and harmless (non-B3) (SK.183/MenLHK/Setjend/PSLB.3/3/2016). The results of the Toxicity Characyeristic Leachate Procedure (TCLP) test show that fly ash and bottom ash contain elements such as magnesium (MgO), calcium (CaO), aluminum (Al2O3) and iron (Fe2O3) as well as silicates (SiO2) which are very much needed. for coral growth. So that it will influence and take advantage of the growing needs of coral life and the speed of the coral and ultimately provide better utilization and rehabilitation efforts for the coastal waters environment [10].

Previous research conducted in Paiton Waters from 2018 to 2022, showed that Acropora formosa had a survival rate of 82.61%; A. Pulchra 80.95%; with growth rates of 1,958 and 1,730 cm/month, respectively. The growth speed of Acopora sp which was transplanted into the FaBa canstine media showed that FABA was very qualified and suitable for use for coral reef rehabilitation".

2. Material and method

This research was conducted from June to September 2022 with a location in Bama, Baluran National Park - East Java. The map of the research location can be seen in Figure 1.

Table 1. Research tools and materials in coral observation.

No.	Tools and Materials	Description
1	SCUBA (*Self-Contained Underwater Breathing Apparatus*) equipment	Diving equipment
2	Underwater camera	As a photo-taking tool
3	waterproof paper	Data recording media
4	Pencil	Observational data writing
5	Ruler	Measurement of the length and width of coral fragments
6	Global Positioning System (GPS)	Determination of the location of the field experiment station
7	Transplant rack	Artificial substrate attachment media
8	Artificial substrate	Place of attachment of coral fragments
9	Coral Fragment	Experimental material of the type *Montipora*
10	Waterproof marker	Coral module numbering
11	Tight Cable	Number binder to the substrate and substrate binder to rack
12	Personal Computer (PC)	Processing of observational data

Station location decisions are made by considering factors that support coral reef growth in the waters of Bama Baluran National Park. Because of this, it is known that these bodies of water are very well located. The current study was based on accessing a nearby survey station location and discovering that one of the factors supporting coral reef growth is sunlight. Sunlight at this point is more optimal than the other side

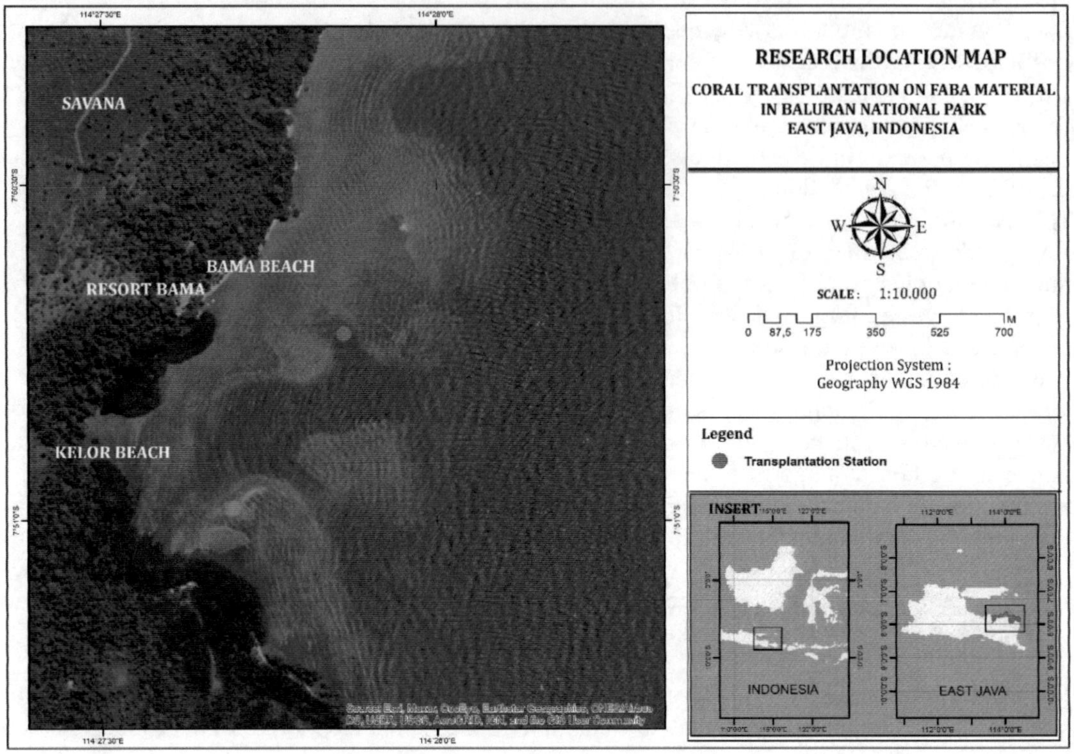

Figure 1. The map of the research location

2.1 Growth observation and survival rate

Computational analysis of observed growth and survival of coral fragments refer to the formula used in [13]. Growth dimension measurements consisted of increased length (longest) and increased height (highest) of coral fragments (Fig. 2).

Measure coral length and height with a ruler. The measurement process is performed directly underwater using scuba diving equipment.

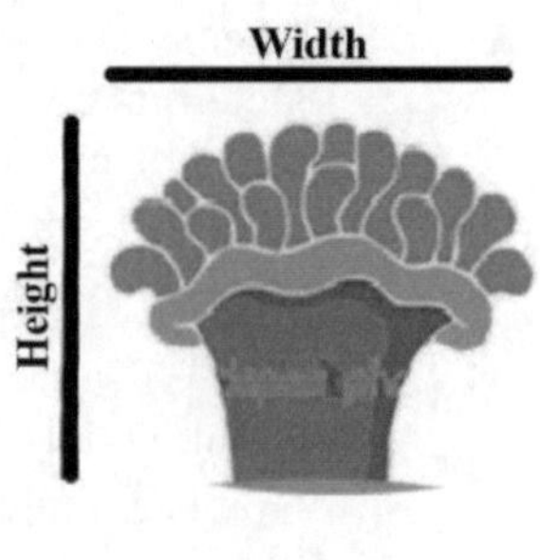

Figure 2. Dimensional measurement of coral fragments.

The growth parameter in the first analysis was the absolute growth, is the total growth from the beginning of the measurement to the end of the measurement. The following formula was used to calculate the absolute growth or growth performance of transplanted Montipora sp coral fragments:

$$\beta = Lt - L0$$

Information:
β = Increase in width/height of coral fragments
L_t = Average width/height of coral fragments after month-t
L_0 = Average width/height of coral fragments at month-0

The second step is to calculate the growth rate of coral fragments transplanted every month using the following equation:

$$\alpha = \frac{L(i+1) - L(i)}{t(i+1) - t(i)}$$

Information:
α = The rate of increase in length or width of transplanted coral fragments
L_{i+1} = Average width or height of fragment at time-
i+1 L_i = Average width or height of fragment at a time- i
t_{i+1} = time- i+1
t_i = time- i

Research coral life (SR) can be determined by comparing the number of live corals (N_t) at the end and the number of transplanted corals (N_0) at the beginning of the observation. Viability (in this case coral fragments) is the level that indicates that some corals are still alive and remain physically and biologically active for a certain period of time. The formula for calculating the survival rate of transplanted corals is:

$$S = Nt/No \times 100\%$$

Information:
SR = Survival Rate
Nt = Number of individuals (fragment) at the end of the study
N0 = Number of individuals (fragment) at the beginning of the study

3. Result and disscusions

3.1 Hydro-Oceanographic conditions of the research environment

The condition of the aquatic environment has an influence on the growth and life of corals. Good and suitable water conditions will determine the growth rate and survival for various types of coral. Parameters of the aquatic environment, both physical and chemical, can also affect the morphology and physiology of corals living in the area. The combination of these environmental parameters will determine the coral growth pattern.

The following is data on the physical and chemical parameters of the Bama Coast waters from June to September 2022 which are presented in Table 2.

Table 2. Physical and chemical parameters of Bama Coastal waters, June–September 2022

Parameters	Unit	2022				Permitted Quality Standards *
		June	July	August	Septembe r	
salinity	psu	32	31	34	33	33-34
temperature	°C	29	28	30	30	28-30
turbidity	NTU	1.2	1.3	1.6	1.5	5
Brightness	%	100	100	100	100	-
current velocity	m/s	0.13	0.07	0.31	0.22	-
Nutrien						
nitrate	mg/l	0.004	0.009	0.026	0.016	0.0080
orthophosphate	mg/l	0.021	0.011	0.008	0.028	0.0150
Ammonia	mg/l	0.067	0.048	0.077	0.189	0.3000
sedimentation	mg/cm^2/day	3.21	3.92	2.81	5.92	-

Source: *Decree of the State Minister of the Environment No. 51 of 2004, concerning Seawater Quality Standards for marine biota.

In general, the water conditions of the research location on the Bama Coast are a suitable environment for the growth of coral communities. has a salinity range of about 31-34 psu, a temperature of 28-30 C, 100% brightness, a turbidity of 1.2-1.6 NTU, a current velocity of 0.07 m/s to 0.31 m/s, and sedimentation ranged from 2.81 mg/cm2/day to 5.92 mg/cm2/day, nitrate 0.004-0.026 mg/l, ammonia 0.048-0.189 mg/l, and orthophosphate 0.008-0.028 mg/l. In general, the water conditions in Bama are still in quite good condition for coral growth [11,12].

3.2 *Absolute growth*

Figure 3 shows a graph of the absolute growth of Montipora corals. Absolute growth of Montipora sp. For about 3 months, observations showed that the width growth of Montipora corals was much higher than the absolute height growth values. This description is consistent with that of Pratiwi et al. [13] He showed that the genus Montipora tends to grow in the same pattern. That is, the transplantation was done at different locations and depths, but the length (width) growth was greater than the height.

Coral species with a leafy colony growth morphology, such as Montipora, have vigorous long growth rather than tall growth. The leafy form of Montipola has a tendency to elongate. In this organism, Montipora paints the skeleton white, and the skeleton is turned sideways instead of upward, following a morphological form. The growth patterns and morphology of Montipora foliosa may also be driven by the physical properties and characteristics of the water bodies in which these corals live. The physical characteristics of coral transplants on the Bama coast are shallow water and strong waves. In shallow areas, coral growth is shortened and dulled when exposed to adequate light supply and large waves [14].

The graph also shows that the best absolute growth of Montipora sp is 6 meters in depth, and the average absolute growth in fragment height and width is 21.31 mm and 60.56 mm in depth, with the lowest absolute growth. . The value is 8 meters deep, with a maximum growth of 9.5 meters and a height and width of 24.25 meters. The average absolute growth height and width of fragments at this depth are 5.33 mm and 15.14 mm, respectively. The dramatic difference in absolute growth values at both depths can be caused by several factors. That means 6 meters deep for optimal light and water circulation.

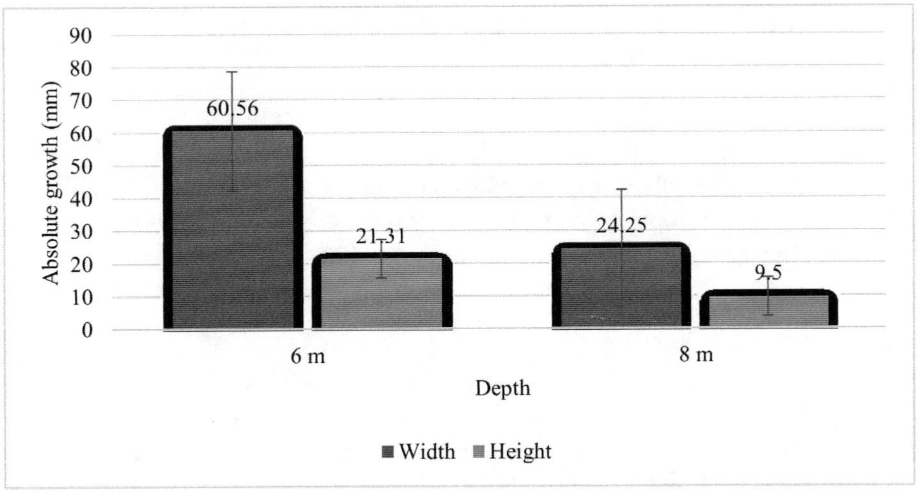

Figure 3. Chart of transplanted Montipora coral absolute growth

Figure 4. Montipora coral transplantation

3.3 Aviability (Survival rate)

SR also pays attention to the accuracy of the method, especially in handling fragments and biological factors such as the type of hard coral planted and their response to environmental conditions [15]. The live data of coral fragments during the observation period is presented in Table 3 as follows.

Table 3. Montipora coral survival in growing media made from FABA

Observation Month	June	July	August	September	
Number of Fragments	92	83	74	73	
SR (%)					**80.5**

Based on this data, the SR of transplanted Montipora corals was 80.5% by the end of the study at both 6m and 8m water depth. This SR value indicates very good viability of the Montiporasp fragment on Bama. According to Harriot and Fisk [16], successful coral transplantation can be judged by a survival rate of 50% or more. This is also supported by Suhendar et al. [17] states that the survival rate of Montipola corals is very high.

Overall, the SR share has shown a downward trend and is expected to decline further after September 2022. We speculate that the main reason for the decrease in SR fraction is the release of coral debris from the grafting base. It was noted that very little algae or dead coral debris was found. According to Clarke and Edwards.[15], one of the factors affecting coral mortality in transplantation is the fragment attachment technique. Jaap too. [18] also mentioned the importance of tying the coral fragments firmly and tightly so that they do not detach from the module [19,20].

4. Conclusions

Based on studies conducted over almost four months, it can be concluded that the growth rate of Montipola corals is 5.3 mm/month average monthly growth rate and 15.1 mm monthly width growth rate at 6 meters water depth /Moon. Minimum coral growth is 8 meters deep, increasing on average 2.4 mm/month in height and 6.1 mm/month in width. Survival rate (SR) at both depths reached 80.5%, which is quite good.

5. References

[1] Haya L O M Y and Fujii 2019 *J. Coast. Conserv.* 23(3) 699–711.
[2] Mora C, Graham N A J, and Nyström M 2016 *Coral Reefs.* 35(4) 1271–1280.
[3] Kordi M G H 2010 *Coral reef ecosystem: Potential, function & management* Jakarta, Indonesia: Rineka Cipta.
[4] Pattiasina T F, Sartimbul A, Herawati E Y, Semedi B, Mulyadi Krey M, and Simatauw F F 2020 *AACL Bioflux. 13*(3) 1522-1534.
[5] Pattiasina T F, Herawati E Y, Semedi B, Sartimbul A, Pelasula D D, and Krey M 2018 *Aquac Aquarium, Conserv Legis.* 11(1) 118-131.
[6] Luthfi O M, Isdianto A, Sirait A P R, Putranto T W, and Affandi M 2020 *Ecol Environ Conserv. 26*(4) 1798-1805.
[7] Luthf O M, Yulianto F, Handayani M, and Soegianto A 2020 *Ann Biol.* 36(2) 288-292.
[8] Anggara D P, Rahardja B S and Suciyono 2022 *IOP Conf Ser Earth Environ Sci.* 1036(1) 012110.
[9] Suciyono, Anggara D P, Azhar M H, Ulkhaq M F, and Fasya A H 2020 *AACL Bioflux*, *13*(1), 241-249.
[10] Khasanah R I 2019 *BIODIVERSITAS.* 20(12) 3593- 3598.
[11] Minister of State for the Environment. 2008. Decree of the State Minister for the Environment No. 51 of 2004 (Attachment 3) concerning Seawater Quality Standards. Jakarta.
[12] Purnomo P W, and Ruswahyuni R 2009 *J Ilm Perikan dan Kelaut.* 1(1), 93-102.
[13] Pratiwi D B, Ramses R, and Efendi Y 2019 *Simbiosa.* 8(1) 10–19
[14] Nybakken and J. Willard 2001 *Marine biology: An ecological approach*, 5th ed. New York: Benjamin Cummings.
[15] Clark S and Edwards A J 1995 *Coral Reefs.* 14(4) 201–213.
[16] Harriott V J and Fisk D A 1988 *Proc 6th Int Coral Reef Symp.* 375–379
[17] Suhendar D B, Johan O, Idris S, Yusri, and Maduppa H H 2020 *Semin Nas Kelaut dan Perikan Ke-VII.* 1–11
[18] Jaap W C 1999 *Ecol. Eng.* 15(2) 345–364
[19] Rizal A, Akbarsyah N, Kdyp P, Permana R, Andhikawati A 2020 *World Sci News.* 144 384-396
[20] Rizal A, Wirawan C A, Pratiwy F M, Pratiwi D Y 2022 *Integrated Development of Marine and Fisheries of Sangihe Islands District, North Sulawesi, Indonesia.* The Institute of Biopaleogeography named under Charles R. Darwin 12. p 1-63

6. Acknowledgement

Our gratitude goes to the Research and Community Service Institute (LPPM) of Universitas Brawijaya, Faculty of Fisheries and Marine Sciences of Universitas Brawijaya, Faculty of Marine Science of State Islamic University Sunan Ampel Surabaya for supporting the procurement of field survey equipment and scientific divers. Thanks also to all those who have helped to conduct this research, PT. Jawa Power - PT. YTL East Java; Baluran National Park; BBKSDA East Java. The author also thanks the reviewers who have provided improvements and suggestions for writing this article.

The effect of temperature differences on oxygen consumption level (OCL) and stress level in the process of spawning vaname shrimp (*Litopenaeus vannamei*)

D Meidiana[1], E D Masithah[2,3] and W Isroni[2]

[1]Program Study Aquaculture, Faculty of Fisheries and Marine Science, Universitas Airlangga. Kampus C, Jl. Dharmahusada Permai No. 330, Mulyorejo, Surabaya 60115, East Java.
[2]Departmen of Aquaculture, Faculty of Fisheries and Marine Science, Universitas Airlangga. Kampus C, Jl. Dharmahusada Permai No.330, Mulyorejo, Surabaya 60115, East Java

[3]Corresponding author: endang_dm@fpk.unair.ac.id

Abstract. Vannamei shrimp *(Litopeaeus vannamei)* is main commodity as well as a mainstay export product in the aquaculture industry program. One of the problems faced in this case of the spawning process caused by poor air quality, especially at temperature thah can increase energy production. The energy released by the vaname shrimp comes from glucose in the *haemocytes*. This study aims to determine the effect of temperature differences on Oxygen Consumption Level (OCL) and stress level in the spawning prcocess of vaname shrimp. This research is experimental using Completely Randomized Design (RAL). This study used 4 treatments, including 28°C as a control, 30°C, 32°C, and 34°C. The results were analyzed using Analysis of Variance (ANOVA) and continued with the DMRT. The result of this study indicate that the difference in temperature affects the Oxygen Consumption Level (OCL) and stress level in the spawning process of vaname shrimp ($p<0,05$). The treatment that has a major influence on the value of the (OCL) and stress level (by measure to blood glucose levels) is at 34°C for OCL with values range from 0,74 – 0,822 mg/O_2/gr/jam and for stress level mostly influences temperature is 32°C and 34°C with values range from 45,6 – 64,8 mg/dL.

1. Introduction

Vannamei shrimp (*Litopenaeus vannamei*) is one of the shrimp species that has high economic value, so that people use it as a fishery product that can benefit foreign exchange for the country. Vannamei shrimp have qualities that can meet market demand, such as being more resistant to disease and water fluctuations, growing relatively fast, and being stocked at high density. Intensification of cultivation of vaname shrimp is carried out by utilizing marine waters, due to the enormous marine potential, relatively high and constant dissolved oxygen of seawater, and higher quality cultured shrimp [1].

Vannamei shrimp (*Litopenaeus vannamei*) is the main commodity as well as the mainstay export product in the aquaculture industry program in addition to tuna, tuna, skipjack, and seaweed. One of the leading commodities that is currently the choice of shrimp farmers is vaname shrimp [2].Based on data taken from the Ministry of PPN/BAPPENAS [3], it is stated that the land used for shrimp cultivation until 2017 has only reached 20% of the total potential, with a national production volume of 886,520 tons. Java Island was the largest contributor at 28.52% (252.813.89 tons) and Sumatra Island at 18.61% (165.020.35 tons). It was recorded from 15 export destination countries that shrimp exports in Indonesia only met 7% of the total world shrimp needs [3].

The development of vaname shrimp cultivation has spread to national shrimp cultivation centers, including North Sumatra, South Sumatra, Lampung, West Java, Central Java, East Java, West Nusa Tenggara, South Sulawesi, North Sulawesi, and Gorontalo. Meanwhile, the highest vaname shrimp product came from East Java with a production of 146,076.46 tons in 2017 [3].

Along with the increasing demand for vaname shrimp production in the community, the aquaculture actors need good quality vaname shrimp broodstock in accordance with SNI. One of the problems faced in this case is the failure of the spawning process caused by poor water quality, especially at temperature. The temperature is determined based on the maximum and minimum levels, which can be tolerated by certain shrimp species and obtain the highest resistance throughout the year in the tropics. In aquaculture, the minimum standard must have a temperature quality that should not be below 12°C, while the highest temperature tolerance is not more than 34.5°C, but care must be given when low water circulation is used [4].

The existence of external factors such as the temperature mentioned above can affect the survival of the parent vaname itself in the present. Therefore, it is necessary to find an appropriate temperature level in the maintenance of laboratory-scale vaname brooders to obtain seeds that are strong against the influence of external factors so that they can help farmers obtain greater profits at a shorter cost and production process time.

One way to find out the degree of temperature that can affect individuals from the parent vaname is to treat the temperature difference and then compare it with the reference of previous studies. According to [5], it is stated that the ideal temperature for rearing white vaname shrimp is 26–32°C, while this temperature treatment method is carried out by adjusting the fluctuations in water temperature so that white vaname shrimp broods are stimulated to spawn due to the effects of these changes.Research on temperature treatment of shrimp has previously been carried out on northern shrimp (*Pandalus borealis*). According to [6], stated that the increase in temperature given to the broodstockoviger shrimp can accelerate the time of egg development, but their survival at a temperature of 8oC decreases as energy reserves are also low.

Giving temperature treatment to white vaname shrimp is a research that has been done by [7], which is a study of female Pandalus borealis broodstock which is commonly used to assess the effect of temperature and feed ratio on female conditions during incubation and determine how to combine the effects of female temperature and condition on egg viability, embryonic development, and larval characteristics. The study showed that an increase in temperature accelerated egg development time, but decreased at 8oC. This study also shows that at high temperatures.

Changes that occur in the environment where vaname shrimp live will be able to increase energy production. One of these changes is the sudden change in temperature that can cause stress to the vaname shrimp that are kept. The energy released by vaname shrimp comes from glucose in the body's hemocytes. According to [8] mentioned that glucose is the main material for producing energy in the body, and glucose levels are usually used to evaluate the metabolic state in aquatic animals by many researchers. [9,25] proposed that increased blood glucose levels in L. Schmitti as an indicator of short-term stress. However, it was reported that the glucose concentration in the hemolymph did not change under the following conditions: 4 weeks at high density.

Based on the description above and considering the importance of the effect of differences in water temperature in the spawning process to measure the level of stress and the level of oxygen consumption (OCL) of white shrimp, the researchers conducted a study entitled "The Effect of Temperature Differences on the Level of Oxygen Consumption (OCL) and Stress Levels in Spawning Process of Vaname Shrimp (Litopenaeusvannamei)".

2. Material and methods

2.1 Place and time of research

This research was conducted at the Anatomy and Aquaculture Laboratory, Faculty of Fisheries and Marine Affairs, Airlangga University, Surabaya from October 2021 to January 2022.

2.2 Material

The material used as the research sample was vaname shrimp broodstock with the age of 6 months consisting of male broodstock with an average weight of 35 grams and female broodstock with an average weight of 45 grams. The mother prawns have not yet spawned and are taken at the same level of gonad maturity. The vaname shrimp broodstock was taken from UD. Kesatria Mas Harvest, Hatchery &Benur Shrimp Tuban, East Java as many as 40 pairs with a ratio of female to male parents of 1:1, water with a salinity of 32 ppt. Another material used in this research is shrimp commercial feed. The materials used to check blood glucose levels are tissue, aquades, and a glucose test script.

2.3 Research Tools

The tools used during this research process are tools for rearing vaname shrimp, namely 20 aquariums measuring 40 x 20 x 20 cm, tank tanks, aquarium filters, aeration hoses, siphon hoses, aeration stones, aerator pumps, heaters, seser, DO (Dissolved Oxygen) meter, refractometer, pH meter, and thermometer. The tools used to calculate the stress level in white shrimp broodstock are the Sinocare Safe-Accu Glucometer, 1ml syringe, and pH meter. The tool used to calculate the TKO (Oxygen Demand Level) is the DO (Dissolved Oxygen) meter.

2.4 Method and Research Design

The method used in this research is experimental or experimental. The research design used was Completely Randomized Design (CRD) using 4 treatments with each treatment getting 5 replications. The treatment given was the difference in the spawning temperature of vaname shrimp (Litopenaeusvannamei). The following treatments will be tested:

P Factor

P1treatment : Maintenance with temperature 28oC

P2 treatment : Maintenance with temperature 30oC
P3 treatment : Maintenance with temperature 32oC
P4 treatment : Maintenance with a temperature of 34oC

The pattern of treatment placement and replication was done randomly. The layout of the placement of each treatment and replication can be seen in the following table.

2.5 Research design

The research variables used during the research process consisted of three types of variables including independent variables, dependent variables, and control variables.
The independent variable in this study is the treatment of temperature differences.
The dependent variable in this study is the level of oxygen consumption and blood glucose levels.
The control variable in this study was the type of sample, namely vaname shrimp aged 6 months and water qualiqualit

2.6 Research Parameters

The parameters in this study consisted of two parameters, namely the main parameters and the main parameters and supporting parameters.

2.7 Main Parameters

The parameters in this study consisted of two parameters, namely the main parameters and supporting parameters. The main parameters observed were blood glucose levels and the level of oxygen consumption (TKO) of white vaname shrimp.

2.8 Support Parameters

The supporting parameters in this study are water quality which includes pH, temperature, and salinity.

2.9Procedure

Preparation in this study began with sterilization using chlorine at a dose of 0.3 ppm for 20 liters of water in a rearing aquarium with a size of 40 x 20 x 20 cm as a research container, preparation of a series of rearing containers, preparation of seawater with a salinity of 35 ppt, and preparation of Parent shrimp 7 months old with a male size of 35 grams and a female 40 grams which will be used for rearing. Sterilization of equipment including maintenance aquariums, hoses, and aerators using chlorine is then left for one night and then cleaned using clean water until the chlorine smell disappears and dried again. The arrangement of a series of maintenance containers includes determining the location of the tank profile reservoir, rearing aquarium, aquarium filter, and aerator, then the rearing aquarium is filled with seawater.

2.10 Sampling and Acclimatization

The sample of vaname shrimp used was 7 months old which was taken from then transported closed using a plastic bag that had been given oxygen and placed into styrofoam that had been given ice cubes with a travel time of 8 hours.

Preparation of vaname shrimp begins with acclimatization. According to [2] acclimatization is done by changing the environment slowly so that the fish are able to adapt themselves to the new environment.

IOP Conf. Series: Earth and Environmental Science 1273 (2023) 012069 doi:10.1088/1755-1315/1273/1/012069

Based on this explanation, the purpose of acclimatization is to prevent high mortality in shrimp during the stocking process. The acclimatization process is carried out in two stages, first by acclimatizing to temperature, namely by immersing a plastic bag containing a sample of shrimp into the rearing water for 30-60 minutes until steam appears in the plastic bag which indicates the water temperature between the rearing container and the water in the plastic bag. Been the same.

2.11 Temperature Treatment and Stocking

The temperature treatment was carried out when the shrimp had stabilized, namely after the acclimatization period in the rearing container. Treatments with temperatures of 28oC, 30oC, 32oC, and 34oC were given into a 20 liter volume maintenance container containing 2 vanamebroodstock per aquarium. A total of 40 shrimp were placed in each experimental unit (aquarium). The treatment temperature setting is done by heating the aquarium water using a heater to the required temperature with the help of a thermometer. During the stocking period, the shrimp were fed ad libitum with a natural diet consisting of the same feed 3 times a week [4].

*2.12 Vaname Shrim*p

Vannamei shrimp rearing was carried out for 28 days starting from the temperature treatment given to white vaname shrimp broodstock. Vannamei shrimp maintenance includes feeding, siphoning shrimp rearing tanks, and measuring water quality. Commercial feed was given four times a day at 08.00, 11.00, 14.00, 17.00, and 21.00 WIB with a dose of 5% of the shrimp body weight. Siphoning or changing the rearing water is carried out to clean the remaining feed and feces from the shrimp to maintain water quality. Water changes are carried out as much as 30% of the total volume of water in the morning. Measurement of water quality is carried out once a day, namely in the morning which includes pH, Dissolved Oxygen (DO), salinity, and temperature.

2.13 Inspection and Data Collection

Measurement of Vaname Shrimp Blood Glucose Level. Sampling of shrimp from each rearing tank was one sample, so that 20 samples were obtained from each treatment. Shrimp hemolymph was taken using a 0.1 mL syringe. Haemolymph was taken from the base of the 5th leg of the shrimp net. The collected haemolymph is then dripped onto the end of the test script which has been inserted into the SINOCARE Safe-Accu digital blood glucose test kit. The results of blood glucose levels will be displayed on the device screen after 10 seconds, which is a number with units of mg/dL. Then, recording or recording the results shown on the screen.

2.14 Measurement of Vaname Shrimp's Oxygen Consumption Level

Oxygen Consumption Rate is one of the main parameters. The rate of oxygen consumption is also a variable used to determine metabolic rate, and is also related to growth. Data from the whole was taken on the 28th day. A pair of vaname shrimp broodstock for each treatment was placed in an aquarium containing treatment media. Before being given treatment, the parent was fasted for 24 hours, then weighed first. The vaname mother was then put into the treatment medium aquarium which was covered with a lid and was fully aerated. After reaching oxygen saturation, aeration was stopped. The measurement of the decrease in DO content in the media can be carried out at the beginning (0 minutes) and 60 minutes later using a DO meter [12].

3. Ressult and discussion

The level of oxygen consumption is calculated based on the following formula information :

OC : Oxygen Consumption Rate (mg/O2/g/hour)
V : Volume of water in the container (L)
DO0 : Dissolved oxygen concentration at the beginning of observation (mg/L)
Dot : Dissolved oxygen concentration at the end of observation (mg/L)
T : Time
W : Weight of test fish (g)

Where OC is the level of oxygen consumption (mg/O2/g/hour), V as the volume of water in the container (L), DO0 dissolved oxygen concentration at the beginning of the observation (mg/L), dissolved oxygen DO concentration in time t (mg/L), W as the weight of the tested shrimp (g) , T as the observation period (hours). As for what was done before the measurement, the shrimp were weighed using an analytical balance. Then measured DO at the beginning of the observation and DO at the end at 1 hour using a DO meter.

If there is a significant difference from the One Way ANOVA test data, it will be continued with Duncan's Multiple Range Test to find out the best treatment, according to the design used, namely a completely randomized design (CRD) [24]. Meanwhile, other parameters, namely water quality, were analyzed descriptively.

Understanding the physiological mechanisms that support tolerance to temperature is critical for predicting the impact of present and future increases on global temperature [21]. To support the growth of vaname shrimp in the environment, by paying attention to water quality parameters, one of which is dissolved oxygen. Dissolved Oxygen / DO is an important component needed by all living bodies for respiration, metabolic processes or the exchange of substances which then produce energy for growth and reproduction [35]. The main source of oxygen in waters comes from a diffusion process from free air and the results of photosynthesis of organisms that live in these waters [37].

Based on the results of the analysis shows that there is an effect of temperature on the level of oxygen consumption and stress levels. The higher the effect of the given temperature, the higher the level of oxygen consumption and stress levels. Based on the results of the study, it was shown that both male and female vaname brooders, the results of treatment P4 (34oC) gave the maximum effect in increasing the level of oxygen consumption compared to treatments P1 (28oC), P2 (30oC), and P3 (32oC).C). This can be shown by the higher the treatment temperature, the higher the oxygen consumption level. According to [37] states that an increase in temperature in a water can trigger an increase in oxygen consumption in aquatic organisms, but on the other hand it will result in a decrease in the solubility of oxygen in the water. Aquatic organisms are often unable to meet dissolved oxygen levels for metabolism and respiration in these conditions [37].

However, there were significant differences in all treatments in the second week due to a decrease in the TKO of males at P1 which initially ranged from 0.33 mg/O2/gr/hour to 0.302 mg/O2/gr/hour; P2 decreased to 0.69 mg/O2/gr/hour; P3 decreased to 0.94 mg/O2/gr/hour; and P4 decreased to 1,289 mg/O2/gr/hour. Meanwhile, the TKO value in female vanamebroodstock also decreased in the second week, namely at P1 to 0.37 mg/O2/gr/hour; P2 decreased to 0.732 mg/O2/gr/hour; P3 decreased to 1.024

mg/O2/gr/hour; and P4 decreased to 1,394 mg/O2/gr/hour. Increased oxygen consumption at all temperature treatments can occur because if the spread of oxygen into the shrimp body does not meet metabolic needs, then this will cause disturbances in the shrimp body system. This often occurs at high metabolic rates caused by high temperatures because metabolism is accelerated and because oxygen is more likely to be limiting under the hottest conditions when its solubility in water is lower [19]. According to [42] the highest level of normal oxygen consumption is 0.450 mg/O2/gr/hour and the lowest is 0.002 mg/O2/gr/hour. The rate of oxygen consumption in white vaname shrimp tends to decrease because it is in line with the increase in body weight of white vaname shrimp because its TKG increases [33]. This often occurs at high metabolic rates caused by high temperatures because metabolism is accelerated and because oxygen is more likely to be limiting under the hottest conditions when its solubility in water is lower [19]. According to [42] the highest level of normal oxygen consumption is 0.450 mg/O2/gr/hour and the lowest is 0.002 mg/O2/gr/hour. The rate of oxygen consumption in white vaname shrimp tends to decrease because it is in line with the increase in body weight of white vaname shrimp because its TKG increases [33] This often occurs at high metabolic rates caused by high temperatures because metabolism is accelerated and because oxygen is more likely to be limiting under the hottest conditions when its solubility in water is lower [19]. According to [42] the highest level of normal oxygen consumption is 0.450 mg/O2/gr/hour and the lowest is 0.002 mg/O2/gr/hour. The rate of oxygen consumption in white vaname shrimp tends to decrease because it is in line with the increase in body weight of white vaname shrimp because its TKG increases [33].

Stress is a non-specific response in the body to many needs due to exposure to stressors. Stress is also considered as an effort to maintain the stability of the internal environment through changes in the form of active adaptive processes through the production of various mediators such as adrenal steroids, catecholamines, cytokines, tissue mediators, and genes [43].

Glucose is the main material for producing energy in the body, and glucose levels are usually used to assess the state of metabolism in aquatic animals by researchers. According to [25] stated that the increase in glucose levels in his observations of L. Schmitti was an indicator of short-term stress. Vannamei shrimp can face environmental conditions that cause stress by utilizing energy from the glycolysis process which is regulated by Crustacean Hyperglycemic Hormone (CHH). CHH functions to increase glucose levels and lactate concentrations in the mobilization of intracellular glycogen reserves, lipid and carbohydrate metabolism, molting, reproductive processes, and osmoregulation [16].

Based on the results of statistical tests showed that the measurement of blood glucose levels in both female and male broodstock, which ranged from 24.8 to 77.6 mg/dL, for 28 days showed significantly different results in each treatment. The blood glucose levels of male and female vanamebroodstock at P1 (28oC) and P2 (30oC) are in the range of levels 24.8 – 37.6 mg/dL and 26.2 – 37 mg/dL where these values are normal or can be said At the temperature treatment, the vaname shrimp broodstock was not in a stressful condition or in other words the two treatments were the best results. In contrast to treatments P3 (32oC) and P4 (34oC) increased glucose levels with a range of values in male vanamebroodstock, which ranged from 45.6 to 51.8 mg/dL and 67.2 to 74.8 mg/dL, while in females namely 51.2 – 54.6 mg/dL and 74.6 – 77, 6 mg/dL which indicates that in both treatments the mother shrimp was stressed. Stress conditions in crustaceans are indicated by the increase in the highest blood glucose levels, such as in L. vannamei shrimp by 45 mg/dL [38] and *P. monodon* by 80 mg/dL [34]. The results of each treatment had significantly different results, where each temperature treatment each week experienced an increase in blood glucose levels. This is because increased blood glucose levels are associated with physiological responses to stress in shrimp, in accordance with previous research conducted by [18] explaining that increased blood glucose levels in *septia*by open air, reduced DO, or increased CO2 dissolved, can be an indication of potential stress factors in shrimp. In this regard, when viewed from other water qualities, there is a decrease in pH and accompanied by an increase in salinity which results in the quality of the cultivation media increasing in decreasing level, so that it becomes a source of stress [8].

The level of oxygen consumption and glucose levels is not only influenced by differences in exposure temperature but also by water quality. Measurement of water quality parameters in this study include dissolved oxygen, temperature, pH and salinity. The results of water quality measurements during treatment and maintenance are closely related to the level of oxygen consumption and blood glucose levels. During the measurement at the time of the study, the DO content in the rearing water of vaname shrimp was in the range from 5.68 to 7.64 mg/dL. This is in accordance with SNI 8037.1:2014 which states that the optimal dissolved oxygen level for rearing white vaname shrimp is more than 4 mg/dL. The results of temperature measurements in the maintenance media water during the study were 28o–34oC. This shows that the temperature in the media is optimal but not at 34oC. The temperature range that is said to be good for the life of vaname shrimp broodstock according to SNI 8037.1:2014 is 28o – 33oC. The results of pH measurements during maintenance ranged from 5.8 to 8.2 which was not in accordance with SNI 8037.1:2014. This can occur due to changes in pH values that vary depending on seawater temperature, dissolved oxygen concentration and the presence of anions and cations [8]. While the results of the measurement of salinity in rearing water for vaname shrimp, which ranges from a value of 32 - 34 ppt where this value is not in accordance with SNI 8037.1:2014, especially at a salinity of 34 ppt. According to [39] states that the high salinity value is caused by the high rate of water evaporation (water evaporation) due to the high temperature coupled with the low level of mixing (mixing) of the new water, but it can also adversely affect the osmoregulation of shrimp. Vannamei shrimp can live in the salinity range (0-31) ppt, but still grows well at (15-25) ppt and optimally at salinity (25-30) ppt [45].

4. Conclusion

Differences in temperature affect the level of oxygen consumption of white shrimp (*Litopenaeus vannamei*) ($p<0.05$). Based on this, H1 is accepted. The difference in temperature affects the stress level of vaname shrimp broodstock (*Litopenaeus vannamei*) ($p<0.05$). Based on this, H2 is accepted.

5. References

[1] Atikah D I, Hartinah, and Wahidah 2018 *Proceedings of the National Seminar*. 1 1-2.
[2] Augusta T S 2012 *J of Trop Animal Sci*. 1(2): 78-82.
[3] National Standardization Agency (BSN) 2014 *Vaname Shrimp (Litopenaeus vannamei, Boone 1931) Part 1: Indoor Model Master Production* 8037.1:2014 : 1-7 SNI.
[4] Brillon S, L Yvan and D Julian 2005 *Mar Biol.* 147: 895-911.
[5] Budiardi T, T Batara and D Wahjuningrum 2005 *Ind J of Aqua.* 4 (1): 89-91.
[6] CABI 2018 *Litopenaeus vannamei (Whiteleg Shrimp)* Center for Agriculture and Bioscience International Datasheet. (Accessed November 12, 2021).
[7] Campbell N A, J B Reece and L G Mitchell 2004 *Biology* (Wasnen Manalu Translation Erlangga Publisher: Jakarta).
[8] Kumar K P, Jetani S I, A N Yusuzai, Sayani, Shabir A D, M A Rather 2012 *Adv in Sci Res App.* 3(4): 2033-2041.

[9] Kannan D P, Thirunavukkarasu K, Jagadeesan N, Shettu and K Aswini 2015 *Fish and Aqua J.* 6: 1-4.
[10] Directorate General of Aquaculture 2019 *Business Opportunities and Vaname Shrimp Investment* Pages: 7-23.
[11] Directorate General of Aquaculture 2006 *Semi-Intensive Vaname Shrimp Cultivation* (Ministry of Marine Affairs and Fisheries Jakarta).
[12] Dugassa H and D G Gaetan 2018 *World J of Fish and Mar Sci.* 10: 5-7.
[13] Edward N, P Iskandar and Rusliadi 2015 *J of the Fac of Fish and Mar Sci*. 1.
[14] Effendi I 2016 *Intensive Cultivation of Vannamei Litopenaeus vannamei Shrimp in the Sea: Study of Location, Physiology and Bichemistry* (dissertation Bogor: Bogor Agricultural University)
[15] Fatchurizal R P and M Abdul *Sci J of Fish and Mar Aff* 6(2): 138.
[16] Fendjalan S N M, T Budiardi, E Supiyono and I Effendi 2016 *J of Trop Mar Sci and Technol.* 8(1): 201-214.
[17] Gou B, W Fang, D Shuanglin, D.Yunwei and T Xiangli 2010 *Aquaculture Intern.* 18:921-932.
[18] Hall M R and Ham V E H 1998 *J of The World Aqua Soc*. 29(3): 294.
[19] Hoefnagel K N and Verberk W C 2015 *J of Therm Biol.* 54(56) : 65.
[20] Holthuis L B 1980 FAO *Fisheries Synopsis* 1 (125): 271.
[21] Huey R B, Kearnye M R, Krockenberger A, Holtum J A M, Jess M and Williams S E 2012 *Philosoph Trans of the Rocay Soc.* 367:1665-1679.
[22] Ministry of National Development Planning/BAPPENAS 2019 *Workshop on Sustainable Aquaculture Develop.* 15-26.
[23] Kordi G K and Tancung A B 2007.*Management of Water Quality in Aquaculture* (Rineka Cipta: Jakarta).
[24] Kusriningrum R S 2012 *Experimental Design* (Airlangga University Pres. Surabaya) pp 44-62.
[25] Lourdes P J, Tania R R, Laida R and Yesel G 2006 *Aquaculture J.* 252 591-597.
[26] Malik A 2018 *Thesis* Subject: 8.
[27] Manan A and R P Fatchurizal 2014 *Sci J of Fish and Mar Aff* 6(2): 138.
[28] Marcelo G G, Noe A E, Glenda L S, Gabriel H R, David V M and Pedro H S 2022 *Latin American J of Aquatic Res.* 50(2): 301-302.
[29] Metin K, T Serhat and K Mehmet 2010 *J of Therm Biol.* 35(6): 305-308.
[30] Mercier L, R Illie, Y P Gloria and M Adriana 2009 *Aqua Res.* 40(16): 1849-1863.
[31] Midhitama A, Subandiyono and A H C Haditomo 2018 *J of Trop Aqua Sci.* 2(2): 12-17.
[32] Panjaitan 2012 *Thesis* Graduate Program, Open University.
[33] Pujianti P, Suminto and Diana R 2014 *J of Aqua Manag And Technol.* 3 (4): 158 – 165.
[34] Rustam H, K Jusoff, S T Hadijah and Ilmiah 2013 *World App Sci J.* 26:82-88.
[35] Salmin 2005 *Oceana.* XXX (3): 21 – 26. ISSN 0216-1877.
[36] Santos E A and R Keller 1993 *J of Comp Physiol.* 163:374-379.
[37] Saputri A, Johnny M T S and Dian R 2017*Journal of the University of Pontianak*
[38] Simon A B, M Pinon, R Racotta and I S Racotta 2010 *Aquaculture.* 289 : 308-314.
[39] Supono, R R Putra and D Hermon 2017 *E-Journal of Aqua Engine and Technol.* 6(1).
[40] Supratno K P and Kasnadi 2002 *Cultivation of Rat Grouper (Cromileptes altivelis) in Ponds Through Improvement of Nutrition and Environment* (Jepara Brackish Water Cultivation Center. Directorate General of Cultivation. Ministry of Marine Affairs and Fisheries).
[41] Stentiford G D, E S Chang, S A Chang and D M Neil 2001*Gen and Comp Endoc.* 121:13-22.
[42] Wafi A, Ariadi H, Muqsith A, Mahmudi M, and Fadjar M 2021 *J of Aqua and Fish Health.* 10(1): 17 – 22.
[43] W Aan F, Pantjara B, Adhiyudanto N B and Rachmansyah 2011 *J of Aqua Res.* 6(2):235.

[44] Wyban J A and Sweeney J 1991 *Internship Shrimp Production Technology the Oceanic* (Institute Shrimp Manual the Oceanic Institute, Honolulu, HI, USA). 158 pp.
[45] Y Hirono 1992 *Current Practices of Water Quality Management in Shrimp Farming and Their Limitations* (Proceedings of The Special Session on Shrimp Farming. USA: World Aquaculture Society).
[46] Zhuo X, Qin Y, He P, Wei P, Zhang B, Chen X and Peng J 2020 *Gene.* 145090.
[47] Zonneveld N E, Husiman and J H Bond 1991 *Principles of Fish Cultivation* (Jakarta: Gramedia Pusataka Utama)

6. Acknowledgement

The authors are grateful to the Faculty of Fisheries and Marine, Universitas Airlangga which had provided the space and laboratory to conduct this research.

IOP Conf. Series: Earth and Environmental Science 1273 (2023) 012070 doi:10.1088/1755-1315/1273/1/012070

Nutrients retention function indications in the water quality of the floodplain area on the lower Batanghari River

G S Ajie[1,2], R L Toruan[1], Yustiawati[1]

[1]Research Center for Limnology and Water Resources, National Research and Innovation Agency (BRIN), Jl. Raya Bogor Km. 46 Cibinong 16911 Bogor Indonesia

[2]Corresponding authors: guruh.satria.ajie@gmail.com

Abstract. The Lower Batanghari River often experiences mass deaths of fish. This could be a symptom of eutrophication caused by nutrient enrichment in the water bodies. Floodplain areas are important in reducing parts of these nutrient loadings. This study aimed to examine water qualities related to the nutrient retention function within the Batanghari River floodplain ecosystem. A sampling expedition was conducted in April-May 2021, during a dry season transition period. Sampling was conducted at 20 sampling points stretching over ±200 km long within the river and its tributaries. These sampling points were categorized as main river channels, tributaries, and floodplain swamps and lakes. The water quality parameters measured were temperature, pH, conductivity, dissolved oxygen (DO), total nitrogen (TN), total phosphorus (TP), and total organic matter (TOM). The main river TP concentrations ranged from 0.21-0.33 mg/L, suitable for fishery activities. The overall sampling point groups have distinct characteristics regarding their DO, TP, and TOM values. The average TP in floodplain lakes was found significantly lower ($t(13)=8.636$, $p<0.05$) than in the river. Meanwhile, a decrease in TN up to 90% was also found in its lower main river channel. These lower nutrient concentrations along the river floodplain gradients indicate a nutrient retention function. Thus, to reduce eutrophication impacts, floodplain conservation is becoming important.

1. Introduction

Poisonous algae blooms and hypoxic conditions that cause mass fish deaths are increasingly occurring in river and estuary areas. These are symptoms of eutrophication of the water bodies. This problem is happened due to the high nutrient loadings that exceed the capacity of the aquatic ecosystem to recover [1,2,3,4,5]. The damage would then threaten the sustainability of the environmental resources and services provided by these aquatic ecosystems. In the lower Batanghari River within the study area, mass fish deaths are often reported. The most recent one in February 2021 occurred in Muara Bulian with an estimated death of 2 tons of fish. Previously, mass fish deaths also occurred in floating net cage aquacultures (FCA) in Muara Duren in 2019. Several mass deaths of up to thousands of fish, with an almost annual frequency, have also been reported in the Kumpeh area.

The nutrient input that causes eutrophication can be divided into 2 main sources, namely point sources and non-point sources. Point source loading can be estimated by measuring the nutrient water concentration from the sewer and tributaries into the main river. Usually, this source comes from household and industrial waste. The wastewater is then discharged into pipelines that are connected directly or indirectly to the main river While major non-point sources came from fertilizer in agricultural soil runoff from surrounding river catchment areas [6,7] Point source loadings can be reduced by installing a wastewater treatment plant, while non-point sources can be handled by conserving natural floodplains in the river section. Floodplains could then function as a natural buffer for nutrient retention from wastewater inputs.

Floodplains are defined as the area around a waterway that was formed from alluvial sediment deposited during a flood. Many ecological concepts ignore how important the nutrient biogeochemical cycles that occur in this area are. The characteristics of a typical floodplain arise from the constant dynamics of the hydrological conditions, resulting in a dynamic interaction process between the morphological, hydrological, biological, and chemical aspects of this ecosystem. The ability of this area to hold some of the flood water has led to changes in the river morphology to form an environment that would then support its rich biodiversity. Floodplain is one of the most productive ecosystems on earth with production levels exceeding that of terrestrial and aquatic ecosystems [8,9,10] Aside from other important ecosystem functions, the nutrient retention function of the floodplain as a buffer and a natural filter for the river water quality, especially in the lower river, would play a very important role to overcome the eutrophication that is currently happening in many rivers and estuaries. Meanwhile, studies on the nutrient retention function of the lower Batanghari River floodplain have yet to be conducted. This study aims to examine the water quality related to nutrient retention function in the lower Batanghari River floodplain. Not only to contribute scientifically, but the research is also important for the conservation of the floodplain areas along the lower Batanghari.

2. Material and methods

A sampling expedition has been carried out in April-May 2021. Sampling was commenced at 20 sampling points stretching from Muara Tembesi to Muara Kumpeh (Figure 1 & Table 1) at the lower Batanghari River. The sampling points consisted of the main river channel (SB), its tributaries (AS), and surrounding floodplain lakes and swamps (DR). Table 1 contains descriptions of the environmental conditions from each sampling point. The research area was divided into 3 sections based on the floodplain area characteristics. Section A of the Batanghari River, which stretches between sampling points SB.1-SB.3.1, has a meandering character and is surrounded by floodplain swamps and lakes. Meanwhile, section B, located between points SB.3-SB.4, is part of the Batanghari River that passes through Jambi city center to the busy industrial port area. This section is characterized by the surrounding floodplains and developed urban areas point SB.3, FCA activities were found on the river bank. Also in this section, floodplain lakes of Lake Sipin and Lake Teluk were sampled. Meanwhile, Jambi Port, industrial areas, and large oil palm plantations are located on the banks of the river along section C (SB.4-SB.6). This section is the most affected by human activities, with a relatively straight river channel without any surrounding floodplains. The last 3 points, namely SK.1, SK.2, and SK.3 are on the Kumpeh River. This river is a branch from the Batanghari River that would then flow through a large area of peat swamp before returning to the main channel (SB.6).

A composite surface water sampling and in situ water quality measurements were carried out at each sampling point. Water quality was measured using a calibrated Horriba® water quality checker (WQC) with parameters of temperature, dissolved oxygen (DO), acidity (pH), and conductivity. Water samples obtained were then directly cooled and preserved using H_2SO_4 for further laboratory analysis of total nutrients. In the laboratory, TOM was measured using the titrimetric method of $KMnO_4$. While, for TN and TP, the spectrophotometric method was applied. Before the measurement, samples were oxidized with potassium persulfate. As for determining the TN content, which has been oxidized into NO_3^-, the brucine method was used. While TP was analyzed as an oxidized form of PO_4^{3-} using the blue molybdate method of the standard method of the American Public Health Association (APHA, 2012). Statistical data processing was carried out using SPSS 13. The analysis was carried out for the descriptive statistics, principal component analysis (PCA), and the mean differences test using an independent t-test.

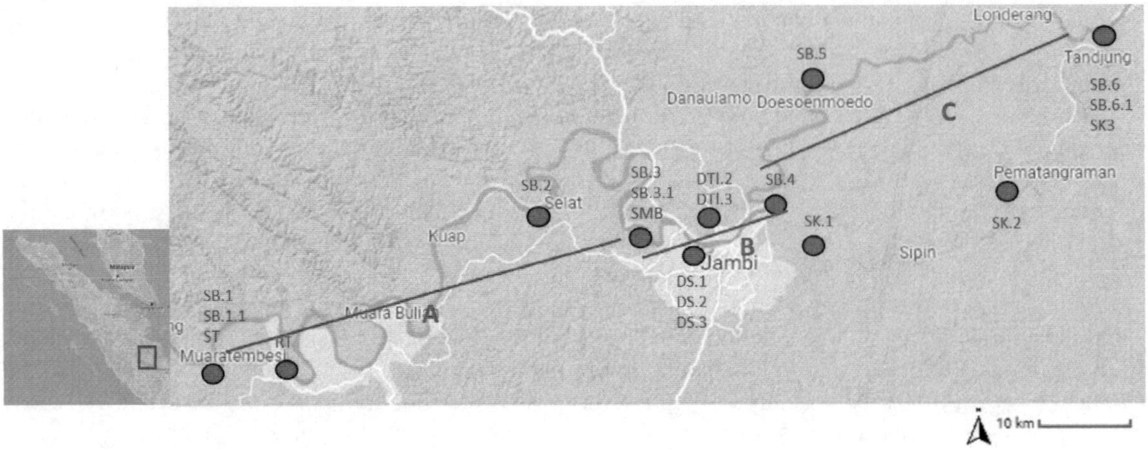

Figure 1. Sampling points along the lower Batanghari River gradient

Table 1. Description of the Sampling points along the lower Batanghari River

No	Sampling Points	Area Annotation
	Batanghari River (SB)	
1	SB.1	Located after the Tembesi River mouth
2	SB.1.1	Located before the Tembesi River mouth
3	SB.2	River island background in the location
4	SB.3	Before entering Jambi City, areas of FCA activities in the river bodies located after Duren River Mouth
5	SB.3.1	Located before Duren River Mouth
6	SB.4	Port Jambi, surrounding the industrial area of the Jambi city
7	SB.5	Palm oil plantation area at river banks
8	SB.6	Located after the Kumpeh River mouth
9	SB.6.1	Located before the Kumpeh River mouth
	Tributaries (AS)	
10	SK.1	Inlet Kumpeh from Batanghari River
11	SK.2	Peat area of Kumpeh River
12	SK.3	Kumpeh River mouth
13	ST	Tembesi River mouth
14	SMB	Duren River mouth
	Floodplain Lakes and Swamp (DR)	
15	DS.1	Inlet water from Batanghari River to the Lake Sipin
16	DS.2	Middle of the Lake Sipin
17	DS.3	Tourism area of the lake, the furthest point from the inlet
18	D.Tl 1	Bay area of the Lake Teluk
19	D.Tl 2	Middle of the Lake Teluk
20	RT	Swamp area after Tembesi River mouth

3. Results and Discussions

This research studied water quality dynamics, especially on the parameters of total nutrients, in the lower reaches of the Batanghari River. Sampling was carried out during a dry season transition period. Therefore, some sampling points, such as Teluk Lake and Terusan Swamp, were no longer connected to the main river channel. The total length of the Batanghari River and its tributaries that was sampled reaches approximately 200 km. The results of measuring water quality on temperature, DO, pH, and conductivity parameters along sampling points groups of SB, AS and DR are shown in table 2. Meanwhile, the results from the TN, TOM, and TP parameters from each sampling point were listed in table 3.

Table 2. Results water quality parameters of temperature, DO, pH, and conductivity

Sampling Points	Temperature (^{o}C)	DO (mg/L)	pH	Conductivity (mS/cm)
Batanghari River (SB)				
SB.1				
SB.1.1	27.8	6.68	8.21	0.046
SB.2	28.34	6.93	8.03	0.046
SB.3	29	6.42	7.71	0.039
SB.3.1	29.16	6.07	7.03	0.038
SB.4	28.9	6.41	6.9	0.038
SB.5	29.56	6.57	7.25	0.044
SB.6	29.64	6.46	7.39	0.042
SB.6.1	28.9	6.29	7.37	0.042
Tributaries (AS)	28.61	6.32	7.21	0.041
SK.1				
SK.2	28.46	4.21	6.35	0.067
SK.3	28.83	2.58	5.37	0.04
ST	30.13	2.52	5.78	0.037
SMB	28.46	6.47	7.82	0.036
Floodplain Lakes and	28.04	2.85	6.06	0.028
Swamp (DR)				
DS.1				
DS.2	29.24	3.21	7.04	0.093
DS.3	31.29	3.82	6.91	0.098
DTl.1	28.7	4.43	7.16	0.098
DTl.2	26.1	4.43	7.61	0.092
RT	26.2	4.51	7.15	0.091
	30.6	4.85	7.72	0.054

The temperature in the lower Batanghari River sampling area ranges from 26-31^{0}C. The average pH was observed neutral at a value of 7.1, with the most acidic condition found in the Kumpeh River whose value was 5.3. The SB sampling points were found with the highest DO average of 6.5 mg/L. While in the DR group, the average DO was found only at 4.2 mg/L. However, this group has the highest average conductivity of 0.0877 mS/cm which is found to range from 0.05-0.1 mS/cm.

Overall, the range of TN concentrations was observed between 0.343-5.012 mg/L with the highest values found in Sipin Lake. The Kumpeh River peat area showed the largest organic content with a TOM concentration of 27.48 mg/L. While for TP, the overall range of the obtained value was 0.051-0.359 mg/L with SB as the highest group mean value.

Table 3. Results of water quality parameters of TN, TP, and TOM

Sampling Points	TN (mg/L)	TP (mg/L)	TOM (mg/L)
Batanghari River (SB)			
SB.1	3.322	0.326	13.37
SB.1.1	0.569	0.256	13.36
SB.2	0.343	0.239	13.35
SB.3	3.552	0.243	15.70
SB.3.1	0.645	0.298	16.49
SB.4	0.352	0.248	11.78
SB.5	3.256	0.253	12.57
SB.6	4.501	0.292	14.93
SB.6.1	4.142	0.214	21.99
Tributaries (AS)			
SK.1	3.486	0.197	14.93
SK.2	4.278	0.108	27.48
SK.3	2.884	0.103	14.15
ST	0.935	0.359	18.07
SMB	2.960	0.086	15.72
Floodplain Lakes and Swamp (DR)			
DS.1			
DS.2	4.298	0.088	6.31
DS.3	4.988	0.128	11.80
D.Tl 1	5.012	0.098	12.58
D.Tl 2	3.572	0.053	9.45
RT	1.555	0.051	10.22
	4.168	0.135	15.72

The results of the PCA are presented in Tables 4 and 5. The two main principal components (PC 1 and PC 2) represent more than 78% of the total variance obtained from the main measured parameters. PC 1 is strongly correlated with the DO and TP, while PC 2 is strongly negatively correlated with measured values of TOM. It was also found that PC 2 correlates moderately with pH and conductivity values. Using these two principal components, the measurement results of all the sampling points are presented in Figure 2.

Table 4. Total variance explained for each principal component result from PCA

Principal Component (PC)	Initial Eigenvalues		
	Total	% Variances	Cumulative (%)
1	2.915	48.583	48.583
2	1.811	30.188	78.772
3	0.763	12.719	91.491
4	0.265	4.411	95.902
5	0.174	2.896	98.798
6	0.072	1.202	100

Table 5. Correlation matrix of 2 main principal components of PCA with measured water quality parameters

Water Quality Parameter	PC 1	PC 2
DO	0.932	0,214
pH	0.651	0,663
Conductivity	-0.593	0,716
TN	-0.654	-0.14
TP	0.912	-0.135
TOM	0,105	-0.891

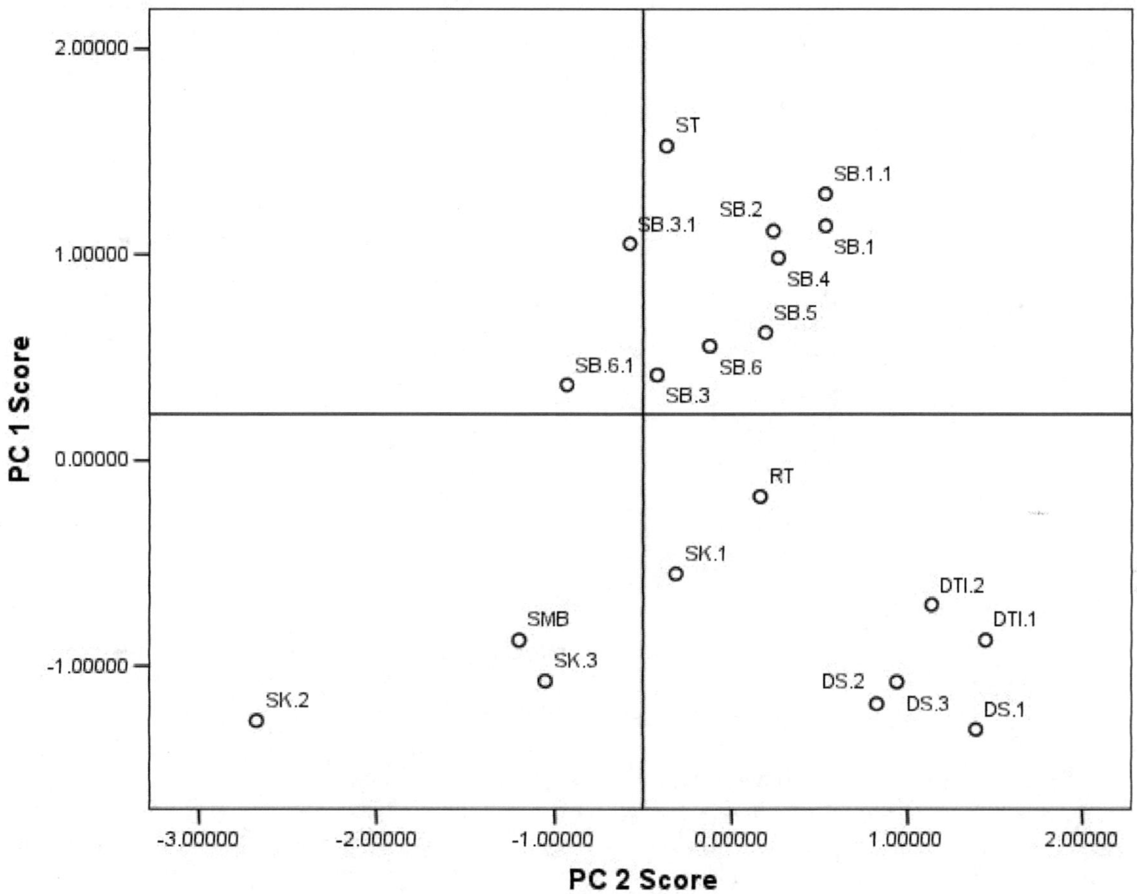

Figure 2. PC 1 and PC 2 scores of PCA from the Lower Batanghari Sampling Points

The overall results of the measured parameters, with exception of ST, showed distinct characteristics of the lower Batanghari River sampling points within groups of SB, DR, and AS, Where the water quality in SB has characteristics of high TP and DO with relatively neutral pH values. Meanwhile, the DR has lower TP and DO concentrations with the lowest TOM values than the 2 other groups. In AS group, sampling points with the highest organic material and the lowest pH were found.

The pH measurement results in AS were found significantly lower than those in SB and DR, $t(12)=3.25$, $p=0.047$, and $t(9)=-2.411$, $p=0.037$ consecutively. This could be possible due to the influence of the low pH in the peaty area of the Kumpeh River. In addition, higher pH values in the DR group could be attributed to the photosynthesis in the floodplain lakes. The process would consume species of dissolved carbon dioxide, e.g., carbonic acid, causing the lake water to be slightly more alkaline during the day [11,12,13]

DO in the SB was significantly higher than DR, $t(12)=4.948$, $p<0.05$. Flowing water within the river would result in turbulence, therefore allowing the surface area of the river water to constantly be exposed to oxygen in the air increasing the river water DO [14,15] The higher DO could also be attributed to the oxygen-generated photosynthesis of the river phytoplankton, and periphyton community [14,15] Meanwhile, the net supply of DO from photosynthesis in the DR water column was not as high as the constant supply of oxygen input found within the river.

The boxplot graphs comparing the range of values from the parameters of total nutrients from each sampling point group are shown in Figures 3 and 4. The DR's TN average value was found higher than SB and AS. In contrast, TP values were found lower in floodplain lakes compared to what was found in the river water.

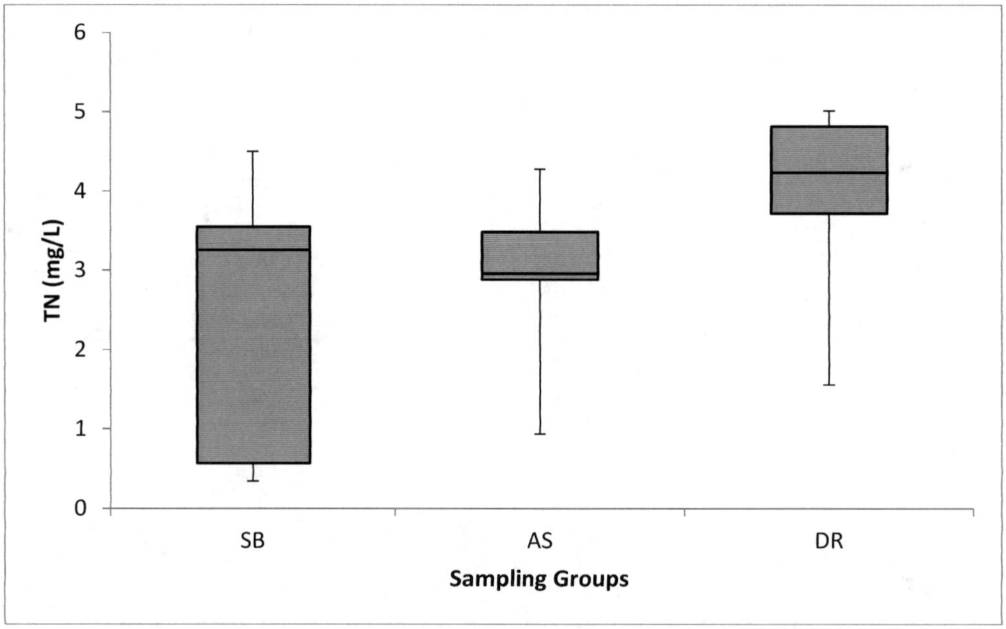

Figure 3. Boxplot Diagram of TN from the sampling points groups

TN within SB sampling points has a relatively wide range which can then be categorized into two groups. The first group is sampling points with relatively high TN content, namely at sampling points of SB.6, SB.6.1, SB.5, SB.3, and SB.1. Meanwhile, the second group consists of sampling points of SB.1.1, SB.3.1, SB.2 and SB.4 which have low TN content. It was observed that the starting points of main river segments A and B, namely SB.1 and SB.3, contain higher TN of 3.322 mg/L and 3.552 mg/L, respectively. The concentration would then become lower at the endpoints within each segment of A and B namely, point SB. 3.1 and SB.4 of 0.645 mg/L and 0.352 mg/L, respectively. With floodplain areas of swamps and lakes found around these two sections, indicates that this floodplain-river ecosystem is effective in reducing the TN in the water. Nitrogen could be removed efficiently from the river water column by nitrification and denitrification processes in this ecosystem [16,17]. Meanwhile, the midpoint of section A namely, point SB.2 has a lower TN content of 0.343 mg/L compared to SB 3.1. The higher TN concentration in SB.3.1 could be attributed to FCA activities in this area. FCA disposes of fish leftover feed and feces directly into river water bodies, which would contaminate TN into water and sediment in the area [18,19]. On the other hand, in section C, where there is no floodplain, an increase in TN concentration from 0.352 mg/L at SB.4 to 3.256 mg/L and 4.142 mg/L at SB.5 and SB.6.1 respectively was observed. Fertilizers spilled from palm oil plantations, sewage that came from industrial areas around the main river canal, and leaks of oil and fuel from boats and tankers crossing this river section could potentially be sourced from the TN contamination of river water [20]

Although not significant, the larger TN concentration in the DR group could be related to the receding period of flooding in the river. During this period DR area was cut off from the water in the main river. The water supply to DR suddenly decreased, while evaporation would continue to occur. This would result in TN becoming more concentrated in the lake water column. In addition, a decrease in DO at the bottom sediment of the lakes due to decomposition could also trigger the release of TN back into the lake water column [21]

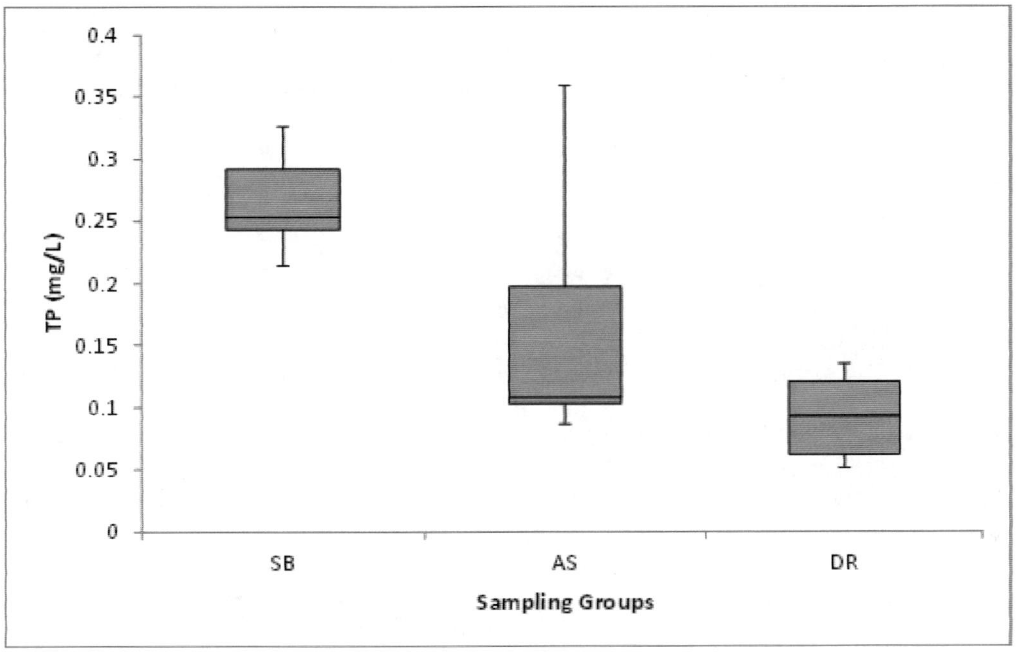

Figure 4. Boxplot Diagram of TP from the sampling points groups

The observed TP content in the main channel of the Batanghari River water generally falls in the class 3 category of water quality utilization according to Indonesian government regulation (PP) No.82.Tahun 2001 [22]. This regulation recommends water uses for fisheries activities. In fact, Batanghari River water has a wide range of uses from its local community, including clean water, irrigation for agricultural farmland, and, as well as for tourism and aquaculture.

The decrease in TP concentration in the main Batanghari River canal was lower than in TN. Where in the A segment the decrease in TP concentration was observed from 0.326 mg/L to 0.243 mg/L. Similar to what was found in the TN, the middle area of this section (SB.2), has a smaller TP content compared to SB.3.1. This also indicates a TP contamination due to the FCA activities. While in the B segment, the TP content is relatively constant at a value of 0.24 mg/L. The overall TP decrease in sections A and B, where floodplains are found, was observed at only 24%. This value is much lower than the 90% decrease in TN in this area. The TP retention in the main river channel with floodplain area is less efficient than what is found in floodplain lakes. This could happen due to the TP removal in the river channel is mainly contributed by the deposition of the TP particulate phase. However, when a strong current occurs, part of this sedimented particulate phase could then be swept back into the river water. In addition to that, lower productivity within the flowing water (*lentic*) compared to floodplain lake ecosystems (*lotic*) is often observed [23,24]. Thus, the removal of the bioavailable dissolved TP in the water column would also be less efficient. On the other hand, in the C segment of Batanghari River, where there is no floodplain, the concentration of TP was found to slightly increase from 0.248 mg/L at point SB.4 to 0.253 mg/L and 0.292 mg/L at sampling points of SB.5 and SB.6, respectively.

For the group comparison, it was also found that there was a significant difference in TP concentration between SB and DR groups, $t(13)=8{,}636$, $p<0.05$. The larger average value in the SB and AS indicates that river water is the main source of TP. While floodplain lakes would then function as an effective TP retention from the river water. Parts of the TP are removed through the deposition from the particulate fraction into lake sediments [25]. While in the dissolved phase, bioavailable P will be used for lake productivity to build new biomass [26,27] Phosphorus was found as the limiting factor for the Batanghari floodplain lakes productivity with a TN: TP ratio found of more than 100.

4. Conclusion

The water quality data indicate nutrient retention function in the ecosystem of the lower Batanghari River floodplain. Thus, conservation of the floodplain area is important to overcome the existing eutrophication problem.

5. References

[1] Dodds W K 2006 *Limnol Oceanogr* 51 671–680.
[2] Smith V H, and Schindler D W 2009 *Trends Ecol Evol* 24 201–207.
[3] Schmidt S, Bernard C, Escalier J M, Etcheber H, Lamouroux, M 2017 *Environ Sci Pollut Res* 24 3251–3259.
[4] O'Hare M T, Pedersen A B, Baumgarte I, Freeman A, Gunn D M., Lázár A N, Sinclair R, Wade A J, and Bowes M J 2018 *Plant Sci* 9 451.
[5] Tang X, Li R, Han D, and Scholz M 2020 *Water* 12 1634.
[6] Liu Y, Li H, Cui G, Cao Y 2020 *Sci. Rep* 10 3012.
[7] Surbakti J A, Dewi I A L, Alamsjah M A dan Lamid M 2019 *IOP Conf Ser Earth Environ Sci.* 236 (1) 12020.
[8] Jakubínský J, Prokopov M, Raška P, Salvati L, Bezak N, Cudlín O, Cudlín P, Purkyt J, Vezza P, Camporeale C, Daneˇk J, Pastor M, Lepeška T 2021 *WIR Wat* 8 1545
[9] Imron M F dan Titah H S 2018 *Environ Engine Rese* 23 (4) 374-382.
[10] Dewi A N dan Abidin Z 2021 *J Aqua and Fis Heal* 10 (1) 46-55.
[11] American Public Health Association (APHA) 2012 Standard Methods for the examination of water and wastewater New York the USA
[12] Hammer K J, Kragh T, Jensen K S 2019 *Freshw Biol* 64 1956–1970.
[13] Bogard M J, Finlay K, Waiser M J, Tumber V P, Donald B D, Wiik E, Simpson G L, del Giorgio PA, Leavitt, P R 2017 *PLoS ONE* 12
[14] Rosmanida R, Fauziyah S dan Pranoto A P 2020 *Jourl of Trop Biodiv and Biotec* 5 (2) 100-107.
[15] Song J, Hou C, Liu Q, Wu X, Wang Y, Yi Y 2020 *J Clean Prod* 261.
[16] Atapaththu K, Asaeda T, Yamamuro M, Kamiya H 2017 *Ekoló* 36 1- 9.
[17] Salem T A 2021 r *Egypt J Aquat Res* 47.
[18] Cahyonugroho O H, Hariyanto S and Supriyanto G 2022 *Glob J of Environ Scie and Manag* 8 (1) 59-74.
[19] Adame M F, Franklin H, Waltham N J, Rodriguez S, Kavehei E, Turschwell M P, Balcombe S R, Kaniewska P, Burford M A, Ronan M 2019 *Mar Freshw Res* 70 1513-1521
[20] Li S and Twilley R R 2021 44 1899–1915.
[21] Anas P, Jubaedah I, Sudino D 2017 *J Penyuluh Perikan Kelaut* 11 35-47.

[22] Ronny R, Bun B Y, Jasmana J, Rusli R and Notobroto H B 2020 *Intern Journ on Adva Sci, Engine and Inform Techno* 10 (5) 2103-2110.
[23] Weigelhofer G, Hein T, Bondar-Kunze E 2018 *Riv Ecos Manag Spri Ope*187-202.
[24] Peraturan Pemerintah (PP) No 82 2001 *Pengelolaan Kualitas Air.*
[25] Sharma P and Giri A 2018 *MOJ Eco Environ Sci* 5 311-317.
[26] Spespatri J A, Rahardja B S and Abdillah A A 2019 *IOP Conf Ser Earth Environ Sci*. 236 (1) 12057.
[27] Yu H, Xu S, Li W, Liu J 2021 *Environ Sci Pollut Res* 28 49588–49601.
[28] Lepori F and Capelli C 2020 *Hydrob* 848 613–629.

6. Acknowledgements

The authors expres many thanks to the Rumah Program Manajemen Sumber Daya Air dan Danau Prioritas BRIN for funding this research. The authors also thanks to the research team, editors and reviewers for their suggestions and input on our manuscript

IOP Conf. Series: Earth and Environmental Science 1273 (2023) 012071 doi:10.1088/1755-1315/1273/1/012071

The biological and ecological impact of the ship grounding and ore nickel spill of the barge ship in Morowali Waters, Central Sulawesi, Indonesia

S Bahri[1], Y Jiliansyah[1], and O Johan[2,3]

[1]Directorate General of Marine and Fisheries Resources Surveillance. Jl. Medan Merdeka Timur No. 16, Jakarta.
[2]Research Center for Conservation of Marine and Inland Water Resources, National Research and Innovation Agency. Jln. Raya Bogor Km 46, Cibinong 16911, Bogor, Jawa Barat, Indonesia.

[3]Corresponding author's e-mail : ofri.johan@brin.go.id

Abstract. This study's purpose is to know the condition of corals in locations that are under pressure due to high sedimentation and coral damage due to ship aground. The research was carried out in August 2021 by collecting data related to the coral reef's condition. The method used the Underwater Photo Transect (UPT) at 2 location points as controls which were assumed not to be affected by the ship aground and 2 location points where the ship ran aground. The study was successful to obtain data on the affected corals on the BG Financia 22 ship at 13.13%, while at the control location, live coral cover was obtained by 17.16%, grouped in very damaged condition. Observations on the BG Intan Kelana 3 ship from a total observation area of 127,600 cm^2, obtained a live coral cover of 24.70%, this condition is grouped in very damaged coral conditions. Types of coral were found in 17 genera from 247 colonies observed during the study. Coverage at affected locations is 4.20%. Observations in this study can be concluded at the location where the condition of the coral is very damaged plus the pressure due to the ship aground, thus worsening conditions in the area.

1. Introduction

The coral reefs of Indonesia are a global concern because they are mega biodiversity of the world, including the coral triangle [1,2]. In addition, coral reef ecosystems as a source of fishery production play an important role in obtaining animal protein food needs [3], other physical functions such as coastal protection from waves and tsunamis, and other important biological and ecological functions [23,24].

Central Sulawesi province is the largest province on Sulawesi Island with a land area of 63.330 km^2 and a sea area of 189.480 km^2. The extent of the sea area allows this area to have the largest coral reef area in the province of Central Sulawesi. However, there are not many reports on the coral reef's condition in this location.

One of the research results informs that the coral reefs condition facing the mainland waters of Central Sulawesi (West-Northwest) generally had a poor to fair condition, while those facing the open waters of Tolo Bay (East-Southeast) had a fair to very good. One of the locations reported to be in very good condition was found in the Taka Pakangkah Alonya area 77.24%-72.16% (IM 0.08-0.17)[4]. Other locations in the western part of Maputi island found 33.3% live coral cover and 19% in the eastern part in 2013 [5]. Damage to coral reefs has been widely reported, including the use of destructive fishing gear, sedimentation, tourism, ornamental corals and including ship grounding, and other marine accidents [6,25,29].

Coral damage due to ship grounding is currently one of the government's concerns to obtain compensation for the impact of the damage by rehabilitating the perpetrators of the damage. The compensation value is obtained from the verification process, clarification directly to the field where the damage occurred so that damage data is obtained in the form of the damaged area, types of corals affected, and all these losses will be calculated for economic value so that the overall replacement is requested from the shipowner.

Several cases of ship aground have been negotiated for the replacement of their economic value through out-of-court negotiations. Among these cases occurred in Raja Ampat, Komodo, Bangka Belitung, Thousand Islands, Karimun Java, and East Nusa Tenggara [6,26].

The growth rate of the affected corals is a concern, especially coral species with massive growth forms. This group of corals has a slow growth rate, which there is 1 cm or less per year. In this condition, it is important to propose coral reef rehabilitation in the ship grounding area. As for the hope, compensation for the damage to coral reefs can increase the awareness of all parties to be responsible for the conservation value in all areas in Indonesia.

This study which carried out the verification activities of the barges BG Finacia and BG Intan Kelana 3 is intended to obtain data on the impact of ship aground and nickel ore soil spills on biological and ecological coral damage, obtain data on the types of corals affected around control locations and locations, and obtain data on the coral cover in affected locations and control locations.

2. Material and methods

2.1 Time and sites

Verification has been carried out in Morowali waters (Figure 1) on 09 – 13 November 2021. The location of the ship aground on the Barges BG Finacia 22 (Tb Entebe Star 10) and BG Intan Kelana 3 (Tb Intan Megah 17) is shown in Figure 1, and data collection was affected at the BG Finacia location, namely on the reef flat, reef slope and coral conditions as a control located on the west side of Lemo Island (2°53'8.31"S and 122°12'44.45"E). Meanwhile, the location of BG Intan Kelana is at the coordinates of 8.4780 South Latitude - 119.5560 East Longitude.

The area of the ship grounding was close to the river's mouth and the mangrove area. There is also a place to dock for ships, about 5 km from the location of the shipwrecked BG Finacia or at the location of the shipwreck of the BG Intan Kelana 3. Both of these areas still have stretches of coral reefs, although conditions have been and are currently undergoing degradation due to high sedimentation.

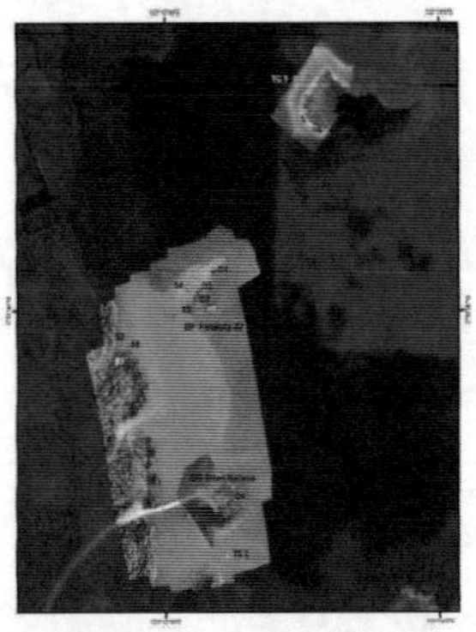

Figure 1. The ship grounding and data collection location from the two ships that ran aground BG Finacia (above) and BG Intan Kelana 3 (below).

2.2 Techniques of collecting data on reefs habitat

The data was collected by using diving apparatus for affected locations and control locations of the BG Finacia ship. However, for the location of the ship affected by the Intan Kelana 3 BG, time diving was carried out to estimate live coral cover due to limited visibility and the use of underwater cameras due to the location being on the ship's dock and having high turbidity.

Data observation was carried out as long as a 50 m transect tape line. Data include the coral condition with types of substrates used which analyzing from 30 samples of each photo. Underwater Photo Transect was used to get photos by using a digital camera Canon G16 on a standard frame and procedure [7, 8, 9]. This method is carried out at reef slopes and control locations. Meanwhile, at the reef flat location, direct visual observation was used using a 58cmx44 cm frame. Data on live coral, rubble, dead coral, algae, and sand are recorded on waterproof bottom paper.

The belt transect and enumeration methods were used to get species of corals on each location [10], where the data collector swam in a transect point on 2 x 50 m to record and took documentation of each type of coral for identifying species names.

Other data including profile, substrate/damage type, water brightness, depth, length, width, and current direction are recorded to an underwater sheet synchronized with photo/video documentation. The Geographic Positioning System was recorded to clarify the observation on the sites where the accident happened.

2.3 Data Analyses

Coral Point Count with Excel extensions (CPCe) software was used to analyze the data from the underwater photo transect [11]. The benthic cover was analyzed to compare with the previous studies [10] by using the formula:

$$\text{Cover categories percentage} = \frac{\text{Number categories - i}}{\text{Lots of random points}}$$

Assessment of the coral reef's condition generally refers to the criteria standard for the degradation of coral reefs based on the State Minister of the Environment's Decree no. 4 in 2001. Substrate degradation type and form were identified and analyzed to compare the damaged and control areas. The areas directly affected are classified into three general categories [6], and indirectly, namely the death of live coral due to buried nickel ore soil, sand, and a fragment of dead coral which hit by ship grounding. The other impact of corals was carried out as descriptively 'in situ' through direct observations about the coral reef's condition and the presence of benthos biota. Specifically, observations of corals followed the last study [6]. Dead coral was counted from the number of dead colonies, including the colonies that tend to be 100 percent white, white or covered by algae) [12, 13, 14, 15].

3. Result and discussion

3.1 Condition of corals at affected sites and BG Finacia control sites

3.1.1 Impacted area

Table 1. Live coral cover and other substrate criteria at affected sites and control sites from the BG Finacia and BG Intan Kelana 3. FI (Finacia), IA (Impacted Area), CA (Control Area), IK (Intan Kelana 3) ship.

Criteria	FI IA 1	FI IA2	FI CA	IK IA	IK CA
Coral (HC)	13.13	4.3	24.7	4.20	17.16
Recent Dead Coral (DC)	0.00	2.1	0.2		
Dead Coral with Algae (DCA)	21.28	1.4	1.4		27.45
Coral Millepora (CME)			3.2		3.20

Soft Coral (SC)	0.15	0.1		
Sponge (SP)	0.75	1.7	2.7	1.96
Fleshy Seaweed (FS)	0.00			1.08
Other Biota (OT)	0.10			0.10
Macro Algae (MA)		1.2	0.3	
Rubble (R)	20.05	1		19.61
Sand (S)	38.86	0.2	64	18.73
Silt (Si)	0.00			
Rock (RCK)	5.67	2.5	3.6	13.92

Live coral cover (HC) on the reef flat was 4.3%, including the severely damaged group because the live coral cover was below 25% based on the decision of the Ministry of Environment and Forestry [16]. Dead Coral (DC) is mostly by sand indicating that coral death is caused by sand, in this case, it is due to nickel ore soil. This death, DCS, is indicated to be new which is different from Dead Coral with Algae (DCA). DCA criteria can appear if the death of DCS at a certain time will be overgrown by algae (Table 1).

Other data in Table 1 shows that live coral cover is 13.13% which is grouped in very damaged coral conditions [16]. Other variables showed DCA 21.28%, coral rubble (20.5%), and Rock (RCK) 5.67%, where this variable came from live coral but due to certain conditions in this case caused by nickel ore soil so that corals died and with increasing when it will be attached by algae, the fracture turns into a small size (rubble) and there is still a fixed size (rock). The complete data is in Table 1.

The location is affected by nickel ore soil on the reef flat and upper reef slope, the coral condition is in very damaged condition due to the coral cover. The main factor for the low live coral cover apart from the nickel ore soil pile, was the low visibility of the waters and a large amount of sediment from nickel ore soils and sediments covering the coral colonies, as shown in Figure 2.

Figure 2. Corals condition were covered by ore nickel, affected by ship aground in locations.

Sediment cover can interfere with coral growth, some even cause coral death. The growth of *Pachyseris* sp coral was disturbed where an abnormal colony shape was seen where several sides were degraded due to the presence of sediment cover in the substrate (Figure 2a). Corals died (Figure 2a,b,c) causing the growth of new colonies to be hampered due to the condition of the sedimentary substrate. The corals from the massive group were still small in competition for survival with the sediments adjacent to the dead coral that had been covered by algae and sponges (Figure 2b). Branched coral *Acropora* sp also experienced some of its branches due to nickel ore soil and sediments, some branches were covered by these sediments (Figure 2d).

3.1.2 Condition of coral at BG Finacia's control site

Locations as a comparison at locations adjacent to affected locations in the same waters were carried out on the west side of Lemo Island, Site 2 on the map.

Live coral cover (HC) was obtained at 17.16%, on the other hand, data that supported the damage obtained data on dead coral covered with algae (DCA) at 27.45%, coral fracture (Rubble, R) of 19.61%, and Rock (RCK) of 13.92%. This live coral cover is classified in very poor condition because it is under the 25% criteria based on the decision of the Ministry of Environment [16]. This low can be proven by 3 recorded parameters, namely DCA, R, and RCK with a total of 60.98%. The low coral cover in this control location is inseparable from the low visibility due to sedimentation that occurs in the area.

Coral cover at control sites (17.16%) minus coral cover at affected locations from coral cover on reef flat and upper reef slope (4.3% and 13.13%) with an average of 8.72%, the difference between control and affected areas is 8.44%.

3.1.3 Diversity of coral species

The results of direct observations found 36 species of coral from 33 different genera. Other biotas recorded at this time are from the soft coral and ascidian groups. The number of coral genera as many as 33 genera is included in the high group and is almost the same compared to several locations in the Thousand Islands 2007 data [17], although the visibility of the waters is low due to sedimentation due to mining activities and especially affected by nickel ore soil spills.

Among the coral species that enter the slow growth rate are *Diploastrea heliopore, Favia* sp, *Favites* sp, *Goniastrea* sp, *Lobophyllia* sp, *Montastrea* sp, *Pavona* sp, *Platygyra* sp, and *Porites* sp. Coral damage of this type will certainly take longer than the recovery time for corals with fast growth rates, as shown in Table 2.

Table 2. Types of coral found at affected locations and control locations of the BG Finacia and BG Intan Kelana 3. Several genera from the slow growth rate of corals (*)

Jenis karang	Finacia	Intan Kelana 3
Acropora falifera	√	-
Acropora sp	√	√
Anacropora puertogalerae	√	-
Anacropora sp	√	-
Astreopora sp	√	-
Coscinarae sp	-	√
Cyphastrea sp	√	√
*Diploastrea heliopora**	√	-
Echinopora sp	-	√
Euphyllia ancora	√	-
*Favia sp**	√	√
*Favites sp**	√	√
Fungia sp	√	√
Galaxea archelia	√	-

*Goniastrea sp**	√	√
Heliopora sp	√	-
Leptoseries sp	-	√
*Lobophyllia sp**	√	-
Millepora sp	√	-
*Montastrea colemani**	√	-
Montipora verrucosa	√	
Montipora sp	-	√
Mycedium elephantotus	√	-
Oulastrea sp	-	√
Oxypora spp	√	-
Pachyseris rugosa	√	-
Pavona sp	√	√
Pectinia sp	√	√
*Platygyra sp**	√	√
Pocillopora sp	√	-
Podabacia sp	√	-
Porites cylindrica	√	-
Porites lutea *	√	-
Porites rus	√	-
*Porites sp**	√	√
Psammocora sp	√	√
Seriatopora hystrix	√	-
Symphyllia sp	-	√
Sarcophyton sp	√	-
Ascidian	√	-

3.2 Condition of coral at the affected location and the control location of BG Intan Kelana

The area observed using 50 frames along the transect is 127,600 cm^2. Live coral cover at a depth of ± 1 m obtained 24.70%, classified in poor coral condition because the cover was below 25% [16]. The turbid waters near the mouth of the river have a major impact on the condition of the corals at the site where the ship ran aground. Coral reefs have also been found in the same turbid waters as in the Gosong Gabuo area and Pisang Island in the waters of Padang City, West Sumatra [18]. Other locations such as in the waters of Karanggeneng in Rembang, Central Java [19].

In the area of the observation area, 17 genera of coral were found, with the highest number of colonies being Favites (79 colonies), then Porites (73 colonies), Favia (40 colonies), and Pavona (18 colonies) in an area of 127,600 cm². Several types of coral included in the slow growth are Favia, Favites, Goniastrea, Pavona, Platygyra, Porites, and Symphyllia as shown in Figure 3.

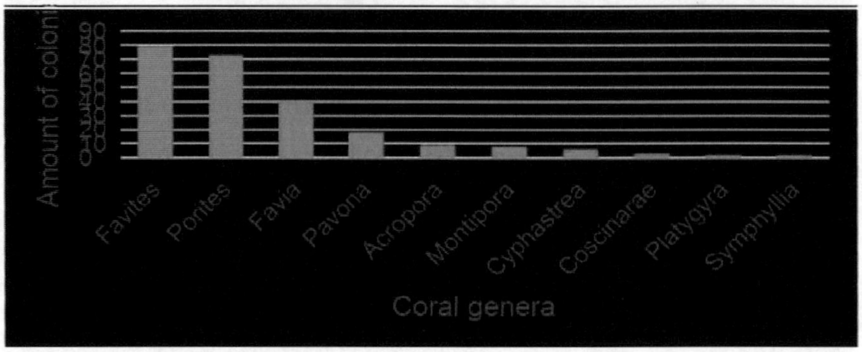

Figure 3. Number of coral colonies found at control sites in the highest ten genera

3.2.1 Coral cover in the affected area
Coral conditions at the affected locations were compared with coral cover in waters that were similar to other locations [27]. The data obtained on the reef flat becomes a benchmark for estimating coral cover at the ship aground location.

Visual estimates directly found coral at the ship aground location with conditions already buried by nickel ore soil and an estimated live coral cover of around 8% (Sadarun, personal communication). The second method is by making comparisons with other locations that have water characteristics that are close to the same, namely waters near river mouths and cloudy waters with low visibility.

The waters of Karangeneng, Masaran (Central Java) have murky waters, located in front of the river about 1 km with underwater visibility of about 50 cm, found various types of coral (17 genera). Live coral cover was obtained at 26.23% on the reef flat (1-3 m) and 23.28 % on the reef crest (3-7 m). Live coral cover on the reef slope (8 - 12 m) was obtained by comparison with the location on Pisang Island (Padang), also in front of the river mouth (Muaro Padang) about 5 km, turbid water conditions obtained coral cover in the reef crest 25% and a reef slope of 14% [18].

Another, more extreme location is in Gosong Gabuo (behind the Bung Hatta Padang campus), close to the Air Tawar river, about 1 km from Padang. Live coral cover at the reef crest is 7% and the reef slope is 0% [18]. The cloudy water conditions greatly affect coral cover in the reef slope area, such as at the location of the ship aground. Thus the difference between the reef crest and reef slope is 11% difference. The difference between reef flat and reef slope at Karangeneng is 2.95% in conditions not at the pier where loading and unloading activities are carried out, so in the case of the ship aground BG Intan Kelana 3 which occurred at the pier and the river mouth, it is estimated that the difference is 6%. Thus, the affected location on the BG Intan Kelana 3 ship is 7.70%. Comparative data on the cover of several locations are shown in Table 1.

3.2.2 The biology and ecology of coral impact
Turbid waters and nickel ore soil spills in the waters have contributed to adding to the turbidity [27] (Figure 4b) and these sediments have an impact on coral colonies, there is sediment attachment to algae that also attach to coral colonies in the sea thus disturbing coral growth (Figure 4b,c) and some colonies experienced a death near the site of the ship aground (Figure 4d).

In addition to causing death, the colony will experience a whitish discoloration due to the obstruction of light penetration and a decrease in the number of zooxanthellae [28]. Another study added that although there was a slight accumulation of sediment, it still had an impact on the recruitment process for new corals at the site [20]. Sedimentation can cause a decrease in coral abundance and diversity that affects coral physiological processes [21]. Long-term observations for 6 years (2004-2010) in sediment-affected areas obtained 20.47-27.87% cover in the Western Gulf of Thailand [22].

Figure 4. Condition of coral colonies in the control area

4. Conclusions

The BG Finacia ship found a difference in coral cover at the control location and the affected location by 8.44%. Coral diversity at the site, observations found 33 different genera of hard corals. Among the coral species that enter the slow growth rate are *Diploastrea heliopora*, *Favia* sp, *Favites* sp, *Goniastrea* sp, *Lobophyllia* sp, *Montastrea* spp, *Pavona* sp, *Platygyra* sp, and *Porites* sp. At the location of the BG Intan Kelana 3 ship, it was found that the difference in coral cover at the control location and the affected location was (24.70% – 7.70%) = 17%. Coral diversity at the observation site found 17 different genera of hard corals. Among the coral species that enter the slow growth rate are *Favia* sp, *Favites* sp, *Goniastrea* sp, *Platygyra* sp, *Porites* spp, and *Symphyllia* sp.

5. Acknowledgment

Thank you to the Ministry of Marine and Fisheries, local fisheries government, local non-government, TNI AL, POLAIRUD, and all experts and staff who support and assist in the collection of data on this survey. Our gratitude is also extended to Isai Yusidarta who assisted in the collection of coral condition data by using a frame and underwater camera. All authors are the main contributor to this paper who collaborate to collect the coral condition, data analysis of Image-J, documentation of photos underwater, paper writing, and coordination among experts and teams for the survey carried out effectively based on the target.

6. References

[1] Lasabuda R 2013 *J Ilmiah Platax* 1-2: 92-101.
[2] Lestari M M 2013. *J Selat 1(1):* 8-12.
[3] Djunaidah IS 2017 *J Pen Perikan Kel 11(1):* 12-24.
[4] Tis'in, M., & Mansyur K 2016 *Condition of coral reefs in the main coral cluster of the Toili oil and gas block, Central Sulawesi.17*(4): 34–46.
[5] Ndobe S, Moore A and Wayhudi D 2014 *Coral reef ecosystem on Maputi Island, Donggala Regency, Province. March 2016.*
[6] Idris, Johan O, Fakhrurrozi, Widodo M P S, Yusri S, Sianipar O 2021 *J Symp on Aquatic Sci Res Manag.* 744-012031.
[7] Giyanto, Iskandar B H, Soedharma D, and Suharsono 2010 *Ocean and Limnol.* 36(1): 111-130.
[8] Giyanto. 2012 *Osean dan Limnol Indones.* 38(1): 1-18
[9] Giyanto 2013 *Oseana, 28*(1), 47–61.
[10] English S, Wilkinson C and Baker V 1997 *Survey Manual for Tropical Marine Resources* (2nd Edition). Australian Institute of Marine Science, Australia: x + 390 p.

[11] Kohler, K. E: M. Gill.2006 *Comput Geosci* 32(9):1259-1269.
[12] Precht W F, 1998 *Geotimes* 43(1): 16–20.
[13] Jaap WC 2000 *Ecol Engineer* 15, 345–364.
[14] Precht WF, Aronson R.B, and Swanson DW 2001 *Bull Mar Sci.* 69, 1001–1012.
[15] Precht W F and Robbart M 2005 *Coral Reef Restoration: The Rehabilitation of an Ecosystem Under Siege* ed W E Precht (Florida: CRC Press LLC) pp 1–24.
[16] Hadi T A, Abrar M., Giyanto, Prayudha B, Johan O, Budiyanto A, Dzumalek A R, Alifatri L O, Sulha S and Suharsono 2020 *The status of Indonesian coral reefs 2019* (RHM Secretariat, Research Center for Oceanography)
[17] Estradivari, Setyawan and E Yusri S 2009 *Jakarta's coral reefs: Long-term observations of the Thousand Islands' coral reefs (2003-2007) TERANGI Foundation Jakarta.*Viii+102 p.
[18] Steffen J H *2001 Patterns of coral reef degradation along the west coast of Sumatra, Indonesia* Center for Tropical Marine Ecology. Dissertation. 75 p.
[19] Johan O, Timotius S, and Soekarno 2002 *Ornamental coral stock assessment in Rembang and Lampung, Indonesia.Indonesian Coral Reef Foundation* Traffic-Europe and AKKII 13 p.
[20] Babcock R and Smith L 2000 Effects of sedimentation on coral settlement and survivorship. Marine Biology, I (October), 1–4. http://www.coremap.or.id/downloads/ICRS9th-BabcockEffectsofsedimentationcameraready-corrected.pdf.
[21] Hadi T A 2019 Impacts of Sedimentation on Stony Corals. *Oseana*, 42(2), 45–58.
[22] Yeemin, T, Pengsakun S, Yucharoen M, Klinthong W, Sangmanee K, and Sutthacheep M 2013 *Deep-Sea Research Part II: Topical Studies in Oceanography* 32–40.

Quantity and quality of ephippia from moina (*Moina macrocopa*) culture with bran suspension feed with Lemuru fish oil

A A Mabruroh[1], **A S Mubarak**[1,2], **W Tjahjaningsih**[1]

[1]Faculty of Fisheries and Marine. Universitas Airlangga. Surabaya. East Java. Indonesia

[2]Corresponding author: shofy.ua@gmail.com

Abstract. Epipia *Moina macrocopa* is a product of sexual reproduction in the form of dormant egg cells that can be stored for a long time. The epipia can hatch at any time similar to the Artemia cyst. The quality and quantity of ephippia is influenced by the quality and quantity of the feed. The study used a completely randomized design (CRD) consisting of five treatments and four replications. The treatments consisted of a control without the addition of fish oil (P0). suspension of bran with the addition of 1.5 ml of fish oil (P1). suspension of bran with the addition of 3 ml of fish oil (P2). suspension of bran with the addition of 4.5 ml of fish oil (P3). and suspension of bran with the addition of 6 ml fish oil (P4). The main parameters are the quality and quantity of epihipia. The concentration of lemuru fish oil 6 ml/100 gr of bran can increase the quantity of ephippia as much as 426 ± 60.53 grains/L with the percentage of ephippia containing eggs as much as 98%.3 grains/L with the percentage of ephippia containing eggs as much as 98%.

1. Introduction

Moina macrocopa is a potential natural food for fish larvae because it has high protein, fat, and carbohydrates and is very suitable for use as larval feed because its size (500–1000 μm) corresponds to the mouth opening of fish larvae [1]. In the aquaculture industry, the availability of Moina is very important for feeding larvae, fry, and fingerlings [2]. *M. macrocopa* can be bred and cultivated using feed from agricultural, livestock and food industry waste [3,4]. The potential possessed by *M. macrocopa* is that it can replace Artemia's natural feed, it is still less attractive to farmers because it is not practical. Cultivators prefer natural food *Artemia* sp. which are easily obtained from hatching cysts of *Artemia* sp. Ephippia *M. macrocopa* is a product of sexual reproduction containing eggs/dormant embryos that can be stored for a long time and hatched at any time [4,5] similar to Artemia cysts so that the availability of *M. macrocopa* ephippia can be optimal as natural feed.

Ephippia *M. macrocopa* can be stored for a long time and can be hatched at any time [4,5]. The production of ephippia *M. macrocopa* begins with sexual reproduction which produces sexual male and female offspring [4]. Sexual females are females that do not reproduce partogenesis but produce eggs from gametogenesis which will be fertilized by males [6]. Ephippia *M. macrocopa* produced from sexual reproduction can be induced by regulating the density, quality, and quantity of feed [7].

The quality and quantity of ephippia produced can be influenced by the quality and quantity of the feed [8,9]. The quality of ephippia is determined by the number of ephippia containing two eggs or completely fertilized [10], while the quantity of ephippia is determined by the number of ephippia produced by *M. macrocopa*. Feed containing EPA. and DHA can improve the quality of ephippia [11] and quantity, because eggs in ephippia (ephippium) require higher concentrations of EPA and DHA than their parthenogenetic eggs [12].

In the culture of *M. macrocopa* with bran suspension feed, the quality and quantity of epiphia were quite low (360±81 grains/L, hatching rate: 20%) because it contained omega-3 eicosapentanoic acid (EPA) and low docosahexanoic acid (DHA). Mubarak *et al.* [7], the content of EPA and DHA in feed can increase the quality and quantity of ephippia [12], so it is necessary to add fish oil to the bran suspension, including lemuru fish oil. Lemuru fish contain 9.6% EPA and 10.09% DHA [13]. Lemuru fish oil contains omega-3 fatty acids of 34.9% of the total fatty acids in fish oil [14]. This study aims to determine the effect of lemuru fish oil mixed into bran suspension to improve the quality and quantity of ephippia *M. macrocopa*.

2. Materials and methods

2.1. Materials

The main research material is Moina of the same age obtained from hatching ephippia and getting bran suspension. and lemuru fish oil.

2.2 Methods

2.2.1. Research Design

The study used a completely randomized design (CRD) consisting of five treatments and four replications. The treatment consisted of control without the addition of lemuru fish oil (P0), bran suspension plus 1.5 mL lemuru fish oil (P1), bran suspension plus 3 mL lemuru fish oil (P2), bran suspension plus 4.5 mL lemuru fish oil (P3), and suspension of bran plus 6 mL of lemuru fish oil (P4). The fish oil treatment dose refers to the fat requirement of aquatic organisms of 10-16% [15]. The fat content in the bran is 10.62% [7] so the addition of fat required is 6% of the total fat.

2.2.2 Preparation of Cultivation Media

The culture medium for *M. macrocopa* uses fresh water that has been accommodated in a 1000 L fiber tank. The water that is put into the fiber tub is aerated for at least three days before being used. Water from the fiber bath was filtered using a 40 μm sieve before being put into the experimental tank to remove competing zooplantons [7].

2.2.3 Feed M. macrocopa

The bran was prepared as much as 100 grams for each treatment and then each was mixed with lemuru fish oil according to the treatment. Furthermore, the mixed bran and fish oil were suspended with 500 ml of water using a blender for 5 minutes twice. After being allowed to stand for 30 minutes, it was then resuspended for another 5 minutes. Furthermore, the suspension results were filtered using a filter with a size of 2 mm and 40 m. The suspension that passes the filter is added with water so that the volume becomes 500 mL [7].

2.2.4 Cultivation of Moina macrocopa

Moina at a density of 20 ind/L with a water volume of 10 L was adapted to bran suspension feed. Moina's children aged less than 24 hours from cultivation were then used as inoculants in this study. Moina cultivation uses a glass container with a water volume of 300 mL [7]. Moina is cultivated for 12 days. During the cultivation period, the container is changed every two days, while the water change is carried out from the fifth day to the eleventh day. During cultivation, the container is aerated [7].

Feed is given twice a day at 08.00-09.00 WIB and 16.00-17.00 WIB, each as much as 50% of the daily concentration, the following is the feeding that will be given [7].

Table 1. Dosage of bran suspension for Moina [7]

Days to	0	1	2	3	4	5	6	7	8	9	10	11	12
Dosage of bran suspension (mL/10 L)	3	3	3.5	4.1	5	6	7.2	86	10	10	10	10	10

2.2.5 Water Quality
Observations of water quality were carried out every day before the first feeding in the morning. Measurement of water temperature using a thermometer, pH was measured with a pH meter, and dissolved oxygen was measured with a DO meter.

2.2.6 Research Parameters
The main parameters observed in this study were the quality and quantity of ephippia. The quality of ephippia can be determined from the number of ephippia containing two eggs or completely fertilized. The quantity of ephippia was determined based on the amount of ephippia produced by *M. macrocopa*. Ephippia collection was carried out on day 3 to day 12. The percentage of ephippia to the total female in mating and the percentage of ephippia containing n eggs (number of eggs in ephippia) were calculated using the equation below [7].

1. Ephippia contains (n) eggs (%) = $\frac{\text{Number of epipphia containing (n) eggs}}{\text{Number of ephippia}}$ x 100%

2. Ephippia per total population (%) = $\frac{\text{Number of ephippia Moina}}{\text{Total population}}$ x 100%

2.2.7 Analysis of Data
Data were analyzed using Analyze of Variance (ANOVA). If there is an effect of treatment, the calculation is continued with Duncan's Multiple Range Test [16].

3. Result and discussion

3.1. Result
The bran suspension without fish oil (P0) had a water content of 92.74% with a fat content of 0.70%. Giving lemuru fish oil in bran suspension can increase the fat content in the bran suspension as feed for *M. macrocopa* (Table 2)

Table 2. Precentage of protein and fat content in each treatment

Treatment	Protein (%)	Lemak (%)
P0	1.50	0.70
P1	1.49	1.02
P2	1.48	1.35
P3	1.47	1.68
P4	1.46	2.01

Description: P0 = without lemuru fish oil, P1 = 1.5% lemuru fish oil, P2 = 3% lemuru fish oil, P3 = 4.5% lemuru fish oil, P4 = 6% lemuru fish oil.

Ephippia began to be produced on the seventh day in all treatments with the addition of lemuru fish oil (Table 3). Cultivation of *M. macrocopa* using bran suspension without the addition of lemuru fish oil. epihippia begins to be produced on the eighth day.

Table 3. Daily production of ephippia (grains/L) fed bran suspension with the addition of lemuru fish oil with different concentrations

Day of culture	Production Ephippia	Lemuru Fish Oil Concentration				
		0 mL/100 g	1.5 mL/100g	3 mL/100 g	4.5 mL/100 g	6 mL/100 g
1	0	-	-	-	-	-
2	0	-	-	-	-	-
3	0	-	-	-	-	-
4	0	-	-	-	-	-
5	0	-	-	-	-	-
6	0	-	-	-	-	-
7	Ephippia - 1	00.00 ± 00.00^a	9.37 ± 0.37^b	15.42 ± 0.86^c	19.4 ± 0.56^d	27.52 ± 0.87^e
8	Ephippia - 2	28.22 ± 1.15^a	34.7 ± 2.47^b	48.7 ± 0.96^c	53.4 ± 1.76^d	54.62 ± 0.54^d
9	Ephippia - 3	88.02 ± 1.18^a	95.9 ± 1.36^b	105.72 ± 3.62^c	157.32 ± 2.22^d	198.95 ± 0.58^e
10	Ephippia - 4	145.9 ± 28.77^a	208.02 ± 1.36^a	286.57 ± 6.42^b	316.17 ± 81.03^b	425.8 ± 60.53^c
11	Ephippia - 5	91.97 ± 1.07^a	104.37 ± 1.98^b	214.67 ± 0.65^c	296.15 ± 1.65^d	383.92 ± 0.33^e
12	Ephippia - 6	54.07 ± 1.04^a	62.57 ± 1.10^b	73.05 ± 2.56^c	107 ± 2.71^d	207.47 ± 2.87^e
13	Ephippia - 7	46 ± 1.40^a	52.32 ± 5.42^a	64.42 ± 1.75^b	84.95 ± 6.36^c	125.87 ± 8.93^d
14	Ephippia - 8	34.35 ± 0.86^a	43.92 ± 2.17^b	55.2 ± 0.80^c	62.17 ± 5.86^d	77.4 ± 2.34^e
15	Ephippia - 9	25.72 ± 4.52^a	37.37 ± 2.62^b	43.95 ± 1.36^c	56.9 ± 2.8^d	64.25 ± 3.92^e

The ephippia produced by *M. macrocopa* consisted of three groups, namely ephippia containing 2 eggs, ephippia containing 1 egg and ephippia without eggs. The quality of the ephippia was determined by the presence of eggs in the ephippia (Figure 1).

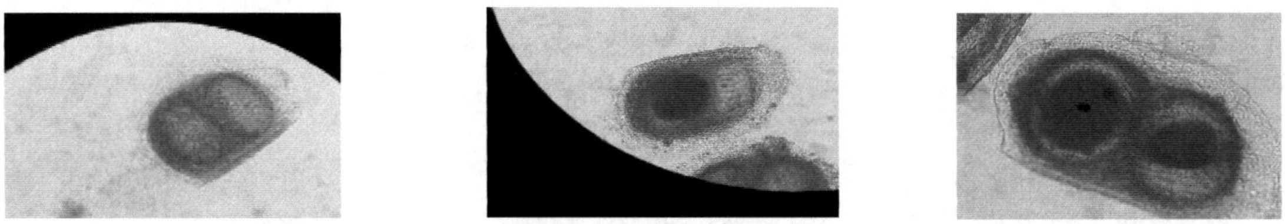

A B C

Figure 1. (A) Ephippia without eggs (B) Ephippia containing 1 egg (C) Ephippia containing 2 eggs

Culture of *M. macrocopa* using bran suspension feed with the addition of lemuru fish oil in various concentrations had a significant effect ($p<0.05$) on the number of *M. macrocopa* ephippia containing eggs (Table 4).

Table 4. Total number of ephippia containing eggs daily (grain/L)

Day of culture	Production Ephippia	Lemuru Fish Oil Concentration				
		0 mL/100 g	1.5 mL/100g	3 mL/100 g	4.5 mL/100 g	6 mL/100 g
1	0	-	-	-	-	-
2	0	-	-	-	-	-
3	0	-	-	-	-	-
4	0	-	-	-	-	-
5	0	-	-	-	-	-
6	0	-	-	-	-	-
7	Ephippia - 1	-	8.82 ± 0.29^b	14.62 ± 0.74^c	18.65 ± 0.59^d	26.67 ± 0.47^e
8	Ephippia - 2	26.0 ± 1.0^a	30.47 ± 2.98^b	46.35 ± 0.75^c	51.25 ± 1.70^d	53.55 ± 0.46^d
9	Ephippia - 3	80.45 ± 1.95^a	90.7 ± 1.32^b	100.45 ± 3.43^c	151 ± 2.15^d	193.45 ± 2.68^e
10	Ephippia - 4	134.22 ± 26.45^a	195.52 ± 1.25^a	272.22 ± 6.05^b	303.52 ± 77.76^b	417.32 ± 107.01^c
11	Ephippia - 5	88.62 ± 3.15^a	100.65 ± 3.68^b	203.95 ± 0.45^c	284.75 ± 2.05^d	376.22 ± 0.33^e
12	Ephippia - 6	49.42 ± 0.92^a	58.72 ± 0.89^b	69.15 ± 2.38^c	103.27 ± 2.97^d	202.12 ± 3.10^e

13	Ephippia - 7	42.2 ± 1.23[a]	49.15 ± 5.06[a]	61.22 ± 1.56[b]	81.65 ± 6.27[c]	123.15 ± 8.55[d]
14	Ephippia - 8	31.55 ± 0.75[a]	41.3 ± 2.05[b]	52.45 ± 0.61[c]	59.67 ± 5.73[d]	76.02 ± 2.31[e]
15	Ephippia - 9	23.75 ± 4.11[a]	35.2 ± 2.32[b]	41.67 ± 1.25[c]	54.75 ± 2.79[d]	62.92 ± 3.82[e]

The production of the first ephippia to the ninth ephippia has obtained the highest percentage of ephippia containing eggs, namely at a fish oil concentration of 6 mL/100 g bran by 98% ephippia containing eggs from the total number of ephippia, and the lowest yield was obtained in feed without the addition of fish oil by 82% ephippia contained eggs from the total number of ephippia (Figure 2).

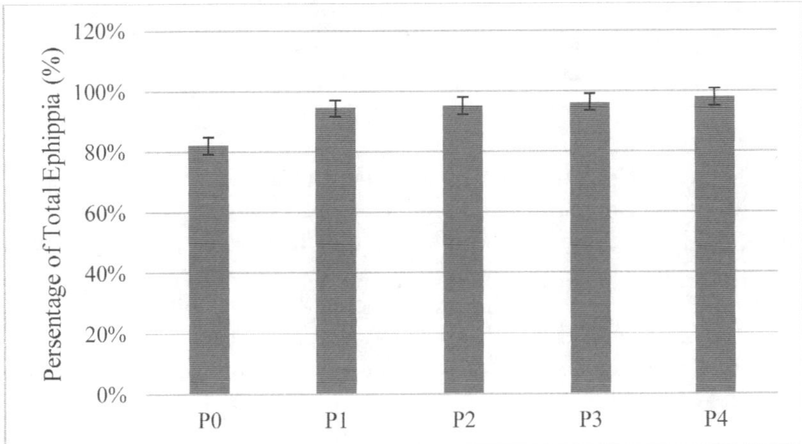

Figure 2. Percentage of total ephippia containing eggs. Description: P0 = without lemuru fish oil, P1 = 1.5% lemuru fish oil, P2 = 3% lemuru fish oil, P3 = 4.5% lemuru fish oil, P4 = 6% lemuru fish oil.

The culture media used for the culture of *M. macrocopa* has a DO range of 5.5-5.8 mg/L, with a pH value of 8.0-8.2. Water temperature during culture ranged from 27.8-28.4°C (Table 5).

Table 5. Value of water quality parameters of M. macrocopa media in each treatment

Parameter	Lemuru Fish Oil Concentration				
	0 mL/100 g	1.5 mL/100g	3 mL/100 g	4.5 mL/100 g	6 mL/100 g
Temperature (°C)	27.8 – 28.2	27.9 – 28.2	27.6 – 28.2	27.9 – 28.2	27.9 – 28.4
pH	8.0 – 8.2	8.0 – 8.1	8.0 – 8.1	8.0 – 8.1	8.0 – 8.2
DO (mg/L)	5.5 – 6.0	5.3 – 5.8	5.7 – 5.8	5.5 – 6.0	5.4 – 5.8

3.2. Discussion
The addition of lemuru fish oil in bran suspension has been able to increase the fat content in the bran suspension, especially omega-3 fatty acids, because lemuru fish oil contains high omega-3 fatty acids including EPA as much as 8.97% and DHA as much as 6.56% [17]. The concentration of fat in the diet of cladosera (*D. magna*) affects the allocation of energy from metabolism, and cholesterol is a hormone-forming precursor [18]. In Cladoscera PUFAs play a role in reproduction [19], improve growth, and survival [20]. The availability of EPA fatty acids in the diet of Copepod species affects the number of eggs produced and the quality of eggs produced [11]. The production of ephippia *M. macrocopa* is also limited by omega-3 fatty acids, EPA, and DHA [12].

Water quality of culture media. *M. macrocopa* in this study were the same between treatments and the water quality conditions of the media were suitable for the culture of *M. macrocopa*, such as DO of 5.3-3.9 mg/L [7]. temperature ranged from 26-31°C [21] and pH 7.0-8.0, so the difference in quantity and quality of ephippia produced was only due to the effect of treatment.

Bran contains high -6 such as linoleic acid 6.35-6.85%, but with a low content of omega-3 fatty acids [22]. Lemuru fish oil contains 38.15% SFA, MUFA by 32.18%, and PUFA by 28.58%, and has a high content of EPA and DHA compared to other fish [17]. The addition of lemuru fish oil as much as 6 mL/100 g of bran as feed for *M. macrocopa* induced the highest ephippia production of 425.8 ± 60.53 eggs/L with ephippia containing eggs of 417.32 ± 107.01 eggs/L. This proves that the Copepod species. Egg production and egg quality are affected by the availability of EPA, DHA in feed [11], which increased with the addition of lemuru fish oil. Giving lemuru fish oil to bran suspension as much as 6 mL/100 g of bran has been able to increase the percentage of ephippia *M. macrocopa* containing the highest eggs by 98%, while *M. macrocopa* with bran suspension feed without the addition of lemuru fish oil has only resulted in the percentage of ephippia *M. macrocopa* contains 82% eggs.

4. Conclusion

The addition of lemuru fish oil into the bran suspension as feed for *M. macrocopa* can affect the quantity and quality of ephippia. The optimal concentration of lemuru fish oil in bran suspension that can increase the quantity of ephippia is 6 mL/100 g of bran with an average ephippia production of 425.8 grains/L with a percentage of ephippia containing eggs as much as 98%. The results of this study have been able to become a solution to increase the quantity and quality of production of ephippia *M. macrocopa* which is currently still low. Success in producing ephippia *M. macrocopa* will be able to maximize the utilization of *M. macrocopa* as natural feed to replace the use of Artemia which is still imported from abroad

5. References

[1] Mudjiman A 2008 *Makanan Ikan* (Jakarta: Penebar Swadaya) [in Indonesian].
[2] Sofian M F, Yusuf H, Kamal A H M, and Karim N U 2021 *JIPK.* 13(2) 121-132.
[3] Patil S S, Ward A J, Kumar M S, and Ball A S 2010 *Bioresour Technol.* 101 3371-3378.
[4] Dodson S, Caceras C, and Rogers C 2010 *Ecology and Classification of North America Freshwater Invertebrates. Chapter 20: Cladosera and Other Branchiopoda.* 3th Ed. (San Diego, California: Academic Press).
[5] Alekseev V R, Stasio B T, and Gilbert J J (Eds) 2007 *Diapause In Aquatic Invertebrates: Theory and Human Use* (Netherland: Spinger Science & Business Media).
[6] Rottmann R W, Graves J S, Watson C, and Yanong R P E 2003 *Culture Techniques* of *Moina: The Ideal Daphnia for Feeding Freshwater Fish Fry* (Florida: University of Floria. Florida).
[7] Mubarak A S, Dedi J, Muhammad Z J., and Muhammad A S 2017 *JAI.* 16(2), 223-233.
[8] Hakima B, Khémissa C, and Boudjéma S 2013 *Sci. J Biol Sci.* 5, 25-31.
[9] Koch U, Creuzburg D, Grossart P, and Straile D 2011 *Oecologia* 67, 981-989.
[10] Choi J, Kim S, Hwan L G, Chang K, Kim D, Jeon K, Park M, Joo J, Kim H, and Jeong K 2016 *Ecol. Evol.* 6, 2817- 2832.
[11] Jonasdottir S H, Visser A, and Jespersen C 2009 *Mar Ecol Prog Ser.* 382, 139-150.
[12] Abrusan G, Fink P, and Lampert W 2007 *Limnol Ocean.* 52, 1724-1728.
[13] Estiasih T, and Ahmadi K 2012 *JTP.* 5, 116-128 [in Indonesian].
[14] Supadmo 1997 *Effects of Chitin Sources and Carnitine Precursors and Lemuru Fish Oil on Fat and Cholesterol Levels and Omega-3 Fatty Acids in Broiler Dissertation (Bogor: Graduate Program of IPB) [in Indonesian].*
[15] Hastuti S, and Subandiyono 2016 *Textbook of Fish Nutrition (Semarang: Institute for Educational Quality Development and Assurance, Dipenegoro University, Semarang) [in Indonesian].*
[16] Kusriningrum R S 2012 Experiment Design (Surabaya: Airlangga University Press) [in Indonesian].
[17] Maulana I T, Sukraso, and Sophi D. 2014 *JITKT.* 6(1), 121-130 [in Indonesian].
[18] Nagaraju G 2011 *J Exp Biol.* 214, 3-16.

[19] Wacker A, and Creuzburg M D 2007 *Funct Ecol.* 21, 738-747.
[20] Fereidouni A E, Fathi N, and Khalesi M K 2013 *Turkish J Fish Aqua Sci.* 13, 119-126.
[21] Tan Q G, and Wang W X 2010 *Limnol Oceano*. 55, 1426-1434.
[22] Faria S, Bassinello P, and Penteado M 2012 *Braz J Pharm Sci.* 48, 35-45.

6. Acknowledgment

The authors are grateful to the Faculty of Fisheries and Marine, Universitas Airlangga which had provided the space and laboratory to conduct this research.

IOP Conf. Series: Earth and Environmental Science 1273 (2023) 012073 doi:10.1088/1755-1315/1273/1/012073

Morphology and cytochrome b gene identification of *Cempedik* fish for ascertainment of domestication broodstock source in Belitung Island

Kurniawan A[1,2], A F Syarif[1] and A Kurniawan[1]

[1]Aquaculture Departement, Universitas Bangka Belitung

[2] Coresponding email : ardian_turen@yahoo.co.id

Abstract. *Cempedik*, a native fish of Belitung Island, has economic potential due to its dependence on natural supplies in local rivers. Domestication efforts are needed for this species to continue its availability and protect the sustainibility of its natural population. However, there is little information regarding this freshwater fish belonging to *Osteochilus spilurus* and the emergence of phenotypic variations in fish with the same local name from different rivers. This study identified morphology and cytochrome b genes from *Cempedik* fish obtained from Gantung, East Belitung and Membalong, West Belitung. The fin membranes of fish from West Belitung are reddish in color, while the color is brownish gray in fish from East Belitung. Meristic variation is found in scales on lateral line. Fish from both locations were in different clusters in the phylogenetic tree construction with large genetic distances. It is possible that there are cryptic conditions or species differences in the two fish samples.

1. Introduction

Osteochilus spilurus is one of the most popular native freshwater fish in Belitung Island. Local people, especially in East Belitung, like this fish as fish consumption with the majority obtained from purchases. This species, that called as *Cempedik* fish in Belitung people, were consumed and traded by 79.07 – 87.93% East Belitung citizen [1]. The trading of this species is only found in Belitung Island and there is no indication of the same condition in other region of Indonesia [2]. Various local wisdoms emerged regarding this species, ranging from how to handle it to reduce the bitter taste to its use as a local batik motif [3]. There is local knowledge that provides understanding to most people about this fish.

The high demand for consumption and the potential impact on the exploitation of local rivers is feared to put pressure on the natural population. The fishing season that occurs in the rainy season has the potential to disrupt the reproductive cycle of fish because the majority of fish caught are gonadally mature. More than half of the fish caught are at level 4 maturity [4]. Efforts to domesticate this species have been made. The first level of domestication has been achieved by water recirculation treatment [5, 6].

Continued of domestication is still ongoing today. The Fish Seed Center of East Belitung Regency and the Hatchery of Bangka Belitung University are currently trying to breed them in an artificial environment.

The domestication process requires accurate species justification for broodstock because it is obtained from natural catches [7]. There are several variations regarding the captivity of *O. spilurus* in East and West Belitung. The fishing season in West Belitung is in the dry season, while in East Belitung it is in the rainy season. These distinct patterns require identification to prevent misapplication of the parent from nature habitat. We used a morphological approach and the cytochrome b gene to identify the species similarity between the two.

2. Material and methods

Sampling of *Cempedik* fish with the character of *Osteochilus spilurus* on Belitung Island was carried out at Air Tering river, Perpat Village, Membalong Sub District, Belitung Regency (3°02'38" S, 107°40'13" E) and Lenggang river, Lintang Village, Gantung Sub District , East Belitung Regency (2°55'09"S 108°06'35"E) (Table 1). The location and condition of the location of the fish sample is shown in Figure 1.

Table 1. Sampling location of *Cempedik* fish in Belitung Island

No	Island	Location	River
1	Belitung	Perpat Village, Membalong Sub District, Belitung Regency (3°02'38" S, 107°40'13" E)	Air Tering
2	Belitung	Lintang Village, Gantung Sub Distric, East Belitung Regency (2°55'09"S 108°06'35"E)	Lenggang

Figure 1. Sampling location of *Cempedik* fish in Belitung Island; A) Map of location at West Belitung (1) and East Belitung (2), B) Condition of Air Tering River in West Belitung, C) Lenggang River, East Belitung

IOP Conf. Series: Earth and Environmental Science 1273 (2023) 012073 doi:10.1088/1755-1315/1273/1/012073

Sample collection was carried out randomly using fishing gear in accordance with the tools used by local communities. Samples of fish were killed in cold water. Meristic observations included dorsal fin rays, anal fin rays, ventral fin rays, pectoral fin rays, linea lateral scales, pre dorsal scales, scales above linea lateral, and scales under linea lateral (Figure 2). The results of the observations were compared with the descriptions of Webber and Beaufort [8].

Figure 2. Meristic characters on *Osteochilus spilurus* . (1. Dorsal fin rays, 2. Anal fin rays, 3. Pectoral fin rays, 4. Pre dorsal scales, 5. Linea lateral scales, 6. Scales on caudal penducle, 7. scales above and under linea lateral).

Fish samples for genetic analysis were preserved in 96% alcohol for molecular analysis. After being submerged in 96% alcohol for at least 24 hours and the sample was completely fixed with a change in color, the fish were wrapped in tissue paper moistened with 96% alcohol and stored at -20°C [6]. Isolation of DNA in this study using the wizard Promega kit. The primers used were LA-cyp 5'-ATGGCAAGCCTACGAAAAAC-3' for forward and HA-cyp 5'-TCGGATTACAAGACCGATGCTT - 3' for reverse [9]. The PCR process uses 2% forward and reverse primers, 50% PCR mix, 36% ddH2O, and 10% template DNA. The PCR settings used were pre-PCR for five minutes on 94°C, denaturation for 30 seconds on 94°C, annealing or attachment (52°C for 30 seconds), extension or elongation (72°C for 30 seconds), and post-PCR (72°C for five minutes) repeated 35 cycles. The nucleotide sequences of the PCR results were then read using Applied Biosystems on the 1st Base. The resulting sequences were cut using Chromas 2.6.6, Consensus sequences using Ugene 1.32 and alignment using Mesquite. The phylogenetic tree was constructed using the Mega X application. The comparison sequences used are as shown in Table 2. The phylogenetic tree was constructed using the 2-parameter neighbor joining model with 500 bootstrap replications.

Table 2. Comparative sequences of O. *spilurus* phylogenetic constructs using the cyt b gene

Spesies	Country	Location	NCBI Accesion Number
Osteochilus hasseltii	Indonesian	Indonesian markets	KR007710.1
Osteochilus hasseltii	Viet Nam	Dak Lak	MN541329.1
Osteochilus waandersii	Viet Nam	Dak Lak	MN541330.1
Osteochilus waandersii	Aquarium [10]	-	JX074267.1
Osteochilus lini	Thailand [10]	-	JX074265.1
Osteochilus sp.	Indonesian	Pasar Batraja, Sumatera Selatan	JX074233.1
Osteochilus vittatus	Cambodia	Landing port, Kampong	JX074226.1

		Chhnang	
Osteochilus melanopleurus	Indonesian	Siak River, Riau	MK036876.1
Osteochilus melanopleurus	Indonesian	Siak River, Riau	MK036875.1
Osteochilus longidorsalis	Aquarium	-	KX227549.1
Osteochilus salsburyi	Cina	-	MH500676.1
Osteochilus hasseltii	Indonesian	Indonesian market	KR007710.1
Osteochilus spilurus	Malaysia	Balung River, Tawau, Sabah	DQ366162.1
Osteochilus sp.	Malaysia	Layar River, Betong, Sarawak	DQ366158.1
Osteochilus salsburyi	China	Rong'an, Guangxi	GU086539.1
Osteochilus nashii	India	Tungabhadra	KP998075.1
Bangana sp	Myanmar	Salween River, eastern Myanmar	JX074231.1
Bangana dero	India	-	MH973621.1
Labeo gonius	Aquarium	-	KU577402.1
Labeo barbatulus	Laos	-	KC631289.1

3. Result and Discussion

All samples obtained from all sampling locations have similar characteristics with the descriptions of *Osteochilus spilurus* of Weber and Beaufort [8] and Karnasuta [11]. In general the samples showed a silvery color and turned brown when prepared using 96% alcohol (Figure 3 and 4). All fish samples had black blots on the caudal penducle. The phenotypic variation was found in the color of the fin membranes, which was brownish in samples from East Belitung and reddish in samples from West Belitung. The difference in body shape between the two sampling locations may be due to phenotypic plasticity as has been reported in this species from different habitat conditions [7]. Fish behavior and diet choice affect variations in body shape [12]. However, geographical isolation that limits the relationship between populations allows for morphological variations caused by genetic variations [13].

Figure 3. Fresh *Cempedik* fish samples from East Belitung (left) and West Belitung

Figure 4. *Cempedik* fish with alcohol preparation from East Belitung (left) and West Belitung

Meristic characters obtained from all samples show variations in the number of linea lateral scales. Fish samples from East Belitung had 29 – 30 scales, while fish from West Belitung had 30-31 scales (Table 3). Webber and Beaufort [8] noted the number of scales attached to the linelateralis of *O. spilurus* was 29-30 scales, while Gunther [14] stated 30 scales, and Karnasuta [11] showed 28-29 scales. *O. Spilurus* found on Bangka Island also exhibits lateral scales of 29-30 units [7, 12]. Variations in meristic characteristics can separate freshwater fish in different clusters, but still require genetic analysis to confirm [15].

Table 3. The results of the meristic calculation of *Cempedik* fish samples from Belitung Island.

No	Meristic Characters		East Belitung	West Belitung
1	Dorsal fin rays	soft hardened	3	3
		soft	11-12	11-12
2.	Anal fin rays		6	6
3.	Ventral fin rays		9	9
4.	Pectoral fin rays		8-9	8
5.	Linea lateral		29-30	30-31
6.	Pre dorsal scales		8	8
6.	Scales above linea lateral		4	4
7.	Scales under linea lateral		4	4

The phylogenetic tree show that *Cempedik* fish from East Belitung is in a different cluster to the West Belitung sample (Figure 5). East Belitung sample has similarity with O. *spilurus* in the gene bank. This happened because the gene bank data for the species O. *spilurus* also came from the same river, the Lenggang river in Gantung Subdistrict [6]. The similarity is not high because the gene bank data uses the gene cyt b 408 bp, while the length of the sequence in this study is more than 1100 bp.

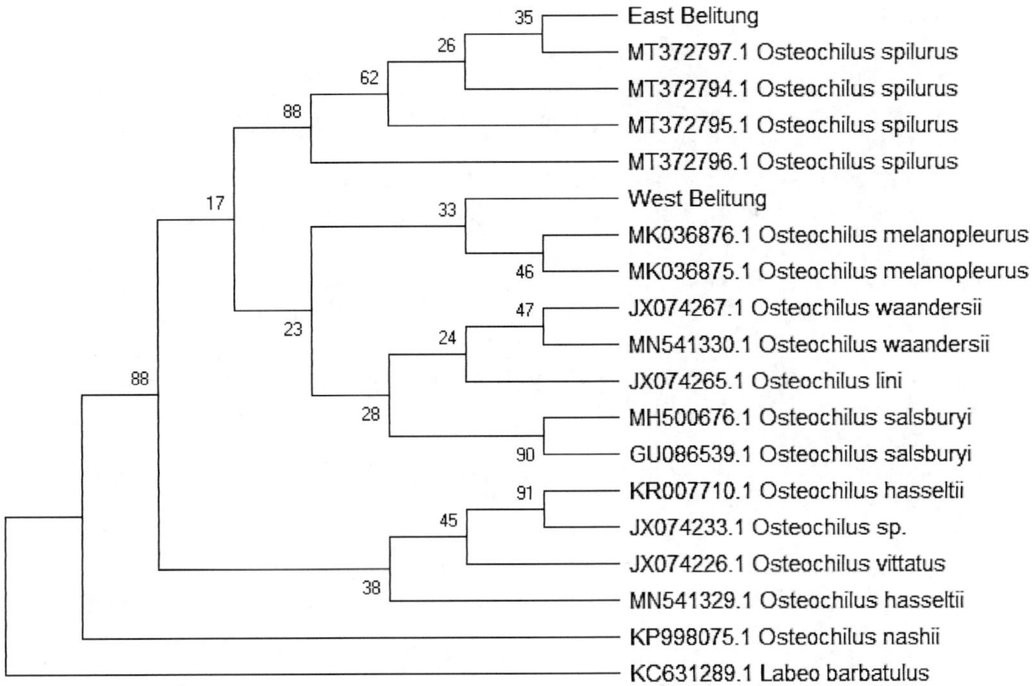

Figure 5. Phylogenetic tree of *Cempedik* fish from East and West Belitung using cyt b gene

Samples from west Belitung are more closely related to O. *melanopleurus*. This condition indicates the existence of criptic species, which are morphologically close to O. *spilurus* but are closely related to O. *melanopleurus* genetically. Morphological variations allow for the existence of cryptic species or different species [16]. The genetic distance between samples from East Belitung and West Belitung was at 6.66%. The intraspecific limitation of freshwater fish associations was generally at 3% [17]. A genetic distance of more than 3% in the family Cyprinidae qualifies as a new species [18]. This morphological similarity has the potential to cause errors when exploring fish broodstock for domestication. We recommend choosing fish from East Belitung or fish with morphological characters like *Cempedik* from East Belitung.

4. Conclusion

Fish sample with morphological characteristics of *O. spilurus* in East Belitung and West Belitung showed differences in the number of linae lateral scales. Samples from East Belitung are closely similarity to *O. spilurus*, samples from West Belitung closer to *O. melanopleurus* and potential to become new species in the genus Osteochilus. The selection of *O. spilurus* broodstock for domestication is recommended to choose fish from East Belitung and avoid fish that have reddish fins.

5. Reference

[1] Kurniawan A and Triswiyana I 2019 *Ecsofim.* 7(01) 109-119.
[2] Pramono D Y, Indrayati A, Hermanto H, and Triswiyana I 2020 *Asian J of Ethnobiol.* 3(2).

[3] Kurniawan A, Azhari M and Prasetiyono E 2019 *IOP Conf Series: Earth and Environ Sci*. 230 1 012116.
[4] Kurniawan A, Hariati AM, Rizkika N, and Wiadnya DGR 2020 *IOP Conf Series: Earth and Environ Sci.* 441(1) 012099
[5] Kurniawan A, Azhari M, and Prasetiyono E 2019 *IOP Conf Series: Earth and Environ Sci.* 230(1) 012116
[6] Kurniawan A, Hariati A M and Wiadnya D G R 2021 *Biodiv J of Biol Div*. 22(2).
[7] Kurniawan A, A M Hariati, Kurniawan A, Haryono H, and G R Wiadnya 2020 *Egyp J Aquatic Biol Fish*. 24 835-845.
[8] Weber M W C and de-Beaufort L F 1916 *The Fishes of the Indo-Australian Archip.* 3
[9] Kurniawan A., A Kurniawan, A M Hariati, N Kurniawan, TW Nugroho, E Bidayani., A F Syarif, and D G R Wiadnya 2022 *Biodiv J of Biol Div.* 23(9).
[10] Tang K L, M K Agnew, M V Hirt, T Sado, L M Schneider, J Freyhof, and K Saitoh 2010 *Mol Phylogene and Evol.* 57(1) 189-214
[11] Karnasuta J 1993 *J Fish and Environ* 19 1-105
[12] Kurniawan A, Hariati A M, Darmawan A, and Wiadnya D G R 2020 *IOP ConfSeries: Earth and Environ Sci.*493(1): 012024
[13] Svanbäck R and Eklöv P 2006 *Evol Ecol Res.* 8(1): 37-49.
[14] Gunther Albert 1868 *Cataloque of the Fishes in the British Museum* Vol 7. The Trustees London.
[15] Sarower-E-Mahfuj M, Rahman M M, Islam M, Samad M A, Paul A K, and Adhikary R K 2019 *J Adv Vet Animal Res.* 6(1): 117.
[16] Kottelat M and Tan H H 2012 *Ichth Explor of Freshwater.* 23(1) 37.
[17] Aminan A W, Kit L L W, Hui C H and Sulaiman B 2020 *Trop life sci res.* 31(2): 33-49
[18] Ng C K C and Tan J 2021 *J Fish Biol* 99(4): 1256-1273

6. Acknowledgment

We would like to thank Universitas Bangka Belitung and Directorate of Research and Community Service (DRPM) from the Indonesian Ministry of Research, Technology and Higher Education for research funding and the laboratory facilities.

IOP Conf. Series: Earth and Environmental Science 1273 (2023) 012074 doi:10.1088/1755-1315/1273/1/012074

The value of the nutritional composition of fresh fish and flour-anchovy

C Litaay[1,4], Indriawati [2], L Nuraini[1], C E W Anggara[1], Sriharti[1], A Triyono[1], T Hidayat [3]

[1]Research Center for Appropriate Technology-BRIN, Subang 41213, Indonesia
[2]Indonesia Directorate of Laboratory Management, Research Facilities, and Science and Technology Park – BRIN, Cibinong
[3]Research Center for Agroindustry-BRIN, Serpong, Indonesia

[4]Corresponding author: christina_litaay@yahoo.com

Abstract. Anchovy provides benefits in increasing bone density and preventing bone loss (osteoporosis) because it contains a lot of calcium and phosphorus. The aim of this study was to determine the value nutritional composition of anchovy after the immersion method. The fish was immersion in different methods of fresh anchovy, flour-anchovy water method, and flour-anchovy alkaline for 1 hour. The value of nutritional composition showed that protein, ash, fat, water, phosphate content, and calcium content were affected by water or alkaline immersion methods. Anchovy flour produced by immersion water for 1 hour showed the best composition value with an ash content of 17.45%, protein content of 74.08%, phosphorus 33948.87 mg/kg, and calcium 6466.51 mg/100g.

1. Introduction

Fish is a food with high nutritional content because it provides complete and essential amino acids for the body. Fish has relevant nutritional value due to the presence of minerals such as phosphorus, iron, and calcium as well as high protein content [1]. Fish is a very good source of animal protein because it has a high nutritional content such as vitamins, essential amino acids, minerals, and polyunsaturated fatty acids so beneficial for health [2,3,4]. The chemical composition of fish provides many health benefits, so consuming fish is very important [5]. Fish has nutritional value so that apart from being consumed directly, it can also be used as additional protein to enrich various food products [6].

In Indonesia, several areas have very low levels of fish consumption. This is due to the high price of fish species, and the low social value of fish so they are not in demand by the public even though the fish has a high nutritional content. Fish consumption depends on the availability of good quality fish and the level of consumer purchasing power [7]. Low fish consumption is related to consumption habits due to impractical fish processing and marketing and distribution problems.

Anchovy is a type of fish that has a high economic value and is used as a side dish of food. Types of fish with small sizes consumed whole significantly contributed to the increase in protein nutrition and micronutrient levels. Fish is one of the animal proteins that have the best protein intake, with excellent absorption of iron [8,9]. Phosphorus and calcium are important in the process of bone mineralization so if the nutritional intake of phosphorus and calcium is lacking in toddlers it can affect bone growth, and in the long term, it can lead to stunting [10]. Anchovy which is rich in calcium, phosphorus, and protein can be used as a solution to reducing the prevalence of stunting [11,12].

Anchovy is a fishery product that is easily damaged and decomposed (*perishable*). To overcome this problem, a preservation process is needed so that it can increase shelf life. Fish meal is one of fish preservation because the quality of fish deteriorates very quickly [13]. Fish meal is an excellent source of protein because they can increase food and feed consumption. The chemical composition of the fish meal includes 60% crude protein, 1.2% ash content, 2.5% water content, 2.54% fat, and 1.2% ash content [14].

Fish meal has benefits, namely high nutritional content, as a fortification ingredient in food products that are low in protein, easy to store and have a long shelf life, and can be processed into various food products. The process of making fresh anchovy into flour-anchovy resulted in changes to the value of the nutritional composition. Analysis of nutritional composition, namely protein content, ash, water, fat, phosphorus content, and calcium content. The purpose of this study was to determine the value of the nutritional composition of fresh fish and flour-anchovy (*Stolephorus* sp).

2. Materials and methods

2.1 Sample preparation

The main ingredients used in this study were fresh anchovies (*Stolephorus* sp.). The total length of the fish ranged from 3.7-9.5 cm with a weight of 1.0-6.0 g. The fish were obtained from Blanakan Subang, West Java. The fish were then put into an ice Styrofoam box, and brought to the laboratory. The fish were washed clean of dirt, drained, then carried out the immersion process using mineral water and alkali (sodium bicarbonate 0.8%) method for 1 hour.

2.2 Making anchovy flour process

The process of making anchovy flour is carried out in the pilot plan laboratory (Research Center for Appropriate Technology - BRIN Subang, West Java). First, the process of boiling anchovies at a temperature of 80°C for 10 minutes. Second, pressing the fish for 10 minutes. Third, drying at a temperature of 150˚C, for 60 minutes. Next is flouring and packaging.

2.3 *Proximate composition*

The nutritional composition consisted of protein value, fat, ash, and moisture. Protein value by the Dumas (DuMaster Buchi D480, Swiss), fat using the Soxhlet method [15], ash content, and moisture content using the gravimetric method [16].

2.4 Phosphorus content

Calculations of phosphorus were analyzed using the ICP OES method regarding 18-13-1/MU/ SMM-SIG [17]

2.5 Calcium content

Calculations of calcium were analyzed using the ICP OES method regarding 18-13-1/MU/ SMM-SIG [17].

2.6 Statistical analysis

Experimental data were analyzed using the IBM SPSS version 25 program. One-way analysis of variance (ANOVA) was used. The advanced test of Duncan performed when ANOVA on treatment had a significant effect ($p<0.05$).

3. Results and discussion

3.1 Proximate composition

The value of the proximate composition of fresh anchovy, flour-anchovy water method, and flour-anchovy alkaline (sodium bicarbonate 0.8%) method can be seen in Figure 1.

The test results showed that the protein of fresh anchovy increased from 10.3% to 74.08% in the manufacture of flour-fish with the alkaline method. The ash and fat value of anchovy increased by

16,86% and 3.03%, respectively. Fresh anchovy has a high moisture value of 80.0% and decreased by 1,92%. Statistical analysis showed that fresh fish anchovy and flour-anchovy with different immersions had a significant effect ($p<0.05$) on the value of protein, fat, ash, and moisture.

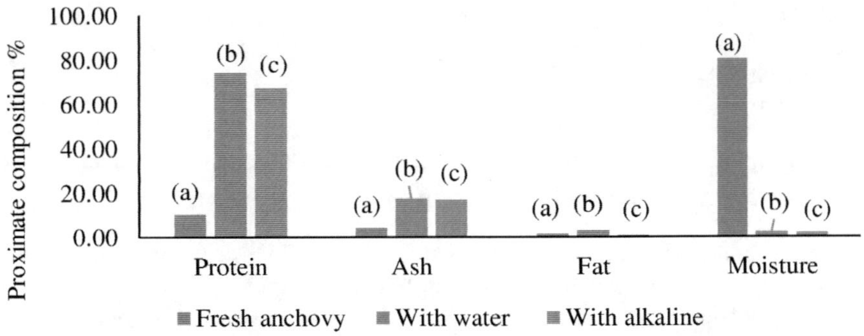

Figure 1. Figure 2. Proximate composition of fresh anchovy, flour-anchovy water method, and flour-anchovy alkaline. The numbers followed by different superscript letters (a,b,c) showed a significant difference ($p<0.05$) in the immersion method.

The results of this study are almost the same as those of skipjack tuna fish meal (*Katsuwonus pelamis*) by 71,46% [18]. The fish flour with protein content between 60% and 72% is of good quality. Marketed fish meal has a crude protein of 65%, but can vary from 57-77% depending on the fish species used [19]. The use of alkaline sodium bicarbonate can reduce fat value. Acids and alkalis can minimize fat or eliminate fat, besides that alkali can open the matrix bond structure in bones so that fat and other compounds are easily released [20].

The ash value of fish flour is increased because all parts of the anchovy body are consumed from the head, and meat to bones. In contrast to the research of [21], the value of fish ash by immersion in water media for 30 minutes was 4.82% and [22] immersion in lime juice 0.2-0.3% for 1 hour was 7.76%. The moisture value tends to decrease in the alkaline immersion method. The presence of sodium bicarbonate concentration affects the decrease in moisture content.

3.2 Phosphorus content

The results of the phosphorus content which is obtained from fresh anchovy, flour-anchovy water method, and flour-anchovy alkaline are presented in figure 3. The phosphorus content of fresh anchovy, flour-anchovy water method, and flour-anchovy alkaline show a significant difference ($p<0.05$).

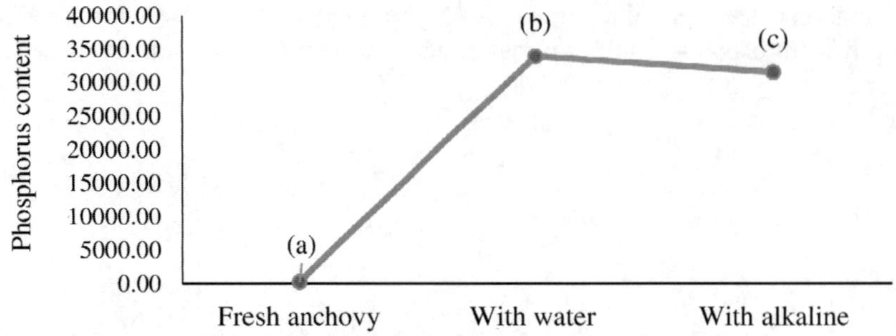

Figure 2. Phosphorus content of fresh anchovy, flour-anchovy water method, and flour-anchovy alkaline. The numbers followed by different superscript letters (a,b,c) showed a significant difference ($p<0.05$) in the immersion method.

The flour-anchovy water and alkaline method resulted in a significant increase ($p<0.05$) in the phosphorus content compared with fresh anchovy, which corresponded to 0,75% and 0,80%, respectively. Based on the nutritional composition of Indonesian food [23], anchovy flour has a phosphorus content of 1200 mg/kg, and research by [24] on catfish bone meal was 8.91%-11.78% lower than the results of the study. Under alkaline conditions, calcium and phosphorus in the bone can form crystals of minerals that are insoluble in alkali or alkaline pH. Consumption of foods containing phosphorus can increase the bioavailability of calcium.

3.3 Calcium content

The results of the calcium content which is obtained from the fresh anchovy, flour-anchovy water method, and flour-anchovy alkaline are presented in figure 4.

The calcium content of fresh anchovy, flour-anchovy water method, and flour-anchovy alkaline show a significant difference ($p<0.05$). The flour-anchovy water method and flour-anchovy alkaline resulted increase compared with fresh anchovy, which corresponded to 15,0 % and 15,4%, respectively.

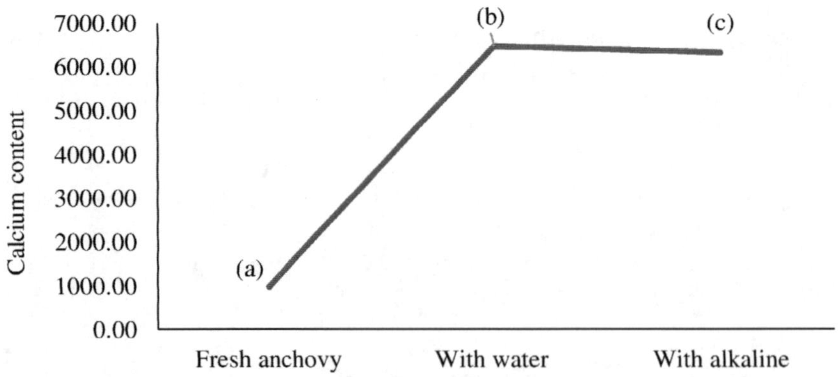

Figure 3. Calcium content of fresh anchovy, flour-anchovy water method, and flour-anchovy alkaline. The numbers followed by different superscript letters (a,b,c) showed a significant difference ($p<0.05$) in the immersion method.

IOP Conf. Series: Earth and Environmental Science 1273 (2023) 012074 doi:10.1088/1755-1315/1273/1/012074

The calcium value of anchovy meal detected without immersion was lower by 4608 mg/100g [23]. The results of the study found by [22] showed a calcium value of 504 mg upon immersion in 0.2-0.3% lime juice for 1 hour, and [25] on anchovy flour resulting from the drying process of 645,37 mg.

The increase in calcium levels of anchovy flour is caused by the use of water and alkaline sodium bicarbonate. The sodium bicarbonate compound contains sodium which is a mineral component. Anchovy flour has a high calcium value because the size of the fish is very small and is eaten whole with the bones [26]. High calcium nutrition is very important for the community, especially for toddlers and pregnant women because it is related to the growth and development of children's teeth and bones [27].

4. Conclusion

The value of nutritional composition showed that protein, ash, fat, water, phosphate content, and calcium content were affected by water or alkaline immersion methods. Anchovy flour produced by immersion water for 1 hour showed the best composition value with an ash content of 17.45%, the protein content of 74.08%, phosphorus 33948.87 mg/kg, and calcium 6466.51 mg/100g.

5. References

[1] Dos E S, Goes R *et al.*, 2015 *FST.* 35(4) 683–689.
[2] Rebolé A *et al.*, 2015 *Food Chem.* 174 614–620.
[3] Litaay C, Indriati A, and Mayasti N K I 2022 *FST.* 42 1-7.
[4] Prastica E E, Masithah E D, and Pursetyo K T 2021 *World's Vet. J.* 11(1) 119-123.
[5] De Brito B M, Lira G M, Pinheiro A G A, da C M A, Santana S, and Amaral I L 2019 *FST.* 39 640-645.
[6] Corapci B, and Guneri N 2020 *FST.* 40 163-169.
[7] Da E S, and Maciel 2013 *FST* . 33(3) 451-456.
[8] Litaay C *et al.*, 2022 *FST.* 42.
[9] Mahmudiono T, Nindya T S, Rachmah Q, Segalita C, and Wiradnyani L A A 2020 *Intern J. of Env Research and Public Healt.* 17(19) 1-15.
[10] Mahmudiono T, Ardianti B P, Kombih M F, Amira K A, and Indriani D 2020 *Systematic Review in Farm.* 11(8) 400-405.
[11] Fadhilah R N, Suhartini, and Rahardyan P 2013 *Insisiva Dent. J.* 2(1) 19-26.
[12] Litaay C *et al.*, 2022 *IOP EES.* 1033(1).
[13] Saputra E, Andriyono S, and Isroni W 2022 *IOP EES.* 1036 (1) 012009.
[14] Jassim J M 2010 *Int. J. Poult. Sci.* 9(12), 1097-1099.
[15] Indonesian National Standard 1992 Method of Testing Food and Beverages.
[16] AOAC 2005 *North Frederick Venue Gaithersburg, Mary Land* 20877-2417.
[17] PT Saraswanti Indogenetech 2021 *Annex To Accreditation Laboratory Certificate NO. LP-184-IDN-SNI ISO/IEC 17025:2017 (ISO/IEC 17025:2017).*
[18] Litaay C, and Santoso J 2013 *JITK.* 5(1) 85-92.
[19] Maigualema M A, and Gernat A G 2003 *Intern J. Poultry Sci.* 2(3) 195-199.
[20] Nolsøe H, and Undeland I 2009 *FBT.* 2(1) 1-27.
[21] Canti M, Hadi T C, and Lestari D 2021 *J. Gizi Pangan.* 16 37-44.
[22] Dewi T K, Budyghifari L, and Adam A *Proximate Characteristics and Nutritional Value of White Anchovy Flour.*
[23] Ministry of Health T R Ministry of Health Directorate General of Public Health Table of Indonesian Food Composition, and K. R. Health, Catalog in.
[24] Lekahena V, Faridah D N, Syarief R, and Peranginangin R 2014 *JTIP.* 25(1) 57-64.
[25] Nurani F P, Rosida J, and Susanti R A 2018 *ICST.*
[26] Rahmi Y, Widya N, Anugerah P N, and Tanuwijaya L K *Nutrire Diaita.* 10(1) 34-44.
[27] Litaay C, Indriati A, Mayasti N K I, Anggara C E W, and Astro H M 2021 *J. Pengolah Has Perikan Indones.* 24(2) 148-159.

6. Acknowledgments

The authors would like to thank the National Research and Innovation Agency - Education Fund Management Institute (BRIN - LPDP) which has provided research funds on the Utilization of Anchovy Flour (Stolephorus sp.) in Sago Starch-Based Noodle Products to Increase Protein, Phosphorus, and Calcium Content Implications for Increasing Nutritional Intake in Children with Stunting through the Research and Innovation for Advanced Indonesia (RIIM) program to author Dr. Christina Litaay, S.Pi., M.Si. The author would also like to thank BRIN – Research Center for Appropriate Technology, and the entire research team who have participated in research activities.

Potency of formulated diets with different protein on feed consumption, feed conversion ratio, feed efficiency and nutrient retention of scalloped spiny lobster (*Panulirus homarus*)

S D Maharani[1], M Amin[2,4], M Lamid[3]

[1]Study Program of Aquaculture, Surabaya Campus, Faculty of Fisheries and Marine, Universitas Airlangga, Indonesia
[2]Department of Aquaculture, Faculty of Fisheries and Marine, Airlangga University, Surabaya 60115 Indonesia
[3]Department of Animal Husbandry, Faculty of Veterinary Medicine, Airlangga University, Surabaya 60115 Indonesia

[4]Corresponding author: muhamad.amin@fpk.unair.ac.id

Abstract. Scalloped spiny lobster (*Panulirus homarus*) is a commodity due to its high economic value, and it increases the exploitation of lobster. This increase needs to be balanced with good cultivation, which one is feed. Currently, the main feed given by the cultivators still depend on trash fish. The purpose of this study was to increase formulated diets with different protein on feed consumption, feed conversion ratio, feed efficiency and nutrient retention.This study was conducted using Completely Randomized Design with 5 treatments and 4 replications, namely P0 : trash fish (control), P1 : protein feed 30 %, P2 : protein feed 40 %, P3 : protein feed 50 %, and P4 : protein feed 60 %. Lobster rearing is carried out for 30 days at the Faculty of Fisheries and Marine Airlangga University. Analysis of research data used Analysis of Variance (ANOVA) and continued with Duncan's Multiple Range Test with a level of 5% to determine the best results.The result showed that potency of formulated diets with different protein had a significant effect ($p < 0,05$) on feed consumption and nutrient retention, but had no significant effect ($p > 0,05$) on feed convertion ratio and feed efficiency.

1. Introduction

Scalloped spiny lobster (*Panulirus homarus*) is a commodity due to its high economic value [1]. The high selling price of lobster makes exploitatiom in nature very intensive. Based on data from the central statistics agency, the export of lobster seed in 2019 was 1.633.220 tons or US$ 33.189.390 million and in 2020 it increased to 2.150.420 tons or US$ 76.106.250 million [2]. Efforts that can be made to maintain the sustainability of lobster resources and long-term stock sustainability are by cultivating lobsters.

In the aspect of cultivation, feed is one of the most important factors. Currently, the main feed given by cultivators still relies on trash fish. Trash fish has a protein content of 57.46%, fat 7.40%,

IOP Conf. Series: Earth and Environmental Science 1273 (2023) 012075 doi:10.1088/1755-1315/1273/1/012075

carbohydrates 1.14%, water 13.20%, and ash 20.80% [3]. However, the feeding of trash fish has a weakness, namely easily deteriorate, high FCR, and depend on the season [4]. Therefore, to support the growth of lobsters can be made by development formulated diets is carried out based on the nutritional needs of fish protein content 42-47 % and the quality of raw materials that content high protein, namely snail flour and wild shrimp flour [5].

Snail flour and wild shrimp flour are relatively high sources of animal protein so that they can be used to replace fish meal in feed. Snail flour content protein nutrients 54,29 -64,14 % [6]. and protein wild shrimp flour 59,4 % [7]. The use of snail flour and wild shrimp flour in feed can support optimal growth. The optimal growth can be seen from the ability of fish to utilize feed nutrients. The utilization of feed nutrients can be seen from the level of feed consumption, feed conversion ratio, protein retention, fat retention, and energy retention in scalloped spiny lobster (*Panulirus homarus*).

2. Materials and methods

2.1 Place and time

The research was conducted in June to July 2022. The feed manufacture of treatment and maintenance was carried out at the Laboratory of Aquaculture and Anatomy, Faculty of Fisheries and Marine, Universitas Airlangga, Surabaya. Proximate analysis of feed ingredients was carried out at the Animal Feed Laboratory, Faculty of Veterinary Medicine, Airlangga University, Surabaya. Proximate Analysis of Animal Test Treatments at the Laboratory of Chemistry and Fish Food, Faculty of Fisheries and Marine Affairs, Universitas Airlangga

2.2 Tools and materials

The tools that used in this research are 20 aquariums with a size of 40 x 30 x 30 cm3, aerator pump, shelter pipe, aerator stone, aerator hose, tray, label paper, spoon, ruler, digital scale, thermometer, refractometer, pH meter, DO (Dissolved Oxygen) meter, stationery, test kit (nitrate, nitrite, and ammonia). The materials that used in this research are scalloped spiny lobster (*Panulirus homarus*) puerulus phase with a length of 1.56 and an average weight of 0.20 grams, snail flour, wild shrimp flour, corn flour, agar powder, fish oil, premix and kuniran trash fish.

2.3 Work procedures

First step before making feed is the ingredients in the proximate to determine the nutritional content. Then compile the pellet making formula with a linear programming method using solver software. The preparation of the formula in this study focuses on different protein content which refers to the protein needs of lobsters, which range from 47-53 % [5]. The preparation of the formula can be seen in table 1.

Table 1. Formulation of Feed

No	Materials	Proportion (%)				
		T0	**T1**	**T2**	**T3**	**T4**
1.	Trash fish	100	-	-	-	-
2.	Snail flour	-	9,87	24	37,35	82
3.	Wild shrimp flour	-	41	45	50	8
4.	Corn flour	-	38,13	20	3,00	3
5.	Agar powder	-	4,00	4,00	4,00	4,00
6.	Fish oil	-	4,00	4,00	4,00	4,00
7.	Vitamin Mix	-	1,50	1,50	1,50	1,50
8.	Mineral Mix	-	1,50	1,50	1,50	1,50
TOTAL			**100**	**100**	**100,85**	**104**

Protein content (%)	**62,06**	**30**	**40**	**50**	**60**
Fat content (%)	**11,98**	**8,15**	**9,45**	**10,82**	**10,34**
Carbohydrate content (%)	**9,41**	**30,75**	**19,51**	**8,64**	**14,69**
ME (Kcal/kg)	**3193,12**	**2839,58**	**2778,31**	**2737,97**	**3274,72**

Second step is to sterilize equipment, maintenance media, and seawater using chlorine. Next, the test animal used in this study was the scalloped spiny lobster (*Panulirus homarus*) puerulus phase. Lobster rearing was carried out for 30 days by feeding 15% of the biomass per day. Feeding was carried out 2 times a day at 09.00 and 16.00.

2.4 Research parameters and data analysis

The observed parameters were feed consumption, feed conversion ratio, feed efficiency, protein retention, fat retention and energy retention of scalloped spiny lobster. The data obtained were analysis of research data using Analysis of Variance (ANOVA) and followed by Duncan's Multiple Range Test with a level of 5% to determine the best results.

3. Results and discussion

3.1 Result

The results showed that the potency of artificial feed with different protein content had a significant effect on feed consumption and nutrient retention ($p < 0.05$), but had no significant effect on feed conversion ratio and feed efficiency ($p > 0.05$).

Table 2. The calculation result of Feed Consumption (FC), Feed Conversion Ratio (FCR), Feed Efficiency (FE), Protein Retention (PR), Fat Retention (FR), and Energy Retention (ER) of scalloped spiny lobster reared for 30 days.

Parameter	**Treatments**				
	T0	**T1**	**T2**	**T3**	**T4**
FC (gr)	$3,16 \pm 0,33^{b}$	$2,20 \pm 0,39^{a}$	$2,69 \pm 0,43^{abc}$	$2,91 \pm 0,47^{bc}$	$2,50 \pm 0,40^{ab}$
FCR	$2,56 \pm 0,40^{a}$	$2,59 \pm 0,25^{a}$	$2,62 \pm 0,55^{a}$	$2,36 \pm 0,13^{a}$	$2,55 \pm 0,22^{a}$
FE (%)	$39,88 \pm 6,99^{a}$	$38,92 \pm 3,63^{a}$	$39,50 \pm 8,28^{a}$	$42,50 \pm 2,35^{a}$	$39,43 \pm 3,77^{a}$
PR (%)	$38,36 \pm 6,49^{b}$	$23,05 \pm 2,37^{b}$	$16,35 \pm 3,44^{a}$	$13,25 \pm 0,69^{a}$	$12,82 \pm 1,16^{a}$
FR (%)	$77,03 \pm 11,70^{d}$	$23,93 \pm 2,48^{a}$	$21,54 \pm 4,58^{a}$	$19,19 \pm 0,90^{a}$	$18,56 \pm 1,60^{a}$
ER (%)	$51,34 \pm 8,37^{c}$	$11,29 \pm 1,18^{a}$	$11,55 \pm 2,68^{a}$	$12,71 \pm 0,64^{a}$	$11,46 \pm 0,98^{a}$

Values (mean ± SD) with different superscripts in the same row are significantly different ($p < 0,05$).

3.2 Discussion

3.2.1 Feed Consumption

Feed consumption rate is the final amount of feed consumed by fish during rearing. The high and low levels of feed consumption will affect growth [8]. Feed consumption in this study range from 2,20-3,16 grams while the research of Trisnasari.,2020 resulted in feed consumption between 13,15-16,54 grams [9]. The highest feed consumption rate is found in P0 with a value of 3.16 grams. This is due to the high feed consumption affected by the nutrients contained in the feed. Trash fish is one of the natural foods that has a protein content of 25-75%, rich in nutrients such as amino acids, energy, fatty acids, minerals and contains attractants [10]. Complete nutritional content and good feed palatability (texture, taste, size, and smell) favored by fish results in high feed consumption [11]. While the value of feed consumption decreased in protein (30% : 2.20 grams). The decrease is due to the high composition of ingredients such as carbohydrates, causing the fish to be full quickly and the process of absorption of nutrients is short [12].

IOP Conf. Series: Earth and Environmental Science 1273 (2023) 012075 doi:10.1088/1755-1315/1273/1/012075

3.2.2 Feed Conversion Ratio

Feed conversion ratio is an indicator to determine the utilization of feed given to the weight gain of fish during maintenance [13].The results showed that different proteins did not affect the feed conversion ratio (p>0.05). The highest feed conversion ratio value was in protein feed is 40 % (P2: 2.62) and the lowest was in protein is 50 % (P3:2.36). Protein is a nutrient that greatly affects the growth of fish weight. The amount of protein content in the feed will affect the high and low feed conversion ratio. Factors affect protein intake are protein consumption and protein digestibility, the higher the protein consumption and protein digestibility will cause a low feed conversion ratio [14].

Feed conversion ratio in this study range from of 2.36-2.62, while the result of research conducted by Nugraha *et al.,* 2019 show the value of the feed conversionratio ranging from 1,78-2,22 [15], and the research of Jones *et al* 2013 show the value ranging 3,3-4,5 [16]. This indicates that the value of feed conversion ratio in this study is relatively good because lobster can utilize feed to the maximum [15].

3.2.3 Feed Efficiency

Feed efficiency is the ratio of weight gain to the amount of feed consumed during maintenance which is expressed as a percent [17]. The results showed that different proteins did not affect feed efficiency (p>0.05). Feed efficiency in this study range from of 38.92-42.50%, these results was higher than research Anggraini *et al.,*2018 where the value of feed efficiency range from 15,27-27,83 % [18]. The highest feed efficiency value at 50% protein treatment (P3: 42.50%), the increase in efficiency value indicates that the feed consumed could be utilized properly, so that the protein was remodeled to meet the energy needs and the rest for growth. Factors that affect feed efficiency are protein quality, feed sources, and amino acid content [19]. The decrease in value of feed efficiency in the protein (30% : 38,92 %) is due to the small amount of feed consumed, so that the feed used to produce more weight results in less efficiency.

3.2.4 Protein Retention

Protein Retention is a comparison between the amount of protein stored and utilized in the form of tissue in the fish body with the amount of protein in the feed consumed during maintenance [20]. The results of average protein retention experienced an increase in the P0 (trash fish) treatment, which was 38.36%. High retention is caused by increased feed consumption and the amount of protein absorbed in the body [21].

The high retention value can be see from its amino acid composition which is suitable for biota needs. The treatment (P0: trash fish) contains a complete amino acid composition, namely arginine 9.34%; threonine 6.76%; isoleucine 5.00%; phenylalalin 4.06%; histidine 3.11%, lysine 21.22%; leucine 11.06%; valine 6.90%; alanine 12.52%; glycine 10.08%; proline 7.02%; serine 6.11%; and tyrosin 2.60 % [22]. The amino acid composition is in accordance with the amino acid needs of lobster, namely arginine 7.35%; threonine 5.52%; isoleucine 3.36%; phenylalalin 3.98%; histidine 4.42%, lysine 5.92%; leucine 8.29%; valine 3.15%; alanine 5.02%; glycine 11.02%; proline 5.62%; serine 5.30%; and tyrosin 2.41 % [23].

The value of protein retention that decreased in the P4 treatment was 12.82%. This causes the excess protein content in the feed given is not all used for protein synthesis, but the excess protein content is catabolized and excreted into ammonia. [24] high protein content does not always increase fish growth. An increase in feed protein without being followed by a balance of non-protein energy sources will cause protein to be used as an energy source. Excess protein intake in the feed results in protein being overhauled into energy, but not used for growth.

Protein retention in this study range from of 12,82-38,36 %, these results are almost the same as research Selva, 2008 where the protein retention values range from 9,08-12,93 % [25] and the research of Minarni *et al.,* 2017 show the value ranging 15,25-26,69 % [26]. The high value of protein retention is due to the suitability of the amino acd composition with the needs of the body biota, so that the feed can be digested into protein [25]

3.2.5 Fat Retention
Fat retention is the amount of fat stored and utilized in the form of tissue by the body of the fish divided by the amount of feed fat consumed. Fat is needed by fish as a source of energy and to maintain the shape and function of tissues [27]. Fat retention in this study range from 18.56-77.01%. The results of average fat retention increased in the P0 treatment (trash fish) which was 77.01%. The high value of fat retention is due to an excess of nutrients in the form of carbohydrates. Excess carbohydrate energy will be converted into body fat [28]. Sufficient non-protein energy content available in the treatment of trash fish causes protein to be utilized optimally for growth [10].

The average value of P4 treatment feed resulted in the lowest fat retention at 18.56%. The decrease in body fat retention of P4 treatment feed is caused because the feed consumed has a protein and non-protein balance that meets the needs of fish, so that the use of fat is more efficient as energy and as a result the fat deposited in the body is not high [27]. The low value of fat retention is caused because fat has been absorbed and used as an energy source which causes low fat contained in lobster meat [29].

3.2.6 Energy Retention
Energy retention is a comparison of the amount of energy stored in the fish body in the form of tissues with the amount of feed energy consumed The amount of feed consumption can affect of the contribution feed energy contribution [30]. The results showed that different proteins affect energy retention ($p<0.05$). Energy retention is in the range of 11.29-51.34%, the results were higher than research Selva, 2008 where the value range from 4,45-7,05 % [25]. The results of average energy retention experienced an increase in the P0 treatment (trash fish) which was 51.34%. The result of high energy retention means that feeding can be utilized properly by fish for growth. High energy retention occurs due to the balance of protein and energy levels. Energy needs are met, so proteins are not broken down and used as energy sources [31].

The results of average energy retention value in the P1 treatment decreased by 11.29%. This is because lack of energy can reduce the growth rate. Feed with a low energy content causes the use of part of the protein to meet energy needs, so that the amount of protein used for growth willdecrease. Therefore, the fulfillment of energy sources (protein and energy content) must be balanced [32].

4. Conclusions

The potency of formulated diets with different protein affects the level feed consumption with the value of the level feed consumption treatment of trash fish (P0) did not significantly different from protein treatment 50% (P3) and protein treatment 40% (P2), also the nutrient retention of scalloped spiny lobster with the highest value in the trash fish treatment (P0). While the formulated diets with different protein did not affect the feed conversion ratio with an average value ranging from 2.36-2.62 and the efficiency of scalloped spiny lobster with an average value ranging from 38.92-42.50 %.

5. References

[1] Asvin M R, Restu I W and Ekawaty R. 2019 *Curr Trends Aquat Sci.* 24 108-114.
[2] Ministry of Maritime Affairs and Fisheries 2020 Export-import data from the Ministry of Maritime Affairs and Fisheries Jakarta.
[3] Atifah Y 2016 *J Eksakta.* 1 42-49.
[4] Suharyanto 2012 *Maj Ilm Biol Biosf A Sci J.* 29 93-101.
[5] Radhakrishnan E V, Philips B P, and Achmveetil G 2019 *Lobster : biology, fisheries and aquaculture* (Singapore: Springer).
[6] Nasution H, Deliani W, Isnaniar, and Wahyunungsih 2017 *Phot J Sains dan Kesehat.* 7 115-123
[7] Hidayat M R 2014 *J Penelit dan Pengemb Borneo Akcaya.* 1 37-45

[8] Abidin Z, Junaidi M, Paryono, Cokrowati N and Yuniarti S 2015 *Depik* 4 33-39
[9] Trisnasari V, Subandiyono, and Hastuti S 2020 *J Sains akuakultur Trop.* 4 19-30
[10] Utomo N B P, Susan, and Setiawati M 2013 *J Akuakultur Indones.* 12 158-168
[11] Sumanteri A, Sukarti K, and Nikhlani A 2021 *J Sains dan Teknol Akuakultur.* 7 10-19
[12] Manganang Y A P, and Mose N I 2019 *J MIPA.* 8 116-121
[13] Mustofa A, Hastuti S and Rachmawati D 2018 *PENA Akuatika* 17 41-58
[14] Radona D, Subagja J and Kusmini I I 2017 *Media Akuakultur* 12 27-33
[15] Nugraha M D, Setyowati D N and Waspodo S 2019 *J Perikan.* 9 153-159
[16] Jones C and Anh L 2013 *Development of Land-Based Lobster Production Systems in Vietnam and Australia.* p 285 1-22
[17] Arief M, Fitriani.N and Subekti S 2014 *J Ilm Perikan dan Kelaut.* 6 49-53
[18] Anggraini W, Abidin Z, and Waspodo S 2018 *J Perikan.* 8 20–29
[19] Hutabarat G M, Rachmawati D and Pinandoyo 2015 *J Aquac Manag Technol.* 4 10-18
[20] Khalida A, Agustono A, and Lokapirnasari W P 2017 *J Ilm Perikan dan Kelaut.* 9 98-106
[21] Hasyim A Y R, Arief M, and Rahardja B S 2017 *J Aquac Fish Heal.* 7 1-9
[22] Zilda D S, Patantis G, Fawzya Y N, and Martosuyono P 2021 *JPB Kelaut dan Perikan.* 16 151-161
[23] Arumugam A, Dineshkumar R, and Rasheeq A A 2020. *Agric Sci Dig.* 40 411-417
[24] Tahapari E and Darmawan J *J Ris Akuakultur.* 13 47-56
[25] Selva D A 2008 *Effect of different artificial feed protein levels on protein retention and energy retention in freshwater crayfish (Cherax quadricarinatus)* (Malang : Brawijaya University)
[26] Minarni E, Yudha I G, and Suparmono S 2017 *Proc Semin Nas Perikan dan Ilmu Kelaut.* 45–58
[27] Subekti S, Prawesti M, and Arief M 2011 *J Mar Sci Technol.* 4 90-95
[28] Suryaningrum L H, Mulyasari, and Samsudin R 2017 *Ber Biol.* 16 157-165
[29] Kurniawan L A, Arief M, Manan A and Nindarwi D D2016 *J Aquac Fish Heal.* 6 32-40
[30] Yudiarto S, Arief M, and Agustono 2012 2017 *J Ilm Perikan dan Kelaut.* 4 135-140
[31] Ariyandra R, Agustono, and Satyantini W H 2017 *J Aquac Sci.* 1 10-18
[32] Tantri A F, Rahardja B S and Agustono 2016 *J Aquac Fish Heal.* 5 36-42

6. Acknowledgement

The author would like to thank Faculty of Fisheries and Marine, Airlangga University, for conducting the seminar.

Detection of Tilapia lake virus (TiLV) using semi-nested RT-PCR method in farmed Tilapia (*Oreochromis niloticus*) from ponds East Java, Indonesia

S R Dita[1], N F Laela[1], D W S Putri[2], Laminem[3], T W Anita[3], R Laili, E Yohan[3], Rozi[2],[3]

[1]Program of Study of Aquaculture, Faculty of Fisheries and Marine, Universitas Airlangga, Surabaya 60115, Indonesia
[2]Department of Aquaculture, Faculty of Fisheries and Marine, Universitas Airlangga
[3]Balai Karantina Ikan Pengendalian Mutu dan Keamanan Hasil Perikanan Surabaya II, East Java, Indonesia

[3]Corresponding author: rozi@fpk.unair.ac.id

Abstract. Tilapia lake virus disease is an emerging viral disease causing mortality up to 90% worldwide of aquaculture production. It has been recorded as a virus attacting wild and cultured tilapias in Asia, Africa, and Latin America. There is the urgent need for rapid Diagnostic techniques for viral diseases to identify infected tilapia to control the outspread in individual farms. This study aims to identify the presence of TiLV with the semi-nested Reverse Transcriptase-Polymerase Chain Reaction method allowing the rapid detection of TiLV in fish organs. A total of 4 samples of tilapia were isolated from different locations and then used a silica extraction kit to obtain RNA from the TiLV virus. Afterwards amplification step was carried out by semi-nested RT-PCR test. The primers used in PCR target the 250-bp fragment in the electrophoresis. The results showed that the 250 bp amplicon of TiLV RNA was detected at 1/4 tested Tilapia. PCR clearly identified infective TiLV viral RNA in tilapia that produced a cytopathic effect with protein similarity (weak homology) in the conserved Polymerase Sub Unit PB1 Influenza-C and the genome of Orthomyxoviruses causing brain, nervous system and liver damage in fish and may have applications as rapid species-specific virulence test.

1. Introduction

In the past decade, the production of fish acuaqulture in Indonesia has decreased significantly due to outbreaks of infectious viral diseases. Viral disease is an important disease that has the potential to threaten death in the priority commodities aquaculture production such as Tilapia lake virus (Tilapia), Koi Herpesvirus (carp and koi), Infectious Myonecrosis Virus, Whitespot Baculovirus (*Litopenaeus vannamei*), Viral Nervous Necrosis (Groupers) and others [1-6]. However, Tilapia Lake Virus is the dominant pathogenic infects Tilapia and has caused disease and mass death of populations of this species both in culture and in the wild.

The tilapia lake virus (-ssRNA) is a particular virus for tilapia (*Oreochromis niloticus*), and it has an icosahedron with a diameter of 55–100 nm [1], consisting of ten genomic fragments, ranging in size from 465 to 1641 nucleotides, make up this organism. With a total length of 10,323 kb [8] and weak similarity to the influenza C virus subunit PB1 polymerase [7], the first segment encodes a protein. Based on the physical similarities of its first segment [8,9], this virus was initially categorized as a novel Orthomyxo-like virus from the family Orthomyxoviridae. It also has a round to oval and filamentous/tubular envelope [10]. The only species of the genus Tilapinevirus and the only member of the new family Amnoonviridae, Tilapia tilpinevirus, is the official classification as of right now [7,8,10,11]. With a peak mass fatality at 30°C, this virus is active in infecting fish at water temperatures > 25°C [12]. The histological and clinical symptoms of an infected tilapia include stomach enlargement, severe anemia, bilateral exophthalmia, anorexia, poor body condition, aberrant swimming, skin erosiveness and congestion, and protruding scales [9]. Tilapia target organs for TiLV virus detection include the brain, eyes, spleen, kidneys, and liver [13].

The global tilapia industry is at risk from the spread of this virus, which can kill large numbers of fish within a month of being moved from hatcheries to rearing facilities [15]. Other fish species, such as wild carp (*Barbonymus schwanenfeldii*) [16] and giant gourami (*Osphronemus goramy*) [17,18] or healthy fish through water in two to three days with a mortality rate of 50%, have also tested positive for TiLV. The death rate of tilapia as a result of this virus typically varies from 20 to 90% [18,19]. The TiLV virus is a member of the HPIK 1 group, according to KEPMEN-KP number 17 of 2021, and as a result, TiLV surveillance must be started in an effort to stop the spread of the virus and its unfavorable effects [8]. Therefore, quick action is required, including the destruction of diseased stocks, the implementation of quarantine measures, trade prohibitions, and the control of potential vectors using the semi-nested reverse transcription-PCR (RT-PCR) assay method [1].

2. Methodology

2.1. Preparation Sampel

A collection of 4 samples of tilapia taken from the necropsy room of the Fish Quarantine, Quality Control and Safety of Fishery Products Surabaya II. Target organs taken from tilapia for TiLV test were gills and kidneys. then put it in a plastic sample that has been given a sample code.

2.2. Extraction of TiLV virus RNA from Tilapia samples(Extraction using a silica extraction kit)

Extraction was started by inserting the sample into a 1.5 ml microtube containing 900µl GT Buffer. Puree the sample with a pestle. then centrifuged at 12,000 rpm for 3 minutes. After that, add 40 l of silica which has been previously vortexed into a new microtube. then transferred 600 l of supernatant into a microtube containing silica and centrifuged at 12,000 rpm for 15 seconds. then discard the supernatant. After that, washing the silica with 500µl GT Buffer and vortexing until all pellets are mixed then centrifuged at 12,000 rpm for 15 seconds and discarding the supernatant. After that, add 1 ml of 70% ethanol to wash the silica pellets and vortex again until the pellets are all mixed. After that, centrifuge at 12,000 rpm for 15 seconds and discard the supernatant. The next step, add 1 ml of DEPC ddH2O to dissolve the silica pellet, vortex again until all the pellets are mixed. Then incubate the microtube at 55°C for 10 minutes. vortex then centrifuge at 12,000 rpm for 2 minutes. Transferred 500µl of supernatant to a new microtube. and ready-to-use templates for advanced applications.

2.3. Amplification

Amplification was carried out in a double step (semi nested), starting with coding for each microtube according to the sample code, then in each sample tester using one microtube each for TiLV control (positive sample standard) and one microtube for TiLV negative control (ddH2O/Nuclease Free Water). then fill each tube for First Step PCR, with a total reaction of 25µl consisting of 12.5µl PCR 2x mastermix solution, 1µl Primer First Forward (Nested ext-1): 5'-TAT GCA GTA CTT TCC CTG CC-3' , 1µl primer MEI : 5'- GTT GGG CAC AAG GCA TCC TA-3', 2µl RNA template, 0.5 AMV Reverse Transcriptase Enzyme, 8µl ddH2O (add up to a total volume of 25µl). After that, ensure that all liquid is at the bottom of the tube and minimized the presence of bubbles or fluid buildup on the microtube wall. Then put the sample in the thermal cycler and run according to the program (TiLV F). First step RT-PCR amplification temperature settings: 1 cycle (50°C-30 minutes), 1 cycle (94°C-2 minutes), 30 cycles (94°C-30 seconds; 60°C-30 seconds; 72° C-30 seconds), 1 cycle (72°C-5 minutes), 1 cycle (12°C-∞). Semi nested PCR, for a total of 25µl semi nested PCR reactions consisting of 12.5µl PCR 2x master mix solution 1µl Primer Nested Forward (7450/150R/ME2) : 5'- TAT CAC GTG CGT ACT CGT TCA GT-3', 1µl, Primer First Reverse (MEI) :5'- GTT GGG CAC AAG GCA TCC TA-3'. Then ensure that all liquid is at the bottom of the tube and minimize the presence of bubbles or fluid buildup on the walls of the microtube. After that, put it in the thermal cycler and run it according to the program (TiLV N). Semi-nested amplification temperature settings: 1 cycle (94°C-2 minutes), 25 cycles (4°C-30 seconds; 50°C-30 seconds; 72°C-30 seconds), 1 cycle (72°C-5 minutes), 1 cycle (4°C-∞). The final result of the semi-nested amplification was obtained as much as 10 µl of amplicon volume.

2.4. Electrophoresis
The results of the amplification were electrophoresed using 1.5% agarose gel. Agarose gel by dissolving 3 grams of agarose into 200 ml of TAE 1x (Tris Acid EDTA) and made until dry/homogeneous. Once warm, enter Sybr Safe 4 µl and print in agarose. As much as 10 l of the amplification was put into the agarose well. For each test, 100 bp marker was added as a DNA marker by inserting 5 µl of agarose into the well. Electrophoresis was performed at a minute power of 115 V (for ± 45) and then observed using a UV transilluminator.

3. Result and discussion
3.1. Result
3.1.1 Clinical signs
In Indonesia, in recent years there have been cases of mass deaths in tilapia cultivation, especially in Sumatra, Java, Bali and Lombok. Although the cause of death is not yet known, when viewed from the clinical symptoms seen, externally, tilapia showed wounds on the skin, change the color of the fish to dark, pale gills, hemorrhage (bleeding) on leptomeninges, spleen and kidney congestion it may be caused by infection with the TiLV (Figure 1).

Figure 1. Blood loss on the leptomeninges and body discolouration were the symptoms seen in infected tilapia (bold arrow).

3.1.2 Detection of TiLV gene fragments
TiLV RNA PCR results in five samples of tilapia using the Semi-Nested RT-PCR method identified positive TiLV fish and revealed that 25% of the samples exhibited weak homology (similarity in protein structure) and weighed 250 thing (Figure 2).

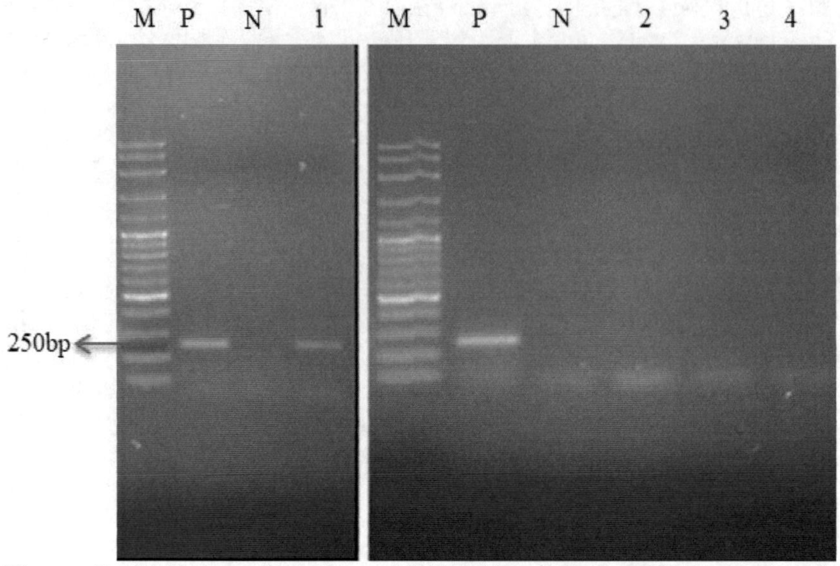

Figure 2. TiLV gene detection and identification through electroforesis of PCR test fragments. M stands for the 100-bp ladder and is used as a marker. Lanes P and N represent the positive and negative controls, respectively. Lanes 1 and 2 represent the sampely NT1 and NT2, while Lanes 5 and 6 represent sampely NT3 and NT4, respectively.

According to do figure above, a band parallel to TiLV positive control was formed on sequence 250 bp which signifies that the tilapia fish was positive with TiLV albeit only mildly. This result is in accordance to Dong [19] who stated that minor infection may induce DNA band on 250 bp. On the other hand, severely TiLV infected tilapia will form a band on 450 bp.

3.1.3 Etiology
The genome of the ssRNA virus tilapia lake virus has ten segments that are between 55 and 100 nm in diameter and 10.323 kb in length. Ten segments revealed an improbable gene connection with various viruses from the Orthomyxoviridae family. Parts 1, 2, 4, and 5 all displayed similarities to the dhori virus. The influenza virus was comparable to segments 6, 7, and 8. Segment 3 only occasionally shared weak homologies with the salmon anemia virus [20].

3.1.4 Analysis on the extraction process of TiLV virus examination
Extraction is a purification method to separate DNA/RNA from cell membrane, protein, and other cellular components through methods such as organic ion extraction (phenol-chloroform method), non-organic (salting and K proteination treatment), as well as adsorbtion (silicate gel membrane). In BKIPM Surabaya II molecular biology laboratory, the preferred method was the silicate gel membrane method, this is due to the high quality and quantity of DNA it produced, ease of use, economical, and could bind both DNA and RNA [21]. As for the process of extraction, it's begun by sampling a little bit of the target organ and muddling it in a pestle in order to make it homogenous, increasing the number of lysed cells, as well as to extract the nucleic acid from the tissue/organ. It was then added with 900 µl of GT Buffer which functions as a separator of the nucleic acid and cell lysis, due to its

hypotonic nature. Next, the mixture was centrifuged for 3 minutes on 12,000 rpm to separate the nucleic acid and continued with the preparation of silicate kit as much as 40 μl pada in a clean microtube. As much as 600 μl of sample supernatant were added, the silicate immediately binds the DNA and RNA, effectively extracting it, as well as separating it from lipids, protein, and other contaminants. The mixture was then centrifuged, the supernatant was disposed and the remaining sediment was added with 500 μl GT Buffer in order to wash the silicate kit. Back to the centrifuge, the supernatant was disposed of again and the sediment was added with 1 ml of ethanol in order to inactivate nuclease that could denature/damage DNA/RNA. The mixture was then centrifuged again, supernatant disposed, and sediment added with DEPC ddH2O which serves to re-dissolve the DNA pellets. The mixture was then incubated on a thermoblock for 10 minutes on 55°C to help with DNA pellet dissolving and the inactivation of nuclease. The final result of the extraction is a DNA/RNA template ready to be amplified.

3.1.5 Analysis on the amplification and electrophoresis process of TiLV virus examination
Amplification is a technique to duplicate DNA sequence from extraction components for the purpose of clearer and easier analysis [22]. TiLV is an RNA virus which requires the Reverse Transcriptase-Polymerase Chain Reaction (RT-PCR). RT-PCR is done through a reversed transcription of nRNA molecules into complementary DNA (cDNA) molecules. This specific process requires various components such as RNA Access quick master mix go taq green, primer, AMV Reverse Transcriptase enzyme, NFW (*Nuclease Free Water*), and RNA *template.* Master mix is a reagent mixture composed of DNA taq polymerase, dNTPs (dATP, dGTP, dCTP, dTTP), $MgCl_2$ as well as loading dye. DNA taq polymerase serves as an enzyme to synthesize new cDNA; dNTPS and $MgCl_2$ buffer supports the chemical environment needed for the process. AMV Reverse Transcriptase enzyme reverse transcribes the mRNA into complementary DNA (*cDNA*). TiLV uses the semi double step method because one of the primer pairs used was *Nested ext-2* [23].

3.1.6 Analysis on the electrophoresis process of TiLV virus examination
Electrophoresis is used to study the properties of single loaded species and as a separation technique. The usage of agarose gel is effective in separating DNA/RNA fragment with size from 100 bp up to 25 kb. 2.5% agarose gel used in this study was diluted with TAE Buffer 1x solution. TAE Buffer is a standard buffer containing weak acids, acetic acid with neutral and anionic ions. As a pH buffer, it keeps the conductivity of the media and supports the DNA fragment resolution excellently [24].

3.2. Discussion
The results of molecular analysis using semi-nested Real Time PCR showed that the death of tilapia (*Oreochromis niloticus*) that occurred suspiciously was caused by many factors and not necessarily due to TiLV infection. The target molecular weight of 250 bp indicates that tilapia is positively infected with TiLV. The second parameter observed was the clinical symptoms and behavior of the fish samples. The observed samples of tilapia showed symptoms of wounds on the skin, change the color of the fish to dark, pale gills, hemorrhage (bleeding) on leptomeninges, spleen and kidney congestion. This is in accordance with the general characteristics if fish are indicated to be infected with Tialpia Lake Virus (TiLV) such as showed infections in fish, includes body discoloration, exophthalmic or blocked eyes, loss of scales, and a pale and enlarged liver [16,25].

The development of TiLV is strongly influenced by the environment, where Indonesia has a tropical climate which is very conducive to the development of TiLV. However, the triggering factors for TiLV disease outbreaks are water temperature, poor water quality, and high density. The temperature range is in accordance with the standard temperature for rearing tilapia of 28°C. An increase in temperature will increase the need for maintenance energy and fish will be more active in looking for food. The pH value that can be tolerated by tilapia ranges from 6 - 9, but that causes outbreaks to occur above 25^{0}C, temperatures that do not cause death are below 20^{0}C, and the peak of mass mortality occurs at 30^{0}C [12]. Outbreaks of TiLV in various countries Clinically have been reported during the summer months, i.e. May to October (at water temperature 22^{0}C to 32^{0}C) in Israel [1], June to October (≥25^{0}C) in Egypt [26] to November (25^{0}C to 27^{0}C) in Ecuador [27]. In Thailand, several samples that led to positive TiLV detection were gathered between October and May [15].

Poor water quality can encourage the spread of infections in fish and the environment, which can harm fish physically and induce behavioral and physiological abnormalities as well as immune system suppression. Because it occurs in the summer, the high mortality rates reported in Israel [1], Egypt [26], and Ecuador [27] were referred to as the "summer death syndrome," which suggests that rising water temperatures may be a significant factor. In addition, there is no statistically significant relationship between monthly water temperature and the presence of TiLV. On the other hand, there is a strong link between tilapia infections with various Aeromonas spp. and poor water quality. Multiple infections, which happen when the host is infected by two different pathogens, are rather common in fish.

4. Conclussion

According to the study's findings, 1/4 of tilapia (*Oreochromis niloticus*) were TiLV-positive using the PCR method, which revealed an amplicon of 250 bp. This is a serious concern considering that Indonesia is the second largest aquaculture in the world.

5. References

[1] Eyngor M, Zamostiano R, Tsofack J E, Berkowitz A, Bercovier H, Tinman S, Lev M, Hurvitz A, Galeotti M, Bacharach E, and Eldar A 2014 *J Clin Microbiol.* 52 4137–4146.
[2] Pradaan M S , Suprapto H, and Suwarno S 2015 *JIPK.* 7(1): 39-45.
[3] Baladrat N K, Nurhudah M and Utari H B 2022 *JIPK.* 14(1): 83–92.
[4] Rani Y, Nanik R B, and Handian F E P 2018 *JIPK.* 10(2): 124-130.
[5] Fitriatin E, and Abdul Manan 2015 *JIPK.* 7(2): 149-152.
[6] Nurani F R, Suprapto H, and Suwarno S 2014 *JIPK.* 6(2): 102-115.
[7] Bacharach E, Mishra N, Briese T, Zody M C, Tsofack J E, Zamostiano R, Berkowitz A, Ng J, Nitido A, Corvelo A, Toussaint N C, Abel Nielsen S C, Hornig M, Del Pozo J, Bloom T, Ferguson H, Eldar A, and Lipkin W I 2016 *M Bio.* 7 e00431–00416
[8] Debnath P P, Delamare-Deboutteville J, Jansen M D, Kornsunee P, Afsana D, Abir H, Saengchan S, Chadag V M, Ha Thanh D, and Channarong R 2020 *J Fish Dis.* 43 1381–1389.
[9] Tattiyaponga, Worawan D, and Win S 2017 *Vet Micro.* 207 170–177.
[10] Del-Pozo J, Mishra N, Kabuusu R, Cheetham S, Eldar A, Bacharach E, Lipkin W I, and Ferguson H W 2017 *Vet Pathol.* 54(1): 164-170.
[11] ICTV Master Species List 2018bv1 2018.
[12] Tsofack J E K, Rachel Z, Salsabeel W, Asaf B, Ezra R, Nischay M, Thomas B, W Ian L, Richard M K, Hugh F, Jorge del P, Avi E, and Eran, B 2016 *J Cli Microbiol.* 55: 759–767.
[13] Husna N, Hapsari K, Mohammad F U, Arif H F, and Dwi L 2020 *JAS.* 5(2): 77-87.
[14] Jansen M D, Dong H T, and Mohan, C V 2018 *Rev in Aqua*. 11: 1–15.
[15] Surachetpong W, Janetanakit T, Nonthabenjawan N, Tattiyapong P, Sirikanchana K, Amonsin A 2017 *Emerg Infect Dis.* 23(6): 1031–1033.
[16] Abdullah A, Ramly R, Ridzwan M S M, Sudirwan F, Abas A, Ahmad K, Murni M, and Kua B C 2018 *J. of Fish Dis.* 41(9), 1459-1462.
[17] Chiamkunakorn C, Machimbirike V I, Senapin S, Khunrae P, Dong H T, and Rattanarojpong T 2019 *J. of Fish Dis.* 42(11): 1629-1636.
[18] Piamsomboon P, and Wongtavatchai J 2021 *JFAC.* 21(4): 205-209.
[19] Dong H T, Siriro S, Meemetta W, Santimanawong W, Gangnonngiw W, Pirarat N, Khunrae K, Rattanarojpong T, Vanichviriyakit R, and Senapin S 2017 *Aquaculture* 476, 111–118.
[20] Aich N, Paul A, Choudhury T G, and Saha H 2021 *Aquaculture and fisheries* 7(1) pp. 7-17 ISSN:2468-550X.

[21] Ariyanti Y, and Sianturi S 2019 *J. Sci & App Tech.* 3(1): 40-4.
[22] Rahmadhan D 2019 *JMFK. UNTAN* 4(1).
[23] Koesharyani I, Gardenia L, Widowati Z, Khumaira K, and Rustianti D 2018 *JRA.* 13(1): 85-92
[24] Kurniawan E 2022 *MMLS.* 6(1): 16-22.
[25] Jamal K, Rahmatia F, and Dhewantara Y L 2019 *JSM.* 4(2): 115-121
[26] Nicholson P, Fathi M A, Mohan C, Schieck E, Mishra N, Heinimann A, Frey J, Wieland B and Jores J 2017 *J Fish Dis.* 40(12): 1925-1928.
[27] Ferguson H W, Kabuusu R, Beltran S, Reyes E, Lince J A, and del Pozo, J 2014 *A case report J Fish Dis.* 37(6): 583-589.

6. Acknowledgments

For their invaluable assistance with fieldwork and sample collection, we are grateful to our team of researchers, which includes both laboratory technicians.

The potential of *Moringa oleifera* extract as a phytobiotic candidate for the growth of *Lactobacillus fermentum* bacteria by in vitro

D F Arumdani[1], W P Lokapirnasari[2], S Sarudji[2], M Lamid[2], S Chusniati[2], I S Hamid[2]

[1] Magister of Veterinary Agribusiness, Airlangga University, Indonesia
[2] Faculty of Veterinary Medicine, Airlangga University, Indonesia

[3] Corresponding author: widya-p-l@fkh.unair.ac.id

Abstract. *Moringa oleifera* is used to explore the biodiversity of the ability of herbal plants as a candidate phytobiotic. The total carbohydrate content in the ethanol extract is 42% consisting of monosaccharides, disaccharides, and oligosaccharides which are gut micro-flora nutrient. The study aimed to discover the effect of moringa leaf extracts with different dosages to improve *Lactobacillus fermentum* bacteria's growth. The study employed moringa leaf extracts and *Lactobacillus fermentum* bacteria then divided them into four groups, i.e., control group (P0), treatment P1 with moringa leaf extracts of 0.1% dosage, treatment P2 with moringa leaf extracts of 0.2%, and treatment P3 with moringa leaf extracts of 0.3% dosage. At the end of the treatment, each colony was calculated using the Total Plate Count method. Data obtained were analyzed using ANOVA and subjected to the Duncan test. The study results show that moringa leaf extracts with dosages of 0.1%, 0.2%, and 0.3% could improve *Lactobacillus fermentum* bacteria's growth. The average growth log of *Lactobacillus fermentum* bacteria on treatment P0 was 3,532 log CFU/ml, 4,320 log CFU/ml for treatment P1, 4,368 log CFU/ml for treatment P2, and 4,480 log CFU/ml for treatment P3 as the highest value.

1. Introduction

Growing world population is expected to increase the demand for fishery in domestic and international markets [1]. In some cases cheap and easy fish are available and the demand for fish has intensified aquaculture activities. Infection is a major problem in the robbery and trade of aquatic products causing huge losses in the aquaculture industry. The expansion of intensive aquaculture leads to higher levels in different types of fish to identify or understand diseases [2]. It has been demonstrated that the incidence of fish diseases is related to several factors related to conservation and environmental conditions. As a result, farmed fish can become more susceptible to not only pathogens but also opportunistic bacteria [2,3].

The importance of food security follows the increased community awareness of quality aquatic feed. Besides its high nutritional value, the product should be free of microbe contaminants, chemicals, or other materials that harm the health. Following this reason, Antibiotic Growth Promotor (AGP) utilization is prohibited. Hence, other alternatives are necessary to replace AGP utilization. Phytobiotic as feed additives could be used for antibiotic reduced in fisheries [4].

Phytobiotics are plants that contain chemical compounds that are beneficial to living organisms. In addition phytobiotics play a dual role as feed additives in feed formulations to increase productivity. Food additives are substances that are not naturally present in feed and are added for a specific purpose. Phytobiotics are appetite stimulants and growth promoters that generally stimulate the secretion of digestive enzymes and directly affect the intestinal flora. Plant extracts have been shown to improve nutrient digestibility and bioavailability resulting in improved feed conversion improved protein synthesis and improved fish growth [5]. Raw materials for phytobiotics can be herbal extracts or plant parts that have the highest concentration of biologically active substances [6].

The Lactic Acid Bacteria (LAB) group, particularly the Lactobacillus genus can found in fish digestive tract. The lactic acid bacteria that live in the intestines of fish can stimulate the immune system and protect against disease [7]. Lactobacillus can produce lactic acid and inhibit the growth of pathogenic bacteria by creating acidic conditions [8]. An acidic environment has a beneficial effect on good bacteria and a toxic effect on pathogenic bacteria.

Moringa plants are used to explore the biodiversity of herbal plants as the phytobiotic candidate. Moringa plant (*Moringa oleifera* Lam.) is one of the plants with high carbohydrate concentration. The total carbohydrate concentration in the ethanol extract is 42%, consisting of monosaccharides, disaccharides, and oligosaccharides, i.e., gut micro-flora nutrient [9]. Phytochemical compounds in *Moringa oleifera* are glycosides, triterpenes, flavonoids, luteolin, quercetin, tannins [10], also has proteins, fats, fibers, calcium, magnesium, phosphorus, potassium, copper, iron, vitamins, and essential amino acids [11].

The addition of living microorganisms and phytobiotic candidate for bacterial growth, such as *Lactobacillus fermentum* with *Moringa oleifera* extract, will benefit from this combination because it improves the life expectancy of that bacteria since the specific substrate is available for fermentation. The purpose of this study was to confirm the efficacy leaf extract of *Moringa oleifera* as a phytobiotic candidate for in vitro bacterial growth of *Lactobacillus fermentum.*

2. Material and methods

The study design was Complete Randomized Design, which comprised four groups, i.e., P0, P1, P2, and P3, with five repetitions. The treatment in this study was different dosages of moringa leaf extracts on each growth medium of *Lactobacillus fermentum* (collected by WP Lokapirnasari).

2.1. *Preparation of Moringa Leaf Extract*

Preparation of moringa leaf extract was acquired from powder moringa leaf samples for 2 kg. The powder was processed with the maceration method by inserting it into an Erlenmeyer flask and adding ethanol solution 70% for 6 liters into the flask. Leaf extraction was carried out using ethanol solution 70% to obtain oligosaccharide substances. The maceration was performed for 3 x 24 hours and stirred once per day. Then, filtration was carried out using the Buchner funnel that contains a filtering paper, obtained macerates. The macerates were put into a rotary evaporator of 60°C with a 50 rpm until all solvents and water content was separated, generating concentrated *Moringa oleifera* leaf extracts.

2.2. *Initial Calculation of Lactobacillus fermentum Bacteria Cells*

The initial calculation of *Lactobacillus fermentum* bacterial cells was carried out using the Mc Farland standard method. A bacterial suspension tube containing colonies of *Lactobacillus fermentum* bacteria and aquadest 10 ml were compared with the Mc Farland standard with a 0.5 scale and cell density estimation of $1{,}5 \times 10^8$ cell/ml. The bacterial suspension was formulated by mixing aquadest of 10 ml and colonies of *Lactobacillus fermentum* bacteria taken using a sterile needle to obtain the same turbidity as the Mc Farland standard with a 0.5 scale. After reaching the same turbidity, dilution was conducted by adding 5 ml aquadest to obtain bacterial suspension with a density estimate of 10^8, then conducting a gradual dilution. A 10^{-1} dilution was carried out by taking 1 ml from the 10^8 tubes and put it into another tube with 9 ml aquadest, then retook another 1 ml of the 10-1 dilution tube to perform the 10^{-2} dilution in the tube with 9 ml aquadest. One ml was taken from the 10^{-2} tube containing 9 ml aquadest and performing the same treatments on 10^{-4}, 10^{-5}, 10^{-6}, and 10^{-7} dilutions. On the 10^{-7} dilution, the cell density estimate was 10^1 cell/ml, then 1 ml of *Lactobacillus fermentum* bacteria was put into 9 ml of MRS Broth media.

2.3. *Treatment*

After culturing isolate in MRS Broth media, leaf extracts of *Moringa oleifera* were added as the phytobiotic candidate with concentrations as stated by the treatment. P0 was in the absence of *Moringa oleifera* leaf extracts, P1 was added with 0.1% of Moringa extracts, P2 was added with 0.2% of Moringa extracts, and P3 was added with 0.3% of Moringa extracts. The incubation of tubes containing MRS Broth, isolate bacteria, and Moringa extracts was carried out for 12 hours.

2.4. *Calculation of Bacteria using the Total Plate Count (TPC) Method*

Calculate isolate bacteria was conducted by a pour plate method in MRSA by taking 1 ml of bacterial suspension and putting it into an empty petri dish sterile, adding liquid MRS media incubating it for 12 hours on 37°C. Calculation of isolate bacterial colonies with the TPC method was carried out on the 12 hours. Isolate bacteria growing after 12 hours were calculated using the Standard Plate Count (SPC), where colonies were calculated in the petri dish containing 30 to 300 colonies. The number of colonies growing on the petri dish was multiplied with a dilution factor.

3. Result and discussion

Test results using the Analysis of Variance (ANOVA) for the growth of *Lactobacillus fermentum* bacteria was $p<0.05$; hence, it can be stated that each treatment had a real influential result and should be followed with the Duncan follow-up test with a 5% significance level to observe the differences of each treatment group. The Duncan follow-up test demonstrated significant differences in the growth of *Lactobacillus fermentum* bacteria on each treatment. At 12 hours incubation showed that the addition of 0.3% *Moringa oleifera* extract indicated the highest growth than the control and the addition of 0.1% and 0.2% *Moringa oleifera* extract.

3.1. *The growth of Lactobacillus fermentum*

Table 1. Average log of the growth calculation result of *Lactobacillus fermentum* bacteria on each treatment with a 10^{-2} dilution factor on the MRS medium on 37°C within 12 hours of incubation.

Treatment	Average log TPC ± SD (log CFU/ml), 12 hours of incubation
P0	$3.532^{a} \pm 0.033$
P1	$4.320^{b} \pm 0.035$
P2	$4.368^{c} \pm 0.011$
P3	$4.480^{d} \pm 0.000$

Note: Superscripts a, b, c, and d on the same column shows real differences on each treatment

The average growth log of Lactobacillus fermentum in the data can be seen that the lowest growth in P0 was 3.532a ± 0.033 log CFU/ml, followed by P1 at 4.320b ± 0.035 log CFU/ml, and P2 at 4.368c ± 0.011 log CFU/ml. Then P3 treatment with 4.480d ± 0.000 log CFU/ml was the highest number observed. In Figure 1, it can be seen that the more the dose of the extract increases, the bacterial growth will increase as well.

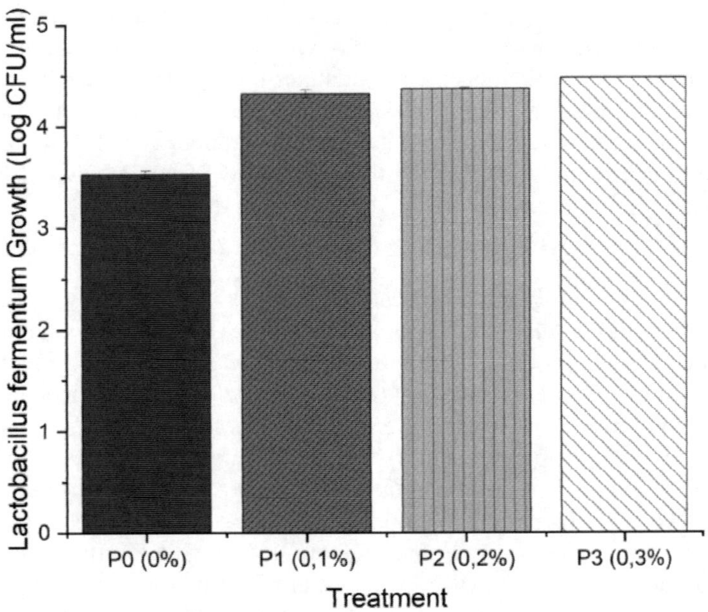

Figure 1. The growth of Lactobacillus fermentum bacteria on P0, P1, P2, and P3 treatment with a 10^{-2} dilution factor on the MRS medium on 37°C within 12 hours of incubation.

Moringa leaf extract concentration added into the MRS Broth media influenced *Lactobacillus fermentum* bacteria's growth, illustrated by the increased colony growth during the same incubation time, i.e., 12 hours. The study results show significant differences in *Lactobacillus fermentum* bacteria's growth in each treatment group, i.e., P0, P1, P2, and P3. *Lactobacillus fermentum* bacteria's average growth shows the lowest growth on group P0 since it was incubated without moringa leaf extracts. Treatment group P1 showed growth improvement of *Lactobacillus fermentum* bacteria. However, it was lower than treatment groups P2 and P3. Treatment group P3 showed the highest growth improvement of *Lactobacillus fermentum* bacteria than treatment groups P1 and P2.

In this study, P0 was without Moringa oleifera leaf extract, P1 was given 0.1% Moringa oleifera leaf extract, P2 was given 0.2%, Moringa oleifera leaf extract and P3 was given 0.3% Moringa oleifera leaf extract. MRS Broth media contains several compositions that support the growth of Lactobacillus fermentum bacteria. Yeast extract contained in MRS Broth media have substance amino acids, peptides and vitamins which are very beneficial for growth. Peptone and triammonium citrate in MRS Broth media are sources of nitrogen while glucose and sodium acetate are sources of carbon. Then, the functions of Lab lemco powder in MRS Broth media as a source of vitamins B for microorganisms that promote growth for cell proliferation. MRS broth media also contains KH_2PO_4. The function of KH_2PO_4 as a buffer to balance pH. The potassium and phosphate content in KH_2PO_4 is used for nourishing microorganism cells as well as a key element in controlling cell metabolism, transport processes, and is needed in carbohydrate metabolism. KH_2PO_4 contains high K elements and as a source of P. Magnesium sulfate functions as a source of S which plays a role in ribosome stabilization, membrane stabilization and cell walls, while sorbitan monooleate, also known as tween 80, acts as an emulsifying agent. The nutrients contained in the MRS Broth media caused the Lactobacillus fermentum bacteria to grow in all treatments, including the control treatment which did not add Moringa extract.

The difference in the number of colonies in each treatment results from the addition Moringa extract with different concentrations. The difference in the added dose affects the nutrients contained in the MRS Broth media. The leaf extract of *Moringa oleifera* contains carbohydrates as a source of carbon, protein as a source of nitrogen, and various vitamins, including vitamin A, vitamin B, and vitamin C. That leaf extract also contains calcium, magnesium, phosphorus, sulfur, and potassium as a mineral source. The addition leaf extract of *Moringa oleifera* to P1, P2, and P3 resulted in different nutrients in each treatment because the dose leaf extract that added was also different. The increase in the number of colonies occurred due to the addition of nutrients from *Moringa oleifera* extract which functions as a source of carbon, nitrogen, vitamins, and minerals that support the growth of *Lactobacillus fermentum* bacteria.

3.2. Moringa oleifera as a phytobiotic candidate
In a study by Anudeep and Radha [9], *Moringa oleifera* leaf extracts contained oligosaccharides. Fructooligosaccharides and malto-oligosaccharides also have a significant effect of stimulating the growth of BAL organisms [12]. *Lactobacillus fermentum* bacteria is a lactic acid bacterium, that can catalyze carbohydrate polymers and utilize them as an energy source. Carbohydrates as a carbon source will be degraded into simple substances, such as disaccharides and monosaccharides. Then, monosaccharides will be metabolized by *Lactobacillus fermentum* bacteria to be organic acids such as lactic acid or acetic acid. Carbohydrate metabolism by lactic acid bacteria occurs by glycosidic-breaking enzyme aid, e.g., glycosyl hydrolase [13]. According to Azmi et al [14], organic acid formulation from microbe metabolism will lower fermentation media's pH value following increased bacterial biomass. This study indicated that substances from *Moringa oleifera* could stimulate the growth of bacteria. *Moringa oleifera* leaves contain various amino acids and micronutrients, including the B vitamins [15, 16]. Most Lactic Acid Bacteria (LAB) species, including *Lactobacillus fermentum*, require various amino acids and vitamins as nutritions.

4. Conclusion
The conclusion of this research was *Moringa oleifera* Lam. whose leaves have been extracted with dosages of 0.1% to 0.3% added into the MRS Broth media could increase the number of *Lactobacillus fermentum* bacteria. Therefore, it is potential as a phytobiotic candidate.

5. References

[1] Compagno L J V 1988 *Sharks of the Order Carcharhiniformes* (Princeton: Princeton University Press) p 570.
[2] Boyd C E, McNevin A A, and Davis R P 2022 *Food Security*. 14 805–827.
[3] Mavraganis T, Constantina C, Kolygas M, Vidalis K, and Nathanailides C 2020 *J. of Aqua Biol and Fisheries*. 24(2) 441-450.
[4] Deekshit V K, Maiti B, Krishna K, Kotian A, Pinto G, Bondad-Reantaso M G, and Karunasagar I 2022 *Antimicrobial resistance in fish pathogens and alternative risk mitigation strategies Reviews in Aquaculture.*
[5] Madhupriya V, Shamsudeen P, Manohar G R, Senthilkumar S, Soundarapandiyan V, and Moorthy M 2018 *Intern J. of Sci. Environment.* 7 815-22.
[6] Citarasu T 2010 *Aquaculture International*. 18(3) 403-414.
[7] Ulrikh E V, Bagno O A, Izhmulkina E A, and Grachev S Y 2019 *Prensa Medica Argentina.* 105(9) 607-612.
[8] He S, Ran C, Qin C, Li S, Zhang H, and de Vos W M 2017 *Sci Rep.* 7 13195.
[9] Lokapirnasari W P, Maslachah L, Sahidu A M, Soepranianondo K, Afikasari D, Yulianto A B, Pribadi T B, and Hariyati I 2018 *Philippine J. of Vet Med.* 5(SI) 73-78.
[10] Anudeep S, Radha C 2018 *Carbohydrates of Moringa oleifera seeds* (Mysore India: Central Food Technological Research Institute) 5: 103-108.

[11] Castillo-Lopez R I, Leon-Felix J, Angulo-Escalante M A, Gutierrez-Dorado R, Muy-Rangel M D, and Heredia J B 2017 *Pak J Bot.* 49 161-168.
[12] Mensah J K, Ikhajiagbe B, Edema N E, and Emokhor J 2012 *JNPR.* 2 107-112.
[13] Endo A, Nakamura S, Konishi K, Nakagawa J, and Tochio T 2016 *Intern J. of Food Science and nutrition.* 67 125-132.
[14] Ganzle M G, and Follador R 2012 *Frontier Microbiology.* 3 1-15.
[15] Azmi A F, Mustafa S, Hashim D M, and Manap Y A 2012 *Molecules.* 17 1635-1651.
[16] Olagbemide P T, and Philip C N A 2014 *IISTE.* 24 92-99.
[17] Olusanya R N, Kolanisi U, Van Onselen A, Ngobese N Z, and Siwela M 2020 *S. Afr. J. Bot.* 129 175-180.

6. Acknowledgement

The research team was grateful to the Directorate General of Research Enhancement and Development, the Ministry of Research Technology and Higher Education, then Chair of the Research and Innovation Institute, and also the Rector Universitas Airlangga for funding this research (PDUPT).

IOP Conf. Series: Earth and Environmental Science 1273 (2023) 012078 doi:10.1088/1755-1315/1273/1/012078

Improving the quality of shellfish products with boiler-based steaming machines in Banjar Kemuning Village, Sedati, Sidoarjo

W Tjahjaningsih[1], **A S Mubarak**[1], **E Saputra**[1], **K T Pursetyo**[1,2]

[1] Faculty of Fisheries and Marine, Universitas Airlangga, Surabaya, Indonesia

[2] Corresponding author: kustiawan@fpk.unair.ac.id

Abstract. Shellfish are filter feeder animals that filter food from the mud of their habitat and are used as bioindicators of water pollution. This has an impact on food safety if it does not meet the quality standards of shellfish sanitation. To provide a sense of security to consumers is to reduce some of the hazardous materials either from biological, chemical, or physical hazards. Banjar Kemuning Village is located in Sedati District, Sidoarjo Regency, East Java has five Joint Business Groups (KUB) which are engaged in catching shellfish. Shellfish taken from the waters should be handled related to sanitation processes. One solution for processing shellfish to meet quality standards is the method of steaming shellfish based on a boiler machine. Activities include (1) counseling through face-to-face meetings and discussions, (2) demonstration plots on the application of mussel steaming technology using a steamer boiler, and (3) monitoring and evaluation of post-catch shellfish handling related to sanitation and hygiene. The application of appropriate technology with the boiler-based shellfish steaming method can increase the production and quality of shellfish products.

1. Introduction

Sidoarjo is one of the districts in East Java that has great fishery potential because it has a beach with a length of 27 km, which is part of the northern coastal waters of Java. The development of mussel fishing production in Sidoarjo Regency in 2014 - 2017 tends to increase. Shellfish catch production in 2017 increased by 59.75% compared to 2016 [1]. Increased market demand both at local and international levels is an opportunity for shellfish fishermen to meet market demand. However, food quality and safety are very important because they are related to public health.

The type of shellfish caught by shellfish fishermen in Banjar Kemuning Village and Gisik Cemandi Village is the blood clam (*Anadara* sp.) which is known by the local community as kukur clam and feather clam. In addition to these types of shells, there are also batik shells (*Paratapes undulatus*), ax shells (*Atrina* sp.), and a type of sea snail commonly called tiger snail (*Babylonia spirata*). Banjar Kemuning Village has five Joint Business Groups (KUB) which are engaged in catching shellfish. The average production of shellfish produced by fishermen is 100 kg/day with a varied composition of shellfish catches. Since 2015, KUB Anadara Bahari and Lestari (ABATA) and Sari Laut have received orders for shellfish from Fish'n Blues (www.fishnblues.com), a partner of the World Wild Foundation (WWF) to be sent to Jakarta. The ordering process carried out by partners is flexible, so it cannot be expected to be sustainable. The problem faced is the problem of cleaning the shells sent by KUB. Cleanliness of the caught shellfish is related to food quality and safety to ensure that shellfish products are safe for human consumption.

Generally, fishermen sell shellfish in a processed form that has been boiled and peeled. The boiling process is still conventional, namely by using a large skillet. The rest of the cooking water is reused to boil the next shellfish, so it is very susceptible to contamination with *Escherichia coli* and *Salmonella* sp. The Directorate of Food Product Standardization [2] explained that pathogenic *E. coli* strains can produce endotoxins that cause severe cases of diarrhea in all age groups. Food products contaminated with *Salmonella* sp. can cause intestinal infection followed by diarrhea, nausea, chills, and headache.

Fishermen use well water in their area to wash and boil shellfish. This has resulted in the non-fulfillment of shellfish sanitation standards set by the Decree of the Minister of Marine and Fisheries number KEP.17/MEN/2004 concerning Shellfish Sanitation System. The decision explains how the process of catching clams from the location of catching to the process of shipping mussels to reach consumers. One of the requirements in shellfish sanitation is the content of polluting microorganisms, namely *E. coli* and *Salmonella*. Research Mustahal *et al.* [3] detected Coliform (8.3 - >1600 APN/g), *E. coli* (0.18 -130 APM/g), and Vibrio (0.082-25 x10^4 CFU/mL) in green mussel cultured in Tanjung waters. Kait, Tangerang, Banten. According to Gourmelon *et al.* [4], shellfish collected from coastal environments may serve as the transmission of Shiga toxin-producing *E. coli.*

Contamination of shellfish products is the result of 1) handling/processing from the stages of catching, processing, packaging, transportation, storage, or preparation, and 2) pollution from the environment. The microbial contaminants are most likely caused by poor hygiene processing. If both types of microbes are contained in shellfish products in amounts exceeding the sanitation requirements of shellfish, it is possible that they can cause foodborne disease. According to Rippey [5] and Scoging [6], post-harvest contamination is one of the causes of foodborne illnesses associated with shellfish consumption.

The presence of these microbial contaminants can be reduced by the steaming method by flowing superheated steam from the steamer boiler machine. The steaming method uses a pressure of 1 atm with a temperature of 121.1°C for ± 5 minutes. High temperatures in the steaming method can destroy microorganisms, but toxins may not be affected by heat, therefore the handling of caught shellfish products that meet shellfish sanitation requirements needs to be a concern for fishermen.

The fishing communities of the Joint Business Group (KUB) ABATA and KUB Sari Laut Banjar Kemuning Village, Sedati-Sidoarjo require technology to steam large amounts of clams. The steaming equipment that is owned is only able to accommodate 8 kg of clams and takes ± 10 minutes. This becomes an obstacle when the shellfish catch season is abundant. Therefore we need a steamer boiler with a capacity of 100 kg, which is easy to apply as an appropriate technology. The use of a steamer boiler to process shellfish is expected to meet the quality requirements of food products set by the National Standardization Agency.

2. Materials and methods

2.1. Materials

Steamer boiler (separate model) with specifications: 1) capacity of 100 kg, 2) dimensions of 70 x 120 x 110 cm, 3) dimensions of steamer tube diameter of 70x120 cm, 4) dimensions of boiler 30 x 90 cm, 5) stainless steel steamer material, and 6) mild steel boiler material.

Figure 1. Steamer boiler with a capacity of 100 kg

2.2. Methods

Post-harvest handling of shellfish using the steam boiler method is the Community Partnership Program at the Joint Business Group for Shellfish Fishermen. This program is intended for shellfish fishing groups under KUB ABATA and KUB Sari Laut Banjar Kemuning Village.

Handling of shellfish raw materials after capture begins with the shellfish washing process. The washing uses a high-pressure pump to remove the remaining mud attached to the shells. Furthermore, a demonstration plot of the application of steaming shellfish technology was carried out using a steam boiler. Previously, the boiling process was carried out conventionally so that the resulting product did not meet the standard of shellfish sanitation.

3. Result and discussion

The selection of a separate model steamer boiler machine because the product produced is better than the combined model and the mussel meat is not dry. If the mussel meat is too dry, fishermen feel disadvantaged because the weight is reduced. Boiler is a closed vessel-shaped device that is used to heat water to produce steam (steam). The heat from the combustion of fuel (Liquid Petroleum Gas) in the boiler will be transferred to the water medium flowing in the pipe. When the water temperature has reached a certain temperature, evaporation will occur. Furthermore, the hot steam is flowed into a steamer which functions as a steamer.

Using a steamer with a high temperature for 5 minutes will produce mussels that are easy to open from their shells so there is no need to pick them with nails, the mussel meat is soft and of good quality. The principle of using a steamer such as a pressure cooker is the process of heating food at a high temperature (121.1°C or 250°F) with a specified time aiming to kill the target microbe until it reaches a certain level.

The support from the Banjar Kemuning Village Head is to provide a building location for the operation of a 100 kg steamer boiler machine so that it is safe from rain and heat. The location is 1.5 km from the Village Hall. The Community Service Team proposed that the building for the placement of the steamer boiler also has sanitation facilities. The availability of sanitation facilities is very important to ensure the quality of shellfish products produced by shellfish fishing groups. This is because the shellfish fishermen in the Banjar Kemuning Village area do not apply sanitation and hygiene, even though they have received counseling on the sanitation of shellfish processing. The team conducted interviews with the Head of KUB and one of the fishermen's wives who is also a member of the Indonesian Women's Coalition and a member of the Banjar Kemuning PKK (Family Welfare Development). The interviews were related to sanitation and hygiene practices. From the results of these interviews, it can be concluded that:

1. Before processing shellfish, fishermen rarely wash their hands first.
2. Fishermen washing shells do not use tools such as gloves.

3. There are still some clam processors using a scallop chopper made of rusty nails.
4. Work with footwear.
5. When working, do not use jewelry and do not smoke.
6. Cleanliness of equipment after use is not considered.
7. Fishermen use clean and running water to wash shellfish.
8. After leaving the bathroom, wash your hands in running water using soap.
9. Clean work clothes when processing shellfish are not noticed by fishermen.

Food contamination is an ingredient or compound that is accidentally added to a food product that can affect the quality and safety of food because it is related to consumer health. To minimize food contamination, sanitation and hygiene in food processing must be considered. Steaming clams using a steamer boiler can reduce the number of microbes in clam meat so that it meets the standard of less than 1×10^5 colonies/g for the Total Plate Number (ALT) test parameter. According to the Directorate of Food Product Standardization [2], ALT indicates the number of microbes in a product. Another term for ALT is the Total Plate Count (TPC), which indicates the number of mesophilic aerobic microbes per gram or milliliter of the sample determined by standard methods.

Unclean processing and poor personal hygiene can cause cross-contamination of shellfish products. This is evident from the research of Khasanah *et al.* [7] which showed *E. coli* contamination in samples of blood clam satay sold in traditional stalls. An understanding of sanitation and hygiene practices in shellfish processing is very important because it is related to food safety, so it must be carried out continuously until there is an improvement in the values of the shellfish fishing community. The US Food and Agriculture Agency (US-FDA) ranks the threats to food safety as (1) microbiological contamination, (2) malnutrition, (3) hazardous chemical contamination, (4) the dangers of natural toxic compounds in food, (5) pesticide residues, and (6) incorrect use of food additives [8].

The Chairperson of the KUB Banjar Kemuning Village must be a role model for its members to utilize the technology of steamed clams with a steamer boiler machine as well as the application of sanitation and hygiene in the processing of shellfish to packaging. Thus, the production of shellfish in Banjar Kemuning Village can increase 10 times, so that it can fulfill orders from WWF. Cross-contamination that can occur from post-catch shellfish, depuration process and shellfish washing to boiler-based steaming can also be minimized. For the operational sustainability of these tools, the role of Village-Owned Enterprises (BUMDes) is needed as an organization that accommodates government assistance programs, maintains program assets, and protects village assets themselves. The establishment of BUMDes is regulated by the Government Regulation of the Republic of Indonesia Number 11 of 2021. The role of BUMDes is the involvement of local governments in managing the sustainability of shellfish processing in Banjar Kemuning Village. According to Dewinta and Ma'aruf [9], increasing government capacity, implementing technology and infrastructure, increasing the role of the processing industry, increasing the quality and quantity of resources are management strategies that must be prepared.

4. Conclusion

The method of steaming shellfish based on a boiler using a steamer boiler can be applied by a group of fishermen in Banjar Kemuning Village, Sedati District, Sidoarjo Regency. A separate model steamer boiler with a capacity of 100 kg can improve the quality of shellfish products that are safe for consumers. The role of BUMDes Banjar Kemuning as operational control of the steamer boiler is necessary for the sustainability of the production of quality processed shellfish.

5. References

[1] Data from the 2019 Dynamic Regional Development Planning Agency for Sidoarjo Regency Quarter 1-2019 (Sidoarjo: Regional Development Planning Agency for the Sidoarjo Regency Government).

[2] Directorate of Food Product Standardization 2012 Guidelines for Contamination Criteria in Ready to Serve Food and Home Industry Food (Jakarta: Directorate of Food Product Standardization. Deputy for Supervision of Food Safety and Hazardous Substances. Food and Drug Supervisory Agency. Republic of Indonesia) [in Indonesian].
[3] Mustahal, Vauziah S, Indaryanto F R, and Soraya S D 2019 *JPK.* 9(2) 194-201.
[4] Gourmelon M, Montet M P, Lozach S, Le Mennec C, Pommepuy M, Beutin L, and Vernozy-Rozand C 2006 *J. Appl. Microbiol.* 100 85-97.
[5] Rippey S R 1994 *Clin Microbiol Rev.* 7 419-425.
[6] Scoging A C 1992 *Can Comm Dis Rep* 18, 19-23.
[7] Khasanah U, Mahasri G, and Kusdarwati K 2021 *World Vet J.* 11(1) 79-84.
[8] Hariyadi P 2010 (October) Realizing Food Security for Regional Superior Products Proceedings of the 2010 National Seminar on the Role of Food Security for Regional Superior Products in Supporting Food Security and Suppressing the Purwokerto Inflation Rate [in Indonesian].
[9] Dewinta A F, and Ma'aruf W F 2020 *JIPK.* 12(1) 81-96.

6. Acknowledgment

The authors are grateful to the Faculty of Fisheries and Marine, Universitas Airlangga which had provided funds for community service activities in Banjar Kemuning Sedati Village, Sidoarjo, East Java.

Crude protein and crude fiber content of fish silage fermented with kombucha tea

A W W Pratama[1], W H Satyantini[2], and H Kenconojati[1,3*]
[1]Study Program of Aquaculture, School of Health and Life Sciences, Universitas Airlangga, Banyuwangi, Indonesia
[2]Department of Aquaculture, Faculty of Fisheries and Marine, Universitas Airlangga, Surabaya, Indonesia

[3]Corresponding author: hapsari@fpk.unair.ac.id

Abstract. Fish silage production using mineral acid is considered to have a negative effect because it is highly corrosive. As an alternative, kombucha tea which contains several organic acids and microorganisms can be used as a starter for fish silage fermentation. This study aims to determine the effect of different doses and duration of fermentation on the crude protein and crude fiber content of fish silage. The research method used was a factorial completely randomized design (CRD) method with 2 factors, namely factor A (dose) and factor B (fermentation time). The results showed that the difference in the dose of kombucha tea was not significantly different ($p>0.05$) on the increase in crude protein, while the difference in fermentation time was considerably different ($p<0.05$). Meanwhile, the difference in kombucha tea dose was not significantly different ($p>0.05$) on the decrease in crude fiber, but the difference in fermentation time was significantly different ($p<0.05$). Based on these results, there was an interaction between the dosage and the length of fermentation time on the crude protein content, while there was no interaction in crude fiber content. In conclusion, fermentation of fish silage using kombucha tea increased crude protein content while decreasing crude fiber content.

1. Introduction

Fish feed is one of the aspects that determine the success of an aquaculture activity because it contains nutrients to promote growth and maintain fish survival [1]. Generally, the main component in fish feed is fish meal as the source of protein which is relatively high price [2]. Therefore, one alternative that can be done to overcome this problem is to replace fish meal by utilizing fishery waste in the form of fish silage.

Fish silage is a feed ingredient or ration with high water content which is fermented under anaerobic conditions for a certain time, based on the literature, the manufacture of fish silage is known in two ways, namely chemically using mineral acids and biologically using lactic acid bacteria [2]. Silage made using mineral acid is very corrosive so it needs to be neutralized before use [3]. This neutralization process requires additional time and costs so it is less effective, so there is a need for alternative materials that can replace the use of mineral acids and organic acids. One alternative that can be done is the use of Kombucha Tea.

Kombucha tea is a traditional fermented tea made from a solution of tea and sugar by mixing a starter culture in the form of SCOBY (Symbiotic Culture of Bacteria and Yeasts) containing bacteria and yeast (yeast) [4]. The bacteria found in kombucha tea are lactic acid bacteria that produce protease enzymes, and cellulase, a protease enzyme whose role is to break down proteins (complex molecules) into peptides or amino acids (simpler molecules). Meanwhile, cellulase enzymes play a role in degrading cellulose into simpler molecules, including glucose [5].

Research conducted by Wahidah *et al.* [6] on the manufacture of trash fish silage using lactic acid bacteria with different doses showed different results, while the research of Erfanto *et al.* [7] stated that to obtain a high protein content, silage was made biologically for 14 days to produce the highest protein. Based on this, this study aims to determine the effect of different fermentation times and different doses on the crude protein and crude fiber content of fish silage and interaction between these two factors.

2. Material and methods

2.1. Time and location

This study was carried out in Maret-April 2022 at Laboratory School of Health and Life Science Airlangga University Banyuwangi.

2.2. Tools and Materials

The research tools used included 18 plastic 600 gram jars, blender, knife, cutting board, digital scale, 50 ml syringe, measuring cup, camera, analytical balance, basin. The material used in this study was 1.8 kg of lemuru fish taken at a fish auction, commercial kombucha tea.

2.3. Experimental design

The research method used is a laboratory experimental method. The variables used in this study include: independent variables, namely the dose of kombucha tea on fish silage and the length of fermentation time. The dependent variable is crude protein and crude fiber, the control variable is the anaerobic condition. The research design used was a factorial Completely Randomized Design (CRD) with 2 factors, namely factor A, which consisted of A1 = 15% kombucha tea dose, and A2 = 20% kombucha tea dose [8]. Meanwhile, factor B, namely the length of fermentation, consisted of B1 = day 3, B2 = day 7 and B3 = day 14 [9].

2.4. Preparation of tools and materials

Prepare tools and materials to be used such as 18 pieces of 600 gram plastic, blender, knife, cutting board, digital scale, 50 ml syringe, measuring cup, camera, analytical scale, basin, 1.8 kg of trash fish taken at the auction place fish, and commercial kombucha tea.

2.5. Making fermented fish silage

The lemuru fish are cleaned first, then chopped/grinded as finely as possible. Then lemuru fish that has been in the form of silage is added with kombucha tea as much as 15%, and 20% (according to the treatment given) of the weight of the fish. Then the lemuru fish silage that has been mixed with kombucha tea is stirred until evenly distributed. The mixture was then put into plastic, then stored for 3, 7, and 14 days (according to the treatment given) to be fermented under anaerobic conditions. After the fermentation process ended, the fish silage was then tested using the feed proximate test to determine the levels of crude protein, crude fiber and organoleptic tests.

2.6. Statistical analysis

Processing and analysis of research data consisting of crude protein and crude fiber, then analyzed using Two Way Analysis of Variance (ANOVA) to determine whether there is an effect of treatment

IOP Conf. Series: Earth and Environmental Science 1273 (2023) 012079 doi:10.1088/1755-1315/1273/1/012079

on the results of the study. Furthermore, the Duncan Multiple Range Test was carried out with a 5% significance level to determine the difference between one treatment and another [10].

Processing and analyzing data from organoleptic test results, then analyzed descriptively by reading and explaining tables and graphs from the resulting data. Then, the resulting data is associated with the quality of fish silage to determine the quality of the resulting fish silage.

3. Results and discussion

3.1. Proximate analysis of lemuru fish meat

The results of the proximate analysis (crude protein and crude fiber) of lemuru fish meat are presented in table 1.

Table 1. Value of crude protein and crude fiber of lemuru fish meat

Treatment	Crude Protein (%)	Crude Fiber (%)
A0	51,14	3,02

3.2. Crude protein

Crude protein content of lemuru silage from fermentation treatment using kombucha tea is presented in table 2.

Table 2. Crude protein value from fermented lemuru silage using kombucha tea with different doses and duration of fermentation

Treatment	Crude Protein (%)
A1B1	$54,72^{b} \pm 0,71$
A1B2	$54,28^{ab} \pm 0,64$
A1B3	$54,63^{b} \pm 0,50$
A2B1	$55,02^{b} \pm 0,39$
A2B2	$55,62^{b} \pm 0,37$
A2B3	$53,16^{a} \pm 1,19$

Different superscripts in the same column showed a significant difference ($p \leq 0.05$)

The results showed an increase in crude protein, it was shown in the difference in the higher crude protein content in the treatment of kombucha tea (Table 2) compared to the crude protein of lemuru fish meat (Table 1). Kombucha tea treatment on fish silage showed a significant effect ($p<0.05$) on the crude protein content of fish silage. The administration of different doses of kombucha tea in fermented fish silage was not significantly different ($p>0.05$) on crude protein content, while the difference in duration of fermentation of fish silage using kombucha tea was significantly different ($p<0.05$) on crude protein content and showed there is an interaction ($p<0.05$) between the two factors.

3.3. Crude fiber

Crude fiber content of lemuru silage from fermentation treatment using kombucha tea is presented in table 3.

Table 3. Crude fiber value from fermented lemuru silage using kombucha tea with different doses and duration of fermentation

Treatment	Crude Fiber (%)
A1B1	$6{,}74^{c} \pm 0{,}50$
A1B2	$5{,}56^{b} \pm 0{,}82$
A1B3	$3{,}93^{a} \pm 0{,}27$
A2B1	$8{,}00^{d} \pm 0{,}40$
A2B2	$6{,}32^{bc} \pm 0{,}32$
A2B3	$5{,}83^{bc} \pm 0{,}40$

Different superscripts in the same column showed a significant difference ($p \leq 0.05$)

The results showed that crude fiber had increased, it was shown in the difference in crude fiber content which was higher in the treatment of kombucha tea (Table 3) than the crude fiber of lemuru fish meat (Table 1). Kombucha tea treatment on fish silage showed a significant effect ($p < 0.05$) on crude fiber content. The administration of different doses of kombucha tea in fermented fish silage was significantly different ($p<0.05$) on crude fiber content, while the difference in duration of fermentation of fish silage using kombucha tea was significantly different ($p<0.05$) on crude fiber content and showed no interaction ($p> 0.05$) between the two factors.

The increase in crude protein was thought to be due to the presence of *Lactobacillus* sp. in kombucha tea which produces protease enzymes, protease enzymes will remodel proteins into polypeptides, then into simple peptides, then the peptides will be remodeled into amino acids [11]. While the decrease in crude fiber occurs due to the cellulase enzyme produced by Lactobacillus sp. bacteria, the cellulase enzyme plays a role in degrading cellulose into simpler molecules, including glucose [12].

Protease enzymes are naturally produced by microbes to hydrolyze polypeptides in the media into peptides and amino acids [13]. Amino acids are then used by bacteria for growth, especially nitrogen compounds that will enter bacterial cells [14]. Cellulase enzymes play a role in degrading cellulose into simpler molecules, including glucose [15]. Glucose is used as an energy source for bacterial growth [16]. So that the crude fiber produced by *Acetobacter xylinum* is degraded by the cellulase enzyme into glucose, while the crude protein is degraded by the protease enzyme into amino acids and nitrogen. Nitrogen and glucose are used as energy sources for bacterial growth, amino acids are used by bacteria to multiply. Amino acids will be used by microbes to multiply, increasing the number of microbial colonies during the fermentation process can indirectly increase the crude protein content because microbes are a source of single cell protein [17], so that crude protein has increased and crude fiber has decreased.

The administration of different doses of kombucha tea on fermented fish silage was not significantly different ($p>0.05$) on crude protein content. This is presumably due to the dose of kombucha tea which is not significantly different so that the crude protein does not experience a significant difference in crude protein content, this is in accordance with research conducted by [19], which states that the administration of kombucha tea with different doses no significant effect on the increase in crude protein fermented water hyacinth using kombucha tea.

The administration of different doses of kombucha tea on fermented fish silage showed a significant difference ($p<0.05$) on the crude fiber content. This is thought to be due to the combination of *Lactobacillus* sp. and Saccharomyces sp bacteria which are effective in degrading crude fiber. *Lactobacillus* sp. and *Saccharomyces* sp. produce enzymes that have the ability to degrade crude fiber (cellulose or hemicellulose) into simpler carbohydrates. [18].

The difference in the length of fermentation of fish silage using kombucha tea was significantly different ($p<0.05$) on the content of crude protein and crude fiber. This is thought to be due to the length of fermentation in line with the growth phase of the bacteria. According to [19] the longer the fermentation time, the more time lactic acid bacteria can use to remodel the complex compounds contained in them.

Overall, administration of kombucha tea with different doses and duration of fermentation led to an increase in crude protein content and a decrease in crude fiber from fermented fish silage as indicated by an increase in crude protein content ($p<0.05$; Table 2) and a decrease in crude fiber content ($p<0.05$; Table 2) and a decrease in crude fiber content ($p <0.05$; Table 3). In addition, there was an interaction between different doses and duration of fermentation on crude protein content, whereas in crude fiber there was no interaction between these two factors. The interaction of treatment with different doses and duration of fermentation is shown in the factors that influence the decrease in crude protein including the number of bacteria, bacterial growth phase, nutrients, competition between microbes and the environment.

4. Conclusion

Based on the present study, The difference in dosage in the manufacture of fish silage with kombucha tea did not significantly increase the crude protein content, but had a significant effect on the decrease in crude fiber content. The difference in the length of fermentation for making fish silage with kombucha tea has a significant effect on increasing crude protein content and decreasing crude fiber content. So there was an interaction between the difference in dosage and duration of fermentation on the increase in crude protein, but there was no interaction between the difference in dosage and the length of time for fermentation of kombucha tea on the decrease in crude fiber.

5. Bibliography

[1] Muntafiah I 2020 *Analys. J. Rese.* 4 35-39.
[2] Handajani H, Dwi H S 2013 *Use. J. Depi.* 2 126-132.
[3] Akhirany N 2011 *Fish.* Book. *UPTD. PSP3.*
[4] Wistiana D, E Zubaidah 2015 *Chara. J. Pangan* 4 1446-1457.
[5] Lokapirnasari W P, Widodo S O, Emy K S 2018 *Potent. J. Fisher* 10 75-82.
[6] Wahidah S, Idris A P S, Nawawi 2018 *Utili. J. Agro* 17 18-23.
[7] Erfanto F, J Hutabarat, Endang A 2013 *Effe. J. Aqua* 2 26-36.
[8] Wahidah S, Idris A P S, Nawawi 2018 *Utili. J. Agro* 17 18-23.
[9] Handajani H, Dwi H S 2013 *Use. J. Depi.* 2 126-132.
[10] Kusriningrum 2008 *Base.Book. Facul. Airla.* 21.
[11] Agustono, Hidayat S, Widya P L 2010 *Effec. J. Fisher.* 2 179-183.
[12] Lokapirnasari W P, Widodo S O, Emy K S 2018 *Potent. J. Fisher.* 10 75-82.
[13] Wilson P, Z Remigio 2012 *Produc. Int. J. Afri.* 6 5542-5551.
[14] Setyati W A, Subagiyo 2012 *Isola. J. Marin.* 17 164- 168.
[15] Lokapirnasari W P, Widodo S O, Emy K S 2018 *Potent. J. Fisher.* 10 75-82.
[16] Suciati P 2016 *Activi. J. Fishe.* 8.
[17] Agustono, Hidayat S, Widya P L 2010 *Effec. J. Fisher.* 2 179-183.
[18] Suryani Y, Hernaman I, Ningsih 2017 *Effec. J. Integr.* 5 13-17.
[19] Yunus Y, Zubaidah E 2015 *Effec. J. Food.* 2 303-3012.

6. Acknowledgement
The authors would like to acknowledge Dr Woro Hastuti Satyantini Ir., M.Si. and Hapsari Kenconojati S.Si., M.Si. for his advice in starting this study and School of Health and Life Science Universitas Airlangga for supporting facilities during this study.

IOP Conf. Series: Earth and Environmental Science 1273 (2023) 012080 doi:10.1088/1755-1315/1273/1/012080

Analysis of fisherman's income affected by the Covid-19 pandemic in Kranji fish landing bases, Paciran, Lamongan District

A S Magdalena[1], A M Sahidu[2,4], W Isroni[3]

[1] Program Fishery Products Technology, Faculty of Fisheries and Marine Universitas Airlangga, Kampus C Unair, Jalan Mulyorejo Surabaya 60115, Indonesia.
[2] Department of Marine, Faculty of Fisheries and Marine Universitas Airlangga, Kampus C Unair, Jalan Mulyorejo Surabaya 60115, Indonesia.
[2] Department of Fish Health Management and Aquaculture, Faculty of Fisheries and Marine Universitas Airlangga, Kampus C Unair, Jalan Mulyorejo Surabaya 60115, Indonesia.

[4] Corresponding author: adriana-m-s@fpk.unair.ac.id

Abstract. Fisherman income was influenced by various factors. One of the factors that take affect income fishermen is the pandemic COVID-19. This study aims to analyze the income level of fishermen before and during the Covid-19 pandemic and to determine the effect of the Covid-19 pandemic on fishermen's income. The method used in this study was survey method, which describes the problem according to the facts that have just taken place, then determination of respondents by means of Simple Random Sampling. The research data were statistically tested using the Different T-Test. The average income of fishermen before the Covid-19 pandemic was Rp. 1,225,531.92 while during the Covid-19 pandemic it was Rp. 1,216,893.62. The results of the Paired Sample Test on the income of fishermen in Kranji Landing Fish Base (PPI) before and during the Covid-19 pandemic showed a p value> 0.05 so it can be concluded that there was no significant difference between the income of fishermen before and during the Covid-19 pandemic or the Covid-19 pandemic has no significant effect on the income of fishermen in Kranji Fish Landing Bases (PPI). The main factor that affects the income of fishermen in Kranji Fish Landing Bases (PPI) is the weather or season.

1. Introduction

Paciran is a sub-district located in the northern part of Lamongan Regency. The north area of Paciran is directly adjacent to the Java Sea. The site is a contributor to the large-scale capture fisheries sector. The fisheries sector itself has a large contribution to the Gross Regional Domestic Product (GRDP) of Lamongan Regency, namely 13.84% in 2014 and 16.98% in 2018 [1]. In 2020, the production of marine-caught fish reached 76,692.96 tons, with a total production value of Rp. 1,188,671,626,220 [2].

One of the livelihoods of the Paciran people is as a fisherman. Paciran fishing activities are supported by 2 PPI units, one of which is the Kranji PPI in Kranji Village, Paciran District, Lamongan Regency. Capture fisheries production at PPI Kranji in 2020 amounted to 4,758.89 tons with a value of 41,043,296.32 million [2]. The problem currently being faced by fishermen at PPI Kranji is the spread of Covid-19.

Corona virus disease 2019 (Covid-19) is an infectious disease caused by acute respiratory syndrome coronavirus 2 (Sars-CoV-2) [3]. Covid-19 was first identified in December 2019 in Wuhan, China. The global status of the Covid-19 pandemic has been established by the World Health Organization (WHO) after this virus spread in various parts of the world.

The spread of the Covid-19 pandemic has caused panic among people around the world [4]. Several countries have implemented various policies to overcome this, one of which is Indonesia which has imposed Large-Scale Social Restrictions (PSBB), work from home, physical distancing as stated in RI government regulation No. 21 of 2020.

The Covid-19 pandemic has affected changes in human life and has resulted in changes in the economies of various countries. Changes in consumer consumption patterns, many restaurants that have stopped operating and delays in sending fish due to the implementation of the lockdown have caused the price of fish caught by fishermen at PPI Kranji to decrease. Mubarok and Fajar [5] explained that the impact of the Covid-19 pandemic that most fishermen feel is the drastic reduction in fish prices up to 50%, but this is not comparable to the operational costs of fishing which have not decreased. The Covid-19 factor is believed to affect fishermen's income. If this goes on for a long time, the fishing business will become a business that is not feasible to continue [6]. Seeing the current situation, income analysis is needed to determine the effect of Covid-19 on fishermen's income at PPI Kranji.

2. Material and methods

2.1. Material

The tools and materials used in the research included stationery, laptops, questionnaires, mobile phones, respondents.

2.2. Research methods

This study uses the interview method. Sampling of respondents using probability sampling with simple random sampling method. The main parameter is the total income of fishermen before and during the Covid-19 pandemic.

2.3. Data analysis

Examination data is presented in the form of tables and figures, the analysis will be explained using a *paired t-test.*

3. Result and discussion

3.1. Result

3.1.1. Overview of research locations

Kranji is the name of a village in the Paciran District, Lamongan, and East Java and includes the coastal area of the northern coast (North Coast). Kranji Village has an area of ± 484,107 ha/m^2 which includes ± 8,114 ha/m^2 of mountainous or hilly land and ± 475,993 ha/m^2 of plains. Kranji Village is a village located on the sea coast with a land elevation of 2 m/asl.

Kranji Village is one lane with the main road leading to Paciran District, so this village is easily accessible. Kranji Village is divided into two parts, namely the residential part and the ocean part whichis directly bounded by the main road.

Kranji Village has the following territorial boundaries:

North : Java Sea
South : Dagan Village, Payaman, Solokuro District
West : Tunggul Village, Sendangagung Village, Paciran District
East : Banjarwati village, Drajat village, Paciran district

The location of Kranji Village which is directly adjacent to the sea makes the majority of the Kranjipeople work as fishermen. The fishing activities of the Kranji community are supported by the development of facilities and infrastructure to increase business and production efficiency. One of the facilities and infrastructure that provides convenience for fishermen in production and marketing activities is the fish landing base (PPI).

3.1.2. Respondent characteristics
Respondent characteristics are data from a respondent collected to determine or find out the profile of the respondent being studied [7]. The research respondents were Kranji PPI fishermen, where the sample of respondents used was 94 people who were considered to represent the fisherman population. Interviews with research respondents were carried out by distributing questionnaires directly to fishermen and assisting fishermen in filling out questionnaires who did not understand the questions asked while maintaining health protocols during the Covid-19 pandemic. The characteristics of the respondents in this study were classified based on age, education, and experience or length of time working as fishermen

3.1.2.1. Characteristics of Respondents by Age
Age is an independent variable that is expected to influence the level of welfare and supporting factors for success in business activities. People who are productive in making decisions will be faster towards new innovations [8]. The characteristics of the age distribution of fishermen in the Kranji PPI area can be seen in the following table:

Table 1. Characteristics of Fisher Respondents in the Kranji Fish Landing Base (PPI) by Age.

Age range	Frequency (Person)	Percentage (%)
< 20 years	4	4,3
20 – 30 years	13	13,8
31 – 40 years	19	20,2
41 – 50 years	35	37,2
> 50 years	23	24,5
Total	94	100,0

Table 1. shows that fishermen in the Kranji PPI area are mostly aged 41-50 years with a percentage of 37.2 percent. Fishermen aged > 50 years are in second place with a percentage of 24.5 percent. Fishermen aged 31-40 years have a percentage of 20.2 percent. Fishermen aged 20-30 years have a percentage of 13.8 percent, and the lowest is in the <20 years age group with a percentage of 4.3 percent.

3.1.2.2. Characteristics of Respondents Based on Education
Education is believed to be able to increase the expertise and knowledge of fishing communities in managing marine potential [9]. This means that the higher the level of education, fishermen are expected to have a rational mindset and tend to be more responsive to new things, so that they can easily accept new technologies to increase their production and enable increased income. The distribution of educational levels of PPI Kranji fishermen can be seen in Table 2.

Table 2. Characteristics of Fisher Respondents in the Kranji PPI Fish Landing Base Area byEducation.

Education	Frequency (Person)	Percentage (%)
Elementary School	42	44,7
Junior High School	11	11,7
Senior High School	14	14,9
Other	27	28,7
Total	94	100,0

Table 2 shows that 44.7 percent or 42 fishermen in the Kranji PPI area have an elementary school education. Fishermen with junior high school education have a percentage of 11.7 percent or as many as 11 respondents. Fishermen with high school education have a percentage of 14.9 percent or as many as 14 respondents. Fishermen who answered others had a percentage of 28.7 percent or as many as 27 respondents. 3.

3.1.2.3. Characteristics of Respondents Based on Experience

Experience is the best teacher, so experience in any activity is indispensable. That is, a person learns from the experience he has. The level of experience is a measure of the time or length of work spent by an individual to understand and perform well on the tasks of a job [10]. The percentage of experience or length of time working for PPI Kranji fishermen can be seen in Table 3.

Table 3. Characteristics of Respondent Fishermen in the Kranji Fish Landing Base (PPI) Area basedon experience or length of work as fishermen

Experience	Frequency (Person)	Percentage (%)
< 1 years	0	0
1 – 2 years	3	3,2
3 – 4 years	15	16,0
≥ 5 years	76	80,9
Total	94	100,0

Table 3. shows, based on the results of interviews, fishermen with a working period of 1-2 years only amounted to 3 respondents or around 3.2 percent of the total fisherman population in the area. Fishermen with 3-4 years of work totaled 15 respondents or 16 percent. The remaining 76 respondents or around 80.9 percent had the longest working period as fishermen, namely ≥ 5 years.

3.1.3. The Impact of the Covid-19 Pandemic on Fishermen's Income

To find out the impact of the Covid-19 pandemic on the income of fishermen at the Kranji Fish Landing Base (PPI), a Paired T-test was carried out, namely by looking at the difference in the average income of fishermen before the Covid-19 pandemic and during the Covid-19 pandemic whether there was a significant difference. Paired T-test can be done if the classical assumption test is fulfilled.

3.1.3.1. Classical Assumption Test

The classic assumption test is a requirement that must be met before the paired t test is carried out. The classical assumption test is carried out to gain confidence that the data obtained can be further processed. This study uses the classical assumption test which includes the normality test and homogeneity test.

3.1.3.1.1. Normality test

Ikasari *et al.* [11] stated that the normality test is a prerequisite test in the use of parametric statistics using the assumption that data from each variable must be normally distributed. If the data is not normally distributed, the parametric test cannot be used. The sample required in the Kolmogorov-Smirnov test for normality is at least 5 samples [12]. Testing the normality of a data usingthe Lilliefors method as follows:

Hypothesis:

Ho : The population of statistical test scores is normally distributed H1: The population of statistical test scores is not normally distributedLillifors Critical Value:

Value α = significance level = 5% = 0.05

Based on the normality test table in Appendix 1, the significance value in the Kolmogorov Smirnov test before the Covid-19 pandemic was 0.116, and the significance value during the Covid-19 pandemic was 0.173. This value is greater than the expected significance value of 0.05, which means that it can be concluded that fishermen's income before and during the Covid-19 pandemic was normally distributed.

3.1.3.1.2. Homogeneity Test

Homogeneity test is used to ensure that the data set studied in the analytical process comes from a population with a diversity that is not much different. This test is used as a condition for t-test analysis and analysis of variance in parametric statistics [13]. The requirements for the homogeneity test are that the data must be normally distributed, the data is on a ratio or interval scale, the data for all samples must be independent, all samples must have the same variance.

Based on the Test of Homogeneity of Variance table in Appendix 1, it shows that the results of a significance value of 0.522 are greater than 0.05, so it can be concluded that the data is homogeneous or the data comes from a normally distributed population.

3.1.3.2. Paired T-Test

Paired T-Test is a parametric test that can be used on two paired data. The purpose of this test is to see whether there is an average difference between two samples that are interconnected or in pairs. Because it is paired, the data from the two samples must have the same amount or come from the same source.

The Paired Sample Statistics test table in Appendix 2 shows the results of the statistical tests for the two samples, namely fishermen's income data before the Covid-19 pandemic and during the Covid-19 pandemic. It can be seen that the fishermen's income data before the Covid-19 pandemic had an average value of Rp. 1,225,531.92 with N as many as 94 respondents. Fishermen's income during the Covid-19 pandemic has an average of IDR 1,216,893.62 with N as many as 94 respondents.

The Paired Sample Correlations table in Appendix 2 illustrates the significance value of the paired samples correlation test of 0.000, where the sig value is less than 0.05, it can be concluded that fishermen's income data before the Covid-19 pandemic and during the Covid-19 pandemic were correlated.

The Paired Sample Test table in Appendix 2 shows the significance value of the Paired Sample Test is 0.105, which means it is greater than 0.05, then Ho is accepted and Ha is rejected. It can be concluded that there is no significant difference between the results of PPI Kranji fishermen's income before and during the pandemic Covid-19.

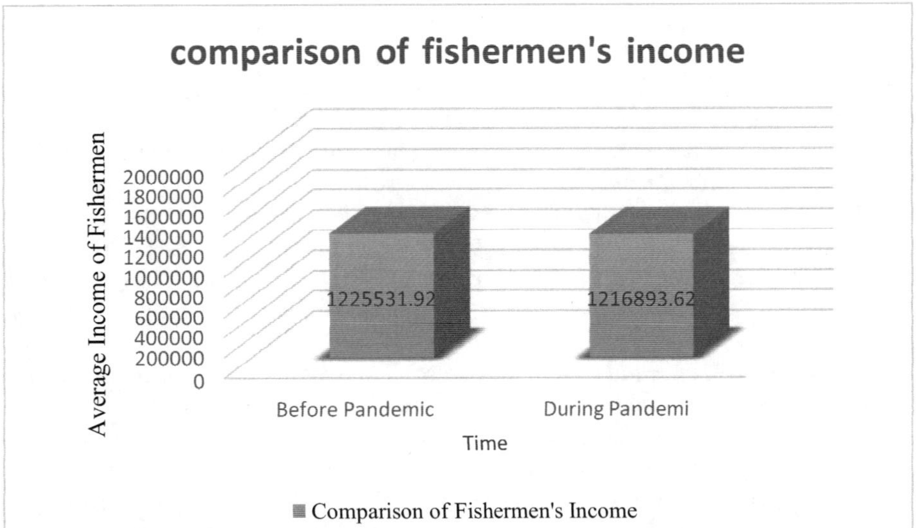

Figure 1. Graph of Fishermen's Income Before and During the Covid- 19 Pandemic.

3.2. Discussion

Based on data on the characteristics of fishermen who were respondents at the Kranji Fish Landing Base (PPI) the average Kranji Fish Landing Base (PPI) fishermen were of productive age, with relatively low education, but had longer experience or more than 5 years as fishermen. This shows that productive age and the amount of experience that fishermen have an effect on fishermen's income, but education has no significant effect on fishermen's income.

Most of the Kranji Fish Landing Base (PPI) fishermen are aged 41-50 years. At that age can be categorized into a productive age. According to Tjiptoherijayanto [14], the age structure of the population is divided into three groups, namely the young age group (under 15 years), the productive age group (age 15-64 years), and the old age group (over 65 years). Fishermen's income decreases with increasing age of fishermen. The increasing age of fishermen makes their physical condition decrease so that there is a decrease in productivity at sea and also makes income decrease, and vice versa. According to Ariska and Prayitno [15] a person's physical strength to carry out activities is very closely related to age because when a person's age has passed the productive period, then his physical strength decreases so that his productivity also decreases and his income also decreases.

Education in this study is formal education. The highest percentage of fisherman's education as respondents is still having their last education at the elementary school level (SD) while in this study what is needed is non-formal education including experience and skills such as the time to go to sea and the month that is suitable for going to sea. The longer the fisherman's education, the income generated has no effect on increasing or decreasing [15]. Fishermen's formal education hasno effect on fishermen's income.

The largest percentage of fishermen who were respondents in the Kranji Fish Landing Base (PPI) area chose to work as fishermen for 5 years or more. Experience is needed in every job. Experience is needed by fishermen to produce fish. Searching for fish requires time, equipment, fish location, and the right fish bait. This can be known by fishermen based on the passage of time. The longer the experience, the more sensitive fishermen are to the position of the fish, making it easier for fishermen to catch fish in the sea [16]. According to Gitosudarmo and Mulyono [17] due to increased experience in doing a job or producing an item, it can reduce the average cost per unit of item. This is logical because with increasing one's experience in doing the job, of course, lessons will be obtained to do it better and more efficiently. Mistakes that have been made can be known and henceforth not repeated again for the same mistakes. If the experience goes to sea, there will be a significant reduction in production costs per unit so that it can reduce the amount of production expenditure incurred by fishermen.

The spread of Covid-19 in Indonesia made the government implement a policy to break the chain of distribution, namely by imposing a PSBB or lockdown in several areas. The implementation of the lockdown has caused problems in and out of the distribution of fish caught in several areas. This did not have a significant effect on the income of fishermen at PPI Kranji, which can be seen from the results of the average income of fishermen before the Covid-19 pandemic and during the Covid-19 pandemic (Figure 1.) which had no significant difference. The flow of marketing or distribution of caught fish requires several chains or marketing actors to reach the hands of consumers. The fishermen themselves are one of the first chains or producers, and the fishermen's catch is only sold at TPI Kranji. Marketing of caught fish outside the area is carried out by distributors or collectors from TPI to the areas that are the destination. The Covid-19 pandemic has affected distributors who are constrained by access to entry and exit in marketing fish to a number of areas.

Fishermen's income is also determined by the price of fish. The Covid-19 pandemic has no impact on fish prices at PPI Kranji. Determination of the rise and fall of fish prices at PPI Kranji is influenced by consumer demand and the number of catches. A decrease in consumer demand with abundant catches results in a decrease in fish prices, and vice versa if consumer demand increases, fish prices also increase. This is in accordance with the Theory of Demand Determination which states that "if demand increases, relative prices will rise, conversely if demand falls, relative prices will fall [18].

The main factors affecting the income of PPI Kranji fishermen come from natural factors such as season/weather. The occurrence of big waves and high rainfall make fishing activities dangerous, so it is estimated that there will be a decrease in the productivity of fishermen by 60% [19].Some fishermen in the Kranji PPI area choose to switch professions during bad weather conditions. However, there are also those who work entirely as fishermen. Kholis *et al.*, [20] stated that seasonaland weather factors have a significant effect on fishermen's length of time at sea and have an impact on catches. Possible solutions can be applied to these obstacles or problems, one of which is by providing assistance in the form of cold storage. Cold storage can be used to store abundant catches during the peak season so that it can be used to deal with when the lean season arrives. Empowerment of fishing communities is also needed to increase fishermen's knowledge and skills, such as how to process fish tomarketing. Fish processing which aims to store fish for a long time so that the price of fish can be maintained and can maintain the stability of fish stocks.

4. Conclusion

The average income of PPI Kranji fishermen before the Covid-19 pandemic was Rp. 1,225,531.92 while the average income of fishermen during the Covid-19 pandemic was Rp. 1,216,893.62. There was no significant difference between the income of fishermen before and during the Covid-19 pandemic or the Covid-19 pandemic had no significant effect on the income of PPI Kranji fishermen. The main factor that has the most influence on the income of PPI Kranji fishermen is natural factors such as weather or season

5. References

[1] Wardono, B. dan T. Muslih 2020 *J Kebijakan Sosek KP* 10 (1) : 25-38.
[2] Department of Fisheries and Maritime Affairs of Lamongan Regency. 2020. *Fisheries Profile. Ministry of Maritime Affairs and Fisheries*. Lamongan.
[3] Supriatna, E. 2020 *J Sos Budaya Syar-i*. 7 (6) : 555-564.
[4] Azamfirei, R 2020 *J Crit Care Med*. 6 (1) : p3-4.
[5] Mubarok F. dan J. Fajar 2020 https://www.mongabay.co.id/2020/04/02/dampak-Covid19-harga-tangkapan-ikan-nelayan-turun-drastis/. 07 November 2020
[6] Setyawan, E. Y. 2020 Maritime and Fisheries Research and HR Agency: https://kkp.go.id/brsdm/sosek/artikel/22843-riset-analisis-kebijakan-pemanfaatansumber-daya-perikanan-tangkap-dalam-pandemic-Covid-19
[7] Sari, A. R. P. dan N. N. K. Yasa 2020 *Lakeisha* Jawa Tengah.
[8] Suroyya, A. N., I. Triarso, dan B. A. Wibowo 2017 *J Fish Resour Util Manag Technol*. 6 (4): 30-39.
[9] Baharudin 2014 *J Jur Pendidik IPS Ekon*. 6 : 57-67.
[10] Foster, B 2001 PPM. Jakarta.
[11] Ikasari, I. H., T. Hidayati, dan I. Handayani 2019 *Pena Persada*. Purwokerto.
[12] Widhiarso, W. 2012 Fakultas Hukum UGM. Yogyakarta.
[13] Budiwanto, S 2017 UM Press. Malang.
[14] Tjiptoherijanto, P. 2000 *Development Planning Magazine*. Jakarta.
[15] Ariska, P. E. Dan B. Prayitno 2019 *J Ekon*. 1(1) : 38-47.
[16] Hendra 2019 Faculty of Islamic Economics and Business Sulthan Thaha Saifuddin State Islamic University. Jambi.
[17] Gitosudarmo, I. dan A. Mulyono 2012 BPFE. Yogyakarta.
[18] Febianti, Y. N. 2014 *J Edunomic*. 2 (1) : 16-24.
[19] Iry, N. dan D. S. Rain 2020 STIE Jambatan Bulan. Papua. Pp. 1 – 27.
[20] Kholis, M. N., Fraternesi, dan L. O. Wahidin 2020 *Albacore* 4 (1) : 001-011.

IOP Conf. Series: Earth and Environmental Science 1273 (2023) 012081 doi:10.1088/1755-1315/1273/1/012081

Evaluation of larval Beautiful Groupers (*Epinephelus fuscoguttatus* x *E. microdon*) using *Chlorella sp.* and *Artemia sp.*

F Ramadhany[1], Sumito[2], L A Sari[3,4], D D Nindarwi[3], dan S Arsad[4]

[1]Aquaculture Study Program, Faculty of Fisheries and Marine Sciences, Universitas Airlangga, Campus C Jalan Mulyorejo, Surabaya 60115 East Java, Indonesia.
[2]Department of Aquaculture, Faculty of Fisheries and Marine Sciences, Universitas Airlangga, Campus C Jalan Mulyorejo, Surabaya 60115 East Java, Indonesia.
[3] Unit of Technical Implementation of Brackishwater and Marine Aquaculture, Brackishwater Aquaculture Development Center Situbondo. East Java 68351. Indonesia.
[4]Department of Aquatic Resources Management, Faculty of Fisheries and Marine Sciences, Universitas Brawijaya, Jalan Veteran Malang 65145 East Java, Indonesia.

[4]Corresponding author: luthfianaas@fpk.unair.ac.id

Abstract. Grouper is one of the fishery commodities having a fairly high price and high economic value at the cultivator level. Due to the increasing market demand in the world, the larval rearing of the beautiful groupers (*Epinephelus fuscoguttatus* x *E. microdon*) has been carried out in order to obtain quality seeds. The purpose of this study was to determine the results of the evaluation of larval rearing of beautiful groupers (*Epinephelus fuscoguttatus* x *E. microdon*) using *Chlorella sp.* and *Artemia sp.* The working methods used in this study included preparation of tanks and water maintenance of culture media, management of feed, stocking and hatching of eggs, frequency and timing of feeding larvae/seeds, water quality management, pest and disease control, and grading and harvesting. Seed harvesting could be done after the seeds were aged 45 to 50 days old with a size of 3 cm. Water quality parameters measured were temperature at 28-32°C, pH at 7-8, salinity at 29-33 ppt, and DO at 6.25-7.3 mg/L. Survival rate (SR) in larval rearing was 57%, and the hatching rate (HR) was 60%.

1. Introduction

Grouper is one of the fishery commodities having a fairly high price and high economic value at the cultivator level. Due to the high price of groupers, the cultivation of groupers is increasing rapidly along with the development of hatchery technology which makes the demand for seeds for several types of groupers start to be fulfilled. In addition to fulfilling the integrity of the sea in Indonesia, seed stocks are also sent abroad including Vietnam, Thailand, Malaysia, Singapore, China, Taiwan, and Hong Kong [1]. Fish hatchery is a science that studies the activities of breeding / multiplying / breeding fish naturally, semi-artificially and artificially. Apart from producing fish seeds, fish hatchery activities will also avoid the extinction of certain fish commodities by continuously breeding cultured fish. Fish hatchery begins with proper broodstock management and broodstock selection in accordance with the Good Management of Fish Hatchery (CPIB) standards and the brood spawning process, egg hatching, larval rearing and seed stocking in accordance with production needs so that optimal cultivation results can be obtained.

Grouper hatchery has become a promising business for community and can absorb a lot of labor [2]. One of the grouper fish that is in great demand is the beautiful groupers, a hybrid between the male batik groupers (*Epinephelus microdon*) and the female tiger groupers (*Epinephelus fuscoguttatus*). Batik groupers have resistance to disease but they have slow growth, and until now it is difficult to breed them despite their high demand for consumption. On the other hand, the tiger groupers are easy to hatch and have faster growth [3].

One of the efforts that support the cultivation of beautiful groupers is the larval rearing stage to obtain quality seeds. However, there are obstacles in the larval rearing process, namely low larval survival [4]. According to Sudewi *et al.* [5], the low survival rate of the larvae is due to the small mouth opening of the fish larvae, which requires feed that is sized according to the mouth opening of the fish larvae. Phytoplankton is the main component in the food chain in the waters [6]. Phytoplankton are organisms that float or hover in waters whose movements are relatively passive. Phytoplankton is also an aquatic organism which plays an important role in influencing primary productivity in waters [7]. The presence of microorganisms can be used as bioindicators of water conditions because phytoplankton have tolerance limits to certain substances [8]. Phytoplankton are organisms that are sensitive to environmental changes so that the number of certain phytoplankton species can be used as an indicator of water pollution. Phytoplankton is a floating organism whose life is influenced by currents and is commonly used as an indicator of biological changes in the waters because this group of aquatic biota is generally very sensitive to environmental changes, and its life cycle is relatively short [9]. Based on the description above, a study was carried out regarding the evaluation of larval rearing of beautiful groupers (*Epinephelus fuscoguttatus* x E. *microdon*) using *Chlorella sp*. and *Artemia sp.*

2. Material and methods

2.1. Feed management for beautiful grouper larvae and seeds

The feed used for larval rearing activities is divided into two types, namely natural feed and artificial feed. Natural feed is classified into two, which are phytoplankton and zooplankton. Examples of natural food for phytoplankton are chlorella, and for zooplankton are rotifers, rebon shrimp and artemia. Chlorella is a microalgae belonging to the class Chlorophyceae which has good nutritional value and is easy to digest by larvae [10]. Rotifers are used because they contain sufficient EPA and DHA as a source of essential fatty acids for fish larvae [11]. The use of Artemia as feed for beautiful grouper larvae/seeds has advantages such as practicality in its use [12]. Rebi shrimp is also used as feed for beautiful groupers because it has a high nutritional value of protein [13].

Culture of *Chlorella sp.* carried out before the stocking of eggs aimed to provide basic feed from mass cultured seeds/larvae. The advantages of using *Chlorella sp.* include the fact that some of them can breed quickly, are easy to cultivate, do not depend on the season, and do not require a long time to grow [14,15,16]. The culture was carried out in a 4 x 2 x 1.25 m3 concrete tank which had been cleaned first by scrubbing it with 200 ppm chlorine to deactivate other organisms before rinsed and dried for about 2-3 days until it was odorless and ready for use.

The culture tank that had been prepared was filled with seawater which had been filtered with a filter bag as much as 30% of the volume of the tank. Afterwards, the culture tank was treated using 120 ppm chlorine and aerated strongly for 8 hours so that it is evenly distributed after previously having been neutralized using 60 ppm sodium thiosulfate. *Chlorella* sp. seeds used were as much as 20% of the total volume of water [17]. Moreover, fertilization was done by using bulk fertilizer with 15 ppm Urea, 30 ppm TSP, and 30 ppm ZA fertilizer. Harvesting was done after the culture was 6-8 days old to be later distributed using a pump. Rotifer culture was carried out when *Chlorella sp.* had been prepared beforehand. The advantages of rotifers as natural food for fish larvae/seeds include the fact that rotifers swim so slowly that it can be easily preyed upon, easy to digest, and easy to breed [18].

The culture tank used for the culture of rotifers was a round fiber tank measuring d = 2 m and t = 1.5 m. The tank was filled with *Chlorella sp.* that were ready to be harvested (minimum age of four days) as much as 30% of the size of the volume of the tank. The tank was then filled with seawater as much as 40% of the size of the volume of the tank. After that, the tank was stocked with rotifer seeds with an initial density of about 30 individuals/mL. Every day, the culture tank of rotifers was added with *Chlorella sp.* as much as 10% - 20% of the volume of the tank as feed. Moreover, harvesting was done using the daily method, where harvesting was carried out as much as 20% - 30% of the size of the tank volume using a plankton net with mesh size 400 after a minimum culture age was four days or after the density reached 100-150 individuals/mL. After the harvesting process, the culture tank of rotifers was added with *Chlorella sp.* as much as the volume at harvest. The harvesting process also occured with the use of *Artemia sp.* or *brine shrimp* as it is one of the natural food for fish larvae widely used in fish hatchery activities. There are advantages of using *Artemia sp.* as feed, among which are high nutritional content and practical use [12].

According to Panggabean [19], decapsulation is a stripping technique to open the outer layer of the cyst of *Artemia sp.* by thinning the cyst shell of *Artemia sp.* to make it easier for the embryo to come out. The process of thinning the cyst shell of *Artemia sp.* can be done by immersion using 40% NaOH, 5.25% NaOCl, and chlorine. The decapsulation steps could be carried out as follows:

1) The *Artemia* cysts were weighed as needed.
2) The cysts were filtered and rinsed with water to clean the dirt that sticked and later were drained.
3) After drained, the cysts were soaked in 1 liter of seawater, while stirred and added with 40% NaOH solution and 5.25% NaOCl little by little so as not to make the temperature reach 40°C because it could kill the embryo.
4) Stirring was carried out until the color changed from brown to white and finally to orange, and then cysts were filtered and washed with water until the dirt and chlorine smell disappeared.
5) The eggs were put into 1 liter of seawater and aerated for 3-4 hours; the eggs that did not contain cysts would settle to the bottom and the ones that were still intact would be on the surface.
6) The cysts could be taken and stored in the refrigerator for reprocessing.

The process of preparing *Artemia* as natural food for beautiful groupers was carried out by hatching eggs (cysts) of *Artemia sp.* by means of decapsulation. After the decapsulation process, Artemia culture could be carried out by washing the decapsulated eggs using sea water until clean before 0.025-1 g/L eggs were put into a hatching container filled with sea water. After that, Artemia eggs were strongly aerated for 24 hours until they hatched. Newly hatched *Artemia* would swim to the surface of the water, while those that settled to the bottom did not contain cysts. *Artemia* could be harvested using a sieve and then fed to beautiful grouper larvae.

Rebon shrimps were also used as feed for beautiful groupers. An advantage of using rebon shrimps as feed is that they have a high protein content. According to Lazaren [13], although the nutritional content of protein and fat in rebon shrimps based on the proximate test results has a high protein content compared to rough fish, in order to avoid malnutrition in beautiful grouper larvae, artificial feeding can be done. However, artificial feed can also have a negative impact on water quality due to the decomposition of uneaten feed residue. Artificial feeding was carried out when the larvae were aged D12. During the larval growth period, artificial feed was given based on the size of the mouth opening. The type of artificial feed used was Otohime brand [20].

2.2. *Egg stocking and hatching*

The process of egg stocking was carried out after the preparation of the larval rearing tank in the room (indoor) by cleaning the tank. Placement of the tank for egg hatching and larval rearing indoor aimed to control the conditions [21]. The cleaned larval rearing tank was then filled with seawater as much as 70% of the volume of the tank to be later filtered using a filter bag. The process of egg stocking was carried out in the morning as many as 200,000 eggs. Prior to stocking, the eggs were subjected to an acclimatization process for 10-15 minutes until the plastic bag condensed as a sign that the temperature in the bag and pond was the same. Then, the egg stocking was done slowly, and the eggs would hatch within 18-20 hours. The characteristics of a good and developing egg cell are being transparent or floating [22]. Meanwhile, unfertilized eggs have the characteristics of being white in color and settling at the bottom. Unfertilized eggs can be removed by siphoning. According to Melianawati and Astuti [23], the characteristics of fertile eggs are clear in color and during selection are on the surface or float, yet unfertilized eggs will be white and sink to the bottom.

2.3. Frequency and timing of feeding

Beautiful grouper larvae that had just hatched would begin to be fed after 2-3 days because after hatching beautiful grouper larvae would have innate energy reserves in the form of yolk eggs and oil granules with a larval life span of 7 days [24]. According to Ismi *et al.* Beautiful grouper larvae began to open their mouths and fed on D3 and were active in swimming after developing dorsal and pectoral fins on D4. The first feed for beautiful grouper larvae aged D2 was *Chlorella* at a dose of 100-200 thousand cells/mL and rotifers as much as 3-5 ind/mL in the evening. The administration of *Chlorella* and rotifers was continued until D30 with the frequency of administering *Chlorella* once a day (morning) and rotifers three times a day (morning, afternoon, and evening). After the administration of *Chlorella* and rotifers, the tank was moderately aerated.

3. Results and discussion

3.1. Hatching rate

The hatching rate obtained was 60% with a media temperature of 31°C. According to Simbolan *et al,*. [25] the best hatchability of beautiful grouper eggs at a temperature of 29-30°C was 91%. The temperature of the media affects the hatching rate of fish eggs that are incubated; setting eggs at a higher temperature will accelerate the hatching process, and conversely low temperatures will slow down the hatching process [26]. After the eggs hatched, preventive measures were taken where the hatchery tank was given squid oil by dripping it on the surface of the water until it was evenly distributed. The provision of squid oil was to prevent the death of beautiful grouper larvae that floated due to being trapped from the surface tension of the water as well as to serve as additional nutrient intake. According to Prakosa *et al.* [27], newly hatched fish larvae have a habit of up and down movement; if the fish larvae rise to the surface of the water, it will trap the larvae from the water surface so that when the larvae will dive they will be stuck. The beautiful grouper larvae will be treated from the process of egg hatching to the age of D23 to be later subjected to the first grading process with the purpose of avoiding cannibalism. After the grading process, the beautiful grouper seeds would be transferred to the rearing tank.

3.2. Feeding

The dose of feeding would increase day by day according to the growth of the larvae or seeds of the beautiful groupers. When the larvae were aged D2-D6, the dose of rotifers was 3-5 ind/mL added with *Chlorella* as much as 100-200 thousand cells/mL, and at D7-20 the dose of *Chlorella* was 500 thousand-1 million cells/mL and *Rotifers* was 10-15 ind/mL, at D21-D30 the dose of *Chlorella* was 500 thousand cells/mL and rotifers was 10-15 ind/mL. Artificial feed with Otohime brand was also given to fish larvae at the age of D11. In order to be eaten by larvae, artificial feed was adapted to the mouth opening of the beautiful grouper larvae. There were various sizes of artificial feed used, which consisted of A1 (250 μm), B1 (360 μm), B2 (360-580 μm), C1 (580-910 μm), S1 (1.0 mm), S2 (1.4 mm), EP1 (1.5 mm), and EP2 (2.3 mm).

According to Slamet *et al.*, [28] each type of fish has a different stage of larval development so that the start of artificial feeding is also different. Feeding artificial feed too early will cause growth to slow down, and on the other hand administering artificial feeding too late will result in mortality rates due to a lack of nutrients [29]. Provision of artificial feeds A1, B1, and B2 was done when the fish larvae were aged D11-D20 with a dose of At satiation (as needed). Artificial feed C1 was given when the larvae were aged D20-D30, which was mixed with artificial feed S1 to train the mouth opening of fish seeds. When the seeds were aged D30-D39, a mixture of S1, S2, and EP1 pellets was given. After the seeds were aged D40, EP2 pellets were given until approaching the harvesting process. Artificial feeding was done by sowing the feed little by little with the size of the artificial feed that had been adjusted to the mouth opening of the larvae/seeds. Provision of artificial feed was carried out with sufficient doses (At satiation) with a frequency of four times a day (at 10.00, 12.00, 14.00, and 16.00). Artificial feed reduction could also be done if at the time of feeding there were still sufficient amounts of *Chlorella*, Rotifers, and *Artemia* to avoid negative impacts on water quality due to spoilage of uneaten feed residues [30,31].

Artemia feeding for larvae/seeds of beautiful groupers was carried out at the age of D14-D40. Giving *Artemia* has an effect on the weight and body length of grouper seeds [32]. The frequency of feeding *Artemia* was twice a day when the larvae/seeds were aged D14-D29 with a dose of 1-3 ind/mL and three times a day when the seeds/larvae were aged D30-D40 at a dose of 3-10 ind/mL. The amount of *Artemia* as feed cannot be determined with certainty because feeding *Artemia* in one hour must be arranged so that the larvae/seeds consumed all of the *Artemia*. The larvae/seeds of beautiful groupers aged D30-D45 were given ad-libitum amount of rebon shrimp. The administration of feeding for the larvae of beautiful groupers at UPT BAPL can be seen in Table 1.

Table 1. Feeding for the larvae of beautiful groupers

Fish Age	Feed Type	Dose	Note
DO-D1	*Yolk egg*	-	-
D2-D6	*Chlorella*	100-200 thousand cells/mL	1x/ day
	Rotifer	3-5 ind/mL	1x/ day
D7-D20	*Chlorella*	500 thousand – 1 million cells /mL	1x/ day
	Rotifer	10 – 15 ind/mL	3x /day
	Artemia	1-3 ind/mL	2x /day
	Artificial Feeding	*at satiation*	4 x/ day (A1,B1,B2)
D21-D30	*Chlorella*	500 thousand cells/mL	1x/ day
	Rotifer	10-15 ind/mL	3x /day
	Artemia	1-3 ind/mL	3x /day
	Artificial Feeding	*at satiation*	4x /day (C1,S1)
D31-D40	*Artemia*	3-10 ind/mL	3x /day
	Artificial Feeding	*at satiation*	4x /day (S1,S2,EP1)
	Rebon Shrimp	*ad libitum*	1x /day
D40-D50	Artificial Feeding	*at satiation*	4x /day (EP1,EP2)
	Rebon Shrimp	*ad libitum*	1x /day

3.3. Water Quality

In general, groupers live in stable waters, so the need for water changes is very important to maintain stable water quality during the maintenance of grouper seeds in controlled containers [33]. The water change process was carried out when the beautiful grouper larvae had reached the age of D8 because the larvae's immune system was strong enough to accept the water change process. Water change used water that had been sterilized in an indoor water reservoir. The rate of water change in the larvae/seeds of beautiful groupers varied according to the age of the larvae/seeds. At the age of D8-D20, 10% -20% of the total volume of the maintenance tank was carried out, and at the age of D21-D30 the water change could be increased by 20% - 50% of the total volume of the maintenance tank. Starting at the age of D31-D45 water changes were carried out as much as 50% - 75% and at the age of D46-D50 water changes could be carried out as much as 75% - 100%. According to Aslianti [34], grouper fish in nature generally live in coral waters, are solitary and prefer clean habitats. Anticipating the natural nature of the groupers, the process of changing water is an important factor in maintaining the stability of water quality.

The siphoning process was carried out when the beautiful grouper larvae were aged D20, it could be done by looking at the the basic condition of the larval rearing tank whether it was dirty or not. Siphoning was carried out by placing a siphon pipe slowly to the bottom of the rearing tank to clean the remnants of metabolism such as feces and feed residues so that the water quality of the media was maintained [35]. The siphoning process was carried out when the larvae were aged D20 because the larval body's resistance could be said to be strong and not stressed against siphoning. According to Mas'ud and Rahayu [38], siphoning is done to reduce toxic ammonia in the water, but there is a risk to the siphoning process as it can stress fish larvae/seeds so that the fish's appetite will decrease. Water quality management carried out in this study was done by checking water quality parameters and changing water every day. Parameters examined included temperature, pH, dissolved oxygen, and salinity. Water quality parameters in the rearing tanks of beautiful groupers had an average temperature of 30°C - 31°C, dissolved oxygen of 7.7 mg/L, salinity of 30 - 32 ppt, and pH of 7.78 - 8.11. According to Ismi *et al.* [36], the range of water quality during the rearing process for beautiful grouper larvae had an average temperature between 28,40 - 30,10°C, dissolved oxygen between 4.80 - 6.50 mg/L, salinity between 33 - 34 ppt, and a pH between 8.14 – 8.19. The water quality parameter values can be seen in Table 2.

Tabel 2. Average water quality measurement results for one month

Water Quality Parameter	Week 1	Week 2	Week 3	Week 4
Ph	7.46 – 8.01	7.78 – 8.11	7.65 – 7.99	7.55 -7.99
Dissolved oxygen (mg/L)	6.6 – 7.1	6.25 - 7	6.66 – 7.3	6.65 – 7.21
Temperature (°C)	28-30	29. 5 - 32	27 - 30	29 – 30
Salinity (ppt)	30 - 32	30 - 33	29 - 32	29 – 30

3.4. Grading and harvesting

Beautiful grouper seeds could be graded at the age of D35 to D40. Siphoning was done before the grading process so that during the grading the maintenance media was not dirty as it could interfere with the grading process. Catching the seeds when grading was not carefully conducted would cause the seeds to be easily stressed. The grading activity aimed to uniform the seeds placed in the rearing tank according to the size of the seed body and also to reduce the cannibalism of beautiful grouper seeds.

Afterwards, harvesting activities were carried out when the seeds were aged D45-D50. The harvesting process was carried out by reducing the volume of water in the maintenance tank until 10% of the total volume of the rearing tank remained. Then, the seeds of the beautiful groupers were caught using a fish net and transferred to a basin filled with aerated seawater. After the seeds were transferred, grading activities and the counting of the number of seeds were carried out. Harvesting was done by reducing the volume of the water in the tank until it reached a height of 15cm, and then the seeds were immediately scooped up and collected in one container. The survival rate of beautiful grouper larvae/seeds was very good, reaching 57%, which means the number of beautiful grouper seeds obtained was 68,400 individuals.

4. Conclusion

Evaluation of larval rearing of beautiful groupers (*Epinephelus fuscoguttatus* x E. *microdon*) using *Chlorella sp.* and *Artemia sp.* included preparation of tanks and water maintenance of culture media, management of feed, stocking and hatching of eggs, frequency and timing of feeding, water quality management, pest and disease control, and grading and harvesting. Seed harvesting could be done after the seeds were aged 45 to 50 days old with a size of 3 cm. Water quality parameters measured were temperature at 28-32°C, pH at 7-8, salinity at 29-33 ppt, and DO at 6.25-7.3 mg/L. Survival rate (SR) in larval rearing was 57%, and the hatching rate (HR) was 60%.

5. References

[1] Ismi S 2010 *ICAI* and *ICOSA*. 1147-1150.
[2] Siar S V, Johnston W L, and Sim S Y 2002 *Col. APEC*. 21p.
[3] Giri NA 2001 *War. Penelitian Per. Indo.* 7(1) p3.
[4] Rimmer M A, Mcbride S, and Will K C 2004 *ACIAR Monograph.* 110 pp 137.
[5] Sudewi Y N, Asih, dan Nasukha A 2020 *JFMR*. 4(1) 87-93.
[6] Sari L A, Pursetyo K T, Arsad S, Masithah E D, Setiawan E, and Affandi M 2019 *Poll. Research*. 38 S27-S32.
[7] Diana N Z A, Sari L A, Arsad S, Pursetyo K T, and Cahyoko Y 2021 *JEE*. 22(1) 29-35.
[8] Hutami W W, Sari L A, Masithah E D, Sahidu A M, and Pursetyo K T 2020 *Asian J. of Water, Env. and Poll.* 17(3) 81-86.
[9] Sari L A, Sari P D W, Nindarwi D D, Arsad S, and Affandi M 2019 *EEC*. 25 S26-S31.
[10] Novianti T, Zainuri M, dan Widowati I 2017 *JBPB*. 1(2) 1-8.
[11] Purba, dan Resmayeti 1995 *Oseana*. 20(3) 21-27.
[12] Purba C Y 2012 *JAMT*. 1(1) 102-115.
[13] Lazaren C C, Karang I W G A, dan Faiqoh E 2018 *JAMS*. 4(1) 86-95.
[14] Bariyyah, Siti K, Fasya A G, Abidin M, dan Hanapi A 2013 *Alchemy*. 2(3) 150-204.
[15] Arsad S, Daryanto A O, Sari L A, Saputra D K, dan Pratiwi F D 2021 *JEMT*. 12(3) 797-807.
[16] Arsad S, Kholifah S N, Prabawati E, Sari L A, Kadim M K, dan Kilawati Y 2020 *Asian J. of Water Env and Poll*. 17(3) 13-17.
[17] Arsad S, Sari L A, Suherman S P, Cahyani Devi, Nadhira Tara, Yulinda E N, Musa Muhammad, Lusiana E D, and Prasetiya F S 2020 *AACL Bioflux*. 13(5) 2878-2885.
[18] Iksan M J, dan Mukhlis A 2016 *JBT*. 16(1) 1-9.
[19] Panggabean M G L 1984 *Oseana*. 9(2) 57-65.
[20] Ghufron M, Rahardja B S, dan Sari L A 2020 *AACL Bioflux*. 13(3) 1328-1335.
[21] Satyani D, dan Priono B 2012 *Med. Akua*. 7(1) 14-19.
[22] Melati V H, Sari L A, Cahyoko Y, Arsad S, Pursetyo K T, Dewi N N, and Idris M. 2021 *Eco. Eng. and Env. Tech.* 22(3) 82–90.
[23] Melianawati R, dan Astuti N W W 2012 *JRA*. 7(3) 421-428.
[24] Sinjal dan Hengky 2014 *JBP*. 2(1) 14-21.
[25] Simbolon A J, Sutarjo G A, and Hariyadi H 2019 *IJOTA* .2(1) 16-24.
[26] Muslim M, Fitriani M, Afrianto A M 2018 *Aqua. Ind.* 19 90-94.
[27] Slamet B, Aslianti T, Setiawati K M, Andriyanto W, and Nasukha A *JRA*. 10(4) 531-540.
[28] Kawahara S, Setiadi I, Ismi S, Tridjoko, dan Sugama K 2000 *Lolitkanta-JICA Booklet*. 11 16.
[29] Musa M, Arsad S, Kilawati Y, Hartiono R F, Fajrin I., Sari L A, dan Prasetiya F S 2020 *EPE,* 46(4) 69-76.
[30] Musa M, Arsad S, Sari L A, Lusiana E D, Kasitowati R D, Yulinda E N, and Cahyani D 2021 *JEE*. 22(2) 70-76.
[31] Mudiarti L, and Zainuddin M 2016 *JFST*. 12(1) 7-11.
[32] Nurhidayat K 2001 Skripsi. Bogor: Institut Pertanian Bogor.
[33] Aslianti T 2010 *JITK*. 2(2) 26-33.
[34] Raharjo E I, and Erdiansyah M 2014 *Ruaya*.3(1). 21-25.
[35] Mas'ud F, dan Rahayu A P 2018 *Grouper*. 9 (1) 17-21.
[36] Ismi S, and Asih Y N, 2014 *JITKT* . 6(2) 403-414.

6. Acknowledgement

Authors wishing to acknowledge assistance or encouragement from colleagues, special work by technical staff or financial support from organizations should do so in an unnumbered Acknowledgments section immediately following the last numbered section of the paper.

Feed conversion ratio and feed efficiency of giant gouramy (*Osphronemus goramy*) seed with different densities on biofloc media

R R Wiyono[1], N N Dewi[2,4], M Amin[2] , A L Fauzan[3]

[1]Program Study of Aquaculture, Faculty of Fisheries and Marine, Universitas Airlangga, Surabaya 60115 Indonesia
[2]Department of Aquaculture, Faculty of Fisheries and Marine, Universitas Airlangga, Surabaya 60115 Indonesia
[3]Omah Koi Farm Indonesia, Banyuwangi East Java Indonesia

[4]Corresponding author: ninanurmaliadewi@fpk.unair.ac.id

Abstract. This study aims to determine the effect of different densities on the feed conversion ratio and feed efficiency of giant gourami on biofloc media. This study used an experimental method using a completely randomized design. Giant gourami seeds were reared for 42 days with four treatments and three replications. K1 (control, non-biofloc, density of 100 fish/pond), K2 (control, non-biofloc, density of 150 fish/pond), P1 (treatment of biofloc with a density of 100 fish/pond), and P2 (treatment of biofloc with a density of 150 fish/pond)). The results showed that different densities had no significant effect on the feed conversion ratio and feed efficiency of giant gourami seeds on biofloc media ($p>0.05$). However, the lowest feed conversion ratio value was found in treatment P2 (1.17 ± 0.01), then followed by treatment P1 (1.28 ± 0.2), K1 (1.99 ± 1.02), and the highest value was in treatment K2. (2.11 ± 0.83). The highest feed utilization efficiency value was found in treatment P2 (85.78 ± 0.01%), then followed by treatment P1 (79.10 ± 0.11%), K1 (58.58 ± 0.24%), and the lowest value was found in treatment K2 (52.02 ± 0.18%).

1. Introduction

Gouramy (*Osphronemus goramy*) is a type of freshwater fish native to Indonesia that is widely preferred with high market demand, and selling price so that production per year increases [1,2]. Gouramy is very popular with the community, but its production is still quite low. Gouramis production in 2010 – 2013 experienced a positive increase, with an average increase of 15.74% per year [3]. An increase in production also occurred in 2013 – 2016, in 2016 the production of gouramy reached 149,553 tons [4]. However, the amount of gouramy production is still relatively low when compared to the production of other freshwater fish consumption. This is because the maintenance period of gouramy is quite long. One of the efforts that can be done to increase production is to carry out intensive hatchery efforts [1].

Intensification is done by increasing the density, limiting land use, good environmental management, and the use of artificial feed is one alternative way to increase the production of gouramy. Intensification of cultivation carried out can cause a decrease in the quality of the cultivation environment. The decline in environmental quality is caused by organic waste from feed residues and manure which are generally dominated by toxic inorganic nitrogen compounds [5].

Biofloc technology can be an alternative solution to the problem of aquaculture waste. This technology can provide higher production yields with lower adverse impacts [6]. The use of biofloc technology can reduce stress levels and fish death due to deteriorating water quality because of the high density. Inorganic nitrogen control can be done by manipulating the C/N ratio. This method is a practical and inexpensive approach because it only requires the addition of carbohydrates as nutrients or food for bacteria. The relationship between carbohydrate addition, ammonium reduction, and microbial protein production depends on the microbial conversion coefficient, the C/N ratio in the microbial biomass, and the carbon content of the additive [7]. Optimal density in maintenance needs to be known to get the optimal value of growth rate, feed conversion ratio, and feed efficiency.

The purpose of this study was to determine the effect of biofloc media and maintenance with different densities on the feed conversion ratio, and feed efficiency of gouramy fry and to determine the optimal density in gouramy seeds culture in biofloc media.

Depending on the results of this study, it is expected to be able to determine the effect of biofloc media and maintenance with different densities on the feed conversion ratio and efficiency of gouramy fry and can be useful information for gourami farmers so that they can accelerate aquaculture activities, reduce operational costs and maintain fish seed stocks.

2. Materials and methods

2.1. Time and Place

This research was conducted in August – September 2020 at the Omah Koi Farm experimental pond, Curah Ketangi Village, Genteng, Banyuwangi, East Java.

2.2. Tools and Materials

This study used 12 round tarpaulin ponds with a diameter of 2 meters and a height of 1 meter, aerators, pipes, hoses, oxygen regulators, aeration stones, hapa, plastic tubs, nets, digital scales, rulers, millimeter block paper, DO meters, pH paper, test kit (ammonia, nitrite, and nitrate), thermometer and stationery.

The materials used in this study were gouramy seeds with a length of 2-4 cm as many as 1500 fish, molasses, commercial fish feed PF-800 with a protein content of 39-41%, fat 5%, fiber 6%, ash 12% and high levels of 10% water and salt.

2.3. Experimental Design

The method in this research is an experimental research type. Experimental research can be interpreted as a research method used to find the effect of certain treatments on others under controlled conditions [8]. This study used a completely randomized design (CRD) with 4 treatments and each had 3 replications. The treatments provided include:

K1 = Control treatment with a density of 100 fish/treatment without biofloc media.
K2 = Control treatment with a density of 150 fish/treatment without biofloc media.
P1 = Treatment density of 100 fish/treatment with biofloc media.
P2 = Treatment density of 150 fish/treatment with biofloc media.

The tested fish were kept for 42 days and sampling was done once a week with the parameters of the length and weight of the fish. Giving molasses is also done once a week. Water quality measurements for temperature and pH were carried out every day (morning and evening), while for DO, nitrite, nitrate, and ammonia 3 times during the maintenance period.

2.4. *Pond Preparation*

Preparation is done by washing the pond using potassium permanganate and then rinsing it clean. The pond is drained and after drying it will be filled with fresh water as much as 1884 liters. 1 kg of salt is added to each pond for disease prevention. Aeration is added by 4 points in each pond.

2.5. *Biofloc Media Preparation*

Floc formation was carried out by giving probiotic containing 3 species of bacteria (*Bacillus subtilis, Bacillus lincheniformis*, and *Bacillus fumillus*). This product also contains molasses as a carbon source. The water that already contains bacteria and molasses will be allowed to stand for 2 days [9], while still being aerated to grow the floc [10]. The frequency of giving molasses is once a week [11]. The amount of organic carbon added to the maintenance medium was calculated by the formula [12] :
The assumptions used in this research include:

1. Feed protein content of 40%
2. Nitrogen Content of 16%
3. The nitrogen content of feed that is wasted into the cultivation media is 75%
4. Target C/N Ratio 12 (Rosmawati et al., 2017)
5. Carbon content in molasses (% C) 53.78% [13].

2.6 *Feeding*

The feed used in this study was commercial feed PF - 800 with a dose of 5% of the fish's body weight. Feed is given 2 times a day, every morning (07.00) and afternoon (16.00).

2.7 *Research Parameters*

The test parameters used in this study were feed conversion ratio and feed efficiency. The calculation formula used is as follows:

A. *Feed Convertion Ratio*

The feed conversion ratio is calculated by the formula [14] :

$$FCR = \frac{F}{(Wt + D) - Wo}$$

Description :
FCR = Feed conversion ratio (feed conversion ratio).
F = Weight of feed consumed (g).
Wt = Biomass of the tested animals at the end of rearing (g).
D = Weight of dead fish (g).
Wo = Biomass of the tested animals at the beginning of rearing (g).

B. *Feed Efficiency*

Feed efficiency can be calculated by the formula [14] :

$$FE = \frac{Bt - Bo - Bd}{F} \times 100\%$$

Description :

FE = feed efficiency (%).
Bo = initial fish weight (kg).
Bt = final fish weight (kg).
Bd = dead fish weight (kg).
F = amount of feed (kg).

2.8 *Data Analysis*
The data obtained from the results of the study were observed using the analysis of variance (ANOVA) at a significant level (α) 0.05. If there are a difference in the effect on the treatment, the calculation is continued with Duncan Multiple Range Test (DMRT).

3. Result and Discussion
3.1 Feed Convertion Ratio
The results of the ANOVA test in this study showed that different densities had no significant effect on the gouramy seed feed convertion ratio($p<0.05$). In the treatment with biofloc media, the feed conversion ratio value was lower when compared to the treatment without biofloc media. An overview of the average feed conversion ratio values during the study can be seen in Figure 1 below:

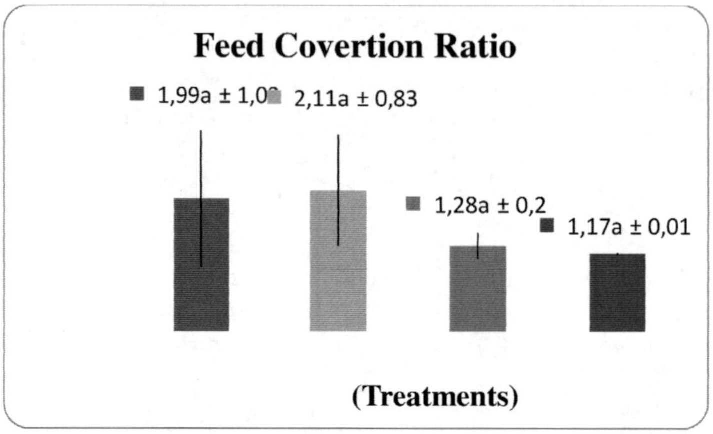

Figure 1. Average value of feed conversion ratio of gouramy seed

The results of the ANOVA test showed that there were no significant differences in all treatments. The FCR value data obtained respectively in K1, K2, P1, and P2 treatments were 1.99, 2.11, 1.28, and 1.17. The best FCR value is found in the P2 treatment. All biofloc treatments had a lower FCR value when compared to the control treatment, the difference in FCR values in the biofloc treatment and the control treatment was also quite significant. P2 has a lower FCR value when compared to P1, but K1 has a lower FCR value when compared to K2. This means that the density treatment given does not significantly affect the FCR value.

3.2 *Feed Efficiency*
The results of the ANOVA test in this study showed that different densities had no significant effect on the feed efficiency of gouramy seeds ($p<0.05$). In the treatment with biofloc media, the feed efficiency value was higher than in the treatment without biofloc media. An overview of the average feed efficiency values during the study can be seen in Figure 2.

The results of the ANOVA test showed that there were no significant differences in all treatments. The data on feed efficiency values obtained respectively in the K1, K2, P1, and P2 treatments were 58.58%, 52.02%, 79.10%, and 85.78%. The best feed efficiency value was found in the P2 treatment. All biofloc treatments had a higher feed efficiency value when compared to the control treatment, the difference in feed efficiency values in the biofloc treatment and the control treatment was also quite significant.

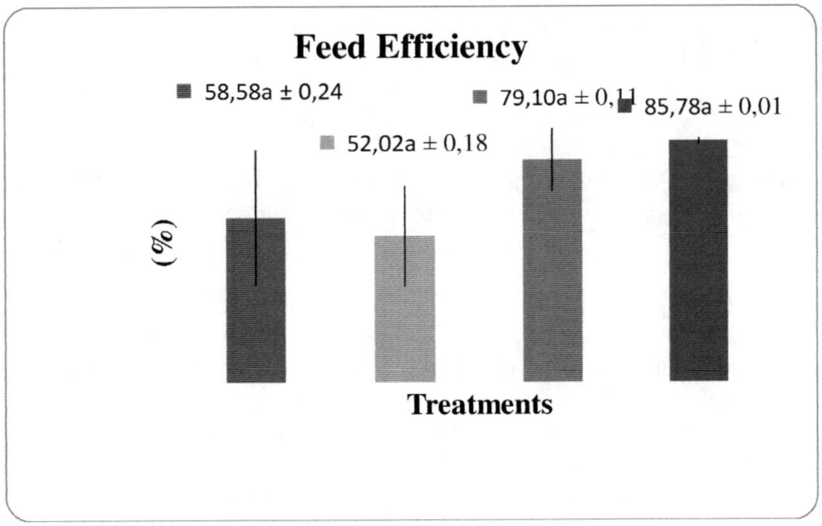

Figure 2. Average value of feed efficiency of gouramy seed

3.3 Discussion

The feed conversion ratio is the ratio between the total amount of feed given to the total weight of the fish produced [11]. Research in the field of aquaculture discusses many ways to reduce the value of the feed conversion ratio. Feed is an expensive and important component in aquaculture because the feed has the function to meet the energy needs needed by fish. Thus it is very necessary to carry out feed management in aquaculture activities. Feeding management is an effort to maximize the use of feed for growth, with feeding management it is hoped that the feed given can be used by fish effectively and efficiently to produce optimal fish growth [15]. Biofloc technology is known to reduce feed conversion ratio by minimizing feed intake. Compared to traditional aquaculture techniques, biofloc technology provides a more sustainable approach with minimal water exchange along with reduced feed intake and transforms it into a low-cost sustainable technology for sustainable aquaculture development. That's because biofloc media contains bacterial biomass that can be used as a source of protein for cultured organisms.

The best feed conversion ratio value was found in the P2 (biofloc) treatment with a value of 1.17 ± 0.01 and the lowest value was found in the K2 (non-biofloc) treatment with a value of 2.11a ± 0.83. This value is better than the previous research [16], with the best feed conversion ratio value of 1.31. The value of the feed conversion ratio in aquaculture with the application of a biofloc system will be slightly lower due to an increase in floc biomass as a source of nutrition or additional food for cultured organisms [17]. The carbon source added to the culture media will be converted into an energy source by heterotrophic bacteria to produce proteinaceous bacterial biomass [18]. This bacterial biomass can be used as an alternative food source for cultivators and can increase the growth rate of cultured organisms [19].

The best feed efficiency value was found in the P2 (biofloc) treatment with a value of 85.78% and the lowest value was found in the K2 (non-biofloc) treatment with a value of 52.02%. This value is better than the previous research [16], with the best feed efficiency value of 76.29%. The protein content contained in the floc reached 42% in dry weight [20]. The high protein content in the floc plays an important role in increasing the feed efficiency value of cultured organisms [21]. With the application of biofloc technology, the value of feed efficiency will be higher due to an increase in biofloc biomass which can be used as a source of nutrition or additional food for cultivated organism [17].

Bacteria added as probiotics into the culture media can improve the digestion of organisms so that more feed will be absorbed into the fish body [21]. Furthermore, bacteria in the digestive tract will secrete digestive enzymes such as proteases and amylase [22]. *Bacillus* sp. can increase feed absorption by increasing protease enzymes in the digestive tract, increasing growth, and reducing the number of pathogenic bacteria contained in the digestive tract of organisms [23].

4. Conclusion
The results of the ANOVA test in this study showed that different densities did not significantly affect the feed conversion ratio and feed efficiency of gouramy fry ($p<0.05$). However, the resulting value in all treatments with biofloc media was higher than the control treatment. Treatment with a density of 150 fish/treatment (P2) gave the best value on the feed conversion ratio and feed efficiency of gouramy seeds.

5. References
[1] Handayani D L, Sumaimi M E, Ashari T A and Jumadi 2020 *J Ilm Perikan dan Kelaut.* 12 (2) : 302 – 307.
[2] Pranata A, Raharjo E I and Farida 2017 *J Ruaya.* 5 (1) 1-6.
[3] General Directorate of Aquaculture Indonesia 2013 *Annual Report of the Directorate of Production of the Ministry of Maritime Affairs and Fisheries p 45.*
[4] Hardaningsih I 2018 *Gouramy Cultivation for Food Security and Poverty Eradication in Rural Areas. 15th Annual National Seminar* (Yogyakarta : Faculty of Agriculture Gadjah Mada University)
[5] Hermawan T E A, Sudaryono A and Prayitno S B 2014 *J Aquac Manag Technol.* 3 (3) 35 – 42.
[6] Bossier P and Ekasari J 2017 *Microb Technol.* 10 (5) : 1012 – 1016.
[7] Avnimelech Y 1999 *Aquaculture* 176 : 227 – 235.
[8] Koriaty S 2016 *J Edukasi.* 14 (1) : 132-142.
[9] Afifi I 2014 *Utilization of Biofloc in Dumbo Catfish (Clarias sp.) Cultivation with Different Stocking Densities Against Growth Rate and Survival Rate (SR). Undergraduate Thesis.* (Surabaya : Faculty of Fisheries and Marine Universitas Airlangga)
[10] Ma'ruf I 2016 *Societa* 5 (2) : 82 – 86.
[11] Zaman Y R, Wiguno T A A, Dewi N N and Agustono, Wulansari P D, Prayogo, Fauzan A L 2021 *Eco. Env. & Cons.* 27 (2) : 752 – 758.
[12] De Schryver P, Crab R, Defoirdt T, Boon N and Verstraete W 2008 *Aquaculture* 277 : 125 – 137.
[13] Widanarni, J Ekasari and S Maryam 2012 *Hayati J Biosci.* 19 (2) : 73 – 80.
[14] Effendie M I 1997 *Fisheries Biology* (Yogyakarta : Yayasan Pustaka Nusantara) p 163.
[15] Nurhad S, Murniasih S. and Dewi N N 2021 *IOP Conf. Series : Earth and Environmental Science* 718 (2021) : 1 – 9.
[16] Rosmawati and Muarif 2017 *Int J Sci Basic Appl Res.* 36 (6) : 47 – 59.
[17] Widanarni, Yuniasari D, Sukenda and Ekasari J 2010 *Hayati J Biosci.*17 (3) : 115 – 119.
[18] Husain N, Putri B dan Supono 2014 *e J Rekayasa dan Teknol Budid Perair.* 3 (1) : 343 – 350.
[19] Khanjani M H and Sharifinia M 2020 *Rev Aquac.* 12(3) : 1836-1850.
[20] Crab R, Avnimelech Y, Defoirdt T, Bossier P and Vestraete W 2007 *Aquaculture* 270 : 1 – 14.
[21] Simanjuntak I C B H, Suminto and Sudaryono A 2016 *J Aquaqulture Manag Technol.* 5(2): 1-8.
[22] Irianto A 2003 *Aquaculture Probiotics* (Yogyakarta : Gadjah Mada University Press) p 125.
[23] Kennedy SB, Tucker Jr, Neidic J W, Carole L, Cooper G K, Jarrel J L and Sennett D G 1998 *Bull Mar Sci.* 62 : 573 – 558.

IOP Conf. Series: Earth and Environmental Science 1273 (2023) 012083 doi:10.1088/1755-1315/1273/1/012083

Histopathological condition of *Litopenaeus vannamei* infected with Infectious Myonecrosis Virus (IMNV) on different days of culture collected from Aceh, Sidoarjo, Probolinggo and Lampung Indonesia

I Puspitasari[1,2], D N Suseno[1], IGP G R Yudana and M K Anwar[1],

[1] Politeknik Kelautan dan Perikanan Sidoarjo

[2] Corresponding author: indah.p@gmail.com

Abstract. Histopathological detection is one of many methods that has been used on shrimp diseases detection. A study on histopathology of *Litopenaeus vannamei* infected by infectious Myonecrosis Virus (IMNV) has been conducted in Sidoarjo, East Java. The purpose of this study was to give a clear view of the histopathological condition of shrimp infected by IMNV which is necessary to examine the destructive level of the disease within the shrimp cells and hopefully can contribute to adding more information about the disease. About 16 samples of shrimp from different days of cultured that shown an IMNV infection clinically sign has been examined on their muscle tissue, lymphoid organ and hepatopancreas condition. The result has shown that shrimp infected by IMNV has a necrosis and haemocyte infiltration on muscle, spheroids on lymphoid organ and shown congestions and haemocyte infiltrations on hepatopancreas. There were significant differences ($P<0.05$, Kruskal-Wallis) on muscle and lymphoid organ condition between shrimp before and after 2 months cultured. But there was no significant difference in shrimp hepatopancreas condition.

1. Introduction

In 1996, there was a decline in tiger shrimp production due to disease attacks and a decrease in environmental quality in Indonesia. This prompted the government to make vannamei shrimp as an alternative in aquaculture. The demand for vannamei shrimp has increased from year to year. However, in 2009 the production of vannamei shrimp decreased due to the attack of INMV (Infectious Myo Necrosis Virus) [1]. Until now, IMNV has still become one of the diseases that has been a threat to vannamei culture. The fast mortality of shrimp due to IMNV infection has made the disease is needed to be studied. Morphological and behavioural symptoms on infected shrimp have been reported in previous studies. On the other hand, a few histopathology analyses on infected shrimp have been studied.

Histopathology is a microscopic analysis of disease wherein histopathological observation the information obtained is in the form of an overview of organ/tissue changes [2]. This is following the previous study, that histopathological analysis can be used as a biomarker to determine the health condition of shrimp through structural changes that occur in organs that are the main targets of

IOP Conf. Series: Earth and Environmental Science 1273 (2023) 012083 doi:10.1088/1755-1315/1273/1/012083

contaminants such as gills, lymphoid, hepatopancreas, intestines and so on [3]. The cause of infection can be identified by histopathological examination. There were a lot of previous studies on histopathology analysis on shrimp infected by IMNV, but there was no study related to histopathology conditions on different days of culture. The importance of histopathological examination in disease diagnosis may be able to identify the cause and level of infection. Therefore, a study on histopathological analysis of shrimp infected with IMNV in different days of culture is necessary to add information about knowing the cell damage caused by the virus.

2. Materials and methods

This study has been conducted from March until May 2021 in PT. Central Proteina Prima Sidoarjo, East Java. There were 16 samples of shrimp *Litopenaeus vannamei* collected from Aceh, Sidoarjo, Probolinggo and Lampung Indonesia, have been observed. The samples that have been observed were on 27, 36, 47, 70, 85 and 94 days of culture. All samples have shown IMNV symptoms such as white colour on the muscle.

2.1. Histopathology procedures

2.1.1. Fixation. The samples were soaked with Davidson's solution for 24 hours in room temperature, in order to make the tissue well preserved and stiff enough to be able to sliced as thin as possible. This is necessary to make coloring process to be successful.

2.1.2. Trimming. The preserved tissue samples were trimmed into smaller size and each were sliced with ± 0,5 cm thickness on the head and 1st, 3rd and 6th abdomen segments and put into the labeled cassettes. The cassettes then soaked in a flowing water for 2 hours.

2.1.3. Tissue processing. In this step, the paraffin was infiltrated into the tissue. The tissue was dehydrated by soaked into ethanol 70%, 90%, 96%, ethanol reintz I, II and ethanol absolute I, II. After that the tissue was cleared by xylene I and II. This processing was done by the Automatic Tissue Tek Processor for 90 minutes.

2.1.4. Blocking and Sectioning. The processed tissues were embedded into the paraffin by using Embedding Tissue Tek machine. Then, the blocks were storage at -20°C for 1 hour and sectioned by using microtome rotary until it reached 3-5 μm thickness.

2.1.5. Staining and Mounting. After blocking and sectioning steps, the sample was stained with Haematoxylin solution by using Tissue Tek DRS machine. Samples on the object glass then mounted by entellan.

2.2. *Data analysis.* Data were analyzed by using Kruskal Wallis test with significancy level P<0,05 in Minitab version 16.

3. Results and Discussion

3.1. Histopathology on muscle

In the earlier studies, vaname shrimp infected with IMNV will experience shrimp muscle transparency where the muscle will look white, the white color is usually caused by necrosis of skeletal muscle due to infection with the IMNV virus [4], [5], [6], [7] as well as other clinical symptoms of IMNV can be seen through the histology of muscle tissue, lymphoid organs, and hepatopancreas organs with hematoxylin-eosin (HE) staining [8].

Histopathological observation of muscle using a microscope with a magnification of 20 m found that the vannamei shrimp muscle has shown necrosis and hemocytes infiltration as shown in figure 1a. Whereas on normal vannamei shrimp samples, the muscle organs did not show the necrosis (figure 1b).

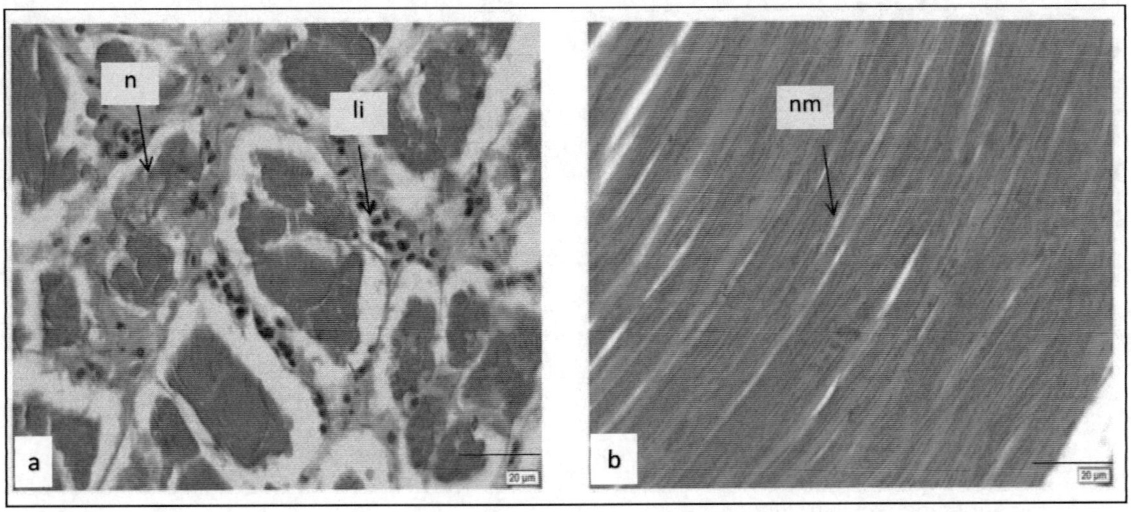

Figure 1. Histopathology of muscle cells in IMNV infected vannamei. a. IMNV infected. nm = normal muscle; li= lympocyte infiltration; n= necrosis; b. normal cells (mn) (H&E, bar scale = 20 µm).

Muscle is a tissue that is commonly used as a sample for cell scale testing for IMNV, it is explained that the clinical symptoms of shrimp infected with IMNV are when the infection starts to get worse, the white belly and tail turn reddish due to necrosis and the shrimp will experience cramps in the abdomen. muscle tissue or tissue damage and hemocytes infiltration when infected with IMNV [9]. Necrosis is a morphological change in which cells and tissues experience decreased activity and eventually die [10]. Cells that undergo necrosis look destroyed and will be lost or die. In general, shrimp infected with the IMNV virus show skeletal muscle lesions, coagulation necrosis including multifocal necrosis, hemocyte congestion, fibrocytes inflammation, phagocytosis and cytoplasmic inclusion bodies and hemocyte infiltration [9,11].

3.2. Histopathology on lymphoid organ

The lymphoid organ is an organ that functions as a producer of lymph in the process of shrimp immunity, this is what causes a very rapid decrease in shrimp immunity when exposed to IMNV disease [12]. Histopathological observation results of lymphoid organs of vannamei shrimp infected with IMNV by using a microscope with a magnification of 10 m and 20 m has shown the presence of spheroids cells. Spheroids are events when part of the organ loses blood loss (lysis) so that the organ has a colour that is not bright and even there is no colour (figure 2a) normal lymphoid organs in Vannamei shrimp samples did not show cells with necrosis (figure 2b).

The formation of the lymphoid organ spheroid (LOS) cells has been observed in many naturally or experimentally infected penaeid prawns with viral diseases. LOS cells have the characteristics of exocytosis, granular hemocytes that have phagocytosed foreign material, particularly viruses [13]. It has been suggested that these spheroid cells constitute a major site of viral degradation [14]. In the research of [15] which stated that vanamei shrimp infected with the IMNV virus found accumulation of lymphoid organ spheroids (LOS) which is a hypertrophy of lymphoid cells after being infected with the virus. Diseases that attack shrimp generally attack the immune system where in shrimp, the immune system is played by the lymphoid organ [16].

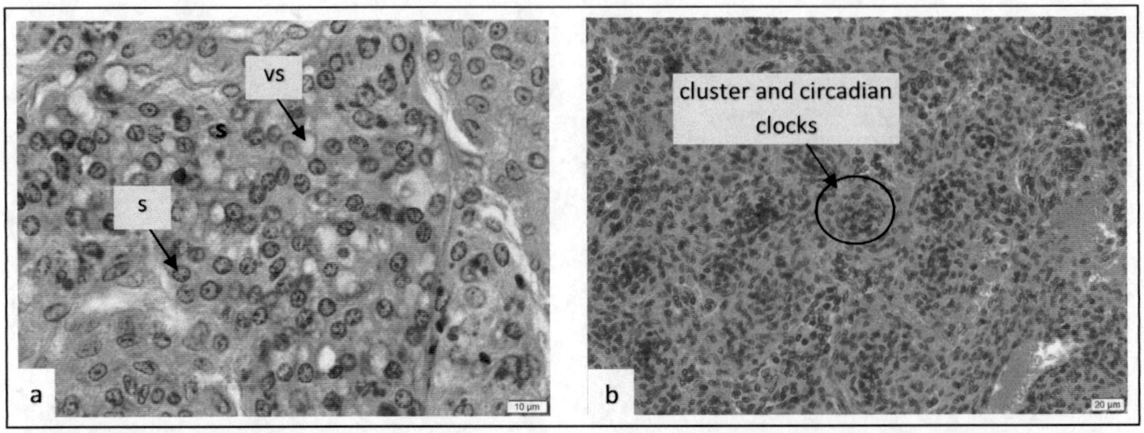

Figure 2. Histopathology of lymphoid organ in shrimp vannamei; a. IMNV infected cells, has shown spheroid cells (s); vacuole spheroid cells (vs); b. normal cells shown an organized, cluster and circadian clocks of lymphoid cells (H&E, bar scale = 10 µm ; 20 µm).

3.3. Histopathology on hepatopancreas

Histopathological observation of shrimp sample hepatopancreas has done by using a microscope with a magnification of 10 m and 20 m. It has shown a vacuole degeneration, congestion, hemocytes infiltration, hemorrhage, and necrosis (figure 3a).

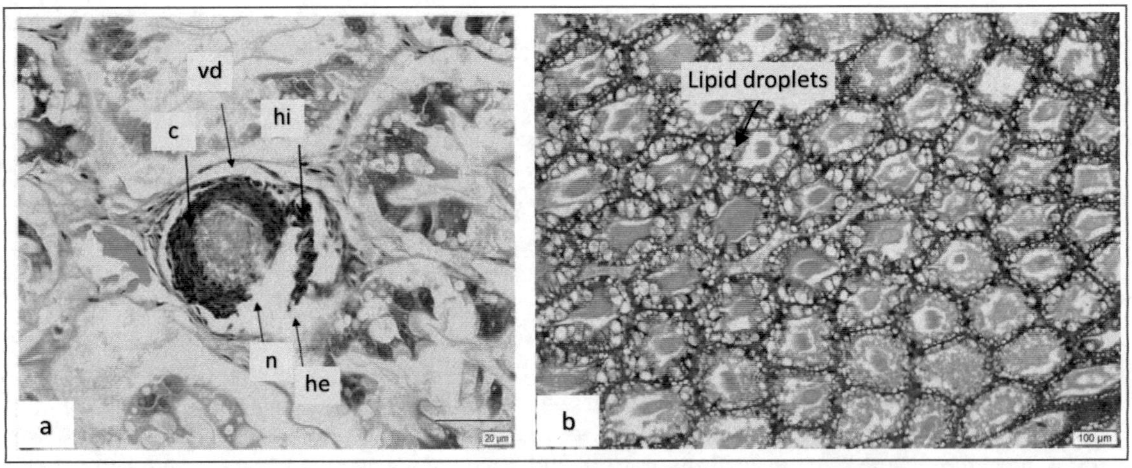

Figure 3. Histopathology of vannamei hepatopancreas; a. IMNV infected hepatopancreas cells of vannamei, c= congestion; vd=vacuole degradation; hi= hemocyte infiltration; n=necrosis; he= hemorrhage, b. normal cells, (H&E, bar scale = 20 µm, 100 µm).

Vacuole degeneration is an early stage of hepatopancreas damage. Vacuole degeneration is reversible (cells can return to normal), while necrosis is irreversible which cannot be cured which results in decreased tissue activity due to the loss of several cell parts one by one and can even cause death [17], [18]. Damping is characterized by the accumulation of very dense red blood cells in blood vessels, which will continue to become clogged blood vessels (congestion or dilation of secondary filaments that occur due to blood accumulation (bleeding in blood vessels) [19], [20]. Congestion can usually occur preceded by degeneration of the vacuole where the enlarged liver cells are filled with erythrocytes which causes the sinusoids to widen resulting in blood damming and haemorrhage. Haemorrhage is the release of red blood cells from the central vein [21], [22] and cell infiltration is an immune response due to disease that enters the tissue cells [23]. The hepatopancreas undergoes necrosis or cell damage, congestion and

IOP Conf. Series: Earth and Environmental Science 1273 (2023) 012083 doi:10.1088/1755-1315/1273/1/012083

infiltration of hemocytes due to pathological changes in the hepatopancreas due to IMNV virus attack [24].

Hepatopancreas in crustaceans is a digestive organ, which has an important function, especially as an absorption organ which is characterized by the presence of microvilli cells that show an absorption and secretory function of enzymes [25]. This is also supported by previous research [26]; [27], [28] that the hepatopancreas functions as a place of metabolism, storage of nutrients, a place for vitellogenin synthesis in addition to the ovaries and detoxification. A normal hepatopancreas cells will show a lots of lipid droplets due to its function in digestive system of a shrimp (figure 3b). Hepatopancreas also being the principal organ for storage of nutrients (mainly lipids) [29]. Therefore, the amount of lipid droplet showing the amount of nutrient absorbed in shrimp.

3.4. Histopathological condition on different days of culture

Shrimp affected by IMNV usually have clinical symptoms that can be seen visually. Shrimp usually move passively, decrease appetite and have bruising on body tissues. In a statement from the Directorate of Fish and Environmental Health [30], it was explained that at the beginning of IMNV infection the shrimp would still eat well. However, 1 week after IMNV infection, the shrimp will show a drastic decrease in appetite. In this study, the analysis was conducted on shrimp infected with IMNV according to the condition of the muscle, lymphoid organ and hepatopancreas with different days of culture (table 1).

Table 1. histopathological observation result on Muscle, Lymphoid organ, and Hepatopancreas including its lipid droplets in IMNV infected vannamei shrimps

DOC	Muscle	Lymphoid organ	Hepatopancreas	Lipid droplets
27	-	-	-	70%
27	+	-	-	60%
27	+	-	-	80%
27	-	-	-	50%
36	-	-	+	40%
36	-	-	-	70%
36	++	-	-	80%
47	-	-	-	80%
47	+	-	+	30%
70	+++	+	-	60%
70	++	-	-	80%
83	+++	++	-	60%
85	-	-	-	80%
85	+++	+	+	60%
94	+++	++	-	70%
94	+++	+++	-	70%

Notes :
- = no damage cells
+ = a few damage cells (10-30% of observation field)
++ = moderate damage cells (40-60% of observation field)
+++ = many damage cells (70-100% of observation field)

The samples were divided into two groups, before and after 2 months of cultures and analyzed. The muscle and lymphoid organ condition have shown a significantly ($P<0,05$, Kruskal Wallis) higher level of damage on the tissue, yet no significant difference has found in shrimp hepatopancreas histopathology. This result is confirmed that muscle and lymphoid organ are the principal target tissue in IMNV infection on shrimp, as mentioned in the previous study that the principal target tissues for IMNV include the striated muscles (skeletal and less often cardiac), connective tissues, hemocytes, and the lymphoid organ parenchymal cells [31]. Whereas hepatopancreas histopathological observation also gives an important information related with the shrimp health condition that support the diagnose from muscle and lymphoid organ.

4. Conclusion

The result has shown that shrimp infected by IMNV has a necrosis and haemocyte infiltration on muscle, spheroids on lymphoid organ and shown congestions and haemocyte infiltrations on hepatopancreas. There were significant differences ($P<0.05$, Kruskal-Wallis) on muscle and lymphoid organ condition between shrimp before and after 2 months cultured. But there was no significant difference in shrimp hepatopancreas condition. It is confirmed that that muscle and lymphoid organ are the principal target tissue in IMNV infection on shrimp.

5. References

[1] Fitriyani W 2014 *Infectious Myo Necrosis Virus* (Universitas Lampung).
[2] Pazra D F 2008 Histopathological Description of Gills, Muscles and Intestines in Catfish (Clarias spp.) Originally from the Bogor Region Thesis (Bogor: Faculty of Veterinary Medicine, Bogor Agricultural University) p 74.
[3] Setyowati A, Awik D H and Nurlita 2012 *J Ristek Akua* 2 22-29.
[4] Taukhid and Nur'aini L 2009 *SEAFDEC International Workshop on Emerging Fish Diseases in Asia* 61 255-262.
[5] Poulos B T and Lightner D V 2006 *Dis. Aquat. Org.* 73 69-72.
[6] Poulos B T, Tang K F J, Pantoja C R, Bonami J R, and Lightner D V 2006 *J. Gen. Virol.* 87 987–996.
[7] Sukenda, Nuryani S, and Sari I R 2011 *J Akua Indonesia* 10 192-202.
[8] Hasan A, Sukenda and Winarni 2011 *Co-infection of Infectious Myonecrosis Virus (IMNV) and Vibrio harveyi in Pacific White Shrimp (Litopenaeus vannamei).* Thesis (Bogor: Post Graduate School, Bogor Agricultural Institute) p 49.
[9] Tang K F, Pantoja C R, Poulos B T, Redman R M, Lightner D V 2005 *Dis. Aquat. Org.* 63 261–265.
[10] Sari R R B, Srjito, Alfabbetian H C H 2015 *J Aqua Manag Tech* 4 26-32.
[11] Takashima F, Hibiya T 1995 *An Atlas of Fish Histology: Second Edition* (Tokyo: Kodansha) pp 789-848.
[12] Widodo D S 2002 Pathogenicity Test of White Spot Virus on Tiger Shrimp (Penaeus monodon Fabricius) with 30, 60 and 90 minutes of soaking time with a concentration of 100 μg/ml Thesis (Bogor: Faculty of Fisheries and Marine Sciences. Bogor Agricultural University) p 87.
[13] Anggraeni M S and Owens L 2000 *Dis Aquat Org.* 40 85-92.
[14] Rusaini, Owens L 2010 *Fish Shellfish Immunol.* 29 367-77.
[15] Andrade T P D, Redman R M, Lightner D V 2008 *Aquaculture* 278 179–183.
[16] Kondo M, Itami T, Takahashi Y, Fujii R, and Tomonaga S 1994 *Dev Comp Immunol* 18 S109.
[17] Rizki M, Rostiana S M T, Damanik B 2009 *Uji histopatologi organ ren, insang, ginjal, intestinum dan hepar ikan mas.*
[18] Sofiana E D, Widjiati, Luqman E M, Samik A 2015 *Veterinaria Medika* 8 131-136.
[19] Robert R J 2001 *Fish Pathology* (London: Elsevier Health Sciences Harcourt Publishers) p 427.
[20] Yuniar V 2009 Mercury (Hg) toxicity on survival rate, growth, blood appearance and organ damage in tilapia (Oreochromis niloticus) Thesis 1-89.
[21] Fricke N F, Stentiford G D, Feist S W, and Lang T 2012 *Mar Env Res.* 82 1-14.
[22] Lubis F U, Marusin N, Zakaria J I 2014 *J Biologi Universitas Andalas* 3 162-167.

[23] Zhahrah Z, Indriyani N, Kadir S 2016 *Media Akuatik* 1 72-76.
[24] Tang K F J, Pantoja C R, Poulos B T, Redman R M, Lightner D V 2008 *Dis Aqua Org.*
[25] Zeng H, Haihui Y, Shaojing L, and Wang G 2010 *Cell. Dev. Biol Animal* 46 431-437.
[26] Soegianto A, Primarastri N A, Winarni D 2004 Effect of cadmium administration on survival rate and damage to gill structure and hepatopancreas in stretch shrimp (Macrobrachium sintangense De-Man) Biological Research File (Surabaya: Faculty of Mathematics and Natural Sciences, Airlangga University) p 10.
[27] Sousa L G, Cuartas E I, and Petriella A M 2005 *Biocell* 29 25-31.
[28] Suseno D N, Epy M L, Mirni L, Akhmad T M, Muhammad A S 2020 *Asian Pacific J Repro* 9 37-43.
[29] Cervellione F, Mcgruk C,Van den Broek W 2017 *J Crustacean Biol.* 37 6.
[30] Directorate of Fish Health and Environment 2010 IMNV Disease Control Technical Guidelines *(Infectious myonecrosis virus)* (Jakarta) p 22.
[31] Lightner DV, Pantoja C R, Poulos B T, Tang K F J, Redman R M, Pasos De Andrade T, and Bonami J R 2004 *Global Aqua Advocate* 7 85.

Rob Flood as impact of sea level rise around Kenjeran Beach Tourism Surabaya

N A Agustina [1,2], Supartono[1] and V D Prasita[1]

[1] Department of Marine Engineering, Faculty of Engineering and Marine Science, Hang Tuah University, Jl. Arief Rahman Hakim no. 150, Surabaya, 60111, East Java, Indonesia

[2]Corresponding author: nur.asyiah@hangtuah.ac.id

Abstract. Kenjeran Beach is located in the coastal area of the East Coast of Surabaya (Pamurbaya) where most of the ecosystems are mangroves, so this area has the potential to be inundated with water (flood). This study aims to analyze tidal flooding as the impact of sea level rise and formulate appropriate adaptation & mitigation strategies so that tourism utilization can be carried out in a sustainable manner. The method used in this study was the tidal calculation which produces sea level rise predictions and the Geographic Information System method which produces a map of the vulnerability of the tidal flood area. The map results were analyzed to determine adaptation and mitigation strategies from the tidal flood. Based on the analysis of sea level rise, the area inundated due to tidal flooding was around 42 cm in the next 50 years. In an effort to adapt and mitigate sea level rise disasters, it was carried out with structural (physical or technical) mitigation and non-structural (non-technical) mitigation.

1. Introduction

The coastal area of the East Coast of Surabaya (Pamurbaya) is mostly mangrove ecosystem and has various uses, including: beach tourism, settlements, and mangrove conservation. In addition, the coastal area is affected by sea level rise [29,33,34] Coastal areas have relatively flat land elevations with relatively low altitudes relative to sea level, making coastal areas vulnerable to natural phenomena, one of which is sea level rise.

Handiani *et al* [12] explained that the issue of sea level rise due to global warming is a separate threat to coastal areas in Indonesia. The Tanjung Perak Maritime BMKG estimated that every year the coastal area of Surabaya experiences tidal flooding due to sea level rise, and in one year there was about four to five times the danger of flooding above the average sea level. The potential impact of sea level rise on densely populated islands, lowlands and coastal areas indicates that this was an urgent problem that requires deep-rooted countermeasures [18, 32].

Surabaya is one of the second big cities in Indonesia which has a long beach tourism bordering the island of Madura. Mayor of Surabaya Tri Rismaharini considered that Surabaya could be seen as a maritime city other than a metropolitan city. This assumption was influenced by the presence of beaches in the North and East areas of the City of Heroes, namely, in the North area of Surabaya, from West to East lies a quite beautiful beach, namely Kenjeran Beach. The Surabaya city government was trying to take advantage of this natural advantage into a very profitable tourist destination. The Surabaya City Culture and Tourism Office has developed Kenjeran Beach to increase tourist attraction through infrastructure development programs and facilities & infrastructure [24, 30]. Kenjeran Beach is one of the interesting tourist attractions with its stage pier that extends into the middle of the sea, as well as Kenjeran Park and Ria Kenjeran Beach which have a variety of artificial tourist objects. These tourist objects and attractions can be used as targets for recreational tourism destinations for tourists who want to visit. The concept of adopting the Industrial - Minimalism design style that elevated Surabaya as an industrial city was applied in such a way that it presents a distinctive tourist attraction [23,31].

However, in a development, especially development in the tourism sector, it was necessary to pay attention continuously in terms of security in the area. One of the safety factors in this case was the state of the tourism environment in the future. According to Ritphring, et al., 2021. The city of Surabaya, especially the coastal area of North Surabaya, was identified as an area prone to waterlogging from overflows of river and channel water discharge during the rainy season and tidal flooding due to the impact of rising sea levels. Global sea level rise affected coastal areas, providing a number of benefits for the tourism sector. By looking at the conditions and impacts of sea level rise in the city of Surabaya, it would directly affect the lowland areas located in the coastal area. Therefore, the restriction area that would be reviewed further was focused on the coastal area and the tourist area [19].

In the Surabaya City Spatial Plan for 2014-2030 regarding policies and strategies for determining spatial patterns, one of the steps to minimize risk and reduce disaster vulnerability is to manage areas prone to natural disasters such as areas prone to tidal waves, tidal flood prone areas through identification of risk levels. areas prone to natural disasters. According to Arini, et al., 2018, the Tanjung Perak Maritime BMKG estimated that every year the Surabaya Coastal Area experiences tidal flooding and in one year there are about 4 to 5 times the danger of flooding above the average sea level. This makes the urgency of doing this research.

Therefore, it is necessary to review the tidal flood due to further sea level rise, especially in the tourist area of Surabaya Kenjeran Beach and its surroundings. In addition, it is necessary to study the local phenomenon of sea level rise, adaptation and mitigation strategies. The purpose of this study is to analyze the tidal flood as the impact of sea level rise and formulate appropriate adaptation & mitigation strategies so that tourism utilization can be carried out in a sustainable manner.

2. Materials and Methods

The research location was in Kenjeran Waters, Surabaya, East Java as shown in figure 1. Tidal data was obtained from the Geospatial Information Agency (BIG) and direct field observations for 15 days. The observations made by BIG were at the Surabaya station with the station code SRBY, located at the PT. PAL, Semampir, Ujung, Surabaya with coordinates of -7.20006° South Latitude and 112.7406° East Longitude. While tidal observations were carried out directly or field observations were carried out at Nambangan Kenjeran with coordinates -7.217884° South Latitude and 112.785467° East Longitude.

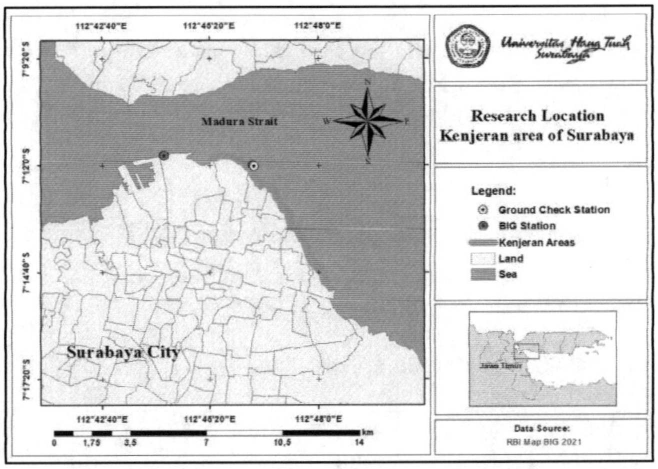

Figure 1. Research Location Map

IOP Conf. Series: Earth and Environmental Science 1273 (2023) 012084 doi:10.1088/1755-1315/1273/1/012084

The stages of this research consisted of several stages, namely data collection, data processing and analysis of regional vulnerability to sea level rise. The stages of the research can be seen in figure

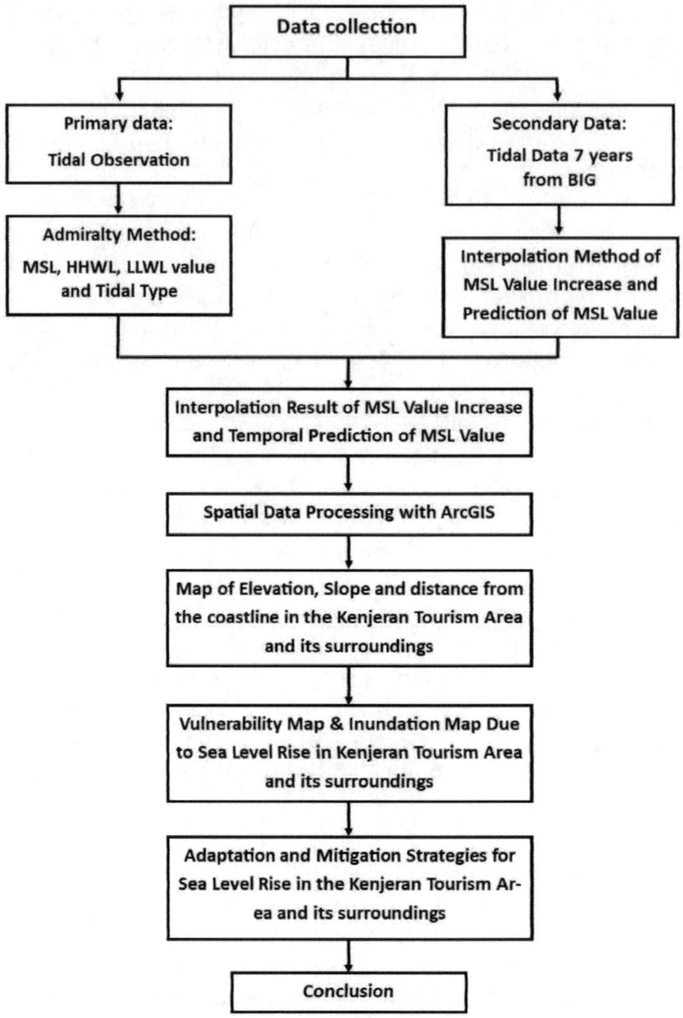

Figure 2. Research Flow Chart

2.1. Data Collection

The tools needed for this research were HP 14s laptop as hardware, and Microsoft Office Excel and ArcMap 10.8 as data processing software. Some data were collected as material for this research, as shown in Table 1. The types of data and their sources, years and uses have been obtained.

Table 1. Material study

No	Data Type	Data Source	Utility	Year
1.	Tides	Information Agency Geospatial (BIG)	Predicting sea level rise due to tides	2016 - 2021
3.	Tides	Observation field in Kenjeran	Finding the Formzhal Value to determine the tidal type	August 2022
4.	DEMNAS	https://tanahair.indonesia.go.id/demnas/#/demnas	Get the land height value (elevation)	2021
5.	RBI Map	https://tanahair.indonesia.go.id/portal-web/download/perregion	Base Map for mapping processing materials	2021

2.2. *Data Processing*

2.2.1. *Sea Level Rise*

The admiralty method is a tidal calculation method used to calculate two components/harmonic constants, namely amplitude and phase difference. The process of calculating the Admiralty method was calculated with the help of tables, where for the time of observation that was not tabled, an approach and interpolation with the help of tables must be made. The process of calculating the harmonic analysis of the Admiralty method was carried out by developing a formula system calculation with the help of Excel software, which would produce the values of several parameters that were tabled so that the calculations in this method would be efficient and have high accuracy and were flexible at any time [9].

The data were analyzed using the 15-day Admiralty method so that the final result would be known about the components of tidal induction in the Kenjeran area of Surabaya which could be used to determine the value of MSL, LLWL, HHWL and types of tides. The amplitude value (A) was used to determine the Formzhal value (F) to obtain the tidal type at the study site (Westplat, et al., 2017). The results of the admiralty method processing were 9 (nine) values of the main tidal constants, namely: M2, S2, N2, K1, O1, M4, MS4, K2 and P1. The next process, to determine the type of sea tide, Formzal numbers could be used, namely by calculating the Formzahl number calculation formula according to Fitriana et al., 2022, with equation 1 as follows:

$$F = \frac{A(K_1)+A(O_1)}{A(M_2)+A(S_2)} \quad [1]$$

Where, F is the formzhal number, A(K1) the amplitude of the tidal generating element K1, A(O1) the amplitude of the tidal generating element O1, A(M2) the amplitude of the tidal generating element M2, A(S2) the amplitude of the tidal generating element tidal generator S2. While the other constants are N2: Harmonic constant which is affected by changes in the distance of the moon, K2: Harmonic constant which is affected by changes in the distance of the sun, O1: Harmonic constant which is affected by the declination of the moon, P1: Harmonic constant which is affected by the declination of the sun K1: Harmonic constant which is affected by the declination of the sun affected by the declination of the sun and the moon M4: Harmonic constants affected by multiple effects MS4: Harmonic constants affected by the interaction between M2 and S2.

Where the results of the F value determines the type of tide, with the classification according to the F value resulting from the calculation of equation 1 above, the classification was in accordance with Table 2 as follows:

Table 2. Tidal Classification

Number Formzal	Tidal Type
0 < F 0.25	Ups and downs daily double (semi-diurnal)
0.25 < F 1.5	Ups and downs mixture leaning daily double
1.5 < F 3	Ups and downs mixture leaning daily single
F >3	Ups and downs daily single (diurnal)

2.2.2. *Regional Vulnerability Mapping*

Spatial data was processed using ArcGIS 10.8 software to produce a vulnerability map of the Kenjeran beach tourism area, Surabaya. The parameters used in mapping the vulnerability of the area were elevation, slope, and distance from the shoreline (coastal proximity). The parameters were reclassified in order to get a class that fits the predetermined criteria. The three parameters were given a vulnerability class, namely: very vulnerable, vulnerable, moderately vulnerable, less vulnerable, and not vulnerable [15]. It can be seen in table 3 for the vulnerability class and also the scoring of the three parameters.

IOP Conf. Series: Earth and Environmental Science 1273 (2023) 012084 doi:10.1088/1755-1315/1273/1/012084

Table 3. Classification of Elevation and Slope Parameters

Elevation (m)	Slope (%)	Distance from Beach (m)	Class Vulnerability	Score
0 - 2.5	0 – 5	0 - 500	Very Vulnerable	5
2.5 – 5	5 – 10	500 – 1000	Susceptible	4
5 – 7.5	10 – 15	1000 – 1500	Enough Susceptible	3
7.5 – 10	15 – 20	1500 – 2000	Less Vulnerable	2
>10	>20	>2000	Not Susceptible	1

Source: Guidelines Preparation of Rehabilitation Patterns Land and Soil Conservation 1986, Ilhami 2014 with modification author.

Scoring and weighting were carried out after the reclassify or categorization process. This process was carried out by assigning a value to each parameter that causes tidal flooding and then calculating it by considering the largest factor that causes tidal flooding in the area. In the processing of tidal flood-prone areas in the Surabaya area by considering the influencing factors, the weighting was divided based on the parameters that most influence getting the largest percentage of weight [3].

Table 4. Weighting Every Parameter

No	Parameter	Weight	Percentage (%)
1	Elevation / Altitude Land	0.25	25
2	Slope	0.35	35
3	Beach Distance	0.30	30
	Total	1	100

Source : Kultsum et al., 2016 with modification author .

In determining the level of potential threat of tidal flooding, an overlay of several variables determining tidal flood-prone areas was carried out using the scoring method, namely giving weights and scores to each parameter. From the results of the overlay, the area that has the highest score was an area that was prone to potential tidal flooding, which was symbolized by the red color (prone). Then after getting the weight of each parameter of the tidal flood threat, the final score was obtained by multiplying the weight and classification of each parameter.

To determine the vulnerability of an area to flooding, it was necessary to divide classes at the level of vulnerability. According to Kingma 1991, to determine the length of the interval in each class was done by dividing the same number of values obtained by the number of class intervals used by using the following formula:

$$Ki = \frac{Xt - Xr}{k}$$

Description: Ki: Interval Class, Xt: Highest Value, Xr: Lowest Value, k: Number of desired classes with class interval value: Highest value= 5 Lowest score= 1 Number of Classes= 5.

After knowing the final score for each parameter, classification was carried out by looking for the value of the tidal flood threat class interval by subtracting the highest score from the lowest score divided by 5 because it has 5 classes, the results of the tidal flood threat class could be seen in Table 5.

Table 5. Classification Vulnerability Flood Rob.

No	Interval	Category Threat	Symbol
1	1 – 1.80	Very Low	
2	1.80 – 2.6	Low	
3	2.60 – 3.40	Currently	
4	3.40 – 4.20	Tall	
5	4.20 - 5	Very high	

After the reclassification of the parameters, an analysis would be carried out using a Weighted Overlay Analysis which shows the distribution of vulnerabilities in the area based on the class that was determined from each parameter [15].

2.3. Disaster Vulnerability Analysis, Strategy and Adaptation

Analysis of the level of vulnerability based on the General Guidelines for Disaster Risk Assessment with weighted parameters (Perkab BNPB No. 2 of 2012) as follows: (1) Social Vulnerability, with parameters of population density, sex ratio, poverty ratio, ratio of disabled people and age group ratio. Social vulnerability index, (2) Physical Vulnerability, with parameters of house density (permanent, semi-permanent and non-permanent), availability of buildings/public facilities and availability of critical facilities, (3) Economic Vulnerability, with parameters of productive land, (4) Environmental Vulnerability, with land cover parameters [27].

The choice of strategies that was carried out for disaster adaptation and mitigation would be determined based on the conditions of the Kenjeran Tourism Area. The strategies and mitigations [25] were as follows. The Adaptation Strategy includes: constructing/maintaining coastal structures to prevent abrasion and coastal erosion, repairing road construction to prevent abrasion, placing new construction in areas that are safer from sea water inundation, handling drainage in coastal areas affected by sea level rise. Mitigation Strategy included : minimizing carbon emissions and encouraging the application of the principle of "Clean Development Mechanism", moving roads along the coast that were affected by rising sea levels, organizing green open spaces in the urban environment.

3. Results and Discussion

3.1. Sea Level Rise

3.1.1. Observation Data Field

The results of the calculation of the tidal harmonic constant using the admiralty method and the tide chart in August 2022, can be seen in Table 6.

Table 6. *Calculation of the tidal harmonic constant*

	S0	M2	S2	N2	K1	K2	O1	M4	MS4	P1
H	54.44	15.38	86.61	-0.28	21.88	23.39	11.09	0.71	2.77	7.22
g°	-	326.54	332.82	97.17	49.09	332.82	206.54	29.87	51.57	49.09

The Formzhal value was obtained from the calculation of the tidal harmonic constant, namely the components AK1, AO1, AM2 and AS2. AK1 constant amplitude; AO1 and AM2; AS2 which were worth 21.88; 11.09 and 15.38; 86.61. The results of the amplitude of this constant produced a Formzhal (F) value of 0.32 in the Kenjeran area of Surabaya, as shown in Table 1, the Formzhal number with a range of 0.25> F F < 1.50 was included in the category of mixed tide type tend to the daily double. In one day there are two high tides and two low tides with very different heights and periods.

It can be seen from figure 3. The hourly chart shows fluctuations in one day there are two highs and lows, but fluctuations in the highest and lowest tides occur around 10.00-17.00, while at 00.00-08.00 the fluctuations in tides and lows are not so significant. Likewise, it is shown in Figure 6. Where the daily fluctuation graph clearly shows that the tides in the Kenjeran area occur twice, each with a different level of fluctuation.

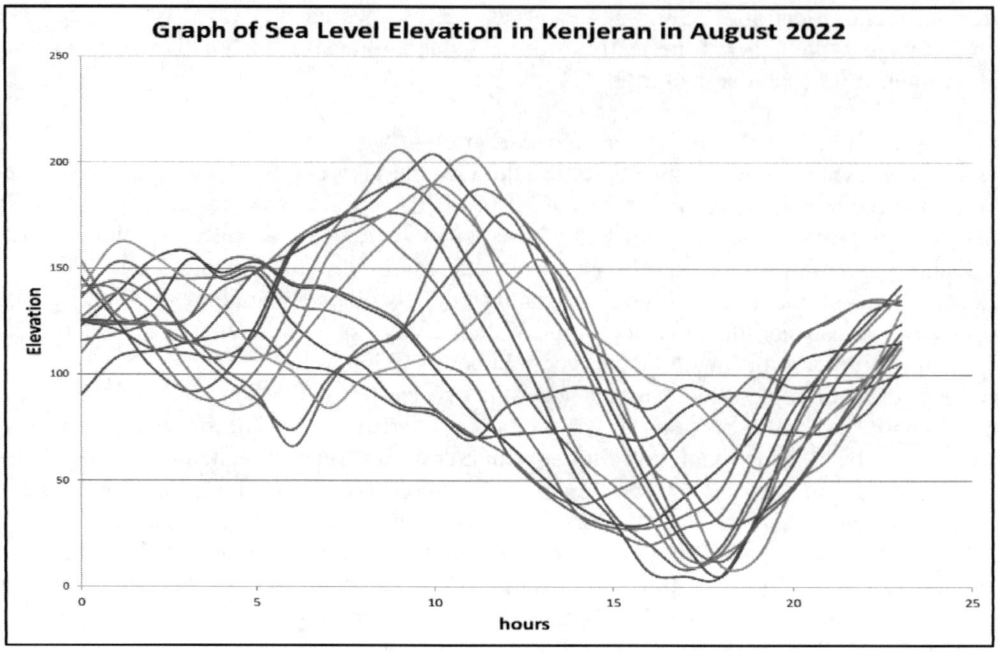

Figure 3. Hourly Tidal Chart for 15 days

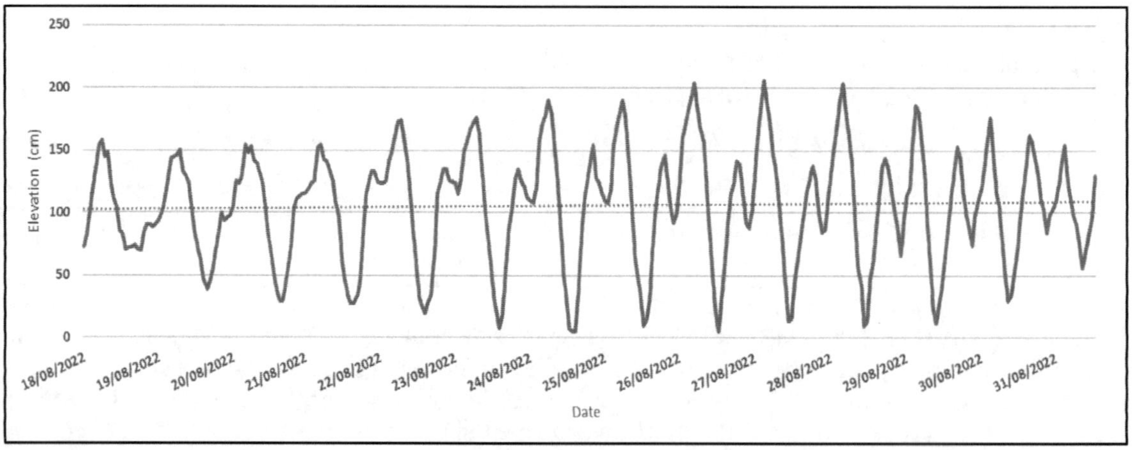

Figure 4. Daily Tidal Chart

The graph in figure 4. It clearly illustrates that the fluctuations in one day have two highs and two lows, but with very different interval values. The highest value or HHWL of tidal fluctuations in August 2022 reached 206 cm while the lowest value or LLWL reached 5 cm which occurred on August 25, 2021 with an MSL position value of 105.26 cm.

3.1.2. Geospatial Information Agency Data

The monthly and yearly tide charts in the Kenjeran Region are very volatile. The graph of the rate of sea level rise based on the annual MSL value is presented in figure 5.

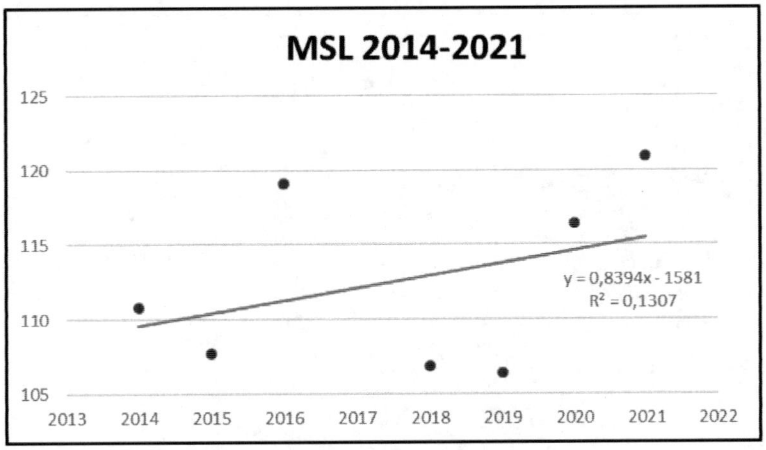

Figure 5. Rate of Sea Level Rise in 7 years

Figure 5 shows the trend of the annual average sea level rise rate following a linear pattern with the equation y = 0.8394 x - 1581 with R2 = 0.1307, where y = MSL (cm) and x = year, thus the rate of sea level rise sea water in 1 year that is equal to 0.839cm in a span of 7 years. The results of this sea level rise that have been obtained can then be predicted in the next 50 years in 2071, this sea level rise will reach 42cm.

3.2. Rob Flood Vulnerability Due to Sea Level Rise

The area's vulnerability to sea level rise is determined based on the parameters of land slope (slope), land elevation (elevation) and distance from the shoreline (coastal proximity). Data processing was carried out for each parameter which was reclassified and analyzed by using the overlay technique to obtain the distribution of the vulnerability of the study area to sea level rise.

The results of processing Figure 6 produce an elevation map at the study location which shows that the Kenjeran area has a dominant low elevation value so that the area is an area that has the potential for flooding because the lower the elevation value in an area, the more prone the area is to flooding. and vice versa if an area has a high elevation value, the area is safer from the potential for flooding.

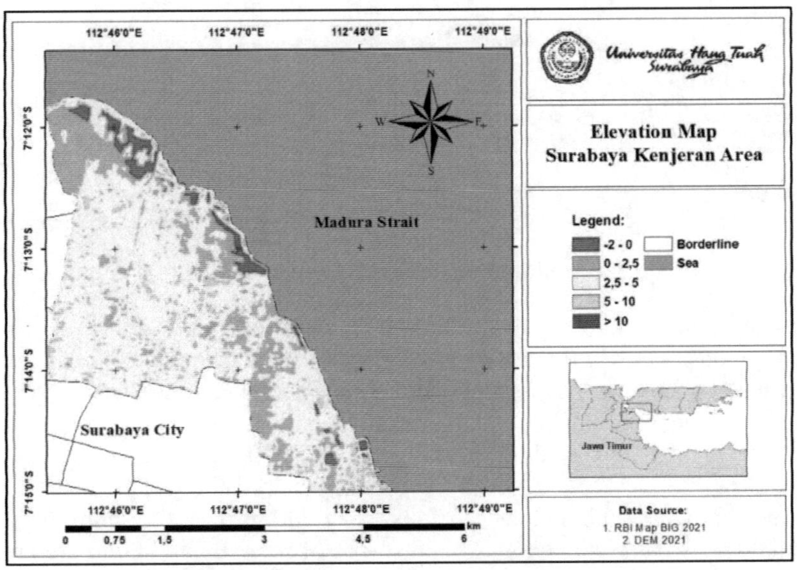

Figure 6. Map of Land Altitude (Elevation)

IOP Conf. Series: Earth and Environmental Science 1273 (2023) 012084 doi:10.1088/1755-1315/1273/1/012084

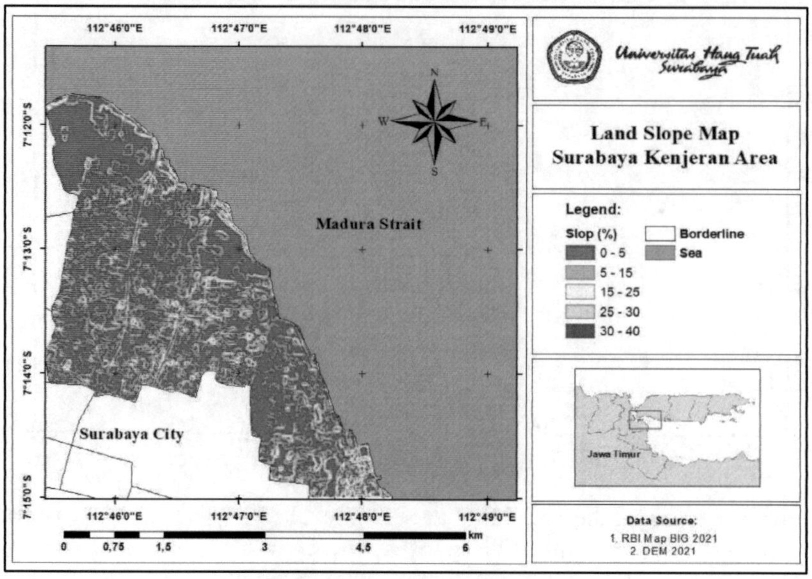

Figure 7. Land Slope Map

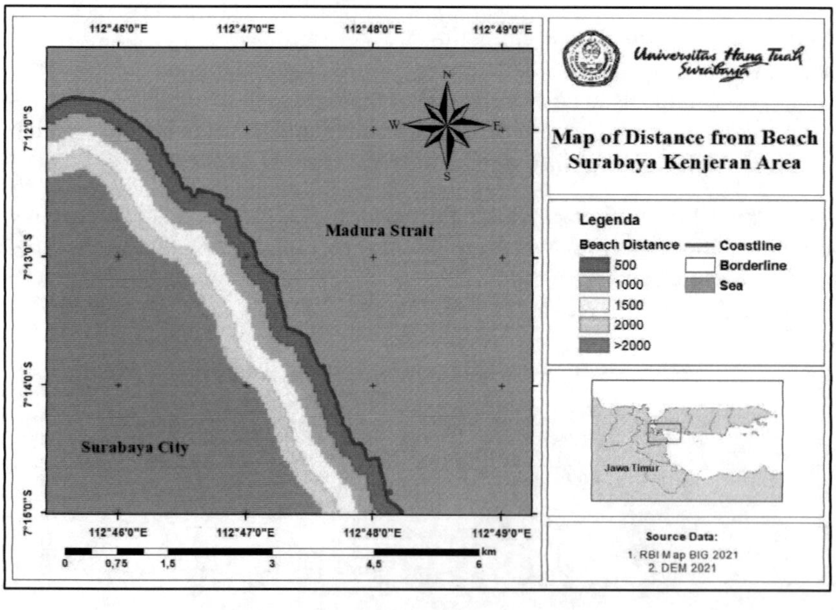

Figure 8. Map of Distance from Coastline

Figure 7 is the result of processing the area data showing that most or dominating the study location has a slope of 0 – 5% in the flat or low category. This shows that the area has the potential for flooding because in a flat area the water will become inundated. Meanwhile, the area that has a very steep area is in the southern part of the study location where the morphology of hills and mountains has less potential for flooding. In this case, the slope is the main factor causing flooding [8].

Figure 8 shows that areas with a distance of less than 500 m from the shoreline are areas that are very prone to tidal flooding, while areas with a distance of more than 2000 m are areas that are very safe against tidal flooding in the region. Similarly, research by Kultsum et al., 2016 shows that the distance from the coastal border is very influential in determining the level of tidal flood vulnerability, where a very close distance certainly has a very high level of vulnerability.

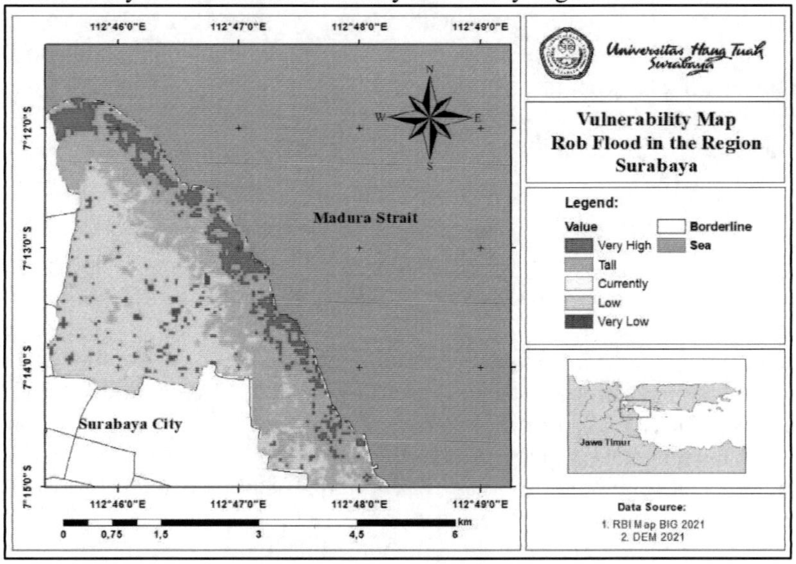

Figure 9. Map of Rob Flood Vulnerability Due to Sea Level Rise

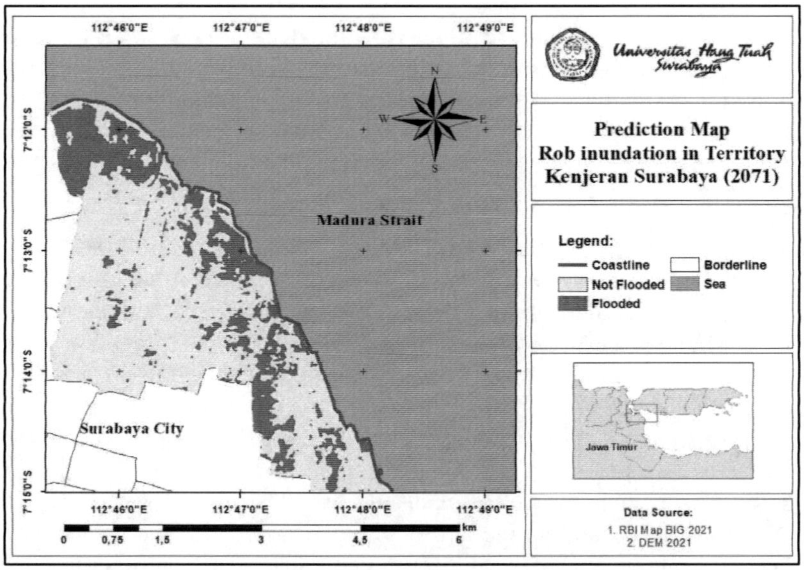

Figure 10. Map of Innudation in Territory Kenjeran Surabaya (2071)

From these results, it is obtained that the level of tidal flood vulnerability in each area in the Kenjeran Region, while the level of threat / vulnerability is divided into 5 classes, namely very low, low, medium, high and very high threats as shown in figure 9. The results of processing with the overlay method from The three parameters result in tidal flood vulnerability in this area being predominantly low, but in some coastal areas of Kenjeran the vulnerability level is very high with an area of 158.8 around the coast.

Based on the tidal analysis of the north coast of Surabaya from 2014-2021, almost every year there is an increase of 0.84cm. so it is possible that there will be an increase in the area of tidal flood inundation in the Kenjeran area of Surabaya. The results of inundation processing in the next 50 years due to sea level rise in this area indicate that the Kenjeran area will experience an inundation of 516.5 hectares with a height of 0.41 meters. The following is the distribution of tidal inundation due to sea level rise in 2071 which is presented with a map layout in Figure 10. The area which has the potential to be affected by tidal inundation if there is a sea level rise is due to the area directly adjacent to the sea and the land height is quite low so it is easy for sea water to enter the land.

3.3. Adaptation and Mitigation

Responses to sea level rise are broadly divided into two categories, mitigation and adaptation. There is a synergistic effect between some of these responses, even from the two different categories, and the positive effect should be recognized and appreciated [28].

3.3.1. Adaptation

Adaptation is one form of response made by the community to climate change and the disasters they face. According to Aldrian et al (2011) adaptation actions are related to the efforts made by the community to suppress the negative impacts caused by natural phenomena such as climate change or natural disasters and take advantage of these circumstances. In this case, adaptation tends to be an effort to manage something that cannot be avoided [26].

From a sustainable adaptation perspective, the implementation of an effective transformative action, such as a new coastal model involves gradual implementation. Long-lasting responses require an adaptive approach over time with better monitoring [11]. Lawrence et al., 2021 suggest a logical chain of actions (monitor, review, update, anticipate) that focuses on the role of information and continuity of follow-up. Because the elaboration of an adaptive approach from time to time depends on the institutional characteristics of the country [11].

The concept of adaptation strategy can be divided into 3 things, namely: (1) Protection, namely the protection carried out by the community affected by the disaster to adapt to the impacts caused by building coastal protective buildings such as APO, groynes, sea dikes; (2) Accommodation, namely adjustments made by the community to the environment such as filling the land, elevating the house building; and (3) Retreat or retreat, namely leaving the coastal area by moving to an area that is safe from natural disasters [26].

3.3.2. Mitigation

Disaster Mitigation is an effort to reduce disaster risk, either structurally or physically through natural and/or artificial physical development as well as non-structural or non-physical through increasing the ability to face disaster threats in Coastal Areas and Small Islands (Law No. 27 of 2007 concerning Regional Management). Coastal and Small Islands). SLR mitigation in coastal areas by taking into account several determinant variables in shaping mitigation in coastal areas, namely risk and vulnerability factors.

Risk can be interpreted as a possibility that can cause harm in the form of material, loss of life, environmental damage, or in general it can be interpreted as a possibility that can damage social, community and environmental structures caused by the interaction between threats and vulnerabilities. Vulnerability Factors Vulnerability can be interpreted as a condition that determines when a natural hazard (natural hazard) that occurs can cause a natural disaster (natural disaster). Vulnerability shows the value of potential losses in an area due to natural disasters, be it environmental values, materials, casualties, social arrangements and others [13].

The rate or trend of sea level rise seems to have increased every year. This is evidenced that the graph of sea level rise based on the annual mean sea level (MSL) continues to increase. The rate or trend of sea level rise in this study is sourced from the annual MSL value obtained based on the average monthly MSL value during the 2014-202 range.

The results of the identification of the tsunami vulnerability variable indicate that the slope value of 0-5% is classified as vulnerable, the elevation of the Kenjeran area <5m is also classified as vulnerable and the distance from the coastline (coastal proximity) the vulnerability is very large the closer it is to the coast. Based on the results of the Weighted Overlay Analysis carried out on each parameter that has been classified as vulnerability class, the vulnerability of the Kenjeran coastal area to tidal flooding is in the vulnerable to very vulnerable category with a land area of 675.3 ha.

The East Java Provincial Spatial Plan for 2011–2031 provides direction for the management of flood-prone areas including: beach reclamation, construction of breakwaters, arrangement of buildings around the coast, development of mangrove forest areas and construction of wave retaining walls.

According to Azuga, [4] Structural Mitigation Efforts Structural efforts are physical or technical efforts aimed at protecting the coastal environment and KDP which are very vulnerable to disasters. The physical countermeasures include: 1. Creating a dynamic coastal protection system such as mangrove planting and static, such as: making a row of groynes, embankments, breakwaters, seawalls. 2. Elevate buildings and roads. 3. Move all forms of facilities and coastal land to land that is far from the reach of sea water. 4. Adjustment of drainage system. Non-Structural Mitigation Efforts Non-structural efforts are non-technical efforts related to adjustments and regulations regarding human activities so that they are in line and in accordance with structural mitigation efforts. Some of the activities that are included in non-structural mitigation efforts are as follows: Formulating policies for the government related to stakeholders regarding the creation of coastal protection systems. 2. Preparation of regulations governing building permits for land affected by erosion due to sea level rise. 3. Build an early warning system for sea level rise by installing monitoring tools for sea level rise and observations of sea tides that are placed at every port or residential area of coastal communities. 4. Make a map of the vulnerability of areas prone to disasters.

4. Conclusion

Admiralti's calculations show that the Kenjeran Region has mixed types of mixed tides with a double daily trend. Where, in one day there are two high tides and two low tides with very different heights and periods. The MSL value generated from the Geospatial Information Agency produces a rate of sea level rise in 1 year which is 0.839cm. Based on the results of mapping and analysis of sea level rise, the area inundated due to tidal flooding is around 42 cm in the next 50 years. In an effort to adapt and mitigate sea level rise disasters, it is carried out with structural (physical or technical) mitigation and non-structural (non-technical) mitigation.

5. References

[1] Aldrian, E., Karmini, M., and Budiman 2011 Climate Change Adaptation and Mitigation in Indonesia. Jakarta: Center for Climate Change and Air Quality BMKG.
[2] Arini D, Setiadi, Priyantini D, Yuliastuti C, Harfika M and Farida I 2018 *J Pengabdian Kesehatan* 1 109-23.
[3] Ariyora Y K S, Budisusanto Y and Prasasti I 2015 *J. Geoid* 10 137-46.
[4] Azuga N A 2021 Vulnerability of Coastal Areas to Sea Level Rise in Indonesia) Earth Science Study Program, Bandung Institute of Technology.
[5] Geospatial Information Agency. Map of DEMNAS 2021. https://tanahair.indonesia.go.id/demnas/#/demnas
[6] Geospatial Information Agency. Earth Map 2021. https://tanahair.indonesia.go.id/portal-web/download/perwilayah
[7] Bapeko Surabaya. Spatial Plan and Surabaya Area 2014-2030.https://bappeko.surabaya.go.id/images/File%20Upload/RPJMD-2016-2021.pdf.
[8] Darmawan K 2017 *J Geode Undip* 6 278–287.
[9] Dewi A M, Efendy M, and Dwi A S 2009 *J Kelautan* 02
[10] Fitriana D, Petala MP and Kusratmoko E 2022 *J of Geosp Informat Scie and Engin* 5

[11] Haasnoot M, Brown S, Scussolini P, Jimenez J A, Vafeidis A T and Nicholls R J 2019 *Environ Res Commun* 1 071006.
[12] Handiani D N, Darmawan S, Heriati A and Aditya Y D 2019 *J Kelau Nasio* 14 145-54.
[13] Hidayat A, 2012 *J Lingku Bina Indo* 1 87.
[14] IPCC 2018 Summary for Policymakers. In: Global Warming of 1.5°C. An IPCC Special Report on the impacts of global warming of 1.5°C above preindustrial levels and related global greenhouse gas emission pathways, in the context of strengthening the global response to the threat of climate change, sustainable development, and efforts to eradicate *World Meteorological Organization, Geneva, Switzerland* 32.
[15] Isdianto A, Kurniasari D, Subagiyo A, Haykal M F and Supriyadi 2021 *J Permu* 16 90–100.
[16] Kingma N C 1991 Natural Hazzard: Geomorphological Aspect of Floodhazard, ITC, The Netherland.
[17] Kultsum U, Muhari A and Fuad Z 2016 *Pertemuan Ilmiah Nasional XIII ISOI 2016*, 833–839.
[18] Mucova S A R, Azeiteiro U M, Filho W L, Lopes C L, Dias J M and Pereira M J 2021 *J of Mar Scie and Engin* 9 205.
[19] Prasita V D, and Kisnarti E A. 2013 *Int J Eng Sci* 2 01-09.
[20] East Java Province Spatial Plan for 2011 – 2031.http://bappeda.jatimprov.go.id/dokumen-perencanaan/
[21] Ritphring S, Nidhinarangkoon P, Udo K and Shirakawa H 2021 *J of Mar Scie and Engin* 9 588.
[22] Salim M A dan Siswanto, A B 2018 *J Tekn Sip* 11 295-303.
[23] Setiawan F Y 2016 *J. Intra* **4**.
[24] Taufiq A and Azizah S 2020 *J Anal Sosio* 9 567-581.
[25] Triana Y T and Hidayah Z 2020 *Juve* 1 141-50.
[26] Wacano D, Rif'an A A, Yuniastuti, Daulay R W and Marfai M A 2013 *Adaptation of Coastal Communities in Demak Regency in Facing Climate Change and Coastal Area Disasters. Book Chapter of the Anthology of Environmental Management Emeralds of the Equator.*ISBN: 978-602-7797-25-3.
[27] Westplat M J H, Tandobala L and Makarau V H 2017 *J Unive Sa Ratu* 4.
[28] Wong K V 2015 *The Op Hydrol J* 9 24-27.
[29] Yasin M, and Trilaksana H,. 2020 *Moroc Jour of Chemis* 8 99-104.
[30] Mahmudiono T, Zebadia E, Indriani D, and Setyaningtyas S, 2021 *Jour of Medic Scien 9* 1549-1552.
[31] Paskarini I, Haqi D N, Arini S Y, Dwiyanti E and Alayyannur P A 2020 *System Revi in Pharm* 11(11) 1643-1648.
[32] Prasetya T A E, McNeil R, and Jani R 2020 *Pertan Jour of Scie and Techno* 28(4).
[33] Zakiyah U, Suwanti L T, Koerniawan M D, Suyono E A, Budiman A, and Siregar U J 2020 *Biodi* 21(3) 1149 - 1159.
[34] Alayyannur P A, and Arini S Y, 2021 *Intern marit hea* 72(3) 195-198.

6. Acknowledgments

The author would like to thank the Kemenristek Dikti which has funded this research and to the Geospatial Information Agency Agency which has provided data to support this research as well as friends who have helped, supported and involved in this research.

The authors are grateful for the research grant received from the Ministry of Education, Culture, Research and Technology, Indonesia. The authors also thank to the Geospatial Information Agency Agency which has provided data to support this research. The authors are also grateful to the Supervisor from Hang Tuah University for providing guidance, input, advice, and support as well as the Chancellor for providing the facilities and infrastructure needed. Furthermore, the authors are grateful to the research team of lecturers and students for their cooperation

IOP Conf. Series: Earth and Environmental Science 1273 (2023) 012085 doi:10.1088/1755-1315/1273/1/012085

Effect of microplastic particles on the gills structure of freshwater fish supplemented with probiotics and vitamin C

A Hayati[1,2], W Adriansyah[1], M R Kusuma[1], A R Astri[1], S Maulidah[1], A Maullani[1], A D Wahyuni[1], M F Zubaidi[1], H Soepriandono[1], M Pramudya[1], L Soehargo[1], Sugiharto[1], F R P Dewi[1], F A Nurbani[1]

[1]Department of Biology, Faculty of Science and Technology, Airlangga University, Surabaya 60115, Indonesia

[2]Corresponding Author: alfiah-h@fst.unair.ac.id

Abstract. This study aims to determine the effectiveness of commercial feed with probioticand vitamin C supplements in repairing damage to the gill structure of freshwater fish, tilapia (*Oreochromis niloticus*). Tilapia in this study were divided into control group (commercial feed), probiotic group (commercial feed with probiotic 200 mg/kg of feed), and vitamin C group (commercial feed with vitamin C 100 mg/kg of feed). Each group was givenmicroplastic particles exposure treatment in aqueous media (0, 0.1, 1, 10 mg/L volume of aquarium water) which was carried out twice a day for 14 days. The results showed that the administration of vitamin C as fish feed supplement reduced the damage that occurred in the tilapia gills in the form of edema (22 ± 3.57%), necrosis (33 ± 4.07%), hyperplasia (38 ± 5.91%), epithelial lifting (28 ± 3.43%), and fusion (17 ± 3.49%) of the secondary lamellae. Meanwhile, the administration of probiotics only reduced the histopathological conditions of necrosis (34 ± 4.57%) and fusion (18 ± 3.07%) of secondary lamellae. From all treatments, it was shown that feeding with vitamin C was the most effective in improving the histopathological condition that occurred in the histological structure of tilapia gills.

1. Introduction

Nile Tilapia (*Oreochromis niloticus*) which is quite popular as a food ingredient and aquaculture cultivation object in Indonesia is a species of tilapia originating from areas in North Africa (especially in the waters of the Nile) [1]. Their fast growth rate and quick reproductive cycle, the weight of a mature fish that can reach up to 5 kg, the distinctive taste of its clean, white, free-of-spines meat, as well as a relatively affordable selling price for the community make tilapia one of the fishery commoditieswith very promising aquaculture prospects and a fairly large market share [2,3].

Hence in its practice in aquaculture, the intensification of tilapia cultivation takes place very quickly so that the fish is not only kept in ponds but is also cultivated in "common" freshwater locations such as “kajapung” in reservoirs, brackish water ponds, lakes, ponds and/or artificial pond connected to the river [4]. Freshwater is one of the ecosystems that is most easily contaminated by human activities, especially the problem of microplastics and nanoplastics pollution which is now increasingly concerning. The production and use of plastic polymer-based products has increased dramatically since 1950 until now, reaching 332 million tons in 2015, and is projected to continue to grow up to 500 million tons to date [5]. This is also in line with the increase in plastic debris formed due to the imperfect decomposition experienced by these plastic polymer products [6].

Plastic particles (microplastics and nanoplastics) originating from fishing and snaring equipment, textile industry, single-use plastic waste, and other macro-sized plastic-based products are susceptible to being eaten and accumulated in the bodies of freshwater fish, especially cultured fish such as nile tilapia(*O. niloticus*) [7]. Fish gills as one of the organs that are first affected by the presenceof plastic particles can experience physical friction injuries that cause damage to the gill filaments, blockage of lamellae or filaments on the gills, chronic inflammation and histopathology due to oxidative stress induced by plastic particles, and further wound infection due to microorganisms that may be carriedas biofilms on the surface of plastic particles [8,9].

This study aimed to examine the effect of exposure to microplastics on the histological structure of nile tilapia gills. In addition, to determine and test the effect of probiotic and vitamin C supplements as an alternative solution to reduce and repair damage or histopathological lesions that arise in the histological structure of nile tilapia gills affected by microplastic particles.

2. Materials and methods

2.1 Ethics approval

All procedures including maintenance and treatment carried out on the research object nile tilapiaare in accordance with internationally recognized guidelines for the ethical use of animals. Fish were quarantined and acclimatized prior to treatment. Facilities used in this study which include an aquarium, water quality, and fish keeping were designed with care so as not to cause stress to the fish being kept. Prior to surgery, the fish were first weakened and killed using 0.2 mg/L clove oil containing eugenol.

2.2 Study period and location

This research was conducted for 6 months starting from September 2021 – February 2022. The research was conducted at the Fish Maintenance Laboratory and Histology Laboratory, Department of Biology, Faculty of Science and Technology, Airlangga University, Surabaya.

2.3 Experimental design

Thirty-six adult nile tilapia (*Oreochromis niloticus*) weighing about 200-250 g were obtained from UPBAT Umbulan, Pasuruan. Tilapia were acclimatized for 2 weeks prior to treatment and fed with commercial pellets (30% protein, 3% fat, and 4% fiber, Takari, Sidoarjo, Indonesia). The whole tilapia samples were divided into 12 treatment groups, including one control group (given commercial feed only), three negative groups (exposed to microplastic concentrations of 0; 0.1; 1; and 10 mg/L), four probiotic treatment groups (given commercial feed mixed with probiotics 200 mL/kg 1×108 CFU and exposed to variations of microplastic concentrations), four vitamin C treatment groups (given commercial feed mixed with vitamin C 100 mg/kg 1×108 CFU and exposed to variations of microplastic concentrations). Each treatment group used 3 fish replicates determined using the Federer equation. Variations in the concentration of microplastics used in this study were based on the reference of previous studies by Soares*et al.* [10].

2.4 Fish gills histological analyses

Surgery and removal of gill organs were carried out with the help of scissors and tweezers by cutting open the gill operculum and loosening the area around the fish's mouth and pharynx to take the gill arches. The gills that have been obtained are weighed first, washed a little bit using physiological salt solution and running water, then stored in urine pot containing NBF solution (neutral buffered formalin) as fixative. The histological preparations of tilapia gills were carried out according to the SOP for paraffin wax-based microtechnical procedures by the histology service unit of Department of Biology, Airlangga University. The histological preparations of the gills were then observed using a light microscope at magnifications of 100× and 400× with the help of a digital camera mounted on the eyepiece of the microscope to facilitate image capture and histological analysis of the histological preparations of the gills. Numeration of lamellae with certain histological characteristics (normal or with histopathological lesions such as epithelial lifting, edema, hyperplasia, fusion, and necrosis) especially in the lamellar parts (primary and secondary lamellae) of the gills was carried out using Optilab and ImageJ software, then an average value is taken from several repetitions of observations and numerations in many differing fields of view.

2.5 Statistical analyses
Statistical analysis was performed using the two¬-way MANOVA test and the Tukey-HSD test (each test was performed at a significance level of = 0.05) with the help of the Statistical Package for the Social Sciences (SPSS) software.

3. Results and discussion
3.1 Results
The number of normal secondary lamellae and histopathological lesions observed showed a significant effect of variations in the type of feed used and variations in microplastic exposure to the histological structure of tilapia gills. In the group that was only fed commercial feed without any supplements, histological analysis of the appearance of the gill tissue of *Oreochromis niloticus* (Figure 1) showed a clear trend for epithelial lifting and fusion to occur more frequently in the secondary lamellaeof the gills as the concentration of microplastic (MP) exposure increased. At concentrations of 1 mg/L (Gn3) and 10 mg/L (Gn4) MP, hyperplasia of the interlamellar and secondary lamellae tissue was also seen, also the lamellae in general tends to be either short or shrunken and experienced epithelial lifting.

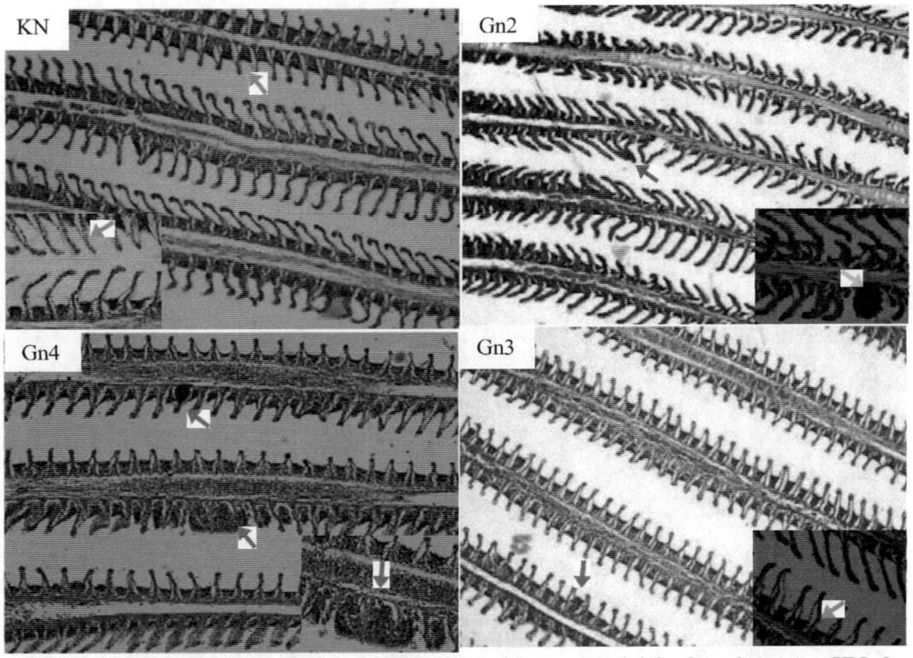

Figure 1. Photomicrograph of *O.niloticus* gill histology in commercial feeds only group. KN. 0 mg/L; Gn2. 0.1 mg/L; Gn3. 1 mg/L; Gn4. 10 mg/L exposure of MP. Hyperplasia (red arrow); epithelial lifting (blue arrow); fusion (green arrow); edema (orange arrow)

Based on the observational data on the histological condition of each secondary lamellae (Table1), it was found that the commercial feed group consistently experienced a drastic decrease in the number of normal secondary lamellae and is significantly different between each MP exposure group. Exposure to high concentrations of MP (10 mg/L) caused a decrease in normal secondary lamellae count and a drastic increase in number of secondary lamellae experiencing epithelial lifting, hyperplasia, and fusion; all of which were significantly different from controls and lower MP exposure. Meanwhile,the number ofobserved edema lesions were not significantly different from the 1 mg/L MP exposure group. The exposure to 0.1 mg/L and 1 mg/L MP in the commercial feed group also did not cause a significant difference in the occurrence of hyperplastic lesions (Tables 1 & 2).

Table 1. Number of histological conditions occurrence observed on Nile Tilapia gill

Type of Feed	Microplastic Concentration (mg/L)	Number of secondary lamellae having certain histological condition (units of lamellae)		
		Normal	Edema	Necrosis
Commercial Feed	0	89 ± 4,23^{b}	11 ± 4,78^{a}	19 ± 2,58^{bc}
	0,1	77 ± 4,03^{d}	9 ± 1,65^{a}	29 ± 4,11^{e}
	1	71 ± 6,58^{e}	18 ± 4,66^{bc}	34 ± 2,95^{f}
	10	57 ± 6,56^{g}	18 ± 4,72^{bc}	48 ± 4,70^{g}
Commercial Feed + Probiotics 200 mg/kg	0	94 ± 5,60^{a}	11 ± 5,64^{a}	18 ± 1,77^{ab}
	0,1	82 ± 6,47^{c}	11 ± 2,55^{a}	22 ± 1,80^{cd}
	1	84 ± 3,97^{c}	18 ± 2,32^{bc}	23 ± 3,33^{d}
	10	70 ± 3,88^{e}	29 ± 3,46^{d}	34 ± 4,57^{f}
Commercial Feed + Vitamin C 100 mg/kg	0	97 ± 4,89^{a}	11 ± 2,06^{a}	15 ± 2,43^{a}
	0,1	83 ± 3,27^{c}	17 ± 3,77^{bc}	22 ± 3,97^{cd}
	1	78 ± 5,24^{cd}	21 ± 4,6^{c}	24 ± 2,18^{d}
	10	64 ± 3,32^{f}	22 ± 3,57^{c}	33 ± 4,07^{f}

Different letters (a, b, c, and d) show significant difference

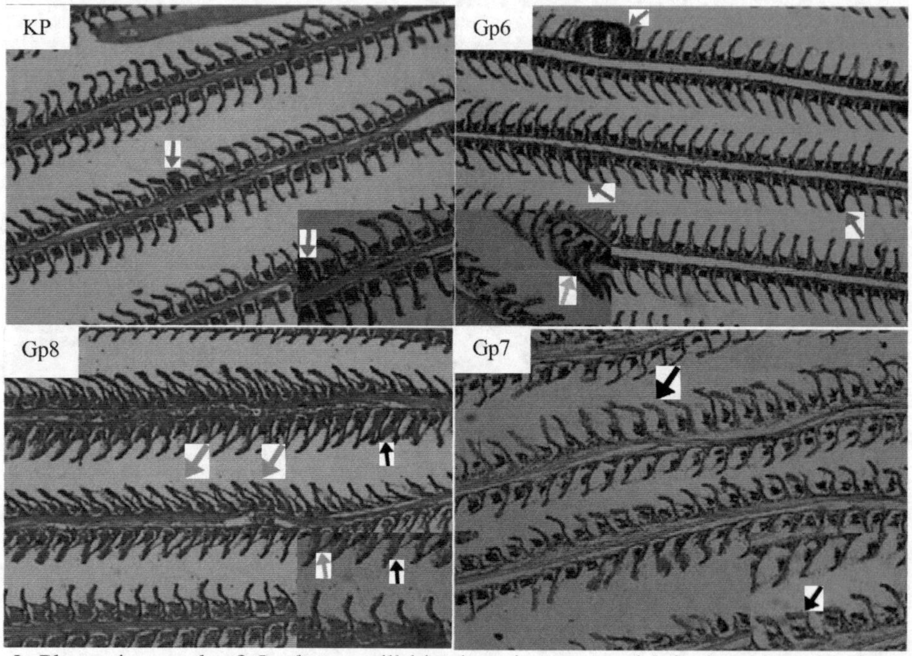

Figure 2. Photomicrograph of *O.niloticus* gill histology in commercial feeds with probiotics group. KP. 0 mg/L; Gp6. 0.1 mg/L; Gp7. 1 mg/L; Gp8. 10 mg/L exposure of MP. Hyperplasia (red arrow); epithelial lifting (blue arrow); fusion (green arrow); necrosis (black arrow)

Then in Figure 2 shows the histological appearance of *O. niloticus* gills fed with commercial feed mixed with 200 mg/kg probiotics. Hyperplasia and epithelial lifting conditions were seen in the secondary lamellae and gill interlamellar tissue which were quite common even in the 0 mg/L MP exposure group and increased along the increase in MP exposure concentration. At the MP concentration of 1 mg/L (Gp7)and 10 mg/L (Gp8), there was a tendency for the secondary lamellae to be damaged and the cells to fade indicating the occurrence of necrosis accompanied by many other secondary lamellae experiencing epithelial lifting.

The number of normal secondary lamellae in the probiotic diet group experienced a significant decrease at 10 mg/L MP exposure, even though the 0.1 and 1 mg/L exposure were not significantly different, both were significantly different from 0 mg/L MP exposure (Table 1). The occurrence of edema,epithelial lifting, hyperplasia, and fusion also increased significantly at 10 mg/L MP exposure compared to lower MP exposure. Between the probiotic diet groups exposed to 0.1 and 1 mg/L MP, the occurrence of fusion and necrosis was not significantly different. Meanwhile, Between 0 mg/L

MP exposure group and 0.1 mg/L MP exposure group, the occurrence of epithelial lifting and edema was not significantlydifferent (Table 2).

Table 2. Number of histological conditions occurrence observed on Nile Tilapia gills (*continued*)

Type of Feed	Microplastic Concentration (mg/L)	Number of secondary lamellae having certain histological condition (units of lamellae)		
		Epithelial Lifting	Hyperplasia	Fusion
Commercial Feed	0	$34 \pm 2{,}98^{bc}$	$22 \pm 3{,}01^{a}$	$11 \pm 3{,}96^{a}$
	0,1	$39 \pm 1{,}89^{d}$	$27 \pm 2{,}97^{b}$	$13 \pm 1{,}90^{b}$
	1	$44 \pm 5{,}66^{e}$	$29 \pm 5{,}64^{bc}$	$16 \pm 3{,}53^{cd}$
	10	$52 \pm 4{,}58^{f}$	$36 \pm 3{,}00^{d}$	$23 \pm 4{,}12^{e}$
Commercial Feed + Probiotics 200 mg/kg	0	$36 \pm 5{,}37^{cd}$	$20 \pm 1{,}98^{a}$	$10 \pm 2{,}92^{a}$
	0,1	$39 \pm 2{,}98^{d}$	$27 \pm 3{,}23^{b}$	$15 \pm 4{,}89^{cd}$
	1	$46 \pm 2{,}76^{e}$	$30 \pm 2{,}7^{c}$	$14 \pm 2{,}04^{bc}$
	10	$51 \pm 6{,}24^{f}$	$40 \pm 2{,}54^{e}$	$18 \pm 3{,}07^{d}$
Commercial Feed + Vitamin C 100 mg/kg	0	$28 \pm 3{,}37^{a}$	$19 \pm 2{,}72^{a}$	$9 \pm 2{,}00^{a}$
	0,1	$30 \pm 2{,}33^{ab}$	$20 \pm 2{,}32^{a}$	$13 \pm 3{,}91^{b}$
	1	$34 \pm 4{,}58^{bc}$	$21 \pm 4{,}08^{a}$	$15 \pm 3{,}72^{c}$
	10	$38 \pm 5{,}91^{d}$	$28 \pm 3{,}43^{bc}$	$17 \pm 3{,}49^{d}$

Different letters (a, b, c, and d) show significant difference

Histological analysis of the appearance of *O. niloticus* gill tissue in the commercial feed group supplemented with 100 mg/kg vitamin C (Figure 3) also showed quite the same trend of increasing occurrence of histopathological lesions in secondary gill lamellae along the increasing concentrations of MP exposure although the rate of increase was not as drastic as in another treatment group. Fish exposed to 0 mg/L MP (KV) showed intact and normal secondary lamellae with some secondary lamellae experiencing mild epithelial lifting. On exposure to 0.1 mg/L MP (Gv10), interlamellar tissue began to experience hyperplasia with some secondary lamellae beginning to undergo necrosis, characterized by damaged, smeared secondary lamellae cells.

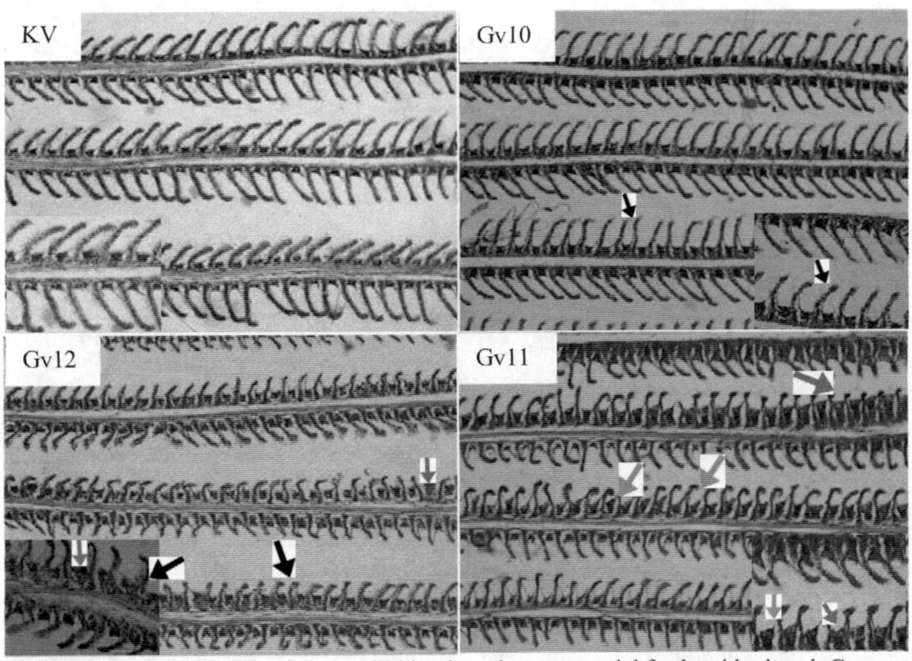

Figure 3. Photomicrograph of *O.niloticus* gill histology in commercial feeds with vitaminC group. KV. 0 mg/L; Gv10. 0.1 mg/L; Gv11. 1 mg/L; Gv12. 10 mg/L exposure of MP. Hyperplasia (red arrow); epithelial lifting (blue arrow); necrosis (black arrow)

Observational data on the vitamin C diet group (Table 2) showed that at 0.1 mg/L MP exposure, the occurrence of epithelial lifting and hyperplasia lesions did not increase significantly, while number ofedema, necrosis, and fusion lesions (Table 1) experienced a significant increase compared to 0 mg/L MPexposure. Then on exposure to 1 mg/L (Gv11) and 10 mg/L MP (Gv12), many secondary lamellae were damaged, folded, and experienced severe epithelial lifting. In some parts, a series of secondary lamellae are also seen fused as a result of the interlamellar tissue undergoing severe hyperplasia, reducing the whole surface area of the secondary lamellae. The number of epithelial lifting and hyperplasia lesions observed in the vitamin C group, although it did not increase significantly at 1 mg/L MP exposure (Table 2), did increased significantly at 10 mg/L MP exposure, as well as other lesions such as edema, necrosis, and fusion which are significantly different at 10 mg/L MP exposure.

3.2 Discussion

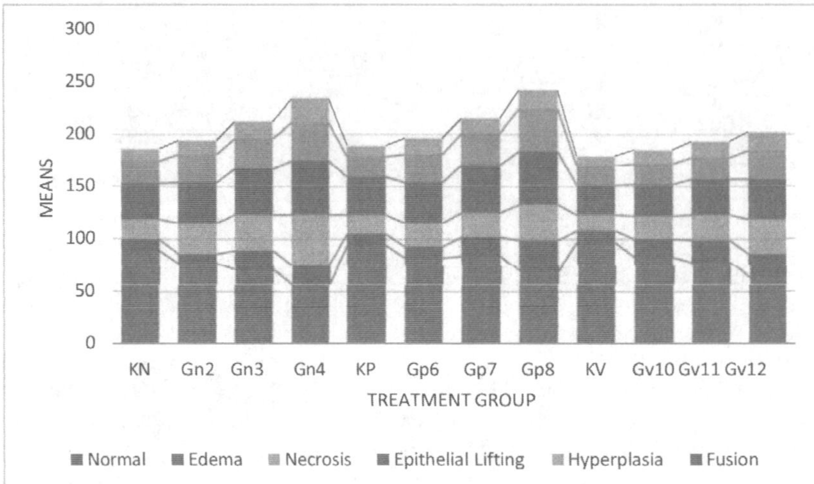

Figure 4. Number of secondary lamellae in various histological conditions in each treatment

Overall, there was a clear trend of increasing number of secondary lamellae with certain histopathological conditions such as edema, necrosis, epithelial lifting, hyperplasia, and fusion as the concentration of microplastics (MP) in the study treatment goes up, i.e. 0 mg/L; 0.1 mg/L; 1 mg/L; up to 10 mg/L (Figure 3). This is in accordance with the research conducted by Hamed *et al.* [11] on juvenile

O. niloticus exposed to MP with concentration of 1 mg/L, 10 mg/L, and 100 mg/L, it was found that at a concentration of 1 mg/L there was vasodilation of blood vessels in the primary lamellae accompanied by several secondary lamellae experiencing epithelial lifting and hyperplastic interlamellar tissue. At higherconcentrations of MP it even causes extensive necrosis and shortening of the lamellar tissue, also a hemorrhage was observed that clots into a hematoma [11]. Research by Almeida [12] also supports this result, with goby fish exposed to 0.2% w/w polystyrene MP showing epithelial lifting of the secondary lamellae and hyperplastic mucus cell in their gill tissue. Consensus from several previousstudies has argued that the histopathological conditions that commonly occur in the gills are generally caused by direct physical trauma from friction and blockage by the MP particles themselves, the presenceof toxic substances such as heavy metals and synthetic chemicals or harmful biological agents/pathogensthat are also carried/adsorbed on the surface of MP particles, and oxidative stress which is suspected to becaused by MP particles which has characteristics as strong oxidants and could overstimulate the immunesystem [13,14,15,16,17].

Microplastic particles (MP) as oxidants generally carry alkyl and peroxyl radicals on the side groups of their polymer long chains, which can lead to an imbalance in the concentration of ROS and antioxidants in gill tissue and cause histopathological lesions associated with oxidative stress [18]. Epithelial lifting lesion tends to be the most common occurring even in the treatment groupthat was not exposed to MP (0 mg/L), and its rate of increasing occurrence along the increasing concentrations of MP exposed was not as fast as other histopathological lesions (Figure 3). This is becauseepithelial lifting is one of the most common tissue adaptation responses found in fish gills and is non- specific and highly sensitive to the fluctuations of environmental condition of the surrounding waters at any time [19].

Meanwhile, edema and necrosis were the types of histopathological lesions that rapidly increasedin numbers along the increasing MP exposure, possibly due to the effects of direct physical trauma with MP particles accumulated in the secondary lamellae of the gills. Other histopathological lesions such as hyperplasia, epithelial lifting, and fusion of secondary lamellae tend to take longer to increase, because they are considered as a form of tissue adaptation/defense against pollutants in water by reducing the surface area of the gills [20,21,22].

MP particles as the activator of a series of cellular physiological events related to cellular oxidative stress have a very complex toxicity mechanism, but some of the most prominent ones happening are attacks to lipid compounds that make up cell plasma membranes by ROS directly from peroxyl radicals in MP polymers which lead to chain reaction of lipid peroxidation which causes damage to the integrity of cell membranes [23]. MP particles could also carry toxic compounds such as PFAS chemicals and heavy metals, which can lead to other toxicity mechanisms that cause cell damage, especially inactivating important cell enzymes. It is also worth noting that most heavy metal ions are ableto undergo Fenton reactions with H2O2 to form more ROS compounds [24].

One of the co-toxicity case of MP with other toxic compounds was reported in the study of Barboza *et al.* [23], which mentions the interaction between MP and mercury bioaccumulation which causes an increase in levels of oxidative stress biomarkers such as SOD, CAT, and GST, as well as lipid peroxidation biomarker LPO in *Dicentrarchus labrax*. Research by Fonte *et al.* [25] also obtained similar results in the cefalexin antibiotic toxicity test with microplastic exposure cofactors and temperature variations. There is a strong positive correlation between cefalexin toxicity with temperature variations and increased exposure to microplastics.

The histomorphological condition of the gill tissue of *O. niloticus* exposed to increasing MP concentration in the group given probiotic supplements was not significantly different from that of commercial feed alone. Even at high MP concentrations (10 mg/L), probiotic administration actually drastically increases the amount of edema that occurs compared to commercial feeding alone and feedingwith vitamin C. This is certainly contradictory to several studies, one of which was conducted by Ghosh *et al.* [26] and Dharmaraj and Dhevendaran [27] on *Xiphophorus helleri*, *X. maculatus*, *Poecilia reticulata*, and *P. sphenops* showed an increase in the rate of growth and development of biomass and body length, as well as easier digestibility of the feed given which was characterized by high specific enzyme activity through measured proteases and amylases in the gastrointestinal tract.

It is suspected that this may occur due to inappropriate doses of probiotics, less synergistic microbial composition used, or the poor quality of the preparation of the probiotics used so that when combined with high MP concentration factors, interactions that have a negative impact on gill tissue occur. Improper use of probiotics also has the potential to increase the amount of organic particulates in the waters, and lead to an imbalance in the microbiome system formed in the gills, supported by the fact that the surface of MP particles can facilitate the growth of biofilms of pathogenic microbes [28,29].

Meanwhile, the administration of vitamin C as a feed supplement under conditions of exposure to increasing MP concentrations was able to significantly reduce the number of secondary lamellae affectedby necrosis, epithelial lifting, hyperplasia, and fusion, consistently while increasing the number of normalsecondary lamellae. This is in line with the results of research by Al-Gabri *et al.* [30] which showed that feed supplementation with vitamin C could increase serum growth hormone levels, hepatic antioxidant activity, interleukin-10 levels, and the percentage of fish survival against infection by *Aeromonas sobria*. In addition, it was reported that the histomorphological conditions of the liver and intestines also improved with the increase in the dose of vitamin C used, up to about 400 mg/kg of givenfeeds.

The positive effect of vitamin C is supported by the fact that vitamin C is also required as a cofactor for lysyl and prolyl hydroxylase which are needed in the process of collagen biosynthesis, one ofthe dominant types of protein in the structural composition of epithelial tissues, such as those found in fishgills [31]. Meanwhile, the histopathological condition of edema tends to increase when vitamin C supplementation is given, most likely due to the effect of vitamin C stimulation on biomarkers of the body's immune system response as reported in research by Xu *et al.* [32] and Al-Gabri *et al.* [30], resulting in vasodilation of lymph fluid and leukocytes into the lamellar tissue of *O. niloticus* gills.

4. Conclusion

Increasing concentration of exposure to microplastics (MP) in the environment has a negative effect on nile tilapia gills, decreasing the number of normal secondary lamellae and increasing the numberof histopathological lesions that arise in the gill tissue. This indicates that MP is toxic to the gill histomorphology of nile tilapia (*Oreochromis niloticus*). The addition of vitamin C as a dietary supplementfor *O. niloticus* could improve the histomorphological condition of the damaged gills of *O. niloticus* consistently at all concentrations of MP exposed. While the positive effect of probiotics provision as a dietary supplement for *O. niloticus* was limited to only reducing the number of secondary lamellae that fused and underwent necrosis. This situation shows the effectiveness of administering vitamin C for healing the histomorphological damage to the gills of *O. niloticus*, while the administration of probiotics has yet to be confirmed to be effective for healing the histomorphological damage to the gills of *O. niloticus*.

5. References

[1] Trewavas, E., 1983. *Tilapiine fishes of the genera Sarotherodon, Oreochromis* and *Danakilia*. London: *British Mus. Nat. Hist*., 583 p.
[2] Eccles, D.H. 1992. *FAO species identification sheets for fishery purposes. Field guide to the freshwater fishes of Tanzania. Prepared and published with the support of the United Nations Development Programme (project URT/87/016)*. Rome: FAO. p 145.
[3] Khairuman, H. & Amri, K. 2013. Tilapia Fish Cultivation. Jakarta: PT AgroMedia Pustaka.
[4] Khairuman, H. & Amri, K. 2008. Intensive Tilapia Cultivation. Jakarta: PT AgroMedia Pustaka.
[5] Lusher A, Hollmann P and Mendoza J 2017 *FAO Fisheries and Aquaculture Technical Paper*. 615 1-126.
[6] Barnes D K A, Morley S, Bell J, Brewin P, Brigden K, Collins M, Glass T, Goodall- Copestake W, Henry L, Laptikhovsky V, Piechaud N, Richardson A, Rose P, Sands C, Schofield A, Shreeve R, Small A, Stamford T & Taylor B 2018 *Curr Biol*. 28 R1137-R1138.
[7] Kasamesiri P 2020 *Int J of GEOMATE*. 18.
[8] Movahedini A, Abtahi B and Bahmani M 2012 *Asian J Anim Vet Adv*. 7 710-717.
[9] Jabeen K, Li B, Chen Q, Su L, Wu C, Hollert H and Shi H 2018 *Chemosphere*. 213.
[10] Soares J, Miguel I, Venâncio C, Lopes I and Oliveira M 2020 *Water*.12.
[11] Hamed M, Soliman H, Badrey A and Osman A 2021 *Tissue and Cell*. 71.
[12] Almeida, D. S. M. 2017. *Histopathological Evaluation of Two Blennius Fishes Exposed to MPs Via Feeding*. Doctoral dissertation.
[13] Jeong C B, Kang H M, Lee M C, Kim D H, Han J, Hwang D, Souissi S, Lee S J, Shin K H, Park H and Lee J S 2017 *Sci Rep*. 7.
[14] Pannetier P, Cachot J, Clérandeau C, van Arkel K, Faure F, de Alencastro L, Sciacca F and Morin B 2017 *Int Conf Microplastics*. 10.1016/B978-0-12-812271-6.00138-1.
[15] Choi J, Jung Y J, Hong N H, Hong S and Park J W 2018 *Marine Pollution Bulletin*, 129, 231-240.
[16] Espinosa-Ruiz C, Beltrán J M, Esteban M and Cuesta A 2018 *Environ Pollut*. 235 30-38.
[17] Imhof H, Rusek J, Thiel M, Wolinska J and Laforsch C 2017 *PLOS ONE*, 12.
[18] Machado A, Kloas W, Zarfl C, Hempel S and Rillig M 2017 *Glob Chang Biol*. *24*(4), 1405-1416
[19] Haaparanta A 1997 *J Fish Biol*. 50 575-591.
[20] Mallatt J 1985 *Can J Fish Aquat Sci*. 42 630-648.
[21] Elshaer F M, Khalaf-Allah H M M, and Bakry S 2013. *World J Fish Marine Sci*, *5*, 693-700.
[22] Monteiro S & Rocha E, Fontaínhas-Fernandes A and Sousa M 2008 *J Fish Biol*. 73 1376 - 1392.
[23] Barboza L, Vieira L, Branco V, Carvalho C and Guilhermino L 2018 *Sci Rep*. 8 15655.
[24] Mittler R 2017 *Trends in Plant Sci*. 22 11 – 19.
[25] Fonte E, Ferreira P and Guilhermino L 2016 *Aquat Toxicol*. 180 173-185.
[26] Ghosh S, Sinha A and Sahu C 2008 *Aquac Nutr*. 14 289 - 299.
[27] Dharmaraj S and Dhevendaran K 2010 *Food Technol Biotechnol*. 48 497 – 504.
[28] Lobelle D and Cunliff M 2011 *Mar Pol Bul*. 62 197-200.
[29] Cruz P, Ibañez A, Monroy O and Ramírez-Saad H 2012 *ISRN Microbiol.*
[30] Al-Gabri N, Ibrahim R, Ahmed S, Amer S, Al-Gabri N, Ahmed A, Abdelwarith A & Younis E 2020 *Aquac Reports*. 18 100545.

[31] Englard S and Seifter S 1986 *Annu Rev Nutr.*, 6, 365-406.
[32] Xu H J, Jiang W D, Feng L, Liu Y, Wu P, Jiang J, Kuang S Y, Tang L and Tang W N , Zhang Y A and Zhou X Q 2016 *Fish Shellfish Immunol*. 58 177-192.

6. Acknowledgement
The author would like to thank Faculty of Fisheries and Marine, Airlangga University, for conducting the seminar.

IOP Conf. Series: Earth and Environmental Science 1273 (2023) 012086 doi:10.1088/1755-1315/1273/1/012086

Antibacterial activity of gel hand sanitizer from seaweed extract (*Kappaphycus alvarezii*)

W K Putri[1], D Y Pujiastuti[2,3], E Saputra[2]

[1]Fisheries Product Technology Study Program, Faculty of Fisheries and Marine, Universitas Airlangga, Kampus C Unair, Jl. Mulyorejo Surabaya 60115, East Java, Indonesia
[2]Department of Marine, Faculty of Fisheries and Marine, Universitas Airlangga, Kampus C Unair, Jl. Mulyorejo Surabaya 60115, East Java, Indonesia

[3]Corresponding Author: dwiyp@fpk.unair.ac.id

Abstract. Seaweed extract *(Kappaphycus alvarezii*) contains flavonoids, triterpenoids, steroids, and alkaloids which have antibacterial activity. The purpose of this study was to determine the potential of seaweed extract (*K. alvarezii*) as a hand sanitizer product and to determine the antibacterial activity of hand sanitizer gel from seaweed extract (*K. alvarezii*) against *S. aureus* and *E. coli* bacteria. This research was conducted by experimental methods Inhibitory Zone Test, Homogeneity Test, Viscosity Test, pH Test, Spreadability Diameter Test, Organoleptic Test and Skin Irritation Test. This study used a completely randomized design (CRD). The data obtained were then analyzed using the ANOVA method and if there was a difference, then the significant difference test was continued using the Duncan Multiple Range Test (DMRT) at a significance level of 5% (α=0.05). The results of this study showed that the seaweed extract of *K. alvarezii* gave good results. The appearance of the hand sanitizer gel in each treatment is homogeneous. The value of gel viscosity is close to standard. The hand sanitizer gel has a pH of 5. The diameter of the spread has a result of 5-6. Does not cause irritation to the skin. Has a zone of inhibition that is classified as strong-very strong.

1. Introduction

Infectious diseases are one of the causes of high morbidity and mortality, which are included in the top ten deadliest diseases in the world [1]. Infectious diseases are caused by bacteria, especially *Escherichia coli* and *Staphylococcus aureus*. These bacteria infect and can cause transmission to humans through the skin, especially on the palms [2]. Prevention of bacterial infections can be done by keeping the palms clean. Health is a very important aspect of life. Various types of viruses, bacteria and fungi can be attached to the hands every day through physical contact. One of the most appropriate prevention efforts to prevent the spread of viruses, bacteria, and fungi is to wash your hands with soapy water and clean water flowing [3]. Washing hands is done because hands are often germ-carrying agents and can cause pathogens to pass from one person to another so that it can cause disease. Prevention can be done by maintaining hand hygiene before eating and drinking by using hand antiseptic gel as a practical way to replace soap and water for washing hands [4].

Hand sanitizer or antiseptic gel is a preparation that has a gel texture and contains antibacterial compounds [5]. Hand sanitizer is a form of technological development to clean hands quickly and practically. But in general, in the market there are hand sanitizers made from chemicals that can cause allergies to the skin because they are often used [6].

One of the plants that have antibacterial activity is seaweed. *Kappaphycus alvarezii* is most commonly found in Indonesia, especially Southeast Sulawesi. *Kappaphycus alvarezii* contains flavonoids, triterpenoids, steroids, and alkaloids that have activity against *Staphylococcus aueus* and *Escherichia coli* bacteria [7]. So, with the potential that exists in seaweed (*K. alvarezii*) and the importance of maintaining health in this pandemic era, This research was conducted in order to obtain information related to the use of (*K. alvarezii*) as an antibacterial preparation packaged in a hand sanitizer gel preparation derived from natural active ingredients.

The purpose of this study was to determine the potential antibacterial activity of seaweed extract (*Kappaphycus alvarezii*) as a hand sanitizer product against *Staphylococcus aureus* and *Escherichia coli* bacteria.

2. Materials and methods

The study was conducted in February – March 2021 at the Chemistry & Analysis Laboratory and the Microbiology & Fish Disease Laboratory, Faculty of Fisheries and Marine Affairs, Airlangga University, Surabaya. Seaweed (*K. alvarezii*) samples were obtained from e-commerce. The research method used is an experimental method using a Completely Randomized Design (CRD) consisting of 3 treatments and 5 replications. This treatment consisted of adding seaweed extract (*K. alvarezii*) with concentrations of 0.5%, 1% and 1.5%. Determination of the dose based on research conducted [8], that the concentration of *K. alvarezii* as much as 0.02%, 0.1% and 0.5% resulted in an average inhibition zone of 10.00 mm, respectively. 11.66 mm, and 14.33 mm.

The main parameter in this study is the antibacterial test. While the supporting parameters in this study were homogeneity test, viscosity test, dispersion diameter test, irritation test, pH test and organoleptic test. Data analysis used Analysis of Variance (ANOVA) test and continued with Duncan's Multiple Distance test.

3. Results and discussion

3.1Results

The characteristics of hand sanitizer gel according to national standards include several things, including homogeneity, viscosity, dispersion diameter, organoleptic, and skin irritation. The results of this study indicate that hand sanitizer gel from seaweed extract (*K. alvarezii*) has results that are in accordance with commercial hand sanitizer gel standards. This hand sanitizer gel has a value of homogeneity, viscosity, pH, dispersion diameter, organoleptic, and skin irritation in accordance with the provisions of national standards referring to SNI No. 06-2588:1992 about antiseptic

Homogeneity

The homogeneity test in this study was carried out using the visual sense of sight. Each sample has the same homogeneity. This is shown when applied to transparent glass by rubbing it on the surface of the glass. The results obtained when the sample was smeared on a transparent glass surface did not show any lumps of material mixed in the sample, this indicates that the sample is homogeneous. The results of the homogeneity test of the hand sanitizer gel can be seen in Table 1.

Table 1. Gel Hand Sanitizer Homogeneity Test Results

TREATMENT	RESULT	Homogeneity Standart SNI 06-2588
P1		
P2	Homogeneous	Homogeneous
P3		

Note: The results of homogeneity test in each treatment have the same results, namely homogeneous.

IOP Conf. Series: Earth and Environmental Science 1273 (2023) 012086 doi:10.1088/1755-1315/1273/1/012086

Viscosity
The results of the viscosity value test on hand sanitizer gel with seaweed extract (*K. alvarezii*) can be seen in Table 2.

Table 2. Gel Hand Sanitizer Viscosity Test Results

TREATMENT	RESULT	VISCOSITY STANDART SNI 06-2588
P1	$176{,}367^{a} \pm 1{,}873$	
P2	$181{,}783^{b} \pm 1{,}954$	2000 – 4000 cPs.
P3	$187{,}174^{c} \pm 4{,}948$	

Note: Different superscripts in the same column show significant differences ($p<0.05$).

pH
The pH value obtained from the hand sanitizer gel with seaweed extract (*K. alvarezii*) in this study can be seen in Table 3.

Table 3. pH test value

TREATMENT	RESULT	pH STANDART SNI 06-2588
P1	$5{,}468^{c} \pm 0{,}00447$	
P2	$5{,}328^{b} \pm 0{,}01095$	4,5 – 6,5
P3	$5{,}232^{a} \pm 0{,}00447$	

Note: Different superscripts in the same column show significant differences ($p<0.05$).

Diameter of Spread Test
The calculation of the dispersion value of the hand sanitizer gel with seaweed extract (*K. alvarezii*) was carried out using a weight of 300 g and clear glass. The results of the dispersion diameter test can be seen in Table 4.

Table 4. Results of the Analysis of the Spreadability of the Hand Sanitizer Gel

TREATMENT	RESULT	DIAMETER OF SPREAD STANDART SNI 06-2588
P1	$6{,}18^{c} \pm 0{,}164$	
P2	$5{,}54^{b} \pm 0{,}054$	5 – 7 cm.
P3	$5{,}18^{a} \pm 0{,}109$	

Note: Different superscripts in the same column show significant differences ($p<0.05$).

Organoleptic
The organoleptic test was carried out using 15 panelists by trying to observe the parameters of color, aroma, texture of hand sanitizer gel samples that had been randomly labeled (Figure 1). The results of the organoleptic test analysis can be seen in Table 5.

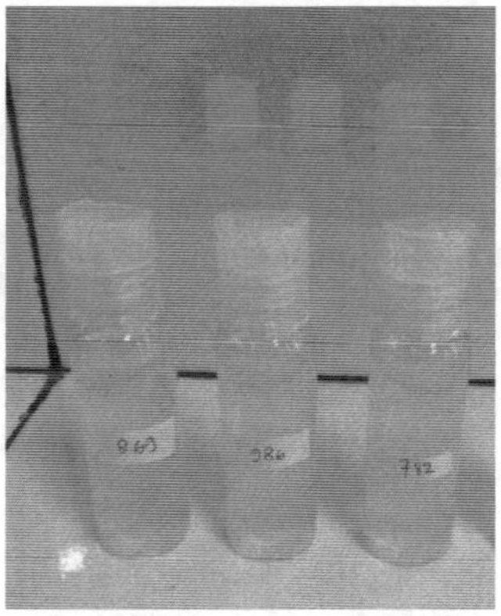

Figure 1. Organoleptic Test Sample

Table 5. Gel Hand Sanitizer Organoleptic Test Results

PARAMETER	RESULT		
	P1	P2	P3
Color	$2{,}67^a \pm 0{,}488$	$2{,}73^a \pm 0{,}458$	$2{,}60^a \pm 0{,}632$
Odor	$2{,}60^a \pm 0{,}507$	$2{,}40^a \pm 0{,}507$	$2{,}60^a \pm 0{,}632$
Texture	$2{,}53^a \pm 0{,}834$	$2{,}40^a \pm 0{,}737$	$2{,}60^a \pm 0{,}632$

Note: The same superscript in the same column shows no significant difference (p>0.05).

Skin Irritation

The skin irritation test in this study was carried out by means of an open test (patch test) with 15 panelists. Panelists who have met the requirements, among others, are in good health, aged between 20-35 years, do not have allergies and are willing to be an irritation test panelist in this study, can perform the test on a predetermined body part. A total of 15 panelists did not show any skin irritation in the designated areas, the results of the skin irritation test can be seen in Table 6.

Table 6. Skin Irritation Test Results on Panelists

TREATMENT	RESULT
P1	
P2	-
P3	

Description : Skin irritation test was performed on 15 volunteers by applying the sample to the inner arm for 3 minutes. The sign (-) indicates that there is no irritation reaction on the skin of the audience who has been treated.

Inhibition Zone

This study used the inhibition zone test to determine the antibacterial activity of the hand sanitizer gel from seaweed extract (*K. alvarezii*) using four treatments, namely (K+ = chloramphenicol, P1 = hand sanitizer gel; 0.5% extract of *K. alvarezii*, P2 = hand sanitizer gel; 1% *K. alvarezii* extract, P3 = hand sanitizer gel; 1.5% *K. alvarezii* extract). There are two types of bacteria used, *namely Staphylococcus aureus* from gram-positive bacteria and *Escherichia coli* from gram-negative bacteria.

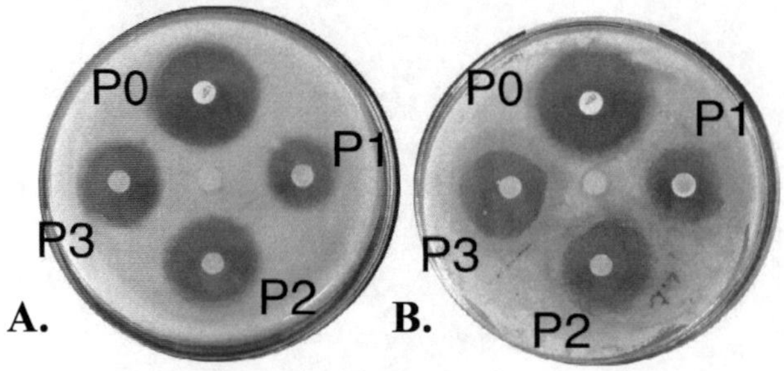

Figure 2. Inhibition Zone Test Results Gel Hand Sanitizer Extract *K. alvarezii*
Description: *S. aureus* (A) *E. coli* (B)

The following is the result of the calculation of the inhibition zone formed by the antibacterial activity of the hand sanitizer gel from seaweed extract (*Kappaphycus alvarezii*) by *S. aureus* bacteria which is presented in Table 7. and by *E. coli* bacteria which is presented in Table 8.

Table 7. Results of Inhibitory Zone Test against *S. aureus* Bacteria

TREATMENT	DOSES	INHIBITION ZONE ± SD
P1	0,5 %	$13{,}96^{a} \pm 2{,}04$
P2	1%	$18{,}06^{b} \pm 3{,}44$
P3	1,5 %	$19{,}00^{b} \pm 5{,}08$

Note: Different superscripts in the same column show significant differences (p<0.05).

Table 8. Results of Inhibitory Zone Test against *E. coli* Bacteria

TREATMENT	DOSES	INHIBITION ZONE ± SD
P1	0,5 %	$15{,}17^{a} \pm 1{,}58$
P2	1%	$20{,}00^{b} \pm 2{,}58$
P3	1,5 %	$20{,}36^{b} \pm 0.96$

Note: Different superscripts in the same column show significant differences (p<0.05).

3.2Discussion

The addition of different concentrations of seaweed extract (*K. alvarezii*) in each treatment in the manufacture of hand sanitizer gel will produce different values of homogeneity, viscosity, pH, dispersion diameter, organoletics, irritation test results and zone of inhibition. The results of each test will be explained as follows:

An indicator contained in the results of mixing product or preparation ingredients is referred to as homogeneity[9]. Homogeneity test is carried out by applying the sample to glass or other suitable transparent material, the preparation must show homogeneous results and not show the presence of coarse grains [10].

The purpose of this homogeneity test is to determine the hand sanitizer gel extract of *K. alvarezii* by looking at the uniformity of the particles in the preparation. The three formulations of hand sanitizer gel preparations of *K. alvarezii* extract had homogeneous results. It is characterized by no parts that are not mixed well.

The data from the homogeneity test results between the formulas can be concluded that there is no difference in the homogeneity of the hand sanitizer gel preparation of *K. alvarezii* extract. So it can be seen that the difference in the concentration of *K. alvarezii* extract as an antibacterial has no effect on the homogeneity of the preparation. Things that affect the homogeneity of a product include the mixing process at the time of product manufacture [11].

Viscosity is a measure to express the viscosity of a fluid. One of the properties of liquids that has a close relationship with the resistance to flow is viscosity [12]. Viscosity can determine the speed at which a fluid flows. Viscosity test is carried out in order to determine the value of the viscosity of a substance. The higher the viscosity value produced, the higher the viscosity of the substance [13].

Based on the test results of the gel hand sanitizer extract of *K. alvarezii*, the highest viscosity results were obtained in the P3 treatment, that is ($187.174^{c} \pm 4{,}948$). This shows that the results obtained do not meet the standards, because a good gel has a standard viscosity of around 2000-4000cPs [14]. One of the factors that can affect the difference in the viscosity of a solute product in a liquid is the concentration [15].

Liquids and gases are two different types and have different viscosity levels. Viscosity is actually the frictional force between the various molecules that make up a fluid. So the various molecules that make up the fluid rub against each other when the fluid flows through each other. The presence of viscosity in liquids occurs because of the cohesive force. Whereas in gases it occurs due to molecular collisions [16].

The measurement of the pH test value has the aim of knowing the acidity value of a product. According [17] one of the important things that can affect the use of cosmetic products is the pH value of the product itself. Based on the measurement results of the pH test value, the lowest average pH value was found in treatment P3 ($5.232^{a} \pm 0.00447$) and the highest pH was in treatment P1 ($5.468^{c} \pm 0.00447$). The pH value that has been obtained is in accordance with the requirements for the pH value in SNI No. 06-2588 which is between 4.5 – 6.5. This shows that the gel hand sanitizer product is safe to use.

The pH value of the P3 treatment tends to be lower. The low pH is caused by the extract which is acidic due to the presence of flavonoids in the extract. The greater the concentration of the extract used, the more acidic the resulting pH. This is in accordance with the research of [18] that extract preparations containing flavonoids tend to have acidic pH values.

The purpose of the dispersion diameter test was to determine the ability of the hand sanitizer gel *K. alvarezii* extract to spread on the skin surface. The easier it is for the sample to be applied to the skin, the greater the contact surface area of the nutritious substance with the skin so that the absorption of the extract will be more optimal [19].

The results of the average dispersion values of various treatments, respectively, were at P1 (6.18), P2 (5.54) and P3 (5.18). Based on the dispersion value data obtained, it can be concluded that the dispersion value meets the requirements of SNI No. 06-2588. A convenient semisolid preparation has a spread of 5 to 7 cm [20].

Things that affect the increase and decrease in spreadability include the consistency of the gel preparation, where this is related to the viscosity value of the preparation. If the viscosity value is high, the dispersion area obtained is low and vice versa. This is due to the high viscosity which makes it difficult for the gel to flow so that the resulting area is small [21].

Organoleptic test is a test carried out by naked eye or direct observation in order to describe the preparation. Organoleptic tests include the shape, color, odor that has been produced by a product. Organoleptic testing on hand sanitizer gel preparations was carried out visually, including color, aroma and texture [22]. This hand sanitizer gel preparation is an external medicine so it is not necessary to test the taste parameters. The results of the organoleptic test analysis on the hand sanitizer gel extract of *K. alvarezii* showed that the color, aroma and texture parameters at P1, P2, and P3 did not show a significant difference ($p>0.05$). There was no difference in the treatment for each parameter because the distance between the concentrations given was too small.

According to research by [23], irritation is a result of skin reactions to chemicals, including strong alkalis, strong acids, solvents and detergents. The reactions given range from hyperemia, edema, vesicles to ulcers. Several types of skin irritation, namely primary irritation that occurs at the site of first touch contact, generally will immediately cause a skin reaction shortly after application to the skin. These irritants occur between a few minutes to an hour after application of the product [24]. Meanwhile, secondary irritation takes several hours to detect a reaction on the skin. Signs that appear include hyperemia, erythema and edema. The skin reaction that is being irritated is local only on the damaged skin area [25].

The results of the skin irritation test obtained in this study showed that there was no skin irritation reaction in a number of panelists who had hand sanitizer gel applied to the forearm. This is because the materials used for the manufacture of hand sanitizer gel preparations do not use alcohol and several other skin irritating ingredients. The use of alcohol in antiseptic preparations is not safe for health because alcohol is an organic solvent that can dissolve the layer of fat and sebum on the skin which has a function to protect against infection with microorganisms. Regular use of alcohol can cause skin irritation [26].

The inhibition zone test aims to determine the ability of the antibacterial activity of the hand sanitizer gel to inhibit the growth of *Staphylococcus aureus* and *Escherichia coli bacteria*. The inhibition zone test was carried out using the disk diffusion method to determine the clear zone created from seaweed extract (*K. alvarezii*). The diffusion method is the process of measuring and observing the diameter of the clear zone formed in the disc area, the measurement process is carried out after being left for 18-24 hours and then measured using a caliper [27].

The test results for the inhibition zone of the gel hand sanitizer which has the largest average inhibition zone are the P3 treatment which is 19.00 mm on *Staphylococcus aureus* bacteria and the P3 treatment which is 20.36 mm on *Escherichia coli* bacteria. This indicates that the zone of inhibition in *Staphylococcus aureus* is strong, while *Escherichia coli* is very strong. The criteria for antibacterial strength are as follows, the diameter of the inhibition zone of less than 5 mm is categorized as lacking, the zone of inhibition between 5-10 mm is categorized as moderate, the zone of inhibition between 11-20 mm is categorized as strong, while the diameter of the inhibition zone of more than 20 mm is categorized as very strong [28].

Antibacterial substances work by inhibiting cell wall synthesis and disrupting membrane function [29]. The difference in the diameter of the inhibition zone at each concentration used was caused by the amount of active substance contained in each concentration. The amount of the active substance component is influenced by the amount of a concentration that can affect the formation of the inhibition zone [30]. Concentrations that are too small generally cannot produce an inhibition zone because the concentration of the extract used greatly affects the antibacterial activityc[31].

Based on the results of the discussion mentioned above, hand sanitizer gel with seaweed extract (*Kappaphycus alvarezii*) is included in hand sanitizer gel with good criteria. This is based on the results of the test values for homogeneity, pH, dispersion diameter, organoleptic and skin irritation which have met the quality requirements in accordance with SNI No. 06-2588 and has an inhibition zone that is classified as strong – very strong.

4. Conclusion

Based on the research that has been done, it can be concluded that hand sanitizer gel from seaweed extract (*Kappaphycus alvarezii*) has antibacterial activity against *Staphylococcus aureus* and *Escherichia coli* bacteria which are classified as strong - very strong. Seaweed extract (*K. alvarezii*) has viscosity test results that do not meet the standards of SNI 06-3588:1992. However, it meets the standards on several test parameters including homogeneity, pH, dispersion diameter, organoleptic, skin irritation and inhibition zone tests.

5. References

[1] WHO 2012 *The Top 10 Causes of Death* [28 Maret 2016].
[2] Shu M 2013 J. Ilmiah Mahasiswa Universitas Sby. 2: 1-14.
[3] Wijaya JI 2013 J. Ilmiah Mahasiswa Universitas Sby. 2(1): 1-14.
[4] Pramita FY 2013 Skripsi Pontianak : Universitas Tanjungpura.

[5] Golin A P, Bhsc D C and Ghahary A 2020 pp 1062 – 1067.
[6] Asngad A and Nopitasari RAB 2018 4(2) pp 61–70.
[7] Andriani Z, A Ghanaim F, Ahmad H 2015 *J. of Chem.* 4: 93-100.
[8] Megawati M, Sari, NI, & Sukmiwati M 2017 Doctoral dissertation, Riau University.
[9] Pujiastuti A dan M Kristiani 2019 *J. Farm Ind.* 16(1) : 42-55.
[10] Ditjen POM 1985 *Indonesian Cosmetics Formulary. Jakarta: Ministry of Health of the Republic of Indonesia* p 22, 29, 83, 97, 356-357, 360.
[11] Rieger M 1994 Emulsi in: Lachman *et al* 1994 T *Theory and Pharmacy Practical Industry Ed ke-2.* Jakarta: UI.
[12] Manurung LS dan Sudrajad H 2018 *J. Pend. Fis.* 6(2), 98-104.
[13] Martin A, Swarbrick J, dan Cammarata A 1993 (Jakarta:Universitas Indonesia) p 1077.
[14] Harimurti S dan Hidayaturahmah R 2016 *FKIK* 1(5) 1-8.
[15] Atkins PW 1996 (Jakarta: Erlangga).
[16] Bird Tony 1993 *Kimia Fisik Untuk Universitas Jakarta* : PT Gramedia.
[17] Wisuda S, D Buchari dan S Loekman 2014 *J. Online Mhs. UNRI.* p 12.
[18] Akib NI, Wulandari IW, Suryani S, dan Hanari H 2019 *Fish Protech. 2*(2), 180-188.
[19] Rohmani S & Kuncoro MA 2019 *Pharm. Sci. and Clinical Res.* 4(1), 16-28.
[20] Garg A, Aggarwal D, Garg S and Sigla AK 2002 *J. Pharm Tech.* 20(2), 84-102.
[21] Arista YN, Paulina VY dan Hamidah S 2013 *J. Ilmiah farm.* 2(2), 18-26.
[22] Ardana, Mirhansyah, Vebry Aeyni, Arsyik Ibrahim 2015 *J Trop Pharm Chem.* 3(2),101 – 108.
[23] Untari EK & Robiyanto 2018 *J. Jamu Indo.* 3(2):55-61.
[24] Sunsmart 1998
[25] Schmitt WH 1996 Blackie Academe and Professional.
[26] Dyer D, Shinder A and Shinder F 2000 *Fam Med.* 32(9) 633-638.
[27] Pratiwi ST 2008 *Microbiologi Pharmacy* Erlangga Yogyakarta.
[28] Davis WW, Stout TR 1971 *J Microbiol.* (4):659-665.
[29] Pelczar MJ, ECS dan Chan 2005 *Basics of Microbiology 1* UI-Press Jakarta.
[30] Brook GF, SB Janet and AM Stephen 2005 Medical Microbiology book 1 Translated by Mudihardi E, Kuntaman, Wasito, Mertaniasih, and Alimsardjono Jakarta: Salemba Medika.
[31] Zuhud AM 2001 *Food Technology and Industry XII.*

6. Acknowledgments

Thank you for the Faculty of Fisheries and Marine Universitas Airlangga for supporting the facilities of the research.

Characteristics of flavour paste from bone milkfish (*Chanos chanos*) with additional of tapioca flour

M Muslimin[1], A M Sahidu[2], D Y Pujiastuti[2,3]

[1]Fisheries Product Technology, Faculty of Fisheries and Marine, Universitas Airlangga, Kampus C Unair, Jl. Mulyorejo Surabaya 60115, East Java, Indonesia
[2]Department of Marine, Faculty of Fisheries and Marine, Universitas Airlangga, Kampus C Unair, Jl. Mulyorejo Surabaya 60115, East Java, Indonesia

[3]Corresponding Author: dwiyp@fpk.unair.ac.id

Abstract. Milkfish bones contains glutamic acid of 7.38%. Glutamic acid is one of the constituents of protein that triggers the umami taste in a dish. Utilization of milkfish bone as a natural flavoring can be obtained through extraction. To make the product in the form of paste, it is necessary to add tapioca flour as a thickener. The purpose of this study was to determine the characteristics of milkfish flavor paste with the addition of different concentrations of tapioca flour. This research using experimental design. The independent variables used were tapioca flour concentrations of 0% (P0), 2.5% (P1), 5% (P2), and 7.5% (P3). Completely randomized design (CRD) was used as the experimental design. Data were analyzed using ANOVA and to find out the best treatment, Duncan test was performed. The addition of tapioca flour affects the protein content with a value (5.38% - 14.78%), fat content with a value (2.45% - 2.53%), moisture content with a value (80.12% - 91.2 %), viscosity with a value (2.11 cSt/s – 558.99 cSt/s), yield with a value (73.69% - 82.4%). While the addition of tapioca flour had no significant effect on the water solubility index of the flavoring paste with a value (98.93% - 99.24%).

1. Introduction

The by-product in the form of fish bones from the processing of boneless milkfish has a fairly large amount. This is in accordance with the statement of Akhmadi et al. [1] that the marine by-products produced are around 30 kg/day and this by-product of bone is only disposed of in landfills, generally. Milkfish bones as a by-product of processed milkfish without the spines has a lot of flesh attached. Based on the statement of Regenstein and Zhou [2] that milkfish bones contain non-essential acids namely glutamic acid of 7.38%. In addition, milkfish meat also has a glutamic acid content of 12.68%. Glutamic acid is one of the constituents of protein that triggers the umami taste in a dish, of course this is a good potential in making added value as a flavor enhancer.

Flavor enhancers on the market are generally in instant form and are given the addition of Mono Sodium Glutamate (MSG) as a flavor enhancer. However, many people are concerned about the use of MSG due to the debate which states that consumption can have a negative impact on the body [3]. This is reinforced by Muntaza and Adi [4] that MSG can damage nerve cells if consumed at high doses (0.5 – 4.0 g/kg body weight).

The use of milkfish bones as a natural flavor enhancer can be obtained through extraction. To make the product in the form of a paste, it is necessary to add tapioca flour as a food thickener when cooked with water. The use of milkfish bones as a flavor enhancer has not been researched yet. Therefore, this

IOP Conf. Series: Earth and Environmental Science 1273 (2023) 012087 doi:10.1088/1755-1315/1273/1/012087

study aims to determine the characteristics of milkfish flavoring paste with the addition of different concentrations of tapioca flour.

2. Materials and methods

The research method used is an experimental method. The independent variables in this study were the concentration of tapioca flour including P0 as a control treatment, P1 with the addition of 2.5% tapioca flour from the total weight of the milkfish bone filtrate, P2 with the addition of 5% tapioca flour from the total weight of the milkfish bone filtrate, P3 with the addition of 7, 5% tapioca flour from the total weight of the milkfish bone filtrate

The working procedures of this research included research preparation, making milkfish bone filtrate, making milkfish flavoring paste with different tapioca flour formulations, testing chemical and physical characteristics which included protein content, fat content, water content, viscosity, water solubility index, yield. Test data were analyzed using ANOVA (Analysis of Variance) with the SPSS application and continued with Duncan's Multiple Distance test

3. Results and discussion

3.1 Protein content

Determination of the protein content in this study has the aim of knowing how much protein content is in the milkfish flavoring paste. Data on protein content results can be seen in Table 1. The protein content of the milkfish flavoring paste produced ranges from 5.38-14.78%.

Table 1. Protein content of milkfish flavoring paste

Treatment	Protein Content (%)	Indonesian Standard 01-4273-1996
P0	$5,38^{d} \pm 0,039$	Minimum 7%
P1	$8,81^{c} \pm 0,087$	
P2	$12,05^{b} \pm 0,137$	
P3	$14,78^{a} \pm 0,318$	

Note: The notation indicated by the same superscript letter in the column indicates that there is no significant difference between treatments (P> 0.05)

3.2 Fat content

Determination of the fat content in this study has the aim of knowing how much fat content is in the milkfish flavoring paste. Data on fat content results can be seen in Table 2. The fat content of the milkfish flavoring paste produced ranges from 2.52-2.58%.

Table 2. Fat content of milkfish flavoring paste

Treatment	Fat Content (%)	Indonesian Standard 01-4273-1996
P0	$2,53^{a} \pm 0,011$	Minimum 2%
P1	$2,50^{b} \pm 0,130$	
P2	$2,48^{c} \pm 0,022$	
P3	$2,45^{d} \pm 0,035$	

Note: The notation indicated by the same superscript letter in the column indicates that there is no significant difference between treatments (P> 0.05)

3.3 Water content

Determination of the water content in this study was to find out how much water content is in the milkfish flavoring paste. Data on water content results can be seen in Table 3. The water content of the milkfish flavoring paste produced ranges from 80.12-91.2%.

Table 3. Water content of milkfish flavoring paste

Treatment	Water Content (%)	Standard
P0	$91,2^{a} \pm 0,022$	63,5% - 74,5% [5]
P1	$87,09^{b} \pm 0,026$	
P2	$83,23^{c} \pm 0,032$	
P3	$80,12^{d} \pm 0,037$	

Note: The notation indicated by the same superscript letter in the column indicates that there is no significant difference between treatments (P> 0.05)

3.4 Viscosity

Viscosity is a test method to assess the thickness of the milkfish flavoring paste. Data on the results of the viscosity value can be seen in Table 4.

Table 4. Viscosity of milkfish flavoring paste

Treatment	Viscosity (cSt/s)	Standard
P0	$2,11^{d} \pm 0,073$	127,67cSt/s – 132,67cSt/s [5]
P1	$4,11^{c} \pm 0,056$	
P2	$93,86^{b} \pm 0,631$	
P3	$558,99^{a} \pm 2,767$	

Note: The notation indicated by the same superscript letter in the column indicates that there is no significant difference between treatments (P> 0.05)

3.5Water Solubility Index

The water solubility index in this study aims to determine the level of solubility in water contained in milkfish flavoring paste. Data on the results of the water solubility index value can be seen in Table 5.

Table 5. Water solubility index of milkfish flavoring paste

Treatment	Water solubility index (%)	Standard
P0	$99,24^{a} \pm 0,081$	98,5% - 99,99% [6]
P1	$99,10^{a} \pm 0,064$	
P2	$99,01^{a} \pm 0,075$	
P3	$98,93^{a} \pm 0,613$	

Note: The notation indicated by the same superscript letter in the column indicates that there is no significant difference between treatments (P> 0.05)

3.6 Yield

The yield in this study serves as a benchmark for further testing. The yield value was obtained from the results of making milkfish flavoring paste. Data on yield values can be seen in Table 6.

Table 6. Yield of milkfish flavoring paste

Treatment	Yield (%)	Standard
P0	$82,40^{a} \pm 2,752$	75,89% - 97,63% [6]
P1	$78,88^{b} \pm 2,467$	
P2	$75,48^{c} \pm 1,844$	
P3	$73,69^{c} \pm 1,670$	

Note: The notation indicated by the same superscript letter in the column indicates that there is no significant difference between treatments (P> 0.05)

3.7 Results

The characterization in this study was conducted to determine the effect of the addition of tapioca flour on the characteristics of milkfish flavoring paste. The protein content in this study increased with each addition of tapioca flour. The minimum limit for protein content in SNI 01-4273-1996 regarding beef seasoning is 7%. The lowest protein content contained in this study was below the minimum SNI limit of 5.38% at P0. Meanwhile, P1, P2, and P3 had protein levels above the minimum SNI limit. The protein content in tapioca flour has an effect on increasing protein in processed food products. This is because tapioca flour itself contains a protein content of 6.98% [7].

The fat content of this study complied with SNI 01-4273-1996 concerning beef seasoning with a minimum limit of 2%. The lowest fat content was obtained by P3 with an average of 2.45%. While the highest fat content was obtained by P0 with an average of 2.53%. This decrease was due to the heating process in the manufacture of milkfish flavoring paste. Based on the statement of Sundari et al., [8] that the decrease in fat content is affected by the damage damage to the fat structure due to the heating process.

The water content produced in this study showed a decrease. The addition of starch causes an increase in amylose content so that the water content of the food decreases due to water binding [9]. The lowest water content results were found in P3 with an average value of 80.12% while the highest value was found in P0 which was the control treatment which was 91.2%. The water content in this study was obtained higher than Rahmi [5] that natural flavoring paste stored at room temperature has an average water content value of 63.5-74.5%. The difference in results is due to the different methods and components in the manufacture of flavoring pastes.

Based on the results of this study, the viscosity value increased with each addition of tapioca flour. The lowest viscosity value is found at P0 with an average value of 2.10 cSt/s, while the highest viscosity value is found at P3 with an average value of 599 cSt/s. The product produced in this study has become thicker with increasing concentration of tapioca flour. Based on the statement of Singh et al., [10] stated that starch suspension in water when heated, the water will penetrate the outer layer of the granules and these granules begin to swell causing gelatinization to occur. During the gelatinization process, the viscosity of the starch-containing material increases when heated with water.

The results of this study indicated a decrease in the water solubility index for each addition of tapioca flour, but this study had no significant effect. The highest average result was obtained by P0 with a value of 99.24%, while the lowest is found in P3 with an average value of 98.94%. This is in accordance with Hartono [6] that the resulting water solubility index values range from 98.5% - 99.99%. The higher the amylopectin in starch, the higher the water absorption. Based on the results of this study, the yield value decreased with each addition of tapioca flour. This is in line with Hartono [6] that the yield results produced have also decreased with value of 97.63% at P0 and 75.89% at P3. The decrease in yield value in this study was due to heating of tapioca flour so that the amorphous regions of the starch granules absorbed water [11]. Water The free-moving particles then become trapped in the starch.

4. Conclusion

Based on the results and discussion of this study, it can be concluded that the addition of tapioca flour affected the protein content with a value (5.38% - 14.78%), fat content with a value (2.45% - 2.53%), water with value (80.12% - 91.2%), viscosity with value (2.11 cSt/s – 558.99 cSt/s), yield with value (73.69% - 82.4%). While the addition of tapioca flour did not significantly affect the water solubility index of the flavoring paste with a value (98.93% - 99.24%).

5. References

[1] Akhmadi MF, Imra and Maulinawati 2019 *J Ilm Perikan dan Kelaut.* 11(1): 49-54.
[2] Regenstein JM and Zhou P 2007 (England: Woodhead Publishing Limited) P 279-303.
[3] Malichati AR and Adi 2018 *J Amerta Nutr.* 2(1): 74-82.

[4] Muntaza Y and Adi 2020 *J Amerta Nutr.* 4(1): 72-78.
[5] Rahmi DA, Dien and Kaparang 2018 *J Media Teknol Has Perikan.* 6(2): 42-47.
[6] Hartono TNA 2020 Utilization of By-Products of White Shrimp (Penaeus Indicus) As Flavor Paste With Addition Of Corn Flour. Airlangga University Fisheries Product Technology Thesis.
[7] Imanningsih N 2012 *J Panel Gizi Makan*.35(1): 13-22.
[8] Sundari D, Almasyhuri and Lamid 2015 *Media Litbangkes*.25(4): 235-242.
[9] Meitta A P, Rosyidi and Widyastuti 2013 *J Ilmu-Ilmu Peternak.* 23(3): 17-26.
[10] Singh J, Kaur L, Sodhi, NS and Gill 2003 *J. Food Chem* 81:219-231
[11] Hariyanti P, Setyawati and Wicaksono 2014 *J Agritech*, 34(3): 308-315

6. Acknowledgments
Thank you for the Faculty of Fisheries and Marine Universitas Airlangga for supporting the facilities of the research.

Size distribution of bilih (*Mystacoleucus padangensis* Blkr.) as an endemic fish spatially and temporally using dipp nets in Lake Singkarak, West Sumatra, Indonesia

S H Nasution[1,2], I Akhdiana[1], M S Syawal[1]

[1]Research Center for Limnology and Water Resources, National Research and Innovation Agency (BRIN), Republic of Indonesia

[2]Corresponding author: syah002@brin.go.id

Abstract. Lake Singkarak has endemic fish resources of only one species in the world that has economic value, namely Bilih (*Mystacoleucus padangensis*, Blkr.). The rapid rate of catching using dipp nets fishing gear in this lake is expected to result in a decline in the Bilih population. Dipp nets is a fishing gear that is not environmentally friendly. Bilih fish samples were obtained using dipp nets fishing gear at two locations, namely Intake PLTA (hydropower intake) and Batu Taba for seven (7) months from June to December 2021. The number and size distribution of Bilih were calculated spatially and temporally. The abundance of fish in the Intake PLTA and Batu Taba locations were 472 fish and 716 fish, respectively. As spatially the size range of fish in Intake PLTA is 45-104 mm and in Batu Taba 40-130 mm. The highest fish size is in the 65-72 mm class range, both spatially and temporally (June). Average water quality at Intake PLTA location (temperature 27.87 °C; pH 8.6; DO 4.68 mg/L; Conductivity 0.143 mS/cm; TDS 0.092 mg/L) at a depth of 14.15 m and at Batu Taba location (temperature 28.14 °C, pH 9.24; DO 4.68 mg/L; Conductivity 0.231 mS/cm; TDS 0.092 mg/L) at a depth of 14.15 m.

1. Introduction

Each lake has its own peculiarities, such as Lake Singkarak which is one of the priority lakes and is a tectonic lake formed from geological processes in the form of faults in the earth's plates. This lake has one type of endemic fish, namely Bilih fish (*Mystacoleucus padangensis*, Blkr.) which has economic value and is the only one found in the world [1], [2], [3]. The population is the most dominant and has important economic value [4].

The rapid increase in fishing activities using dipp nets fishing gear in this lake is estimated to have resulted in a 60% decline in the Bilih fish population within two (2) decades. Dipp nets fishing gear since 2014 has been used to catch Bilih fish using small nets. Dipp nets is a fishing gear that is not environmentally friendly equipped with lights. Fish approach the lamp, both small and large size fish. The number of these fishing gear in 2017 was 373 units.

In Tanah Datar District and Solok Regency, West Sumatra Province, Lake Singkarak is located at an altitude of 361 m above sea level, is a tectonic lake [5], [6]. The area of the lake is recorded at around 12,200 hectares [7] with a maximum depth of 271 m [8].

Water resources in Lake Singkarak are used for various activities, including hydroelectric power (PLTA), drinking water, fisheries, tourism and agriculture (irrigation), with the economic value contribution of 88.5%, 4.41%, respectively. 4.19%, 2.39% and 0.44% of the total economic value of 174.992 billion [9].

The production of total fish catches in Lake Singkarak in the last five years (2016 - 2020) has decreased by 73% [10] as well as fish size. In a 24-year period (1997 - 2021) the size of Bilih fish decreased by 60% [11], [12]. The Bilih fish population in Lake Singkarak is starting to be threatened due to the decreasing catch production and the smaller size of the Bilih fish caught.

The spatial and temporal distribution of the size of the Bilih endemic fish using charts in Lake Singkarak was studied to determine the purpose of this study. In order for Bilih's fish resources and their ecosystem to be sustainable, it is hoped that the results of this research can become a reference for local government policies in an effort to develop a Bilih fisheries management plan.

2. Materials and methods

2.1. Research location and data collection

The sampling locations at the Hydropower Intake (Intake PLTA) and Batu Taba were chosen because of the large number of dipp nets found in Singkarak Lake (figure 1, figure 2 and figure 3). The study was conducted for seven months from June to December 2021. The method used in this research is the analytical survey method. At the research location the samples were analyzed.

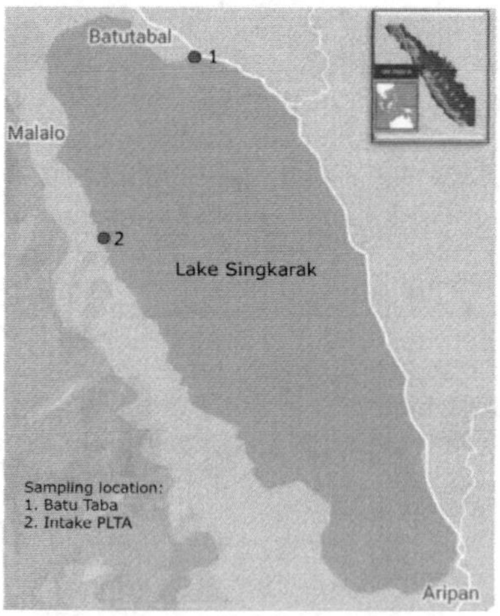

Figure 1. Bilih fish sampling location using dipp nets (Scale = 1 cm : 500 m)

Figure 2. At the PLTA Intake location (hydroelectric power plant) there are shady trees and agricultural/plantation areas located in Nagari Guguak Malalo, Tanah Datar Regency. The bottom of the waters has a sand substrate and clear surface water, but there is a lot of garbage found at the bottom of the waters. Many dipp nets are found in this location.

Figure 3. The location of Batu Taba is on the shores of Lake Singkarak, there is a tourist location, namely Tanjung Mutiara and is located in Batipuh Selatan District, Tanah Datar Regency. At a depth of 30 meters from the edge of the lake on the west side, many dipp nets are found.

2.2 Research Focus
The distribution of the number and size of Bilih is calculated spatially and temporally which is captured using a dipp nets is the aim of this study, the spatial and temporal distribution of the length of Bilih fish. Water quality parameters and depth at the sampling location were also measured.

2.3. Sampling methods
2.3.1. Fish sampling
Bilih fish samples (figure 4) were obtained using dipp nets fishing gear at two locations, namely Intake PLTA (hydropower intake) and Batu Taba for seven (7) months from June to December 2021 in Lake Singkarak. Bilih fish samples were obtained from the catch of the dipp nets then measuring unit mm is used to measure the total length of Bilih fish in a fresh condition.

Figure 5 shows how Bilih fish are obtained using dipp nets. Bagan as fishing gear is not environmentally friendly and using lights will stimulate fish to approach lights (positive phototaxis) of various sizes. The bottom support of the chart is a drum to float on the water. Bagan is equipped with a fine-sized net (such as a mosquito net) [13]. During operation, the owner of the dipp nets goes to the location by boat (generally the distance of the dipp nets from the edge of the lake is about 20-30 m), then the net is lifted using a pulley so that the fish can be caught. Fish are taken using a scoop and separated. Bilih fish are measured from the tip of the head to the tip of the last tail to the nearest unit below it to obtain the total length [14] or to the nearest unit [15].

Figure 4. Bilih (*Mystacoleucus padangensis*, Blkr)
Source: (Nasution, 2022)

Figure 5. How to get bilih fish using dipp nets in Lake Singkarak Source: (Nasution and Syawal, 2022)

2.3.2. Water quality parameter

Direct measurements of water quality parameters at the research site are dissolved oxygen (DO), temperature, pH, total suspended solids (TDS) and conductivity. The tool used is the Water Quality Checker-Horiba U 20 and a depth at the dipp nets location using a scaled rope [9].

3. Results and Discussion

3.1. The spatial frequency of the length distribution of *Mistacoleucus padangensis*

The size structure of fish communities in water bodies is influenced by fishing activities and fishing gear used [16]. Dipp nets that are not environmentally friendly are expected to reduce the number of fish populations in Lake Singkarak.

The high demand for Bilih fish has resulted in high catches, especially using a chart. Bilih fish in fresh form costs Rp. 85,000/kg and Rp. 75,000/kg and processed fish in dry form is Rp. 300,000 and Rp. 200,000/kg (Personal interviews Nasution in May 2021 and Syandri with fishermen in October 2021).

The abundance of fish in the Intake PLTA and Batu Taba locations were 472 fish and 716 fish, respectively. As spatially the size range of fish in Intake PLTA is 40-104 mm (32,9%) and in Batu Taba 40-130 mm (21,9%) (figure 6). The highest fish size is in the 65-72 mm class range, both spatially and temporally (June) (figure 7). The total length of Bilih fish caught using experimental gill nets ranged from 40 – 95 mm and the highest length was 80.3 mm [12]. These results show that the size range of fish caught in the Intake PLTA and Batu Taba is longer than using the experimental gill net, as well as the highest length. The size of the Bilih fish caught using the dipp nets is larger than the Bilih fish caught using experimental gill net fishing gear in Lake Singkarak.

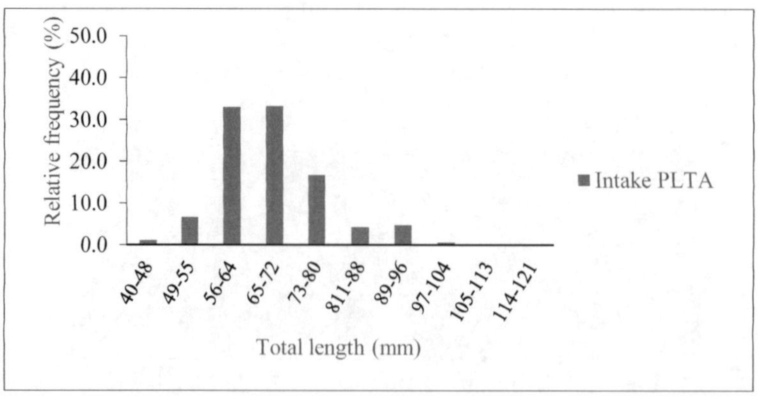

Figure 6. Spatial distribution of the relative length of Mistacoleucus padangensis

The total length mode of Bilih fish in 2003 in Lake Singkarak was 65 mm [17], while in Lake Toba it was 65 mm and 125 mm in 2005. Mode of total length of Bilih fish in 2008 in Lake Toba increased to 135 mm (caught using 1.25 inch gill nets) and 185 mm caught using 1.5 inch gill nets). Natural food factors for Bilih fish are estimated to be quite available and the level of exploitation is not yet intensive in Lake Toba.

3.2. The temporal frequency of the length distribution of *Mistacoleucus padangensis*

Bilih fish as an endemic fish have been successfully introduced to Lake Toba from Lake Singkarak in its native area [17]. However, the population in Lake Toba has decreased drastically. In July and September 2019 catching Bilih fish using experimental gill nets with a mesh size of 5/8; ¾; 1.0 and 1¼ inches in Sumpur Village and Singkarak Village. The total length of female and male Bilih fish caught ranged between 50-109 mm and 50-99 mm [18]. The length of Bilih fish observed using dipp nets in July and September 2021 ranged from 42-100 mm and 65-130 mm, respectively.

The mode of total length of Bilih fish in June was 69 mm, July 65 mm, August 64 mm, September 65 mm, October 60 mm, November 60 mm, and December 60 mm. The average total length of Bilih fish in June is 76.5 mm, July is 66.7 mm, August is 67.9 mm, September is 79.9 mm, October is 74.6 mm, November is 76.8 mm, and December is 74.6 mm. Min-max total length Bilih in June ranged from 60-92 mm, July 42-100 mm, August 40-100 mm, September 65-130 mm, October 50-130 mm, November 60-130 mm, and December was 50 -120 mm (figure 7).

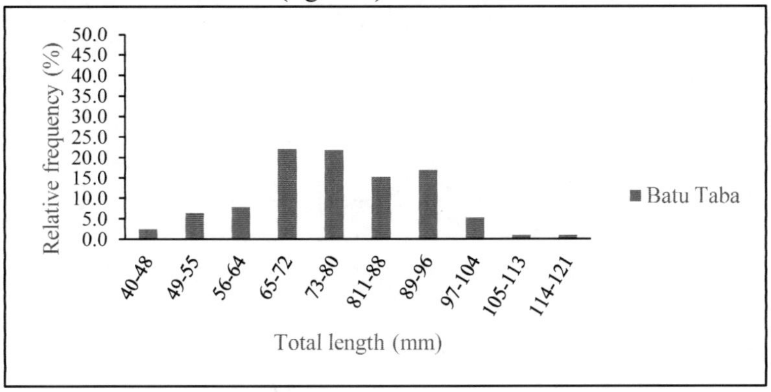

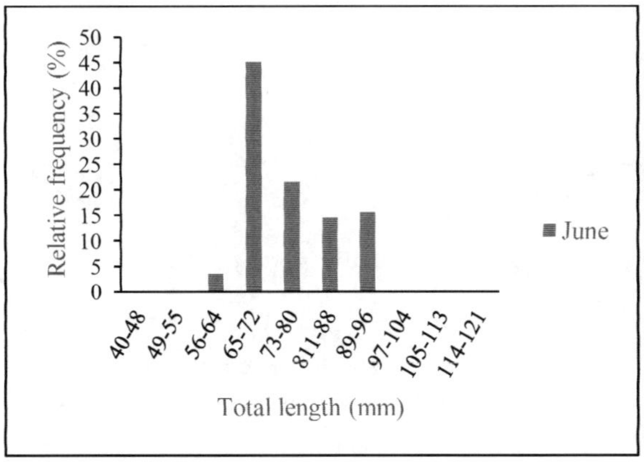

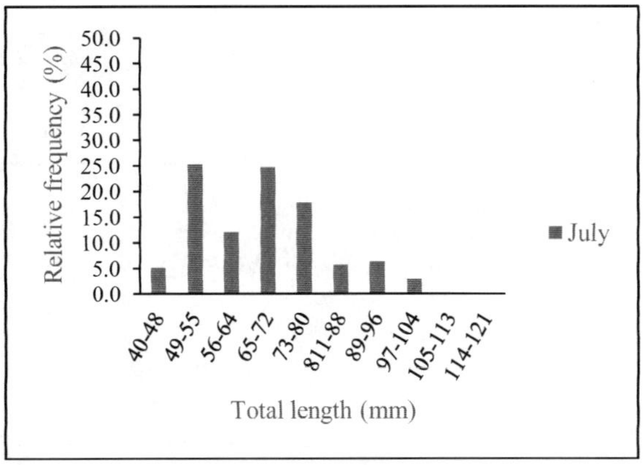

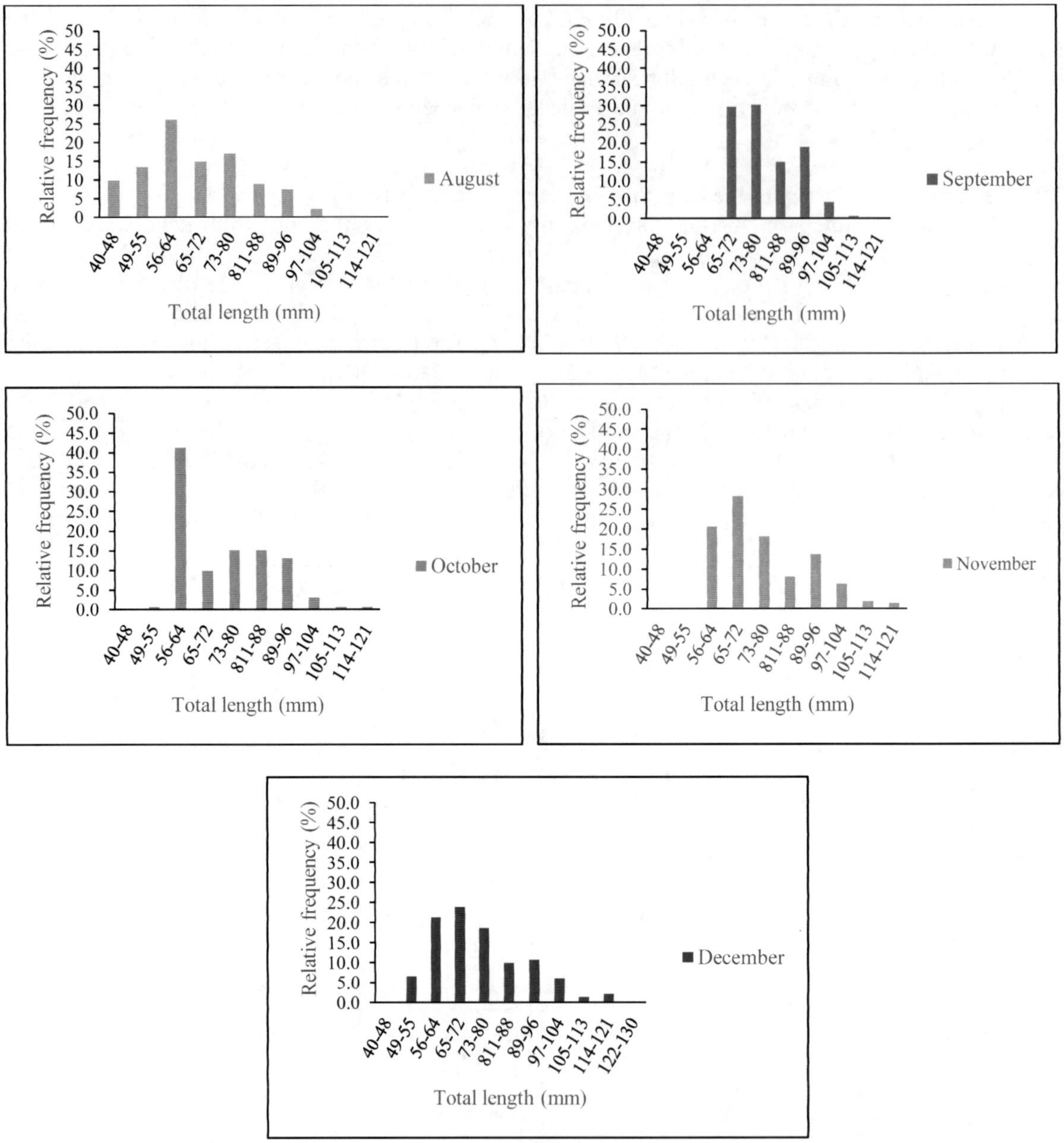

Figure 7. The temporal distribution of the relative length of *Mistacoleucus padangensis*

3.3. The condition of the water environment at the dipp nets location of Lake Singkarak

At the research location the Intake hydropower (Intake PLTA) and Batu Taba have relatively good water quality. At the bottom of the water, the DO value is 4.68 mg/L and the temperature is 27.87 °C lower than other locations. With increasing depth, the dissolved oxygen content will decrease. Organic matter can decompose at the bottom of the waters, it will require a lot of oxygen. According to [19] toxic gases such as H_2S, NH_3, and CH_4 will be produced if the decomposition occurs under anaerobic conditions because the dissolved oxygen content in the hypolimnion layer is very small and even reaches zero. Conductivity 0.143 mS/cm; TDS 0.092 mg/L at a depth of 14.15 m (figure 8). In the Batu Taba location, the condition of the aquatic environment in general is still relatively good. The temperature ranges from

26.97 – 28.93 °C, but at Secchi depth (3.28 m) the pH value is higher than at the Intake PLTA location. This condition is estimated by the entry of nutrients into the waters of the lake by the flow of water directly due to fertilization from agricultural activities carried out by the people around the lake. DO 4.68 mg/L; Conductivity 0.231 mS/cm; TDS 0.092 mg/L) at a depth of 14.15 m (figure 8).

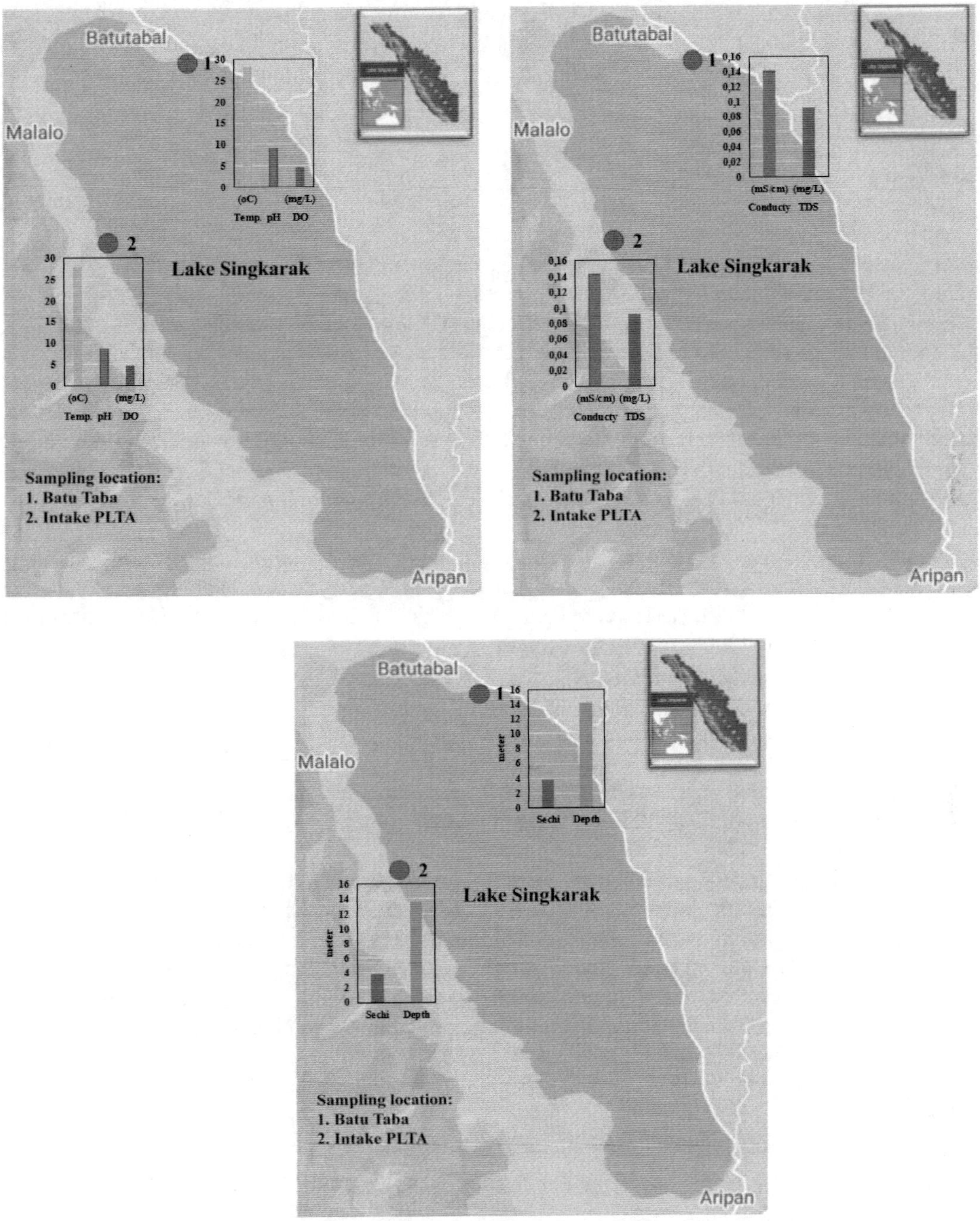

Figure 8. Water quality and depth of the dipp nets sampling location in Lake Singkarak

IOP Conf. Series: Earth and Environmental Science 1273 (2023) 012088 doi:10.1088/1755-1315/1273/1/012088

4. Conclusion

The abundance of fish in the intake PLTA and Batu Taba locations were 472 fish and 716 fish, respectively. As spatially the size range of fish in intake PLTA is 45-104 mm and in Batu Taba 40-130 mm. The highest fish size is in the 65-72 mm class range, both spatially and temporally (June). The mode of total length of Bilih fish in June was 69 mm, July 65 mm, August 64 mm, September 65 mm, October 60 mm, November 60 mm, and December 60 mm. Average water quality at intake PLTA location (temperature 27.87 ^{o}C; pH 8.6; DO 4.68 mg/L; Conductivity 0.143 mS/cm; TDS 0.092 mg/L) at a depth of 14.15 m. Average water quality at Batu Taba location (temperature 28.14 ^{o}C, pH 9.24; DO 4.68 mg/L; Conductivity 0.231 mS/cm; TDS 0.092 mg/L) at a depth of 14.15 m.

5. References

[1] Weber M and KLF de Beaufort 1916 *The fisheries of Indo-Australia archipelago* Vol. III EJ Brill. Ltd., Leiden: 455 pp

[2] Kottelat M, AJ Whitten, SN Kartikasari and S Wirjoatmodjo 1993 *Freshwater fishes of Western Indonesia and Sulawesi*. Periplus Editions, Hong Kong. 221 p

[3] Fishbase 2021. https://fishbase.mnhn.fr/summary/Mystacoleucus- padangensis.html

[4] Panudju L 2010 Kajian ecologis habitat dan sumberdaya bagi konservasi ikan Bilih (*Mystacoleucus padangensis*) di Danau Singkarak, Sumatera Barat. (Skripsi Sarjana). Institut Pertanian Bogor

[5] Lehmusluoto P 1995 Expedition Indodanau, P.O.Box 717, FIN-00101 Helsinki, Finland; Contact by E-mail:(pasi.Lehmusluoto@kolumbus.fi)

[6] Anonimus 2012 Grand Design Penyelamatan Ekosistem Danau Indonesia. KLHK. 174 hal.

[7] PUTL Department 2007

[8] Syandri H 2008 Ancaman terhadap plasma nutfah ikan bilih (*Mystacoleucus padangensis*) dan upaya pelestariannya di Danau Singkarak. Dalam: Pidato yang disampaikan pada Upacara Pengukuhan sebagai Guru BesarTetap Fakultas Perikanan dan Ilmu Kelautan, Universitas Bung HattaPadang, 12 Juli 2008. Indonesia. 35 hlm

[9] Idris 2013 *Jurn Bum Lest* 13 (2) 355-65

[10] Food and Fisheries Office of Tanah Datar Regency, 2021

[11] Syandri H & MI Effendie 1997 *Terubuk* 67 (XVIII) 2-16

[12] Nasution SH, Akhdiana I, Syawal MS and Nurhidayat A 2021 Length-weight relationship and condition factor of endemic fish Bilih (*Mystacoleucus padangensis* Blkr.) in Lake Singkarak, West Sumatra,Indonesia. E3S Web of Conferences Volume 322, 01036 (e-ISSN: 2267- 1242; SCOPUS Indexed).

[13] Bukhari and Eriza M 2014 "Pemetaan daerah penangkapan ikan bilih (*Mystacolecus padangensis*) di Danau Singkarak Sumatera Barat," In A. Isnansetyo, A Husni, D Fransiska, Djumanto, E Setyobudi, H Saksono, & Rustadi (Eds.), Prosiding Seminar Nasional Tahunan XI Hasil Penelitian Perikanan dan Kelautan Yogyakarta: Universitas Gadjah Mada. hal. 619–624s

[14] Sparre P and Venema SC 1999 Introduksi Pengkajian Stok Ikan Tropis. Puslitbangkan, penerjemah. Jakarta: Pusat Penelitian dan Pengembangan Perikanan. Terjemahan dari : *Introduction to Tropical Fish Stock Assessment*

[15] King M 1995 *Fisheries Biology, Assessment, and Management* United Kingdom: Fishing News Books

[16] Silvano RAM, Ramires M and Zuanon J 2009 Reser *Ecol of Fresh Fis* 18 156–66

[17] Kartamihardja ES and Sarnita AS 2008 Populasi ikan Bilih di Danau Toba: Keberhasilan introduksi ikan, implikasi pengelolaan dan prospek masa depan. Pusat Riet Perikanan Tangkap, Badan Riset Kelautan dan Perikanan, DepartemenKelautam dan Perikanan. 62 hal

[18] Warsa A, Hendrawan ALS, Krismono 2020 *Jour of Fish and Mari Resea* 4(1) 178-86

[19] Jones B 2011 Oxygen-the most important water quality parameter? *Spring* 23 1

6. Acknowledgement
The author would like to thank the enumerators who have assisted in this research activity. The authors also thank the editors and reviewers for their suggestions and input on our manuscript.

IOP Conf. Series: Earth and Environmental Science 1273 (2023) 012089 doi:10.1088/1755-1315/1273/1/012089

Analysis of high protein feed on the maintenance of tilapia (*Oreochromis niloticus*) broodstock in brackishwater culture

N Chairany[1], L A Sari[2], and S Arshad[3]

[1]Aquaculture Study Program, Faculty of Fisheries and Marine Sciences, Universitas Airlangga, Surabaya 60115, Indonesia
[2]Department of Aquaculture, Faculty of Fisheries and Marine Sciences, Universitas Airlangga, Surabaya 60115, Indonesia
[3]Department of Aquatic Resources Management, Faculty of Fisheries and Marine Sciences, Universitas Brawijaya, Jalan Veteran Malang 65145 East Java, Indonesia.

[4]Corresponding author : luthfianaas@fpk.unair.ac.id

Abstract. Tilapia (*Oreochromis niloticus*) is a fish native to Indonesian waters which has spread to Southeast Asia and China. Tilapia is generally cultivated in fresh waters, but due to the high demand adaptation of tilapia cultivation in brackishwater is also conducted. The purpose of the study was to determine the analysis of high protein feed in rearing Tilapia (*Oreochromis niloticus*) broodstock in brackishwater culture. The research method included the data collection of specific daily growth rate parameters and survival values. The content of protein used in feed was 35%. The results included the value of the specific daily growth rate of tilapia (*Oreochromis niloticus*) broodstock of 0.0102% and a survival value of 100%. The water quality was in the temperature range of 27-33oC, pH of 6-8, DO of 3.26-6.38 mg/L, and salinity of 0-5 ppt. The study was influenced by several factors including pond preparation, broodstock preparation, broodstock selection, broodstock feeding, water quality management, and pest and disease control.

1. Introduction

Fisheries development is at present directed at increasing the contribution of the fisheries sub-sector in supporting advanced, efficient, and resilient agriculture. Morever, fishery development aims to achieve balanced economic stability between industry and agriculture, as well as to support rural development and improve the living standard of fish entrepreneurs. The need for fish for the community is very high, so the development of fishery businesses should be highly supported [1].

Tilapia (*Oreochromis niloticus*) is a fish native to Indonesian waters which has spread to Southeast Asia and China. This fish is a commodity which is widely developed by farmers. Tilapia is commonly cultivated in fresh waters, but recently the area for freshwater fish cultivation is getting narrower since competition for land use by various types of freshwater fish increases. On the other hand, the availability of ponds is still widely available. This encourages efforts to develop tilapia cultivation in brackish waters (ponds) and the sea or better known as nile tilapia (*Oreochromis niloticus*). Tilapia cultivation has been developed in various parts of Indonesia, namely Aceh, Central Java, West Java, North Sumatra and Lampung. Several studies on salinity and physiological studies of tilapia have also been carried out. The results of Mege's [2] research showed that the growth rate of tilapia was higher when they were reared at salinity > 5 ppt. A similar result was reported by Darwisito [3], stating that at a salinity of 10 ppt the body resistance of tilapia became better. and it was also the best environmental condition for tilapia.

One of the main factors in rearing tilapia broodstock is high protein feed. Tilapia feeds in its natural habitat are in the form of plankton, periphyton, and soft plants, such as Hydrilla and algae. Tilapia is classified as an omnivorous animal (eating everything/animals and plants) which tends to be herbivores. In the maintenance period, tilapia could be given artificial feed (pellets) containing protein of 20–25% [4]. During the maintenance period, tilapia was very responsive to artificial feed (pellets), both floating pellets and sinking pellets [5]. Feeding for tilapia fry was carried out 3-4 times a day, namely in the morning, afternoon, evening, and night. The amount of feed given to fry measuring 5-7 cm was 4-6% of the total body weight of the fish [4]. Therefore, it is necessary to conduct a study to determine the analysis of high protein feed in rearing the tilapia (*Oreochromis niloticus*) broodstock in brackishwater culture.

2. Material and methods

Survival rate is the main parameter in aquaculture which can indicate the success of the production. The standard value of fish survival rate is 75% [6]. The survival rate was calculated based on Effendi's [7] formula. Death of fish in aquaculture can be caused by narrow space for movement, competition in obtaining food (natural or artificial feed) and stress on fish [8]. Growth is influenced by internal and external factors. Internal factors heavily depend on the condition of the fish's body, such as the fish's ability to utilize the remaining energy and protein after metabolism for its growth. On the other hand, external factors such as environmental factors and feed are very influential on fish growth. Tilapia growth is also largely influenced by several factors, one of which is density. The stocking density setting was carried out to minimize competition between individual fish, namely competition for space.

Moreover, natural feed is one of the important factors that influence the success of fish farming [9]. Most of the natural feed for fish is plankton, namely phytoplankton and zooplankton [10]. Natural feed for fish larvae or fry has several advantages, including its relatively small size, size in accordance with the mouth opening of larvae and fish fry, high nutritional value, potential for cultivation, its movement which can stimulate fish to eat it, the ability to breed quickly so that its availability can be guaranteed, and the cost of cultivation which is relatively inexpensive [11][12]. Feed is the most important element in supporting growth and survival.

2.1. Maintenance techniques for tilapia broodstock

Maintenance techniques for Tilapia broodstock included pond preparation, broodstock preparation, broodstock selection, broodstock feeding, water quality management, and pest and disease control [13][14]. Tilapia can be produced in different farming systems and scales with higher or lower dependence on inputs, especially feed. Efficient management of cultivation systems requires monitoring of growth parameters [15].

2.2. Pond preparation

Preparation of the pond was carried out before being used for the broodstock maintenance activities. The pond used for rearing tilapia broodstock was a concrete tank. There were eleven broodstock rearing tanks which had an area of 5x5x8 m2, a wall thickness of the tank of 15cm, and a bottom of the pool in the form of concrete. Before carrying out broodstock maintenance, the pool had been cleaned first by reducing the water volume in the tank. The water level was reduced by opening the pipe at the outlet. The volume of water was reduced until it completely ran out along with the sludge or feed residue that was pushed out through the outlet hole; this aimed to reduce the organic matter present in the tank. Cleaning the tank was done merely by brushing it using a brush and net until the tank was free of dirt. After the tank was drained, the tank was filled with water. The process of filling the broodstock rearing tank with water was done by inserting a 1.5-inch hose placed at the corner of the tank. Then, the tank was filled with water with a height of 50-60 cm. According to Bardach [16], the water depth for rearing freshwater fish ranged from 0.50 to 0.80 cm. The regular filling and changing of water aimed to drain phytoplankton from the tank in order to prevent the excessive development phytoplankton, eliminate harmful metabolic wastes such as ammonia (NH3), and serve as a substitute for aeration. Sources of fresh water came from the groundwater well pumps with a pump size of 1 inch, totaling 3 pumps located near the unit which were then distributed to the cultivation tanks. Water sources in good quantity and quality play an important role in boosting freshwater fishery production. Water sources for freshwater fish farming must be free from contaminants such as household waste, agricultural waste (insecticides), and industrial waste.

2.3. Broodstock preparation

Tilapia broodstock came from BPTPB Cangkringan and BBPBAT Sukabumi. The broodstock of red Tilapia came from Cangkringan, and the broodstock of black tilapia came from Sukabumi. Broodstock used was the broodstock that had superior and good quality. Zulkhairiansyah [17] stated that fulfilling the requirement of a good broodstock is the first step in an effort to produce good seeds.

2.4. Broodstock selection

Broodstock selection aimed to obtain superior seeds, which would not hinder the fish rearing process. This selection was to find out the level of development of gonad maturation in male and female tilapia broodstock. The conditions for the broodstock selected were fast growth, absence of disability and healthy condition, complete limbs, ready size and age to spawn, and state of being free from diseases. According to Arie [18], the characteristics of the prospective broodstock of tilapia used in the spawning process included stocky body, large and flat scales, not dull body color, healthy condition and absence of disability. The selected male and female Tilapia broodstock were aged 8 months when they first experienced gonad maturity (Table 1).

Table 1. Sample weight and length of 8-month-old tilapia broodstock

No	November 2018		January 2019	
	Weight	Length	Weight	Length
1	150 grams	23.9 cm	300 grams	29.8 cm
2	250 grams	23 cm	450 grams	30 cm
3	200 grams	22.5 cm	350 grams	29.6 cm
4	190 grams	22.8 cm	380 grams	30.2 cm
5	200 grams	23.5 cm	350 grams	29 cm
Average	198 grams	23.14 cm	366 grams	29.72 cm

This broodstock selection activity for tilapia had to be carried out in the process of broodstock rearing activities, so that when the broodstock was spawned it could produce good quality and quantity larvae that were resistant to diseases and became superior seeds. This selection activity could improve fish characteristics for productivity such as growth speed, fish resistance to diseases, and feed consumption levels. Tilapia could start to be used as broodstock when it started to weigh approximately 0.4 kg, both female and male tilapia broodstock. According to Khairuman and Amri [19], the productive phase in spawning was 1.5-2 years old with a weight above 500 g/fish. Meanwhile, according to SNI [20], the requirement age for the reproductive characteristics of fish was 6 months, with the standard length of 22 cm and body weight of 200 grams.

On the genitals of adult male fish, the genitals have protrusions that are tapered, and there are two holes in the genitals. Meanwhile, the genitals in females are round and reddish in color, and the abdomen is bulging and wide. Morphologically, the gonadal maturity level (TKG) is defined as a change in shape, length and weight, color, egg diameter, and the development of gonadal contents which is a certain stage of gonadal development before and after fish spawning [21]. Tilapia has a white color on the belly and a bit black on the dorsal. The male parent has the characteristics of rough pectoral fins, has a smaller body size than the female parent, and has genitals that are slightly pale in color and tapered inward. The difference between male and female Tilapia can be seen in Table 2.

Table 2. Differences between male and female tilapia broodstock

No	Fish	Characteristics
1	 Female Parent	- genitals are rounded and reddish - Stomach is bulged or slightly dilated - Body color is black or dark red - Movement is slow - If the stomach is stripped, it does not release liquid but eggs
2	 Male Parent	- Genitals are tapered with a pure white color - Belly color is dark and blackish - Movement is agile and aggressive - If the stomach is stripped, it secretes sperm fluid.

The description of Table 2 above is complemented by the opinion of Kordi [22], which states that the characteristics of gonadal mature parents include the conditions where the tips of the fins on the female parent are pale reddish, while those of the male parent is bright red. The female parent has a lighter body color than the male parent, and if the female's stomach is stripped, eggs will come out while the male will release a clear liquid, namely sperm.

3. Results and discussion

3.1. Broodstock feeding

During the rearing of tilapia broodstock, feed given was round pellet, in the form of floating fish feed. Food enrichment was also made by the technicians with the aim of accelerating the maturity of the parent gonads. The dose of pellet feed was 2% of the weight of fish biomass. Regarding the dose of broodstock feed, Khairruman and Amri [23] stated that broodstock was given pellets at a dose of 3-5% per day of the total broodstock kept.

Feeding with pellets was done 2 times per day, specifically in the morning and evening. Feed pellets were given in the broodstock pond, in the form of BD – IN floating fish pellets with a size of 2-3 mm. The protein content used in rearing the broodstock was in accordance with Setiawati and Supriadi [24], who stated that the tilapia broodstock feed was an artificial feed in the form of pellets with a protein content of 28–35% and a fat content of not more than 3%. This feed was given to the mother tilapia to stimulate the maturity level of the gonads and support the early to late growth period of fish cultivation. Nutrient content in BD – IN floating fish pellets can be seen in Table 3.

Tabel 3. Nutrient contents in BD – IN floating fish pellets

No	Content	Dose
1	Protein	35%
2	Fat	3%
3	Ash	13%
4	Coarse Fiber	6%
5	Water content	10%

(Source: PT. Matahari Sakti)

The feed pellets given to each pond of tilapia broodstock containing 50 male parents with an average weight of 350 g and 150 female parents with an average weight of 200 g every day were:

- Amount of feed per day = average weight x number of male parents x 2%
  = 350 x 50 x 2%
  = 350 gr
- Amount of feed per day = average weight x number of female parents x 2%
  = 200 x 150 x 2%
  = 600 gr

Because the maintenance of male and female parents was mixed in the same tank, the feeding of males and females was ± 1 kg per day. Moreover, because the feeding system was done twice a day, thus it took as much as 500 grams of pellets per tank each time of feeding.

3.2. Specific growth rate (SGR)

The specific growth rate was the % of the difference between the final weight and the initial weight, divided by the length of maintenance time. The percentage of the specific growth rate of tilapia broodstock was 0.0102%/day. This value was lower when compared to the results of Basuki's [25] study which obtained an optimal relative weight growth of 0.05%. Calculations were obtained from the formula Zenneveld et al., [26]; Loekman et al., [27]; Oktaviani et al., [28].

3.3. Survival rate (SR) of tilapia

Survival rate is the degree of survival obtained from the ratio of the number of live fish at the end of the maintenance with the number of fish at the beginning of the maintenance. During the Field Work activities, no mortality was found in tilapia broodstock, thus the SR value for tilapia broodstock was 100%. The percentage of fish survival could be calculated by the formula from Nugroho [29].

3.4. Water quality management

Water as a medium for fish to live is very influential on the life and growth of fish [30][31]. Therefore, the water used for cultivation must have optimal conditions both in terms of quality and quantity. Zulkhairiansyah [17] explained that good water quality is the most important thing for aquatic organisms because it will determine the production of the cultivation business carried out. Water in aquaculture must be free from pollution, both production waste and household waste. Water quality measurements in this study for temperature and DO were carried out using the YSI 550A DO meter. Measurements were taken twice a week in the morning and evening. The table of the average results of water quality observations is in Table 4.

Table 4. Average results of water quality observations

Parameter	Average of Measurement Results
Temperature	27 – 33 °C
pH	6 – 8
DO	3.26 – 6.38 mg/l
Salinity	0 – 5 ppt

3.4.1. Temperature

The results of measuring the water temperature in the rearing tank for tilapia broodstock was in the range of 27–33 oC; that temperature was still in the optimal temperature for tilapia. According to BSNI [6], the optimal temperature for tilapia growth in calm water ponds is between 25-32 0C. The lowest temperature value of 27 oC occurred in the morning, while the highest value in the afternoon was 33 oC. Temperatures that experience rapid changes can cause death in some fish species.

3.4.2. Acidity (pH)

The results of measuring the acidity (pH) in rearing tanks for tilapia broodstock were 6-8. The pH measurement was carried out using a pH pen, Milwaukee pH 56. The relatively neutral pH was a good habitat for tilapia. In general, the pH values obtained during the study were included in the optimal category for fish growth. According to Sugama [32], the ideal pH value for the life of aquatic organisms ranged from 6 to 8.5. This is in accordance with SNI [6] which states that the optimal pH for tilapia is 6.5-8.5. Water conditions that are too acidic or too alkaline will endanger the survival of organisms because it causes metabolic and respiratory disorders. High pH values will cause the balance between ammonium and ammonia in the water to be disturbed, while ammonia concentrations that are too high are toxic to organisms [33].

3.4.3. Dissolved oxygen (DO)

The content of dissolved oxygen (DO) in the water during broodstock rearing was between 3.26 – 6.38 mg/l. From the measurement results, the dissolved oxygen content during the study was still within the tolerance range. This is in accordance with SNI [6] which states that the optimal dissolved oxygen level for tilapia is more than 3 mg/L. The factor that affects the difference in dissolved oxygen is the effect of activity in the pond so that it is easy for oxygen to diffuse from the air to the water. Furthermore, dissolved oxygen is also influenced by the abundance of phytoplankton. Oxygen is one of the limiting factors, so if its availability in the water is not sufficient for the needs of aquatic organisms, all activities of the organisms will be hindered. Sources of dissolved oxygen can come from the diffusion of oxygen in the atmosphere and photosynthetic activity by aquatic plants and phytoplankton [34].

3.4.4. Salinity

The results of salinity measurements in rearing tanks for tilapia broodstock were 0-5 ppt. Salinity measurements were carried out using a salinity refractometer (ATC). In general, the salinity values obtained during the study were included in the optimal category for the growth of tilapia. According to Kordi [35], tilapia could live in waters with a salinity of 0 ppt to 35 ppt, yet the salinity suitable for tilapia to live optimally was 0 ppt to 30 ppt. Salinity is one of the environmental parameters that influence the biological processes of an organism and directly affect the life of the organism including the growth rate, the amount of food consumed (food conversion) and survival [36].

4. Conclusion

The conclusion of the study on the analysis of high protein feed in Tilapia (*Oreochromis niloticus*) broodstock rearing in brackishwater culture showed rearing techniques for tilapia broodstock which included pond preparation (pond cleaning and pond water filling), broodstock preparation, broodstock selection, broodstock feeding, water quality management, and pest and disease control. The results of the study showed that the specific daily growth rate (SGR) of tilapia broodstock was 0.0102%/day. The survival value (SR) of tilapia broodstock was 100%, which means that there were no deaths during maintenance. Moreover, the water quality in the broodstock pond was good, making it suitable for fish farming as evidenced by the water temperature range of 27-33oC, pH of 6-8, DO of 3.26-6.38 mg/l, and salinity of 0-5 ppt.

5. References

[1] Murtidjo B A 2005 *Several Freshwater Fish Hatchery Methods* (Yogyakarta : Kanisius) p 22-24.
[2] Mege R A 1993 Tesis. Bogor : Bogor Agricultural Institute. 116 hal.
[3] Darwisto S 2006 Doctoral Dissertation. Bogor : Bogor Agricultural Institute. 79 hal.
[4] Ghufran M, dan Kordik K 2009 *Aquaculture* (Bandung : PT. Citra Aditya Bakti) p 88-89..
[5] Cholik F R P, dan Jauzi A 2005 *Aquaculture: The Foundation of Hope for the Nation's Future* (Jakarta: Masyarakat Perikanan Nusantara) p 415.

[6] BSNI 2009 SNI No.7550:2009. *Production of Tilapia Fish (Oreochromis niloticus Bleeker) Growing Class in Still Water Ponds. (Jakarta : National Standardization Body)* p 12.
[7] Effendi 1979 *Fisheries biology* (Yogyakarta : Yayasan Pustaka Nusantama) p 163.
[8] Nasution Z, Sari Y D, dan Huda H M 2012 *J. Kebijakan Sos. Eko. Kel. dan Per.*1(1) 19-31.
[9] Azmi K A, Arsad S, and Sari L A 2020 *IOP Conf. Series: EES.*441 (1) 012039 .
[10] Sari L A, Pursetyo K T, Arsad S, Masithah E D, Setiawan E, and Affandi M 2019 *Pollution Research* . 38: S27-S32.
[11] Damayanti K Y, Mubarak A S, and Sari L A 2020 *IOP Conf. Series: EES.* 441(1) 012156.
[12] Holy N H., and Sari L A 2020 *IOP Conf. Series: EES.* 441(1) 012057.
[13] Nindarwi D D, Sari L A, Wulansari P D, Samara S H, and Santanumurti M B 2020 *IOP Conf. Series: EES.* 441(1) 012014.
[14] Dwiardan K H, Sari L A, Sari P D W, Nindarwi D D, and Arsad S 2020 *IOP Conf. Series: EES.* 441(1) 012015.
[15] Olurin K B, and Aderibigbe O A 2006 *World J.Zool.* 1(513) 82-85.
[16] Bardach J E, Ryther J H, and Mc Larney W O 1972 *Aquaculture The Farming and Husbandry of Freshwater and Marine Organisms* (New York : John Wiley and Sons Inc) p 868.
[17] Zulkhairiansyah M 2011 *Fisheries Potential Must Advance* (Jakarta : CV. Simplex) p 18.
[18] Arie U 2001 *Tilapia Gift Breeding and Breeding* (Jakarta : Penebar Suiadaya) p 64.
[19] Khairuman dan Amri K 2007 Business Opportunities and Mass Reproduction Techniques for Toddler Fish (Jakarta: Gramedia Pustaka Utama).
[20] BSNI 2009 SNI No.6138:2009 *Black tilapia broodstock (Oreochromis niloticus Bleeker) Principal Parent Class.*(Jakarta : Badan Standardisasi Nasional) p 12.
[21] Sutanto D 2012 *Tilapia Cultivation* (Yogyakarta : Pustaka Baru Press) p 33-34.
[22] Kordi M G H K, dan Tancung A B 2007 *Water Quality Management in Aquaculture.* (Jakarta : Rineka Cipta) p 210.
[23] Khairuman H, dan Amri K, 2011 2.5 Months Tilapia Harvest (Jakarta Selatan: AgroMedia).
[24] Setiawati M, dan Supriyadi M A 2003 *JAI.* 2 27 – 30.
[25] Basuki F, dan Rejeki S 2015 *Proc. Env. Sci.* 23 142-147.
[26] Zenneveld N, Huisman E A, dan Boon J H 1991 *The principles of fish farming* (Jakarta : PT. Gramedia Pustaka Umum) p 37.
[27] Loekman N A, Manan A, Arief M, and Prayogo P 2018 *JAFH.* 7 78-83.
[28] Oktaviani R, Mubarak A S, and Sudarno S 2020 *JAFH.* 9 20-29.
[29] Nugroho A 2006 *Water Quality Bioindicator* (Jakarta : Universitas Trisakti) p 7.
[30] Izzah N 2020 *JAFH.* 9 81-85.
[31] Artha O A, Sudarno H, Pramono, and Sari L A 2019 *IOP Conf. Series: EES.* 236 (1) 012003.
[32] Sugama K, Tridjoko B, Slamet S. Ismi E, Setiadi, dan Kawahara S 2001 *Technical Instructions for the Seed Production of Cromileptes Altivelis Duck Grouper Fish (Bali: Gondol Marine Cultivation Research Center) p 40.*
[33] Setijaningsih L, dan Suryaningrum L H 2015 *J. Berita Bio.* 14 72.
[34] Novonty V, and Olem H 1994 *Water Quality, Prevention, Identification, and Management of Difusse Pollution* (New York : Van Nostrans Reinhold) p 192 .
[35] Kordi M G H K, dan Tancung A B 2007 *Water Quality Management in Aquaculture.* (Jakarta : Rineka Cipta) p 210.
[36] Andrianto T 2005 *Practical Guidelines for Cultivating Tiger Grouper Fish.* (Yogyakarta : Absolut) p 18-31.

6. Acknowledgements
The author is very grateful to Universitas Airlangga for organizing InCoFims 2021.

AUTHOR INDEX